DEVELOPMENTAL MATHEMATICS

fourth edition

C. L. JOHNSTON

ALDEN T. WILLIS

GALE M. HUGHES
Cerritos College

PWS PUBLISHING COMPANY

I(T)P • **An International Thomson Publishing Company**

*Boston • Albany • Bonn • Cincinnati • Detroit • London • Madrid • Melbourne • Mexico City
New York • Paris • San Francisco • Singapore • Tokyo • Toronto • Washington*

This book is dedicated to our students,
who inspired us to do our best
to produce a book worthy of their time.

PWS PUBLISHING COMPANY
20 Park Plaza, Boston, MA 02116-4324

I(T)P™
International Thomson Publishing
The trademark ITP is used under license.

For more information, contact:

PWS Publishing Company
20 Park Plaza
Boston, MA 02116

International Thomson Publishing Europe
Berkshire House 168-173
High Holborn
London WC1V 7AA
England

Thomas Nelson Australia
102 Dodds Street
South Melbourne, 3205
Victoria, Australia

Nelson Canada
1120 Birchmont Road
Scarborough, Ontario
Canada M1K 5G4

International Thomson Editores
Campos Eliseos 385, Piso 7
Col. Polanco
11560 Mexico D.F., Mexico

International Thomson Publishing GmbH
Königswinterer Strasse 418
53227 Bonn, Germany

International Thomson Publishing Asia
221 Henderson Road
#05-10 Henderson Building
Singapore 0315

International Thomson Publishing Japan
Hirakawacho Kyowa Building, 31
2-2-1 Hirakawacho
Chiyoda-ku, Tokyo 102
Japan

Library of Congress Cataloging-in-Publication Data
Johnston, C. L. (Carol Lee).
    Developmental mathematics / C. L. Johnston, Alden T. Willis, Gale M. Hughes -- 4th ed.
        p.   cm.
    4th ed.
    Includes index.
    Hughes, Gale M.
    ISBN 0534-94500-7
    1. Algebra.   I. Willis, Alden T.   II. Title
QA152.2.J628   1994   94-22941
512'.1--dc20   CIP

Sponsoring Editor   *Susan McCulley Gay*
Editorial Assistant   *Hattie Schroeder*
Developmental Editor   *Maureen Brooks/Elizabeth R. Deck*
Production Coordinator   *Elise S. Kaiser*
Marketing Manager   *Marianne C. P. Rutter*
Manufacturing Coordinator   *Marcia A. Locke*
Production   *Lifland et al., Bookmakers*
Interior/Cover Designer   *Elise S. Kaiser*
Interior Illustrator   *Scientific Illustrators*

Cover Photo   *Dag Sundberg/© The Image Bank*
Compositor   *Beacon Graphics Corporation*
Cover Printer   *New England Book Components, Inc.*
Text Printer and Binder   *Courier/Westford*
Printed and bound in the United States of America.
99—10   9 8 7

CONTENTS

Chapter 1   **Whole Numbers**   1

Chapter 2   **Fractions**   45

$4\frac{1}{2}$ in. $3\frac{3}{4}$ in.

$5\frac{5}{6}$ in.

Chapter 3   Decimal Fractions   87

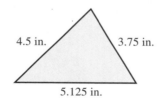

4.5 in.   3.75 in.

5.125 in.

Chapter 4   Ratio, Proportion, and Percent   123

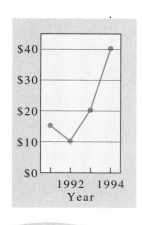

Chapter 11   **Factoring**   303

Chapter 12   **Algebraic Fractions**   331

CONTENTS

DEVELOPMENTAL MATHEMATICS, Fourth Edition, can be used with a lecture format, in a learning laboratory setting, or for self-study. Chapters 1–5 provide an arithmetic review for those students who need to improve their skills. The instructor may choose to use all or part of the arithmetic review, depending on the students' needs and abilities. Chapters 6–16 cover the topics traditionally taught in a beginning algebra course.

Geometry is included in an increasing number of developmental mathematics courses and on competency tests and entry-level examinations for many colleges. Chapter 17 can be used to introduce students to the topics of geometry.

Features of This Book

The major features of this book include the following:

1. This book uses a one-step, one-concept-at-a-time approach. That is, major topics are divided into small sections, each with its own examples and exercises. This approach allows students to master each section before proceeding confidently to the next section.

2. Each topic is explained carefully and in detail, and many concrete, annotated examples illustrate the general mathematical principles covered in each section.

3. Important concepts and algorithms are enclosed in boxes for easy identification and reference.

4. Over 200 "Critical Thinking and Writing Problems" at the ends of the chapters motivate students to think and write about mathematical concepts.

5. In "Word of Caution" screened boxes, students are warned against common mathematical errors.

6. Interspersed throughout the text and the examples are "Notes" that provide students with additional information regarding the topic or example under discussion. These notes are easily identified with the pointing-finger icon: ☞

7. The new edition makes strong, pedagogical use of the second color, with shading, icons, and annotations to guide students through the examples and to clearly differentiate elements in figures and examples.

8. The solutions for many of the examples are explicitly checked in the text. The importance of checking solutions is emphasized throughout the book.

9. The book contains over 6,000 exercises. Answers to all odd-numbered exercises, along with **completely worked-out solutions** for many selected problems from these odd exercises, are provided in the back of the text. The answers to the even-numbered exercises are given in the *Annotated Instructor's Edition* and in the printed Test Bank. In most cases, the odd-numbered exercises provide practice on problems analogous to those in the even-numbered exercises.

10. Special attention is given to the operations with zero in a single section of Chapter 6.

11. A detailed method is provided for translating word statements into algebraic equations for solving word problems. Students are asked to use the following steps to help them set up word problems and organize their work:

   a. State what their variables represent.

   b. Set up an equation and solve it.

   c. Answer the question.

12. Many of the difficulties students have with percent problems are explicitly addressed in the section on percent proportion, which has aids to help students identify the numbers in percent problems.

13. The examples and exercises that lend themselves to the use of a scientific calculator are designated with the calculator icon:

14. A diagnostic test at the end of each chapter can be used for study and review or as a pretest. Complete solutions to all the problems in these diagnostic tests, together with section references, appear in the answer section at the back of the book.

15. A comprehensive summary and a set of review exercises are included at the end of each chapter. The answers, as well as solutions for many selected exercises, are in the answer section at the back of the book.

16. A set of cumulative review exercises is provided after Chapters 5, 10, and 17. The answers are in the answer section at the back of the book.

17. A comprehensive treatment of the metric and English systems of measurement is included in Appendix B.

Changes in the Fourth Edition

The fourth edition incorporates changes that resulted from many helpful comments from reviewers and users of the first three editions. The major changes in the fourth edition include the following:

1. Critical thinking and writing problems have been added at the end of each chapter. We believe that writing about mathematics helps students develop a deeper understanding of the subject. Because communication is one of the essential goals of mathematics, we think that students should be encouraged to express mathematical concepts in writing. Other exercises in this section require students to use the definitions and concepts learned in the chapter to draw further conclusions or to identify common errors.

2. Objectives for every section are listed in the printed Test Bank.

3. In Chapter 1 (Whole Numbers), the section on division now includes an explanation of short division for one-digit divisors.

4. To prepare the student for rounding whole numbers and decimals, Exercises 1.2 and 3.1 have been expanded to include twelve problems requiring the student to identify the digit in a given place value.

5. In Chapter 2 (Fractions), Sections 2.11 and 2.12 have been rewritten to include a second method for adding and subtracting mixed numbers, in which each mixed number is converted to an improper fraction before the operation is performed.

6. A new section on Powers and Roots of Fractions has been added to Chapter 2 (Fractions).

7. A new section on Powers and Roots of Decimals has been added to Chapter 3 (Decimals).

8. Chapter 4 (Ratio, Proportion, and Percent) now includes a discussion of rates and unit cost.

9. In Chapter 5 (Graphs and Statistics), the graphs have been updated to reflect current statistics, and applied problems involving ratio and percent of increase or decrease have been added to the examples and the exercise sets.

10. Chapter 6 (Signed Numbers) now includes a discussion of and exercise problems on additive inverse (opposites) and multiplicative inverse (reciprocals).

11. In Chapter 7 (Evaluating Expressions), the examples and exercise sets have been rewritten and expanded; the difficulty of some of the problems has been increased to give students a higher level of skill in evaluating expressions.

12. In Chapter 8 (Polynomials), the definition of a polynomial has been moved from Section 8.1 to 8.5A to improve the flow of the text, and the discussion of zero as an exponent has been moved from Section 15.1 to 8.2 to make the coverage more complete. Division of a polynomial by a monomial is now introduced using the rule for addition of fractions. More problems have been added to the exercise sets throughout Chapter 8.

13. In Chapter 9 (Equations and Inequalities), the symmetric property of equality is now introduced in Section 9.3, and a corresponding "Word of Caution" showing that inequalities are not symmetric has been added to 9.6. The exercise sets have been expanded with additional problems.

14. In Chapter 10 (Word Problems), more explanation is provided about translating a word statement into an algebraic expression.

15. In Chapter 11 (Factoring), a "Word of Caution" has been added to Section 11.6 showing that the sum of squares cannot be factored. Several exercise sets have been expanded, including the diagnostic test.

16. In Chapter 12 (Algebraic Fractions), the discussion of finding excluded values in Section 12.1 has been rewritten for greater clarity, and excluded values of each problem in Exercises 12.7 are found before the equations are solved. A "Word of Caution" in Section 12.5 warns about canceling across an addition or subtraction sign, and another "Word of Caution" in Section 12.7 compares how the LCD is used in adding fractions, in simplifying complex fractions, and in solving equations.

17. In Chapter 13 (Graphing), the definition of slope has been rewritten as $\dfrac{\text{rise}}{\text{run}}$, a form more familiar and accessible to students. Section 13.4 on writing the equation of a line has been clarified.

18. More exercises have been added to Section 15.1 (Negative Exponents).

19. A new Chapter 16 (Quadratic Equations) covers General Form of a Quadratic Equation (Section 16.1), Solving Quadratic Equations by Factoring (Section 16.2), Incomplete Quadratic Equations (Section 16.3), The Quadratic Formula (Section 16.4), and Word Problems Involving Quadratic Equations (Section 16.5).

20. The chapter on geometry has been moved to the end of the book; it is now Chapter 17.

Major topics are divided into small manageable sections, as part of the book's one-step, one-concept-at-a-time approach

"Notes" to students provide additional information or point out problem-solving hints

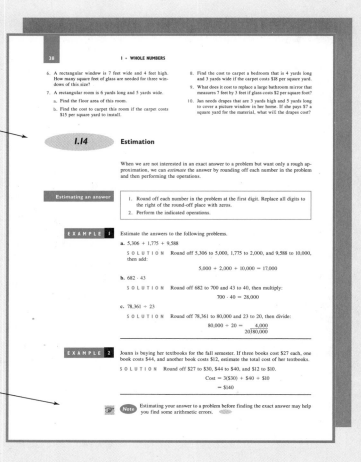

In special screened boxes labeled "A Word of Caution," students are warned against common errors

Boxes enclose important concepts and algorithms, for easy identification and reference

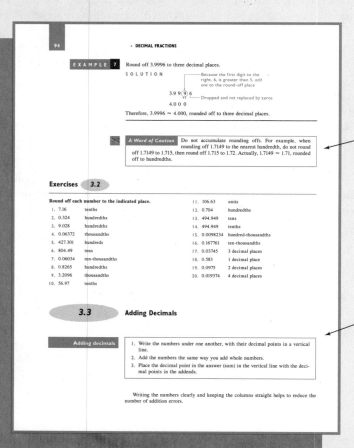

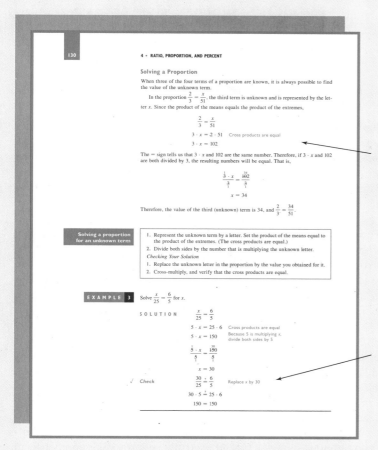

Solving a Proportion

When three of the four terms of a proportion are known, it is always possible to find the value of the unknown term.

In the proportion $\frac{2}{3} = \frac{x}{51}$, the third term is unknown and is represented by the letter x. Since the product of the means equals the product of the extremes,

$$\frac{2}{3} = \frac{x}{51}$$
$$3 \cdot x = 2 \cdot 51 \quad \text{Cross products are equal}$$
$$3 \cdot x = 102$$

The = sign tells us that $3 \cdot x$ and 102 are the same number. Therefore, if $3 \cdot x$ and 102 are both divided by 3, the resulting numbers will be equal. That is,

$$\frac{\overset{1}{\cancel{3}} \cdot x}{\cancel{3}} = \frac{\overset{34}{\cancel{102}}}{\cancel{3}}$$
$$x = 34$$

Therefore, the value of the third (unknown) term is 34, and $\frac{2}{3} = \frac{34}{51}$.

Solving a proportion for an unknown term	1. Represent the unknown term by a letter. Set the product of the means equal to the product of the extremes. (The cross products are equal.) 2. Divide both sides by the number that is multiplying the unknown letter. *Checking Your Solution* 1. Replace the unknown letter in the proportion by the value you obtained for it. 2. Cross-multiply, and verify that the cross products are equal.

EXAMPLE 3   Solve $\frac{x}{25} = \frac{6}{5}$ for x.

SOLUTION
$$\frac{x}{25} = \frac{6}{5}$$
$$5 \cdot x = 25 \cdot 6 \quad \text{Cross products are equal}$$
$$5 \cdot x = 150 \quad \text{Because 5 is multiplying } x,\text{ divide both sides by 5}$$
$$\frac{\overset{1}{\cancel{5}} \cdot x}{\cancel{5}} = \frac{\overset{30}{\cancel{150}}}{\cancel{5}}$$
$$x = 30$$

√ Check
$$\frac{30}{25} \overset{?}{=} \frac{6}{5} \quad \text{Replace } x \text{ by 30}$$
$$30 \cdot 5 \overset{?}{=} 25 \cdot 6$$
$$150 = 150$$

Fully worked out solutions are given, with careful one-step-at-a-time explanation provided and problem-solving points highlighted

The importance of checking solutions is reiterated throughout the book

---

In Exercises 25–36, use the pictograph below.

Daily Sales

Sunday
Monday
Tuesday
Wednesday
Thursday
Friday
Saturday

= 1,000

Figure for Exercises 25–36

25. What day of the week had the highest sales?
26. What day of the week had the lowest sales?
27. What was the amount of sales on Friday?

28. What was the amount of sales on Thursday?
29. On what day of the week did the sales equal $6,000?
30. On what day of the week did the sales equal $4,500?
31. How much higher were sales on Saturday than on Wednesday?
32. What were the total sales for the week?
33. What percent of the total week's sales were made on Saturday?
34. What percent of the total week's sales were made on Wednesday?
35. What is the ratio of Wednesday's sales to Saturday's sales?
36. What is the ratio of Thursday's sales to Saturday's sales?

In Exercises 37–40, use the following set of numbers: {8, 6, 5, 3, 6, 2}.
37. Find the mean.   38. Find the median.
39. Find the mode.   40. Find the range.

In Exercises 41–44, use the line graph on page 174 for Exercises 1–10. Find the following.
41. Find the mean temperature for the year.
42. Find the median temperature for the year.
43. Find the temperature mode.
44. Find the range of the temperatures.

Chapter 5 Critical Thinking and Writing Problems

Answer Problems 1 and 2 in your own words, using complete sentences.

1. Explain how to find the median of a set of numbers.
2. Explain how to find the range of a set of numbers.
3. From the chart below make a bar graph showing the number of gallons of water used for each activity. Be sure to label the horizontal and vertical scales and to give a title to your graph.

Average Family's Water Usage Per Day

Lawn and Pool	180 gallons
Toilets	100 gallons
Showers and Baths	52 gallons
Laundry	48 gallons
Cooking	20 gallons
Total =	400 gallons

4. Use the chart in Problem 3 to find the percent of water used for each activity. Then sketch a circle graph *showing* the percent of water used for each activity.

5. Is the following word problem set up correctly? Explain your answer.

Births at Community Hospital

Use the bar graph to find the percent of decrease in births from 1991 to 1992.

$$\frac{1,200}{1,500} = \frac{P}{100}$$

All new artwork enhances students' ability to visualize problems and examples

Calculator icons designate examples and exercises appropriate for solving with a scientific or graphics calculator

"Critical Thinking and Writing Problems" have been added at the end of every chapter

Using This Book

As mentioned earlier, *Developmental Mathematics,* Fourth Edition, can be used in three types of instructional programs: lecture, laboratory, and self-study.

The Conventional Lecture Course   This book has been class-tested and used successfully in conventional lecture courses by many instructors. It is not a workbook, and therefore it contains enough material to stimulate classroom discussion. Examinations for each chapter are provided in the Test Bank, and computer software for DOS, Windows, and Macintosh enables instructors to create their own tests. Tutorial software is available to help students who require extra assistance.

The Learning Laboratory Class   This text has also been used successfully in many learning labs. The format of explanation, example, and exercises in each section of the book and the tutorial software make the book easy to use in laboratories. Students can use the diagnostic test at the end of each chapter as a pretest or for review and diagnostic purposes. The five forms of each chapter test available in the Test Bank and the computerized test generators allow students to review and then be re-tested on material as needed.

Self-Study   This book lends itself to self-study because each new topic is introduced in an informal, readable manner, and each topic is short enough for students to master before continuing. The numerous examples and completely solved exercises show students exactly how to proceed. Students can use the diagnostic test at the end of each chapter to determine which parts of that chapter they need to study and can thus concentrate on their weak areas. The MathQUEST™ tutorial software extends the usefulness of the new edition in laboratory and self-study settings by providing another form of self-paced review and practice.

In addition to assigning the critical thinking and writing problems that appear in the text, instructors might also encourage students to keep a journal and/or to do other writing about their feelings about mathematics.

Ancillaries

The following ancillaries are available with this text:

- *Annotated Instructor's Edition*   includes the complete student text along with answers to most of the exercises printed adjacent to each exercise in blue. Also included are five teaching essays, written by leading mathematics educators, offering ideas to new and experienced instructors on teaching developmental mathematics.

- *Test Bank with Answers to Even-Numbered Exercises*   includes five different exams for each chapter, an arithmetic pretest, two forms for each of the three unit exams, and two forms for the final exam. All of these exams are prepared with adequate workspace and can be easily removed and duplicated for class use. Answer keys, with complete solutions for all problems on these exams, are provided in this manual. This manual also contains the answers to the even-numbered exercises and the answers to many critical thinking and writing problems. The objectives for each section of each chapter are listed in the front of this manual.

- *Computerized Testing Programs*   are available. **EXPTest** is available in both Windows and DOS versions for the IBM PC and compatibles; **ExamBuilder** is available for the Macintosh. Questions are multiple choice, true-false, and open-ended. Instructors can add to, delete, or change existing questions and produce individual tests. Demos are available.

- *MathQUEST ™ Tutorial Software* is a text-specific, intuitive tutorial, which runs on Microsoft Windows and Macintosh platforms. It allows students to practice the skills taught in the textbook section by section. The tutorial provides hints at the first wrong answer and a detailed solution at the second wrong answer. Students can also view the detailed solution after working the problem correctly. The format is mainly fill-in-the-blank, with some multiple choice. This diagnostic program keeps track of right and wrong responses and can report on the student's progress to the instructor. A demo is available.

- *DOS Tutorial Software* is a text-specific tutoring software system available for IBM PC and compatibles with DOS. It uses an interactive format to tailor lessons to the specific learning problems of students. The multiple-choice questions help students master mathematical ideas and procedures. Demos are available.

- *Videotape Series* teaches key topics from the text and features a professional math instructor. Students (and instructors) may purchase the **Video Guidebook**, which provides exercises that reinforce connections between the content of the text and the content of the videos. Carefully modeled on the style and mathematical presentation in the text, each lesson helps students focus on the essential concepts presented in the videos.

Acknowledgments

We wish to thank the members of the editorial and production staff at PWS Publishing Company for their help with this edition; special thanks go to Susan McCulley Gay, Elise Kaiser, Maureen Brooks, Judith Mustacchia Shoemaker, and Bess Rogerson Deck. Thanks, too, to Sally Lifland from Lifland et al., Bookmakers, for her help in the production of this book.

We are grateful to Laurel Technical Services of Redwood City, California, for checking the solutions to the problems. We would also like to thank the following reviewers for their helpful comments and significant contributions:

Jo Battaglia
Penn State University

Michael E. Bennett
Bronx Community College

Bernice Bowdoin
Bristol Community College

Doris S. Edwards
Motlow State Community College

Philip A. Glynn
Naugatuck Valley Community/Technical College

Shantha Krishnamachari
Borough of Manhattan Community College

Gerald LePage
Bristol Community College

Vincent McGarry
Austin Community College

Lynda D. MacLeod
Robeson Community College

Joseph Payez
Paul D. Camp Community College

Vera Preston
Austin Community College

Mark Swetnam
Ashland Community College

Emmanuel E. Usen
Northeast College of Houston Community College System

John Vangor
Housatonic Community College

To the Student

Mathematics is not a spectator sport! Mathematics is learned by *doing,* not by watching. (The fact that you "understand everything the teacher does on the board" does not mean that you have learned the material!) No textbook can teach mathematics to you, and no instructor can teach mathematics to you; the best that both can do is to help *you* learn.

If you have not been successful in mathematics courses previously and if you really want to succeed now, we strongly recommend that you obtain and study *Mastering Mathematics: How to Be a Great Math Student*, Second Edition, by Richard Manning Smith (Belmont, CA: Wadsworth Publishing Company, 1994). This little book contains *many* fine suggestions for studying, a few of which are listed below:

- Attend all classes. Missing even one class can put you behind in the course by at least two classes. To quote Mr. Smith on this topic:

  > My experience provides irrefutable evidence that you are much more likely to be successful if you attend *all* classes, always arriving on time or a little early. If you are even a few minutes late, or if you miss a class entirely, you risk feeling lost in class when you finally do get there. Not only have you wasted the class time you missed, but you may spend most or all of the first class after your absence trying to catch up. (p. 46)

- Learn as much as you can during class time.
- Sit in the front of the classroom.
- Take complete class notes.
- Ask questions about class notes.
- Solve homework problems on time (*before* the next class).
- Ask questions that deal with course material.
- Organize your notebook.
- Read other textbooks.
- Don't make excuses.

Mastering Mathematics: How to Be a Great Math Student also gives specific suggestions about getting ready for a math course even before the course begins, preparing for tests (including the final exam), coping with a bad teacher, improving your attitude toward math, using class time effectively, avoiding "mental blocks," and so forth.

Two Final Notes (1) If you see a word in this book with which you are not familiar, *look it up in the dictionary.* Learning to use a dictionary effectively is a major part of your education. (2) **It is impossible to overemphasize the value of doing homework!**

It is important for you to realize that, to a very great extent, you can take charge of and have control over your success in mathematics. We believe that studying from *Developmental Mathematics,* Fourth Edition, can help you be successful.

Whole Numbers

CHAPTER

1

n Chapter 1 we show how to perform the basic operations on whole numbers, and we discuss the order of operations used for these operations. Many applied problems are covered, along with an introduction to using a calculator.

1.1   Basic Definitions

In this section we introduce the names and definitions of some numbers and number relations.

Sets   A **set** is a collection of objects or things. Sets are usually represented by listing their members, separated by commas, within braces { }.

Natural Numbers   The numbers

$$1, 2, 3, 4, 5, 6, 7, 8, 9, 10, 11, 12, \text{ and so on}$$

are called the **natural numbers** (or **counting numbers**). These were probably the first numbers invented by humans and used to count their possessions, such as sheep and goats. We call the set of natural numbers N, so

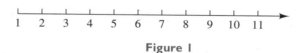

$$N = \{1, 2, 3, 4, \ldots\}$$

which is read

     "N equals the set whose members are $1, 2, 3, 4,$ and so on ."

The three dots to the right of the number 4 indicate that the remaining numbers are found by counting in the same way we have begun: namely, by adding 1 to the preceding number to find the next number.

     The smallest natural number is 1. The largest natural number can never be found, because no matter how far we count there are always larger natural numbers.

Number Line   Natural numbers can be represented by numbered points equally spaced along a straight line (Figure 1). Such a line is called a **number line**. The arrow on the right of the line indicates that the number line continues forever.

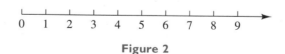

Figure 1

Whole Numbers   When 0 is included with the natural numbers, we have the set of numbers known as **whole numbers** (Figure 2).

Figure 2

We call the set of whole numbers W, so

$$W = \{0, 1, 2, 3, \ldots\}$$

which is read

     "W equals the set whose members are $0, 1, 2, 3,$ and so on."

Digits   In our number system a **digit** is any one of the first ten whole numbers: $\{0, 1, 2, 3, 4, 5, 6, 7, 8, 9\}$. Any number can be written by using a combination of these digits. For this reason, the digits are sometimes called the building blocks of our number system. Numbers are often referred to as *one-digit* numbers, *two-digit* numbers, *three-digit* numbers, and so on.

EXAMPLE 1
a. 35 is a two-digit number.
c. 275 is a three-digit number.
e. The first digit of 785 is 7.
g. The third digit of 785 is 5.

b. 7 is a one-digit number.
d. 100 is a three-digit number.
f. The second digit of 785 is 8.

We show the natural numbers, whole numbers, and digits on the number line in Figure 3.

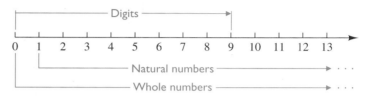

Figure 3

Inequality Symbols   The symbol <, read "less than," and the symbol >, read "greater than," are called **inequality symbols**. They are used to compare numbers that are not equal. Numbers get larger as we move to the right on the number line and smaller as we move to the left.

EXAMPLE 2   **a.** 2 < 7

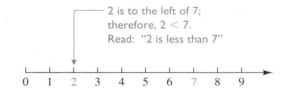

b. 6 > 3

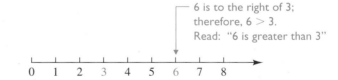

An easy way to remember the meaning of the symbol is to notice that the wide part of the symbol is next to the larger number:

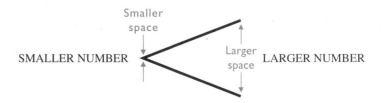

Some people like to think of the symbols > and < as arrowheads that point toward the smaller number.

EXAMPLE 3   **a.** 7 > 6 is read "7 is greater than 6."
b. 7 > 1 is read "7 is greater than 1."
c. 5 < 10 is read "5 is less than 10."

Note that $7 > 6$ and $6 < 7$ give the same information even though they are read differently.

Another inequality symbol is $\neq$. A slash line drawn through a symbol puts a "not" in the meaning of the symbol:

$=$ is read "is equal to."

$\neq$ is read "is *not* equal to."

$<$ is read "is less than."

$\not<$ is read "is *not* less than."

$>$ is read "is greater than."

$\not>$ is read "is *not* greater than."

EXAMPLE 4

a. $4 \neq 5$ is read "4 is not equal to 5."

b. $3 \not< 2$ is read "3 is not less than 2."

c. $5 \not> 6$ is read "5 is not greater than 6."

Exercises 1.1

1. What is the second digit of the number 2,478?

2. What is the fourth digit of the number 1,975?

3. What is the smallest digit?

4. What is the smallest whole number?

5. What is the smallest natural number?

6. What is the largest digit?

7. What is the largest whole number?

8. What is the largest natural number?

9. What is the largest two-digit number?

10. What is the smallest two-digit number?

11. What is the smallest three-digit number?

12. What is the largest three-digit number?

13. Is 12 a digit?

14. Is 12 a whole number?

15. Is 12 a natural number?

16. Is 0 a natural number?

17. Write all the natural numbers < 6.

18. Write all the whole numbers < 6.

In Exercises 19–24, determine which symbol, $>$ or $<$, should be used to make each statement true.

19. 5 ? 8   20. 0 ? 5   21. 7 ? 0   22. 2 ? 6

23. 13 ? 43   24. 27 ? 16

1.2 Reading and Writing Whole Numbers

Place-Value   Our number system is a **place-value** system. This means that the value of each digit in a written number is determined by its place in the number. Whenever a digit is moved one place to the left, its value becomes 10 times larger.

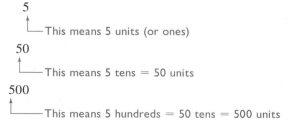

The value of the number 235 is determined as follows:

$$2 \quad 3 \quad 5$$

$$= \quad 5 \text{ units}$$
$$3 \text{ tens} = \quad 30 \text{ units}$$
$$2 \text{ hundreds} = 200 \text{ units}$$

(A more complete discussion of the idea of place-value and our number system is given later.)

Separating Digits into Groups   Commas are used to separate large numbers into smaller groups of three digits. In placing the commas, we count from the right, as shown in Figure 4. Notice that the first group on the left may have one, two, or three digits. Each group of three has a name: The first group on the right is the units group; the second group from the right is the thousands group; the third group from the right is the millions group; and so on (see Figure 4).

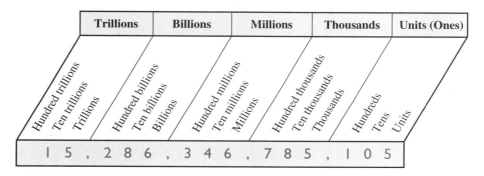

Figure 4

Reading Numbers   To read a number, we read the number formed by the digits in the left group, then say the name of that group. Then we read the number formed by the digits in the next group and say the name of that group. We continue in this way until all groups in the number have been read. The name of the units group is usually omitted when reading the number. Study Figure 4.

Writing Numbers in Words   In writing a number in words, we use commas in the same places as when we write the number in digits. For example, in words, we write the number shown in Figure 4 as follows: fifteen trillion, two hundred eighty-six billion, three hundred forty-six million, seven hundred eighty-five thousand, one hundred five. The following examples demonstrate the reading and writing of numbers:

EXAMPLE 1

a. 8,025 is read "eight thousand, twenty-five."
b. 37,045,200 is read "thirty-seven million, forty-five thousand, two hundred."

Notice in parts a and b that hyphens are used in writing two-digit numbers such as 25 (twenty-five), 37 (thirty-seven), and 45 (forty-five).

c. 275,000,000 is read "two hundred seventy-five million."
d. 9,000,605,000 is read "nine billion, six hundred five thousand."

Notice in parts c and d that when a group is made up of all zeros it is not read.

Note   The word *and* should not be used in reading or writing whole numbers: 605 is read "six hundred five," *not* "six hundred and five."

Numbers Larger Than Trillions   Numbers larger than trillions have names. For those students who want to know the names of larger numbers, we include Figure 5. You will not be expected to know the names of numbers larger than trillions.

12, 315, 218, 557, 314, 708, 515, 900, 000, 304, 017, 708

Decillion Nonillion Octillion Septillion Sextillion Quintillion Quadrillion Trillion Billion Million Thousand Units

Figure 5

Large numbers such as billions and trillions do not have much meaning for most of us. For example, one billion miles is farther than 40,000 times the distance around the earth. We often see large numbers written in newspapers and magazines. For example, the gross national product of the United States is over six trillion dollars; the distance light travels in one year is almost six trillion miles. The word *zillion* means a very, very large number, but it has no exact meaning.

Using Digits to Write Numbers   When we write a number using digits, we have to remember that each group after the first group on the left must have three digits in it. Also, if a group is not mentioned, we need three zeros as placeholders for that group.

EXAMPLE 2   Use digits to write the following numbers.

a. Forty-eight thousand, five hundred

  SOLUTION   48,500

b. Nine million, three hundred twenty

  SOLUTION   9,000,320

There are no thousands; therefore, we need three zeros as placeholders

c. Seventeen million, six thousand, nine hundred

  SOLUTION   17,006,900

Each group after the first group on the left must have three digits; therefore, we will write six thousand as 006 in the thousands group

d. Five billion, fifty thousand

  SOLUTION   5,000,050,000

Three zeros as placeholders in the units group

Three digits in the thousands group

Three zeros as placeholders in the millions group

Exercises 1.2

1. In the number 576:

  a. The 6 represents how many units?

  b. The 7 represents how many units?

  c. The 5 represents how many units?

2. In the number 914:

  a. The 4 represents how many units?

  b. The 1 represents how many units?

  c. The 9 represents how many units?

3. In the number 348:

   a. The 4 represents how many tens?

   b. The 4 represents how many units?

   c. The 3 represents how many hundreds?

   d. The 3 represents how many tens?

4. In the number 862:

   a. The 6 represents how many tens?

   b. The 6 represents how many units?

   c. The 8 represents how many hundreds?

   d. The 8 represents how many tens?

5. In the number 123,456:

   a. What digit is in the tens place?

   b. What digit is in the thousands place?

   c. What digit is in the ten thousands place?

6. In the number 987,503:

   a. What digit is in the hundreds place?

   b. What digit is in the thousands place?

   c. What digit is in the hundred thousands place?

7. In the number 806,239,145:

   a. What digit is in the hundreds place?

   b. What digit is in the hundred thousands place?

   c. What digit is in the millions place?

8. In the number 192,837,064:

   a. What digit is in the ten thousands place?

   b. What digit is in the millions place?

   c. What digit is in the ten millions place?

In Exercises 9–16, write the numbers in words.

9. 16,346

10. 125,006

11. 1,070,200

12. 309,005,000

13. 49,000,070

14. 2,400,000,000

15. 8,000,725,000

16. 73,006,000,900

In Exercises 17–24, mark off with commas, then write the numbers in words.

17. 16601

18. 230023

19. 71040

20. 7100400

21. 710004000

22. 71000040000

23. 5206000000

24. 9000900909

In Exercises 25–36, use digits to write the numbers.

25. Twenty thousand, five

26. Two hundred thousand, fifty

27. Eight million, eight thousand, eight hundred eight

28. Twelve million, six hundred sixty thousand

29. Seven million, seven

30. Forty-five million, sixteen thousand

31. Ten billion, one hundred thousand, ten

32. Two billion, thirty million

33. One hundred million, twenty-nine thousand, six

34. Four hundred fifty million, nine hundred

35. One light-year is the distance light travels in one year. This is about five trillion, eight hundred eighty billion miles.

36. Sirius, the brightest star in our winter sky, is fifty trillion, five hundred sixty billion miles from the earth.

1.3 Rounding Off Whole Numbers

Numbers are often expressed to the nearest million, or to the nearest thousand, or to the nearest hundred, and so on. When we say the earth is 93,000,000 miles from the sun, it is understood that 93,000,000 has been rounded off to the nearest million. The distance around the earth at the equator is 24,902 miles, but in speaking of this distance it is more common to say 25,000 miles. We say that 24,902 miles has been "rounded off to the nearest thousand" miles. When a whole number is rounded off, we must say what place it has been rounded off to.

The symbol ≈, read "is approximately equal to," is often used when numbers have been rounded off.

<table>
<tr><td>Rounding off a whole number</td><td>

1. Locate the digit in the round-off place. (We will circle the digit.)
2. If the first digit to the right of the round-off place is less than 5, the digit in the round-off place is *unchanged*.

   If the first digit to the right of the round-off place is 5 or more, the digit in the round-off place is *increased by 1*.
3. Digits to the *left* of the round-off place are unchanged (special case: Examples 3 and 4).
4. Digits to the *right* of the round-off place are replaced by zeros.

</td></tr>
</table>

EXAMPLE 1   Round off 243,280 to the nearest thousand.

SOLUTION

3 is *unchanged* because the first digit to its right (2) is less than 5

Unchanged ——
Replaced by zeros

2 4 ③, 2 8 0

2 4 3 , 0 0 0

Therefore, 243,280 ≈ 243,000, rounded off to thousands.

EXAMPLE 2   Round off 90,671 to the nearest hundred.

SOLUTION

Add one because the first digit to the right (7) is greater than 5

Unchanged ——
Replaced by zeros

9 0, ⑥ 7 1
     +1

9 0, 7 0 0

Therefore, 90,671 ≈ 90,700, rounded off to hundreds.

EXAMPLE 3   Round off 31,962 to the nearest hundred.

SOLUTION

Since the first digit to the right (6) is greater than 5, *add one* to the round-off place; this changes the 319 to 320

3 1, ⑨ 6 2
      +1

3 2, 0 0 0

Therefore, 31,962 ≈ 32,000, rounded off to hundreds.

EXAMPLE 4   Round off 999,507 to the nearest thousand.

SOLUTION

Since the first digit to the right is a 5, *add one* to the round-off place

9 9 ⑨, 5 0 7
      +1

1, 0 0 0, 0 0 0

Therefore, 999,507 ≈ 1,000,000, rounded off to thousands.

Exercises 1.3

Round off the following numbers to the indicated place.

1. 4,728   Nearest hundred
2. 256,491   Nearest thousand
3. 926   Nearest ten
4. 28,619,000   Nearest million
5. 753   Nearest hundred
6. 485   Nearest ten
7. 63,195   Nearest thousand
8. 28,232   Nearest hundred
9. 798   Nearest ten
10. 629,653   Nearest thousand

11. 19,500,000   Nearest million
12. 26,500   Nearest thousand
13. 52,461,000   Nearest million
14. 853   Nearest ten
15. 3,472   Nearest hundred
16. 78,415   Nearest ten thousand
17. 45,429,444   Nearest ten thousand
18. 45,429,444   Nearest hundred thousand
19. 1,648,723   Nearest hundred thousand
20. 1,648,723   Nearest thousand

1.4 Addition of Whole Numbers

When whole numbers are added, the numbers being added are called the **addends,** and the answer is called the **sum.**

$$
\begin{array}{r}
2 \\
+\ 3 \\
\hline
5
\end{array}
\quad
\begin{array}{l}
\text{Addend} \\
\text{Addend} \\
\text{Sum}
\end{array}
$$

Commutative Property of Addition   Reversing the *order* of two whole numbers in an addition problem does not change their sum.

EXAMPLE 1

a. $4 + 5 = 5 + 4 = 9$
b. $1 + 3 = 3 + 1 = 4$
c. $10 + 2 = 2 + 10 = 12$

The important property of whole numbers demonstrated in Example 1 is called the **commutative property of addition**.

Commutative property of addition

If a and b represent any numbers, then
$$a + b = b + a$$

Associative Property of Addition   The sum of three numbers is unchanged no matter how we *group* (or *associate*) the numbers. It is assumed that this property of addition is true when any three whole numbers are added. Thus, when adding three numbers, we can do either of the following:

1. Find the sum of the first two numbers, then add that sum to the third number.
2. Find the sum of the last two numbers, then add that sum to the first number.

We use parentheses ( ) as grouping symbols to show which two numbers are being added first. Other grouping symbols such as brackets [ ] or braces { } can be used. When grouping symbols are used, the operations within those grouping symbols must be done first.

EXAMPLE 2

Add $2 + 3 + 4$.

SOLUTION

a. $(2 + 3) + 4$
$= \quad 5 \quad + 4 = 9$ The parentheses mean that the 2 and 3 are to be added first
b. $2 + (3 + 4)$
$= 2 + \quad 7 \quad = 9$ Here, the parentheses mean that the 3 and 4 are to be added first

Therefore, $(2 + 3) + 4 = 2 + (3 + 4)$.

The property demonstrated in Example 2 is called the **associative property of addition**.

Associative property of addition
If a, b, and c represent any numbers, then $$(a + b) + c = a + (b + c)$$

Additive Identity   Adding 0 to a number gives the identical number for the sum.

EXAMPLE 3

a. $8 + 0 = 8$ **b.** $0 + 6 = 6$

For this reason, zero is called the **additive identity element.**

Additive identity property
If a represents any number, then $$a + 0 = 0 + a = a$$

Addition of Whole Numbers

When we add numbers, we first arrange the numbers in vertical columns, with all the units digits in the same vertical line, all the tens digits in the same vertical line, all the hundreds digits in the same vertical line, and so on. We then add the digits in each vertical line, starting at the right (the units column) and moving left. Writing the numbers clearly and keeping the columns straight will help reduce the number of addition errors.

EXAMPLE 4

Find the sum $587 + 265$.

SOLUTION

$$\begin{array}{r} \overset{1}{5}\;\overset{1}{8}\;7 \\ +\,2\;6\;5 \\ \hline 5\;2 \end{array}$$

1 ten + 8 tens + 6 tens =
15 tens = 1 hundred + 5 tens

$$\begin{array}{r} \overset{1}{5}\;\overset{1}{8}\;7 \\ +\,2\;6\;5 \\ \hline 8\;5\;2 \end{array}$$

1 hundred + 5 hundreds + 2 hundreds = 8 hundreds

EXAMPLE 5

Add 85 + 354 + 6 + 5,871.

SOLUTION

$$\begin{array}{r} \overset{1}{}\overset{2}{}\overset{1}{}\\ 85 \\ 354 \\ 6 \\ +\,5{,}871 \\ \hline 6{,}316 \end{array}$$

The commutative and associative properties guarantee that when we add a long column of numbers, we will obtain the same sum no matter what order or grouping of the numbers is used. Therefore, we can shorten the work by adding groups of 10.

EXAMPLE 6

Add 46 + 18 + 74 + 32.

SOLUTION

$$\begin{array}{r} \overset{2}{}\\ 46 \\ 18 \\ 74 \\ +\,32 \\ \hline 170 \end{array}$$
10   10   10

To improve your speed and accuracy with addition, see Appendix A for drill and practice with addition facts.

Exercises 1.4

In Exercises 1–18, find the sums.

1. 354 +283	**2.** 678 +254	**3.** 987 +715	**4.** 817 +209				

5. 987 +807	**6.** 483 +967	**7.** 308 25 +691	**8.** 755 9 +307

9. 908 92 +818	**10.** 562 267 +48	**11.** 506 117 +284	**12.** 274 638 +752

13.
5,840
218
21
+ 3,002

14.
75
3,594
315
+ 4,121

15.
1,263
522
4,817
+ 63,548

16.
26
47
64
13
+ 91

17.
28
75
97
42
+ 15

18.
203
576
524
837
+ 641

In Exercises 19–26, arrange the numbers in a vertical column and add.

19. 75,386 + 77 + 105,706,035 + 880,755,009 + 28,388,406

20. 359 + 40,286 + 17 + 284,000,189 + 70,096,347

21. 275 + 80 + 9 + 786,410,075 + 3,000,000 + 259,715,306

22. 7,409 + 701,093,005 + 43 + 806,240 + 576

23. 885,209,734 + 42,076 + 68 + 7,090,300 + 9,004

24. 823 + 10,090 + 520,007,380 + 79 + 64,082

25. 1,723 + 72 + 391,400,082 + 905 + 605,210 + 8

26. 14 + 43,050,908 + 20,809 + 100,926 + 804

In Exercises 27 and 28, use digits to write the numbers, then arrange them in a vertical column and find their sum.

27. Thirty thousand, six. Seventy-five million, one hundred. Two billion, five hundred. Fifty million, one hundred thousand, ten.

28. Ten thousand, forty-seven. Twenty-three million, five thousand. Four billion, six million, seventy-three thousand, forty-two. Two hundred million, one hundred fifty-six thousand, six.

29. Find the sum of the digits.

30. Find the sum of the natural numbers less than 20.

31. The greatest known depth of our oceans is 36,198 feet, and the highest point on the earth is the top of Mt. Everest, at 29,028 feet. What is the vertical distance from the lowest point to the highest point on the earth?

32. The populations of some of the nations of the world are as follows: China, over 1,169,619,000; India, 886,362,000; Russia, 149,527,000; United States, 256,561,000. Find the combined population of China, India, Russia, and the United States.

33. Luis has $75, Jim has $18, and Marie has $12 more than Luis and Jim together. Find the total amount of money the three have together.

34. Mary has $34, Jane has $15, and Helen has $27 more than Mary and Jane together. Find the total amount of money the three girls have together.

1.5 Subtraction of Whole Numbers

Subtraction is the *inverse* operation of addition; that is, subtraction undoes, or cancels out, an addition. Every subtraction problem is related to an addition problem.

$$9 - 4 = 5 \quad \text{because} \quad 5 + 4 = 9$$

$$\begin{array}{rl} 9 & \text{Minuend} \\ -\ 4 & \text{Subtrahend} \\ \hline 5 & \text{Difference} \end{array}$$

Subtraction Is Not Commutative We learned earlier that the operation of addition is commutative. That is,

$$a + b = b + a$$

This means that interchanging the order of the numbers does not change the sum. Can this be done with subtraction? We need only take a single example to see that **subtraction is not commutative**.

EXAMPLE 1

$7 - 4 \neq 4 - 7$

Recall that the symbol $\neq$ is read "is not equal to." $7 - 4 = 3$, whereas $4 - 7$ does not represent a whole number.

Subtraction Is Not Associative We also learned that addition is associative. That is,

$$(a + b) + c = a + (b + c)$$

This means that changing the way the numbers are grouped does not change the sum. Is this also true for subtraction? Again, we need only a single example to show that **subtraction is not associative**.

EXAMPLE 2

$$(9 - 5) - 2 \neq 9 - (5 - 2)$$

because $(9 - 5) - 2 = 4 - 2 = 2$

whereas $9 - (5 - 2) = 9 - 3 = 6$

$\qquad 2 \neq 6$

Subtraction of Whole Numbers

EXAMPLE 3 Subtract 2,502 from 78,514.

SOLUTION Always write units digits below units digits, tens digits below tens digits, hundreds digits below hundreds digits, and so on. Then subtract the digits in each vertical column, starting at the right (the units column) and moving left.

$$
\begin{array}{r}
78,514 \quad \text{Minuend} \\
- \ 2,502 \quad \text{Subtrahend} \\
\hline
76,012 \quad \text{Difference}
\end{array}
$$

EXAMPLE 4 Find $\begin{array}{r} 692 \\ - 456 \end{array}$

SOLUTION Because, in the units column, 6 cannot be subtracted from 2, we must borrow 1 ten from the tens column.

$$
\begin{array}{r}
\overset{8 \ 12}{69\!\!\!\diagup 2} \\
- 456 \\
\hline
236
\end{array}
$$

We borrow 1 ten from 9 tens, leaving 8 tens; the borrowed 1 ten (= 10 units) is added to the 2 units we already have, making 12 units. Then 12 − 6 = 6 units and   8 − 5 = 3 tens and   6 − 4 = 2 hundreds

EXAMPLE 5 Find $\begin{array}{r} 58,067 \\ - \ 4,193 \end{array}$

SOLUTION
$$
\begin{array}{r}
\overset{7 \ \ \overset{9}{10} \ 16}{5\,8,0\,6\,7} \\
- \ 4,1\,9\,3 \\
\hline
5\,3,8\,7\,4
\end{array}
$$

We borrow 1 thousand from 8 thousand, leaving 7 thousand; the borrowed 1 thousand (= 10 hundreds) is added to the 0 hundreds:
10 + 0 = 10 hundreds
Then we borrow 1 hundred from 10 hundreds, leaving 9 hundreds; the borrowed 1 hundred (= 10 tens) is added to the 6 tens:
10 + 6 = 16 tens
Now we can subtract

Checking Subtraction   Subtraction is checked using the related addition problem:

Subtrahend + difference = minuend

EXAMPLE 6

$$
\begin{array}{r}
6 \quad \text{Minuend} \\
-\ 2 \leftarrow \text{Subtrahend} \rightarrow 2 \\
\hline
4 \leftarrow \text{Difference} \rightarrow +\ 4 \\
\hline
6 \quad \text{Minuend}
\end{array}
$$

This same check can be done as follows:

$$
\text{Add} \begin{bmatrix} 6 \\ -\ 2 \\ \hline 4 \\ \hline 6 \end{bmatrix} \text{Check}
$$

EXAMPLE 7 Subtract 1,324 from 3,010, and check.

SOLUTION

$$
\text{Add} \begin{bmatrix} 3,010 \\ -\ 1,324 \\ \hline 1,686 \\ \hline 3,010 \end{bmatrix} \text{Check}
$$

Exercises 1.5

In Exercises 1–20, find the differences.

1. 8,907 − 702	2. 2,806 − 1,502	3. 5,568 − 2,563	4. 7,186 − 7,136
5. 98,765 − 4,565	6. 642 − 218	7. 615 − 287	8. 833 − 295
9. 308 − 179	10. 740 − 256	11. 2,108 − 1,896	12. 6,914 − 6,057
13. 3,300 − 684	14. 2,804 − 1,926	15. 60,701 − 10,808	16. 90,084 − 40,529
17. 693,421 − 355,818	18. 173,041 − 88,350		
19. 7,060,500 − 6,829,711	20. 5,200,365 − 2,843,942		

21. Subtract 4,506 from 7,392.

22. Subtract 10,362 from 19,217.

23. The average distance to the sun is about 92,889,000 miles. The average distance to the moon is about 239,000 miles. How much farther is it to the sun than to the moon?

24. If the population of China is 1,169,619,000 and the population of the United States is 256,561,000, how many more people are there in China than in the United States?

25. If at the beginning of a trip your odometer read 67,856 miles and at the end of the trip it read 71,304 miles, how many miles did you drive?

26. A used car had a list price of $4,120. Ms. Wang bought the car and was given a discount of $455. What did she pay for the car?

27. At the beginning of the month Mr. Hanson's checking account balance was $356. He made deposits of $225, $57, and $375. He wrote checks for $56, $135, $157, $38, and $417. What was his balance at the end of the month?

28. The height of Mt. Everest is twenty-nine thousand, twenty-eight feet. The height of Mt. Whitney is fourteen thousand, four hundred ninety-five feet. How much higher is Mt. Everest than Mt. Whitney?

29. Assim has $75, and Joe has $57. Jamal has $17 more than Assim, and Mike has $13 less than Assim and Joe together. How much money do all four boys have together?

30. Nine years ago the Martinez family bought a house for $76,500. Today the value of the house is $125,000. How much did the value of the house increase in the last 9 years?

1.6   Multiplication of Whole Numbers

Multiplication can be thought of as repeated addition of the same number.

$$
\left.\begin{array}{r} 5 \\ 5 \\ 5 \\ + \ 5 \end{array}\right\} \text{Four 5's} \qquad \text{and} \qquad 4 \times 5 = 20
$$
$$
\overline{\quad 20 \quad}
$$

In the expression $5 \times 6 = 30$, the number we are multiplying, 5, is called the **multiplicand**, and the number we are multiplying by, 6, is called the **multiplier**. The answer, 30, is called the **product**. The numbers 5 and 6 are also called **factors** of 30. That is, the multiplicand and the multiplier are factors of the product. In algebra the word *factor* is more commonly used for the numbers in a product than the words *multiplier* and *multiplicand*.

$$
\begin{array}{ll}
5 & \text{Multiplicand} \\
\underline{\times \ 6} & \text{Multiplier} \\
30 & \text{Product}
\end{array}
\qquad
\underset{\text{Factor}}{5} \ \times \ \underset{\text{Factor}}{6} \ = \ \underset{\text{Product}}{30}
$$

In the expression $3 \times 4 = 12$, 3 and 4 are factors of 12.

$$
\underset{\text{Factor}}{3} \ \times \ \underset{\text{Factor}}{4} \ = \ \underset{\text{Product}}{12}
$$

Multiplication may be indicated in several different ways.

EXAMPLE 1

a. $3 \times 2 = 6$ In algebra we avoid using the times sign ($\times$) because it may be confused with the letter *x*

b. $3 \cdot 2 = 6$ The multiplication dot · is written a little higher than a decimal point

c. $3(2) = 6$ When two numbers are written next to each other in this way, with no

d. $(3)2 = 6$ symbol of operation, it is understood that they are to be multiplied

e. $(3)(2) = 6$

Commutative Property of Multiplication   When we reverse the *order* of the factors, the product remains the same.

EXAMPLE 2

a. $7 \cdot 8 = 8 \cdot 7 = 56$ **b.** $4 \cdot 5 = 5 \cdot 4 = 20$

The property of whole numbers demonstrated in Example 2 is called the **commutative property of multiplication**.

Commutative property of multiplication

> If *a* and *b* represent any numbers, then
>
> $$a \cdot b = b \cdot a$$

Associative Property of Multiplication   The product of three numbers is unchanged no matter how we *group* the numbers. It is assumed that this property of multiplication is true when any three numbers are multiplied.

EXAMPLE 3   Multiply $3 \cdot 4 \cdot 2$.

SOLUTION

a. $(3 \cdot 4) \cdot 2$ **b.** $3 \cdot (4 \cdot 2)$
$= \quad 12 \quad \cdot 2 = 24 \qquad = 3 \cdot \quad 8 \quad = 24$

Therefore, $(3 \cdot 4) \cdot 2 = 3 \cdot (4 \cdot 2)$.

The property of whole numbers demonstrated in Example 3 is called the **associative property of multiplication**.

Associative property of multiplication	If a, b, and c represent any numbers, then $$(a \cdot b) \cdot c = a \cdot (b \cdot c)$$

The *commutative* and *associative* properties hold true for both addition and multiplication.

Multiplicative Identity   Multiplying any number by 1 gives the identical number for the product.

EXAMPLE 4   **a.** $8 \cdot 1 = 8$ **b.** $1 \cdot 5 = 5$

For this reason, 1 is called the **multiplicative identity element**.

Multiplicative identity property	If a represents any number, then $$a \cdot 1 = 1 \cdot a = a$$

Multiplication by Zero   Since multiplication is repeated addition of the same number, multiplying a number by 0 gives a product of 0.

EXAMPLE 5   **a.** $3 \cdot 0 = 0 + 0 + 0 = 0$ **b.** $4 \cdot 0 = 0 + 0 + 0 + 0 = 0$

Because of the commutative property of multiplication, it follows that

$$3 \cdot 0 = 0 \cdot 3 = 0$$
$$4 \cdot 0 = 0 \cdot 4 = 0$$

Multiplication property of zero	If a represents any number, then $$a \cdot 0 = 0 \cdot a = 0$$

Multiplication of Whole Numbers

EXAMPLE 6 Find $4 \cdot 36$.

SOLUTION

Hundreds / Tens / Units

$$
\begin{array}{r}
\overset{2}{3}\ 6 \quad \leftarrow \text{Carry number} \\
\times \quad 4 \\
\hline
4 \quad \leftarrow 4 \text{ units} \times 6 \text{ units} = 24 \text{ units} \\
= 2 \text{ tens} + 4 \text{ units}
\end{array}
$$

$$
\begin{array}{r}
\overset{2}{3}\ 6 \\
\times \quad 4 \\
\hline
1\ 4\ 4
\end{array}
$$

4 units × 3 tens = 12 tens

Carry

12 tens + 2 tens = 14 tens
= 1 hundred + 4 tens

EXAMPLE 7 Find $527 \cdot 34$.

SOLUTION

Tens / Units

$$
\begin{array}{r}
5\ 2\ 7 \\
\times \quad 3\ 4 \\
\hline
2\ 1\ 0\ 8 \quad \leftarrow
\end{array}
$$

Step 1. Multiply 527 units by 4 (the units digit of 34):
527 units × 4 = 2108 *units*
For this reason, the 8 in 2108 is placed in the *units* column

First partial product

$$
\begin{array}{r}
5\ 2\ 7 \\
\times \quad 3\ 4 \\
\hline
2\ 1\ 0\ 8 \\
1\ 5\ 8\ 1 \quad \leftarrow \\
\hline
1\ 7\ 9\ 1\ 8
\end{array}
$$

Step 2. Multiply 527 units by 3 (the tens digit of 34):
527 units × 3 tens = 527 units × 30
= 15,810 units
= 1581 tens

The second partial product is 1581 *tens*; therefore, the 1 is placed in the *tens* column

Step 3. Add all partial products

Therefore, the final product is 17,918.

EXAMPLE 8 Find $4{,}385 \cdot 739$.

SOLUTION

Units

$$
\begin{array}{r}
4\ 3\ 8\ 5 \\
\times \quad 7\ 3\ 9 \\
\hline
3\ 9\ 4\ 6\ 5 \\
1\ 3\ 1\ 5\ 5 \\
3\ 0\ 6\ 9\ 5 \\
\hline
3\ 2\ 4\ 0\ 5\ 1\ 5
\end{array}
$$

Tens

$$
\begin{array}{r}
4\ 3\ 8\ 5 \\
\times \quad 7\ 3\ 9 \\
\hline
3\ 9\ 4\ 6\ 5 \\
1\ 3\ 1\ 5\ 5 \\
3\ 0\ 6\ 9\ 5 \\
\hline
3\ 2\ 4\ 0\ 5\ 1\ 5
\end{array}
$$

Hundreds

$$
\begin{array}{r}
4\ 3\ 8\ 5 \\
\times \quad 7\ 3\ 9 \\
\hline
3\ 9\ 4\ 6\ 5 \\
1\ 3\ 1\ 5\ 5 \\
3\ 0\ 6\ 9\ 5 \\
\hline
3\ 2\ 4\ 0\ 5\ 1\ 5
\end{array}
$$

Notice that the right digit of each partial product is directly below the digit of the multiplier that was used to obtain it.

EXAMPLE 9   Find 305 · 1,734.

SOLUTION

$$
\begin{array}{r}
1734 \\
\times\ \ 305 \\
\hline
8670 \\
0000 \\
5202\ \ \ \\
\hline
528870
\end{array}
$$

Since the multiplication by zero results in a row of zeros, we shorten the writing by omitting this row of zeros →

$$
\begin{array}{r}
1734 \\
\times\ \ 305 \\
\hline
8670 \\
5202\ \ \ \\
\hline
528870
\end{array}
$$ = 528,870

EXAMPLE 10   Find 4,006 · 2,314.

SOLUTION

$$
\begin{array}{r}
2\ 3\ 1\ 4 \\
\times\ \ \ 4\ 0\ 0\ 6 \\
\hline
1\ 3\ 8\ 8\ 4 \\
9\ 2\ 5\ 6\ \ \ \ \ \\
\hline
9\ 2\ 6\ 9\ 8\ 8\ 4
\end{array}
$$

Notice that the right digit of each partial product is directly below the digit of the multiplier that was used to obtain it

EXAMPLE 11   Find 2,341 · 18,000.

SOLUTION   When one or both of the numbers end in zeros, we may use a shortcut to multiply the numbers. Rather than write the problem in the form

$$
\begin{array}{r}
2341 \\
\times\ 18000
\end{array}
$$

we can move the multiplier to the right and draw a vertical line to separate the zeros from the 18. Multiply by 18, then attach the zeros later, as shown below:

$$
\begin{array}{r}
2341\,| \\
\times\ \ 18\,|\,000 \\
\hline
18728\,| \\
2341\ \ \ \,| \\
\hline
42138\,|\,000
\end{array}
$$

$$= 42{,}138{,}000$$

This method works because

$$2{,}341 \cdot 18{,}000$$

$$= 2{,}341 \cdot (18 \cdot 1{,}000)$$

$$= (2{,}341 \cdot 18) \cdot 1{,}000$$

$$= 42{,}138 \cdot 1{,}000$$

$$= 42{,}138{,}000$$

EXAMPLE 12   Find $175{,}000 \cdot 3{,}200$.

SOLUTION

$$
\begin{array}{r}
175\,|\,000 \\
\times \quad 32\,|\,00 \\
\hline
350 \\
525 \\
\hline
5600\,|\,00000
\end{array}
$$

3 zeros } to the right
+ 2 zeros } of the line

5 zeros

5 zeros to the right of the line

$= 560{,}000{,}000$

To improve your speed and accuracy with multiplication, see Appendix A for drill and practice with multiplication facts.

Exercises *1.6*

In Exercises 1–42, find the products.

1. 638×4
2. 794×6
3. 962×7
4. 548×9

5. $6{,}497 \times 8$
6. $8{,}697 \times 5$
7. $23{,}106 \times 3$
8. $68{,}075 \times 9$

9. 438×57
10. 897×98
11. $7{,}836 \times 64$
12. $9{,}267 \times 38$

13. $90{,}276 \times 95$
14. $43{,}807 \times 76$
15. 235×416
16. 487×395

17. $7{,}043 \times 642$
18. $8{,}106 \times 387$
19. $4{,}372 \times 5{,}168$
20. $7{,}864 \times 8{,}794$

21. 356×204
22. 725×306
23. $3{,}804 \times 709$
24. $9{,}067 \times 508$

25. $7{,}802 \times 1{,}009$
26. $3{,}085 \times 4{,}007$
27. $60{,}058 \times 9{,}005$

28. $20{,}109 \times 6{,}008$
29. $75{,}009 \times 30{,}007$
30. $820{,}040 \times 90{,}007$

31. $2{,}500 \cdot 376$
32. $12{,}000 \cdot 507$
33. $9{,}200 \cdot 3{,}154$

34. $300 \cdot 7{,}855$
35. $500 \cdot 3{,}751$
36. $2{,}000 \cdot 799$

37. $6{,}600 \cdot 449$
38. $5{,}000 \cdot 7{,}008$
39. $9{,}500 \cdot 7{,}893$

40. $7{,}500 \cdot 3{,}500$
41. $8{,}960 \cdot 5{,}600$
42. $38{,}000 \cdot 7{,}800$

43. A ream of paper contains 500 sheets. A school uses 1,427 reams in a year. How many sheets of paper are used in that time?

44. A town has 7,500 inhabitants who pay property tax. If the average tax paid is $504, what is the town's income from property taxes?

45. Each computer disc costs $2. Find the cost of 8 boxes of computer discs, if each box contains 10 discs.

46. Each tanker in an oil fleet can carry 503,024 gallons of oil. If the fleet has 207 tankers, how many gallons of oil can the fleet transport at once?

47. Debbie earns $850 a month. Her husband earns $275 a week. How much do they earn together in one year? (1 year = 12 months or 52 weeks)

48. If a machine can sort 1,650 checks in 1 minute, how many checks can it sort in an 8-hour day?

1.7   Powers of Whole Numbers

Now that you have learned how to multiply whole numbers, it's possible to consider products in which the same number is repeated as a factor. For example,

$$3 \cdot 3 \cdot 3 \cdot 3 = 3^4 = 81$$

In the symbol 3^4, the 3 is called the **base**; the 4 is called the **exponent** and is written above and to the right of the base 3. The entire symbol 3^4 is called the *fourth* **power** *of three* and is commonly read "three to the fourth power."

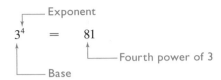

An exponent is used to show repeated multiplication by the same number. Thus 3^4 is a shorthand way to show 3 used as a factor four times. Notice the importance of the position of the 4:

$$3^4 \qquad = 3 \cdot 3 \cdot 3 \cdot 3 = 81$$

——The raised 4 indicates repeated *multiplication*

$$3 \cdot 4 = 3 + 3 + 3 + 3 = 12$$

——This 4 indicates repeated *addition*

EXAMPLE 1

a. $2^3 = 2 \cdot 2 \cdot 2 = 8$ **b.** $4^2 = 4 \cdot 4 = 16$
c. $1^4 = 1 \cdot 1 \cdot 1 \cdot 1 = 1$ **d.** $25^2 = 25 \cdot 25 = 625$
e. $10^1 = 10$ **f.** $10^3 = 10 \cdot 10 \cdot 10 = 1{,}000$

Zero as a Base   When 0 is raised to any whole-number power other than 0, we get 0 (Example 2).

EXAMPLE 2

a. $0^2 = 0 \cdot 0 = 0$ **b.** $0^5 = 0 \cdot 0 \cdot 0 \cdot 0 \cdot 0 = 0$

Zero as an Exponent   When any number (other than 0) is raised to the 0 power, we get 1 (Example 3). The reason for this definition will be covered later.

EXAMPLE 3

a. $2^0 = 1$ **b.** $5^0 = 1$ **c.** $10^0 = 1$

In general,

Zero as an exponent

> If a represents any number except 0,
> $$a^0 = 1$$

Zero as Both Exponent and Base   The symbol 0^0 is not defined or used in this book.

A calculator can be used to find powers of large numbers.

EXAMPLE 4

Find 25^4, using a calculator.

SOLUTION   If your calculator has a power key that looks like $\boxed{y^x}$ (or $\boxed{x^y}$), follow the steps below:

Key in   25 $\boxed{y^x}$ 4 $\boxed{=}$

Answer   390,625

Note Your calculator may require you to use the [INV] or [2nd F] key before the [y^x] key. Consult your instruction manual for the correct order of keys for your calculator.

If your calculator does not have a power key, you will need to use repeated multiplication:

Key in 25 [×] 25 [×] 25 [×] 25 [=]

Answer 390,625

Some calculators will allow you to repeat an operation with the same number by pressing the [=] repeatedly:

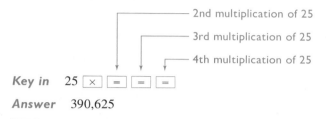

Key in 25 [×] [=] [=] [=]

Answer 390,625

Exercises 1.7

In Exercises 1–4, rewrite each expression using exponents.

1. $5 \cdot 5 \cdot 5$ 2. $8 \cdot 8$ 3. $3 \cdot 3 \cdot 3 \cdot 3 \cdot 3$

4. $10 \cdot 10 \cdot 10 \cdot 10$

In Exercises 5–28, find the value of each expression.

5. 2^3 6. 3^2 7. 5^2 8. 4^3

9. 3^3 10. 2^4 11. 6^2 12. 5^3

13. 10^0 14. 10^1 15. 10^2 16. 10^3

17. 10^4 18. 10^5 19. 10^6 20. 10^7

21. 0^3 22. 1^3 23. 3^0 24. 2^2

25. 4^2 26. 8^2 27. 1^5 28. 2^5

In Exercises 29–36, use a calculator to find the value of each expression.

29. 12^5 30. 36^3 31. 9^6 32. 8^5

33. 245^3 34. 765^2 35. 15^0 36. 0^{15}

1.8 Powers of Ten

Powers in which the base is 10 have many important uses in mathematics and science. We show some powers of 10 below.

Power of 10	Name
$10^0 = 1$	$= 1$ unit
$10^1 = 10$	$= 1$ ten
$10^2 = 100$	$= 1$ hundred
$10^3 = 1,000$	$= 1$ thousand
$10^4 = 10,000$	$= 1$ ten thousand
$10^5 = 100,000$	$= 1$ hundred thousand
$10^6 = 1,000,000$	$= 1$ million

Notice that 10^3 has a value of 1 followed by three zeros $= 1,000$ and is read "one thousand"; 10^5 has a value of 1 followed by five zeros $= 1\,00,000$ and is read "one hundred thousand."

Notice also that the successive names of the powers of 10 correspond exactly to the names of the places when we read or write a number. See Figure 6.

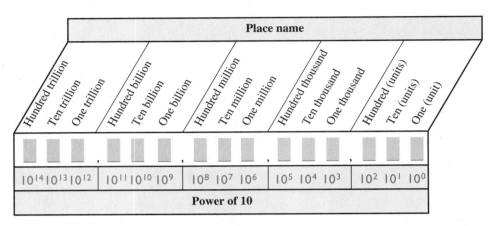

Figure 6

Multiplying a Whole Number by a Power of Ten   Suppose you make $5 a week for mowing and taking care of your neighbor's lawn. As you know, in 10 weeks you would make a total of $50. In 100 weeks you would make a total of $500. That is,

$$1\,0 \cdot 5 = 5\,0$$
$$1\,00 \cdot 5 = 5\,00$$
$$1,\,000 \cdot 5 = 5,\,000$$

Because $10 = 10^1$, $100 = 10^2$, $1,000 = 10^3$, and so on,

$$10^1 \cdot 5 = 5\,0$$
$$10^2 \cdot 5 = 5\,00$$
$$10^3 \cdot 5 = 5,\,000, \text{ and so on}$$

> When a whole number is multiplied by a power of 10, follow the number by as many zeros as the exponent of 10.

EXAMPLE 1

a. $12 \cdot 10^3 = 12,000$ or $12 \cdot 1,000 = 12,000$

b. $275 \cdot 10^4 = 2,750,000$ or $275 \cdot 10,000 = 2,750,000$

c. $4,806 \cdot 10^2 = 480,600$ or $4,806 \cdot 100 = 480,600$

d. $7 \cdot 10^0 = 7 \cdot 1 = 7$

Exercises   *1.8*

Find the value of each expression, without using written calculations.

1. $3 \cdot 100$
2. $75 \cdot 1,000$
3. $3 \cdot 10^2$
4. $75 \cdot 10^3$
5. $10 \cdot 16$
6. $100 \cdot 24$
7. $10^3 \cdot 8$
8. $10^4 \cdot 44$
9. $10 \cdot 10,000$
10. $100 \cdot 10$
11. $4 \cdot 10^0$
12. $10^0 \cdot 44$
13. $20 \cdot 100$
14. $1,000 \cdot 50$
15. $80 \cdot 10^3$
16. $400 \cdot 10^2$
17. $900 \cdot 10$
18. $12,000 \cdot 100$
19. $10^5 \cdot 103$
20. $250 \cdot 10^4$

1.9 Division of Whole Numbers

Division is the *inverse* operation of multiplication; that is, division undoes, or cancels out, a multiplication. Every division problem is related to a multiplication problem.

$$20 \div 5 = 4 \quad \text{because} \quad 4 \cdot 5 = 20$$

Just as multiplication is repeated addition of the same number, division can be thought of as repeated subtraction of the same number.

Multiplication as repeated addition	*Division as repeated subtraction*
When we think	When we think

$$2 + 2 + 2 = 6$$

$$\begin{array}{r} 6 \\ -\ 2 \\ \hline 4 \\ -\ 2 \\ \hline 2 \\ -\ 2 \\ \hline 0 \end{array}$$

← 1st subtraction of 2
← 2nd subtraction of 2
← ③rd subtraction of 2

we say $3 \cdot 2 = 6$. we say $6 \div 2 = ③$.

The **dividend** is the number we are dividing into, the **divisor** is the number we are dividing by, and the **quotient** is the answer to a division problem. If the divisor does not divide evenly into the dividend, the part left over is called the **remainder**.

Division can be written several ways.

$$21 \div 7 = 3 \longleftarrow \text{Quotient} \longrightarrow 3$$
$$\text{Divisor} \longrightarrow 7\overline{)21}$$
$$\text{Dividend}$$

We learned earlier that, since $21 = 3 \cdot 7$, the numbers 3 and 7 are *factors of* 21. We also call 3 and 7 **divisors of** 21; that is, 3 (as well as 7) divides into 21 evenly.

EXAMPLE 1

a. Divisors of 15 are $1, 3, 5, 15$ because $15 = 1 \cdot 15 = 3 \cdot 5$.
b. Divisors of 12 are $1, 2, 3, 4, 6, 12$ because $12 = 1 \cdot 12 = 2 \cdot 6 = 3 \cdot 4$.

Even and Odd Numbers   An **even number** is a number that 2 divides into evenly; that is, 2 is a divisor of the number. An **odd number** is not evenly divisible by 2, or 2 is not a divisor of the number. In Example 1, 12 is an even number because 2 is a divisor of 12. However, 15 is an odd number because 2 is not a divisor of 15.

Division Is Not Commutative   Both addition and multiplication are commutative, but subtraction is not commutative. What is the case with division? A single example will show that **division is not commutative**.

$$6 \div 3 \neq 3 \div 6$$

because $6 \div 3 = 2$, whereas $3 \div 6$ does not represent a whole number.

Division Is Not Associative   Both addition and multiplication are associative, but subtraction is not associative. A single example will show that **division is not associative**.

$$(16 \div 4) \div 2 \neq 16 \div (4 \div 2)$$

because

$$(16 \div 4) \div 2 = 4 \div 2 = 2$$

whereas

$$16 \div (4 \div 2) = 16 \div 2 = 8$$

$$2 \neq 8$$

Division Involving Zero   *Zero divided by a nonzero number is possible, and the quotient is zero:*

$$0 \div 2 = 0 \quad \text{Because } 0 \cdot 2 = 0$$

A nonzero number divided by zero is impossible:

$$4 \div 0 = ? \longleftarrow \text{Suppose the quotient is some number } x$$

then

$$4 \div 0 = x \quad \text{Means } x \cdot 0 = 4, \text{ which is certainly false}$$

$x \cdot 0 \neq 4$ because any number multiplied by zero $= 0$. Therefore, dividing any nonzero number by zero is undefined.

Zero divided by zero cannot be determined:

$$0 \div 0 = 0 \quad \text{Means } 0 \cdot 0 = 0, \text{ which is true}$$

$$0 \div 0 = 1 \quad \text{Means } 1 \cdot 0 = 0, \text{ which is true}$$

$$0 \div 0 = 5 \quad \text{Means } 5 \cdot 0 = 0, \text{ which is true}$$

In other words, $0 \div 0 = 0$, 1, and also 5. In fact, it can be any number. Because there is *no unique* answer, we say that $0 \div 0$ is undefined.

Division involving zero

If a represents any number except 0, then

$$0 \div a = 0$$

$$\left.\begin{array}{l} a \div 0 \text{ is undefined} \\[4pt] 0 \div 0 \text{ is undefined} \end{array}\right\} \text{Cannot divide by } 0$$

Long Division

EXAMPLE 2   Find $73 \div 3$.

SOLUTION   We first work this division by showing that division is repeated subtraction:

$$
\begin{array}{r}
\text{Tens} \\
\text{Units} \\
24 \\
3\,\overline{)\ 73} \\
-\ 30 \\
\hline
43 \\
-\ 30 \\
\hline
13 \\
-\ 3 \\
\hline
10 \\
-\ 3 \\
\hline
7 \\
-\ 3 \\
\hline
4 \\
-\ 3 \\
\hline
1
\end{array}
$$

1st   subtraction of   10   (3)

②nd   subtraction of   10   (3)

1st   subtraction of   1   (3)

2nd   subtraction of   1   (3)

3rd   subtraction of   1   (3)

④th   subtraction of   1   (3)

①← The *remainder* $24 = 10 + 10 + 1 + 1 + 1 + 1$

24 3's have been subtracted

   Note   The *remainder*, the number left over after the divisor has been subtracted as many times as possible, must be smaller than the divisor.

The work shown above is usually shortened as follows:

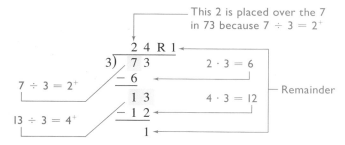

This 2 is placed over the 7 in 73 because $7 \div 3 = 2^+$

$$\begin{array}{r} 2\ 4\ R\ 1 \\ 3\overline{)\ 7\ 3} \\ -\ 6 \\ \hline 1\ 3 \\ -\ 1\ 2 \\ \hline 1 \end{array}$$

$7 \div 3 = 2^+$

$13 \div 3 = 4^+$

$2 \cdot 3 = 6$

$4 \cdot 3 = 12$

Remainder

Therefore, $73 \div 3 = 24$ with a remainder of 1.

Checking Division   You can check division problems with remainders by using the following fact:

> Quotient · divisor + remainder = dividend

EXAMPLE 3   Find $274 \div 6$, and check.

SOLUTION   Because 2 cannot be divided by 6, we divide 27 by 6.

This 4 is placed over the 7 in 27 because $27 \div 6 = 4^+$

$$\begin{array}{r} 45\ R\ 4 \\ 6\overline{)\ 274} \\ -\ 24 \\ \hline 34 \\ -\ 30 \\ \hline 4 \end{array}$$

$4 \cdot 6 = 24$

$5 \cdot 6 = 30$

Check $45 \cdot 6 + 4 \overset{?}{=} 274$

$270 + 4 \overset{?}{=} 274$

$274 = 274$

EXAMPLE 4   Find $9,035 \div 7$, and check.

SOLUTION

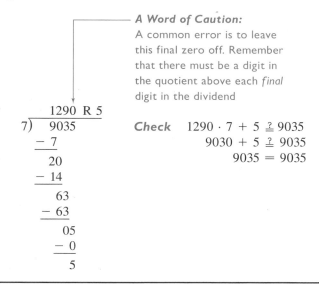

A Word of Caution:
A common error is to leave this final zero off. Remember that there must be a digit in the quotient above each *final* digit in the dividend

$$\begin{array}{r} 1290\ R\ 5 \\ 7\overline{)\ 9035} \\ -\ 7 \\ \hline 20 \\ -\ 14 \\ \hline 63 \\ -\ 63 \\ \hline 05 \\ -\ 0 \\ \hline 5 \end{array}$$

Check $1290 \cdot 7 + 5 \overset{?}{=} 9035$

$9030 + 5 \overset{?}{=} 9035$

$9035 = 9035$

Short Division

When the divisor has only one digit, we can shorten the work by doing the multiplication and subtraction mentally and carrying any remainder of that subtraction to the next column.

EXAMPLE 5   Find $78 \div 4$, using short division.

SOLUTION

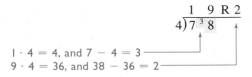

$$\begin{array}{r} 1 \quad 9 \text{ R } 2 \\ 4\overline{)7\,^38} \end{array}$$

$1 \cdot 4 = 4$, and $7 - 4 = 3$ ——

$9 \cdot 4 = 36$, and $38 - 36 = 2$ ——

Notice how the first remainder, 3, is written to the left of 8, making 38 ; then 38 is divided by 4

EXAMPLE 6   Find $2,692 \div 5$, using short division.

SOLUTION

$$\begin{array}{r} 5 \quad 3 \quad 8 \text{ R } 2 \\ 5\overline{)26\,^19\,^42} \end{array}$$

$5 \cdot 5 = 25$, and $26 - 25 = 1$ ——

$3 \cdot 5 = 15$, and $19 - 15 = 4$ ——

$8 \cdot 5 = 40$, and $42 - 40 = 2$ ——

Divisors with Two or More Digits: The Trial Divisor Method

When the divisor has two or more digits, we will use the **trial divisor** method for long division. The *trial divisor* is usually the first digit of the divisor. We use examples to illustrate the method.

EXAMPLE 7   Find $759 \div 31$.

SOLUTION   Use 3 as a trial divisor.

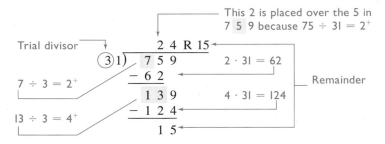

This 2 is placed over the 5 in 7 5 9 because $75 \div 31 = 2^+$

Trial divisor

$$\begin{array}{r} 2 \quad 4 \text{ R } 15 \\ (3\,1)\overline{7\,5\,9} \\ -\,6\,2 \\ \hline 1\,3\,9 \\ -\,1\,2\,4 \\ \hline 1\,5 \end{array}$$

$7 \div 3 = 2^+$

$13 \div 3 = 4^+$

$2 \cdot 31 = 62$

$4 \cdot 31 = 124$

Remainder

In Example 8, instead of using the first digit (3) of the divisor (39) for our trial divisor, we use 4, because 39 is closer to four tens than it is to three tens.

EXAMPLE 8   Find 6,842 ÷ 39.

SOLUTION

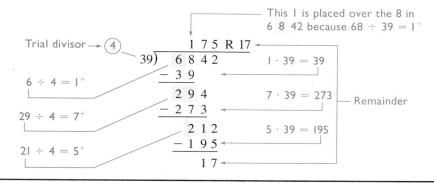

Trial divisor → ④

This 1 is placed over the 8 in
6 8 42 because 68 ÷ 39 = 1⁺

```
              1 7 5  R 17
        39) 6 8 4 2
             − 3 9            1 · 39 = 39
               2 9 4          7 · 39 = 273      Remainder
             − 2 7 3
                 2 1 2        5 · 39 = 195
               − 1 9 5
                   1 7
```

6 ÷ 4 = 1⁺
29 ÷ 4 = 7⁺
21 ÷ 4 = 5⁺

EXAMPLE 9   Find 4,908 ÷ 67.

SOLUTION

Trial divisor → ⑦

This 7 is placed over the 0 in
49 0 8 because 490 ÷ 67 = 7⁺

```
            73  R 17
      67) 4908
         − 469
           218
         − 201
            17
```

EXAMPLE 10   Find 44,490 ÷ 218.

SOLUTION

Trial divisor → ②

This 0 is placed over the 9 in
444 9 0 because 89 ÷ 218 = 0⁺

```
            204  R 18
     218) 44490
          436
           89
            0
          890
          872
           18
```

The trial divisor will help us estimate the quotient, but sometimes we need to adjust our estimate (see Example 11).

EXAMPLE 11   Find 16,605 ÷ 27.

SOLUTION   Using a trial divisor of 3, we have $16 \div 3 = 5^+$. We'll try 5 in the quotient:

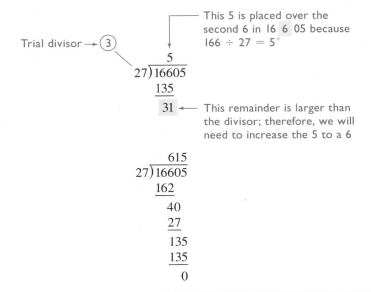

This 5 is placed over the second 6 in 16 6 05 because $166 \div 27 = 5^+$

Trial divisor → ③

$$\begin{array}{r} 5 \\ 27\overline{)16605} \\ 135 \\ \hline 31 \end{array}$$

← This remainder is larger than the divisor; therefore, we will need to increase the 5 to a 6

$$\begin{array}{r} 615 \\ 27\overline{)16605} \\ 162 \\ \hline 40 \\ 27 \\ \hline 135 \\ 135 \\ \hline 0 \end{array}$$

A Word of Caution   Once the long-division process has started, each remaining digit in the dividend must have a *single* digit above it in the quotient.

Exercises *1.9*

In Exercises 1–6, work each problem using short division.

1. 1,235 ÷ 8   **2.** 7,166 ÷ 9   **3.** 5,254 ÷ 5

4. 20,441 ÷ 6   **5.** 65,754 ÷ 9   **6.** 54,432 ÷ 8

In Exercises 7–12, work each problem using long division, and show your check.

7. 847 ÷ 31   **8.** 960 ÷ 15   **9.** 885 ÷ 43

10. 2,415 ÷ 69   **11.** 4,677 ÷ 87   **12.** 3,182 ÷ 63

In Exercises 13–36, work each long-division problem.

13. 8,866 ÷ 31   **14.** 6,870 ÷ 15   **15.** 9,729 ÷ 47

16. 8,904 ÷ 84   **17.** 9,625 ÷ 26   **18.** 22,695 ÷ 27

19. 15,120 ÷ 56   **20.** 31,746 ÷ 78   **21.** 14,450 ÷ 24

22. 25,839 ÷ 41   **23.** 97,414 ÷ 91   **24.** 24,050 ÷ 12

25. 78,800 ÷ 56   **26.** 95,348 ÷ 73   **27.** 185,503 ÷ 89

28. 235,620 ÷ 63   **29.** 65,058 ÷ 185   **30.** 129,596 ÷ 206

31. 421,400 ÷ 301   **32.** 941,200 ÷ 362

33. 565,090 ÷ 715   **34.** 1,160,695 ÷ 144

35. 5,248,749 ÷ 583   **36.** 2,087,490 ÷ 298

37. A man can make 38 machine parts on a lathe in 1 hour. How many hours will he need to make 2,356 parts?

38. Frank figures a tire on his car lasts for 24,000 miles. How many sets of tires will he need to drive his car 96,000 miles?

39. Rosali earns an annual salary of $18,720.

   a. What is her monthly salary?

   b. What is her weekly salary?

40. A family wants to save $5,000 in 4 years to make a trip to Europe. How much must they save each year? If they save $100 each month, will they meet their goal?

41. A soap company produces 96,400 pounds of detergent in one day. How many 16-pound boxes of detergent can be filled in one day?

42. A man pays off a debt of $3,048 in equal monthly payments over a period of two years. Find the amount he pays each month.

43. Eight boys going on a camping trip are trying to divide 100 small boxes of raisins up evenly among their packs. The leader (one of the eight) will take the remainder as well as his share. How many boxes did the leader have to carry?

44. Seven girls pooled their money for a lottery ticket and won $2,500. When they divide the money in dollars, there is a remainder. They draw straws to see who will get the remainder. What will be the total amount received by the winner?

45. Form numbers by arranging the digits 1 through 9 in several different orders. Divide each number so obtained by 9. Compare the remainders in each division.

46. If it takes 6 minutes to saw a log into three pieces, how long will it take to saw the same log into four pieces?

1.10 Square Roots

The symbol $\sqrt{}$ is called a radical sign, and it indicates the **square root** of the number under it. For example, $\sqrt{9}$ is read "square root of 9." Finding the square root of a number is the *inverse* operation of squaring a number. Every square root problem has a related problem involving squaring a number:

$$\sqrt{9} = 3 \qquad \text{because} \qquad 3^2 = 9$$

EXAMPLE 1

a. $\sqrt{4} = 2$ Because $2^2 = 4$
b. $\sqrt{36} = 6$ Because $6^2 = 36$
c. $\sqrt{100} = 10$ Because $10^2 = 100$
d. $\sqrt{0} = 0$ Because $0^2 = 0$
e. $\sqrt{1} = 1$ Because $1^2 = 1$

 A calculator can be used to find the square roots of large numbers.

EXAMPLE 2 Find $\sqrt{625}$ using the square-root key, $\boxed{\sqrt{}}$.

SOLUTION

Key in   625 $\boxed{\sqrt{}}$

Answer   25

Exercises 1.10

In Exercises 1–15, find the indicated square roots.

  1. $\sqrt{25}$ **2.** $\sqrt{49}$ **3.** $\sqrt{64}$ **4.** $\sqrt{4}$ **5.** $\sqrt{81}$

  6. $\sqrt{9}$ **7.** $\sqrt{16}$ **8.** $\sqrt{36}$ **9.** $\sqrt{1}$ **10.** $\sqrt{0}$

11. $\sqrt{144}$ **12.** $\sqrt{121}$ **13.** $\sqrt{100}$ **14.** $\sqrt{400}$ **15.** $\sqrt{10,000}$

 In Exercises 16–20, use a calculator to find the square roots.

16. $\sqrt{225}$ **17.** $\sqrt{529}$ **18.** $\sqrt{961}$

19. $\sqrt{5,625}$ **20.** $\sqrt{18,496}$

I.II   Order of Operations

What does $2 + 3 \cdot 4$ equal? If we add first,

$$\underbrace{2 + 3} \cdot 4$$
$$= \quad 5 \quad \cdot 4 = 20$$

But if we multiply first,

$$2 + \underbrace{3 \cdot 4}$$
$$= 2 + \quad 12 \quad = 14$$

Which answer is correct, 20 or 14? To avoid this confusion, we perform the operations in the following order:

Order of operations

> **I.** Any expressions in parentheses are evaluated first.
>
> **2.** Evaluations are done in this order:
>   *First:* Powers and roots are done.
>   *Second:* Multiplication and division are done in order from *left to right.*
>   *Third:* Addition and subtraction are done in order from *left to right.*

EXAMPLE 1

$$2 + 3 \cdot 4$$ Multiplication before addition
$$= 2 + \quad 12 \quad = 14$$

EXAMPLE 2

$$16 \div 2 \cdot 4$$ Multiplication and division are done *left to right*
$$= \quad 8 \quad \cdot 4 = 32$$

EXAMPLE 3

$$2(20) + 2(15)$$ When two numbers are written next to each other with no symbol of operation, it is understood they are to be multiplied
$$= \quad 40 \quad + \quad 30 \quad = 70$$

EXAMPLE 4

$$4^2 - 2\sqrt{25}$$ Powers and roots are done first
$$= 16 - 2 \cdot 5$$ The understood multiplication of 2 and 5 is done next
$$= 16 - \quad 10 \quad = 6$$ Subtraction is last

EXAMPLE 5

$$2(3)^2 - 4(3) + 3^0$$ Both powers are first
$$= 2 \cdot 9 - 4(3) + 1$$ Both multiplications are next
$$= \quad 18 \quad - \quad 12 \quad + 1$$ Addition and subtraction are done *left to right*
$$= \quad\quad 6 \quad\quad + 1 = 7$$

EXAMPLE 6

$2 \cdot (5 + 6)$ Operations in parentheses are done first

$= 2 \cdot \quad 11 \quad = 22$

EXAMPLE 7

$20 - (4 + 3 \cdot 2)$ Multiplication inside parentheses is first

$= 20 - \quad (4 + 6)$ Addition inside parentheses is next

$= 20 - \quad 10 \quad = 10$ Subtraction is last

Calculators

Calculators use one of three types of logic: algebraic logic, arithmetic logic, and reverse Polish notation (RPN).

Algebraic Logic   For calculators using **algebraic logic**, numbers, symbols of operation, and parentheses can be entered into the calculator in the same order in which they appear in the expression. To find $2 + 3 \times 4$,

Key in   2 ⊞ 3 ⊠ 4 ⊟

Answer   14

Arithmetic Logic   For calculators using **arithmetic logic**, *all* operations are carried out from left to right. If you key in 2 ⊞ 3 ⊠ 4 ⊟, these calculators will display an incorrect answer of 20. Therefore, when using this type of calculator, you must be careful to use the correct order of operations:

Key in   3 ⊠ 4 ⊞ 2 ⊟

Answer   14

RPN   For calculators using **RPN**, there is no ⊟ key. They have an ENTER or SAVE key, and the operations are keyed in after the numbers are entered:

Key in   2 ENTER 3 ENTER 4 ⊠ ⊞

Answer   14

The remaining examples in this book will be performed on a scientific calculator using algebraic logic. Be sure to consult your instruction manual for the correct order of keys for your calculator.

EXAMPLE 8

Find the value of each expression using a calculator.

a. $3^6 + 14 \cdot \sqrt{256}$

SOLUTION

Key in   3 $\boxed{y^x}$ 6 ⊞ 14 ⊠ 256 $\boxed{\sqrt{}}$ ⊟

Answer   953

b. $24 \cdot (75 - 18) \div 19$

SOLUTION

Key in   24 ⊠ ⊏ 75 ⊟ 18 ⊐ ÷ 19 ⊟

Answer   72

Exercises 1.11

In Exercises 1–36, evaluate each expression using the correct order of operations.

1. $10 \div 2 \cdot 5$
2. $36 \div 6 \div 2$
3. $12 - 6 + 2$
4. $8 - 2 - 4$
5. $6 + 4 \cdot 5$
6. $48 - 18 \div 3$
7. $4 \cdot 5 - 15 \div 5$
8. $44 \div 11 + 8 \cdot 7$
9. $15 - 3 \cdot 4 + 6$
10. $16 - 8 \div 2 + 1$
11. $5^2 - 2^3 + \sqrt{9}$
12. $8^2 - \sqrt{16} - 3^3$
13. $4(5) + 3$
14. $4(5 + 3)$
15. $7(9 - 3)$
16. $7(9) - 7(3)$
17. $2(6) + 2(4)$
18. $2(6 + 4)$
19. $14 - (3 - 1)$
20. $14 - 3 - 1$
21. $100 \div 10 \div 2$
22. $100 \div (10 \div 2)$
23. $20 - (2 + 4 \cdot 3)$
24. $20 - 2 + 4 \cdot 3$
25. $12 + 9 - 6 \div 3$
26. $12 + (9 - 6 \div 3)$
27. $10^2 - 10^0 - 1^2$
28. $2^3 + 0^2 + 2^0$
29. $5 \cdot 8 + 2\sqrt{100}$
30. $4 \cdot 5 - 3\sqrt{36}$
31. $24 - 8 \div 4 \cdot 2 + 2^3$
32. $(24 - 8) \div 4 \cdot 2 + 2^3$
33. $24 - 8 \div (4 \cdot 2) + 2^3$
34. $24 - 8 \div 4 \cdot (2 + 2^3)$
35. $(24 - 8 \div 4) \cdot 2 + 2^3$
36. $24 - (8 \div 4 \cdot 2 + 2^3)$

In Exercises 37–48, use a calculator to evaluate each expression.

37. $189 + 560 \div 35$
38. $2{,}400 - 18 \cdot 23$
39. $567 - 6 \cdot 43 + 297$
40. $46 \cdot 90 - 102 \div 17$
41. $560 \div (34 - 18) \cdot 100$
42. $7{,}040 - (9 \cdot 24 - 68)$
43. $(12 + 7) \cdot (48 - 12)$
44. $(234 + 416) \div (89 - 63)$
45. $909 + 8 \cdot 5^4$
46. $19^3 + 56^2$
47. $\sqrt{11{,}025} - 108 \div 12$
48. $\sqrt{9{,}216} - 5\sqrt{324}$

1.12   Average

If a student gets 70 on one test and 90 on another test, you probably know the student's average is 80. Why is 80 called the *average*? It's because

$$70 + 90 = 160$$

and

$$80 + 80 = 160$$

— Same total points

— 80 on *every* test gives the same total points

In other words, the average is the score the student would have to make on *every* test in order to get the same total points. To find the average, we take the sum of all the grades, 160, and divide it by the number of grades, 2.

We can find the average of any kinds of quantities such as grades, weights, speeds, and costs. The average is found by dividing the sum of the quantities by the number of quantities.

Finding the average

1. Find the sum of all the quantities.
2. Divide this sum by the number of quantities.

EXAMPLE 1

Four linemen on a football team weigh 270 lb, 241 lb, 265 lb, and 284 lb. What is their average weight?

SOLUTION

Step 1.

$$\underbrace{270 + 241 + 265 + 284}_{\text{4 weights}} = 1060$$

Step 2.

$$\begin{array}{r} 265 \text{ lb} \longleftarrow \text{Average} \\ 4\overline{)1060} \\ \underline{8} \\ 26 \\ \underline{24} \\ 20 \\ \underline{20} \end{array}$$

EXAMPLE 2

In Mrs. Martinez's quiz section two students scored 5, three students scored 4, and one student scored 2. Find the class average.

SOLUTION Since *two* students scored 5 and *three* students scored 4, the number 5 must be added twice and the number 4 must be added three times.

Step 1.

$$\underbrace{5 + 5 + 4 + 4 + 4 + 2}_{\text{6 students}} = 24$$

Step 2.

$$\begin{array}{r} 4 \longleftarrow \text{Average} \\ 6\overline{)24} \\ \underline{24} \end{array}$$

EXAMPLE 3

Tom's exam scores were 73, 84, 91, and 70. What score will he need on the fifth test so that his test average will be 80 for all tests?

SOLUTION To average 80, Tom must have the same total points as he would if he scored 80 on each of the five tests. Therefore, he needs $80 \cdot 5 = 400$ total points.

$$73 + 84 + 91 + 70 = 318 \longleftarrow \text{Total points on 4 tests}$$

$$400 - 318 = 82 \longleftarrow \text{Points needed on the fifth test}$$

EXAMPLE 4

The sales figures for three months were \$54,230, \$86,045, and \$76,775. Find the average sales per month.

SOLUTION For calculators using algebraic logic, the $\boxed{=}$ key must be pressed at the end of the sum to get the total, which is then divided by 3:

Key in   54230 $\boxed{+}$ 86045 $\boxed{+}$ 76775 $\boxed{=}$ $\boxed{\div}$ 3 $\boxed{=}$

Answer   \$72,350

ALTERNATIVE SOLUTION If your calculator has parentheses keys, $\boxed{(}$ and $\boxed{)}$, they can be used instead of the $\boxed{=}$ to get the total:

Key in $\boxed{(}$ 54230 $\boxed{+}$ 86045 $\boxed{+}$ 76775 $\boxed{)}$ $\boxed{\div}$ 3 $\boxed{=}$

Answer   \$72,350

At this time we only consider problems in which the average is a whole number. Problems in which the average is not a whole number will be discussed in later chapters.

Exercises *1.12*

In Exercises 1–8, find the average of each set of numbers.

1. $\{2, 7, 9\}$ 2. $\{3, 5, 7\}$ 3. $\{6, 8, 9, 5\}$ 4. $\{9, 0, 6, 9\}$

5. $\{21, 24, 33\}$ 6. $\{7, 10, 8, 5, 11, 7\}$

7. $\{74, 88, 85, 69\}$ 8. $\{96, 92, 95, 89, 88\}$

9. Maria's examination scores during the semester were 75, 83, 74, 86, 95, and 61. What was her average score?

10. Mrs. Lindstrom recorded her weight each Monday morning for 6 weeks. The weights were 155 pounds, 150 pounds, 149 pounds, 148 pounds, 150 pounds, and 142 pounds. What is her average weight?

11. Five basketball players have heights of 76 inches, 78 inches, 84 inches, 72 inches, and 75 inches. What is the average height of the team?

12. Traveling across the country, a family stayed in motels. The cost per night for the motels was as follows: $54, $76, $60, $54, $33, and $47. Find the average cost per night for motels.

13. Find the class average on a history test if two students scored 90, one student scored 85, four students scored 70, and one student scored 55.

14. The manager earns $25,000 a year, the assistant manager earns $17,000 a year, and the four clerks each earn $12,000 a year. What is the average income per year?

15. Rosa's exam scores were 61, 73, 48, and 81. What grade will she need to get on her fifth test so that her average will be 70 for all five tests?

16. Two groups of five students took a test. The students in group A made scores of 78, 85, 97, 76, and 84. Those in group B made scores of 95, 87, 78, 80, and 55. Which group made the higher average, and by how much?

17. The monthly rainfall for a city was 17 inches for January, 14 inches for February, 19 inches for March, 15 inches for April, 7 inches for May, 2 inches for June, 1 inch for July, 2 inches for August, 4 inches for September, 7 inches for October, 11 inches for November, and 9 inches for December. What is the city's average monthly rainfall?

18. The high temperatures for six days were 79°, 87°, 96°, 92°, 98°, and 102°. What temperature must occur on the seventh day in order for the average high for the week to be 93°?

1.13 Perimeter and Area

1.13A Perimeter

To measure the length of a line segment, we see how many times a unit of length divides into it. In Figure 7, since the unit of length (inch) fits three times into the length to be measured, we say the length is 3 inches.

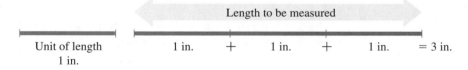

Length to be measured

Unit of length
1 in.

1 in. + 1 in. + 1 in. = 3 in.

Figure 7

The word **perimeter** means "the distance around a figure." To find the perimeter of a geometric figure, we determine the sum of the lengths of all its sides.

EXAMPLE I   Find the perimeter of the figure below.

SOLUTION   To find the perimeter we must add the four sides:

$$\text{Perimeter} = 20 + 15 + 12 + 9$$

$$= 56 \text{ ft}$$

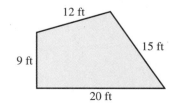

Triangle   A **triangle** has three sides. The perimeter of a triangle is the sum of its three sides: Perimeter = $a + b + c$.

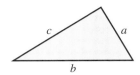

Triangle
$P = a + b + c$

Rectangle   A **rectangle** has four sides and four right angles (square corners). The opposite sides of a rectangle are equal. If we label the sides l for length and w for width, then the perimeter $= 2l + 2w$.

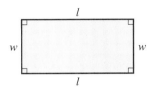

Rectangle
$P = 2l + 2w$

E X A M P L E 2

Find the perimeter of a rectangle with length 10 inches and width 6 inches.

S O L U T I O N   Perimeter $= 2l + 2w$

$$= 2 \cdot 10 + 2 \cdot 6$$

$$= 20 + 12$$

$$= 32 \text{ in.}$$

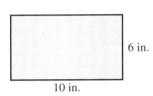

Square   A **square** is a rectangle with all sides equal. If we let s represent the length of one side, then the perimeter $= s + s + s + s = 4s$.

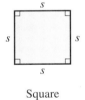

Square
$P = 4s$

E X A M P L E 3

Find the cost to fence in a square playground if the length of each side is 20 feet and the fencing costs $12 per foot.

S O L U T I O N   Perimeter $= 4s$

$$= 4 \cdot 20$$

$$= 80 \text{ ft}$$

Cost $= 80 \cdot 12 = \$960$

20 ft

Exercises *1.13A*

In Exercises 1–8, find the perimeter of each figure.

1.

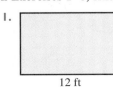

8 ft

12 ft

2.

8 cm

3.

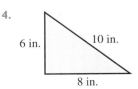

14 in.   10 in.

18 in.

4.

6 in.   10 in.

8 in.

5.

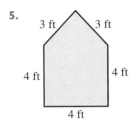

6.

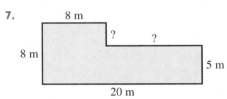

7.

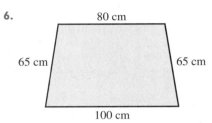

8.

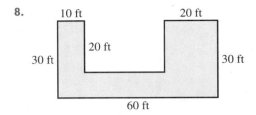

9. Find the perimeter of a square if the length of each side is 15 inches.

10. Find the perimeter of a rectangle with length 25 meters and width 18 meters.

11. What is the total cost to fence a rectangular garden that is 8 feet long and 4 feet wide if the wire fence costs $3 per foot?

12. A door measures 3 feet wide and 6 feet high. Find the cost to replace the molding around the door if the molding costs $2 per foot. (Note: There is no molding at the bottom of the door.)

1.13B Area

The **area** of a geometric figure is the space inside the lines. Area is measured in square units: square feet, square inches, square meters, and so on. To measure the area of a rectangle, we see how many times a unit of area fits into it. In Figure 8, since the unit of area (1 sq. in.) fits into the space six times, the area of the rectangle is 6 sq. in.

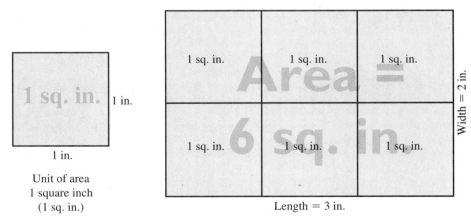

Unit of area
1 square inch
(1 sq. in.)

Figure 8

We can also find the area of the rectangle by multiplying the length times the width.

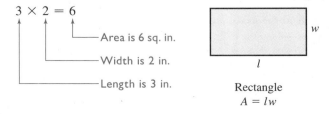

$3 \times 2 = 6$

Area is 6 sq. in.

Width is 2 in.

Length is 3 in.

Rectangle
$A = lw$

Since a square is a rectangle in which the length and width are equal, if we let s represent the length of a side, then the area of a square $= s \cdot s = s^2$.

Square
$A = s^2$

EXAMPLE 4   Find the area of a rectangle with length 9 inches and width 6 inches.

SOLUTION   Area $= lw$

$\quad\quad = 9 \cdot 6$

$\quad\quad = 54$ sq. in.

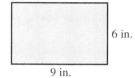

6 in.

9 in.

EXAMPLE 5   Find the area of a square playground if the length of each side is 20 feet.

SOLUTION   Area $= s^2$

$\quad\quad = 20^2$

$\quad\quad = 400$ sq. ft

20 ft

EXAMPLE 6   What is the cost to tile a kitchen that is 12 feet long and 8 feet wide if the ceramic tile costs $3 per square foot?

SOLUTION   Area $= lw$

$\quad\quad = 12 \cdot 8$

$\quad\quad = 96$ sq. ft

Cost $= 96 \cdot 3 = \$288$

8 ft

12 ft

Exercises 1.13B

In Exercises 1–3, find the area of each figure.

1.

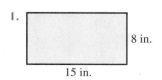

8 in.

15 in.

2.

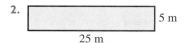

5 m

25 m

3.

6 in.

4. Find the area of a square if the length of a side is 15 centimeters.

5. The floor of a square balcony is 8 feet long. How many square feet of material are needed to cover this floor?

6. A rectangular window is 7 feet wide and 4 feet high. How many square feet of glass are needed for three windows of this size?

7. A rectangular room is 6 yards long and 5 yards wide.

   a. Find the floor area of this room.

   b. Find the cost to carpet this room if the carpet costs $15 per square yard to install.

8. Find the cost to carpet a bedroom that is 4 yards long and 3 yards wide if the carpet costs $18 per square yard.

9. What does it cost to replace a large bathroom mirror that measures 7 feet by 3 feet if glass costs $2 per square foot?

10. Jan needs drapes that are 3 yards high and 5 yards long to cover a picture window in her home. If she pays $7 a square yard for the material, what will the drapes cost?

1.14   Estimation

When we are not interested in an exact answer to a problem but want only a rough approximation, we can *estimate* the answer by rounding off each number in the problem and then performing the operations.

Estimating an answer

> **1.** Round off each number in the problem at the first digit. Replace all digits to the right of the round-off place with zeros.
>
> **2.** Perform the indicated operations.

EXAMPLE 1   Estimate the answers to the following problems.

a. 5,306 + 1,775 + 9,588

   SOLUTION   Round off 5,306 to 5,000, 1,775 to 2,000, and 9,588 to 10,000, then add:

$$5,000 + 2,000 + 10,000 = 17,000$$

b. 682 · 43

   SOLUTION   Round off 682 to 700 and 43 to 40, then multiply:

$$700 \cdot 40 = 28,000$$

c. 78,361 ÷ 23

   SOLUTION   Round off 78,361 to 80,000 and 23 to 20, then divide:

$$80,000 \div 20 = \frac{4,000}{20)\overline{80,000}}$$

EXAMPLE 2   Joann is buying her textbooks for the fall semester. If three books cost $27 each, one book costs $44, and another book costs $12, estimate the total cost of her textbooks.

SOLUTION   Round off $27 to $30, $44 to $40, and $12 to $10.

$$\text{Cost} = 3(\$30) + \$40 + \$10$$

$$= \$140$$

  Note   Estimating your answer to a problem before finding the exact answer may help you find some arithmetic errors.

Exercises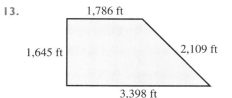

In Exercises 1–10, estimate the answers to the following problems.

1. 6,723 + 8,908 + 3,215

2. 569 + 135 + 409 + 982

3. 92,750 − 18,060

4. 875,210 − 558,008

5. 6,623 · 327

6. 947 · 86

7. 8,274 ÷ 42

8. 319,286 ÷ 631

9. 54(296 + 676)

10. 81 · 45 − 2,765

11. Dan had two hamburgers, an order of french fries, and a Coke for lunch. Each hamburger contained 420 calories, the french fries contained 185 calories, and the Coke contained 267 calories.

   a. Estimate the number of calories in his lunch.

   b. Find the exact number of calories in his lunch.

12. Yuki earns $289 a week. She also receives a stock dividend of $1,225 once a year.

   a. Estimate her total yearly income.

   b. Find the exact amount of her yearly income.

13.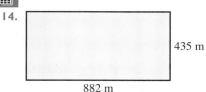

   a. Estimate the perimeter of the figure.

    b. Find the exact perimeter of the figure.

14.

   a. Estimate the area of the rectangle.

    b. Find the exact area of the rectangle.

Chapter 1 REVIEW

Numbers
1.1

Natural numbers are the counting numbers that start with 1 and continue on forever. When 0 is included with the natural numbers, we have the set of *whole numbers*. The *digits* are the first ten whole numbers. Digits are the building blocks used in writing all numbers. We show the natural numbers, whole numbers, and digits on the number line below.

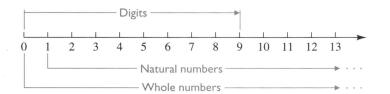

Reading Whole Numbers
1.2

1. Starting at the right end of the number, separate the digits into groups of three by means of commas.

2. Starting at the left end of the number, read the first group and follow it by that group's name. Continue in this way until all groups have been read.

Writing Whole Numbers in Words
1.2

Write the number the same way it is read, putting commas in the same places they appear when the number is written in digits.

Addition
1.4

Addition is repeated counting; its inverse is subtraction.

$$\begin{array}{rl} 5 & \text{Addend} \\ +\ 4 & \text{Addend} \\ \hline 9 & \text{Sum} \end{array}$$

Commutative property of addition:

$$a + b = b + a$$

Associative property of addition:

$$(a + b) + c = a + (b + c)$$

Additive identity property:

$$a + 0 = 0 + a = a$$

Subtraction
1.5

Subtraction is the inverse of addition.

$$\begin{array}{rl} 9 & \text{Minuend} \\ -\ 4 & \text{Subtrahend} \\ \hline 5 & \text{Difference} \end{array}$$

Multiplication
1.6

Multiplication is repeated addition; its inverse is division.

$$\underset{\text{Factor}\quad\text{Factor}\quad\text{Product}}{4 \quad \cdot \quad 5 \quad = \quad 20}$$

Commutative property of multiplication:

$$a \cdot b = b \cdot a$$

Associative property of multiplication:

$$(a \cdot b) \cdot c = a \cdot (b \cdot c)$$

Multiplicative identity property:

$$a \cdot 1 = 1 \cdot a = a$$

Multiplication property of zero:

$$a \cdot 0 = 0 \cdot a = 0$$

Division
1.9

Division is repeated subtraction; its inverse is multiplication.

$$\begin{array}{r} 3 \quad \text{Quotient} \\ \text{Divisor} \quad 12\overline{)40} \quad \text{Dividend} \\ \underline{36} \quad\quad\ \\ 4 \quad \text{Remainder} \end{array}$$

To check a division problem, use

$$\text{quotient} \cdot \text{divisor} + \text{remainder} = \text{dividend}$$
$$3 \quad \cdot \quad 12 \quad + \quad 4 \quad = \quad 40$$

Raising a Number to a Power
1.7

Raising a number to a power is repeated multiplication.

$$\underset{\text{Base}}{2^{\overset{\text{Exponent}}{3}}} = 2 \cdot 2 \cdot 2 = \underset{\text{Power}}{8}$$

Square Roots
1.10

Finding the square root of a number is the inverse of squaring a number.

$$\sqrt{16} = 4 \quad \text{Because } 4^2 = 16$$

Order of Operations
1.11

1. Any expressions in parentheses are evaluated first.

2. Evaluations are done in this order:

First: Powers and roots are done.
Second: Multiplication and division are done in order from left to right.
Third: Addition and subtraction are done in order from left to right.

Finding the Average
1.12

1. Find the sum of all the quantities.

2. Divide this sum by the number of quantities.

Perimeter and Area
1.13

Perimeter is the distance around a figure.

Area measures the space inside a figure and is measured in square units.

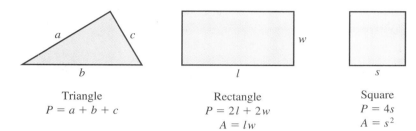

Triangle	Rectangle	Square
$P = a + b + c$	$P = 2l + 2w$	$P = 4s$
	$A = lw$	$A = s^2$

Estimating an Answer
1.14

1. Round off each number in the problem at the first digit. Replace all digits to the right of the round-off place with zeros.

2. Perform the indicated operations.

Chapter 1   REVIEW EXERCISES

1. Write the number 3,075,600,008 in words.

2. Write "five million, seventy-two thousand, six" using digits.

3. Write all the digits less than 5.

4. Write the smallest two-digit natural number.

5. Which symbol, $<$ or $>$, should be used to make each statement true?

    a. 17 ____?____ 12   **b.** 0 ____?____ 19

6. Which of the operations—addition, subtraction, multiplication, division—are neither associative nor commutative?

7. Since $5 \cdot 3 = 15$, 5 and 3 are _____ of 15.

8. Division of whole numbers can be considered repeated _____.

9. In many cases, the first digit of the divisor is used as the _____.

10. Finding the square root of a number is the _____ of squaring a number.

11. Write the term used for each part in the following subtraction example:

$$\begin{array}{r} 8 \leftarrow \\ -\ 2 \leftarrow \\ \hline 6 \leftarrow \end{array}$$

12. Write the term used for each part in the following division example:

$$\begin{array}{r} 14 \\ 17\overline{)243} \\ \underline{17} \\ 73 \\ \underline{68} \\ 5 \end{array}$$

13. Round off the following numbers to the indicated place.

    a. 284,791   Nearest ten thousand

    b. 9,685   Nearest hundred

    c. 42,568,499   Nearest million

14. Find the following quotients. If division cannot be performed, write "undefined."

    a. $6 \div 0$ **b.** $0 \div 3$

    c. $4 \div 1$ **d.** $0 \div 0$

In Exercises 15–26, perform the indicated operations.

15.
$$\begin{array}{r} 1,045 \\ +\ 896 \\ \hline \end{array}$$

16.
$$\begin{array}{r} 785 \\ 396 \\ +\ 29 \\ \hline \end{array}$$

17.
$$\begin{array}{r} 78,619 \\ 788 \\ +\ 6,907 \\ \hline \end{array}$$

18.
$$\begin{array}{r} 8,247 \\ -\ 358 \\ \hline \end{array}$$

19.
$$\begin{array}{r} 7,906 \\ -\ 2,789 \\ \hline \end{array}$$

20.
$$\begin{array}{r} 700,051 \\ -\ 20,893 \\ \hline \end{array}$$

21.
$$\begin{array}{r} 786 \\ \times\ 35 \\ \hline \end{array}$$

22.
$$\begin{array}{r} 9,207 \\ \times\ 704 \\ \hline \end{array}$$

23.
$$\begin{array}{r} 3,967 \\ \times\ 867 \\ \hline \end{array}$$

24. $7\overline{)28,602}$ 25. $38\overline{)2,180}$ 26. $609\overline{)292,320}$

In Exercises 27–38, evaluate each expression.

27. 5^2 28. 2^4 29. 3^0 30. $\sqrt{36}$

31. $\sqrt{81}$ 32. $75 \cdot 100$ 33. $8 \cdot 10^4$ 34. $30 \cdot 10^3$

35. $10 + 20 \div 2 \cdot 5$ 36. $4\sqrt{100} - 6 + 3 \cdot 2^3$

37. $2(3)^2 - 4(3) + 5$ 38. $10 - (8 - 3 \cdot 2) - 6 \div 2$

39. Mr. Washington earns an annual salary of \$9,096. What is his monthly salary?

40. Fencing costs \$7 per foot. What will 256 feet of fencing cost?

41. Arrange the following numbers in a vertical column, then add:
7,825 + 84 + 900 + 45,788 + 9 + 2,000,085

42. How many 3-ounce bottles can a druggist fill from a bottle that contains 16 ounces of peroxide?

43. A man pays off a debt of $3,048 in equal monthly payments, over a period of two years. Find the amount he pays each month.

44. Ms. Chang pays the property tax on her home in equal monthly payments. If her yearly property tax is $576, how much does she pay each month?

45. The average page in a particular textbook has about 703 words. If there are 356 pages in the book, how many words would you estimate there are in the book?

46. Irma has $45. Anita has $23. Felisa has $12 more than Irma and Anita together. Find the total amount of money the three have together.

47. Lee makes monthly payments to pay off a $1,950 debt. He makes 23 payments of $82 each and then one final payment to pay off the balance of the debt. How much is the final payment?

48. Two hundred seventy-five families are invited to a neighborhood picnic. The planning committee estimates the average family size to be four. How many people should they plan on providing refreshments for?

49. After trading his car in for a new car, Joe had a balance of $8,400 due. What equal monthly payments must he make to pay this off in five years?

50. In air, light travels 983,584,800 feet per second and sound travels 1,129 feet per second. The speed of light is how many times the speed of sound?

51. A man owed $2,365 on his car. After making 35 payments of $67 each, how much did he have left to be paid?

52. Suppose the gasoline tank of your car holds 22 gallons, and you average 16 miles to the gallon. How far can you drive on a tankful of gasoline?

53. The full price, including tax and financing, of a certain color television set was $456. Mrs. Stein bought the set. She agreed to pay for it in 24 equal monthly payments.

    a. Find the amount of her monthly payment.

    b. After making 17 payments, how much does she still owe?

54. To get a gasoline mileage check, Mr. Perez filled his gasoline tank and wrote down the odometer reading, which was 53,408 miles. The next time he got gasoline, his tank took 19 gallons and his odometer reading was 53,731 miles. How many miles did he get per gallon?

55. In the last three weeks a cashier worked 20 hours, 24 hours, and 34 hours. How many hours must she work next week in order to average 25 hours per week?

56. Find the class average on a test if one student scored 85, two students scored 80, one student scored 75, and two students scored 65.

57. A rectangle is 12 inches long and 8 inches wide.

    a. Find the perimeter of the rectangle.

    b. Find the area of the rectangle.

58. The length of a side of a square is 9 feet.

    a. Find the perimeter of the square.

    b. Find the area of the square.

59. Estimate the answers to the following.

    a. 3,094 · 575   b. 9,880 ÷ 52

60. Debbie wrote checks for the following amounts: $77, $12, $54, $48, and $192. Estimate the total amount of the checks.

Chapter I Critical Thinking and Writing Problems

Answer Problems 1–12 in your own words, using complete sentences.

1. Explain why the number 17 is not a digit.

2. Explain why the number 0 is not a natural number.

3. Explain how to determine whether 2 + (3 + 4) = 2 + (4 + 3) uses the associative property of addition or the commutative property of addition.

4. Explain how to determine whether 2 · (3 · 4) = (2 · 3) · 4 uses the associative property of multiplication or the commutative property of multiplication.

5. Explain how to determine whether 7 + 0 = 7 illustrates the additive identity property or the multiplication property of zero.

6. Explain how to determine whether 7 · 0 = 0 illustrates the additive identity property or the multiplication property of zero.

7. Explain how to check a subtraction problem.

8. Explain how to check a division problem.

9. Explain why 5 ÷ 0 is undefined.

10. Explain why 0 ÷ 0 is undefined.

11. Explain why $2^3 \neq 6$.

12. Explain what $\sqrt{36}$ means.

13. Find the mystery [?] number.
    Clue 1: The number is a digit greater than 2.
    Clue 2: It is a factor of 12.
    Clue 3: It is a divisor of 20.

14. Find the mystery [?] number.
    Clue 1: It is a two-digit number less than 25.
    Clue 2: When the number is divided by 5, the remainder is 3.
    Clue 3: It is an even number.

Each of the following problems has an error. Find the error and, in your own words, explain why it's wrong. Then work the problem correctly.

15. $3^3 - 2\sqrt{25}$
    $= 27 - 2 + 5$
    $= \quad 25 + 5 = 30$

16. $30 + 20 \div 5 \cdot 2$
    $= 30 + 20 \div 10$
    $= 30 + 2 = 32$

Chapter 1   DIAGNOSTIC TEST

Allow yourself about 50 minutes to do these problems. Complete solutions for all problems, together with section references, are given in the answer section at the end of the book.

1. Light travels about 5,879,200,000,000 miles in 1 year. Write this number in words.

2. Use digits to write the number fifty-four billion, seven million, five hundred six thousand, eighty.

In Problems 3–10, perform the indicated operations.

3. $\begin{array}{r} 5,843 \\ 209 \\ + 6,027 \end{array}$
4. $\begin{array}{r} 946 \\ 7,328 \\ 407 \\ + \quad 24 \end{array}$
5. $\begin{array}{r} 3,564 \\ - \quad 782 \end{array}$
6. $\begin{array}{r} 50,406 \\ - 35,008 \end{array}$

7. $\begin{array}{r} 576 \\ \times \quad 89 \end{array}$
8. $\begin{array}{r} 3084 \\ \times \quad 706 \end{array}$
9. $63\overline{)5,055}$
10. $495\overline{)349,470}$

11. Find the indicated powers.
    a. 2^3 b. 8^2

12. Find the indicated square roots.
    a. $\sqrt{16}$ b. $\sqrt{100}$

13. Just look at the problem, then write your answer. Do not use written calculations.
    a. $40 \cdot 100$ b. $16 \cdot 10^4$

14. Round off the following numbers to the indicated place.
    a. 78,603   Nearest thousand
    b. 3,749   Nearest hundred

15. If at the beginning of a trip your odometer reading was 67,856 miles and at the end of the trip it was 71,304 miles, how many miles did you drive?

16. After trading his car in on a new car, Joe had a balance of $2,016 due. What equal monthly payments must he make to pay this off in 3 years?

17. Enrique has a job paying $168 per week. How much does he earn in one year (52 weeks)?

18. Yelena makes monthly payments to pay off a $1,350 debt. She makes 23 payments of $58 each and then a final payment to pay off the balance. How much is the final payment?

19. Find the perimeter of a rectangle 18 inches long and 12 inches wide.

20. What is the area of a square whose side is 10 centimeters?

21. Estimate the answer. $77,840 \cdot 216$

22. Wanda's test scores during the semester were 76, 84, 92, 63, and 70. What was her average score?

In Problems 23–25, evaluate each expression.

23. $6 + 18 \div 3 \cdot 2$ 24. $30 - 4^2 + 4\sqrt{9}$

25. $5 + 2(10 - 2 \cdot 3)$

Fractions

CHAPTER

2

o far we have been concerned mainly with whole numbers and the operations that can be performed on them. In this chapter we introduce a new kind of number, *fractions*, and show how to perform the basic operations on fractions.

2.1   The Meaning of Fraction

A **fraction** is written $\dfrac{a}{b}$, and a and b are called the *terms* of the fraction. We call b the **denominator**. It tells us into how many equal parts the whole has been divided. We call a the **numerator**. It tells us how many of those parts we have.

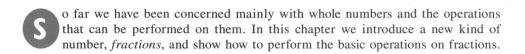

Numerator (any whole number)

Terms of the fraction — $\dfrac{a}{b}$ ← Fraction bar

Denominator (cannot be 0)

EXAMPLE 1

Examples of fractions:

a. If we divide a whole rectangle into two equal parts, then the fraction $\dfrac{1}{2}$, read "one-half," means we have one of the two equal parts of the whole.

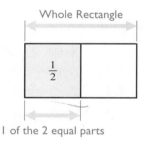

Whole Rectangle

$\frac{1}{2}$

1 of the 2 equal parts

b. If we divide a whole circle into three equal parts, then the fraction $\dfrac{2}{3}$, read "two-thirds," means we have two of the three equal parts.

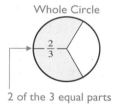

Whole Circle

$\frac{2}{3}$

2 of the 3 equal parts

c. If we divide a whole square into four equal parts, then the fraction $\dfrac{4}{4}$, read "four-fourths," means we have four of the four equal parts.

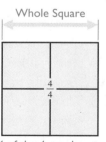

Whole Square

$\frac{4}{4}$

4 of the 4 equal parts

Notice, in Example 1c, that $\dfrac{4}{4} = 1$ whole square. Similarly, $\dfrac{1}{1} = 1, \dfrac{2}{2} = 1, \dfrac{3}{3} = 1$, and so on. In general,

If a represents any number, except 0, then

$$\frac{a}{a} = 1$$

Any Fraction Is Equivalent to a Division Problem Figure 1 shows that $\frac{6}{3} = 2$.

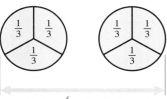

six-thirds $= \frac{6}{3} = 2$ whole circles

Figure 1

We also know that $6 \div 3 = 2$. Because both $\frac{6}{3}$ and $6 \div 3$ represent the same thing (in this case, 2), we can say:

$$\frac{6}{3} = 6 \div 3 = 2$$

Any fraction is equivalent to a division problem

If a and b represent any numbers ($b \neq 0$), then

$$\frac{a}{b} = a \div b$$

EXAMPLE 2 **a.** $\frac{8}{2} = 8 \div 2 = 2\overline{)8} = 4$ **b.** $\frac{36}{4} = 36 \div 4 = 4\overline{)36} = 9$

c. $\frac{39}{11} = 39 \div 11 = 11\overline{)39}$ ← Here the answer is not a whole number; divisions of this type will be discussed in a later section

Proper Fraction A **proper fraction** is a fraction whose numerator is less than its denominator. Any proper fraction has a value less than one unit. Examples of proper fractions are

$$\frac{1}{2}, \quad \frac{2}{3}, \quad \frac{3}{4}, \quad \frac{4}{5}, \quad \frac{3}{8}, \quad \frac{18}{35}$$

Improper Fraction An **improper fraction** is a fraction whose numerator is larger than or equal to its denominator. Any improper fraction has a value greater than or equal to one. Examples of improper fractions are

$$\frac{4}{3}, \quad \frac{5}{4}, \quad \frac{7}{2}, \quad \frac{11}{11}, \quad \frac{131}{17}$$

Exercises 2.1

1. If we divide a whole into eight equal parts and take three of them, what fraction represents the part taken?

2. If you cut a pie into five equal pieces and serve two pieces, what fractional part of the pie is left?

3. If we divide a class into six equal groups and take five of the groups, what fractional part of the class is *not* taken?

4. In a football game, three of the eleven first-string players were injured. What fractional part of the team's first-string players were injured?

5. What fraction represents the shaded portion of the rectangle?

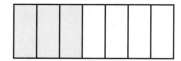

6. What fraction represents the shaded portion of the circle?

For Exercises 7–11, use the following list:

$$\frac{5}{11}, \quad \frac{28}{13}, \quad \frac{17}{22}, \quad \frac{8}{8}, \quad \frac{1}{2}, \quad \frac{98}{107}, \quad \frac{316}{219}, \quad \frac{1}{31}, \quad \frac{4}{4}$$

7. Which fractions in the list are proper fractions?

8. Which fractions in the list are improper fractions?

9. What is the numerator of the first fraction?

10. What is the denominator of the second fraction?

11. Name all of the fractions in the list that are equal to 1.

12. Is $\frac{4}{0}$ a fraction? Give a reason for your answer.

In Exercises 13–22, change the fractions to whole numbers.

13. $\frac{10}{2}$　14. $\frac{9}{3}$　15. $\frac{24}{6}$　16. $\frac{42}{7}$　17. $\frac{35}{35}$

18. $\frac{84}{7}$　19. $\frac{144}{16}$　20. $\frac{751}{751}$　21. $\frac{1,940}{97}$　22. $\frac{33,060}{551}$

2.2　Multiplication of Fractions: An Introduction

When discussing whole numbers, we introduced addition first; our discussions of all other operations on whole numbers were based on an understanding of addition. With fractions, we introduce multiplication first, and all other operations follow.

A recipe for cookies calls for $\frac{1}{3}$ pound of butter. If we want to make only half a batch of cookies, we will need to find $\frac{1}{2}$ of each ingredient. To find the amount of butter, we need to find $\frac{1}{2} \cdot \frac{1}{3}$. Figure 2 shows that we need $\frac{1}{6}$ pound of butter; that is, $\frac{1}{2} \cdot \frac{1}{3} = \frac{1}{6}$.

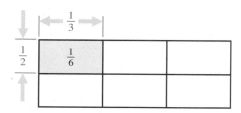

Figure 2

This illustration leads to the following definition of multiplication of fractions:

Multiplying two fractions	$\dfrac{\text{product of}}{\text{two fractions}} = \dfrac{\text{product of their numerators}}{\text{product of their denominators}}$
	In symbols, $\quad \dfrac{a}{b} \cdot \dfrac{c}{d} = \dfrac{a \cdot c}{b \cdot d}$

EXAMPLE 1   **a.** $\dfrac{3}{5} \cdot \dfrac{2}{7} = \dfrac{3 \cdot 2}{5 \cdot 7} = \dfrac{6}{35}$ $\qquad$ **b.** $\dfrac{13}{6} \cdot \dfrac{1}{8} = \dfrac{13 \cdot 1}{6 \cdot 8} = \dfrac{13}{48}$

c. $\dfrac{2}{5} \cdot \dfrac{4}{7} \cdot \dfrac{1}{3} = \dfrac{2 \cdot 4 \cdot 1}{5 \cdot 7 \cdot 3} = \dfrac{8}{105}$

Whole Numbers Written as Fractions Because $\dfrac{8}{1} = 8 \div 1 = 8$, in general, $\dfrac{a}{1} = a \div 1 = a$. This means that any whole number can be written as a fraction by writing it over 1. That is, the whole number becomes the numerator of a fraction whose denominator is 1.

If a is any whole number, then
$$a = \dfrac{a}{1}$$

EXAMPLE 2   **a.** $5 = \dfrac{5}{1}$ $\quad$ **b.** $29 = \dfrac{29}{1}$ $\quad$ **c.** $117 = \dfrac{117}{1}$

This fact makes it possible to multiply fractions by whole numbers.

EXAMPLE 3   Examples of multiplying fractions by whole numbers:

a. $2 \cdot \dfrac{4}{9} = \dfrac{2}{1} \cdot \dfrac{4}{9} = \dfrac{2 \cdot 4}{1 \cdot 9} = \dfrac{8}{9}$ $\quad$ **b.** $\dfrac{3}{4} \cdot 17 = \dfrac{3}{4} \cdot \dfrac{17}{1} = \dfrac{3 \cdot 17}{4 \cdot 1} = \dfrac{51}{4}$

Exercises 2.2

Find the products.

1. $\dfrac{2}{3} \cdot \dfrac{5}{7}$ $\quad$ 2. $\dfrac{1}{2} \cdot \dfrac{5}{3}$ $\quad$ 3. $\dfrac{3}{4} \cdot \dfrac{7}{8}$ $\quad$ 4. $\dfrac{5}{8} \cdot \dfrac{3}{2}$ $\quad$ 5. $\dfrac{4}{5} \cdot \dfrac{3}{5}$

6. $\dfrac{5}{6} \cdot \dfrac{7}{8}$ $\quad$ 7. $\dfrac{4}{9} \cdot \dfrac{2}{3}$ $\quad$ 8. $\dfrac{7}{3} \cdot \dfrac{5}{9}$ $\quad$ 9. $\dfrac{3}{4} \cdot \dfrac{3}{4}$ $\quad$ 10. $\dfrac{6}{7} \cdot \dfrac{6}{7}$

11. $\dfrac{5}{12} \cdot \dfrac{5}{8}$ $\quad$ 12. $\dfrac{3}{8} \cdot \dfrac{5}{16}$ $\quad$ 13. $\dfrac{11}{32} \cdot \dfrac{3}{2}$ $\quad$ 14. $\dfrac{7}{12} \cdot \dfrac{13}{15}$ $\quad$ 15. $\dfrac{7}{8} \cdot \dfrac{11}{13}$

16. $\dfrac{37}{16} \cdot \dfrac{15}{43}$ $\qquad$ 17. $\dfrac{1}{2} \cdot \dfrac{3}{4} \cdot \dfrac{1}{5}$ $\qquad$ 18. $\dfrac{1}{3} \cdot \dfrac{2}{5} \cdot \dfrac{4}{3}$

19. $\dfrac{2}{5} \cdot \dfrac{3}{5} \cdot \dfrac{1}{7}$ 20. $\dfrac{3}{4} \cdot \dfrac{5}{8} \cdot \dfrac{5}{7}$ 21. $2 \cdot \dfrac{1}{3}$ 25. $3 \cdot \dfrac{5}{2}$ 26. $6 \cdot \dfrac{3}{5}$ 27. $\dfrac{7}{12} \cdot 7$

22. $3 \cdot \dfrac{2}{7}$ 23. $\dfrac{1}{5} \cdot 4$ 24. $\dfrac{1}{8} \cdot 5$ 28. $5 \cdot \dfrac{3}{16}$ 29. $5 \cdot \dfrac{1}{12} \cdot \dfrac{5}{4}$ 30. $\dfrac{3}{5} \cdot 7 \cdot \dfrac{1}{8}$

2.3 Changing an Improper Fraction to a Mixed Number

A number such as $1\dfrac{2}{3}$ is called a **mixed number**. A mixed number is the *sum* of a whole number and a proper fraction. Although a mixed number is a *sum*, the $+$ sign is omitted in writing mixed numbers.

> **A Word of Caution**   The mixed number $2\dfrac{1}{2}$ means $2 + \dfrac{1}{2}$.
>
> It does *not* mean $2 \cdot \dfrac{1}{2}$.

EXAMPLE 1   Examples of mixed numbers:

$$2\frac{1}{2}, \quad 3\frac{5}{8}, \quad 5\frac{1}{4}, \quad 12\frac{3}{16}$$

Consider the improper fraction $\dfrac{5}{3}$: Each unit is 3 thirds, so 5 thirds = 3 thirds + 2 thirds = 1 unit + 2 thirds = $1\dfrac{2}{3}$ (see Figure 3).

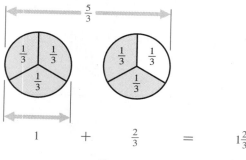

Figure 3

Since a fraction is equivalent to a division, we can change an improper fraction to a mixed number by dividing:

$$\frac{5}{3} = 5 \div 3 = 3\overline{)5}^{\,1\ \text{R}\ 2} = 1\frac{2}{3}$$

Changing an improper fraction to a mixed number	1. Divide the numerator by the denominator.
	2. Write the quotient followed by the fraction: remainder over divisor.

$$\text{quotient} \ \frac{\text{remainder}}{\text{divisor}}$$

EXAMPLE 2

a. $\dfrac{3}{2} = 3 \div 2 = 2\overline{)3}^{\,1\,R\,1} = 1\dfrac{1}{2}$

b. $\dfrac{15}{11} = 15 \div 11 = 11\overline{)15}^{\,1\,R\,4} = 1\dfrac{4}{11}$

c. $\dfrac{37}{5} = 37 \div 5 = 5\overline{)37}^{\,7\,R\,2} = 7\dfrac{2}{5}$

EXAMPLE 3

Mark's test scores are 78, 82, 90, and 75. Find his test average.

SOLUTION   Recall that to find the average, we first find the sum of all the tests, then divide by the number of tests.

$$\text{Average} = \frac{78 + 82 + 90 + 75}{4} = \frac{325}{4} = 81\frac{1}{4}$$

Exercises   2.3

In Exercises 1–18, change the improper fractions to mixed numbers.

1. $\dfrac{5}{3}$ 2. $\dfrac{7}{4}$ 3. $\dfrac{9}{5}$ 4. $\dfrac{11}{4}$ 5. $\dfrac{13}{5}$ 6. $\dfrac{11}{3}$

7. $\dfrac{15}{4}$ 8. $\dfrac{35}{2}$ 9. $\dfrac{23}{6}$ 10. $\dfrac{86}{9}$ 11. $\dfrac{16}{13}$ 12. $\dfrac{32}{23}$

13. $\dfrac{20}{7}$ 14. $\dfrac{56}{15}$ 15. $\dfrac{207}{19}$ 16. $\dfrac{37}{21}$ 17. $\dfrac{54}{17}$ 18. $\dfrac{87}{31}$

In Exercises 19–22, find the average of each set of numbers.

19. $\{7, 8, 3, 5\}$

20. $\{75, 89, 81\}$

21. $(12, 10, 15, 23, 12\}$

22. $\{9, 5, 2, 0, 8, 5\}$

23. On an English quiz, three students scored 5, five students scored 4, and two students scored 2. Find the class average.

24. Tokie made five long distance calls. They were for 15 minutes, 20 minutes, 3 minutes, and two for 5 minutes each. Find the average time of the calls.

2.4 Changing a Mixed Number to an Improper Fraction

Consider the mixed number $2\dfrac{1}{3}$. This means $2 + \dfrac{1}{3} = 2$ units $+ 1$ third. Each unit is 3 thirds; therefore, 2 units $= 2 \cdot 3 = 6$ thirds, so $2\dfrac{1}{3} = 6$ thirds $+ 1$ third $= 7$ thirds $= \dfrac{7}{3}$ (see Figure 4).

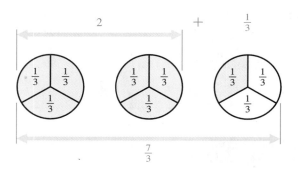

Figure 4

Changing a mixed number to an improper fraction	1. Multiply the whole number by the denominator of the fraction.
	2. Add the numerator of the fraction to the product found in step 1.
	3. Write that sum found in step 2 over the denominator of the fraction.

EXAMPLE 1

a. $3\dfrac{1}{2} = \dfrac{3 \cdot 2 + 1}{2} = \dfrac{6 + 1}{2} = \dfrac{7}{2}$

b. $7\dfrac{2}{5} = \dfrac{7 \cdot 5 + 2}{5} = \dfrac{35 + 2}{5} = \dfrac{37}{5}$

c. $4\dfrac{7}{13} = \dfrac{4 \cdot 13 + 7}{13} = \dfrac{52 + 7}{13} = \dfrac{59}{13}$

Exercises   2.4

Change the mixed numbers to improper fractions.

1. $1\dfrac{1}{2}$ 2. $2\dfrac{3}{5}$ 3. $3\dfrac{1}{4}$ 4. $2\dfrac{5}{8}$ 5. $4\dfrac{5}{6}$

6. $4\dfrac{1}{2}$ 7. $3\dfrac{7}{10}$ 8. $3\dfrac{5}{16}$ 9. $5\dfrac{7}{12}$ 10. $6\dfrac{2}{7}$

11. $12\dfrac{2}{3}$ 12. $3\dfrac{9}{13}$ 13. $6\dfrac{3}{4}$ 14. $4\dfrac{5}{11}$ 15. $3\dfrac{7}{15}$

16. $1\dfrac{8}{17}$ 17. $15\dfrac{3}{4}$ 18. $21\dfrac{2}{3}$ 19. $2\dfrac{7}{10}$ 20. $8\dfrac{3}{100}$

2.5   Equivalent Fractions

Equivalent fractions are equal—they are different ways of writing the same number. An examination of Figure 5 will convince you that $\dfrac{3}{4}$ and $\dfrac{9}{12}$ are equivalent fractions because they represent the same thing.

Recall that the number 1 is the multiplicative identity. That is, multiplying any number by 1 does not change the number's value. Since $1 = \dfrac{1}{1} = \dfrac{2}{2} = \dfrac{3}{3} = \dfrac{4}{4}$, and so on,

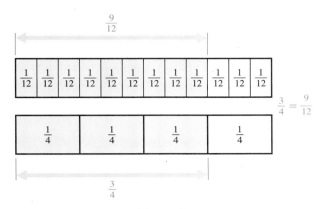

Figure 5

multiplying any fraction by $\dfrac{2}{2}$, or $\dfrac{3}{3}$, or any fraction whose value is 1, will form an equivalent fraction. Another way to see that $\dfrac{3}{4} = \dfrac{9}{12}$ is

$$\frac{3}{4} = \frac{3}{4} \cdot 1 = \frac{3}{4} \cdot \frac{3}{3} = \frac{3 \cdot 3}{4 \cdot 3} = \frac{9}{12}$$

The value of this fraction is 1

Hence, $\dfrac{3}{4}$ and $\dfrac{9}{12}$ are equivalent fractions $\left(\text{or, simply, } \dfrac{3}{4} = \dfrac{9}{12}\right)$. This leads to the following rule for forming equivalent fractions:

Multiplying both numerator and denominator by the same nonzero number gives an equivalent fraction.

$$\frac{a}{b} = \frac{a \cdot c}{b \cdot c}$$

Equivalent fractions

EXAMPLE 1

Examples of finding a particular fraction equivalent to a given fraction:

a. $\dfrac{3}{4} = \dfrac{?}{8}$

Since the denominator 8 is twice the denominator 4, *multiply* the numerator and denominator by 2

$$\frac{3}{4} = \frac{3 \cdot 2}{4 \cdot 2} = \frac{6}{8}$$

The 2 can be found by dividing 8 by 4; since the denominator 4 must be multiplied by 2 to give 8, we must also multiply the numerator by 2

b. $\dfrac{7}{9} = \dfrac{?}{45}$

Since the denominator 45 is five times the denominator 9, multiply the numerator and denominator by 5

$$\frac{7}{9} = \frac{7 \cdot 5}{9 \cdot 5} = \frac{35}{45}$$

$45 \div 9 = 5$

We have already shown that $\dfrac{3}{4}$ and $\dfrac{9}{12}$ are equivalent fractions, but we can look at these two fractions in another way:

$$\frac{9}{12} = \frac{9 \div 3}{12 \div 3} = \frac{3}{4}$$

In general,

> Dividing both numerator and denominator by the same nonzero number that is a divisor of both gives an equivalent fraction.
>
> $$\frac{a}{b} = \frac{a \div c}{b \div c}$$
>
> Equivalent fractions

EXAMPLE 2   Examples of finding a particular fraction equivalent to a given fraction:

a. $\dfrac{15}{20} = \dfrac{?}{4}$ Since the denominator 4 is less than the denominator 20, *divide* the numerator and denominator by 5

$$\frac{15}{20} = \frac{15 \div \boxed{5}}{20 \div \boxed{5}} = \frac{3}{4}$$ The $\boxed{5}$ can be found by dividing 20 by 4; since the denominator 20 must be divided by 5 to give 4, we must also divide the numerator by 5

b. $\dfrac{14}{35} = \dfrac{?}{5}$ Since the denominator 5 is less than the denominator 35, divide the numerator and denominator by 7

$$\frac{14}{35} = \frac{14 \div \boxed{7}}{35 \div \boxed{7}} = \frac{2}{5}$$ $35 \div 5 = \boxed{7}$

Forming equivalent fractions is necessary when reducing fractions and when adding unlike fractions.

Exercises   2.5

Replace the question mark with a number that makes the fractions equivalent (have equal value).

1. $\dfrac{1}{2} = \dfrac{?}{10}$ 2. $\dfrac{1}{3} = \dfrac{?}{6}$ 3. $\dfrac{3}{6} = \dfrac{?}{2}$ 4. $\dfrac{6}{8} = \dfrac{?}{4}$

5. $\dfrac{2}{3} = \dfrac{?}{6}$ 6. $\dfrac{3}{4} = \dfrac{?}{40}$ 7. $\dfrac{6}{10} = \dfrac{?}{5}$ 8. $\dfrac{6}{16} = \dfrac{?}{8}$

9. $\dfrac{3}{5} = \dfrac{?}{15}$ 10. $\dfrac{5}{6} = \dfrac{?}{12}$ 11. $\dfrac{14}{20} = \dfrac{?}{10}$ 12. $\dfrac{18}{45} = \dfrac{?}{15}$

13. $\dfrac{9}{13} = \dfrac{?}{52}$ 14. $\dfrac{15}{18} = \dfrac{?}{36}$ 15. $\dfrac{36}{54} = \dfrac{?}{9}$ 16. $\dfrac{72}{24} = \dfrac{?}{3}$

17. $\dfrac{8}{11} = \dfrac{?}{55}$ 18. $\dfrac{20}{26} = \dfrac{?}{52}$ 19. $\dfrac{85}{34} = \dfrac{?}{2}$ 20. $\dfrac{33}{77} = \dfrac{?}{7}$

2.6   Prime Factorization

A **prime number** is a natural number greater than 1 that can be divided evenly (the remainder is zero) only by itself and 1. A prime number has *no* factors other than itself and 1.

A **composite number** is a natural number greater than 1 that can be divided evenly by some natural number other than itself and 1. A composite number *has* factors other than itself and 1.

Note that 1 is neither prime nor composite.

EXAMPLE 1

a. 9 is composite because $3 \cdot 3 = 9$, so 9 has a factor other than itself and 1.
b. 17 is prime because 1 and 17 are the only factors of 17.
c. 45 is composite because it has factors 3, 5, 9, and 15, along with 45 and 1.
d. 31 is prime because 1 and 31 are the only factors of 31.

A partial list of prime numbers is $2, 3, 5, 7, 11, 13, 17, 19, 23, 29, \ldots$.

The **prime factorization** of a natural number is the product of all its factors that are themselves prime numbers. For example, the prime factorization of 6 is $2 \cdot 3$, because 2 and 3 are both prime numbers that are factors of 6. A factorization that is not a prime factorization is $1 \cdot 6$, because neither 1 nor 6 is a prime number.

Method for Finding the Prime Factorization

The smallest prime is 2; the next smallest is 3; the next smallest is 5, and so on. To find the prime factorization of 24, first try to divide 24 by the smallest prime, 2. Two does divide 24 and gives a quotient of 12. Next try to divide 12 by the smallest prime, 2. Two does divide 12 and gives a quotient of 6. Next try to divide 6 by the smallest prime, 2. Two does divide 6 and gives a quotient of 3. This process ends here because the final quotient 3 is itself a prime. The work of finding the prime factorization of a number can be conveniently arranged as follows:

$$
\begin{array}{r|l}
2 & 24 \\
\hline
2 & 12 \\
\hline
2 & 6 \\
\hline
 & 3
\end{array}
$$

The prime factorization is the product of these numbers $\longrightarrow$

Therefore, $24 = 2 \cdot 2 \cdot 2 \cdot 3 = 2^3 \cdot 3$, where $2^3 \cdot 3$ is the prime factorization of 24.

EXAMPLE 2

a. Find the prime factorization of 30.

SOLUTION

$$
\begin{array}{r|l}
2 & 30 \\
\hline
3 & 15 \\
\hline
 & 5
\end{array}
$$

Prime factorization of $30 = 2 \cdot 3 \cdot 5$

b. Find the prime factorization of 20.

SOLUTION

$$
\begin{array}{r|l}
2 & 20 \\
\hline
2 & 10 \\
\hline
 & 5
\end{array}
$$

Prime factorization of $20 = 2 \cdot 2 \cdot 5$
$= 2^2 \cdot 5$

c. Find the prime factorization of 36.

SOLUTION

$$
\begin{array}{r|l}
2 & 36 \\
\hline
2 & 18 \\
\hline
3 & 9 \\
\hline
 & 3
\end{array}
$$

Prime factorization of $36 = 2 \cdot 2 \cdot 3 \cdot 3$
$= 2^2 \cdot 3^2$

d. Find the prime factorization of 315.

SOLUTION

$$
\begin{array}{r|l}
3 & 315 \\
\hline
3 & 105 \\
\hline
5 & 35 \\
\hline
 & 7
\end{array}
$$

Prime factorization of $315 = 3 \cdot 3 \cdot 5 \cdot 7$
$= 3^2 \cdot 5 \cdot 7$

When we're trying to find a prime factor of a number, we need not try any prime whose square is greater than that number (see Example 3).

EXAMPLE 3   Find the prime factorization of 97.

S O L U T I O N   ┌──── Primes in order of size

    2 does not divide 97.
    3 does not divide 97.
    5 does not divide 97.
    7 does not divide 97.
    11 is not possible because $11^2 = 121$, which is greater than 97.

Therefore, 97 is prime.

Tests for Divisibility

When finding the prime factorization, use the following rules to tell whether a number is divisible by the prime 2, 3, or 5. We have listed only those rules that will be most useful.

Divisible by 2   A number is **divisible by 2** if its *last* digit is $0, 2, 4, 6, 8$.

EXAMPLE 4   1 2 , 30 0 , 203 4 , and 57 8 are divisible by 2, but 9 1 is not.

Divisible by 3   A number is **divisible by 3** if the *sum* of its digits is divisible by 3.

EXAMPLE 5   **a.** 210 is divisible by 3 because $2 + 1 + 0 = 3$, which is divisible by 3.
b. 5162 is not divisible by 3 because $5 + 1 + 6 + 2 = 14$, which is not divisible by 3.

Divisible by 5   A number is **divisible by 5** if its *last* digit is 0 or 5.

EXAMPLE 6   25 0 and 75 5 are both divisible by 5, but 112 7 is not.

We have not included tests of divisibility by larger primes because the tests are usually longer than the actual trial division.

Exercises   2.6

In Exercises 1–16, state whether each of the numbers is prime (P) or composite (C). To justify your answer, give the set of all factors for each number.

1. 5
2. 11
3. 8
4. 15
5. 21
6. 10
7. 13
8. 19
9. 25
10. 49
11. 23
12. 29
13. 41
14. 51
15. 20
16. 75

In Exercises 17–32, find the prime factorization of each number.

17. 14
18. 33
19. 70
20. 42
21. 16
22. 27
23. 18
24. 75
25. 28
26. 24
27. 81
28. 64
29. 250
30. 117
31. 144
32. 360

2.7 Reducing Fractions to Lowest Terms

A fraction is reduced to **lowest terms** when there is no whole number (greater than 1) that divides both numerator and denominator evenly.

Earlier in this chapter we divided numerator and denominator by a whole number (greater than 1) that is a divisor of both to obtain an equivalent fraction:

$$\frac{12}{18} = \frac{12 \div 2}{18 \div 2} = \frac{6}{9} \quad \text{and} \quad \frac{12}{18} = \frac{12 \div 3}{18 \div 3} = \frac{4}{6}$$

Since we have reduced the fraction $\frac{12}{18}$ to two different fractions having *lower* terms, the question arises, How can we reduce $\frac{12}{18}$ to a fraction having the *lowest* terms?

2.7A Reducing Fractions

Reducing a fraction to lowest terms

1. Divide numerator and denominator by any whole number (greater than 1) that you can see is a divisor of both.
2. When you cannot see any more common divisors, write the prime factorization of the numerator and the denominator.
3. Divide numerator and denominator by all factors common to both.

EXAMPLE 1 Reduce $\frac{12}{18}$ to lowest terms.

SOLUTION $\frac{12}{18} = \frac{12 \div 2}{18 \div 2} = \frac{6}{9}$ You can see that 2 is a common divisor

Then $\frac{6}{9} = \frac{6 \div 3}{9 \div 3} = \frac{2}{3}$ You can see that 3 is a common divisor

Therefore, $\frac{12}{18} = \frac{2}{3}$ Numerator and denominator are in prime-factored form and have no common factors

Dividing numerator and denominator by a common factor is often written as follows:

$$\frac{12}{18} = \frac{\overset{2}{\cancel{12}}}{\underset{3}{\cancel{18}}} = \frac{2}{3} \quad \text{Both numerator and denominator were divided by 6}$$

EXAMPLE 2 Reduce $\frac{30}{42}$ to lowest terms.

SOLUTION $\frac{30}{42} = \frac{30 \div 2}{42 \div 2} = \frac{15}{21}$ You can see that 2 is a common divisor

Then $\frac{15}{21} = \frac{15 \div 3}{21 \div 3} = \frac{5}{7}$ You can see that 3 is a common divisor

Therefore, $\qquad \dfrac{30}{42} = \dfrac{5}{7}$ Numerator and denominator are in prime-factored form and have no common factor

Another way of writing this solution is

$$\dfrac{\overset{\overset{5}{15}}{\cancel{30}}}{\underset{\underset{7}{21}}{\cancel{42}}} = \dfrac{5}{7}$$ We first divide 30 and 42 by 2; then we divide 15 and 21 by 3

EXAMPLE 3   Reduce $\dfrac{150}{280}$ to lowest terms.

SOLUTION $\dfrac{\overset{15}{\cancel{150}}}{\underset{28}{\cancel{280}}} = \dfrac{15}{28}$ We first divide 150 and 280 by 10

$\dfrac{15}{28} = \dfrac{3 \cdot 5}{2 \cdot 2 \cdot 7}$ Write numerator and denominator in prime-factored form

Therefore, $\qquad \dfrac{150}{280} = \dfrac{15}{28}$ In lowest terms because numerator and denominator have no common factor

Sometimes numerator and denominator can be divided by the same number more than once. See Example 4.

EXAMPLE 4   Reduce $\dfrac{18}{45}$ to lowest terms.

SOLUTION $\dfrac{\overset{\overset{2}{6}}{\cancel{18}}}{\underset{\underset{5}{15}}{\cancel{45}}} = \dfrac{2}{5}$ We first divide 18 and 45 by 3, then we divide 6 and 15 by 3

Another way to reduce fractions to lowest terms is to *start* with the prime factorization of the numerator and denominator.

Reducing fractions by prime factorization

1. Write the prime factorization of both numerator and denominator.
2. Divide numerator and denominator by all factors common to both.

EXAMPLE 5   Reduce $\dfrac{34}{51}$ to lowest terms.

SOLUTION   If we prime factor both numerator and denominator, we will be able to determine the common factor. The prime factorization of 34 is $2 \cdot 17$. The prime factorization of 51 is $3 \cdot 17$. Therefore,

17 is a factor of the numerator

$$\dfrac{34}{51} = \dfrac{2 \cdot 17}{3 \cdot 17} = \dfrac{2 \cdot \overset{1}{\cancel{17}}}{3 \cdot \underset{1}{\cancel{17}}} = \dfrac{2}{3}$$ Both numerator and denominator are divided by 17

17 is a factor of the denominator

EXAMPLE 6 Reduce $\dfrac{6}{20}$ to lowest terms.

SOLUTION $\quad \dfrac{6}{20} = \dfrac{\overset{1}{\cancel{2}} \cdot 3}{\underset{1}{\cancel{2}} \cdot 2 \cdot 5} = \dfrac{3}{10}$ Both numerator and denominator are divided by 2

In reducing fractions, it's helpful to write every factor rather than using powers: $2 \cdot 2 \cdot 5$ instead of $2^2 \cdot 5$

EXAMPLE 7 Reduce $\dfrac{28}{56}$ to lowest terms.

SOLUTION $\quad \dfrac{28}{56} = \dfrac{\overset{1}{\cancel{2}} \cdot \overset{1}{\cancel{2}} \cdot \overset{1}{\cancel{7}}}{\underset{1}{\cancel{2}} \cdot \underset{1}{\cancel{2}} \cdot 2 \cdot \underset{1}{\cancel{7}}} = \dfrac{1}{2}$ Both numerator and denominator are divided by $2 \cdot 2 \cdot 7 = 28$

 Note When the numerator is divided by $2 \cdot 2 \cdot 7$, the quotient is 1.

Exercises 2.7A

Reduce each fraction to lowest terms.

1. $\dfrac{6}{9}$ 2. $\dfrac{15}{20}$ 3. $\dfrac{12}{16}$ 4. $\dfrac{14}{21}$ 5. $\dfrac{30}{40}$

6. $\dfrac{12}{15}$ 7. $\dfrac{24}{32}$ 8. $\dfrac{36}{90}$ 9. $\dfrac{10}{18}$ 10. $\dfrac{16}{24}$

11. $\dfrac{32}{40}$ 12. $\dfrac{48}{64}$ 13. $\dfrac{54}{90}$ 14. $\dfrac{135}{315}$ 15. $\dfrac{84}{105}$

16. $\dfrac{42}{84}$ 17. $\dfrac{33}{55}$ 18. $\dfrac{44}{66}$ 19. $\dfrac{36}{60}$ 20. $\dfrac{12}{60}$

21. $\dfrac{39}{51}$ 22. $\dfrac{33}{57}$ 23. $\dfrac{80}{180}$ 24. $\dfrac{210}{270}$ 25. $\dfrac{3,000}{4,200}$

2.7B Shortening Multiplication of Fractions

Because of the definition of multiplication of fractions, we can reduce the fraction to lowest terms before multiplying. When we multiply fractions, a factor in any numerator can be canceled with the same factor in any denominator

EXAMPLE 8

a. $\dfrac{7}{12} \cdot \dfrac{4}{21} = \dfrac{\overset{1}{\cancel{7}} \cdot \overset{1}{\cancel{4}}}{\underset{3}{\cancel{12}} \cdot \underset{3}{\cancel{21}}} = \dfrac{1}{9}$ or $\dfrac{\overset{1}{\cancel{7}}}{\underset{3}{\cancel{12}}} \cdot \dfrac{\overset{1}{\cancel{4}}}{\underset{3}{\cancel{21}}} = \dfrac{1}{9}$

b. $\dfrac{4}{25} \cdot \dfrac{15}{8} = \dfrac{\overset{1}{\cancel{4}}}{\underset{5}{\cancel{25}}} \cdot \dfrac{\overset{3}{\cancel{15}}}{\underset{2}{\cancel{8}}} = \dfrac{3}{10}$

c. $\dfrac{9}{2} \cdot \dfrac{1}{3} \cdot \dfrac{35}{14} = \dfrac{\overset{3}{\cancel{9}}}{2} \cdot \dfrac{1}{\underset{1}{\cancel{3}}} \cdot \dfrac{\overset{5}{\cancel{35}}}{\underset{2}{\cancel{14}}} = \dfrac{15}{4} = 3\dfrac{3}{4}$

Fractional Part of a Number

EXAMPLE 9 Find $\dfrac{2}{3}$ of 18.

SOLUTION To find a fractional part of a number, we need to multiply the fraction times the number. In this case, the word *of* means to multiply.

$$\dfrac{2}{3} \text{ of } 18 = \dfrac{2}{3} \cdot \dfrac{18}{1} = \dfrac{2}{\underset{1}{\cancel{3}}} \cdot \dfrac{\overset{6}{\cancel{18}}}{1} = \dfrac{12}{1} = 12$$

EXAMPLE 10   If $\frac{3}{5}$ of the members of a class are women, how many of the 35 students are women?

SOLUTION $\frac{3}{5}$ of $35 = \frac{3}{5} \cdot \frac{35}{1} = \frac{3}{\underset{1}{5}} \cdot \frac{\overset{7}{35}}{1} = 21$ women

Exercises   2.7B

In Exercises 1–16, find the products. Reduce your answers to lowest terms. Write any improper fraction as a mixed number.

1. $\frac{2}{3} \cdot \frac{6}{7}$

2. $\frac{2}{5} \cdot \frac{5}{8}$

3. $\frac{6}{8} \cdot \frac{4}{9}$

4. $\frac{7}{8} \cdot \frac{12}{7}$

5. $\frac{10}{18} \cdot \frac{9}{5}$

6. $\frac{25}{14} \cdot \frac{21}{10}$

7. $\frac{35}{72} \cdot \frac{18}{56}$

8. $\frac{80}{45} \cdot \frac{27}{60}$

9. $\frac{22}{75} \cdot \frac{20}{44}$

10. $\frac{3}{5} \cdot \frac{5}{8} \cdot \frac{1}{6}$

11. $\frac{5}{7} \cdot \frac{3}{10} \cdot \frac{14}{15}$

12. $\frac{4}{5} \cdot \frac{5}{6} \cdot \frac{10}{15}$

13. $\frac{16}{49} \cdot \frac{14}{8} \cdot \frac{9}{12}$

14. $\frac{40}{14} \cdot \frac{21}{5} \cdot \frac{2}{27}$

15. $\frac{4}{15} \cdot \frac{10}{28} \cdot \frac{21}{10}$

16. $\frac{4}{3} \cdot \frac{5}{14} \cdot \frac{7}{12} \cdot \frac{6}{15}$

17. Find $\frac{3}{4}$ of 12.

18. Find $\frac{5}{6}$ of 24.

19. Find $\frac{4}{5}$ of 300.

20. Find $\frac{5}{8}$ of 320.

21. An astronaut on the moon weighs $\frac{1}{6}$ of his weight on Earth. If he weighs 186 pounds on Earth, how much would he weigh on the moon?

22. If $\frac{2}{5}$ of the college's 25,000 students are over 25 years old, how many students are over 25 years old?

23. John filled his 24-gallon gas tank at the beginning of his business trip. At the end of the trip, the tank was $\frac{3}{8}$ full. How many gallons of gas did he use on the trip?

24. On a math test of 20 problems, Marcia solved $\frac{3}{4}$ of the problems correctly. How many problems did she miss?

2.8   Adding and Subtracting Like Fractions

Like fractions are fractions that have the same denominator.

EXAMPLE 1   Examples of like fractions:

$$\frac{2}{3}, \quad \frac{5}{3}, \quad \frac{1}{3}$$

Same denominator

Unlike fractions are fractions that have different denominators.

EXAMPLE 2   Examples of unlike fractions:

$$\frac{1}{3}, \quad \frac{2}{7}, \quad \frac{3}{8}$$

Different denominators

Adding Like Fractions

Figure 6 illustrates how to add like fractions.

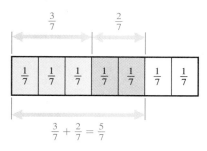

Figure 6

Adding like fractions	1. Add their numerators.

2. Write the sum found in step 1 over the common denominator:

$$\frac{a}{c} + \frac{b}{c} = \frac{a + b}{c}$$

3. Reduce the resulting fraction to lowest terms.
4. Any improper fraction found in step 3 is usually changed to a mixed number.

EXAMPLE 3

a. $\dfrac{11}{23} + \dfrac{5}{23} = \dfrac{11 + 5}{23} = \dfrac{16}{23}$ Already lowest terms

b. $\dfrac{1}{9} + \dfrac{4}{9} + \dfrac{7}{9} = \dfrac{1 + 4 + 7}{9} = \dfrac{12}{9} = \dfrac{\overset{4}{\cancel{12}}}{\underset{3}{\cancel{9}}} = \dfrac{4}{3} = 1\dfrac{1}{3}$

Subtracting Like Fractions

Subtraction of like fractions is done in a similar manner.

Subtracting like fractions	1. Subtract their numerators.

2. Write the difference found in step 1 over the common denominator:

$$\frac{a}{c} - \frac{b}{c} = \frac{a - b}{c}$$

3. Reduce the resulting fraction to lowest terms.
4. Any improper fraction found in step 3 is usually changed to a mixed number.

EXAMPLE 4

a. $\dfrac{13}{30} - \dfrac{7}{30} = \dfrac{13 - 7}{30} = \dfrac{\overset{1}{\cancel{6}}}{\underset{5}{\cancel{30}}} = \dfrac{1}{5}$

b. $\dfrac{25}{8} - \dfrac{11}{8} = \dfrac{25 - 11}{8} = \dfrac{\overset{7}{\cancel{14}}}{\underset{4}{\cancel{8}}} = \dfrac{7}{4} = 1\dfrac{3}{4}$

Exercises 2.8

Perform the indicated operations. Reduce your answers to lowest terms. Write any improper fractions as mixed numbers.

1. $\dfrac{1}{6} + \dfrac{3}{6}$ 2. $\dfrac{3}{8} + \dfrac{1}{8}$ 3. $\dfrac{3}{5} + \dfrac{2}{5}$

4. $\dfrac{3}{4} + \dfrac{1}{4}$ 5. $\dfrac{7}{10} - \dfrac{5}{10}$ 6. $\dfrac{11}{12} - \dfrac{3}{12}$

7. $\dfrac{5}{9} - \dfrac{2}{9}$ 8. $\dfrac{9}{14} - \dfrac{5}{14}$ 9. $\dfrac{5}{8} + \dfrac{7}{8}$

10. $\dfrac{5}{16} + \dfrac{13}{16}$ 11. $\dfrac{11}{24} + \dfrac{4}{24}$ 12. $\dfrac{35}{80} + \dfrac{27}{80}$

13. $\dfrac{27}{35} - \dfrac{6}{35}$ 14. $\dfrac{45}{52} - \dfrac{6}{52}$ 15. $\dfrac{70}{81} - \dfrac{19}{81}$

16. $\dfrac{123}{144} - \dfrac{43}{144}$ 17. $\dfrac{1}{6} + \dfrac{5}{6} + \dfrac{3}{6}$ 18. $\dfrac{5}{12} + \dfrac{2}{12} + \dfrac{9}{12}$

19. $\dfrac{3}{15} + \dfrac{1}{15} + \dfrac{6}{15}$ 20. $\dfrac{29}{45} + \dfrac{16}{45} + \dfrac{3}{45}$

2.9   Lowest Common Denominator (LCD)

Before adding unlike fractions, we ordinarily find their lowest common denominator. The **lowest common denominator (LCD)** of two or more fractions is the smallest number that is exactly divisible by each of their denominators. In this section we show two methods for finding the LCD: inspection and prime factorization.

Finding the LCD by Inspection

By inspection, try to find a number exactly divisible by each denominator. If you cannot find the LCD by inspection, then use the prime factorization method.

EXAMPLE 1   Find the LCD for $\dfrac{1}{2} + \dfrac{3}{4}$.

SOLUTION   By inspection we see that the LCD = 4 because 4 is the smallest number exactly divisible by both denominators, 2 and 4.

EXAMPLE 2   Find the LCD for $\dfrac{5}{2} + \dfrac{1}{3}$.

SOLUTION   By inspection we see that the LCD = 6 because 6 is the smallest number exactly divisible by both denominators, 2 and 3.

Finding the LCD by Prime Factorization

Consider the problem $\dfrac{7}{12} + \dfrac{4}{15}$. Here, the LCD is not easily found by inspection. To find the LCD in this case:

1. Find the prime factorization of each denominator:

$$\begin{array}{c} 2\,\lfloor 12 \\ 2\,\lfloor 6 \\ 3 \end{array} \qquad \begin{array}{c} 3\,\lfloor 15 \\ 5 \end{array}$$

$$12 = 2^2 \cdot 3 \qquad 15 = 3 \cdot 5$$

2. Write down each different factor that appears in the prime factorizations:

$$2, \quad 3, \quad 5$$

3. Raise each factor to the highest power to which it occurs in any denominator:

$$2^2, \quad 3, \quad 5$$

4. The LCD is the product of all the powers found in step 3:

$$\text{LCD} = 2^2 \cdot 3 \cdot 5 = 60$$

The prime factorization method for finding the LCD is summarized in the following box.

Finding the LCD by prime factorization	1. Write the prime factorization of each denominator. Repeated factors should be expressed as powers. 2. Write down each different factor that appears. 3. Raise each factor to the highest power to which it occurs in any denominator. 4. The LCD is the product of all the powers found in step 3.

EXAMPLE 3 Find the LCD for $\dfrac{3}{8} + \dfrac{7}{10}$.

SOLUTION

Step 1. $8 = 2 \cdot 2 \cdot 2 = 2^3$ Prime factorization of 8
$\phantom{\textbf{Step 1.}}\ 10 = 2 \cdot 5$ Prime factorization of 10

Step 2. 2, 5   2 and 5 are the only different prime factors that appear in the prime factorizations

Step 3. $2^3, 5^1$ Highest power of each factor

Step 4. $\text{LCD} = 2^3 \cdot 5^1 = 40$ Product of all the powers found in step 3

2^3 is a *power* of 2; 3 is only the *exponent* of 2

EXAMPLE 4 Find the LCD for $\dfrac{5}{14} + \dfrac{13}{21}$.

SOLUTION

Step 1. $14 = 2 \cdot 7$
$\phantom{\textbf{Step 1.}}\ 21 = 3 \cdot 7$ Prime factorization of each denominator

Step 2. 2, 3, 7   The only different prime factors that appear

Step 3. $2^1, 3^1, 7^1$ Highest power of each factor

Step 4. $\text{LCD} = 2^1 \cdot 3^1 \cdot 7^1 = 42$ Product of all the powers found in step 3

EXAMPLE 5 Find the LCD for $\dfrac{2}{9} + \dfrac{7}{15} + \dfrac{3}{20}$.

SOLUTION

Step 1. $9 = 3 \cdot 3 = 3^2$
$\phantom{\textbf{Step 1.}}\ 15 = 3 \cdot 5$ Prime factorization of each denominator
$\phantom{\textbf{Step 1.}}\ 20 = 2 \cdot 2 \cdot 5 = 2^2 \cdot 5$

Step 2. 2, 3, 5   All the different prime factors

Step 3. $2^2, 3^2, 5^1$ Highest power of each factor
Step 4. LCD $= 2^2 \cdot 3^2 \cdot 5^1$
$= 4 \cdot 9 \cdot 5 = 180$

EXAMPLE 6

Find the LCD for $\dfrac{3}{5} + \dfrac{1}{2} + \dfrac{2}{3}$.

SOLUTION   Since each denominator is a prime number, the LCD is the product of the three denominators:

$$LCD = 5 \cdot 2 \cdot 3 = 30$$

Exercises   2.9

Assume the following sets to be denominators of fractions. In Exercises 1–6, find the LCD by inspection, then check it by prime factoring.

1. $\{2, 3, 4\}$ 2. $\{4, 5, 10\}$ 3. $\{2, 8, 4\}$

4. $\{3, 6, 9\}$ 5. $\{3, 5, 15\}$ 6. $\{7, 2, 14\}$

In Exercises 7–22, find the LCD by either method.

7. $\{14, 10\}$ 8. $\{16, 12\}$ 9. $\{7, 5\}$

10. $\{4, 9\}$ 11. $\{18, 30\}$ 12. $\{20, 24\}$

13. $\{6, 8, 9\}$ 14. $\{4, 15, 18\}$ 15. $\{40, 15, 25\}$

16. $\{3, 7, 5\}$ 17. $\{4, 5, 21\}$ 18. $\{4, 6, 9, 12\}$

19. $\{6, 13, 26\}$ 20. $\{45, 63, 98\}$ 21. $\{66, 33, 132\}$

22. $\{24, 40, 48, 56\}$

2.10   Adding and Subtracting Unlike Fractions

To add $\dfrac{1}{2}$ dollar and $\dfrac{1}{4}$ dollar (one quarter), we change the $\dfrac{1}{2}$ dollar to two $\dfrac{1}{4}$ dollars (two quarters). Then

$$\frac{1}{2} \text{ dollar} + \frac{1}{4} \text{ dollar} = \frac{2}{4} \text{ dollar} + \frac{1}{4} \text{ dollar}$$

$$= \left(\frac{2}{4} + \frac{1}{4} \right) \text{ dollar}$$

$$= \frac{3}{4} \quad \text{dollar}$$

We converted $\dfrac{1}{2}$ into $\dfrac{2}{4}$, which has the same denominator as $\dfrac{1}{4}$. The *lowest common denominator* (LCD) of $\dfrac{1}{2}$ and $\dfrac{1}{4}$ is 4. The method of adding unlike fractions is given in the following box.

Adding or subtracting unlike fractions

1. Find the LCD.
2. Convert each fraction to an equivalent fraction having the LCD as its denominator.
3. Add or subtract the like fractions as before.
4. Reduce the resulting fraction to lowest terms.
5. Any improper fraction found in step 4 is usually changed to a mixed number.

EXAMPLE 1 Add $\dfrac{1}{2} + \dfrac{2}{3}$.

SOLUTION   The LCD is 6, by inspection.

$$\begin{array}{r} \dfrac{1}{2} = \dfrac{3}{6} \\[2mm] +\dfrac{2}{3} = \dfrac{4}{6} \\[2mm] \hline \dfrac{7}{6} = 1\dfrac{1}{6} \end{array}$$

Convert each fraction to an equivalent fraction having the LCD 6 as a denominator; then add the like fractions

The sum

EXAMPLE 2 Subtract $\dfrac{4}{5} - \dfrac{3}{10}$.

SOLUTION   The LCD is 10, by inspection.

$$\begin{array}{r} \dfrac{4}{5} = \dfrac{8}{10} \\[2mm] -\dfrac{3}{10} = \dfrac{3}{10} \\[2mm] \hline \dfrac{5}{10} = \dfrac{\overset{1}{\cancel{5}}}{\underset{2}{\cancel{10}}} = \dfrac{1}{2} \end{array}$$

Reduce

In Examples 3–6, the LCD is found by prime factorization.

EXAMPLE 3 Add $\dfrac{5}{6} + \dfrac{3}{10} + \dfrac{4}{15}$.

SOLUTION

Finding the LCD

$$\begin{aligned} 6 &= 2 \cdot 3 \\ 10 &= 2 \cdot 5 \\ 15 &= 3 \cdot 5 \\ \text{LCD} &= 2 \cdot 3 \cdot 5 \\ &= 30 \end{aligned}$$

Adding like fractions

$$\begin{array}{r} \dfrac{5}{6} = \dfrac{25}{30} \\[2mm] \dfrac{3}{10} = \dfrac{9}{30} \\[2mm] +\dfrac{4}{15} = \dfrac{8}{30} \\[2mm] \hline \dfrac{42}{30} = \dfrac{\overset{7}{\cancel{42}}}{\underset{5}{\cancel{30}}} = \dfrac{7}{5} = 1\dfrac{2}{5} \end{array}$$

EXAMPLE 4 Add $\dfrac{3}{16} + \dfrac{5}{12} + \dfrac{5}{24}$.

SOLUTION

Finding the LCD

$$\begin{aligned} 16 &= 2^4 \\ 12 &= 2^2 \cdot 3 \\ 24 &= 2^3 \cdot 3 \\ \text{LCD} &= 2^4 \cdot 3 \\ &= 48 \end{aligned}$$

Adding like fractions

$$\begin{array}{r} \dfrac{3}{16} = \dfrac{9}{48} \\[2mm] \dfrac{5}{12} = \dfrac{20}{48} \\[2mm] +\dfrac{5}{24} = \dfrac{10}{48} \\[2mm] \hline \dfrac{39}{48} = \dfrac{\overset{13}{\cancel{39}}}{\underset{16}{\cancel{48}}} = \dfrac{13}{16} \end{array}$$

EXAMPLE 5   Subtract $\dfrac{7}{15} - \dfrac{5}{12}$.

SOLUTION

Finding the LCD

$$15 = 3 \cdot 5$$
$$12 = 2^2 \cdot 3$$
$$LCD = 2^2 \cdot 3 \cdot 5$$
$$= 60$$

Subtracting like fractions

$$\dfrac{7}{15} = \dfrac{28}{60}$$
$$-\dfrac{5}{12} = \dfrac{25}{60}$$
$$\dfrac{3}{60} = \dfrac{\overset{1}{3}}{\underset{20}{60}} = \dfrac{1}{20}$$

EXAMPLE 6   Add $\dfrac{8}{48} + \dfrac{5}{12} + \dfrac{4}{18}$.

SOLUTION

Finding the LCD

$$48 = 2^4 \cdot 3$$
$$12 = 2^2 \cdot 3$$
$$18 = 2 \cdot 3^2$$
$$LCD = 2^4 \cdot 3^2$$
$$= 144$$

Adding like fractions

$$\dfrac{8}{48} = \dfrac{24}{144}$$
$$\dfrac{5}{12} = \dfrac{60}{144}$$
$$+\dfrac{4}{18} = \dfrac{32}{144}$$
$$\dfrac{116}{144} = \dfrac{\overset{29}{116}}{\underset{36}{144}} = \dfrac{29}{36}$$

If we had reduced the original fractions to lowest terms before finding the LCD and adding, the work could have been simplified. For example, if we first reduce the fractions to lowest terms,

$$\dfrac{\overset{1}{8}}{\underset{6}{48}} + \dfrac{5}{12} + \dfrac{\overset{2}{4}}{\underset{9}{18}} = \dfrac{1}{6} + \dfrac{5}{12} + \dfrac{2}{9}$$

Finding the LCD

$$6 = 2 \cdot 3$$
$$12 = 2^2 \cdot 3$$
$$9 = 3^2$$
$$LCD = 2^2 \cdot 3^2$$
$$= 36$$

Adding like fractions

$$\dfrac{1}{6} = \dfrac{6}{36}$$
$$\dfrac{5}{12} = \dfrac{15}{36}$$
$$+\dfrac{2}{9} = \dfrac{8}{36}$$
$$\dfrac{29}{36}$$

Note   It is not necessary to use the LCD when adding unlike fractions. *Any* common denominator will work. However, if the LCD is not used, larger numbers appear in the work. In this case, the resulting fraction must be reduced.

EXAMPLE 7 Add $\dfrac{1}{6} + \dfrac{2}{3} + \dfrac{3}{4}$.

SOLUTION A *common denominator* (not the LCD) is 24, because each denominator divides into 24 exactly.

$$\dfrac{1}{6} = \dfrac{4}{24}$$
$$\dfrac{2}{3} = \dfrac{16}{24}$$
$$+ \dfrac{3}{4} = \dfrac{18}{24}$$
$$\dfrac{38}{24} = \dfrac{\overset{19}{38}}{\underset{12}{24}} = \dfrac{19}{12} = 1\dfrac{7}{12}$$

Exercises 2.10

First reduce fractions to lowest terms and find the LCD, then perform the indicated operation. Reduce your answers to lowest terms. Write any improper fractions as mixed numbers.

1. $\dfrac{1}{2} + \dfrac{3}{4}$ 2. $\dfrac{1}{3} + \dfrac{5}{6}$ 3. $\dfrac{3}{5} + \dfrac{3}{10}$ 4. $\dfrac{5}{8} + \dfrac{1}{2}$

5. $\dfrac{2}{3} - \dfrac{1}{6}$ 6. $\dfrac{7}{10} - \dfrac{1}{2}$ 7. $\dfrac{2}{3} - \dfrac{5}{12}$ 8. $\dfrac{9}{15} - \dfrac{2}{5}$

9. $\dfrac{2}{3} + \dfrac{1}{4}$ 10. $\dfrac{2}{5} + \dfrac{1}{3}$ 11. $\dfrac{3}{6} + \dfrac{5}{15}$ 12. $\dfrac{2}{6} + \dfrac{6}{8}$

13. $\dfrac{6}{7} - \dfrac{4}{12}$ 14. $\dfrac{10}{16} - \dfrac{5}{12}$ 15. $\dfrac{12}{30} - \dfrac{1}{5}$ 16. $\dfrac{12}{16} - \dfrac{1}{6}$

17. $\dfrac{25}{32} - \dfrac{3}{4}$ 18. $\dfrac{13}{16} - \dfrac{5}{8}$ 19. $\dfrac{56}{64} - \dfrac{14}{24}$ 20. $\dfrac{14}{35} - \dfrac{5}{20}$

21. $\dfrac{3}{20} + \dfrac{1}{8}$ 22. $\dfrac{4}{15} + \dfrac{7}{12}$ 23. $\dfrac{7}{9} + \dfrac{5}{6}$ 24. $\dfrac{11}{12} + \dfrac{7}{8}$

25. $\dfrac{5}{12} - \dfrac{2}{15}$ 26. $\dfrac{11}{18} - \dfrac{4}{15}$ 27. $\dfrac{19}{35} - \dfrac{5}{14}$ 28. $\dfrac{15}{16} - \dfrac{5}{24}$

29. $\dfrac{2}{3} + \dfrac{5}{6} + \dfrac{1}{2}$ 30. $\dfrac{1}{2} + \dfrac{2}{3} + \dfrac{3}{4}$ 31. $\dfrac{3}{5} + \dfrac{1}{2} + \dfrac{3}{10}$

32. $\dfrac{7}{10} + \dfrac{3}{5} + \dfrac{2}{3}$ 33. $\dfrac{1}{4} + \dfrac{7}{12} + \dfrac{5}{8}$ 34. $\dfrac{5}{6} + \dfrac{4}{12} + \dfrac{3}{8}$

35. $\dfrac{4}{6} + \dfrac{6}{14} + \dfrac{2}{3}$ 36. $\dfrac{6}{9} + \dfrac{8}{12} + \dfrac{9}{10}$ 37. $\dfrac{5}{6} + \dfrac{3}{8} + \dfrac{1}{12}$

38. $\dfrac{1}{3} + \dfrac{3}{5} + \dfrac{2}{11}$ 39. $\dfrac{3}{4} + \dfrac{2}{14} + \dfrac{1}{2}$ 40. $\dfrac{1}{12} + \dfrac{3}{16} + \dfrac{4}{10}$

2.11 Adding Mixed Numbers

Two different methods can be used to add mixed numbers.

Method 1 One way to add mixed numbers is to convert each mixed number to an improper fraction, then perform the addition. If the final answer is an improper fraction, convert it to a mixed number.

EXAMPLE 1

Add $1\frac{5}{6} + 2\frac{1}{4}$.

SOLUTION

$$1\frac{5}{6} = \frac{11}{6} = \frac{22}{12}$$
$$+\ 2\frac{1}{4} = \frac{9}{4} = \frac{27}{12}$$
$$\overline{\qquad\qquad\frac{49}{12} = 4\frac{1}{12}}$$

Method 2   Another way to add mixed numbers is to add the whole number parts and the fractional parts separately. If the sum of the fractional parts is an improper fraction, it must be changed to a mixed number and added to the whole number part (see Examples 3 and 4).

EXAMPLE 2

Add $12\frac{1}{2} + 14\frac{1}{3}$.

SOLUTION

$$12\frac{1}{2} = 12 + \frac{3}{6}$$
$$+\ 14\frac{1}{3} = 14 + \frac{2}{6}$$
$$\overline{\qquad\qquad 26 + \frac{5}{6} = 26\frac{5}{6}}$$

The fraction parts are changed to equivalent fractions having the LCD for their denominators

---

EXAMPLE 3

Add $12\frac{3}{4} + 21\frac{3}{8} + 45\frac{1}{2}$.

SOLUTION

$$12\frac{3}{4} = 12 + \frac{6}{8}$$
$$21\frac{3}{8} = 21 + \frac{3}{8}$$
$$+\ 45\frac{1}{2} = 45 + \frac{4}{8}$$
$$\overline{\qquad\qquad 78 + \frac{13}{8}}$$

Replace $\frac{13}{8}$ by $1\frac{5}{8}$

$$= 78 + 1\frac{5}{8} = 79\frac{5}{8}$$

EXAMPLE 4

Find the perimeter of the triangle.

SOLUTION   To find the perimeter of a figure, we must add the lengths of all its sides.

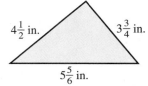

$$4\frac{1}{2} = 4 + \frac{6}{12}$$
$$3\frac{3}{4} = 3 + \frac{9}{12}$$
$$+\ 5\frac{5}{6} = 5 + \frac{10}{12}$$
$$\overline{\qquad\qquad 12 + \frac{25}{12}}$$
$$= 12 + 2\frac{1}{12} = 14\frac{1}{12} \text{ in.}$$

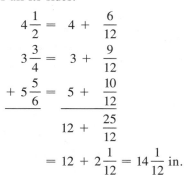

Exercises 2.11

In Exercises 1–28, find the sums. Reduce the answers to lowest terms. Write any improper fractions as mixed numbers.

1. $2\frac{3}{5} + 1\frac{1}{10}$ 2. $1\frac{3}{8} + 2\frac{1}{4}$ 3. $3\frac{1}{3} + 2\frac{1}{2}$

4. $5\frac{3}{6} + 1\frac{1}{2}$ 5. $4\frac{1}{6} + 3\frac{2}{3}$ 6. $2\frac{1}{2} + 3\frac{3}{4}$

7. $1\frac{5}{8} + 2\frac{1}{2}$ 8. $3\frac{1}{8} + 2\frac{1}{5}$ 9. $7\frac{5}{6} + 3\frac{2}{3}$

10. $8\frac{1}{5} + 2\frac{7}{8}$ 11. $2\frac{5}{8} + 3$ 12. $4 + 2\frac{3}{5}$

13. $1\frac{1}{2} + 2\frac{1}{3} + 3\frac{1}{4}$ 14. $2\frac{1}{3} + 1\frac{3}{4} + 3\frac{5}{6}$

15. $5\frac{1}{4} + 3 + 2\frac{3}{8}$ 16. $6 + 2\frac{3}{5} + 1\frac{7}{10}$

17. $27\frac{5}{6} + 9\frac{3}{4}$ 18. $46\frac{7}{9} + 21\frac{1}{6}$

19. $16\frac{3}{4} + 32\frac{7}{10}$ 20. $58\frac{1}{6} + 43\frac{9}{10}$

21. $\begin{array}{r} 72\frac{2}{3} \\ 81\frac{3}{4} \\ + 93\frac{1}{2} \\ \hline \end{array}$ 22. $\begin{array}{r} 68\frac{2}{9} \\ 97\frac{1}{3} \\ + 55\frac{5}{6} \\ \hline \end{array}$ 23. $\begin{array}{r} 17\frac{1}{3} \\ 28\frac{2}{5} \\ + 15\frac{4}{15} \\ \hline \end{array}$

24. $\begin{array}{r} 32\frac{3}{14} \\ 78\frac{19}{28} \\ + 21\frac{5}{7} \\ \hline \end{array}$ 25. $\begin{array}{r} 20\frac{5}{8} \\ 46\frac{1}{16} \\ + 53\frac{3}{4} \\ \hline \end{array}$ 26. $\begin{array}{r} 56\frac{4}{5} \\ 81\frac{7}{15} \\ + 22\frac{3}{10} \\ \hline \end{array}$

27. $\begin{array}{r} 117\frac{5}{6} \\ 28\frac{1}{2} \\ + 232\frac{7}{9} \\ \hline \end{array}$ 28. $\begin{array}{r} 168\frac{7}{16} \\ 312\frac{19}{32} \\ + 406\frac{1}{2} \\ \hline \end{array}$

29. Find the sum of $7\frac{3}{4}$ ounces, $5\frac{7}{8}$ ounces, $8\frac{5}{16}$ ounces, and $10\frac{2}{5}$ ounces.

30. Tony weighs $145\frac{1}{2}$ pounds. Mike weighs $157\frac{3}{4}$ pounds. Yuri weighs $7\frac{3}{4}$ pounds more than Tony. Find the weight of the three men.

31. Find the perimeter of the figure.

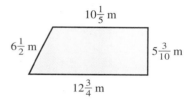

32. Find the perimeter of the triangle.

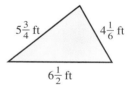

2.12   Subtracting Mixed Numbers

Two different methods can be used to subtract mixed numbers.

Method 1   One way to subtract mixed numbers is to convert each mixed number to an improper fraction, then perform the subtraction. If the final answer is an improper fraction, convert it to a mixed number.

EXAMPLE 1   Subtract $3\frac{1}{2} - 1\frac{2}{5}$.

SOLUTION

$$3\frac{1}{2} = \frac{7}{2} = \frac{35}{10}$$
$$-1\frac{2}{5} = \frac{7}{5} = \frac{14}{10}$$
$$\frac{21}{10} = 2\frac{1}{10}$$

Method 2   Another way to subtract mixed numbers is to subtract the whole number parts and the fractional parts separately. If the fractional part being subtracted is larger than the fractional part we are subtracting from, we must "borrow" 1 from the whole number part (see Examples 3–5).

EXAMPLE 2   Subtract $15\frac{3}{8} - 3\frac{1}{4}$.

SOLUTION

$$15\frac{3}{8} = \quad 15\frac{3}{8}$$
$$-3\frac{1}{4} = - \quad 3\frac{2}{8}$$
$$12\frac{1}{8}$$

EXAMPLE 3   Subtract $12\frac{1}{6} - 4\frac{1}{2}$.

SOLUTION

We cannot subtract $\frac{3}{6}$ from $\frac{1}{6}$

$$12\frac{1}{6} = \quad 12\frac{1}{6} = \overset{11+1}{\cancel{12}} + \frac{1}{6} = 11 + \frac{6}{6} + \frac{1}{6} = 11\frac{7}{6} = \quad 11\frac{7}{6}$$
$$-4\frac{1}{2} = - \quad 4\frac{3}{6} \qquad\qquad\qquad\qquad\qquad\qquad\qquad\qquad - 4\frac{3}{6}$$

1 is *borrowed* from 12, leaving

11; this 1 is changed to $\frac{6}{6}$;

then $12\frac{1}{6} = 11\frac{6}{6} + \frac{1}{6} = 11\frac{7}{6}$

$$7\frac{4}{6} = 7\frac{2}{3}$$

We shorten the work by writing the problem as follows:

$$12\frac{1}{6} = \quad 12\overset{11\ \frac{7}{6}}{\cancel{\frac{1}{6}}}$$
$$-4\frac{1}{2} = - \quad 4\frac{3}{6}$$
$$7\frac{4}{6} = 7\frac{2}{3}$$

EXAMPLE 4   Subtract $58\frac{2}{7} - 25\frac{1}{3}$.

SOLUTION

$$57\frac{27}{21} \leftarrow \text{Borrow } 1 = \frac{21}{21}; \text{ then add } \frac{21}{21} + \frac{6}{21} = \frac{27}{21}$$

$$58\frac{2}{7} = 58\frac{6}{21}$$

$$-25\frac{1}{3} = -25\frac{7}{21}$$

$$32\frac{20}{21}$$

EXAMPLE 5   Subtract $15 - 5\frac{3}{8}$.

SOLUTION

$$14\frac{8}{8} \leftarrow \text{Borrow } 1 = \frac{8}{8}$$

$$\cancel{15}$$

$$-5\frac{3}{8}$$

$$9\frac{5}{8}$$

Exercises 2.12

In Exercises 1–20, find the differences. Reduce the answers to lowest terms. Write any improper fractions as mixed numbers.

1. $14\frac{3}{4}$
   $-10\frac{1}{4}$

2. $21\frac{5}{6}$
   $-18\frac{1}{6}$

3. $17\frac{3}{4}$
   $-5\frac{1}{8}$

4. $19\frac{3}{5}$
   $-6\frac{1}{10}$

5. 8
   $-4\frac{1}{2}$

6. 7
   $-3\frac{1}{3}$

7. 6
   $-2\frac{3}{5}$

8. 4
   $-1\frac{5}{6}$

9. $4\frac{1}{4}$
   $-1\frac{3}{4}$

10. $5\frac{1}{3}$
    $-1\frac{2}{3}$

11. $12\frac{2}{5}$
    $-7\frac{3}{10}$

12. $54\frac{2}{3}$
    $-39\frac{1}{6}$

13. $3\frac{1}{12}$
    $-1\frac{1}{6}$

14. $23\frac{5}{8}$
    $-17\frac{3}{4}$

15. 45
    $-38\frac{2}{3}$

16. 32
    $-28\frac{7}{15}$

17. $68\frac{5}{16}$
    $-53\frac{3}{4}$

18. $107\frac{2}{3}$
    $-99\frac{1}{6}$

19. $234\frac{5}{14}$
    $-157\frac{3}{7}$

20. $7,005\frac{2}{3}$
    $-2,867\frac{2}{3}$

21. When a $1\frac{1}{4}$-pound steak was trimmed of fat, it weighed $\frac{7}{8}$ pound. How much did the fat weigh?

22. Mr. Angelini took $7\frac{1}{2}$ days to paint the interior of his house. A professional painter said he could have done it in $2\frac{1}{3}$ days. How much time would have been saved by having the painter do it?

23. Mrs. Segal has $5\frac{3}{4}$ square yards of carpet left after carpeting her living room. How much more will she need to carpet a bathroom that has a floor area of 7 square yards?

24. Jim wants three boards for shelves that measure $5\frac{3}{4}$ feet, $2\frac{5}{12}$ feet, and $3\frac{1}{2}$ feet. Can he cut all three shelves from a 12-foot board, allowing $\frac{1}{4}$ foot for waste?

2.13   Multiplying Mixed Numbers

To multiply mixed numbers, there is only one method: Change each mixed number to an improper fraction, then multiply.

$$\frac{a}{b} \cdot \frac{c}{d} = \frac{a \cdot c}{b \cdot d}$$

EXAMPLE 1
a. $2\frac{11}{12} \cdot 1\frac{3}{5} = \frac{\overset{7}{35}}{\underset{3}{12}} \cdot \frac{\overset{2}{8}}{\underset{1}{5}} = \frac{14}{3} = 4\frac{2}{3}$

b. $3\frac{1}{8} \cdot 16 = \frac{25}{8} \cdot \frac{\overset{2}{16}}{1} = 50$

c. $2\frac{1}{4} \cdot 4\frac{2}{3} \cdot 1\frac{1}{5} = \frac{\overset{3}{9}}{\underset{\underset{1}{2}}{4}} \cdot \frac{\overset{7}{14}}{\underset{1}{3}} \cdot \frac{\overset{3}{6}}{5} = \frac{63}{5} = 12\frac{3}{5}$

EXAMPLE 2   Find the area of a rectangle with length $2\frac{2}{3}$ yards and width $1\frac{3}{4}$ yards.

SOLUTION   Recall that to find the area of a rectangle we multiply the length times the width:

$$\text{Area} = l \cdot w$$

$$= 2\frac{2}{3} \cdot 1\frac{3}{4} = \frac{\overset{2}{8}}{3} \cdot \frac{7}{\underset{1}{4}} = \frac{14}{3} = 4\frac{2}{3} \text{ sq. yd}$$

Exercises 2.13

In Exercises 1–16, find the products. Reduce the answers to lowest terms. Write any improper fractions as mixed numbers.

1. $1\frac{2}{3} \cdot 2\frac{1}{2}$

2. $1\frac{5}{6} \cdot 3\frac{1}{2}$

3. $2\frac{2}{3} \cdot 2\frac{1}{4}$

4. $2\frac{4}{5} \cdot 2\frac{1}{7}$

5. $8 \cdot 3\frac{3}{4}$

6. $4\frac{2}{3} \cdot 6$

7. $2\frac{5}{8} \cdot 4$

8. $6 \cdot 2\frac{5}{12}$

9. $1\frac{1}{12} \cdot 1\frac{4}{5}$

10. $2\frac{3}{10} \cdot 2\frac{1}{2}$

11. $2\frac{2}{15} \cdot 1\frac{1}{4}$

12. $1\frac{5}{16} \cdot 3\frac{1}{3}$

13. $3\frac{3}{10} \cdot \frac{6}{11} \cdot 1\frac{2}{3}$

14. $1\frac{1}{8} \cdot \frac{4}{9} \cdot 1\frac{5}{6}$

15. $3\frac{3}{4} \cdot 4\frac{2}{5} \cdot 3\frac{1}{2}$

16. $2\frac{1}{6} \cdot 1\frac{5}{7} \cdot 4\frac{2}{3}$

17. Each of eight hikers carries a food pack weighing $2\frac{7}{16}$ pounds. How much food are they carrying in all?

18. A roast is to be cooked for $\frac{1}{3}$ hour for each pound. If the roast weighs 7 pounds, how long will it take to cook?

19. A large truck can carry $10\frac{1}{2}$ tons of lettuce. If the truck is $\frac{4}{5}$ full, how many tons of lettuce is it carrying?

20. Hahn's $7\frac{1}{2}$-gallon fish tank is only $\frac{2}{3}$ full. How many gallons of water does he need to add to fill the tank?

21. Find the area of the rectangle.

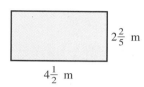

$2\frac{2}{5}$ m

$4\frac{1}{2}$ m

22. Mary is putting tile on her kitchen floor, which measures 8 feet long and $6\frac{1}{2}$ feet wide. She ran out of tiles after completing only $\frac{3}{4}$ of the floor. How many square feet of tile must she buy to finish the job?

2.14   Division of Fractions

If the product of two numbers is 1, they are called **multiplicative inverses**, or **reciprocals**, of each other. To find the multiplicative inverse, or reciprocal, of any nonzero number, we write the number as a fraction, then invert the fraction. The number zero has no reciprocal because 0 times any number is 0 (not 1).

Multiplicative inverse property	If $\dfrac{a}{b}$ represents any fraction except zero, then $$\underset{\underset{\text{Reciprocals}}{\uparrow\qquad\uparrow}}{\frac{a}{b}\cdot\frac{b}{a}=1}$$

EXAMPLE 1

a. The reciprocal of $\dfrac{4}{7}$ is $\dfrac{7}{4}$. $\dfrac{4}{7}\cdot\dfrac{7}{4}=\dfrac{28}{28}=1$

b. The reciprocal of $\dfrac{1}{3}$ is 3. $\dfrac{1}{3}\cdot3=\dfrac{1}{3}\cdot\dfrac{3}{1}=\dfrac{3}{3}=1$

c. The reciprocal of 5 is $\dfrac{1}{5}$. $5\cdot\dfrac{1}{5}=\dfrac{5}{1}\cdot\dfrac{1}{5}=\dfrac{5}{5}=1$

Consider the quotient $\dfrac{3}{5}\div\dfrac{4}{7}$. First we will write this quotient in fraction form. Then we will multiply the numerator and the denominator by the reciprocal of the denominator.

The value of this fraction is 1, and multiplying a number by 1 does not change its value

$$\frac{3}{5}\div\frac{4}{7}=\frac{\frac{3}{5}}{\frac{4}{7}}=\frac{\frac{3}{5}}{\frac{4}{7}}\cdot\frac{\frac{7}{4}}{\frac{7}{4}}=\frac{\frac{3}{5}\cdot\frac{7}{4}}{\frac{4}{7}\cdot\frac{7}{4}}=\frac{\frac{3}{5}\cdot\frac{7}{4}}{1}=\frac{3}{5}\cdot\frac{7}{4}$$

Therefore,

$$\frac{3}{5}\div\frac{4}{7}=\frac{3}{5}\cdot\frac{7}{4}$$

This leads to the rule for division of fractions.

Dividing fractions

> Invert the second fraction and multiply.
>
> $$\frac{a}{b} \div \frac{c}{d} = \frac{a}{b} \cdot \frac{d}{c}$$
>
> First fraction ⟶   ⟵ Second fraction

EXAMPLE 2

a. $\dfrac{3}{5} \div \dfrac{3}{7} = \dfrac{3}{5} \cdot \dfrac{7}{3} = \dfrac{\cancel{3}^{1}}{5} \cdot \dfrac{7}{\cancel{3}_{1}} = \dfrac{7}{5} = 1\dfrac{2}{5}$

b. $\dfrac{5}{6} \div \dfrac{2}{3} = \dfrac{5}{6} \cdot \dfrac{3}{2} = \dfrac{5}{\cancel{6}_{2}} \cdot \dfrac{\cancel{3}^{1}}{2} = \dfrac{5}{4} = 1\dfrac{1}{4}$

c. $\dfrac{15}{32} \div 2 = \dfrac{15}{32} \cdot \dfrac{1}{2} = \dfrac{15}{64}$

d. $7 \div \dfrac{3}{8} = \dfrac{7}{1} \cdot \dfrac{8}{3} = \dfrac{56}{3} = 18\dfrac{2}{3}$

There is only one method of dividing mixed numbers: Change each mixed number to an improper fraction, then invert the second fraction and multiply.

EXAMPLE 3

a. $6\dfrac{4}{5} \div 1\dfrac{7}{10} = \dfrac{34}{5} \div \dfrac{17}{10} = \dfrac{\cancel{34}^{2}}{\cancel{5}_{1}} \cdot \dfrac{\cancel{10}^{2}}{\cancel{17}_{1}} = 4$

b. $12 \div 2\dfrac{2}{3} = \dfrac{12}{1} \div \dfrac{8}{3} = \dfrac{\cancel{12}^{3}}{1} \cdot \dfrac{3}{\cancel{8}_{2}} = \dfrac{9}{2} = 4\dfrac{1}{2}$

Exercises 2.14

In Exercises 1–35, find the quotients. Reduce the answers to lowest terms. Write any improper fractions as mixed numbers.

1. $\dfrac{3}{4} \div \dfrac{1}{2}$ 2. $\dfrac{2}{5} \div \dfrac{1}{3}$ 3. $\dfrac{1}{2} \div \dfrac{2}{3}$ 4. $4 \div \dfrac{2}{5}$

5. $\dfrac{1}{2} \div 5$ 6. $\dfrac{4}{3} \div \dfrac{8}{6}$ 7. $\dfrac{5}{2} \div \dfrac{5}{8}$ 8. $\dfrac{7}{8} \div 7$

9. $1 \div \dfrac{1}{2}$ 10. $\dfrac{100}{150} \div \dfrac{4}{9}$ 11. $\dfrac{3}{5} \div \dfrac{3}{10}$ 12. $\dfrac{7}{12} \div \dfrac{1}{3}$

13. $\dfrac{3}{16} \div \dfrac{9}{20}$ 14. $\dfrac{13}{28} \div 39$ 15. $34 \div \dfrac{17}{56}$ 16. $\dfrac{8}{15} \div \dfrac{24}{10}$

17. $\dfrac{35}{16} \div \dfrac{42}{22}$ 18. $\dfrac{7}{18} \div \dfrac{21}{15}$ 19. $36 \div \dfrac{4}{5}$ 20. $\dfrac{4}{9} \div 36$

21. $\dfrac{6}{35} \div \dfrac{8}{15}$ 22. $\dfrac{11}{84} \div \dfrac{22}{60}$ 23. $\dfrac{14}{24} \div 210$ 24. $22 \div \dfrac{11}{5}$

25. $\dfrac{56}{15} \div \dfrac{28}{90}$ 26. $1\dfrac{3}{7} \div 1\dfrac{1}{4}$ 27. $1\dfrac{7}{9} \div 2\dfrac{2}{3}$

28. $2\dfrac{3}{5} \div 1\dfrac{4}{35}$ 29. $3\dfrac{2}{3} \div 1\dfrac{7}{15}$ 30. $7 \div 4\dfrac{2}{3}$

31. $3\dfrac{4}{5} \div 19$ 32. $3\dfrac{1}{3} \div 5$ 33. $11 \div 3\dfrac{1}{7}$

34. $10\dfrac{1}{2} \div 5\dfrac{5}{6}$ 35. $6\dfrac{2}{3} \div 3\dfrac{5}{9}$

36. A board $5\dfrac{1}{3}$ feet long is cut into 8 pieces. How long is each piece?

37. How many tablets, each containing 3 milligrams of a heart medicine, must be used to make up a $4\dfrac{1}{2}$-milligram dosage?

38. How many pins $2\dfrac{1}{2}$ inches long can be cut from a rod 30 inches long?

39. A bookshelf is 36 inches long. How many books that are $1\dfrac{1}{2}$ inches thick can stand on that shelf?

2.15   **Complex Fractions**

A **simple fraction** has only one fraction line.

EXAMPLE 1   Examples of simple fractions:

$$\frac{2}{3}, \quad \frac{18}{5}, \quad \frac{2}{7-3}, \quad \frac{13+6}{9-2}$$

A **complex fraction** is a fraction that has more than one fraction line.

EXAMPLE 2   Examples of complex fractions:

$$\frac{\frac{2}{5}}{3}, \quad \frac{8}{\frac{1}{2}}, \quad \frac{\frac{4}{7}}{\frac{5}{9}}, \quad \frac{\frac{3}{2}-\frac{2}{5}}{\frac{5}{6}+\frac{3}{4}}$$

Simplifying complex fractions

1. Simplify the numerator and the denominator when possible.
2. Divide the simplified numerator by the simplified denominator.

EXAMPLE 3

a. $\dfrac{\frac{5}{12}}{\frac{8}{15}} = \dfrac{5}{12} \div \dfrac{8}{15} = \dfrac{5}{\cancel{12}_{4}} \cdot \dfrac{\cancel{15}^{5}}{8} = \dfrac{25}{32}$

b. $\dfrac{6}{\frac{3}{5}} = 6 \div \dfrac{3}{5} = \dfrac{\cancel{6}^{2}}{1} \cdot \dfrac{5}{\cancel{3}_{1}} = \dfrac{10}{1} = 10$

c. $\dfrac{\frac{4}{5}}{12} = \dfrac{4}{5} \div \dfrac{12}{1} = \dfrac{\cancel{4}^{1}}{5} \cdot \dfrac{1}{\cancel{12}_{3}} = \dfrac{1}{15}$

d. $\dfrac{2\frac{3}{5}}{1\frac{5}{8}} = \dfrac{\frac{13}{5}}{\frac{13}{8}} = \dfrac{13}{5} \div \dfrac{13}{8} = \dfrac{\cancel{13}^{1}}{5} \cdot \dfrac{8}{\cancel{13}_{1}} = \dfrac{8}{5} = 1\dfrac{3}{5}$

e. $\dfrac{\frac{1}{6}+\frac{2}{3}}{\frac{5}{8}-\frac{1}{4}} = \dfrac{\frac{1}{6}+\frac{4}{6}}{\frac{5}{8}-\frac{2}{8}} = \dfrac{\frac{5}{6}}{\frac{3}{8}} = \dfrac{5}{6} \div \dfrac{3}{8} = \dfrac{5}{\cancel{6}_{3}} \cdot \dfrac{\cancel{8}^{4}}{3} = \dfrac{20}{9} = 2\dfrac{2}{9}$

   Note   It is necessary to know which fraction line is the *main* fraction line of a complex fraction. The main fraction line will be heavier and longer than the other fraction lines.

EXAMPLE 4　Show that

$$\frac{2}{\dfrac{3}{4}} \neq \frac{\dfrac{2}{3}}{4}$$

SOLUTION　If the upper fraction line is the main fraction line, then

$$\frac{2}{\dfrac{3}{4}} = \frac{2}{1} \div \frac{3}{4} = \frac{2}{1} \cdot \frac{4}{3} = \boxed{\frac{8}{3}}$$

If the lower fraction line is the main fraction line, then

$$\frac{\dfrac{2}{3}}{4} = \frac{2}{3} \div \frac{4}{1} = \frac{\overset{1}{2}}{3} \cdot \frac{1}{\underset{2}{4}} = \boxed{\frac{1}{6}}$$

which is different from the first value, $\dfrac{8}{3}$.

Exercises　2.15

In Exercises 1–25, simplify the complex fractions.

1. $\dfrac{\frac{3}{4}}{\frac{1}{6}}$　　2. $\dfrac{\frac{15}{16}}{\frac{12}{5}}$　　3. $\dfrac{\frac{2}{3}}{\frac{1}{2}}$　　4. $\dfrac{\frac{3}{4}}{\frac{7}{8}}$

5. $\dfrac{\frac{3}{5}}{\frac{3}{10}}$　　6. $\dfrac{\frac{7}{16}}{\frac{7}{24}}$　　7. $\dfrac{\frac{3}{8}}{\frac{5}{12}}$　　8. $\dfrac{\frac{5}{7}}{\frac{10}{21}}$

9. $\dfrac{\frac{6}{2}}{3}$　　10. $\dfrac{\frac{20}{2}}{5}$　　11. $\dfrac{12}{\frac{15}{2}}$　　12. $\dfrac{15}{\frac{9}{5}}$

13. $\dfrac{1\frac{2}{3}}{10}$　　14. $\dfrac{8\frac{1}{3}}{100}$　　15. $\dfrac{6\frac{2}{3}}{100}$　　16. $\dfrac{2\frac{1}{2}}{10}$

17. $\dfrac{1\frac{1}{2}}{3\frac{3}{4}}$　　18. $\dfrac{3\frac{5}{9}}{2\frac{2}{3}}$　　19. $\dfrac{\frac{1}{4}+\frac{2}{5}}{\frac{1}{6}}$　　20. $\dfrac{\frac{3}{8}}{\frac{9}{10}-\frac{2}{5}}$

21. $\dfrac{\frac{1}{8}+\frac{3}{4}}{\frac{1}{2}-\frac{1}{3}}$　　22. $\dfrac{\frac{11}{4}-\frac{5}{9}}{\frac{7}{18}+\frac{13}{36}}$　　23. $\dfrac{\frac{1}{7}+\frac{9}{28}}{\frac{13}{14}-\frac{3}{7}}$

24. $\dfrac{\frac{16}{5}-\frac{7}{15}}{\frac{9}{30}+\frac{3}{10}}$　　25. $\dfrac{2+1\frac{3}{4}}{4-2\frac{1}{3}}$

26. The monthly rainfall for a city was $2\frac{1}{2}$ inches in January, $6\frac{3}{4}$ inches in February, 10 inches in March, $7\frac{1}{2}$ inches in April, 3 inches in May, and $1\frac{3}{4}$ inches in June. What was the average monthly rainfall for the first six months?

27. ABC's stock prices for four days were $11\frac{1}{4}$, $12\frac{1}{2}$, $11\frac{5}{8}$, and $9\frac{5}{8}$. What price must the stock reach on the fifth day so that the average price of the stock will be 11 for the week?

2.16 Powers and Roots of Fractions

Powers of Fractions

We can raise a fraction to a power the same way we raised a whole number to a power. Just as

$$2^3 = 2 \cdot 2 \cdot 2 = 8$$

so

$$\left(\frac{1}{2}\right)^3 = \frac{1}{2} \cdot \frac{1}{2} \cdot \frac{1}{2} = \frac{1}{8}$$

The exponent 3 indicates repeated multiplication of the base, $\frac{1}{2}$

EXAMPLE 1

a. $\left(\frac{3}{4}\right)^2 = \frac{3}{4} \cdot \frac{3}{4} = \frac{9}{16}$

b. $\left(\frac{1}{10}\right)^4 = \frac{1}{10} \cdot \frac{1}{10} \cdot \frac{1}{10} \cdot \frac{1}{10} = \frac{1}{10,000}$

c. $\left(2\frac{1}{2}\right)^2 = \frac{5}{2} \cdot \frac{5}{2} = \frac{25}{4} = 6\frac{1}{4}$

d. $\left(\frac{5}{9}\right)^0 = 1$ Any number raised to the zero power is 1

Square Roots of Fractions

Finding the square root of a number is the inverse operation of squaring a number. Recall that

$$\sqrt{9} = 3 \quad \text{Because } 3^2 = 9$$

Likewise,

$$\sqrt{\frac{1}{9}} = \frac{1}{3} \quad \text{Because } \left(\frac{1}{3}\right)^2 = \frac{1}{9}$$

EXAMPLE 2

a. $\sqrt{\frac{4}{25}} = \frac{2}{5}$ Because $\left(\frac{2}{5}\right)^2 = \frac{4}{25}$

b. $\sqrt{\frac{49}{100}} = \frac{7}{10}$ Because $\left(\frac{7}{10}\right)^2 = \frac{49}{100}$

c. $\sqrt{\frac{81}{16}} = \frac{9}{4} = 2\frac{1}{4}$ Because $\left(\frac{9}{4}\right)^2 = \frac{81}{16}$

Exercises 2.16

In Exercises 1–22, find the indicated powers and roots.

1. $\left(\frac{1}{4}\right)^2$

2. $\left(\frac{1}{6}\right)^2$

3. $\left(\frac{2}{3}\right)^2$

4. $\left(\frac{4}{5}\right)^2$

5. $\left(\frac{1}{2}\right)^4$

6. $\left(\frac{1}{10}\right)^5$

7. $\left(\frac{3}{4}\right)^3$

8. $\left(\frac{2}{5}\right)^3$

9. $\left(1\frac{1}{2}\right)^3$

10. $\left(1\frac{3}{4}\right)^2$

11. $\left(\frac{7}{10}\right)^0$

12. $\left(2\frac{2}{3}\right)^0$

13. $\left(\frac{8}{9}\right)^2$

14. $\left(\frac{11}{20}\right)^2$

15. $\sqrt{\frac{9}{100}}$

16. $\sqrt{\frac{1}{36}}$

17. $\sqrt{\frac{16}{81}}$

18. $\sqrt{\frac{64}{49}}$

19. $\sqrt{\frac{4}{81}} + \left(\frac{1}{6}\right)^2$

20. $\sqrt{\frac{25}{64}} + \left(\frac{3}{4}\right)^2$

21. $\left(1\dfrac{1}{2}\right)^2 - \sqrt{\dfrac{49}{144}}$

22. $\left(\dfrac{2}{5}\right)^2 - \sqrt{\dfrac{1}{100}}$

24. Find the cost to carpet a square room that measures $4\dfrac{1}{2}$ yards on each side if the carpet costs \$12 per square yard.

23. Find the area of the square.

$3\dfrac{1}{3}$ ft

2.17   ## Combined Operations

In evaluating expressions with more than one operation, we use the same order of operations with fractions that we used with whole numbers.

Order of operations	
	1. Any expressions in parentheses are evaluated first.
	2. Evaluations are done in this order:
	First: Powers and roots
	Second: Multiplication and division in order from *left to right*
	Third: Addition and subtraction in order from *left to right*

EXAMPLE 1

$$2 - \frac{3}{4} + 3\frac{1}{2}$$

$$= \frac{2}{1} - \frac{3}{4} + \frac{7}{2}$$

$$= \frac{8}{4} - \frac{3}{4} + \frac{14}{4}$$

$$= \frac{5}{4} + \frac{14}{4}$$

$$= \frac{19}{4} = 4\frac{3}{4}$$

EXAMPLE 2

$$2\frac{4}{5} \div 7 \cdot 1\frac{2}{3}$$

$$= \frac{14}{5} \div \frac{7}{1} \cdot \frac{5}{3}$$

$$= \frac{\overset{2}{\cancel{14}}}{\underset{1}{\cancel{5}}} \cdot \frac{1}{\underset{1}{\cancel{7}}} \cdot \frac{\overset{1}{\cancel{5}}}{3} = \frac{2}{3}$$

EXAMPLE 3

$$2\frac{1}{3} + \frac{5}{6} \div 1\frac{3}{4}$$

$$= \frac{7}{3} + \frac{5}{6} \div \frac{7}{4}$$

$$= \frac{7}{3} + \frac{5}{\underset{3}{6}} \cdot \frac{\overset{2}{4}}{7}$$

$$= \frac{7}{3} + \frac{10}{21}$$

$$= \frac{49}{21} + \frac{10}{21}$$

$$\frac{59}{21} = 2\frac{17}{21}$$

EXAMPLE 4

$$8 - \frac{2}{3} \cdot 2\frac{1}{2}$$

$$= \frac{8}{1} - \frac{\overset{1}{2}}{3} \cdot \frac{5}{\underset{1}{2}}$$

$$= \frac{8}{1} - \frac{5}{3}$$

$$= \frac{24}{3} - \frac{5}{3}$$

$$= \frac{19}{3} = 6\frac{1}{3}$$

EXAMPLE 5

$$\left(\frac{3}{4}\right)^2 + 2\frac{4}{5} \cdot 1\frac{1}{4}$$

$$= \frac{3}{4} \cdot \frac{3}{4} + \frac{\overset{7}{14}}{\underset{1}{5}} \cdot \frac{\overset{1}{5}}{\underset{2}{4}}$$

$$= \frac{9}{16} + \frac{7}{2}$$

$$= \frac{9}{16} + \frac{56}{16}$$

$$= \frac{65}{16} = 4\frac{1}{16}$$

EXAMPLE 6

$$\left(1\frac{3}{8} - \frac{1}{2}\right) \div 1\frac{5}{16}$$

$$= \left(\frac{11}{8} - \frac{1}{2}\right) \div \frac{21}{16}$$

$$= \left(\frac{11}{8} - \frac{4}{8}\right) \div \frac{21}{16}$$

$$= \frac{7}{8} \div \frac{21}{16}$$

$$= \frac{\overset{1}{\cancel{7}}}{\underset{1}{\cancel{8}}} \cdot \frac{\overset{2}{\cancel{16}}}{\underset{3}{\cancel{21}}} = \frac{2}{3}$$

EXAMPLE 7   Find the perimeter of a rectangle with length $3\frac{1}{4}$ feet and width $2\frac{1}{2}$ feet.

SOLUTION   Perimeter $= 2l + 2w$

$$= 2 \cdot 3\frac{1}{4} + 2 \cdot 2\frac{1}{2}$$

$$= \frac{\cancel{2}}{1} \cdot \frac{13}{\underset{2}{\cancel{4}}} + \frac{\cancel{2}}{1} \cdot \frac{5}{\underset{1}{\cancel{2}}}$$

$$= \frac{13}{2} + \frac{5}{1}$$

$$= \frac{13}{2} + \frac{10}{2}$$

$$= \frac{23}{2}$$

$$= 11\frac{1}{2} \text{ ft}$$

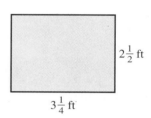

$2\frac{1}{2}$ ft

$3\frac{1}{4}$ ft

Exercises   2.17

In Exercises 1–18, evaluate each expression using the correct order of operations.

1. $3\frac{3}{4} \div \frac{5}{8} \cdot \frac{5}{9}$ **2.** $\frac{4}{5} \cdot 1\frac{2}{3} \div 1\frac{7}{9}$ **3.** $4 - \frac{2}{3} + 1\frac{1}{2}$

4. $2\frac{1}{8} + \frac{3}{4} - 1\frac{1}{2}$ **5.** $\frac{1}{2} \div \frac{1}{3} \div \frac{5}{6}$ **6.** $4\frac{1}{2} \div \frac{1}{4} \div \frac{3}{5}$

7. $\frac{7}{8} + \frac{3}{4} \cdot \frac{5}{6}$ **8.** $\frac{3}{4} + \frac{2}{3} \div \frac{4}{9}$ **9.** $\frac{3}{5} - \frac{1}{6} \div 1\frac{1}{4}$

10. $2\frac{1}{2} - 1\frac{1}{5} \cdot \frac{3}{4}$ **11.** $\left(\frac{2}{3}\right)^2 + 1\frac{2}{3} \cdot \frac{1}{10}$

12. $\left(\frac{1}{5}\right)^2 + \frac{4}{5} \div 1\frac{1}{3}$ **13.** $\left(8\frac{1}{4} - 1\frac{2}{3}\right) \cdot 3$

14. $\left(\frac{2}{3} + \frac{4}{5}\right) \div 2\frac{1}{5}$ **15.** $\left(3 - \frac{3}{10}\right) \div \sqrt{\frac{9}{25}}$

16. $\left(\frac{5}{6} + \frac{1}{4}\right) \cdot \sqrt{\frac{4}{9}}$ **17.** $7\frac{1}{2} \div \frac{3}{5} + 1\frac{7}{8} \cdot 2\frac{2}{5}$

18. $1\frac{1}{4} \cdot \frac{2}{3} - 1\frac{1}{6} \div 1\frac{1}{2}$

19. A room measures $5\frac{1}{3}$ yards long and $3\frac{1}{4}$ yards wide.

   a. Find the perimeter of the room.

   b. What would it cost to put ceiling molding around the room if the molding costs $12 per yard?

20. What is the total length of weather stripping needed to go around all sides of three windows if one window measures $5\frac{1}{2}$ feet by $2\frac{1}{4}$ feet and the other two square windows measure $2\frac{1}{4}$ feet on each side?

2.18 Comparing Fractions

Sometimes we need to recognize which of a group of fractions is largest. We can compare the sizes of fractions by converting all the given fractions to equivalent fractions having the same denominator. It is convenient (but not necessary) to use the LCD for the denominator of all the equivalent fractions.

EXAMPLE 1 Arrange the following fractions in order of size, with the largest first: $\dfrac{3}{8}, \dfrac{1}{3}, \dfrac{5}{12}$.

SOLUTION

Finding the LCD	*Finding the equivalent fractions*
$8 = 2 \cdot 2 \cdot 2 = 2^3$	$\dfrac{3}{8} = \dfrac{9}{24}$
$3 = 3$	
$12 = 2 \cdot 2 \cdot 3 = 2^2 \cdot 3$	$\dfrac{1}{3} = \dfrac{8}{24}$
$\text{LCD} = 2^3 \cdot 3 = 8 \cdot 3 = 24$	$\dfrac{5}{12} = \dfrac{10}{24}$

Arranging the *equivalent* fractions in order of size, we have

$$\frac{10}{24} > \frac{9}{24} > \frac{8}{24}$$

Therefore, the *original* fractions arranged in order of size are

$$\frac{5}{12} > \frac{3}{8} > \frac{1}{3}$$

Exercises 2.18

In Exercises 1–8, arrange the fractions in order of size, largest to smallest.

1. $\dfrac{3}{4}, \dfrac{5}{6}, \dfrac{2}{3}$ 2. $\dfrac{7}{9}, \dfrac{2}{3}, \dfrac{5}{6}$ 3. $\dfrac{5}{8}, \dfrac{3}{4}, \dfrac{11}{16}$ 4. $\dfrac{4}{9}, \dfrac{5}{12}, \dfrac{1}{3}$

5. $\dfrac{9}{14}, \dfrac{5}{7}, \dfrac{3}{4}$ 6. $\dfrac{7}{10}, \dfrac{3}{4}, \dfrac{4}{5}$ 7. $\dfrac{2}{15}, \dfrac{3}{10}, \dfrac{1}{6}$ 8. $\dfrac{19}{32}, \dfrac{5}{8}, \dfrac{9}{16}$

In Exercises 9–14, determine which symbol, > or <, should be used to make each statement true.

9. $5\dfrac{1}{2} \, ? \, 5\dfrac{13}{24}$

10. $3\dfrac{7}{18} \, ? \, 3\dfrac{1}{2}$

11. $10\dfrac{5}{8} \, ? \, 10\dfrac{7}{12}$

12. $25\dfrac{6}{25} \, ? \, 25\dfrac{4}{15}$

13. $\dfrac{1}{20}$ of 15 $? \, \dfrac{5}{16}$ of 2

14. $\dfrac{1}{4}$ of 10 $? \, \dfrac{2}{3}$ of 4

15. A stock price went from $12\dfrac{5}{8}$ to $12\dfrac{3}{4}$. Did it go up or down? By how much?

16. A stock price went from $16\dfrac{3}{8}$ to $16\dfrac{1}{4}$. Did it go up or down? By how much?

17. A salesperson measured the length of a drape to be $45\dfrac{3}{8}$ inches. Is the length closer to 45 or 46 inches?

18. A machinist measured the length of a steel rod to be $9\dfrac{17}{32}$ inches. Is the length closer to 9 or 10 inches?

Chapter 2   REVIEW

Definitions
2.1

A **fraction** is equivalent to a division problem.

$$\underset{\uparrow}{\underset{\text{Numerator}}{}}$$

$$\frac{a}{b} = a \div b$$

Numerator points to a; Denominator (cannot be zero) points to b.

A **proper fraction** is a fraction whose numerator is less than its denominator.

An improper fraction is a fraction whose numerator is larger than or equal to its denominator. Any improper fraction has a value greater than or equal to 1.

2.3   **A mixed number** is the sum of a whole number and a proper fraction.

2.6   **A prime number** is a natural number greater than 1 that can be divided evenly only by itself and 1. A prime number has no factors other than itself and 1.

A **composite number** is a natural number greater than 1 that can be divided evenly by some natural number other than itself and 1. A composite number has factors other than itself and 1.

2.15   **A simple fraction** is a fraction having only one fraction line.

A **complex fraction** is a fraction having more than one fraction line.

Changing Improper Fractions
to Mixed Numbers
2.3

To change an improper fraction to a mixed number, divide the numerator by the denominator. Write the quotient, followed by the fraction: remainder over divisor.

Changing Mixed Numbers
to Improper Fractions
2.4

To change a mixed number to an improper fraction:

1. Multiply the whole number by the denominator of the fraction.

2. Add the numerator of the fraction to the product found in step 1.

3. Write the sum found in step 2 over the denominator of the fraction.

Equivalent Fractions
2.5

To form an equivalent (equal) fraction, either

1. Multiply both numerator and denominator by the same nonzero number, *or*

2. Divide numerator and denominator by the same nonzero number that is a divisor of both.

Reducing Fractions
2.7A

To reduce a fraction to lowest terms:

1. Divide numerator and denominator by any whole number greater than 1 that you can see is a divisor of both.

2. When you cannot see any more common divisors, write the prime factorization of the numerator and the denominator.

3. Divide numerator and denominator by all factors common to both.

Finding a Fractional
Part of a Number
2.7B

To find a fractional part of a number, multiply the fraction times the number.

Finding the LCD
2.9

To find the LCD by prime factorization:

1. Write the prime factorization of each denominator. Repeated factors should be expressed as powers.

2. List each different factor that appears.

3. Raise each factor to the highest power to which it occurs in any denominator.

4. The LCD is the product of all the powers found in step 3.

Adding and Subtracting Fractions
2.10

To add or subtract fractions:

1. Find the LCD.

2. Convert each fraction to an equivalent fraction having the LCD as its denominator.

3. Add or subtract the numerators, and write the result over the common denominator.

4. Reduce the fraction to lowest terms.

$$\frac{a}{c} + \frac{b}{c} = \frac{a+b}{c} \qquad \frac{a}{c} - \frac{b}{c} = \frac{a-b}{c}$$

Multiplying Fractions
2.2, 2.13

To multiply fractions, write the product of the numerators over the product of the denominators:

$$\frac{a}{b} \cdot \frac{c}{d} = \frac{a \cdot c}{b \cdot d}$$

Dividing Fractions
2.14

To divide fractions, invert the second fraction and multiply:

$$\frac{a}{b} \div \frac{c}{d} = \frac{a}{b} \cdot \frac{d}{c}$$

Simplifying Complex Fractions
2.15

To simplify a complex fraction, simplify its numerator and denominator, then divide the simplified numerator by the simplified denominator.

Order of Operations
2.17

1. Any expressions in parentheses are evaluated first.

2. Evaluations are done in this order:
First: Powers and roots
Second: Multiplication and division in order from left to right
Third: Addition and subtraction in order from left to right

Comparing Fractions
2.18

To compare fractions, convert the given fractions to equivalent fractions having the same denominators.

Chapter 2   REVIEW EXERCISES

In Exercises 1–4, change the improper fractions to mixed numbers.

1. $\frac{7}{3}$ 2. $\frac{11}{8}$ 3. $\frac{28}{5}$ 4. $\frac{29}{12}$

In Exercises 5–8, change the mixed numbers to improper fractions.

5. $2\frac{3}{5}$ 6. $3\frac{7}{8}$ 7. $9\frac{2}{11}$ 8. $13\frac{5}{6}$

In Exercises 9–12, reduce to lowest terms.

9. $\frac{42}{54}$ 10. $\frac{45}{105}$ 11. $\frac{28}{57}$ 12. $\frac{200}{250}$

In Exercises 13–30, perform the indicated operations. Reduce answers to lowest terms. Write any improper fractions as mixed numbers.

13. $\frac{1}{2} + \frac{5}{6} + \frac{4}{9}$ 14. $2\frac{2}{3} + 1\frac{3}{5}$ 15. $4\frac{5}{16} + \frac{5}{8}$

16. $153\frac{2}{5} + 135\frac{3}{4}$ 17. $5\frac{4}{5} - 3\frac{7}{10}$ 18. $5\frac{1}{3} - 2\frac{3}{4}$

19. $16 - 5\frac{7}{8}$ 20. $351\frac{2}{3} - 272\frac{7}{9}$ 21. $\frac{4}{5} \cdot \frac{7}{8} \cdot 15$

22. $3\frac{2}{3} \cdot 6\frac{3}{5}$ 23. $1\frac{3}{5} \cdot 3\frac{3}{4}$ 24. $9 \cdot 1\frac{5}{12}$

25. $2\frac{2}{5} \div 1\frac{1}{15}$ 26. $16 \div \frac{8}{13}$ 27. $1\frac{9}{16} \div 10$

28. $\frac{7}{12} + 2\frac{2}{3} \cdot \frac{1}{4}$ 29. $\frac{2}{3} \div 1\frac{1}{4} \cdot \left(\frac{3}{4}\right)^2$

30. $\left(\frac{3}{10} + \frac{3}{4}\right) \cdot \sqrt{\frac{4}{9}} + \left(\frac{5}{6}\right)^0$

In Exercises 31–34, simplify the complex fractions.

31. $\dfrac{\frac{5}{8}}{\frac{5}{6}}$ 32. $\dfrac{\frac{12}{6}}{11}$ 33. $\dfrac{1\frac{2}{3}}{\frac{3}{4} + \frac{1}{2}}$ 34. $\dfrac{\frac{2}{3} + \frac{1}{2}}{\frac{8}{9} - \frac{1}{3}}$

In Exercises 35 and 36, arrange the fractions in order of size, largest to smallest.

35. $\dfrac{2}{5}, \dfrac{1}{4}, \dfrac{3}{10}$ 36. $\dfrac{5}{8}, \dfrac{2}{3}, \dfrac{5}{6}$

37. Which drill bit, $\dfrac{2}{3}$ inch, $\dfrac{5}{8}$ inch, or $\dfrac{11}{16}$ inch, will produce a hole that best fits a $\dfrac{3}{4}$-inch bolt?

38. A bookshelf is 42 inches long. How many books that are $\dfrac{3}{4}$ inch thick can stand on that shelf?

39. In driving across the country, a woman drove $4\dfrac{1}{2}$ hours on Monday, $12\dfrac{3}{4}$ hours on Tuesday, $8\dfrac{1}{3}$ hours on Wednesday, and $15\dfrac{1}{6}$ hours on Thursday. What was her total driving time for the trip?

40. A family used $10\dfrac{1}{2}$ square yards of carpet for the living room and $8\dfrac{3}{4}$ square yards for the hall and bedroom combined.

    a. What was the total amount of carpet used?

    b. If carpet cost $8 per square yard, how much did it cost them to carpet the living room, hall, and bedroom?

41. A satellite travels around the earth in $2\dfrac{1}{3}$ hours. How many trips does it make in 1 week?

42. A board that is 73 inches long is cut into three pieces of equal length. If $\dfrac{1}{8}$ inch is wasted each time the board is sawed, how long is each of the three finished pieces?

43. A man has a 16-foot board he is going to use for shelves. If the shelves are to be $3\dfrac{1}{2}$ feet long:

    a. How many shelves can be cut from the board? $\left(\text{Assume } \dfrac{1}{8} \text{ inch is wasted each time the board is sawed.}\right)$

    b. What length will be left from the original board after he cuts as many shelves from it as he can?

44. A seamstress is making the costumes for the high school drill team, which has 26 members. If $1\dfrac{1}{4}$ yards of material are needed for each skirt and $\dfrac{5}{8}$ yard of material is needed for each vest, how many yards of material should she order for the entire team?

45. A freight car can carry $10\dfrac{2}{3}$ tons of wheat. If the car is $\dfrac{3}{4}$ full, how many tons of wheat is it carrying?

46. A man weighing 200 pounds went on a diet and lost $3\dfrac{1}{2}$ pounds the first week, $2\dfrac{3}{4}$ pounds the second week, and $1\dfrac{5}{8}$ pounds the third week. How much did he weigh at the end of three weeks?

47. Princess had five puppies that weighed $2\dfrac{1}{4}$ pounds, $2\dfrac{1}{2}$ pounds, $1\dfrac{7}{8}$ pounds, $1\dfrac{3}{4}$ pounds, and $2\dfrac{1}{4}$ pounds. Find the average weight of her puppies.

48. Katrina spends $\dfrac{1}{3}$ of her income on rent, $\dfrac{1}{4}$ of her income on food, and $\dfrac{1}{3}$ of her income on car expenses. If she earns $18,000 a year, how much does she have left over after paying for rent, food, and her car?

49. The length of the side of a square is $1\dfrac{1}{4}$ feet.

    a. Find the perimeter of the square.

    b. Find the area of the square.

50. A rectangle is $4\dfrac{1}{2}$ meters long and $2\dfrac{3}{4}$ meters wide.

    a. Find the perimeter of the rectangle.

    b. Find the area of the rectangle.

Chapter 2 Critical Thinking and Writing Problems

Answer Problems 1–6 in your own words, using complete sentences.

1. Explain why 13 is a prime number.

2. Explain why 18 is a composite number.

3. Explain how you can tell whether a number is divisible by 3.

4. Explain how you can tell whether a number is divisible by 2.

5. Explain how to write a whole number as a fraction.

6. Explain how to compare the sizes of unlike fractions.

7. In the problem "find $\frac{5}{6}$ of 30," the word *of* indicates what operation? Find the answer to the problem.

8. True or False: Any proper fraction is less than any improper fraction. Explain your answer.

9. Make up a word problem that involves multiplying by $\frac{3}{4}$.

10. Make up a word problem that involves adding two mixed numbers.

11. Find the mystery $\boxed{\frac{?}{?}}$ fraction.

Clue 1: The numerator is an even prime number.

Clue 2: The denominator is less than 12.

Clue 3: The denominator is an odd composite number.

12. Find the mystery $\boxed{\frac{?}{?}}$ fraction.

Clue 1: The numerator is a digit divisible by 7.

Clue 2: The denominator is a digit divisible by 3.

Clue 3: The fraction is a proper fraction.

Each of the following problems has an error. Find the error, and, in your own words, explain why it's wrong. Then work the problem correctly.

13. $\dfrac{1}{\overset{}{\underset{5}{\cancel{10}}}} + \dfrac{\overset{2}{4}}{5} = \dfrac{3}{5}$

14. $4\dfrac{1}{2} \cdot 5\dfrac{1}{3} = 20\dfrac{1}{6}$

15. $\dfrac{\frac{5}{6}}{\frac{2}{3}} = \dfrac{2}{\underset{1}{3}} \cdot \dfrac{\overset{2}{6}}{5} = \dfrac{4}{5}$

Chapter 2   DIAGNOSTIC TEST

Allow yourself about 50 minutes to do these problems. Complete solutions for all problems, together with section references, are given in the answer section at the end of the book.

1. Change $\dfrac{69}{8}$ to a mixed number.

2. Change $5\dfrac{7}{12}$ to an improper fraction.

3. Reduce $\dfrac{180}{540}$ to lowest terms.

4. Find $\dfrac{7}{8}$ of 120.

In Problems 5–20, perform the indicated operations. Reduce all answers to lowest terms. Write any improper fractions as mixed numbers.

5. $\dfrac{5}{6} + \dfrac{7}{8}$

6. $4\dfrac{2}{3} + 3\dfrac{1}{2}$

7. $\dfrac{5}{6} \cdot \dfrac{3}{20}$

8. $1\dfrac{1}{3} \cdot 42$

9. $\dfrac{3}{8} \div \dfrac{9}{16}$

10. $\dfrac{7}{10} - \dfrac{1}{6}$

11. $8 - 3\dfrac{4}{9}$

12. $2\dfrac{2}{9} \div 3\dfrac{1}{3}$

13. $4\dfrac{1}{5} \cdot 2\dfrac{1}{7}$

14. $5\dfrac{1}{4} - 2\dfrac{5}{6}$

15. $\left(\dfrac{5}{6}\right)^2 - \sqrt{\dfrac{1}{16}}$

16. $\dfrac{5}{12} + \dfrac{3}{8} + \dfrac{5}{6}$

17. $\dfrac{\frac{5}{8}}{\frac{15}{16}}$

18. $\dfrac{\frac{3}{8} + \frac{3}{4}}{7\frac{1}{2}}$

19. $\dfrac{4}{5} \div 2\dfrac{2}{3} \cdot \left(\dfrac{2}{3}\right)^2$

20. $\dfrac{3}{4} + \dfrac{1}{4} \cdot 1\dfrac{3}{5}$

21. When a $1\frac{1}{8}$-pound steak was trimmed of fat, it weighed $\frac{3}{4}$ pound. How much of the steak was fat?

22. How many tablets, each containing 3 milligrams of medicine, must be used to make up a $7\frac{1}{2}$-milligram dosage?

23. Find the area of a rectangle with length $5\frac{1}{4}$ feet and width $3\frac{1}{3}$ feet.

24. Kevin received three shipments that weighed $23\frac{1}{2}$ pounds, $16\frac{1}{4}$ pounds, and $37\frac{7}{8}$ pounds. What was the total weight of the shipments?

25. Arrange in order, largest to smallest: $\frac{8}{15}, \frac{7}{10}, \frac{3}{5}$.

Decimal Fractions

CHAPTER 3

I n this chapter we introduce decimal numbers and show how to perform the basic operations on decimals. We will discuss solving applied problems involving decimal numbers that occur in our daily lives.

3.1   Reading and Writing Decimal Numbers

A **decimal fraction** is a fraction whose denominator is a power of 10.

EXAMPLE 1

a. $\dfrac{3}{10^1} = \dfrac{3}{10}$ Read "three tenths"

b. $\dfrac{3}{10^2} = \dfrac{3}{100}$ Read "three hundredths"

c. $\dfrac{25}{10^3} = \dfrac{25}{1,000}$ Read "twenty-five thousandths"

d. $\dfrac{5}{10^0} = \dfrac{5}{1} = 5$

e. $\dfrac{0}{10^0} = \dfrac{0}{1} = 0$

Examples 1d and 1e show that whole numbers are also decimal fractions.

A decimal fraction can be written in two ways: in fraction form or in decimal form using a decimal point.

EXAMPLE 2

Fraction form		*Decimal form*	
a. $\dfrac{4}{10}$	=	0.4	Read "four tenths"
b. $\dfrac{5}{100}$	=	0.05	Read "five hundredths"
c. $\dfrac{27}{1,000}$	=	0.027	Read "twenty-seven thousandths"

Whole numbers are decimal fractions. When the decimal point is not written in a whole number, it is understood to be to the right of the units digit.

EXAMPLE 3

Whole number		*Decimal form*
a. 4	=	4.
b. 130	=	130.

Although we must never lose sight of the fact that decimal fractions are *fractions*, it's common practice to shorten the term *decimal fraction* to *decimal*. In most cases, we will use the term *fraction* for fractions having denominators other than powers of 10.

In Figure 1 we show place-values of decimals. Note that the decimal point is written between the units place and the tenths place.

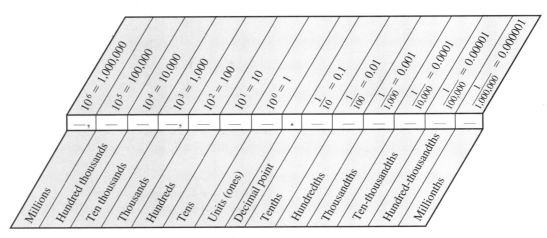

Figure 1 Place-values of decimals

 Note In Figure 1 we can see that the value of each place is one-tenth the value of the first place to its left. In addition, we can see that the place names to the right of the decimal point all end in "ths." The place names to the left of the decimal point do not end in "ths." For example, the second place to the left of the units place is the "hundreds" place. The second place to the right of the units place is the "hundred ths " place.

Reading a decimal

1. Read the number to the left of the decimal point as you read a whole number.
2. Say "and" for the decimal point.
3. Read the number to the right of the decimal point as a whole number, then say the name of the place occupied by the right-hand digit of the number.

Reading and Writing Decimals Less Than One

EXAMPLE 4

a. 0. 5 is read "five tenths."

        └── Tenths place

b. 0.0 6 is read "six hundredths."

        └── Hundredths place

c. 0.00 7 is read "seven thousandths."

        └── Thousandths place

d. 0.5 0 is read "fifty hundredths."

        └── Hundredths place

e. 0.50 0 is read "five hundred thousandths."

        └── Thousandths place

f. 0.56 7 is read "five hundred sixty-seven thousandths."

        └── Thousandths place

In writing decimals less than 1, we usually write a 0 to the left of the decimal point to call attention to the decimal point so that it is not overlooked. For example, seventy-five hundredths is written 0.75. However, both 0.75 and .75 are correct ways of writing seventy-five hundredths.

Reading and Writing Decimals Greater Than One

EXAMPLE 5

a. 6.27 is read "six *and* twenty-seven hundredths."

and indicates the decimal point

b. 175.006 is read "one hundred seventy-five and six thousandths."

c. 8.0000 4 is read "eight and four hundred-thousandths."

Hundred-thousandths place

d. 8.400 is read "eight and four hundred thousandths."

e. 107,060.756 is read "one hundred seven thousand, sixty, and seven hundred fifty-six thousandths."

 Note Examples 5c and 5d show the importance of the hyphen in the word "hundred-thousandths": "Eight and four hundred-thousandths" represents 8.00004, but "eight and four hundred thousandths" represents 8.400. The hyphen in the word "ten-thousandths" is similarly important.

Writing Numbers in Decimal Notation

EXAMPLE 6

a. Twenty and nine tenths

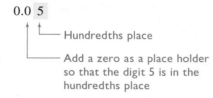

The *and* indicates the decimal point

20. 9

Tenths place

b. Five hundredths

0.0 5

Hundredths place

Add a zero as a place holder so that the digit 5 is in the hundredths place

c. Four hundred six thousandths

0.40 6

Thousandths place

d. Four hundred and six thousandths

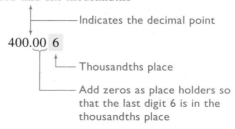

Indicates the decimal point

400.00 6

Thousandths place

Add zeros as place holders so that the last digit 6 is in the thousandths place

An Easy Way to Remember the Name of a Decimal Place   To find the name of a decimal place, we write a 0 for each decimal place in the number, then precede the zero(s) by a 1.

EXAMPLE 7

a. 0.0 *X*
   |
   10 0 ——— Hundredths place

b. 0. *X*
   1 0 ——— Tenths place

c. 0.00 *X*
   | |
   1,00 0 ——— Thousandths place

d. 0.000 00 *X*
   | | | | | |
   1,000,00 0 ——— Millionths place

Exercises 3.1

1. In the number 0.3456:

   a. What digit is in the tenths place?

   b. What digit is in the thousandths place?

   c. What digit is in the ten-thousandths place?

2. In the number 0.15039:

   a. What digit is in the hundredths place?

   b. What digit is in the thousandths place?

   c. What digit is in the hundred-thousandths place?

3. In the number 249.175:

   a. What digit is in the hundreds place?

   b. What digit is in the hundredths place?

   c. What digit is in the thousandths place?

4. In the number 473.28:

   a. What digit is in the tens place?

   b. What digit is in the tenths place?

   c. What digit is in the hundredths place?

In Exercises 5–16, write the numbers in words.

5. 0.35

6. 0.054

7. 3.016

8. 7.08

9. 0.0004

10. 0.070

11. 20.900

12. 860.03

13. 9,000.50

14. 5,006

15. 0.5006

16. 9.00275

In Exercises 17–30, write the numbers in decimal notation.

17. Nine hundredths

18. Eighteen thousandths

19. Two and three thousandths

20. Twelve and six ten-thousandths

21. Four hundred ten-thousandths

22. Five hundred thousandths

23. Sixty and eight hundredths

24. Sixty-eight hundredths

25. Seven hundred twenty thousandths

26. Seven hundred and twenty thousandths

27. Three thousand and fifty-five hundredths

28. Three thousand fifty and five hundredths

29. One hundred and four ten-thousandths

30. One hundred four ten-thousandths

3.2   Rounding Off Decimals

Earlier we explained rounding off whole numbers and why it is done. There is also a need for rounding off decimals. For example, since the smallest coin we have is the 1¢ piece, any calculations done with numbers representing money are usually rounded off to the nearest cent. When figuring federal income taxes, we are permitted to round off our calculations to the nearest dollar to make it easier to figure and pay our taxes. Some measuring instruments are accurate to thousandths of an inch or tenths of a foot. For this reason, we usually round off measurements to the accuracy of the instruments used.

Rounding off a decimal	**1.** Locate the digit in the round-off place. (We will circle the digit.)
	2. If the first digit to the right of the round-off place is less than 5, the digit in the round-off place is *unchanged*.
	If the first digit to the right of the round-off place is 5 or more, the digit in the round-off place is *increased by 1*.
	3. Digits to the *left* of the round-off place are unchanged (special case: Examples 5 and 7).
	4. Digits to the *right* of *both* the round-off place and the decimal point are dropped.
	5. Digits to the *right* of the round-off place and the *left* of the decimal point are replaced by zeros.

E X A M P L E   1   Round off 3.249 to the nearest tenth.

S O L U T I O N

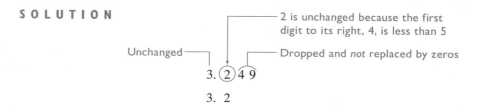

Therefore, 3.249 ≈ 3.2, rounded off to tenths.

 Note   The symbol ≈ means "is approximately equal to." We use this symbol to show that we are rounding off the number.

E X A M P L E   2   Round off 473.28 to the nearest ten.

S O L U T I O N

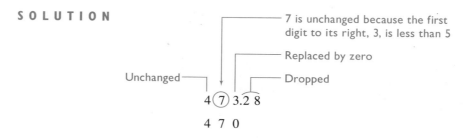

Therefore, 473.28 ≈ 470, rounded off to tens.

EXAMPLE 3 Round off 2.4856 to the nearest hundredth.

SOLUTION

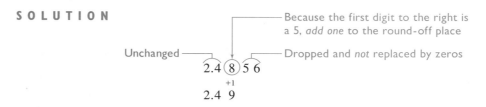

Therefore, 2.4856 ≈ 2.49, rounded off to hundredths.

EXAMPLE 4 Round off 82,674.153 to the nearest hundred.

SOLUTION

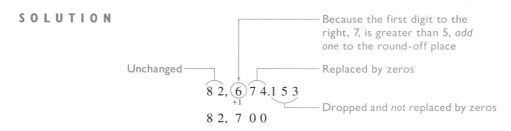

Therefore, 82,674.153 ≈ 82,700, rounded off to hundreds.

EXAMPLE 5 Round off 64.982 to the nearest tenth.

SOLUTION

Because the first digit to the right, 8, is greater than 5, *add one* to the round-off place; this changes 64.9 to 65.0

$$6\,4.\,\textcircled{9}\,8\,2$$
$${}^{+1}$$
$$6\,5.\,0$$

Therefore, 64.982 ≈ 65.0, to the nearest tenth. The zero in the tenths place must be written to show that the number has been rounded to the nearest tenth.

The **number of decimal places** in a number is the number of digits written to the right of the decimal point. A whole number has no decimal places.

EXAMPLE 6
 a. 0.25 has two decimal places.
 b. 0.054 has three decimal places.
 c. 14.5 has one decimal place.
 d. 0.5000 has four decimal places.
 e. 167. has no decimal places.

EXAMPLE 7   Round off 3.9996 to three decimal places.

SOLUTION

Because the first digit to the right, 6, is greater than 5, *add one* to the round-off place

$$3.9\ 9\ \underset{+1}{\textcircled{9}}\ 6$$

Dropped and *not* replaced by zeros

$$4.0\ 0\ 0$$

Therefore, 3.9996 ≈ 4.000, rounded off to three decimal places.

 A Word of Caution   Do not accumulate rounding offs. For example, when rounding off 1.7149 to the nearest hundredth, do not round off 1.7149 to 1.715, then round off 1.715 to 1.72. Actually, 1.7149 ≈ 1.71, rounded off to hundredths.

Exercises 3.2

Round off each number to the indicated place.

1.	7.16	tenths
2.	0.324	hundredths
3.	9.028	hundredths
4.	0.06372	thousandths
5.	427.301	hundreds
6.	804.49	tens
7.	0.06034	ten-thousandths
8.	0.8265	hundredths
9.	3.2096	thousandths
10.	56.97	tenths

11.	106.63	units
12.	0.704	hundredths
13.	494.949	tens
14.	494.949	tenths
15.	0.0098234	hundred-thousandths
16.	0.167761	ten-thousandths
17.	0.03745	3 decimal places
18.	0.583	1 decimal place
19.	0.0975	2 decimal places
20.	0.019374	4 decimal places

3.3   Adding Decimals

Adding decimals

1. Write the numbers under one another, with their decimal points in a vertical line.
2. Add the numbers the same way you add whole numbers.
3. Place the decimal point in the answer (sum) in the vertical line with the decimal points in the addends.

Writing the numbers clearly and keeping the columns straight helps to reduce the number of addition errors.

EXAMPLE 1   Add 75.4 + 186 + 0.056 + 1.207 + 2,350.

SOLUTION

$$
\begin{array}{r}
\text{Thousands} \\
\text{Hundreds} \\
\text{Tens} \\
\text{Units} \\
\leftarrow\text{Decimal point} \\
\text{Tenths} \\
\text{Hundredths} \\
\text{Thousandths}
\end{array}
$$

```
       7 5 . 4
     1 8 6 .
       0 . 0 5 6
       1 . 2 0 7
  + 2,3 5 0 .
  ─────────────
    2,6 1 2 . 6 6 3
```

You may find it easier to keep the columns straight when zeros are written in the open spaces, as shown:

```
   2 1      1
      7 5 . 4 0 0
    1 8 6 . 0 0 0
    0 0 0 . 0 5 6
    0 0 1 . 2 0 7
  + 2,3 5 0 . 0 0 0
  ─────────────────
    2,6 1 2 . 6 6 3
```

The method for adding decimals shown in Example 1 works because *only like things can be added directly*.

EXAMPLE 2   Add 1 apple + 2 pears.

SOLUTION   These cannot be added because they are *not* like things.

EXAMPLE 3   Add 3 apples + 5 apples.

SOLUTION   These can be added because they *are* like things. The sum is 8 apples.

EXAMPLE 4   Add 12.6 + 3.24.

SOLUTION   If the decimal points are not in a vertical line, the addition cannot be done.

```
    1 2 . 6     ← 6 tenths + 4 hundredths
  + 3 . 2 4       These cannot be added directly
        ↑
        └── Notice that the decimal points
            are not in a vertical line
```

But when the decimal points are in a vertical line,

$$
\begin{array}{r}
1\,2.\,\boxed{6} \\
+\quad 3.2\,4 \\
\hline
1\,5.\,8\,4
\end{array}
$$

← 6 tenths + 2 tenths = 8 tenths
These can be added directly

That is why we place the decimal points in a vertical line

Exercises 3.3

In Exercises 1–8, find the indicated sums.

1. $6.5 + 0.66 + 80.75 + 287 + 0.078$

2. $100 + 20 + 7 + 0.6 + 0.09 + 0.008$

3. $\$0.35 + \$24.79 + \$127.50 + \$18.84 + \$96$

4. $\$0.85 + \$286.83 + \$7.89 + \$46 + \$19.95$

5. $75.5 + 3.45 + 180 + 0.0056$

6. $185 + 35.06 + 0.186 + 0.0007$

7. $987.46 + 35.778 + 1{,}750.46 + 706.188 + 7{,}556.189$

8. $75{,}000 + 398.46 + 79.06 + 5.0789 + 186{,}300 + 35.45$

9. Mrs. Ramirez spent the following for lunch: Monday, $5.33; Tuesday, $7.47; Wednesday, $3.89; Thursday, $6.28; and Friday, $4.65.

   a. Estimate the total spent for lunch.

   b. How much did she spend for lunch that week?

10. Frank checked his gasoline credit slips after making a short trip and found that he had used the following amounts of gasoline: 11.2 gallons, 10.8 gallons, 14.1 gallons, 6.7 gallons, and 9.4 gallons. How many gallons did he use for the trip?

11. Find the sum of the following numbers: three thousand, fifty, and thirty-seven hundredths; five and two hundred-thousandths; seventy and one hundred fifty ten-thousandths.

12. a. Estimate the perimeter of the triangle.

    b. Find the exact perimeter of the triangle.

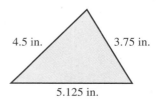

4.5 in.   3.75 in.

5.125 in.

3.4   Subtracting Decimals

Subtracting decimals

1. Write the number being subtracted (*subtrahend*) under the number it is being subtracted from (*minuend*), with their decimal points in the same vertical line. If the minuend has fewer decimal places than the subtrahend, attach zeros to the minuend so that both numbers have the same number of decimal places (see Example 3).

2. Subtract the numbers the same way you subtract whole numbers.

3. Place the decimal point in your answer (*difference*) in the vertical line with the other decimal points.

EXAMPLE 1   Subtract 6.07 from 75.14.

SOLUTION

$$
\begin{array}{r}
{\scriptstyle 6\ \ 15\ \ 0\ 14} \\
7\!\!\!/5.\!\!\!/1\,4 \\
-\quad 6.0\,7 \\
\hline
6\,9.0\,7
\end{array}
$$

EXAMPLE 2 Subtract 121.6 from 304.178.

SOLUTION

$$
\begin{array}{r}
\overset{2}{\cancel{3}}\overset{10}{\cancel{0}}\overset{3}{\cancel{4}}.\overset{11}{\cancel{1}}78 \\
-\ 1\ 2\ 1.6 \\
\hline
1\ 8\ 2.5\ 7\ 8
\end{array}
$$

EXAMPLE 3 Subtract 1.654 from 38.6.

SOLUTION

$$
\begin{array}{r}
\overset{7}{\cancel{3}}\overset{15\ 9}{8}.\overset{5\ 10\ 10}{6\ 0\ 0} \\
-\ 1.6\ 5\ 4 \\
\hline
3\ 6.9\ 4\ 6
\end{array}
$$

← Attach zeros to the right of 6, in the hundredths and thousandths places

The method for subtracting decimals shown in Examples 1, 2, and 3 works because *only like things can be subtracted directly.*

EXAMPLE 4 Subtract 1 apple from 2 pears.

SOLUTION These cannot be subtracted directly because they are *not* like things.

EXAMPLE 5 Subtract 2 apples from 5 apples.

SOLUTION These can be subtracted directly because they *are* like things. The difference is 3 apples.

EXAMPLE 6 Subtract 3.24 from 12.8.

SOLUTION

$$
\begin{array}{r}
\overset{0}{\cancel{1}}\overset{12}{2}.\overset{7\ 10}{8\ 0} \\
-\ 3.2\ 4 \\
\hline
9.5\ 6
\end{array}
$$

← 10 hundredths − 4 hundredths = 6 hundredths

Exercises 3.4

In Exercises 1–10, find the differences.

1. $7.85 - 3.44$

2. $84.07 - 0.66$

3. $208.5 - 7.16$

4. $715.75 - 28.19$

5. $300 - 0.145$

6. $7,000 - 3.68$

7. $81,284.56 - 2,784.8$

8. $2,000,046 - 30,015.8$

9. $5.785 - 0.9665$

10. $6.005 - 0.8476$

11. Subtract 46.8 from 224.

12. Subtract 2.093 from 187.5

13. Mrs. Geller's bank statement showed a balance of $254.39 at the beginning of the month. During the month she made the following deposits: $183.50, $233.75, and $78.86. During the month she wrote the following checks: $27.15, $86.94, $123.47, $167.66, $122.20, and $38.67. Find her balance at the end of the month.

14. When two dragsters raced, the first car's time was 6.05 seconds, and the second car's time was 6.375 seconds. Find the difference in their times.

3.5   Multiplication of Decimals

Multiplying decimals	1. Multiply the numbers as if they were whole numbers.
	2. Add the number of decimal places in the two numbers being multiplied.
	3. Place the decimal point in the answer so that the answer has as many decimal places as the *sum* found in step 2.

In Example 1, we show why the number of decimal places in the product is equal to the sum of the decimal places of the numbers being multiplied.

EXAMPLE 1   Multiply 0.2 by 0.003.

SOLUTION

$$0.2 \times 0.003 = \frac{2}{10} \times \frac{3}{1{,}000} = \frac{6}{10{,}000} = 0.0006$$

1	+	3	= 4
decimal		decimal	decimal
place		places	places

EXAMPLE 2   Multiply 0.035 by 0.25.

SOLUTION

```
        0.2 5        2 decimal places
  ×   0.0 3 5      + 3 decimal places
        1 2 5        5
        7 5
  0.0 0 8 7 5 ←—— 5 decimal places
```

Since changing the order of the numbers being multiplied does not change the product, it is usually easier to use the number with fewer nonzero digits as the multiplier. We show this by working Example 3 two different ways.

EXAMPLE 3   Multiply 4.6 by 3.749.

SOLUTION 1

```
      3.7 4 9
  ×         4.6
    2 2 4 9 4
  1 4 9 9 6
  1 7.2 4 5 4
```

SOLUTION 2

```
          4.6
  ×   3.7 4 9
        4 1 4
      1 8 4
    3 2 2
  1 3 8
  1 7.2 4 5 4
```

EXAMPLE 4   Multiply 0.86 by 18,000.

SOLUTION   We handle final zeros the same way we did earlier.

$$
\begin{array}{r}
0.8\,6 \\
\times\quad 1\,8|0\,0\,0 \\
\hline
6\,8\,8 \\
8\,6 \\
\hline
1\,5{,}4\,8|0.0\,0
\end{array}
$$

2 decimal places
+ 0 decimal places
2
← 2 decimal places

EXAMPLE 5   If gasoline costs 92.9 cents per gallon, (a) estimate the cost of 17.5 gallons of gasoline, and (b) find the exact cost of 17.5 gallons of gasoline.

SOLUTION   First, because the answer to this problem will be in dollars and cents, we will rewrite 92.9 cents as $0.929.

a. To estimate the answer, we round off each number at the first nonzero digit, then perform the operation. Round off $0.929 to $0.9 and 17.5 to 20, then multiply:

$$\$0.9 \times 20 = \$18.00$$

b.
$$
\begin{array}{r}
0.9\,2\,9 \\
\times\quad 1\,7.5 \\
\hline
4\,6\,4\,5 \\
6\,5\,0\,3 \\
9\,2\,9 \\
\hline
1\,6.2\,5\,7\,5
\end{array}
$$

3 decimal places
+ 1 decimal place
4
← 4 decimal places

$\approx \$16.26$ Rounded to the nearest cent

Exercises 3.5

In Exercises 1–16, find the products.

1. 0.03×0.8
2. 0.9×0.007
3. 0.04×0.06
4. 0.005×0.03
5. 42×0.3
6. 0.06×94
7. 0.17×0.8
8. 0.029×0.5
9. 3.6×0.75
10. 8.4×9.5
11. 1.06×0.37
12. 0.41×20.4
13. 500×0.79
14. 0.26×800
15. $0.19 \times 43{,}000$
16. $6{,}800 \times 6.7$

In Exercises 17–20, estimate the products.

17. 0.875×0.39
18. 0.0718×0.56
19. 0.685×42.5
20. 0.00934×206

21. Marcia earns $6.15 an hour.

   a. Estimate her earnings if she worked 26 hours.

   b. How much did she earn if she worked 26 hours?

22. If gasoline sells for 87.9 cents a gallon:

   a. Estimate the cost of 13.2 gallons of gasoline.

   b. Find the cost of 13.2 gallons of gasoline.

23. A rectangular playground measures 12.5 meters long and 8.6 meters wide.

   a. Estimate the area of the playground.

   b. What is the area of the playground?

   c. Find the cost to blacktop the playground if the blacktop costs $20.50 per square meter.

24. Miguel is planning a square garden that measures 5.8 meters on each side.

   a. Estimate the perimeter of the garden.

   b. What is the perimeter of the garden?

   c. Find the cost to fence in the garden if fencing costs $11.90 per meter.

25. If a pipe weighs 3.2 pounds per foot, find the weight of a pipe 16 feet long.

26. Irv makes $12.68 an hour.

    a. What is his salary for a regular 40-hour week?

    b. Irv receives time-and-a-half for each hour over 40 that he works in 1 week. How much does he make for working a 50-hour week?

27. A car rental agency charges $12.00 a day and 15 cents a mile to rent their cars. How much would it cost to rent a car for a week if you drove 875 miles?

28. The Cliffton County Telephone Company charges $8.25 a month plus 18 cents for each call over 30 during the month. What is the monthly phone bill for 57 calls made that month?

3.6 Division of Decimals

3.6A Dividing a Decimal by a Whole Number

Dividing a decimal by a whole number	**1.** Place a decimal point above the quotient line directly above the decimal point in the dividend. **2.** Divide the numbers the same way you divide whole numbers.

EXAMPLE 1

Divide 150.4 by 47.

SOLUTION

$$
\begin{array}{r}
3.2 \\
47\overline{)150.4} \\
\underline{141} \\
94 \\
\underline{94}
\end{array}
$$

EXAMPLE 2

Divide 48.4 by 85. (Round off your answer to two decimal places.)

SOLUTION   When rounding off, we carry out the division to *one more place* than required, then round off.

$$
\begin{array}{r}
.569 \approx 0.57 \\
85\overline{)48.400} \\
\underline{425} \\
590 \\
\underline{510} \\
800 \\
\underline{765} \\
35
\end{array}
$$
The division has been carried out to three decimal places, then rounded off to two decimal places

EXAMPLE 3

Scott swam the 100-meter freestyle in 54.23 seconds, 59.60 seconds, and 57.26 seconds. What was his average time?

SOLUTION

$$
\text{Average} = \frac{54.23 + 59.60 + 57.26}{3}
$$

$$
= \frac{171.09}{3} = 3\overline{)171.09} \quad \begin{array}{r} 57.03 \text{ sec} \end{array}
$$

Exercises 3.6A

In Exercises 1–10, the quotients are exact. Do not round off the quotients.

1. $86.96 \div 8$
2. $249.2 \div 7$
3. $93.6 \div 6$
4. $673.2 \div 9$
5. $33.6 \div 32$
6. $43.26 \div 21$
7. $6.825 \div 39$
8. $28.13 \div 58$
9. $4.977 \div 63$
10. $311.1 \div 85$

In Exercises 11–20, divide, and round off each quotient to the indicated place.

11. $8.56 \div 7$ 2 decimal places
12. $456.7 \div 9$ 1 decimal place
13. $376.3 \div 8$ 3 decimal places
14. $514.7 \div 6$ 3 decimal places
15. $58.6 \div 42$ tenths
16. $75.4 \div 51$ hundredths
17. $3.86 \div 76$ thousandths
18. $5.77 \div 84$ ten-thousandths
19. $76.5 \div 208$ hundredths
20. $90.6 \div 555$ thousandths

21. Find the average of 6.3, 8.4, 10.2, and 9.7.

22. Find the average of 0.12, 0.21, 0.34, 0.15, and 0.22.

23. The rainfall in Smalltown for five weeks was 2 inches, 1.6 inches, 3.1 inches, 4.7 inches, and 0.5 inch. What was the average rainfall for the five weeks? Round off the answer to the nearest tenth of an inch.

24. William ran the 400-meter hurdles in 56.28 seconds, 52.93 seconds, and 54.67 seconds. How fast will he need to run the next race to average 54.25 seconds for all four races?

25. Terry bought three tapes for $8.50 each and two tapes for $6.95. Find the average cost of the tapes.

26. Virginia bought four birthday cards for $1.75, one card for $2.25, and one card for $3.65. What was the average cost of the cards?

3.6B Dividing a Decimal by a Decimal

In our study of fractions, we learned that the value of a fraction is not changed when the numerator and denominator are *multiplied* by the same number.

$$\frac{5}{8} = \frac{5 \times 10}{8 \times 10} = \frac{50}{80}$$

We know that a fraction is equivalent to a division; therefore,

$$0.7\overline{)16.8} = 16.8 \div 0.7 = \frac{16.8 \times 10}{0.7 \times 10} = \frac{168}{7} = 7\overline{)168}$$

This shows that $0.7\overline{)16.8} = 7.\overline{)168}$. Therefore, *the quotient is unchanged when the decimal point in the divisor and dividend are both moved the same number of places to the right.* Using carets ($_\wedge$), we would rewrite the above example as

$$0.7\overline{)1\,6.8} = 0.7_\wedge\overline{)1\,6.8_\wedge}$$

These carets are used to indicate the new positions of the decimal points

$$
\begin{array}{r}
2\,4. \\
0.7_\wedge\overline{)1\,6.8_\wedge} \\
\underline{1\,4} \\
2\,8 \\
\underline{2\,8}
\end{array}
$$

The divisor has become 7 instead of 0.7

It is also true that the value of a fraction is not changed when the numerator and denominator are *divided* by the same number. Therefore, *the quotient is unchanged when the decimal points in the divisor and dividend are both moved the same number of places to the left.*

EXAMPLE 4

Divide 166.4 by 40.

SOLUTION

$$166.4 \div 40 = 40\overline{)166.4} = \frac{166.4}{40} = \frac{166.4 \div 10}{40 \div 10} = \frac{16.64}{4} = 4\overline{)16.64}$$

Both numerator and denominator *divided* by 10

We will write $40\overline{)166.4} = 4_\wedge 0\overline{)16_\wedge 6.4}$

These carets are used to indicate the new positions of the decimal points

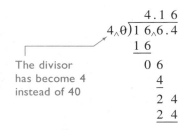

The divisor has become 4 instead of 40

```
          4 . 1 6
  4ᴧ0)1 6ᴧ6 . 4
      1 6
        0 6
        4
        2 4
        2 4
```

Dividing a decimal by a decimal

1. Place a caret (ᴧ) to the right of the last nonzero digit of the divisor.
2. Count the number of places between the decimal point and the caret in the divisor.
3. Place a caret in the dividend the same number of places to the right (or left) of its decimal point as counted in step 2. Attach zeros to the dividend when needed.
4. Place a decimal point in the quotient directly above the caret in the dividend.
5. Divide the numbers the same way you divide whole numbers.

EXAMPLE 5

Divide 2.368 by 0.32.

SOLUTION

```
              7 . 4
  0 . 3 2ᴧ)2 . 3 6ᴧ8
      2 2 4
        1 2 8
        1 2 8
```

EXAMPLE 6

Divide 0.144 by 1.20.

SOLUTION

```
                    . 1 2
  1 . 2ᴧ0)0 . 1ᴧ4 4
              1 2
                2 4
                2 4
```

The divisor has become 12

Example 6 shows that *we place the caret so that we make the divisor the smallest possible whole number.*

EXAMPLE 7 Divide 3.51 by 0.065.

SOLUTION

$$
\begin{array}{r}
5\,4. \\
0.065_\wedge\overline{)3.510_\wedge} \\
\underline{3\,25} \\
2\,6\,0 \\
\underline{2\,6\,0}
\end{array}
$$

← This zero was added so that three places come between caret and decimal point

EXAMPLE 8 Divide 197.2 by 0.29.

SOLUTION

This zero was added to hold the place above the 0 of the dividend

$$
\begin{array}{r}
6\,8\,0. \\
0.29_\wedge\overline{)197.20_\wedge} \\
\underline{1\,7\,4} \\
2\,3\,2 \\
\underline{2\,3\,2}
\end{array}
$$

This zero was added so that two places come between caret and decimal point

EXAMPLE 9 Divide 0.2429 by 5.7, and round off the answer to three decimal places.

SOLUTION

This zero was added to hold the place above the 4

$$
\begin{array}{r}
.0\,4\,2\,6 \approx 0.043 \\
5.7_\wedge\overline{)0.2_\wedge4\,2\,9\,0} \\
\underline{2\,2\,8} \\
1\,4\,9 \\
\underline{1\,1\,4} \\
3\,5\,0 \\
\underline{3\,4\,2}
\end{array}
$$

This zero was added so that we could carry the division out to four decimal places, then round off to three places

EXAMPLE 10 Divide 7 by 4.1, and round off the answer to the nearest hundredth.

SOLUTION

$$
\begin{array}{r}
1.\,7\,0\,7 \approx 1.71 \\
4.1_\wedge\overline{)7.0_\wedge0\,0\,0} \\
\underline{4\,1} \\
2\,9\,0 \\
\underline{2\,8\,7} \\
3\,0\,0 \\
\underline{2\,8\,7}
\end{array}
$$

← These zeros were added so that we could carry the division out to the thousandths place, then round off to hundredths

Exercises 3.6B

In Exercises 1–16, the quotients are exact. Do not round off the quotients.

1. $9.612 \div 2.7$
2. $17.898 \div 3.8$
3. $478.24 \div 6.1$
4. $12.42 \div 9.2$
5. $51.2 \div 800$
6. $40.5 \div 500$
7. $0.0196 \div 1.40$
8. $0.242 \div 1.10$
9. $0.4077 \div 0.45$
10. $5.226 \div 0.65$
11. $2.0349 \div 0.057$
12. $1.6472 \div 0.29$
13. $2,173 \div 2.65$
14. $6,120 \div 8.16$
15. $1.0488 \div 0.368$
16. $7.3143 \div 0.774$

In Exercises 17–26, divide, and round off each quotient to the indicated place.

17. $7.68 \div 2.90$ hundredths
18. $0.0089 \div 0.056$ thousandths

19. $12 \div 0.39$ tenths

20. $29 \div 1.4$ hundredths

21. $0.041 \div 0.0064$ thousandths

22. $0.0023 \div 0.093$ thousandths

23. $0.25 \div 8.07$ 4 decimal places

24. $0.2 \div 35.9$ 5 decimal places

25. $66 \div 0.073$ 1 decimal place

26. $246 \div 0.816$ 2 decimal places

27. The balance owed on a car amounts to $2,467.14. What would the monthly payments have to be in order to pay it off in 3 years? Round off the payment to the nearest cent (two decimal places).

28. At the beginning of a trip, Raul's odometer reading was 65,479 miles. At the end of the trip the reading was 67,784 miles. He used 147 gallons of gasoline. How many miles did he get to the gallon of gasoline? Round off the answer to one decimal place.

29. How many pieces of ribbon 3.5 feet long can be cut from a spool 50 feet long?

30. If Alice paid $13.92 a month on her finance company loan, how long would it take her to pay off a balance of $250.56?

3.7   Multiplying and Dividing Decimals by Powers of Ten

Here are some powers of 10.

$$10^1 = 10, \qquad 10^2 = 100, \qquad 10^3 = 1,000$$

Multiplying by Powers of 10

EXAMPLE 1

a.
$$
\begin{array}{r}
5.3 \\
\times\,1\,0 \\
\hline
0\,0 \\
5\,3 \\
\hline
5\,3.0
\end{array}
$$

b.
$$
\begin{array}{r}
5.3 \\
\times\,1\,0\,0 \\
\hline
0\,0 \\
0\,0 \\
5\,3 \\
\hline
5\,3\,0.0
\end{array}
$$

Notice that when we multiply 5.3 by 10 or 100, the decimal point in the product moves to the right the same number of places as there are zeros in the power of ten.

Multiplying a decimal by a power of ten | Move the decimal point to the *right* as many places as the number of zeros in the power of 10. Attach zeros when needed.

EXAMPLE 2 Examples of multiplying a decimal by a power of ten:

a. $24.7 \times 10 = 24_\wedge 7. = 247$

    1 zero   1 place

b. $24.7 \times 100 = 2,4_\wedge 70. = 2,470$

    2 zeros   2 places

c. $1.0567 \times 1{,}000 = 1{\wedge}056.7 = 1{,}056.7$

3 zeros   3 places

d. $0.0973 \times 10^2 = 0{\wedge}09.73 = 9.73$

Exponent 2   2 places

e. $34 \times 10^4 = 34{\wedge}0000. = 340{,}000$

Exponent 4   4 places

Dividing by Powers of 10

EXAMPLE 3

a. $5.3 \div 10 = 10\overline{)5.30}$

$$\begin{array}{r} .53 \\ \hline 5.30 \\ 5\ 0 \\ \hline 3\ 0 \\ 3\ 0 \\ \hline \end{array}$$

b. $5.3 \div 100 = 100\overline{)5.300}$

$$\begin{array}{r} .053 \\ \hline 5.300 \\ 5\ 00 \\ \hline 3\ 00 \\ 3\ 00 \\ \hline \end{array}$$

When we divide 5.3 by 10 or 100, the decimal point in the quotient moves to the left the same number of places as there are zeros in the power of ten.

Dividing a decimal by a power of ten	Move the decimal point to the *left* as many places as the number of zeros in the power of 10. Attach zeros when needed.

EXAMPLE 4

Examples of dividing a decimal by a power of ten:

a. $395 \div 100 = 3.95_{\wedge} = 3.95$

2 zeros   2 places

b. $75.6 \div 10 = 7.5_{\wedge}6 = 7.56$

1 zero   1 place

c. $\dfrac{0.315}{100} = 0.00_{\wedge}315 = 0.00315$

2 zeros   2 places

d. $\dfrac{4{,}165.2}{10^3} = 4.165_{\wedge}2 = 4.1652$

Exponent 3   3 places

e. $75.6 \div 10^4 = 0.0075_{\wedge}6 = 0.00756$

Exponent 4   4 places

Exercises 3.7

In Exercises 1–16, perform the indicated operations.

1. $\dfrac{95.6}{10}$ 2. $7.98 \div 10$ 3. $573 \div 100$

4. $\dfrac{64.8}{100}$ 5. 27.8×100 6. 8.95×10

7. $1{,}000(0.2094)$ 8. $1{,}000(3.097)$ 9. $\dfrac{750.2}{10^2}$

10. $\dfrac{98.47}{10^2}$ 11. 9.846×10^2 12. 0.0837×10^3

13. $\dfrac{100}{10^3}$ 14. 200×10^3 15. $10^4 \times 27.4$

16. $0.48 \div 10^2$

17. The \$146.35 cost of a party was shared by 10 people. How much did each person have to pay? (Be sure to round the answer off to the nearest cent.)

18. A club charged \$1.50 for each lottery ticket. If 10,000 tickets were sold, how much money did the club raise from this lottery?

19. 537 people attended a \$100-a-plate fund-raising dinner given for a political candidate. How much campaign money did this dinner raise?

20. The \$23,758 cost of putting in electric service along a rural road was shared equally by 100 home owners. How much did each owner have to pay?

3.8 Powers and Roots of Decimals

Powers of Decimals

We can raise a decimal to a power the same way we raised whole numbers and fractions to a power. Recall that the exponent indicates repeated multiplication of the base.

EXAMPLE 1 Examples of finding powers:

a. $(0.2)^3 = 0.2 \times 0.2 \times 0.2 = 0.008$

$$1 + 1 + 1 = 3$$
decimal place + decimal place + decimal place = decimal places

b. $(1.1)^2 = 1.1 \times 1.1 = 1.21$

c. $(3.25)^0 = 1$ Any number raised to the zero power is 1

Square Roots of Decimals

Recall that finding the square root of a number is the inverse operation of squaring a number.

EXAMPLE 2 Examples of finding square roots:

a. $\sqrt{0.25} = 0.5$ Because $(0.5)^2 = 0.5 \times 0.5 = 0.25$

b. $\sqrt{0.01} = 0.1$ Because $(0.1)^2 = 0.1 \times 0.1 = 0.01$

c. $\sqrt{1.21} = 1.1$ Because $(1.1)^2 = 1.1 \times 1.1 = 1.21$

A Word of Caution A common error is to say that

$$\sqrt{0.25} = 0.05 \qquad \text{but} \qquad (0.05)^2 = 0.05 \times 0.05 = 0.0025$$

$$\underset{\substack{\text{decimal} \\ \text{places}}}{2} + \underset{\substack{\text{decimal} \\ \text{places}}}{2} = \underset{\substack{\text{decimal} \\ \text{places}}}{4}$$

The decimal point in a product is located at the *sum* of the decimal places in the numbers being multiplied. To obtain a two-decimal-place product, such as 0.25, we will need two one-decimal-place factors. Therefore,

$$\sqrt{0.25} = 0.5 \qquad \text{because} \qquad (0.5)^2 = 0.5 \times 0.5 = 0.25$$

$$\underset{\substack{\text{decimal} \\ \text{place}}}{1} + \underset{\substack{\text{decimal} \\ \text{place}}}{1} = \underset{\substack{\text{decimal} \\ \text{places}}}{2}$$

EXAMPLE 3

Examples of using a calculator to find square roots and powers:

a. Find $\sqrt{86.49}$, using a calculator.

    Key in   86.49 $\boxed{\sqrt{\;}}$

    Answer   9.3

b. Find $(8.3)^3$, using a calculator.

    Key in   8.3 $\boxed{y^x}$ 3 $\boxed{=}$

    Answer   571.787

Exercises 3.8

In Exercises 1–20, find the indicated powers and roots.

1. $(0.3)^2$ **2.** $(0.2)^2$ **3.** $(0.8)^2$ **4.** $(0.9)^2$

5. $(0.01)^2$ **6.** $(0.03)^2$ **7.** $(1.3)^2$ **8.** $(1.5)^2$

9. $(0.4)^3$ **10.** $(0.1)^4$ **11.** $(2.5)^0$ **12.** $(3.1)^0$

13. $\sqrt{0.04}$ **14.** $\sqrt{0.36}$ **15.** $\sqrt{1.69}$ **16.** $\sqrt{1.44}$

17. $\sqrt{0.0016}$ **18.** $\sqrt{0.0001}$

19. $(0.4)^2 + \sqrt{0.09}$ **20.** $\sqrt{0.64} - (0.6)^2$

21. A square measures 8.2 centimeters on each side.

    a. Estimate the area of the square.

    b. Find the area of the square.

22. A square acre measures approximately 69.6 yards on each side.

    a. Estimate the number of square yards in an acre.

    b. Find the number of square yards in an acre, and round off your answer to the nearest ten.

 In Exercises 23–30, use a calculator to evaluate each expression.

23. $(2.5)^4$ **24.** $(1.66)^3$ **25.** $(1.2)^5$ **26.** $(0.142)^2$

27. $\sqrt{153.76}$ **28.** $\sqrt{47.61}$ **29.** $\sqrt{0.6084}$ **30.** $\sqrt{0.006724}$

3.9   Combined Operations

In evaluating expressions with more than one operation, we use the same order of operations with decimals that we used with whole numbers and fractions.

Order of operations

1. Any expressions in parentheses are evaluated first.
2. Evaluations are done in this order
   First: Powers and roots
   Second: Multiplication and division in order from *left to right*
   Third: Addition and subtraction in order from *left to right*

EXAMPLE 1

$4.6 + 2.3 \times 5.4 - 8.6$ Multiply first

$= 4.6 + \quad 12.42 \quad - 8.6$ Addition and subtraction are done left to right

$= \qquad 17.02 \qquad - 8.6 = 8.42$

EXAMPLE 2

$54 \div 10 - (0.3)^2 \times 5$ Powers first: $(0.3)^2 = 0.3 \times 0.3 = 0.09$

$= 54 \div 10 - \quad 0.09 \quad \times 5$ Multiplication and division are done left to right

$= \quad 5.4 \quad - \quad 0.09 \quad \times 5$

$= \quad 5.4 \quad - \quad 0.45 \quad = 4.95$ Subtraction is last

---

EXAMPLE 3

Estimate the answer to $\dfrac{0.38 \times 63}{0.51}$.

SOLUTION   Round off 0.38 to 0.4, 63 to 60, and 0.51 to 0.5:

$$\frac{0.4 \times 60}{0.5} = \frac{24}{0.5}$$

Now we can round off 24 to 25 (because 25 is a *convenient* number to divide by 0.5), or we can round off 24 to 20 and then divide.

If we round off 24 to 25: $\dfrac{25}{0.5} = 0.5_{\wedge}\overline{)2\,5.0_{\wedge}} \;\; 5\,0.$ Both 50 and 40 are considered good estimates

If we round off 24 to 20: $\dfrac{20}{0.5} = 0.5_{\wedge}\overline{)2\,0.0_{\wedge}} \;\; 4\,0.$

EXAMPLE 4

Find $\dfrac{0.38 \times 63}{0.51}$, using a calculator. Round off the answer to one decimal place.

SOLUTION

Key in   .38 $\boxed{\times}$ 63 $\boxed{\div}$ .51 $\boxed{=}$

Answer $46.941176 \approx 46.9$

EXAMPLE 5

The Southern Electric Co. charges 7.95 cents per kWh for the first 250 kWh used and 12.2 cents for each kWh used over the first 250. Mr. Carson's electric meter read 22,607 on September 1 and 23,039 on October 1. (kWh means kilowatt-hour and is a measure of electrical energy. One kWh of electricity will light a 100-watt bulb for

10 hours.) (a) Find the amount of Mr. Carson's electric bill for September. (b) Find his daily average of kWh used for September (30 days).

S O L U T I O N First, find the number of kWh used by subtracting the meter readings.

Key in 23039 $\boxed{-}$ 22607 $\boxed{=}$ 432 ← kWh used

and $\boxed{-}$ 250 $\boxed{=}$ 182 ← kWh at higher rate

a. To find the cost of Mr. Carson's electric bill,

Key in 250 $\boxed{\times}$ .0795 $\boxed{+}$ 182 $\boxed{\times}$ .122 $\boxed{=}$

Answer 42.079 ≈ $42.08

b. To find the daily average, divide the total number of kWh used by the number of days.

Key in 432 $\boxed{\div}$ 30 $\boxed{=}$

Answer 14.4

Exercises 3.9

In Exercises 1–12, find the value of each expression using the correct order of operations.

1. $2.5 + 4.3 \times 0.8 - 1.5$ **2.** $12 \div 0.2 - 1.8 \times 4$

3. $7.06 - 0.4 \times 0.6 + (0.6)^2$ **4.** $18 + 2.4 \div 6 \times 0.4$

5. $9 \times \sqrt{0.16} + 75 \div 10$

6. $1.42 \times 10^3 - 4.65 \times 10^2$

7. $42 \div 1,000 + 0.057 \times 100$

8. $0.5(2.8 + 12)$

9. $46 - (5.6 - 0.34)$ **10.** $46 - 5.6 - 0.34$

11. $2.3 + 1.6 + 1.2 \div 3$ **12.** $(2.3 + 1.6 + 1.2) \div 3$

In Exercises 13–18, estimate the answer (estimates may vary).

13. $3.8 + 7.1 - 4.3$ **14.** $0.72 \times 0.394 \times 0.53$

15. $2(8.9 + 4.7)$ **16.** $\dfrac{0.405}{7.69}$

17. $\dfrac{2.1 \times 6.3}{0.292}$ **18.** $\dfrac{7.2 \times 4.8}{0.0613}$

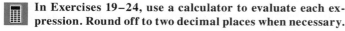 **In Exercises 19–24, use a calculator to evaluate each expression. Round off to two decimal places when necessary.**

19. $6.89 + 0.53 \times 7.34$ **20.** $\sqrt{39.69} - 3.51 \div 0.78$

21. $1,722(1 + 0.055)$ **22.** $750 \times 0.085 \times \dfrac{30}{365}$

23. $\dfrac{4.1 \times (1.5)^2}{0.625}$ **24.** $5,000\left(1 + \dfrac{0.8}{12}\right)$

25. A rectangle measures 8.65 meters long and 5.3 meters wide. What is the perimeter of the rectangle?

26. A door measures 2.5 feet wide and 6.5 feet high. Find the cost to replace the molding around the door if the molding costs $1.80 per foot. (Note that there is no molding at the bottom of the door.)

27. A taxi charges $2 for the first mile and $0.75 for each additional mile. What is the taxi fare for 6 miles?

28. A delivery company charges $12 plus $0.35 per pound. How much would it cost to send a 12-pound package?

 29. The gas company charges $0.365 per therm (a unit of heat) for the first 65 therms used and $0.785 for each therm used over the first 65. The gas meter read 9,876 on November 1 and 9,963 on December 1.

   a. What was the amount of the gas bill for November?

   b. Find the daily average of therms used for November (30 days).

   c. Find the average daily cost of gas for November.

 30. The electric company charges 6.78 cents per kWh for the first 280 kWh used and 10.9 cents for each kWh used over the first 280. The electric meter read 12,865 on May 1 and 13,237 on June 1.

   a. Find the amount of the electric bill for May.

   b. Find the daily average of kWh used for May (31 days).

   c. Find the average daily cost of electricity for May.

3.10   Changing a Fraction to a Decimal

Because

$$\frac{a}{b} = a \div b$$

we can change a fraction to a decimal as follows.

Changing a fraction to a decimal	Divide the numerator by the denominator.

Any fraction can be changed to either a **terminating decimal**, for which the division comes to an end (see Example 1), or a **repeating decimal**, in which a digit or a group of digits will repeat indefinitely (see Examples 2 and 3).

EXAMPLE 1   Change $\frac{5}{8}$ to a decimal.

SOLUTION

$$\frac{5}{8} = 5 \div 8 = 8\overline{)5.000}$$

```
        .6 2 5
  8)5.0 0 0
    4 8
      2 0
      1 6
        4 0
        4 0
```

EXAMPLE 2   Change $\frac{2}{11}$ to a decimal.

SOLUTION

$$\frac{2}{11} = 11\overline{)2.0000}$$

```
         .1 8 1 8...  ←— The three dots indicate that the
  11)2.0 0 0 0              decimal continues forever
     1 1
       9 0
       8 8
         2 0
         1 1
           9 0
           8 8
             2
```

The digits 1 and 8 repeat forever, and we may indicate this with a bar above the 1 and 8.

Therefore,

$$\frac{2}{11} = 0.1818\ldots$$

or

$$\frac{2}{11} = 0.\overline{18}$$

EXAMPLE 3   Change $\dfrac{1}{12}$ to a decimal.

SOLUTION

$$\dfrac{1}{12} = 12)\overline{\begin{array}{l} .0\,8\,3\,3\,3\ldots = 0.08\overline{3} \\ 1.0\,0\,0\,0\,0 \end{array}}$$

                   9 6

                     4 0

                     3 6

                       4 0

                       3 6

                         4 0

                         3 6

                           4

← The bar is over the 3 because only the digit 3 is repeating

Sometimes we are interested in an answer that has been rounded off. For example, numbers representing money are usually rounded off to the nearest cent. When rounding off, carry out the division to one more place than required, then round off.

EXAMPLE 4   One share of stock in a computer company is selling for $5\dfrac{3}{8}$ dollars. Find the cost of one share of stock rounded off to the nearest cent.

SOLUTION 1   Change $\dfrac{3}{8}$ to a decimal, then add its value to 5.

$$8)\overline{\begin{array}{l} .3\,7\,5 \\ 3.0\,0\,0 \end{array}}$$

        2 4

          6 0

          5 6

            4 0

            4 0

Therefore, $\dfrac{3}{8} \approx 0.38$. Then $5\dfrac{3}{8} \approx 5 + 0.38 = \$5.38$.

SOLUTION 2   Change $5\dfrac{3}{8}$ to the improper fraction $\dfrac{43}{8}$. Then change $\dfrac{43}{8}$ to a decimal.

$$5\dfrac{3}{8} = \dfrac{43}{8} = 8)\overline{\begin{array}{l} 5.3\,7\,5 \\ 4\,3.0\,0\,0 \end{array}}$$

          4 0

          3 0

          2 4

            6 0

            5 6

             4 0

             4 0

Therefore, $5\dfrac{3}{8} \approx \$5.38$.

EXAMPLE 5 Change $3\dfrac{7}{24}$ to a decimal, using a calculator. Round off to the nearest thousandth.

 SOLUTION $3\dfrac{7}{24} = 3 + \dfrac{7}{24}$

Key in   3 ⊞ 7 ÷ 24 =

Answer   3.2916666 ≈ 3.292

Exercises   3.10

In Exercises 1–18, change the fractions and mixed numbers to exact decimals.

1. $\dfrac{3}{4}$ 2. $\dfrac{5}{8}$ 3. $2\dfrac{1}{2}$

4. $5\dfrac{1}{4}$ 5. $\dfrac{1}{8}$ 6. $\dfrac{3}{5}$

7. $\dfrac{2}{3}$ 8. $\dfrac{7}{9}$ 9. $\dfrac{3}{11}$

10. $\dfrac{5}{18}$ 11. $3\dfrac{1}{15}$ 12. $2\dfrac{5}{12}$

13. $\dfrac{1}{25}$ 14. $\dfrac{1}{40}$ 15. $\dfrac{5}{16}$

16. $\dfrac{7}{8}$ 17. $6\dfrac{1}{20}$ 18. $3\dfrac{3}{250}$

19. Change $\dfrac{7}{11}$ to a decimal rounded off to two decimal places.

20. Change $4\dfrac{5}{7}$ to a decimal rounded off to three decimal places.

21. Change $\dfrac{2}{9}$ to a decimal rounded off to two decimal places.

22. Change $2\dfrac{7}{12}$ to a decimal rounded off to three decimal places.

23. Change $\dfrac{2}{23}$ to a decimal rounded off to the nearest thousandth.

24. Change $\dfrac{3}{11}$ to a decimal rounded off to the nearest thousandth.

25. Change $6\dfrac{7}{15}$ to a decimal rounded off to the nearest thousandth.

26. Change $9\dfrac{1}{14}$ to a decimal rounded off to the nearest ten-thousandth.

 In Exercises 27–32, use a calculator to change the fractions and mixed numbers to decimals. Round off to four decimal places when necessary.

27. $\dfrac{9}{16}$ 28. $\dfrac{17}{40}$ 29. $2\dfrac{7}{24}$

30. $3\dfrac{8}{27}$ 31. $9\dfrac{5}{13}$ 32. $6\dfrac{13}{21}$

3.11   Changing a Decimal to a Fraction

3.11A   Simple Decimals

Because a decimal is actually a fraction, in this section we write the decimals in fraction form (with numerator and denominator), then reduce the fraction to lowest terms.

| **Changing a decimal to a fraction** | 1. *Read* the decimal, then *write* it in fraction form, with numerator and denominator.
 2. Reduce the fraction to lowest terms. |

EXAMPLE 1 Change 0. 4 to a fraction in lowest terms.

└─── Tenths place

SOLUTION   *Read* the decimal "four tenths."

Write $\dfrac{4}{10} = \dfrac{2}{5}$ Reduced to lowest terms

EXAMPLE 2 Change 0.2 5 to a fraction in lowest terms.

└─── Hundredths place

SOLUTION   *Read* the decimal "twenty-five hundredths."

Write $\dfrac{25}{100} = \dfrac{1}{4}$ Reduced to lowest terms

EXAMPLE 3 Change 7.0 4 to a mixed number in lowest terms.

└─── Hundredths place

SOLUTION   *Read* the decimal "seven and four hundredths."

Write $7\dfrac{4}{100} = 7\dfrac{1}{25}$

EXAMPLE 4 Change 0.00785 to a fraction.

SOLUTION   We show an easy way to write a decimal as a fraction:

$$0.0\ 0\ 7\ 8\ 5$$
$$\text{| | | | | |} \rightarrow \frac{785}{100,000} = \frac{157}{20,000}$$

Write 1 ──→ 1 0 0,0 0 0
to the left
of the zeros

└─── Under each digit to the right of
the decimal point, write a zero

Exercises 3.11A

In Exercises 1–18, change the decimals to fractions or mixed numbers reduced to lowest terms.

1. 0.6	**2.** 0.8	**3.** 0.05	**4.** 0.65
5. 0.075	**6.** 0.750	**7.** 0.66	**8.** 1.8
9. 2.5	**10.** 3.7	**11.** 5.24	**12.** 4.08
13. 0.0625	**14.** 2.125	**15.** 37.5	**16.** 2.1875

17. 0.00012   **18.** 0.000875

19. A machinist must drill a 0.875-inch hole through a metal brace. What fractional size drill should he use?

20. A carpenter must drill a 0.625-inch hole through a rafter. What fractional size drill should she use?

3.11B Complex Decimals

Decimals that include fractional parts, such as $0.33\dfrac{1}{3}$ and $0.67\dfrac{1}{2}$, are called **complex decimals**.

Changing a complex decimal to a fraction

1. Write the complex decimal as a complex fraction.
2. Simplify the complex fraction by dividing the numerator by the denominator.

EXAMPLE 5   Change $0.12\frac{1}{2}$ to a fraction in lowest terms.

SOLUTION

2nd decimal place

$$0.1\,\left(2\frac{1}{2}\right)\longrightarrow\ \frac{12\frac{1}{2}}{100}=\frac{\frac{25}{2}}{100}=\frac{25}{2}\div\frac{100}{1}=\frac{\overset{1}{25}}{2}\cdot\frac{1}{\underset{4}{100}}=\frac{1}{8}$$
$$1\ 0\ 0$$

EXAMPLE 6   Change $0.2\frac{2}{9}$ to a fraction in lowest terms.

SOLUTION

1st decimal place

$$0.\,\left(2\frac{2}{9}\right)\longrightarrow\ \frac{2\frac{2}{9}}{10}=\frac{\frac{20}{9}}{10}=\frac{20}{9}\div\frac{10}{1}=\frac{\overset{2}{20}}{9}\cdot\frac{1}{\underset{1}{10}}=\frac{2}{9}$$
$$1\ \ 0$$

EXAMPLE 7   Change $2.16\frac{2}{3}$ to a fraction in lowest terms.

SOLUTION $2.16\frac{2}{3}=2+.16\frac{2}{3}$

$$0.1\,\left(6\frac{2}{3}\right)\longrightarrow\ \frac{16\frac{2}{3}}{100}=\frac{\frac{50}{3}}{100}=\frac{50}{3}\div\frac{100}{1}=\frac{\overset{1}{50}}{3}\cdot\frac{1}{\underset{2}{100}}=\frac{1}{6}$$
$$1\ 0\ 0$$

Therefore, $2.16\frac{2}{3}=2+\frac{1}{6}=2\frac{1}{6}$.

Exercises   3.11B

Change the complex decimals to fractions or mixed numbers reduced to lowest terms.

9. $0.00\frac{5}{12}$ 10. $2.00\frac{1}{4}$ 11. $0.001\frac{1}{6}$ 12. $0.041\frac{2}{3}$

1. $0.37\frac{1}{2}$ 2. $0.62\frac{1}{2}$ 3. $0.33\frac{1}{3}$ 4. $0.66\frac{2}{3}$

13. $0.8\frac{1}{3}$ 14. $0.2\frac{6}{7}$ 15. $0.13\frac{1}{3}$ 16. $0.16\frac{2}{3}$

5. $0.5\frac{5}{6}$ 6. $0.2\frac{2}{3}$ 7. $1.0\frac{1}{5}$ 8. $0.0\frac{2}{3}$

17. $2.10\frac{5}{7}$ 18. $3.08\frac{8}{9}$

3.12 Operations with Decimals and Fractions

When a problem involves both decimals and fractions, we either change the decimal to its fractional form or change the fraction to its decimal form, then perform the indicated operation.

EXAMPLE 1 Find $3.2 + \dfrac{3}{4}$.

SOLUTION 1 Change the fraction $\dfrac{3}{4}$ to a decimal, then add.

$$\frac{3}{4} = 4\overline{)3.00} \begin{array}{l} .75 \\ 2\,8 \\ \overline{2\,0} \\ 2\,0 \end{array}$$

Therefore,

$$3.2 + \frac{3}{4} = 3.2 + 0.75 = \begin{array}{r} 3.2 \\ +\ 0.75 \\ \hline 3.95 \end{array}$$

SOLUTION 2 Change the decimal 3.2 to a fraction, then add.

$$3.2 = 3\frac{2}{10} = 3\frac{1}{5} \quad \text{Reduce by 2}$$

Therefore,

$$3.2 + \frac{3}{4} = 3\frac{1}{5} + \frac{3}{4} = \begin{array}{r} 3\dfrac{1}{5} = 3\dfrac{4}{20} \\ +\ \dfrac{3}{4} = \dfrac{15}{20} \\ \hline 3\dfrac{19}{20} \end{array}$$

EXAMPLE 2 Find $2\dfrac{1}{12} \times 0.08$. The answer must be exact.

SOLUTION The fraction $\dfrac{1}{12}$ is a repeating decimal (see Section 3.10, Example 3). To obtain an exact answer, we must change the decimal 0.08 to a fraction.

$$0.08 = \frac{8}{100} = \frac{2}{25} \quad \text{Reduce by 4}$$

Then

$$2\frac{1}{12} \times 0.08 = \frac{\overset{1}{25}}{\underset{6}{12}} \times \frac{\overset{1}{2}}{\underset{1}{25}} = \frac{1}{6}$$

EXAMPLE 3   Find $\dfrac{0.76}{\dfrac{4}{5}}$. Write the answer in decimal form.

SOLUTION   Because we want the answer in decimal form, we will change the fraction $\dfrac{4}{5}$ to a decimal:

$$\frac{4}{5} = 5\overline{)\begin{array}{c} .8 \\ 4.0 \end{array}}$$

Then

$$\frac{0.76}{\dfrac{4}{5}} = \frac{0.76}{0.8} = 0.8_\wedge\overline{)\begin{array}{r} .95 \\ 0.7_\wedge60 \\ \underline{72} \\ 40 \\ \underline{40} \end{array}}$$

EXAMPLE 4   Find the cost of $9\dfrac{1}{4}$ feet of copper pipe selling for 98 cents a foot. Round off your answer to the nearest cent.

SOLUTION   Because this problem deals with money rounded to the nearest cent, we will change the fraction $9\dfrac{1}{4}$ to a decimal, then multiply.

$$9\frac{1}{4} = \frac{37}{4} = 4\overline{)\begin{array}{r} 9.25 \\ 37.00 \\ \underline{36} \\ 10 \\ \underline{8} \\ 20 \\ \underline{20} \end{array}}$$

Therefore,

$$\begin{array}{r} 9.25 \\ \times\ 0.98 \\ \hline 7400 \\ 8325\ \ \\ \hline 9.0650 \approx \$9.07 \end{array}$$ Rounded to the nearest cent

Exercises 3.12

In Exercises 1–8, perform the indicated operations. Write the answers in decimal form.

1. $4.62 + \dfrac{7}{10}$ 2. $9.6 - \dfrac{3}{100}$ 3. $2\dfrac{3}{20} - 1.6$ 4. $1\dfrac{6}{25} + 5.9$

5. $\dfrac{5}{8} \times 0.6$ 6. $0.84 \times \dfrac{5}{16}$ 7. $\dfrac{\dfrac{7}{20}}{0.014}$ 8. $\dfrac{7.6}{\dfrac{19}{50}}$

In Exercises 9–16, perform the indicated operations. Write the answers in fractional form.

9. $\dfrac{7}{8} + 0.35$ 10. $3.6 + 4\dfrac{2}{3}$ 11. $6.75 - 2\dfrac{5}{6}$

12. $8\dfrac{1}{12} - 4.125$ 13. $0.56 \times \dfrac{5}{6}$ 14. $1\dfrac{2}{3} \times 2.7$

15. $\dfrac{2\frac{5}{6}}{0.85}$ 16. $\dfrac{1.25}{1\frac{9}{16}}$

26. Find the cost of $2\frac{3}{4}$ yards of silk at \$4.59 a yard.

27. Carmen makes \$6.45 an hour. How much does she earn in a $7\frac{1}{2}$-hour day?

In Exercises 17–20, perform the indicated operations. The answers must be exact.

17. $1\frac{1}{9} \times 0.75$ 18. $\dfrac{5}{6} \times 0.16$

19. $0.4 \times 1\frac{1}{4}$ 20. $0.05 \times \dfrac{4}{5}$

28. Kevin makes \$7.80 an hour on Saturday. How much does he earn for $5\frac{3}{4}$ hours of work at that rate?

29. Suppose you drive 18,000 miles a year, and your car gets $22\frac{1}{2}$ miles per gallon. If the average cost of gasoline is \$1.05 per gallon, how much do you spend on gasoline in one year?

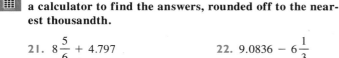 **In Exercises 21–24, (a) estimate the answers, and (b) use a calculator to find the answers, rounded off to the nearest thousandth.**

21. $8\frac{5}{6} + 4.797$ 22. $9.0836 - 6\frac{1}{3}$

23. $7\frac{2}{11} \times 0.1238$ 24. $\dfrac{19\frac{1}{6}}{41.6}$

30. Larry paid \$60.50 for fabric to reupholster his chair. If he bought $3\frac{2}{3}$ yards of fabric, find the cost per yard of the fabric.

In Exercises 25–30, round off your answers to the nearest cent when necessary.

25. SunAi needs $2\frac{3}{8}$ yards of material to make a dress. If she buys a fabric costing \$3.95 a yard, what is the total cost of the material?

3.13 Comparing Decimals

Sometimes we need to recognize which of a group of decimals is largest. *We can compare the sizes of decimals by writing them all with the same number of decimal places.*

EXAMPLE 1 Arrange the following decimals in order of size, with the largest first, using the symbol $>$: 0.27, 0.205, 0.2, 0.250.

SOLUTION Because the largest number of decimal places in any of the given decimals is three, we add final zeros wherever necessary to change all the given decimals to three decimal places:

$$0.27\,0\,, \quad 0.205, \quad 0.2\,00\,, \quad 0.250$$

Now arrange the three-decimal-place numbers in order of size—the largest first:

$$0.270 > 0.250 > 0.205 > 0.200$$

Therefore, the *original* decimals arranged according to size are

$$0.27 > 0.250 > 0.205 > 0.2$$

EXAMPLE 2   Which is larger, $\dfrac{5}{8}$ or 0.65?

SOLUTION   First, we change $\dfrac{5}{8}$ to a decimal:

$$\dfrac{5}{8} = 8\overline{)5.000}^{.625}$$

Since $0.65 = 0.65\ 0$

and $0.650 > 0.625$

we see that
$$0.65 > \dfrac{5}{8}$$

Exercises   3.13

In Exercises 1–8, arrange the decimals in order of size, with the largest first, using the symbol >.

1. 0.409, 0.49, 0.41, 0.4
2. 0.35, 0.3, 0.305, 0.335

3. 3.075, 3.1, 3.05, 3.009
4. 7.0, 7.1, 7.08, 7.099

5. 0.075, 0.07501, 0.0749, 0.07

6. 0.06, 0.1998, 0.6, 0.059

7. 5.05, 5.5, 5.0501, 5, 5.0496

8. 8.0505, 8.051, 8.0695, 8.199, 8

In Exercises 9–12, determine which symbol, > or <, should be used to make each statement true.

9. $\dfrac{3}{4}$? 0.8   10. $\dfrac{3}{8}$? 0.35   11. 0.5 ? $\dfrac{9}{20}$ 12. 0.075 ? $\dfrac{2}{25}$

13. A drill press operator uses a $\dfrac{5}{16}$-inch drill to put a hole through a steel bracket. Would a 0.325-inch pin fit in the hole?

14. Will a $\dfrac{7}{8}$-inch pin fit in a 0.8215-inch hole?

15. Star Fish tuna costs $1.02 for 7 ounces, and Whale of the Sea tuna costs $1.89 for 12 ounces. Which is the better buy, and by how much? (Round off to the nearest tenth of a cent.)

 16. Which brand is the best buy:
Brand A: $1.98 for 24 ounces
Brand B: $1.20 for 16 ounces
Brand C: $0.69 for 7.5 ounces

Chapter 3   REVIEW

Decimal
3.1

A **decimal** is a fraction whose denominator is a power of 10.

Reading a Decimal
3.1

1. Read the number to the left of the decimal point as a whole number.

2. Say "and" for the decimal point.

3. Read the number to the right of the decimal point as a whole number, then say the name of the place occupied by the right-hand digit of the number.

The Number of
Decimal Places
3.2

The number of decimal places in a number is the number of digits written to the right of the decimal point.

Rounding Off a Decimal **3.2**	**1.** Locate the digit in the round-off place.
	2. If the digit to the right of the round-off place is less than 5, the digit in the round-off place is unchanged. If the digit to the right of the round-off place is 5 or more, the digit in the round-off place is increased by 1.
	3. Digits to the left of the round-off place are unchanged.
	4. Digits to the right of both the round-off place and the decimal point are dropped.
	5. Digits to the right of the round-off place and the left of the decimal point are replaced by zeros.
Adding Decimals **3.3**	Write the numbers under one another, with their decimal points in the same vertical line, then add the numbers the way you add whole numbers. Place the decimal point in your answer in the same vertical line as the other decimal points.
Subtracting Decimals **3.4**	Write the number being subtracted under the number it is being subtracted from, with their decimal points in the same vertical line. Subtract the numbers the way you subtract whole numbers. Place the decimal point in the answer in the same vertical line as the other decimal points.
Multiplying Decimals **3.5**	Multiply the numbers the way you multiply whole numbers. Your answer has as many decimal places as the sum of the decimal places in the numbers being multiplied.
Dividing Decimals **3.6**	**1.** Place a caret ($\wedge$) in the divisor to make it the smallest possible whole number.
	2. Place a caret in the dividend the same number of places to the right (or left) of its decimal point as was done in the divisor.
	3. Place the decimal point in the quotient directly above the caret in the dividend.
	4. Divide the numbers the way you divide whole numbers.
Multiplying a Decimal by a Power of 10 **3.7**	To multiply a decimal by a power of 10, move the decimal point to the right as many places as the number of zeros in the power of 10.
Dividing a Decimal by a Power of 10 **3.7**	To divide a decimal by a power of 10, move the decimal point to the left as many places as the number of zeros in the power of 10.
Changing a Fraction to a Decimal **3.10**	To change a fraction to a decimal, divide the numerator by the denominator.
Changing a Decimal to a Fraction **3.11**	**1.** Read the decimal, then write it in fraction form (with numerator and denominator). **2.** Reduce the fraction to lowest terms.
Order of Operations **3.9**	The order of operations for decimals is the same as the order of operations for whole numbers and fractions.
Comparing Decimals **3.13**	To compare decimals, write all of them with the same number of decimal places.

Chapter 3   REVIEW EXERCISES

In Exercises 1 and 2, write the numbers in words.

1. 0.145   2. 250.06

In Exercises 3 and 4, write the numbers using digits.

3. Sixteen ten-thousandths

4. Five hundred and seventy-five hundredths

In Exercises 5–8, round off to the indicated place.

5. 0.83671   thousandths   6. 0.5967   hundredths

7. 24.74   tenths   8. 185.75   tens

In Exercises 9–20, perform the indicated operations.

9. $75.23 + 186.6 + 34,932 + 8.0205$

10. $23 + 5.6 + 0.875 + 0.0016 + 7.29$

11. 7.8×0.64

12. $0.0817 \div 0.95$

13. $509.6 - 41.345$

14. $7.85 \times 1,000$

15. $0.64 \div 10^2$

16. 0.014×0.59

17. $57.4 \div 28$

18. $10^4 \times 0.0056$

19. $476 - 39.4$

20. $9.6 \div 100$

In Exercises 21 and 22, divide, and round off the answer to two decimal places.

21. $50 \div 4.8$

22. $0.262 \div 0.086$

In Exercises 23–26, perform the indicated operations in the correct order.

23. $12.5 + 5.6 \times 10^2 + \sqrt{1.44}$

24. $67 - (9.6 - 1.8)$

25. $72 \div 0.8 \times (0.3)^2$

26. $2(2.7) + 2(1.9)$

In Exercises 27 and 28, change the fraction and mixed number to exact decimals.

27. $\dfrac{7}{8}$

28. $5\dfrac{1}{18}$

In Exercises 29 and 30, change the fraction and mixed number to decimals, rounded off to two decimal places.

29. $\dfrac{5}{6}$

30. $3\dfrac{7}{11}$

In Exercises 31–34, change the decimals to fractions or mixed numbers in lowest terms.

31. 0.68

32. 4.025

33. $0.16\dfrac{2}{3}$

34. $0.4\dfrac{1}{6}$

35. Find $7.06 - \dfrac{5}{8}$. Write the answer in decimal form.

36. Find $\dfrac{9}{10} + 0.24$. Write the answer in fraction form.

37. Find $0.025 \times 3\dfrac{1}{3}$. The answer must be exact.

38. Find $\dfrac{4\dfrac{1}{20}}{0.9}$. The answer must be exact.

In Exercises 39 and 40, estimate the answers.

39. $0.32(6.87 + 4.32)$

40. $\dfrac{0.52 \times 0.47}{7.9}$

In Exercises 41 and 42, arrange the decimals in order of size, with the largest first, using the symbol >.

41. $0.6, 0.603, 0.063, 0.06$

42. $3.89, 3.9, 3.098, 3.908, 3$

43. Find the average of 8.7, 6.2, and 7.6.

44. Find the average of 0.62, 0.57, 0.48, and 0.53.

45. A rectangle has a length of 6.25 meters and a width of 4.3 meters.

    a. Find the perimeter of the rectangle.

    b. Find the area of the rectangle.

46. The length of the side of a square is 3.4 centimeters.

    a. Find the perimeter of the square.

    b. Find the area of the square.

47. How many pieces of wire 1.5 meters long can be cut from a spool 60 meters long?

48. A machinist uses a $\dfrac{11}{16}$-inch drill to put a hole through a metal plate. Would a 0.625-inch pin fit in the hole?

49. Carlos bought one pair of shoes for $19.95, two neckties for $3.95 each, three pairs of socks for $1.25 a pair, and one suit for $89.95. What was his total bill?

50. At the beginning of the month, Jim's bank balance was $275.38. During the month he wrote the following checks: $15.98, $46.75, $87.45, $135.46, and $68. He made deposits of $250 and $350. Find his bank balance at the end of the month.

51. Nora made 18 equal monthly payments on her new stereo set. If the total cost of the set was $355, what was her monthly payment? (Round off to the nearest cent.)

52. Find the cost of $15\dfrac{1}{2}$ yards of electrical wire at 29 cents a yard. (Round off to the nearest cent.)

53. Rudy drove his car 9,600 miles last year. His total car expenses were $625 for the year. Find the average cost per mile. Round off the answer to the nearest tenth of a cent.

54. A car traveled 196 miles. If it gets 14 miles per gallon of gas, how many gallons did it use for the trip? If gas is $1.329 per gallon, how much was spent for gas?

Chapter 3 Critical Thinking and Writing Problems

Answer Problems 1–5 in your own words, using complete sentences.

1. Explain how to round off 5.8362 to the nearest hundredth.

2. Explain the difference between "six hundred" and "six hundredths."

3. Explain the difference between "one hundred twelve thousandths" and "one hundred and twelve thousandths."

4. Explain how to change a fraction to a decimal.

5. Explain how to change a decimal to a fraction.

6. If 25 is multiplied by a number that is less than one, will the answer be larger or smaller than 25?

7. If 25 is divided by a number that is less than one, will the answer be larger or smaller than 25?

8. True or false: $\frac{1}{3} = 0.33$. Explain your answer.

9. True or false: $0.79 > 0.729$. Explain your answer.

10. Find the mystery $\boxed{0.???}$ three-digit decimal.

    Clue 1: The decimal is greater than $\frac{1}{100}$.

    Clue 2: The sum of the digits is 3.

    Clue 3: The decimal is less than $\frac{1}{80}$.

11. Find the mystery $\boxed{0.??}$ two-digit decimal.

    Clue 1: Both digits are the same.

    Clue 2: The decimal is greater than $\frac{2}{3}$.

    Clue 3: Both digits are even.

Each of the following problems has an error. Find the error, and, in your own words, explain why it's wrong. Then work the problem correctly.

12. $1\frac{1}{4} = \frac{5}{4} = 5\overline{)4.0}^{.8}$

13. Subtract 0.974 from 3.05.

$$\begin{array}{r} \overset{9}{\underset{2\ \ \overset{10}{} \ 15}{3.0\ 5}} \\ -\ .9\ 7\ 4 \\ \hline 2.0\ 8\ 4 \end{array}$$

14. Find $0.088 \div 0.7$. Round off the answer to the nearest hundredth.

$$0.7_\wedge\overline{)0.0_\wedge 8\ 8}^{\ .1\ 2}$$

Allow yourself about 50 minutes to do these problems. Complete solutions for all problems, together with section references, are given in the answer section at the end of the book.

1. Write 9.015 in words.

2. Use digits to write four hundred twenty and five hundredths.

In Exercises 3–10, perform the indicated operations.

3. $7.8 + 56 + 0.017 + 500.94$ 4. $40.6 - 3.54$

5. 3.7×0.058 6. $62.79 \div 0.078$

7. 0.46×10^3 8. $6.9 \div 100$

9. $75 \div 10 - 0.2 \times 0.06$ 10. $(0.8)^2 + 3 \times \sqrt{0.25}$

11. Divide, and round off to the nearest tenth. $7.2 \div 0.35$

12. Change $2\frac{3}{16}$ to an exact decimal.

13. Change 0.78 to a fraction in lowest terms.

14. Change $0.1\frac{9}{11}$ to a fraction in lowest terms.

15. Find $1\frac{2}{3} \times 0.35$. The answer must be exact.

16. Estimate the answer to $\dfrac{9.9 \times 2.3}{0.42}$.

17. Arrange the decimals in order of size, with the largest first, using the symbol $>$. 7.48, 7.408, 7.084, 7.4

18. Sheila skated the 500-meters in 45.60 seconds, 42.76 seconds, and 44.02 seconds. What was her average time, rounded to the nearest hundredth of a second?

19. If a beam weighs 24 pounds per foot, what is the weight of a beam 4.8 feet long?

20. How many pieces of pipe 0.75-foot long can be cut from a 12-foot length?

Ratio, Proportion, and Percent

CHAPTER 4

In this chapter we will cover the relationships between percents, decimals, and fractions. We will also study how to solve applied problems involving ratios, rates, proportions, and percents.

4.1 Ratio and Rate Problems

Ratios

A **ratio** is a comparison of two like quantities, usually written in fraction form. In algebra, fractions are called rational (*ratio*nal) numbers.

$$\text{The ratio of } a \text{ to } b \text{ is written } \frac{a}{b}$$

$$\text{The ratio of } b \text{ to } a \text{ is written } \frac{b}{a}$$

where a and b are the *terms of the ratio*. The terms of a ratio can be any kind of number; the only restriction is that the denominator cannot be zero.

> **A Word of Caution**   The order of the terms of a ratio is important. The number that is before the *to* is in the numerator, and the number that is after the *to* is in the denominator. Therefore, the ratio of 5 to 6 is $\frac{5}{6}$, not $\frac{6}{5}$.

EXAMPLE 1   In an arithmetic class there are 13 men and 17 women.

a. Find the ratio of men to women.

   SOLUTION   The ratio of men to women is $\frac{13}{17}$.

b. Find the ratio of women to men.

   SOLUTION   The ratio of women to men is $\frac{17}{13}$.

c. Find the ratio of men to the total number of students in the class.

   SOLUTION   The total number of students in the class is $13 + 17 = 30$. The ratio of men to the total number of students in the class is $\frac{13}{30}$.

Reducing a Ratio to Lowest Terms

Because a ratio is a fraction, we can reduce a ratio to lowest terms using the same methods we used earlier for reducing fractions to lowest terms.

Reducing a ratio to lowest terms	1. Divide the numerator and the denominator by any number that you can see is a divisor of both. 2. When you cannot see any more common divisors, use prime factorization.

EXAMPLE 2

Reduce the ratio of 48 to 30 to lowest terms.

SOLUTION

$$\frac{48}{30} = \frac{\overset{\overset{8}{\cancel{24}}}{\cancel{48}}}{\underset{\underset{5}{\cancel{15}}}{\cancel{30}}} = \frac{8}{5}$$

EXAMPLE 3

Reduce the ratio of 42 to 156 to lowest terms.

SOLUTION   We write the prime factorization of the numerator and the denominator, then divide both by the common factors.

The prime factorization of $42 = 2 \cdot 3 \cdot 7$

The prime factorization of $156 = 2 \cdot 2 \cdot 3 \cdot 13$

$$\frac{42}{156} = \frac{\overset{1}{\cancel{2}} \cdot \overset{1}{\cancel{3}} \cdot 7}{\underset{1}{\cancel{2}} \cdot 2 \cdot \underset{1}{\cancel{3}} \cdot 13} = \frac{7}{26}$$

When we are using a ratio to compare like quantities, the numerator and the denominator should be written in the *same units*. (See Example 4.)

EXAMPLE 4

Reduce the ratio of 6 inches to 1 foot to lowest terms.

SOLUTION

$$\frac{6 \text{ inches}}{1 \text{ foot}} = \frac{6 \text{ inches}}{12 \text{ inches}} = \frac{\overset{1}{\cancel{6}}}{\underset{2}{\cancel{12}}} = \frac{1}{2}$$

Because 1 foot = 12 inches

EXAMPLE 5

Reduce the ratio of 5 cents to 1 dollar to lowest terms.

SOLUTION

$$\frac{5 \text{ cents}}{1 \text{ dollar}} = \frac{5 \text{ cents}}{100 \text{ cents}} = \frac{\overset{1}{\cancel{5}}}{\underset{20}{\cancel{100}}} = \frac{1}{20}$$

Because 1 dollar = 100 cents

EXAMPLE 6

Greg takes a 15-minute break for every 2 hours he works. Find the ratio of work time to break time.

SOLUTION

Because 1 hour = 60 minutes

$$\frac{2 \text{ hours}}{15 \text{ minutes}} = \frac{2 \cdot 60 \text{ minutes}}{15 \text{ minutes}} = \frac{\overset{\overset{8}{\cancel{24}}}{\cancel{120}}}{\underset{\underset{1}{\cancel{3}}}{\cancel{15}}} = \frac{8}{1}$$

Note   Since a ratio is a comparison of two quantities, the 1 should be left in the denominator in Example 6.

Reducing a Ratio with Terms That Are Not Whole Numbers

In all the ratios we have studied so far, the terms have been whole numbers. This is not always the case. The terms of a ratio can be any kind of number, the only restriction being that the denominator cannot be 0.

EXAMPLE 7 Simplify the ratio of $1\frac{3}{8}$ to 11.

SOLUTION

$$\frac{1\frac{3}{8}}{11} = 1\frac{3}{8} \div 11 = \frac{11}{8} \div \frac{11}{1} = \frac{\overset{1}{\cancel{11}}}{8} \cdot \frac{1}{\underset{1}{\cancel{11}}} = \frac{1}{8}$$

EXAMPLE 8 Simplify the ratio of $1\frac{3}{4}$ to $3\frac{1}{2}$.

SOLUTION

$$\frac{1\frac{3}{4}}{3\frac{1}{2}} = 1\frac{3}{4} \div 3\frac{1}{2} = \frac{7}{4} \div \frac{7}{2} = \frac{\overset{1}{\cancel{7}}}{\underset{2}{\cancel{4}}} \cdot \frac{\overset{1}{\cancel{2}}}{\underset{1}{\cancel{7}}} = \frac{1}{2}$$

EXAMPLE 9 Simplify the ratio of 3.6 to 12.

SOLUTION   We can clear the decimals from the fraction by multiplying the numerator and the denominator by 10.

$$\frac{3.6}{12} = \frac{3.6 \times 10}{12 \times 10} = \frac{36}{120} = \frac{\overset{3}{\cancel{36}}}{\underset{10}{\cancel{120}}} = \frac{3}{10}$$

EXAMPLE 10 A 6-foot man casts a $7\frac{1}{2}$-foot shadow. Find the ratio of the man's height to the length of his shadow.

SOLUTION

$$\frac{\text{height of man}}{\text{length of shadow}} = \frac{6 \text{ feet}}{7\frac{1}{2} \text{ feet}} = 6 \div 7\frac{1}{2} = \frac{6}{1} \div \frac{15}{2} = \frac{\overset{2}{\cancel{6}}}{1} \cdot \frac{2}{\underset{5}{\cancel{15}}} = \frac{4}{5}$$

The ratio is $\frac{4}{5}$, which means that the man's height is $\frac{4}{5}$ of the length of his shadow.

Rates

A **rate** is a comparison of two unlike quantities. We often use the word *per* to express a rate. The word *per* indicates a division. For example, if we drove 150 miles in 3 hours, our driving rate is in miles *per* hour and is found as follows:

$$\frac{150 \text{ miles}}{3 \text{ hours}} = \frac{\overset{50}{\cancel{150}} \text{ miles}}{\underset{1}{\cancel{3}} \text{ hours}} = \frac{50 \text{ miles}}{1 \text{ hour}} = 50 \text{ miles per hour}$$

Therefore, we drove 50 miles for every 1 hour in driving time.

EXAMPLE 11 Mario typed 230 words in 5 minutes. Find his rate in words per minute.

SOLUTION $\dfrac{230 \text{ words}}{5 \text{ minutes}} = \dfrac{\overset{46}{\cancel{230}} \text{ words}}{\underset{1}{\cancel{5}} \text{ minutes}} = \dfrac{46 \text{ words}}{1 \text{ minute}} = 46$ words per minute

In daily life, rates are often written in decimal form rather than in fraction form.

EXAMPLE 12 Kim earned $94 for working 8 hours. Find her rate of pay in dollars per hour, in decimal form.

SOLUTION $\dfrac{94 \text{ dollars}}{8 \text{ hours}} = 8\overline{)94.00}^{\,11.75} = \$11.75$ per hour

Unit Cost

The **unit cost** of an item is the cost per 1 unit (1 ounce, 1 pound, 1 serving, and so on).

EXAMPLE 13 A 16-ounce loaf of bread costs $1.39 and has 24 slices.

a. Find the cost per ounce. Round off to the nearest tenth of a cent.

SOLUTION Since $1.39 = 139¢, cost per ounce $= \dfrac{139¢}{16 \text{ oz}}$.

Key in 139 ÷ 16 =

Answer 8.6875 ≈ 8.7 cents per ounce

b. Find the cost per slice. Round off to the nearest tenth of a cent.

SOLUTION Cost per slice $= \dfrac{139¢}{24 \text{ slices}}$.

Key in 139 ÷ 24 =

Answer 5.7916666 ≈ 5.8 cents per slice

Exercises 4.1

In Exercises 1–24, reduce the ratios to lowest terms.

1. 18 to 27
2. 32 to 48
3. 135 to 60

4. 40 to 16
5. 60 to 12
6. 75 to 15

7. 12 feet to 60 feet
8. 9 inches to 45 inches

9. 35 cents to 1 dollar
10. 45 cents to 3 dollars

11. 2 hours to 40 minutes
12. 45 minutes to 4 hours

13. 9 inches to 3 feet
14. 18 inches to 2 feet

15. 6 to $\dfrac{2}{3}$
16. 15 to $\dfrac{3}{5}$

17. $1\dfrac{1}{2}$ to 12
18. $2\dfrac{2}{3}$ to 24

19. $1\dfrac{1}{6}$ to $2\dfrac{1}{3}$
20. $4\dfrac{1}{2}$ to $3\dfrac{3}{4}$

21. 1.8 to 2.4
22. 7.5 to 4.5

23. 1.5 to 6
24. 2.4 to 12

25. A college basketball team won 24 out of 30 games played. There were no tie games.

   a. Find the ratio of wins to games played.

   b. Find the ratio of wins to losses.

   c. Find the ratio of losses to wins.

26. In a history class, 30 of the 48 students are women.

   a. Find the ratio of women to students in the class.

   b. Find the ratio of women to men.

   c. Find the ratio of men to women.

27. A coffee shop received its order of pies: 12 apple pies, 8 cherry pies, and 4 lemon pies.

   a. What is the ratio of apple pies to cherry pies?

   b. What is the ratio of apple pies to lemon pies?

   c. What is the ratio of cherry pies to the total number of pies?

28. A restaurant received its order of steaks: 20 top sirloin steaks, 15 porterhouse, and 10 filet mignon.

   a. What is the ratio of porterhouse to filet mignon?

   b. What is the ratio of top sirloin to porterhouse?

   c. What is the ratio of filet mignon to the total number of steaks?

29. Yuri can do a job in 2 hours 40 minutes. Hal takes 4 hours to do the same job. Find the ratio of Yuri's time to Hal's time.

30. Machine A can produce 100 bushings in 2 hours. Machine B takes 1 hour 15 minutes to produce 100 bushings. What is the ratio of machine A's time to machine B's time?

31. A 15-foot pole casts a $4\frac{1}{2}$-foot shadow. What is the ratio of the length of the shadow to the height of the pole?

32. A room is $4\frac{2}{3}$ yards long and $3\frac{1}{2}$ yards wide. What is the ratio of the length to the width?

33. Yuki drove 144 miles in 3 hours. What was her rate in miles per hour?

34. Luis read 750 words in 6 minutes. What was his rate in words per minute?

35. A laser printer produced 90 pages in 12 minutes. Find the printer's rate in pages per minute, in decimal form.

36. An assembly-line worker completed 150 parts in an 8-hour day. Find his rate in parts per hour, in decimal form.

In Exercises 37–40, round off the unit cost to the nearest tenth of a cent when necessary.

37. An 8-ounce package of sliced cheese costs $2.15 and contains 10 slices.

   a. Find the cost per ounce.

   b. Find the cost per slice.

38. A 16-ounce package of bacon costs $2.49 and contains 20 slices.

   a. Find the cost per ounce.

   b. Find the cost per slice.

 39. If six 12-ounce cans of soda cost $2.99, find the cost per ounce.

 40. If a 4.4-ounce tube of toothpaste is on sale for $1.29, what is the cost per ounce?

4.2   Proportion Problems

A **proportion** is a statement that two ratios (or two rates) are equal. Common notation for a proportion is

$$\frac{a}{b} = \frac{c}{d} \qquad \text{Read "}a\text{ is to }b\text{ as }c\text{ is to }d\text{"} \\ \text{or "}a\text{ over }b\text{ equals }c\text{ over }d\text{"}$$

The **terms of a proportion** are numbered as follows:

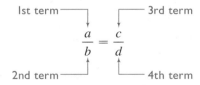

The **means** of a proportion are the second and third terms; the **extremes** are the first and fourth terms.

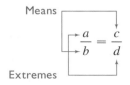

The *means* of the above proportion are *b* and *c*. The *extremes* of the above proportion are *a* and *d*.

EXAMPLE 1 In the proportion $\dfrac{2}{7} = \dfrac{6}{21}$, identify each term, the means, and the extremes.

SOLUTION Read "2 is to 7 as 6 is to 21."

The first term is 2.	The second term is 7.
The third term is 6.	The fourth term is 21.
The means are 7 and 6.	The extremes are 2 and 21.

Product of Means Equals Product of Extremes In any proportion, *the product of the means equals the product of the extremes.* This is sometimes called the *cross-multiplication rule*, and these products are called *cross products.*

In the proportion
$$\frac{2}{3} = \frac{4}{6}$$

$$3 \times 4 \qquad = \qquad 2 \times 6$$

Product of means Product of extremes

The cross-multiplication rule

In the proportion

$$\frac{a}{b} = \frac{c}{d}$$

$$b \cdot c \qquad = \qquad a \cdot d$$

Product of means Product of extremes

In words: The cross products are equal.

We can use the cross-multiplication rule to see whether two ratios form a proportion. If the product of the means equals the product of the extremes, then the ratios form a proportion. If the product of the means does *not* equal the product of the extremes, then the ratios do not form a proportion.

EXAMPLE 2 Do the given ratios form a proportion?

SOLUTION

a. $\dfrac{5}{12} \overset{?}{=} \dfrac{3}{7}$ No, because $5 \cdot 7 \neq 12 \cdot 3$ This is not a proportion

$$35 \neq 36$$

b. $\dfrac{16}{6} \overset{?}{=} \dfrac{8}{3}$ Yes, because $16 \cdot 3 = 6 \cdot 8$ This is a proportion

$$48 = 48$$

c. $\dfrac{\frac{1}{2}}{3} \overset{?}{=} \dfrac{1\frac{1}{3}}{8}$ Yes, because $\dfrac{1}{2} \cdot 8 = 3 \cdot 1\dfrac{1}{3}$ This is a proportion

$$\frac{1}{\underset{1}{\cancel{2}}} \cdot \frac{\overset{4}{\cancel{8}}}{1} = \frac{\overset{1}{\cancel{3}}}{1} \cdot \frac{4}{\underset{1}{\cancel{3}}}$$

$$4 = 4$$

Solving a Proportion

When three of the four terms of a proportion are known, it is always possible to find the value of the unknown term.

In the proportion $\dfrac{2}{3} = \dfrac{x}{51}$, the third term is unknown and is represented by the letter x. Since the product of the means equals the product of the extremes,

$$\frac{2}{3} = \frac{x}{51}$$

$$3 \cdot x = 2 \cdot 51 \qquad \text{Cross products are equal}$$

$$3 \cdot x = 102$$

The = sign tells us that $3 \cdot x$ and 102 are the same number. Therefore, if $3 \cdot x$ and 102 are both divided by 3, the resulting numbers will be equal. That is,

$$\frac{\overset{1}{\cancel{3}} \cdot x}{\underset{1}{\cancel{3}}} = \frac{\overset{34}{\cancel{102}}}{\underset{1}{\cancel{3}}}$$

$$x = 34$$

Therefore, the value of the third (unknown) term is 34, and $\dfrac{2}{3} = \dfrac{34}{51}$.

Solving a proportion for an unknown term	**1.** Represent the unknown term by a letter. Set the product of the means equal to the product of the extremes. (The cross products are equal.) **2.** Divide both sides by the number that is multiplying the unknown letter. *Checking Your Solution* **1.** Replace the unknown letter in the proportion by the value you obtained for it. **2.** Cross-multiply, and verify that the cross products are equal.

EXAMPLE 3

Solve $\dfrac{x}{25} = \dfrac{6}{5}$ for x.

SOLUTION

$$\frac{x}{25} = \frac{6}{5}$$

$$5 \cdot x = 25 \cdot 6 \qquad \text{Cross products are equal}$$

$$5 \cdot x = 150 \qquad \begin{array}{l}\text{Because 5 is multiplying } x, \\ \text{divide both sides by 5}\end{array}$$

$$\frac{\overset{1}{\cancel{5}} \cdot x}{\underset{1}{\cancel{5}}} = \frac{\overset{30}{\cancel{150}}}{\underset{1}{\cancel{5}}}$$

$$x = 30$$

√ **Check**

$$\frac{30}{25} \overset{?}{=} \frac{6}{5} \qquad \text{Replace } x \text{ by 30}$$

$$30 \cdot 5 \overset{?}{=} 25 \cdot 6$$

$$150 = 150$$

EXAMPLE 4   Solve $\dfrac{2}{3} = \dfrac{x}{150}$ for x.

SOLUTION

$$\frac{2}{3} = \frac{x}{150}$$

$$3 \cdot x = 2 \cdot 150 \qquad \text{Cross products are equal}$$

$$\frac{\overset{1}{\cancel{3}} \cdot x}{\underset{1}{\cancel{3}}} = \frac{\overset{100}{\cancel{300}}}{\underset{1}{\cancel{3}}} \qquad \text{Divide both sides by 3}$$

$$x = 100$$

√ **Check**

$$\frac{2}{3} \overset{?}{=} \frac{100}{150} \qquad \text{Replace } x \text{ by 100}$$

$$2 \cdot 150 \overset{?}{=} 3 \cdot 100$$

$$300 = 300$$

EXAMPLE 5   Solve $\dfrac{42}{x} = \dfrac{28}{12}$ for x.

SOLUTION   We can simplify our work by reducing the ratio $\dfrac{28}{12}$ to lowest terms before cross-multiplying.

$$\frac{42}{x} = \frac{28}{12}$$

$$\frac{42}{x} = \frac{7}{3} \qquad \text{Reduce } \frac{28}{12} \text{ to } \frac{7}{3}$$

$$7 \cdot x = 42 \cdot 3 \qquad \text{Cross-multiply}$$

$$\frac{\overset{1}{\cancel{7}} \cdot x}{\underset{1}{\cancel{7}}} = \frac{\overset{18}{\cancel{126}}}{\underset{1}{\cancel{7}}} \qquad \text{Divide both sides by 7}$$

$$x = 18$$

√ **Check**

$$\frac{42}{18} \overset{?}{=} \frac{28}{12} \qquad \text{Replace } x \text{ by 18}$$

$$42 \cdot 12 \overset{?}{=} 18 \cdot 28$$

$$504 = 504$$

EXAMPLE 6   Solve $\dfrac{x}{2} = \dfrac{15}{25}$ for x.

SOLUTION

$$\frac{x}{2} = \frac{15}{25}$$

$$\frac{x}{2} = \frac{3}{5} \qquad \text{Reduce } \frac{15}{25} \text{ to } \frac{3}{5}$$

$$5 \cdot x = 2 \cdot 3 \qquad \text{Cross-multiply}$$

$$\frac{\overset{1}{\cancel{5}} \cdot x}{\underset{1}{\cancel{5}}} = \frac{6}{5} \qquad \text{Divide both sides by 5}$$

$$x = 1\frac{1}{5}$$

√ **Check**
$$\frac{1\frac{1}{5}}{2} \overset{?}{=} \frac{15}{25} \qquad \text{Replace } x \text{ by } 1\frac{1}{5}$$

$$1\frac{1}{5} \cdot 25 \overset{?}{=} 2 \cdot 15$$

$$\frac{6}{\underset{1}{5}} \cdot \frac{\overset{5}{25}}{1} \overset{?}{=} 30$$

$$30 = 30$$

Solving a Proportion with Terms That Are Not Whole Numbers

In all the ratios we have studied so far, the terms have been whole numbers. This is not always the case. The terms of a proportion can be any kind of number, the only restriction being that the denominator in either ratio cannot be 0. Letters other than x are often used to represent the unknown term.

EXAMPLE 7 Solve $\dfrac{P}{3} = \dfrac{\frac{5}{6}}{5}$ for P.

SOLUTION
$$\frac{P}{3} = \frac{\frac{5}{6}}{5}$$

$$5 \cdot P = \frac{\overset{1}{3}}{1} \cdot \frac{5}{\underset{2}{6}}$$

$$5 \cdot P = \frac{5}{2}$$

$$\frac{\overset{1}{5} \cdot P}{\underset{1}{5}} = \frac{\frac{5}{2}}{5}$$

$$P = \frac{5}{2} \div 5 = \frac{\overset{1}{5}}{2} \cdot \frac{1}{\underset{1}{5}} = \frac{1}{2}$$

EXAMPLE 8 Solve $\dfrac{3\frac{1}{2}}{5\frac{1}{4}} = \dfrac{x}{4}$ for x.

SOLUTION
$$\frac{3\frac{1}{2}}{5\frac{1}{4}} = \frac{x}{4}$$

$$5\frac{1}{4} \cdot x = 3\frac{1}{2} \cdot 4$$

$$\frac{21}{4} \cdot x = \frac{7}{2} \cdot \frac{\overset{2}{4}}{1}$$

$$\frac{\frac{\overset{1}{\cancel{21}}}{4} \cdot x}{\frac{\cancel{21}}{\cancel{4}}} = \frac{14}{\frac{21}{4}}$$

$$x = 14 \div \frac{21}{4}$$

$$x = \frac{\overset{2}{14}}{1} \cdot \frac{4}{\underset{3}{\cancel{21}}}$$

$$x = \frac{8}{3} = 2\frac{2}{3}$$

EXAMPLE 9 Solve $\dfrac{1.5}{100} = \dfrac{A}{2.4}$ for A.

SOLUTION

$$\frac{1.5}{100} = \frac{A}{2.4}$$

$$100 \cdot A = 1.5 \times 2.4$$

$$\frac{\overset{1}{\cancel{100}} \cdot A}{\underset{1}{\cancel{100}}} = \frac{3.6}{100}$$

$$A = 0.036$$

Exercises 4.2

1. In the proportion $\dfrac{8}{14} = \dfrac{16}{28}$, find the following:

 a. the first term
 b. the second term
 c. the third term
 d. the fourth term
 e. the means
 f. the extremes

2. In the proportion $\dfrac{3}{5} = \dfrac{x}{20}$, find the following:

 a. the first term
 b. the second term
 c. the third term
 d. the fourth term
 e. the means
 f. the extremes

In Exercises 3–8, do the given ratios form a proportion?

3. $\dfrac{9}{12} \overset{?}{=} \dfrac{6}{8}$

4. $\dfrac{3}{7} \overset{?}{=} \dfrac{4}{9}$

5. $\dfrac{6}{9} \overset{?}{=} \dfrac{10}{16}$

6. $\dfrac{\frac{2}{3}}{8} \overset{?}{=} \dfrac{\frac{3}{4}}{9}$

7. $\dfrac{2\frac{1}{4}}{27} \overset{?}{=} \dfrac{3\frac{1}{3}}{40}$

8. $\dfrac{60}{100} \overset{?}{=} \dfrac{4.2}{0.7}$

In Exercises 9–32, solve for the letter in each proportion.

9. $\dfrac{4}{7} = \dfrac{x}{21}$

10. $\dfrac{5}{12} = \dfrac{25}{x}$

11. $\dfrac{15}{12} = \dfrac{10}{x}$

12. $\dfrac{24}{30} = \dfrac{16}{x}$

13. $\dfrac{P}{100} = \dfrac{30}{40}$

14. $\dfrac{P}{100} = \dfrac{75}{125}$

15. $\dfrac{10}{x} = \dfrac{15}{4}$

16. $\dfrac{6}{5} = \dfrac{9}{x}$

17. $\dfrac{30}{45} = \dfrac{x}{12}$

18. $\dfrac{12}{21} = \dfrac{x}{56}$

19. $\dfrac{7}{100} = \dfrac{A}{5}$

20. $\dfrac{9}{100} = \dfrac{36}{B}$

21. $\dfrac{\frac{3}{4}}{6} = \dfrac{x}{16}$ 22. $\dfrac{\frac{4}{5}}{2} = \dfrac{x}{25}$ 23. $\dfrac{x}{9} = \dfrac{3\frac{1}{3}}{5}$ 27. $\dfrac{\frac{5}{6}}{1\frac{1}{4}} = \dfrac{24}{x}$ 28. $\dfrac{x}{2\frac{1}{2}} = \dfrac{6}{1\frac{2}{3}}$ 29. $\dfrac{6.5}{100} = \dfrac{A}{3.2}$

24. $\dfrac{x}{8} = \dfrac{2\frac{1}{4}}{18}$ 25. $\dfrac{4\frac{1}{2}}{3} = \dfrac{x}{2\frac{2}{3}}$ 26. $\dfrac{3\frac{1}{5}}{x} = \dfrac{4}{2\frac{1}{2}}$ 30. $\dfrac{4}{8.24} = \dfrac{0.5}{y}$ 31. $\dfrac{P}{100} = \dfrac{4.8}{1.5}$ 32. $\dfrac{7.6}{y} = \dfrac{1.9}{5}$

4.3   Word Problems Using Proportions

Solving word problems that lead to proportions

1. Represent the unknown quantity by a letter.
2. Set up a proportion, with the unit of measure next to each number in the proportion.
3. The same units must occupy corresponding positions in the two ratios of your proportion.

   Correct arrangements   *Incorrect arrangements*

   $\dfrac{\text{miles}}{\text{hours}} = \dfrac{\text{miles}}{\text{hours}}$ $\dfrac{\text{dollars}}{\text{weeks}} = \dfrac{\text{weeks}}{\text{dollars}}$

   $\dfrac{\text{hours}}{\text{miles}} = \dfrac{\text{hours}}{\text{miles}}$ $\dfrac{\text{dollars}}{\text{weeks}} = \dfrac{\text{dollars}}{\text{days}}$

   $\dfrac{\text{miles}}{\text{miles}} = \dfrac{\text{hours}}{\text{hours}}$

4. Once the numbers have been correctly entered in the proportion by using the units as a guide, drop the units, then cross-multiply to solve for the unknown.

EXAMPLE 1   A woman used 10 gallons of gas on a 180-mile trip. How many gallons of gas can she expect to use on a 300-mile trip?

SOLUTION   Let $x =$ number of gallons of gas for the 300-mile trip.

10 gallons of gas used for a 180-mile trip	How many gallons of gas for a 300-mile trip?

$$\dfrac{10 \text{ gallons}}{180 \text{ miles}} = \dfrac{x \text{ gallons}}{300 \text{ miles}}$$

Note: The ratios used on each side have gallons in the numerator and miles in the denominator.

$$180 \cdot x = 10 \cdot 300$$

$$\dfrac{\overset{1}{\cancel{180}} \cdot x}{\underset{1}{\cancel{180}}} = \dfrac{\overset{50}{\cancel{3,000}}}{\underset{3}{\cancel{180}}}$$

$$x = \dfrac{50}{3} = 16\dfrac{2}{3} \text{ gallons}$$

EXAMPLE 2   At a soda fountain, 8 quarts of ice cream was used to make 100 milk shakes. How many quarts are needed to make 550 milk shakes?

S O L U T I O N   Let x = number of quarts of ice cream.

8 quarts of ice cream used to make 100 milk shakes	How many quarts are needed to make 550 milk shakes?

$$\frac{8 \text{ quarts}}{100 \text{ milk shakes}} = \frac{x \text{ quarts}}{550 \text{ milk shakes}}$$

$$100 \cdot x = 8 \cdot 550$$

$$\frac{\overset{1}{\cancel{100}} \cdot x}{\underset{1}{\cancel{100}}} = \frac{4{,}400}{100}$$

$$x = 44 \text{ quarts}$$

EXAMPLE 3   If a 6-foot man casts a $4\frac{1}{2}$-foot shadow, how tall is a tree that casts a 30-foot shadow?

S O L U T I O N   Let x = the height of the tree.

6-foot man casts a $4\frac{1}{2}$-foot shadow	How tall is a tree that casts a 30-foot shadow?

$$\frac{6 \text{ feet high (man)}}{4\frac{1}{2} \text{ feet high (shadow)}} = \frac{x \text{ feet high (tree)}}{30 \text{ feet high (shadow)}}$$

$$\frac{6}{4\frac{1}{2}} = \frac{x}{30}$$

$$4\frac{1}{2} \cdot x = 6 \cdot 30$$

$$\frac{9}{2} \cdot x = 180$$

$$\frac{\frac{9}{2} \cdot x}{\frac{9}{2}} = \frac{180}{\frac{9}{2}}$$

$$x = \frac{180}{1} \div \frac{9}{2}$$

$$x = \frac{\overset{20}{\cancel{180}}}{1} \cdot \frac{2}{\underset{1}{\cancel{9}}} = 40 \text{ feet}$$

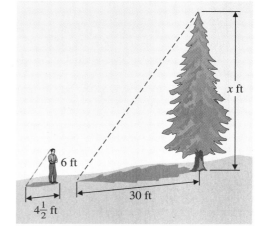

Note: The ground does not need to be horizontal.

Exercises 4.3

Solve the following problems using a proportion.

1. A painter uses about 3 gallons of paint in doing 2 rooms. How many gallons would she need to paint 20 rooms?

2. A man drives 600 miles in $1\frac{1}{2}$ days. How long would it take him to drive 3,000 miles?

3. A 6-foot man has a 4-foot shadow when a tree casts a 20-foot shadow. How tall is the tree?

4. Seven builders finish 10 houses in a month. How many houses could 35 builders finish in the same time?

5. An investment of $3,000 earned $180 for a year. How much would have to be invested to earn $540 in the same time?

6. A store has a bargain price of 85¢ for three jars of grape jelly. How many jars could someone buy for $5.95?

7. The property tax on a $15,000 home is $450. What would be the tax on a $25,000 home?

8. A woman used 12 gallons of gas on a 200-mile trip. How many gallons of gas can she expect to use on a 750-mile trip?

9. The ratio of a woman's weight on earth compared to her weight on the moon is 6 to 1. How much would a 150-pound woman weigh on the moon?

10. The ratio of a man's weight on Mars compared to his weight on earth is 2 to 5. How much would a 180-pound man weigh on Mars?

11. The ratio of the weight of lead to the weight of an equal volume of aluminum is 21 to 5. If an aluminum bar weighs 150 pounds, what would a lead bar of the same size weigh?

12. The ratio of the weight of platinum to the weight of an equal volume of copper is 12 to 5. If a platinum bar weighs 18 pounds, what would a copper bar of the same size weigh?

13. A market is selling three cans of beets for 99 cents. How much will 12 cans cost at the same rate?

14. A baseball team wins seven of its first 12 games. How many will it win out of its first 36 games if the team continues to play with the same degree of success?

15. On a map, $\frac{1}{2}$ inch represents 12 miles. How many miles would $2\frac{1}{4}$ inches represent?

16. An apartment house manager spent 22 hours painting 3 apartments. How long can he expect to take painting the remaining 15 apartments?

17. Ho drove 420 miles in $\frac{3}{4}$ of a day. About how far can he drive in $2\frac{1}{2}$ days?

18. Amalia noticed that her shadow was 4 feet long when that of a 16-foot flagpole was 12 feet long. What is her height in feet and inches?

19. A crew of 10 mechanics takes a week to overhaul 25 trucks in a fleet of 100 trucks. How many mechanics would it take to complete the fleet overhaul in one week?

20. A car burns $2\frac{1}{2}$ quarts of oil on an 1,800-mile trip. How many quarts of oil can the owner expect to use on a 12,000-mile trip?

21. The scale in an architectural drawing is 1 inch equals 8 feet. What are the dimensions of a room that measures $2\frac{1}{2}$ by 3 inches on the drawing? (*Hint:* Find the length and width of the room separately.)

22. Fifteen defective axles were found in 100,000 cars of a particular model. How many defective axles could we expect to find in the 2 million cars made of that same model?

4.4   Percent

Consider a 5-gallon paint can containing 2 gallons of paint (see Figure 1). We can say, "The can is $\frac{2}{5}$ full." Because $\frac{2}{5} = \frac{2 \cdot 20}{5 \cdot 20} = \frac{40}{100}$, we can also say, "The can is $\frac{40}{100}$ full." The decimal representation for $\frac{40}{100}$ is 0.40.

Figure 1

The word **percent** means "per hundred." (*Cent*um means 100 in Latin.) The symbol for percent is %. Since $\dfrac{40}{100}$ is 40 per hundred, we say, "The can is 40% full." We have described the amount of paint in the can in three different ways:

1. The can is $\dfrac{2}{5}$ full.   *Fraction* representation

2. The can is 0.40 full.   *Decimal* representation

3. The can is 40% full.   *Percent* representation

EXAMPLE 1

On a test, a student answered 17 out of 20 problems correctly. Describe the score three ways.

SOLUTION

a. $\dfrac{17}{20}$ correct   Fraction form

b. $\dfrac{17}{20} = 20\overline{)17.00}^{\,.85}$ correct   Decimal form

c. $\dfrac{17}{20} = \dfrac{17 \cdot 5}{20 \cdot 5} = \dfrac{85}{100} = 85\%$ Percent form

As Example 1 shows, $\dfrac{17}{20}$, 0.85, and 85% are three different ways of saying the same thing. We need to know how to change from one form to another.

4.4A Changing a Decimal to a Percent

Because percent means the number of hundredths, to change a decimal to a percent, we determine the number of hundredths in the decimal and then write the percent symbol (%).

EXAMPLE 2

a. 0.25 = 25 hundredths = 25%   **b.** 0.03 = 3 hundredths = 3%
c. 0.5 = 0.50 = 50 hundredths = 50%

Example 2 leads us to the following procedure:

Changing a decimal to a percent

Move the decimal point two places to the *right*, and write the percent symbol (%).

EXAMPLE 3   **a.** $0.136 = 0.13{_\wedge}6 = 13.6\%$

b. $0.07 = 0.07{_\wedge} = 7\%$

c. $0.3 = 0.30{_\wedge} = 30\%$

d. $2 = 2.00{_\wedge} = 200\%$

Exercises 4.4A

Change each decimal to a percent.

1. 0.27   **2.** 0.35   **3.** 0.06   **4.** 0.125   **5.** 1.4

6. 2.05   **7.** 0.186   **8.** 0.015   **9.** 0.075   **10.** 0.175

11. 2.9   **12.** 3.8   **13.** 2.005   **14.** 3.015   **15.** 1.36

16. 2.11   **17.** 4   **18.** 3

4.4B Changing a Percent to a Decimal

Because percent means the number of hundredths, to change a percent to a decimal, we write the percent as a fraction over 100, then change to decimal form.

EXAMPLE 4   **a.** $75\% = 75 \text{ hundredths} = \dfrac{75}{100} = 0.75$

b. $8\% = 8 \text{ hundredths} = \dfrac{8}{100} = 0.08$

c. $20\% = 20 \text{ hundredths} = \dfrac{20}{100} = 0.20 = 0.2$

Example 4 shows us the following procedure:

Changing a percent to a decimal	Move the decimal point two places to the *left*, and remove the percent symbol (%).

EXAMPLE 5   **a.** $38.5\% = {_\wedge}38.5\% = 0.385$

b. $2\% = {_\wedge}02.\% = 0.02$

c. $80\% = {_\wedge}80.\% = 0.8$

d. $400\% = 4{_\wedge}00.\% = 4$

e. $5\dfrac{1}{2}\% = 5.5\% = {_\wedge}05.5\% = 0.055$

First change the mixed number

$5\dfrac{1}{2}$ into the decimal 5.5

Exercises 4.4B

Change each percent to a decimal. If the decimal is not exact, round off to four decimal places.

1. 45% **2.** 78% **3.** 125% **4.** 150% **5.** 6.5%

6. 8.6% **7.** 9% **8.** 7% **9.** $2\frac{1}{2}$% **10.** $4\frac{3}{4}$%

11. $3\frac{1}{4}$% **12.** $5\frac{2}{5}$% **13.** 10.05% **14.** 2.08% **15.** $\frac{3}{4}$%

16. $\frac{1}{2}$% **17.** $66\frac{2}{3}$% **18.** $33\frac{1}{3}$%

4.4C Changing a Fraction to a Percent

Changing a fraction to a percent	1. Change the fraction to a decimal by dividing numerator by denominator. 2. Change the decimal to a percent by moving the decimal point two places to the right and writing the percent symbol.

EXAMPLE 6

Fraction ⟶ Decimal ⟶ Percent

a. $\dfrac{3}{4}$ $= 4\overline{)3.00}^{\,.75} = 0.75_\wedge = 75\%$

b. $\dfrac{3}{5}$ $= 5\overline{)3.0}^{\,.6} = 0.60_\wedge = 60\%$

c. $1\dfrac{1}{8} = \dfrac{9}{8} = 8\overline{)9.000}^{\,1.125} = 1.12_\wedge 5 = 112.5\%$

Exercises 4.4C

Change each fraction to a percent. If an answer is not exact, round it off to two decimal places.

1. $\dfrac{1}{2}$ **2.** $\dfrac{1}{4}$ **3.** $\dfrac{2}{5}$ **4.** $\dfrac{4}{5}$ **5.** $\dfrac{3}{8}$

6. $\dfrac{7}{10}$ **7.** $\dfrac{9}{10}$ **8.** $\dfrac{3}{20}$ **9.** $\dfrac{4}{25}$ **10.** $\dfrac{17}{25}$

11. $\dfrac{1}{3}$ **12.** $\dfrac{1}{6}$ **13.** $\dfrac{5}{6}$ **14.** $\dfrac{5}{3}$ **15.** $\dfrac{7}{16}$

16. $1\dfrac{3}{4}$ **17.** $2\dfrac{5}{8}$ **18.** $1\dfrac{2}{5}$

4.4D Changing a Percent to a Fraction

Because percent means number of hundredths, to change a percent to a fraction, we write the percent as a fraction over 100 and reduce the fraction to lowest terms.

$$25\% = 25 \text{ hundredths} = \frac{25}{100} = \frac{1}{4}$$

└── Reduced form

| **Changing a percent to a fraction** | 1. Write the percent as the numerator of a fraction with a denominator of 100. |
| | 2. Reduce the fraction to lowest terms. |

EXAMPLE 7

a. $6\% = \dfrac{6}{100} = \dfrac{3}{50}$

b. $20\% = \dfrac{20}{100} = \dfrac{1}{5}$

c. $150\% = \dfrac{150}{100} = \dfrac{3}{2} = 1\dfrac{1}{2}$

d. $12\dfrac{1}{2}\% = \dfrac{12\dfrac{1}{2}}{100} = \dfrac{25}{2} \div \dfrac{100}{1} = \dfrac{\overset{1}{25}}{2} \cdot \dfrac{1}{\underset{4}{100}} = \dfrac{1}{8}$

Exercises 4.4D

In Exercises 1–18, change each percent to a proper fraction or mixed number.

1. 75%	2. 50%	3. 10%	4. 30%
5. 35%	6. 65%	7. 80%	8. 60%
9. 5%	10. 4%	11. 250%	12. 300%
13. $\dfrac{1}{2}\%$	14. $\dfrac{3}{4}\%$	15. $2\dfrac{1}{2}\%$	16. $3\dfrac{1}{3}\%$
17. $33\dfrac{1}{3}\%$	18. $66\dfrac{2}{3}\%$		

In Exercises 19–44, change the given form into the missing forms.

	Fraction	Decimal	Percent
19.	$\dfrac{1}{2}$		
20.	$\dfrac{1}{4}$		
21.		0.6	
22.		0.4	
23.			10%
24.			20%
25.	$\dfrac{3}{4}$		

	Fraction	Decimal	Percent
26.		0.36	
27.			6%
28.	$\dfrac{4}{5}$		
29.		0.48	
30.			8%
31.	$1\dfrac{1}{8}$		
32.		0.075	
33.			44%
34.	$2\dfrac{3}{8}$		
35.		0.025	
36.			28%
37.			350%

	Fraction	Decimal	Percent
38.			275%
39.		6.25	
40.			$4\frac{1}{2}\%$
41.			$5\frac{1}{4}\%$
42.		8.75	
43.			$\frac{3}{4}\%$
44.			$\frac{1}{2}\%$

45. Because of illness, Manny was absent from school 10 days out of 40 days.

  a. What fraction of the time was he absent?

  b. What decimal part of the time was he absent?

  c. What percent of the time was he absent?

46. A student took an examination in mathematics and solved 9 problems out of 12.

  a. What fraction of the problems did she solve correctly?

  b. What decimal part of the problems did she solve correctly?

  c. What was her percent grade?

47. The price of a car is discounted 20%. What fraction of the original price must the buyer *pay*?

48. If $\frac{1}{3}$ of Irma's salary is deducted from her paycheck, what percent of her check does she get?

49. Duane spends 25% of his income on rent and 20% of his income on his car. What fraction of his income does he spend on rent and his car?

50. On an exam, 12% of the students got an A, 20% got a B, and 33% got a C. What fraction of the class passed the exam with an A, B, or C?

4.5   Finding a Fractional Part of a Number

Earlier we showed that to find a fractional part of a number, we multiply the fraction times the number. When the fractional part is expressed as a decimal or as a percent, we may multiply using the following rules.

> **Finding a fractional part of a number**
>
> **1.** If the fractional part is expressed as a *fraction*, multiply the fraction times the number.
> **2.** If the fractional part is expressed as a *decimal*, multiply the decimal times the number.
> **3.** If the fractional part is expressed as a *percent*, change the percent to a decimal or fraction, then multiply.

EXAMPLE 1   On a 20-problem test, Sun Ai answered $\frac{4}{5}$ of the problems correctly. How many problems did she have correct?

SOLUTION $\frac{4}{5}$ of $20 = \frac{4}{\underset{1}{5}} \cdot \frac{\overset{4}{20}}{1} = 16$ problems

of means to multiply in problems of this type

EXAMPLE 2   Find 0.3 of 25.

SOLUTION

$$0.3 \text{ of } 25 = \begin{array}{r} 25 \\ \times\ .3 \\ \hline 7.5 \end{array}$$

EXAMPLE 3   Find 75% of 24.

SOLUTION   Change 75% to a fraction.

$$75\% = \frac{75}{100} = \frac{3}{4}$$

Then

$$75\% \text{ of } 24 = \frac{3}{\underset{1}{4}} \cdot \frac{\overset{6}{24}}{1} = 18$$

EXAMPLE 4   Find $3\frac{1}{3}\%$ of 150.

SOLUTION   Change $3\frac{1}{3}\%$ to a fraction.

$$3\frac{1}{3}\% = \frac{3\frac{1}{3}}{100} = 3\frac{1}{3} \div 100 = \frac{\overset{1}{10}}{3} \cdot \frac{1}{\underset{10}{100}} = \frac{1}{30}$$

Then

$$3\frac{1}{3}\% \text{ of } 150 = \frac{1}{\underset{1}{30}} \cdot \frac{\overset{5}{150}}{1} = 5$$

EXAMPLE 5   Tran's weekly salary is $725. His total deductions amount to 23% of his check.

a. How much is deducted from Tran's check?

SOLUTION   Change 23% to a decimal:

$$23\% = 0.23$$

Then

$$23\% \text{ of } \$725 = \begin{array}{r} \$7\ 2\ 5 \\ \times\ \ \ \ .2\ 3 \\ \hline 2\ 1\ 7\ 5 \\ 1\ 4\ 5\ 0\ \ \\ \hline \$\ 1\ 6\ 6.7\ 5 \end{array}$$

Amount deducted
from his paycheck

b. What is Tran's take-home pay?

SOLUTION   To find Tran's take-home pay, subtract the deductions from his salary.

$$\begin{array}{r} \$725.00 \\ -\ \$166.75 \\ \hline \$558.25 \end{array}$$

His take-home pay

EXAMPLE 6 Use a calculator to find the fractional parts.

SOLUTION

a. $\dfrac{17}{32}$ of 36   *Key in*   17 ÷ 32 × 36 =

                 Answer   19.125

b. 0.0375 of 1,250   *Key in*   .0375 × 1250 =

                 Answer   46.875

c. 6.5% of 342   *Key in*   342 × 6.5 %

                 Answer   22.23

Exercises 4.5

1. Find $\dfrac{5}{6}$ of 48.

2. Find $\dfrac{7}{8}$ of 32.

3. Find 0.8 of 15.

4. Find 0.35 of 36.

5. Find 15% of 60.

6. Find 8% of 75.

7. Find 200% of 15.

8. Find 300% of 7.

9. Find $\dfrac{4}{15}$ of 9.

10. Find $\dfrac{5}{12}$ of 32.

11. Find 0.06 of 25.

12. Find 0.05 of 98.

13. Find $13\dfrac{1}{3}\%$ of 600.

14. Find $6\dfrac{2}{3}\%$ of 504.

15. Find $\dfrac{1}{2}\%$ of 300.

16. Find $\dfrac{1}{4}\%$ of 200.

17. Audrey's weekly salary is $375. Her total deductions amount to 27% of her check.

    a. How much is deducted from Audrey's check?

    b. What is her take-home pay?

18. On a mathematics examination of 20 problems, Yoshio solved 85% of the problems correctly.

    a. How many problems did he have correct?

    b. How many problems did he miss?

19. During an 88-day spring semester, a student was absent $\dfrac{1}{11}$ of the time. How many days was he absent?

20. A man's weekly salary is $165. His deductions amount to $\dfrac{1}{5}$ of his check. Find his take-home pay.

21. In purchasing a car, a 15% down payment is required. Find the down payment on a car that sells for $6,150.

22. The present value of a house is 450% of its 1976 cost. If the house sold for $64,000 in 1976, what is its present value?

23. If a student sleeps $\dfrac{1}{3}$ of each 24-hour day, how many hours does she sleep in a week?

24. If $\dfrac{3}{4}$ of the total weight of a steer will produce usable products, how many pounds of usable products can be taken from a 1,250-pound steer?

25. An automobile listed at $2,490 was sold at a 15% discount. What was the selling price?

26. Mrs. Martinez, with a salary of $12,500 a year, was given an 8.5% raise. What was her salary after the raise?

In Exercises 27–36, use a calculator to work the problems.

27. Find $\dfrac{5}{6}$ of 8.7.

28. Find $\dfrac{7}{12}$ of 40.5.

29. Find 0.18 of 565.

30. Find 2.4 of 9.6.

31. Find 29% of 824.

32. Find 6% of 905.

33. Find 4.5% of 3,750.

34. Find $6\dfrac{1}{2}\%$ of 78.

35. Manuel is ordering three items that cost $12.95, $19.50, and $7.95. He must add 4.5% sales tax and $3.50 for postage and handling. What is the total cost of his order? (Round off to the nearest cent.)

36. The total surface area of the earth is 196,940,000 square miles, of which 29% is land and 71% is water. Find the land area and the water area of the earth.

4.6 Percent Problems

Let's take a second look at the example of the paint can, shown again in Figure 2.

5 gal

2 gal

Figure 2

We described that situation in three different ways:

1. The can is $\frac{2}{5}$ full.
2. The can is 0.40 full.
3. The can is 40 percent full.

We have talked about three quantities:

1. 40 *percent* (*P*)
2. 5, which represents the *whole thing* and is called the *base* (*B*)
3. 2, which is called the *amount* (*A*)

These three numbers are related by the proportion

$$\frac{2}{5} = \frac{40}{100}$$

When the numbers are replaced by the corresponding letters, *A*, *B*, and *P*, we have

The percent proportion

$$\frac{A}{B} = \frac{P}{100}$$

where *A* = amount, *B* = base, *P* = percent.

There are three letters in the percent proportion. If any two are known, the proportion can be solved for the remaining letter by the method given earlier.

1. The easiest term to identify is the percent *P*, which is the number written with the word *percent* or the symbol %.
2. Next, we identify the base *B*, which is the number that represents 100 percent, or the *total*. *B* is what we are finding the percent of, and therefore it follows the words *percent of*.
3. *A* is the remaining number after *P* and *B* have been identified. *A* is the *amount*.

Identifying the numbers in a percent problem

1. *P* is the number followed by the word *percent* (or %).
2. *B* is the number that follows the words *percent of*.
3. *A* is the number left after *P* and *B* have been identified.

EXAMPLE 1

Example of identifying the numbers in a percent problem:
What number is 8 percent of 40?

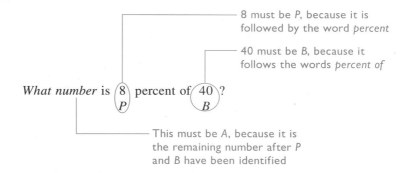

What number is ⑧ percent of ㊵ ?
             P   B

8 must be P, because it is followed by the word *percent*

40 must be B, because it follows the words *percent of*

This must be A, because it is the remaining number after P and B have been identified

---

Solving a percent problem

1. Identify the given numbers as A, B, or P.
2. Substitute the given numbers in the percent proportion.
3. Solve the percent proportion for the unknown.

EXAMPLE 2

What is 15% of 72?

SOLUTION 1 Identify the numbers A, B, and P.

$$\widehat{\text{What}}_{A} \text{ is } \widehat{15}_{P} \% \text{ of } \widehat{72}_{B} ?$$

$$\frac{A}{B} \cdot = \frac{P}{100} \qquad \text{The percent proportion}$$

$$\frac{A}{72} = \frac{15}{100}$$

$$100 \cdot A = 72 \cdot 15$$

$$\frac{\overset{1}{\cancel{100}} \cdot A}{\underset{1}{\cancel{100}}} = \frac{1080}{100}$$

$$A = 10.8$$

Therefore, 10.8 is 15% of 72.

SOLUTION 2 Another way to solve percent problems is to translate the word statement into an equation. The word *is* translates as the $=$ sign, and the word *of* means to multiply. When we use this method, we must change the percent to a fraction or a decimal.

What is 15% of 72?

$$A \quad = \quad 0.15 \quad \times \quad 72$$

$$15\% = \frac{15}{100} = 0.15$$

$$A \quad = \quad 10.8$$

EXAMPLE 3   5 is what percent of 20?

S O L U T I O N   Identify the numbers A, B, and P.

$$\underset{A}{5} \text{ is } \underset{P}{\text{what}} \text{ percent of } \underset{B}{20}?$$

$$\frac{A}{B} = \frac{P}{100} \quad \text{The percent proportion}$$

$$\frac{5}{20} = \frac{P}{100} \quad \text{Reduce } \frac{\overset{1}{5}}{\underset{4}{20}} = \frac{1}{4}$$

$$\frac{1}{4} = \frac{P}{100}$$

$$4 \cdot P = 100$$

$$P = 25$$

Therefore, 5 is 25% of 20.

EXAMPLE 4   30% of what number is 12?

S O L U T I O N $$\underset{P}{30}\% \text{ of } \underset{B}{\text{what number}} \text{ is } \underset{A}{12}?$$

$$\frac{A}{B} = \frac{P}{100}$$

$$\frac{12}{B} = \frac{30}{100} \quad \text{Reduce } \frac{30}{100} = \frac{3}{10}$$

$$\frac{12}{B} = \frac{3}{10}$$

$$3 \cdot B = 120$$

$$B = 40$$

Therefore, 30% of 40 is 12.

Exercises   4.6

Solve the following percent problems.

1. 15 is 30% of what number?

2. 16 is 20% of what number?

3. 115 is what percent of 250?

4. 90 is what percent of 225?

5. What number is 25% of 40?

6. What number is 45% of 65?

7. 15% of what number is 127.5?

8. 32% of what number is 256?

9. What percent of 8 is 18?

10. What percent of 6 is 12?

11. 65% of 48 is what number?

12. 87% of 49 is what number?

13. 750 is 125% of what number?

14. 325 is 130% of what number?

15. 20 is what percent of 16?

16. 30 is what percent of 25?

17. What number is 200% of 12?

18. What number is 300% of 9?

19. 15% of what number is 37.5?

20. What number is 80% of $135?

21. 42 is $66\frac{2}{3}$% of what number?

22. 36 is $16\frac{2}{3}$% of what number?

4.7 Word Problems Using Percent

In the percent proportion

$$\frac{A}{B} = \frac{P}{100}$$

the base B represents 100 percent, or the whole thing. Therefore, B represents the *total* number or the *original* number.

EXAMPLE 1   In an examination a student worked 15 problems correctly. This was 75% of the problems. Find the total number of problems on the examination.

SOLUTION   In this problem we are saying

15 is 75 % of some number
A P B ← *Total* number of problems on exam

$$\frac{A}{B} = \frac{P}{100}$$

$$\frac{15}{B} = \frac{75}{100} \qquad \text{Reduce } \frac{\overset{3}{\cancel{75}}}{\underset{4}{\cancel{100}}} = \frac{3}{4}$$

$$\frac{15}{B} = \frac{3}{4}$$

$$3 \cdot B = 4 \cdot 15$$

$$\frac{\cancel{3} \cdot B}{\cancel{3}} = \frac{60}{3}$$

$$B = 20$$

Therefore, there were 20 problems on the examination.

EXAMPLE 2   Thirty grams (about 1 cup) of a popular breakfast cereal contains 12 grams of sugar. What percent of the cereal is sugar?

SOLUTION   12 is what percent of 30 ?
A P B ← *Total* weight of cereal

$$\frac{A}{B} = \frac{P}{100}$$

$$\frac{12}{30} = \frac{P}{100} \qquad \text{Reduce } \frac{12}{30} = \frac{2}{5}$$

$$\frac{2}{5} = \frac{P}{100}$$

$$5 \cdot P = 200$$

$$P = 40$$

Therefore, the cereal is 40% sugar.

EXAMPLE 3   A CD player that originally sold for $300 is on sale at 25% off.

a. What is the amount of discount?

SOLUTION

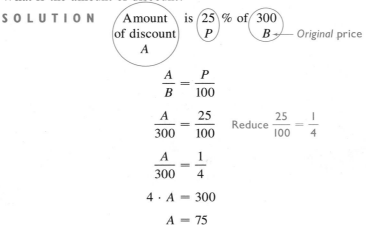

$$\frac{A}{B} = \frac{P}{100}$$

$$\frac{A}{300} = \frac{25}{100} \qquad \text{Reduce } \frac{25}{100} = \frac{1}{4}$$

$$\frac{A}{300} = \frac{1}{4}$$

$$4 \cdot A = 300$$

$$A = 75$$

Therefore, the discount is $75.

b. What is the sale price?

SOLUTION   To find the sale price, subtract the discount from the original price.

> Sale price = original price − discount

Sale price = $300 − $75 = $225

Percent Increase or Decrease   Many changes are stated using the percent of increase or decrease.

Food prices increased 8%.

Enrollment has decreased 12%.

Unemployment has increased 25% this year.

> In the percent proportion
>
> $$\frac{A}{B} = \frac{P}{100}$$
>
> *P* is the *percent of increase* (or decrease).
>
> *A* is the *amount of increase* (or decrease).
>
> *B* is the *original* amount.

EXAMPLE 4   Mr. Lee's salary last year was $25,000. This year his salary has increased 5%.

a. Find the amount of increase.

SOLUTION

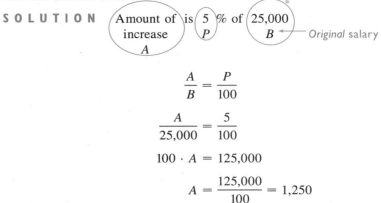

$$\frac{A}{B} = \frac{P}{100}$$

$$\frac{A}{25,000} = \frac{5}{100}$$

$$100 \cdot A = 125,000$$

$$A = \frac{125,000}{100} = 1,250$$

Therefore, the amount of increase is $1,250.

b. Find his new salary.

SOLUTION   To find his new salary, add the amount of increase to his original salary.

New amount = original amount + amount of increase

New salary = $25,000 + $1,250 = $26,250

EXAMPLE 5   The population of a town increased from 20,000 to 25,000.

a. Find the amount of increase.

SOLUTION   To find the amount of increase, subtract the original population from the new population.

Amount of increase = new amount − original amount

Amount of increase = 25,000 − 20,000 = 5,000

b. Find the percent of increase.

SOLUTION   5,000 is what percent of 20,000 ?

Amount of increase → A   P   B ← Original population

$$\frac{A}{B} = \frac{P}{100}$$

$$\frac{5,000}{20,000} = \frac{P}{100} \qquad \text{Reduce } \frac{5,000}{20,000} = \frac{1}{4}$$

$$\frac{1}{4} = \frac{P}{100}$$

$$4 \cdot P = 100$$

$$P = 25$$

Therefore, the population increased 25%.

EXAMPLE 6   Maria weighed 150 pounds. After going on a diet, she weighed 125 pounds. What was the percent of decrease, rounded off to the nearest percent?

SOLUTION   First, find the amount of decrease by subtracting the new weight from the original weight.

Amount of decrease = original − new amount

Amount of decrease = 150 − 125 = 25 pounds

25 is what percent of 150 ?
Amount of decrease— A   P   B — Original weight

$$\frac{A}{B} = \frac{P}{100}$$

$$\frac{25}{150} = \frac{P}{100} \qquad \text{Reduce } \frac{25}{150} = \frac{1}{6}$$

$$\frac{1}{6} = \frac{P}{100}$$

$$6 \cdot P = 100$$

$$\frac{\overset{1}{6} \cdot P}{\underset{1}{6}} = \frac{100}{6}$$

$$P = 16.666\ldots$$

$$P \approx 17 \qquad \text{Rounded to the nearest unit}$$

Therefore, the percent of decrease was about 17%.

Exercises 4.7

1. A team won 80% of its games. If it won 68 games, how many games did it play?

2. On a Spanish test, Eleni answered 17 problems correctly and scored 68%. How many problems were on the test?

3. In a class of 42 students, 7 students received a grade of B. What percent of the class received a grade of B?

4. John's weekly gross pay for his part-time job is $110, but 23% of his check is withheld. How much is withheld?

5. Fifty-four out of 216 civil service applicants passed their exams. What percent of the applicants passed? What percent did *not* pass?

6. A 4,200-pound automobile contains 462 pounds of rubber. What percent of the car's total weight is rubber? What percent of the car's total weight is *not* rubber?

7. Mr. Delgado, a salesman, makes a 6% commission on all items he sells. One week he made $390. What were his gross sales for the week?

8. Mrs. Hidalgo makes a 9% commission on all items she sells. One week she made $369.81. What were her gross sales for the week?

9. An affirmative action committee suggests that 24% of the 1,800 entering freshmen be minority students. How many more than the actual 256 minority freshmen entering would satisfy the committee?

10. Sergio got a $1,800 raise on his $52,000 yearly salary. There is a 4% rate of inflation for the year. How much more (or less) should his raise have been in order for him to keep up with inflation?

11. The evening enrollment at a certain college is 6% less than the day enrollment. If the day enrollment is 2,450, find the evening enrollment.

12. Mr. Edmonson's new contract calls for a 12% raise in salary. His present salary is $36,500. What will his new salary be?

13. Gloria thinks that at least 40% of the 50 teachers at her school should be women. The school board decides to increase the number of women teachers at her school from 16 to 22 (without changing the total number). Will this satisfy Gloria?

14. Seventy percent of the 46,000 burglaries in a city were committed by persons who had previously been convicted at least 3 times for the same crime. If all of these criminals had been kept in jail after the third conviction, how many burglaries could have been prevented?

15. There is 43 grams of sulfuric acid in 500 grams of solution. Find the percent of acid in the solution.

16. 150 grams of a solution contains 7.5 grams of a drug. Find the percent of drug strength.

17. The track team attended 30 meets and won 18. What percent of the meets did they lose?

18. A team wins 105 games. This is 70% of the games played. How many games were played?

19. A computer company surveyed 120 people. If 65% of those surveyed were men, how many women were surveyed?

20. Rosie's weekly salary is $875. If her deductions amount to $280, what percent of her salary is take-home pay?

21. Mrs. Chang's salary was $28,000 a year. At the end of one year, her salary was increased 8%.

  a. Find the amount of increase.

  b. Find her new salary.

22. Last year the Shigetos paid property taxes of $847.50. This year their property taxes will increase 6%. Find their property taxes for this year.

23. Smalltown's population of 13,000 decreased 6%. Find the new population.

24. Sidney Sabot sold $214,000 worth of sailboats in the summer. The company expects its sales to decrease 60% in the winter. Find the expected sales for the winter months.

25. The local university is requesting a $228 tuition increase for next year. If the present tuition is $3,800, find the percent of increase being proposed.

26. A team won 15 games last year. This year they won 3 more games than last year. What is the percent of increase in the number of games won?

27. Last year the average monthly utility bill for a family was $46.00. Today the average utility bill is $80.50.

  a. Find the amount of increase.

  b. Find the percent of increase.

28. When Marisa began selling real estate, the average family home in her city cost $66,000. Today the average family home in her city costs $105,600. Find the percent of increase in the cost of a family home in Marisa's city.

29. When a man bought his new car in 1990, it cost $12,000. The resale value of the car today is $5,400. Find the percent of decrease in the value of the car.

30. After lowering its thermostat from 75°F to 68°F, a family found that its heating bill decreased from $70.60 to $48.10. Find the percent of decrease in the heating bill. (Round off to the nearest tenth of a percent.)

Chapter 4   REVIEW

Ratio 4.1	A **ratio** is a comparison of two like quantities, written in fraction form.
Rate 4.1	A **rate** is a comparison of two unlike quantities, usually written in decimal form.
Proportion 4.2	A **proportion** is a statement that two ratios (or two rates) are equal.
Solving a Proportion for an Unknown Term 4.2	1. First, represent the unknown term by a letter, then set the product of the means equal to the product of the extremes. (The cross products are equal.) 2. Divide both sides by the number that is multiplying the unknown letter.
Checking Your Solution of a Proportion 4.2	1. Replace the unknown letter in the original proportion by the value you obtained for it. 2. Cross-multiply, and verify that the cross products are equal.

Solving a Word Problem by Using a Proportion
4.3

1. Read the word problem completely, then use x to represent the unknown quantity.
2. Set up a proportion with the same units occupying corresponding positions in both ratios of your proportion.
3. Solve your proportion for the unknown x.

Percent
4.4

Percent means per hundred or the number of hundredths.

Changing a Decimal to a Percent
4.4A

Move the decimal point two places to the right, and write the percent symbol (%).

Changing a Percent to a Decimal
4.4B

Move the decimal point two places to the left, and remove the percent symbol (%).

Changing a Fraction to a Percent
4.4C

1. Change the fraction to a decimal by dividing.
2. Change the decimal to a percent by moving the decimal point two places to the right and writing the percent symbol.

Changing a Percent to a Fraction
4.4D

1. Write the percent as the numerator of a fraction with a denominator of 100.
2. Reduce the fraction to lowest terms.

Finding a Fractional Part of a Number
4.5

1. If the fractional part is expressed as a *fraction*, multiply the fraction times the number.
2. If the fractional part is expressed as a *decimal*, multiply the decimal times the number.
3. If the fractional part is expressed as a *percent*, change the percent to a decimal or fraction, then multiply.

Solving a Percent Problem
4.6

1. Identify the given numbers as A, B, or P:
   P is the number followed by the word *percent* (or %).
   B is the number that follows the words *percent of*.
   A is the number left after P and B have been identified.
2. Substitute the given numbers in the percent proportion:

$$\frac{A}{B} = \frac{P}{100}$$

3. Solve the percent proportion for the unknown.

Percent Increase or Decrease
4.7

In the percent proportion

$$\frac{A}{B} = \frac{P}{100}$$

P is the *percent of increase* (or decrease).

A is the *amount of increase* (or decrease).

B is the *original* amount.

Chapter 4   REVIEW EXERCISES

In Exercises 1–3, reduce the ratios to lowest terms.

1. 27 to 12   2. $1\frac{1}{3}$ to 12   3. 2 hours to 20 minutes

In Exercises 4–6, do the given ratios form a proportion?

4. $\frac{5}{12} \overset{?}{=} \frac{125}{300}$ 5. $\frac{46}{32} \overset{?}{=} \frac{3}{2}$ 6. $\frac{1\frac{2}{3}}{3} \overset{?}{=} \frac{5}{9}$

In Exercises 7–12, solve for the unknown letter.

7. $\frac{45}{70} = \frac{x}{56}$ 8. $\frac{39}{x} = \frac{130}{210}$ 9. $\frac{24}{15} = \frac{18}{x}$

10. $\frac{x}{2} = \frac{\frac{9}{16}}{\frac{5}{6}}$ 11. $\frac{2.8}{5.2} = \frac{A}{6.5}$ 12. $\frac{2\frac{1}{2}}{B} = \frac{9}{2\frac{7}{10}}$

13. For a school fund raiser, Ahmed sold 12 packages of popcorn for a profit of $19.20. Find the rate of profit in dollars per package.

14. An 11-ounce bottle of shampoo is on sale for $1.29. What is the cost per ounce? Round off the answer to the nearest tenth of a cent.

15. If 10 pounds of apples costs $6, how much does $2\frac{1}{2}$ pounds cost?

16. Itzik knows his car needs a quart of oil every 1,500 miles. How many quarts will he need for a 12,000-mile trip?

17. Four students finish a class for every five students who begin. For 252 students to finish, how many must have begun the class?

18. If a 6-foot man casts an 8-foot shadow, find the height of a tree that casts a 96-foot shadow.

19. On a map, $\frac{1}{2}$ inch represents 10 miles. How many miles would $3\frac{1}{2}$ inches represent?

20. A machine can finish 80 parts in 2 hours 30 minutes. How many parts can it finish in an 8-hour day?

21. Change to a percent:

    a. 0.7   b. 1.25

22. Change to a decimal:

    a. 4%   b. 65%

23. Change to a percent:

    a. $\frac{3}{5}$ b. $\frac{7}{8}$

24. Change to a reduced fraction:

    a. 24%   b. $37\frac{1}{2}\%$

25. Find $\frac{5}{12}$ of 48.   26. Find 0.4 of 120.

27. Find 45% of 300.   28. What number is 6% of 40?

29. What percent of 20 is 25?

30. 360 is 150% of what number?

31. At a certain college the fall enrollment was 5% more than the spring enrollment. The spring enrollment was 3,560. Find the fall enrollment.

32. If you work 9 problems correctly on an examination of 12 problems, what is your percent grade?

33. A $95 suit is marked down 20%. Find the selling price of the suit.

34. The rent on a $400-a-month apartment is increased $12\frac{1}{2}\%$. Find the new rent.

35. 500 grams of a solution contains 75 grams of a drug. Find the percent of drug.

36. One week a sales representative working on a 15% commission made $255. Find her total sales for the week.

37. Mr. Garcia, with a salary of $26,500, was given a 6% raise. What was his salary after the raise?

38. Last year, the Carters paid property taxes of $875. This year their property taxes are $1,015. Find the percent of increase in taxes.

 39. A credit card company charges 1.75% interest each month on the unpaid balance. Find the interest on an unpaid balance of $369.

 40. A lawn mower that originally sold for $250 is on sale at 25% off. Since Wing works at the store, he gets a 15% employee discount. How much will Wing pay for the lawn mower?

Chapter 4 Critical Thinking and Writing Problems

Answer Problems 1–3 in your own words, using complete sentences.

1. Explain how to change a percent to a decimal.

2. Explain how to change a fraction to a percent.

3. Explain how to solve a proportion.

4. Find the mystery $\boxed{\dfrac{?}{?} = \dfrac{?}{?}}$ proportion.

    Clue 1: The first term is a prime number that is a common factor of 12 and 27.
    Clue 2: The second term is the smallest even number divisible by 5.

Clue 3: The third term is a 1-digit composite number that is a factor of 45.

Are Problems 5–7 set up correctly? Explain your answer.

5. If a machine can thread 9 nuts in 15 seconds, how many nuts can it thread in 10 minutes?

$$\frac{9}{15} = \frac{x}{10}$$

6. What percent of 20 is 25?

$$\frac{20}{25} = \frac{P}{100}$$

7. A car manufacturer reduced the weight of its compact model to 2,000 pounds from 2,500 pounds. What was the percent of decrease in the weight of the car?

$$\frac{500}{2,000} = \frac{P}{100}$$

Problems 8–10 each have an error. Find the error, and in your own words, explain why it's wrong. Then work the problem correctly.

8. Reduce the ratio 1.2 in. to 9 in.

$$\frac{1.2 \text{ in.}}{9 \text{ in.}} = \frac{\overset{4}{\cancel{1.2} \text{ in.}}}{\underset{3}{\cancel{9} \text{ in.}}} = \frac{4}{3}$$

9. $\dfrac{x}{\underset{10}{50}} = \dfrac{\frac{1}{5}}{2}$

$2 \cdot x = 10 \cdot 1$

$\dfrac{2 \cdot x}{2} = \dfrac{10}{2}$

$x = 5$

10. Change $3\frac{3}{4}$ to a percent.

$$3\frac{3}{4} = \frac{15}{4} = 4\overline{)15.00}\ \ 3.75\%$$

Chapter 4   DIAGNOSTIC TEST

Allow yourself about 50 minutes to do these problems. Complete solutions for all problems, together with section references, are given in the answer section at the end of the book.

1. A college basketball team won 15 out of 24 games played. There were no ties.
 a. What is the ratio of wins to games played?
 b. What is the ratio of wins to losses?

2. Jamie typed 260 words in 5 minutes. What was her rate in words per minute?

In Exercises 3–5, solve the proportions.

3. $\dfrac{x}{12} = \dfrac{40}{60}$ **4.** $\dfrac{20}{75} = \dfrac{x}{300}$ **5.** $\dfrac{3\frac{1}{2}}{x} = \dfrac{21}{40}$

6. Change 0.8 to a percent.

7. Change 12.5% to a decimal.

8. Change $\dfrac{3}{40}$ to a percent.

9. Change 76% to a reduced fraction.

10. Find $\dfrac{5}{8}$ of 72.

11. Find 0.7 of 35.

12. 15 is 125% of what number?

13. What percent of 30 is 12?

14. If 5 pounds of strawberries cost $4, what will $7\frac{1}{2}$ pounds cost?

15. If a 6-foot man casts a 4-foot shadow, find the height of a flagpole that casts an 18-foot shadow.

16. Amina's car used 2 quarts of oil on a 1,500-mile trip. How many quarts can she expect to use on a 6,000-mile trip?

17. On a test, Bill answered 34 problems correctly and scored 85%. How many problems were on the test?

18. A bike that originally sold for $96 is on sale at 25% off. What is the sale price?

19. The population of a town increased from 4,000 to 4,600. What was the percent of increase?

20. There are 80 students in the class. If 40% of the students are men, how many women are in the class?

Graphs and Statistics

CHAPTER 5

Graphs are used to present numerical data in an organized way. It is often easier to compare amounts or to see a change if the information is in graph form. Graphs are used by scientists, economists, technicians, business organizations, government agencies, and advertising firms.

5.1   Bar Graphs

Bar graphs are used to compare amounts. The bar graph in Figure 1 shows land area for the seven continents. On the vertical scale on the left, labeled "Area in Millions of Square Miles," each interval represents 1 million square miles. The horizontal scale on the bottom is labeled with the names of the continents. The height of each bar corresponds to the land area of the continent.

Notice that you can estimate the area of a continent from the graph, but you cannot read the area exactly. For example, the area of Europe is approximately 4 million square miles (4,000,000 sq. mi).

It is easy to see from the graph that the largest continent is Asia (because its bar is the tallest) and that the smallest continent is Australia. We can also compare the areas. Africa is how many times larger than Europe? Because the bar representing Africa is about 3 times taller than the bar representing Europe, Africa is about 3 times larger than Europe.

Continental Land Area

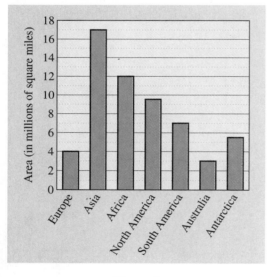

Figure 1

City College Enrollment

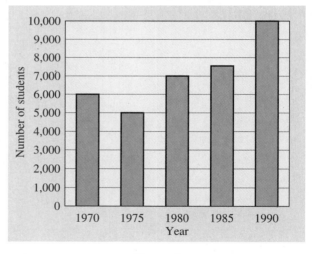

Figure 2

EXAMPLE 1   Examine the bar graph in Figure 2, and answer the following questions.

a. What was the enrollment in 1980?

   SOLUTION   About 7,000 students

b. In what year was the enrollment about 6,000?

   SOLUTION   1970

c. What year had the least enrollment?

   SOLUTION   1975

d. During which years did the number of students increase?

S O L U T I O N   Comparing the heights of the bars in order, we find the enrollment *decreased* from 1970 to 1975 but continually *increased* from 1975 to 1990.

e. Which 5-year period had the largest increase?

S O L U T I O N   The largest difference between the heights of the bars is from 1985 to 1990.

f. How much greater was the enrollment in 1980 than in 1975?

S O L U T I O N   The enrollment was about 7,000 in 1980 and about 5,000 in 1975. Therefore, the difference is $7,000 - 5,000 = 2,000$ more students.

g. What was the percent of increase in enrollment from 1975 to 1980?

S O L U T I O N   2,000 is what percent of 5,000?

Amount   1975
of   enrollment
increase

$$\frac{2,000}{5,000} = \frac{P}{100} \quad \text{Reduce} \quad \frac{2,000}{5,000} = \frac{2}{5}$$

$$\frac{2}{5} = \frac{P}{100}$$

$$5P = 200$$

$$P = 40$$

Therefore, the enrollment increased 40%.

The graphs in Figures 1 and 2 used vertical bars; Figure 3 in Example 2 uses horizontal bars.

E X A M P L E   2   Examine the bar graph in Figure 3, and answer the following questions.

Gasoline Mileage

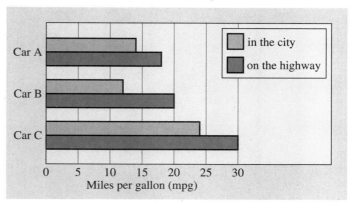

Figure 3

a. What gasoline mileage does car C get in the city?

S O L U T I O N   Different shadings have been used for city driving and highway driving. For city driving car C gets approximately 24 mpg.

b. What gasoline mileage does car C get on the highway?

S O L U T I O N   Approximately 30 mpg

c. What is the ratio of car C's gasoline mileage in the city to its gasoline mileage on the highway?

S O L U T I O N $\dfrac{24 \text{ mpg}}{30 \text{ mpg}} = \dfrac{\overset{4}{\cancel{24}} \text{ mpg}}{\underset{5}{\cancel{30}} \text{ mpg}} = \dfrac{4}{5}$

d. Which car gets the best gasoline mileage in the city?

S O L U T I O N   Car C

e. Which car gets the worst gasoline mileage in the city?

S O L U T I O N   Car B

f. Which car gets the worst gasoline mileage on the highway?

S O L U T I O N   Car A

g. How many times greater is the gasoline mileage for car C than for car B in city driving?

S O L U T I O N   About 2 times

Slanting the Picture

When we read information from a graph, it is important to note the vertical and the horizontal scales. The two graphs in Figure 4 show the same sales figures for the three companies, but the graph on the right distorts the differences between the companies because the vertical scale does not start at 0.

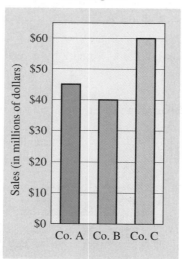
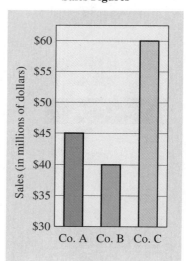

Figure 4

Exercises 5.1

In Exercises 1–10, use the bar graph below.

State Population in 1990

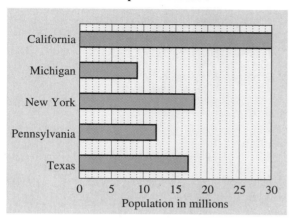

Figure for Exercises 1–10

Store Sales

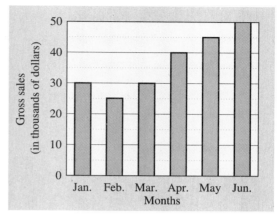

Figure for Exercises 11–20

1. What was the population of California in 1990?

2. What was the population of New York in 1990?

3. Which state had a population of about 17,000,000 in 1990?

4. Which state had a population of about 9,000,000 in 1990?

5. Which state had the largest population in 1990?

6. Which state had the smallest population in 1990?

7. How many more people lived in California than in Texas?

8. How many more people lived in New York than in Pennsylvania?

9. How many times larger was the population of New York than the population of Michigan?

10. What is the ratio of New York's population to Pennsylvania's population?

In Exercises 11–20, use the bar graph at the top of the next column.

11. What were the store sales for April?

12. During which month did the store have sales of $45,000?

13. Which month had the lowest sales?

14. Which month had the highest sales?

15. How much more were the sales in May than in March?

16. During which months did the sales increase?

17. During which months did the sales decrease?

18. Find the total sales for January to June.

19. What was the percent of increase in sales from February to March?

20. What was the percent of increase in sales from April to May?

In Exercises 21–28, use the bar graph in Figure 1 on page 156.

21. What is the land area of Asia?

22. What is the land area of Africa?

23. What is the total area of North America and South America?

24. What is the total area of Europe, Asia, and Africa?

25. How much larger is North America than South America?

26. How much larger is Asia than Europe?

27. What is the ratio of land area of Europe to Africa?

28. What is the ratio of land area of Africa to Australia?

5.2   Pictographs

Pictographs are basically bar graphs that use rows of picture symbols to replace the bars. Each picture symbol represents a certain number of units. Partial symbols represent fractional parts of the given unit.

EXAMPLE I Examine the pictograph in Figure 5, and answer the following questions.

Registered Voters in Small County

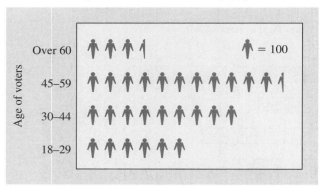

Figure 5

a. How many voters are 18–29 years old?

SOLUTION 6 × 100 = 600 voters

b. How many voters are 45–59 years old?

SOLUTION There are $11\frac{1}{2}$, or 11.5, symbols. Therefore, 11.5 × 100 = 1,150 voters.

c. How many voters are over 44 years old?

SOLUTION (11.5 + 3.5) × 100 = 15 × 100 = 1,500 voters

d. How many registered voters are in Small County?

SOLUTION Total number of symbols = 3.5 + 11.5 + 9 + 6 = 30. Therefore, 30 × 100 = 3,000 voters.

e. What percent of the voters are under 30 years old?

SOLUTION 600 is what percent of 3,000?

Voters under 30   Total voters

$$\frac{600}{3,000} = \frac{P}{100} \quad \text{Reduce} \quad \frac{600}{3,000} = \frac{1}{5}$$

$$\frac{1}{5} = \frac{P}{100}$$

$$5P = 100$$

$$P = 20$$

Therefore, 20% of the voters are under 30 years old.

f. What is the ratio of voters under 30 years old to voters 30 years old or older?

SOLUTION Since 600 voters are under 30 years old, 3,000 − 600 = 2,400 voters are 30 years old or older.

$$\frac{600}{2,400} = \frac{\overset{1}{\cancel{600}}}{\underset{4}{\cancel{2,400}}} = \frac{1}{4}$$

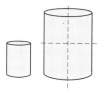

Slanting the Picture

In the hypothetical ad in Figure 6, the height of the large battery is twice the height of the small battery, but to keep the figure proportional, the artist doubled the width also. What the reader sees is an area that is actually 4 times larger.

Our Batteries Last *Twice* as Long

Their Battery   Our Battery

Figure 6

Exercises 5.2

In Exercises 1–8, use the pictograph graph below.

Number of U.S. Farms

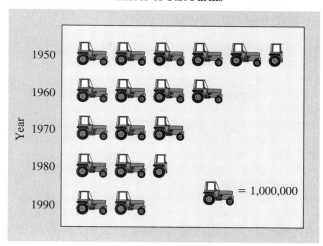

Figure for Exercises 1–8

1. How many farms were in the United States in 1960?

2. How many farms were in the United States in 1980?

3. Did the number of farms increase or decrease from 1980 to 1990?

4. Which 10-year period had the largest decrease?

5. How many more farms were there in 1980 than in 1990?

6. How many more farms were there in 1970 than in 1990?

7. What was the percent of decrease in the number of farms from 1980 to 1990?

8. What was the percent of decrease in the number of farms from 1970 to 1990? (Round off to the nearest percent.)

In Exercises 9–20, use the pictograph below.

College Enrollment

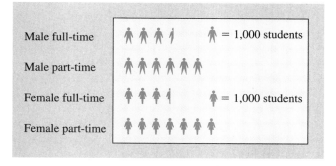

Figure for Exercises 9–20

9. How many male students attend the college full-time?

10. How many female students attend the college part-time?

11. How many students attend the college full-time?

12. How many students altogether attend the college part-time?

13. Estimate the number of male students.

14. Estimate the number of female students.

15. What is the college's total enrollment?

16. What percent of the total enrollment is part-time students?

17. What percent of the total enrollment is full-time students?

18. What percent of the total enrollment is female?

19. What is the ratio of full-time students to part-time students?

20. What is the ratio of male students to female students?

5.3 Line Graphs

Line (or broken-line) graphs are similar to bar graphs, except the data are represented by points instead of bars, then the points are connected in order by line segments. Line graphs are used to show change over a period of time.

EXAMPLE 1 Examine the line graph in Figure 7, and answer the following questions.

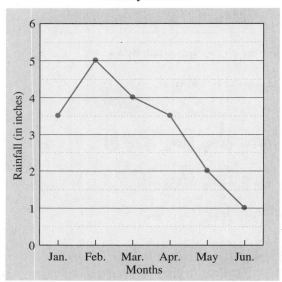

Monthly Rainfall

Figure 7

a. What was the monthly rainfall for January?

SOLUTION   About 3.5 in.

b. What was the monthly rainfall for May?

SOLUTION   About 2 in.

c. What is the ratio of January's rainfall to May's rainfall?

SOLUTION $\dfrac{3.5 \text{ in.}}{2 \text{ in.}} = \dfrac{3.5 \times 10}{2 \times 10} = \dfrac{\overset{7}{\cancel{35}}}{\underset{4}{\cancel{20}}} = \dfrac{7}{4}$

d. Which was the wettest month from January to June?

SOLUTION   February

e. During which months did the rainfall increase?

S O L U T I O N The line segment between January and February is rising; therefore, the rainfall increased from January to February.

f. During which months did the rainfall decrease?

S O L U T I O N All of the line segments between February and June are dropping. This means the rainfall decreased from February to June.

g. What was the percent of decrease in rainfall from February to June?

S O L U T I O N Before we can find the percent of decrease, we must find the *amount* of decrease.

$$5 \text{ in.} - 1 \text{ in.} = 4 \text{ in.} \qquad \text{Amount of decrease}$$

$$4 \text{ is what percent of } 5?$$

Amount of decrease   February's rainfall

$$\frac{4}{5} = \frac{P}{100}$$

$$5P = 400$$

$$P = 80$$

Therefore, the rainfall decreased 80%.

h. Between which two months was the decrease in rainfall the greatest?

S O L U T I O N The greatest decrease in rainfall can be found by looking for the steepest line segment. This occurs between April and May.

EXAMPLE 2   Examine the line graph in Figure 8, and answer the following questions.

Store Sales

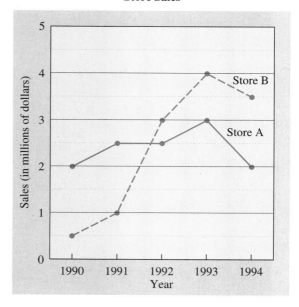

Figure 8

a. What was the amount of sales for Store A in 1992?

S O L U T I O N   The sales for Store A are represented by the solid line. In 1992 Store A's sales were about

$$2.5 \text{ million dollars} = \$2.5 \times 1,000,000$$
$$= \$2,500,000$$

b. Which store had higher sales in 1991?

S O L U T I O N   Store A

c. Which store had higher sales in 1994?

S O L U T I O N   Store B

d. How much more did Store B sell than Store A in 1993?

S O L U T I O N   In 1993 Store A's sales were $3,000,000, and Store B's were $4,000,000. The difference is $4,000,000 − $3,000,000 = $1,000,000 more.

e. Between which two years did sales increase the most for Store B?

S O L U T I O N   Between 1991 and 1992

f. Estimate the total sales for Store A from 1990 to 1994.

S O L U T I O N   In 1990 $ 2,000,000
                        1991   2,500,000
                        1992   2,500,000
                        1993   3,000,000
                        1994   + 2,000,000
                        Total = $12,000,000

g. What percent of the total sales for Store A were its sales in 1993?

S O L U T I O N   3,000,000 is what percent of 12,000,000?

1993 sales   Total sales
for Store A   for Store A

$$\frac{3,000,000}{12,000,000} = \frac{P}{100} \qquad \text{Reduce } \frac{3,000,000}{12,000,000} = \frac{1}{4}$$

$$\frac{1}{4} = \frac{P}{100}$$

$$4P = 100$$

$$P = 25$$

Therefore, 25% of the sales were in 1993.

Slanting the Picture

Both graphs in Figure 9 show the same stock prices, but in the graph on the right the vertical scale was stretched. This makes the line segments rise faster.

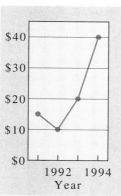

Stock Prices Are Rising

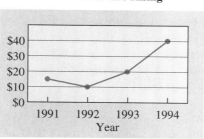

Stock Prices Are Rising

Figure 9

Exercises 5.3

In Exercises 1–10, use the line graph below.

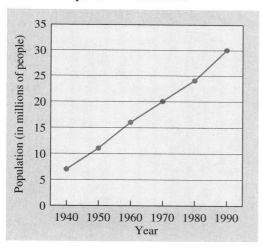

Population of California

Figure for Exercises 1–10

1. What was the population of California in 1970?

2. What was the population of California in 1960?

3. In what year was the population of California about 24,000,000?

4. In what year was the population of California about 7,000,000?

5. How many more people lived in California in 1990 than in 1980?

6. How many more people lived in California in 1990 than in 1970?

7. During which years did the population of California increase?

8. During which years did the population increase the most?

9. What was the percent of increase in population from 1980 to 1990?

10. What was the percent of increase in population from 1970 to 1980?

In Exercises 11–22, use the line graph below.

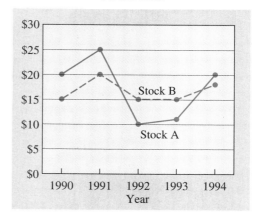

Stock Prices

Figure for Exercises 11–22

11. What was the price of Stock A in 1992?

12. What was the price of Stock B in 1993?

13. What was the highest price for Stock A, and when did it occur?

14. What was the highest price for Stock B, and when did it occur?

15. Which stock had a higher price in 1991?

16. Which stock had a higher price in 1992?

17. Between which two years did the price of Stock A increase the most?

18. Between which two years did the price of Stock B increase the most?

19. How much more was the price of Stock A than that of Stock B in 1991?

20. How much more was the price of Stock A than that of Stock B in 1994?

21. What is the ratio of the price of Stock A to the price of Stock B in 1991?

22. What is the ratio of the price of Stock A to the price of Stock B in 1994?

5.4   Circle Graphs

A circle (or pie) graph is a graph in the form of a circle that is divided into wedges, or slices. The size of each wedge is proportional to the size of the data it represents. Circle graphs are used to compare parts of a whole. The whole circle represents 100%.

EXAMPLE 1   Examine the circle graph in Figure 10, and answer the following questions.

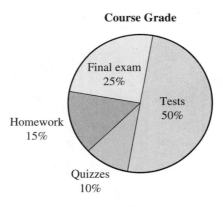

Course Grade

Figure 10

a. What percent of the course grade is based on a student's test scores?

SOLUTION   50% (*Note:* $50\% = \dfrac{50}{100} = \dfrac{1}{2}$ of the whole circle.)

b. What percent of the course grade is based on a student's quiz scores?

SOLUTION   10% (*Note:* $10\% = \dfrac{10}{100} = \dfrac{1}{10}$ of the whole circle.)

c. What percent of the course grade is based on a student's test and quiz scores?

SOLUTION   50% + 10% = 60%

d. Which score will count as $\frac{1}{4}$ of the course grade?

SOLUTION $\frac{1}{4} = 0.25 = 25\%$; therefore, the final exam will count as $\frac{1}{4}$ of the course grade.

e. Which score will count the least?

SOLUTION Since the smallest wedge represents the quizzes, the quiz scores will count the least.

EXAMPLE 2 Examine the circle graph in Figure 11, and answer the following questions.

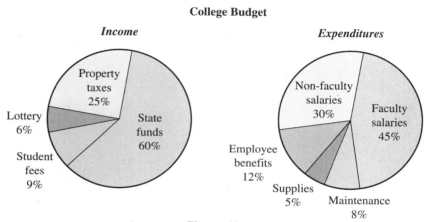

College Budget

Income   *Expenditures*

Figure 11

a. What percent of the college's income comes from property taxes?

SOLUTION 25%

b. What is the largest source of income for the college?

SOLUTION State funds

c. What is the smallest source of income for the college?

SOLUTION Lottery

d. What percent of the college's budget is spent on faculty and non-faculty salaries?

SOLUTION 45% + 30% = 75%

e. If the college's total budget is $50,000,000, how much is spent on supplies?

SOLUTION What is 5% of 50,000,000?

Amount for supplies   Total budget

$$\frac{A}{50,000,000} = \frac{5}{100}$$

$$100A = 250,000,000$$

$$\frac{100A}{100} = \frac{250,000,000}{100}$$

$$A = \$2,500,000 \quad \text{For supplies}$$

f. If the college's total budget is \$50,000,000, how much is spent on faculty salaries?

SOLUTION   What is 45% of 50,000,000?

Faculty   Total
salaries   budget

$$\frac{A}{50,000,000} = \frac{45}{100}$$

$$100A = 2,250,000,000$$

$$\frac{100A}{100} = \frac{2,250,000,000}{100}$$

$$A = \$22,500,000 \quad \text{For faculty salaries}$$

Exercises 5.4

In Exercises 1–12, use the circle graph below.

College Students' Ages

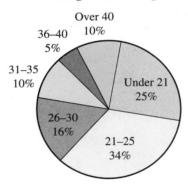

Figure for Exercises 1–12

1. What percent of the students are 21 to 25 years old?

2. What percent of the students are 26 to 30 years old?

3. What percent of the students are over 30 years old?

4. What percent of the students are under 26 years old?

5. What is the largest age group?

6. What is the smallest age group?

7. $\frac{1}{4}$ of the students are in which age group?

8. Approximately $\frac{1}{3}$ of the students are in which age group?

9. If the total number of students is 20,000, how many students are under 21 years old?

10. If the total number of students is 20,000, how many students are 21 to 25 years old?

11. If the college's total enrollment is 30,000 students, how many students are under 21 years old?

12. If the college's total enrollment is 30,000 students, how many students are 21 to 25 years old?

In Exercises 13–24, use the circle graphs below.

Family Budget

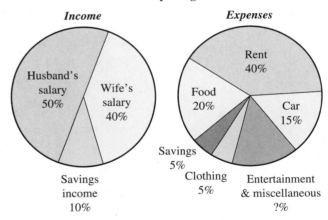

Figure for Exercises 13–24

13. What percent of the family income comes from the wife's salary?

14. What percent of the family income comes from their savings?

15. What percent of the family budget is spent on food?

16. What percent of the family budget is spent on rent?

17. What is the largest source of income?

18. What is the biggest expense for the family?

19. What percent of the family budget is left over for entertainment and miscellaneous expenses?

20. $\frac{1}{5}$ of the family budget is spent for what item?

21. If the total family income after taxes is $30,000 a year, what is the husband's salary?

22. If the total family income after taxes is $30,000 a year, what is the wife's salary?

23. If the total family income after taxes is $30,000 a year, how much is spent on rent each *month*?

24. If the total family income after taxes is $30,000 a year, how much is spent on clothing each *month*?

5.5 Statistics

Statistics helps us to organize and interpret large amounts of data.

Mean

The **mean** of a set of numbers is found by adding all of the numbers and then dividing this sum by the number of numbers. The mean is what we usually refer to as the "average."

$$\text{Mean} = \frac{\text{sum of numbers}}{\text{number of numbers}}$$

EXAMPLE 1 A bowler scored the following in four games: 175, 194, 150, and 165. What is the mean?

SOLUTION $\text{Mean} = \dfrac{175 + 194 + 150 + 165}{4} = \dfrac{684}{4} = 171$

EXAMPLE 2 There are 10 students in a chemistry lab class. On a quiz, 1 student scored 10 points, 4 students scored 9 points, 3 students scored 8 points, and 2 students scored 5 points. Find the class mean.

SOLUTION 1 List all the scores.

$$
\left.
\begin{array}{r}
10 \\
9 \\
9 \\
9 \\
9 \\
8 \\
8 \\
8 \\
5 \\
+\ 5 \\
\hline
80
\end{array}
\right\} \text{10 students}
$$

$\text{Mean} = \dfrac{\text{sum of scores}}{\text{number of students}} = \dfrac{80}{10} = 8$

SOLUTION 2   Group the scores.

Students		*Points*		
1	×	10	=	10
4	×	9	=	36
3	×	8	=	24
+ 2	×	5	=	+ 10
10				80 ← Sum of scores

└─ Number of students

$$\text{Mean} = \frac{\text{sum of scores}}{\text{number of students}} = \frac{80}{10} = 8$$

Median

The **median** of a set of numbers is the *middle* number *after the numbers have been arranged in order of size*. If there are an even number of items, there will be no middle number. In that case we take the average of the two numbers in the middle.

EXAMPLE 3   The daily high temperatures for a week were 75°F, 82°F, 83°F, 85°F, 75°F, 77°F, and 70°F. Find the median.

SOLUTION   To find the median, arrange the temperatures in order of size:

85,   83,   82,   77,   75,   75,   70

The middle number, 77°F, is the median.

EXAMPLE 4   Find the median of the set of numbers $\{19, 16, 12, 15, 13, 9\}$.

SOLUTION   To find the median, arrange the numbers in order of size:

19,   16,   15,   13,   12,   9

$$\text{The median} = \frac{15 + 13}{2} = \frac{28}{2} = 14$$

Mode

The **mode** of a set of numbers is the number that occurs the most often. There may be more than one mode, or there may be no mode.

EXAMPLE 5   Find the mode of the set of numbers $\{3, 3, 4, 6, 6, 6, 7, 9\}$.

SOLUTION   3,   3,   4,   6,   6,   6,   7,   9

Mode

The number 6 occurs the most often; therefore, the mode is 6.

EXAMPLE 6

Find the mode of the set of numbers {24, 17, 19, 24, 26, 17}.

SOLUTION 17, 17, 19, 24, 24, 26

Mode   Mode

Both 17 and 24 occur twice; therefore, there are two modes, 17 and 24.

Range

The **range** of a set of numbers is the difference between the highest and the lowest number.

EXAMPLE 7

Find the range in the following test scores: 78, 95, 86, 64, 72, 52, and 85.

SOLUTION Range = 95 − 52 = 43

Highest − Lowest

EXAMPLE 8

Mrs. Furness paid the following heating bills last winter: $12 in October, $20 in November, $30 in December, $78 in January, $65 in February, and $35 in March. Find the mean, median, mode, and range of the bills.

SOLUTION

$$\text{Mean} = \frac{\$12 + \$20 + \$30 + \$78 + \$65 + \$35}{6} = \frac{\$240}{6} = \$40$$

To find the median, arrange the bills in order of size: $12, $20, $30, $35, $65, $78.

$$\text{Median} = \frac{\$30 + \$35}{2} = \frac{\$65}{2} = \$32.50$$

There is no mode.

$$\text{Range} = \$78 - \$12 = \$66$$

EXAMPLE 9

Compare the salaries paid at Company A with those paid at Company B by finding the mean, median, and range of salaries for each company.

Company A		Company B	
President	$120,000	President	$ 80,000
Vice President	60,000	Vice President	50,000
Manager #1	20,000	Manager #1	30,000
Manager #2	20,000	Manager #2	30,000
Manager #3	20,000	Manager #3	30,000
Manager #4	20,000	Manager #4	30,000
Manager #5	20,000	Manager #5	30,000
Total = $280,000		Total = $280,000	

SOLUTION

Company A	Company B
Mean $= \dfrac{\$280,000}{7} = \$40,000$	Mean $= \dfrac{\$280,000}{7} = \$40,000$
Median $= \$20,000$	Median $= \$30,000$
Range $= \$120,000 - \$20,000$	Range $= \$80,000 - \$30,000$
$= \$100,000$	$= \$50,000$

Although the mean salary for each company is $40,000, the medians are very different. This is because one or two very large numbers, such as the president's salary for Company A, will have a big effect on the mean. We can see the big difference in the salaries for Company A by noting its large salary range. Sometimes the median gives a better idea of the "average" than the mean does.

Exercises   5.5

1. Use the following numbers: $\{6, 4, 3, 4, 7, 9, 2\}$.
   a. Find the mean.
   b. Find the median.
   c. Find the mode.
   d. Find the range.

2. Use the following numbers: $\{3, 7, 9, 3, 4, 6, 9, 1, 3\}$.
   a. Find the mean.
   b. Find the median.
   c. Find the mode.
   d. Find the range.

3. Use the following numbers: $\{22, 26, 29, 21, 20, 32\}$.
   a. Find the mean.
   b. Find the median.
   c. Find the mode.
   d. Find the range.

4. Use the following numbers: $\{98, 74, 83, 81\}$.
   a. Find the mean.
   b. Find the median.
   c. Find the mode.
   d. Find the range.

5. Five people drove a car on the highway to measure the car's fuel consumption in miles per gallon. The results were 24.7 mpg, 18.9 mpg, 20.0 mpg, 16.6 mpg, and 19.8 mpg.
   a. Find the range.
   b. Find the median.
   c. Find the mean.

6. The rainfall for the month was 2.6 inches the first week, 1.7 inches the second week, 0.8 inch the third week, and 1.3 inches the fourth week.
   a. Find the range.
   b. Find the median.
   c. Find the mean.

7. A bowler scored the following in six games: 180, 156, 164, 210, 176, and 170.
   a. Find the median score.
   b. Find the mean score.
   c. Find the range in scores.

8. A golfer scored the following in seven games: 74, 86, 72, 76, 95, 70, and 80.
   a. Find the median score.
   b. Find the mean score.
   c. Find the range in scores.

9. On a math quiz, 2 students scored 10, 1 student scored 9, 2 students scored 8, 3 students 7, and 4 students scored 6.
   a. Find the class mean.
   b. Find the class median.
   c. Find the class mode.
   d. Find the class range.

10. On an English quiz, 1 student scored 20, 3 students scored 14, 4 students scored 12, and 2 students scored 10.

  a. Find the class mean.

  b. Find the class median.

  c. Find the class mode.

  d. Find the class range.

 11. Nine light bulbs were tested to measure the life of a light bulb, in hours. The results are listed below. (Round off answers to the nearest unit when necessary.)

726 hr	764 hr	780 hr
825 hr	801 hr	678 hr
715 hr	690 hr	679 hr

  a. Find the mean.

  b. Find the median.

  c. Find the range.

 12. Mr. Lumus's monthly electric bills for last year are listed below. (Round off answers to the nearest cent when necessary.)

$52.50	$43.43	$51.36
$47.00	$38.00	$45.00
$48.50	$43.95	$61.75
$36.25	$48.27	$70.50

  a. Find the mean.

  b. Find the median.

  c. Find the range.

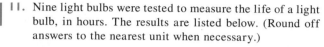

Chapter 5 — REVIEW

Graphs
5.1–5.4
Graphs are used to present data in an organized manner that often makes it easier to compare the data or see a change. Graph formats include bar graphs, pictographs, line graphs, and circle graphs.

Mean
5.5
The **mean** is the sum of the numbers divided by the number of numbers.

$$\text{Mean} = \frac{\text{sum of the numbers}}{\text{number of numbers}}$$

Median
5.5
The **median** is the middle number after the numbers have been arranged in order of size. If there are an even number of numbers, the median is the average of the two numbers in the middle.

Mode
5.5
The **mode** is the number that occurs the most often. There may be more than one mode, or there may be no mode.

Range
5.5
The **range** is the difference between the highest and the lowest number.

Chapter 5   REVIEW EXERCISES

In Exercises 1–10, use the line graph below.

Monthly High Temperatures

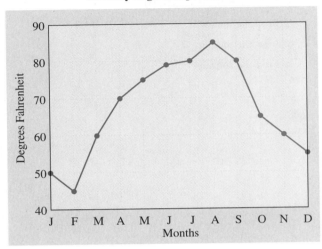

Figure for Exercises 1–10

1. What was the high temperature for March?

2. What was the high temperature for April?

3. Which month was the coldest?

4. Which month was the hottest?

5. During which months did the temperature increase?

6. During which months did the temperature decrease?

7. Between which two months did the temperature increase the most?

8. Between which two months did the temperature decrease the most?

9. What is the percent of decrease in high temperature from January to February?

10. What is the percent of increase in high temperature from February to March?

In Exercises 11–16, use the bar graph at the top of the next column.

11. How many feet does it take to stop a car at 60 mph?

12. How many feet does it take to stop a car at 50 mph?

13. At what speed does it take approximately 40 feet to stop?

14. At what speed does it take approximately 120 feet to stop?

15. If you increase your speed from 20 mph to 40 mph, how many more feet will it take to stop the car?

Stopping Distances for Cars

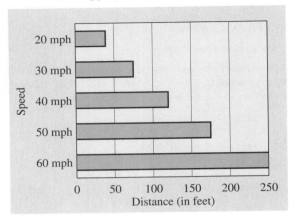

Figure for Exercises 11–16

16. If you increase your speed from 40 mph to 60 mph, how many more feet will it take to stop the car?

In Exercises 17–24, use the circle graph below.

Hair Color

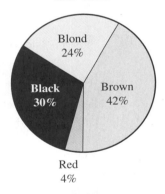

Figure for Exercises 17–24

17. What percent of the people have black hair?

18. What percent of the people have blond hair?

19. What percent of the people do not have red hair?

20. What percent of the people do not have brown hair?

21. Out of 1,000 people, how many would have brown hair?

22. Out of 100,000 people, how many would have brown hair?

23. What is the ratio of the percent of people with blond hair to the percent of people with black hair?

24. What is the ratio of the percent of people with brown hair to the percent of people with black hair?

In Exercises 25–36, use the pictograph below.

Daily Sales

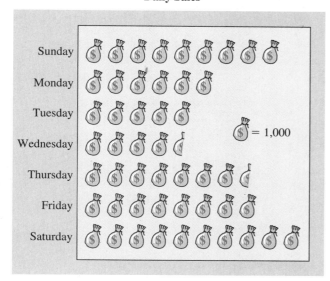

$ = 1,000

Figure for Exercises 25–36

25. What day of the week had the highest sales?

26. What day of the week had the lowest sales?

27. What was the amount of sales on Friday?

28. What was the amount of sales on Thursday?

29. On what day of the week did the sales equal $6,000?

30. On what day of the week did the sales equal $4,500?

31. How much higher were sales on Saturday than on Wednesday?

32. What were the total sales for the week?

33. What percent of the total week's sales were made on Saturday?

34. What percent of the total week's sales were made on Wednesday?

35. What is the ratio of Wednesday's sales to Saturday's sales?

36. What is the ratio of Thursday's sales to Saturday's sales?

In Exercises 37–40, use the following set of numbers: {8, 6, 5, 3, 6, 2}.

37. Find the mean.   38. Find the median.

39. Find the mode.   40. Find the range.

In Exercises 41–44, use the line graph on page 174 for Exercises 1–10. Find the following.

41. Find the mean temperature for the year.

42. Find the median temperature for the year.

43. Find the temperature mode.

44. Find the range of the temperatures.

Chapter 5 Critical Thinking and Writing Problems

Answer Problems 1 and 2 in your own words, using complete sentences.

1. Explain how to find the median of a set of numbers.

2. Explain how to find the range of a set of numbers.

3. From the chart below make a bar graph showing the number of gallons of water used for each activity. Be sure to label the horizontal and vertical scales and to give a title to your graph.

Average Family's Water Usage Per Day

Lawn and Pool	180 gallons
Toilets	100 gallons
Showers and Baths	52 gallons
Laundry	48 gallons
Cooking	20 gallons
Total =	400 gallons

4. Use the chart in Problem 3 to find the percent of water used for each activity. Then sketch a circle graph *showing* the percent of water used for each activity.

5. Is the following word problem set up correctly? Explain your answer.

Births at Community Hospital

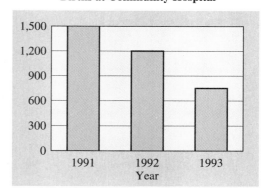

Use the bar graph to find the percent of decrease in births from 1991 to 1992.

$$\frac{1,200}{1,500} = \frac{P}{100}$$

Problems 6 and 7 each have an error. Find the error, and in your own words, explain why it is wrong. Then work the problem correctly.

6. Find the median temperature for the six weeks if recorded temperatures were 84°F, 86°F, 80°F, 72°F, 75°F, and 83°F.

$$84, \quad 86, \quad 80, \quad 72, \quad 75, \quad 83$$

$$\text{Median} = \frac{80 + 72}{2} = \frac{152}{2} = 76°F$$

7. In the ABC school district, 6 schools have 20 computers, 3 schools have 40 computers, 2 schools have 30 computers, and 1 school has 60 computers. Find the mean number of computers at a school in the ABC district.

$$6 \times 20 = 120$$

$$3 \times 40 = 120$$

$$2 \times 30 = 60$$

$$1 \times 60 = 60$$

$$\text{Mean} = \frac{120 + 120 + 60 + 60}{4} = \frac{360}{4} = 90 \text{ computers}$$

Chapter 5   DIAGNOSTIC TEST

Allow yourself about 30 minutes to do these problems. Complete solutions for all problems, together with section references, are given in the answer section at the end of the book.

In Problems 1–5, use the line graph below.

Daytime Temperatures

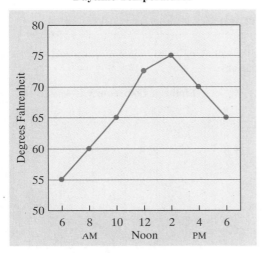

Figure for Problems 1–5

1. What was the temperature at 8 A.M.?

2. At what time was the temperature 75°F?

3. What was the difference in temperature from 4 P.M. to 6 P.M.?

4. During what hours did the temperature increase?

5. What is the percent of increase in temperature from 8 A.M. to 2 P.M.?

In Problems 6–10, use the bar graph below.

Sugar in Cereal

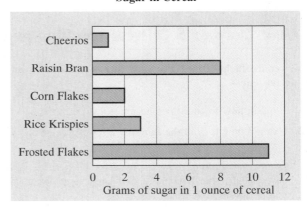

Figure for Problems 6–10

6. Which cereal has the least sugar?

7. How many grams of sugar are in one ounce of Rice Krispies?

8. Which cereal has 8 grams of sugar in 1 ounce of cereal?

9. How many times more sugar is in Raisin Bran than in Corn Flakes?

10. How much more sugar is in one ounce of Frosted Flakes than in one ounce of Rice Krispies?

In Problems 11–15, use the pictograph below.

Car Sales

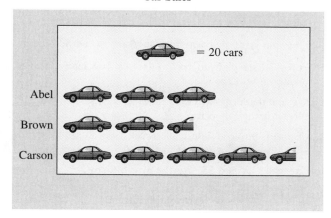

= 20 cars

Figure for Problems 11–15

11. How many cars did Mr. Brown sell?

12. Which sales associate sold the most cars?

13. What is the ratio of cars sold by Ms. Abel to cars sold by Ms. Carson?

14. What were the total sales for all three sales associates?

15. What percent of the total sales did Ms. Abel sell?

In Problems 16–19, use the circle graph below.

Class Grades

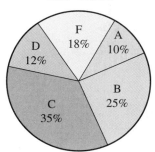

Figure for Problems 16–19

16. What percent of the class got an A?

17. What percent of the class got a C or better?

18. What grade did approximately $\frac{1}{3}$ of the class receive?

19. If there were 48 students in the class, how many students got a B?

In Problems 20 and 21, use the following test scores: 87, 72, 82, 93, and 71.

20. Find the mean test score.

21. Find the median test score.

In Problems 22 and 23, use the following temperatures: 53°F, 66°F, 71°F, and 62°F.

22. Find the temperature range.

23. Find the mean temperature.

In Problems 24 and 25, use the following numbers: 9, 6, 4, 2, 6, 2, 5, 6.

24. Find the median.   25. Find the mode.

Chapters 1–5   CUMULATIVE REVIEW EXERCISES

1. Write, using digits: five thousand, seven hundred and nine hundredths.

In Exercises 2–9, perform the indicated operations. Reduce all fractions to lowest terms, and change any improper fractions to mixed numbers.

2. $\dfrac{2}{9} + \dfrac{5}{6} + \dfrac{1}{2}$

3. $23\dfrac{1}{3} - 10\dfrac{7}{12}$

4. $1\dfrac{1}{6} \cdot 3\dfrac{3}{4}$

5. $\dfrac{\dfrac{1}{2} + \dfrac{2}{3}}{\dfrac{3}{4}}$

6. $54.3 - 5.18$

7. 9.6×0.48

8. $630 \div 1.8$

9. $15 - 12 \div 3 \cdot 2 + 3^2$

10. Arrange in order, largest to smallest: $\dfrac{3}{4}, \dfrac{5}{6}, \dfrac{7}{12}$.

11. Solve for the unknown term: $\dfrac{x}{9} = \dfrac{8}{12}$.

12. Change 0.024 to a fraction in lowest terms.

13. Change $\dfrac{2}{5}$ to a percent.

14. Change 45% to a fraction in lowest terms.

15. 12 is 15% of what number?

16. On a map, $\dfrac{1}{4}$ inch represents 6 miles. How many miles would $2\dfrac{1}{2}$ inches represent?

17. Find the perimeter of a rectangle with length 16 centimeters and width 12 centimeters.

18. A machinist uses a $\dfrac{5}{8}$-inch drill to put a hole through a metal plate. Would a 0.65-inch pin fit through the hole?

19. A woman bought a refrigerator for $665. After making a down payment of $125, she paid the balance in 12 monthly payments. Find the amount of each monthly payment.

20. The rainfall for the first six months was 9.4 inches in January, 10.4 inches in February, 7.8 inches in March, 5.5 inches in April, 2.3 inches in May, and 1.2 inches in June. Find the average rainfall for the first six months.

21. A dressmaker needs $1\dfrac{1}{4}$ yards of material for the top and $3\dfrac{1}{2}$ yards of material for the skirt of a bridesmaid's dress. If she plans to make 4 dresses, how many yards of material will she need?

22. A coat that originally sold for $120 is on sale at 20% off. What is the sale price?

23. Enrollment at a college increased from 8,000 to 9,200. What was the percent of increase?

In Exercises 24 and 25, use the graph below.

Average Monthly Income

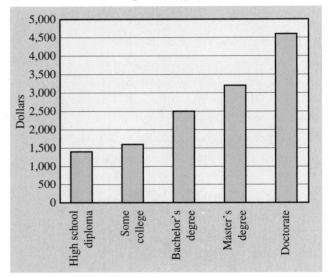

Figure for Exercises 24–25

24. What is the average monthly income with a master's degree?

25. How much more is the average monthly income with a bachelor's degree than with a high school diploma?

Signed Numbers

CHAPTER

6

Arithmetic is calculation with numbers, using fundamental operations such as addition, subtraction, multiplication, and division. Algebra deals with the same fundamental operations with numbers but uses letters to represent some of the numbers.

Before beginning the study of algebra, we review for your benefit a few basic definitions relating to numbers.

6.1   Signed Numbers

Basic Terms

The following basic terms were introduced earlier.

Natural Numbers   The set of **natural numbers** (or *counting numbers*) is

$$N = \{1, 2, 3, 4, 5, \ldots \}$$

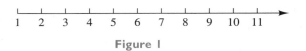

Read "and so on"

The smallest natural number is 1. The largest natural number can never be found because no matter how far we count, there are always larger natural numbers.

Number Line   Natural numbers can be represented by numbered points equally spaced along a straight line (see Figure 1). Such a line is called a **number line**. The arrow on the right of the line indicates that the number line continues forever.

$$1 \quad 2 \quad 3 \quad 4 \quad 5 \quad 6 \quad 7 \quad 8 \quad 9 \quad 10 \quad 11 \longrightarrow$$

Figure 1

Whole Numbers   When 0 is included with the natural numbers, we have the set of numbers known as **whole numbers** (Figure 2).

$$W = \{0, 1, 2, 3, 4, \ldots\}$$

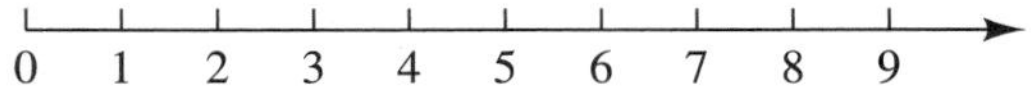

$$0 \quad 1 \quad 2 \quad 3 \quad 4 \quad 5 \quad 6 \quad 7 \quad 8 \quad 9 \longrightarrow$$

Figure 2

Digits   In our number system a **digit** is any one of the first ten whole numbers $\{0, 1, 2, 3, 4, 5, 6, 7, 8, 9\}$. Any number can be written by using a combination of these digits.

Signed Numbers

We now extend the number line to the left and continue with the set of equally spaced points. Numbers used to name the points to the left of 0 on the number line are called *negative numbers*. Numbers used to name the points to the right of 0 on the number line are called *positive numbers*. Zero itself is neither positive nor negative. The positive and negative numbers are referred to as **signed numbers** (see Figure 3).

$$\cdots \; -7 \;\; -6 \;\; -5 \;\; -4 \;\; -3 \;\; -2 \;\; -1 \;\; 0 \;\; +1 \;\; +2 \;\; +3 \;\; +4 \;\; +5 \;\; +6 \;\; +7 \; \cdots$$

Negative numbers   Positive numbers

Figure 3

The numbers used to name the points shown in Figure 3 are called **integers**. The set of integers can be represented in the following way:

$$\{\ldots, -3, -2, -1, 0, +1, +2, +3, \ldots\}$$

We stated above that a largest natural number can never be found because no matter how far we count, there are always larger natural numbers. Similarly, no matter how far we count along the number line to the left of 0, we never reach a smallest integer.

EXAMPLE 1

Examples of reading positive and negative integers:

a. -1 Read "negative one."
b. -575 Read "negative five hundred seventy-five."
c. 25 Read "twenty-five" or "positive twenty-five."

When reading or writing positive numbers, we usually omit the word *positive* and the + sign. Therefore, when there is no sign in front of a number, it is understood to be positive.

EXAMPLE 2

Examples of the use of positive and negative integers:

a. On a cold day in Minnesota, the temperature was $-20°F$. This means that the temperature was 20°F *below* 0.
b. The altitude of Mt. Everest is 29,028 feet. This means that the peak of Mt. Everest is 29,028 ft *above* sea level.
c. The lowest point in Death Valley, California, is -282 feet. This means that the lowest point in Death Valley is 282 ft *below* sea level.

Graphing Points on the Number Line

Many points on the number line are not integers. Decimals, fractions, and mixed numbers are part of the real number system and can be represented by points on the number line (see Figure 4).

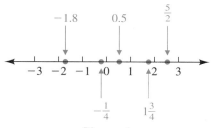

Figure 4

EXAMPLE 3 On the number line below, find the letter that best locates each of the following numbers.

$$A \quad B \qquad C \quad DEF \quad G$$

$$\begin{array}{c} \text{—} \end{array} \qquad -3 \quad -2 \quad -1 \quad 0 \quad 1 \quad 2 \quad 3$$

SOLUTION

a. $1\frac{1}{2}$ $1\frac{1}{2}$ is $1\frac{1}{2}$ units to the right of zero; therefore, $1\frac{1}{2} = G$.

b. $-\frac{1}{2}$ $-\frac{1}{2}$ is $\frac{1}{2}$ unit to the left of zero; therefore, $-\frac{1}{2} = C$.

c. $-2\frac{1}{3}$ $-2\frac{1}{3}$ is $2\frac{1}{3}$ units to the left of zero; therefore, $-2\frac{1}{3} = A$.

d. 0.75   0.75 is $\frac{75}{100} = \frac{3}{4}$ unit to the right of zero; therefore, $0.75 = F$.

Using Inequality Symbols   Recall that numbers get larger as we move to the right on the number line and smaller as we move to the left.

EXAMPLE 4 $-3 > -7$

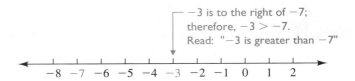

-3 is to the right of -7; therefore, $-3 > -7$. Read: "-3 is greater than -7"

EXAMPLE 5 $-5 < -2$

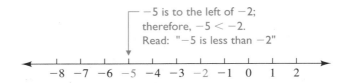

-5 is to the left of -2; therefore, $-5 < -2$. Read: "-5 is less than -2"

An easy way to remember the meaning of the inequality symbol is to notice that the wide part of the symbol is next to the larger number. (See Figure 5.) Some people like to think of the symbols $>$ and $<$ as arrowheads that point toward the smaller number.

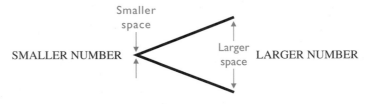

Smaller space

SMALLER NUMBER   Larger space   LARGER NUMBER

Figure 5

EXAMPLE 6 Read the following inequalities aloud, and then verify them by using a number line to determine whether the first number of each pair is to the right or left of the second number of that pair.

SOLUTION

a. $6 > 4$ Read "6 is greater than 4."

b. $-2 > -7$ Read "−2 is greater than −7."

c. $-1 < -\dfrac{1}{2}$ Read "−1 is less than $-\dfrac{1}{2}$."

d. $-5 < 3$ Read "−5 is less than 3."

e. $3 > -5$ Read "3 is greater than −5."

Note that in Examples 6d and 6e, $-5 < 3$ and $3 > -5$ give the same information even though they are read differently.

Exercises 6.1

1. Write −75 in words.

2. Write −49 in words.

3. Use digits to write negative fifty-four.

4. Use digits to write negative one hundred nine.

5. A scuba diver descends to a depth of sixty-two feet. Represent this number by an integer.

6. The temperature in Fairbanks, Alaska, was forty-five degrees Fahrenheit below zero. Represent this number by an integer.

7. Which is larger, −2 or −4?

8. Which is larger, 0 or −10?

9. Which is smaller, −5 or −10?

10. Which is smaller, −1 or −15?

11. What is the largest negative integer?

12. What is the largest integer?

13. What is the smallest integer?

14. What is the smallest whole number?

In Exercises 15–22, find the letter that best locates each number on the number line below.

Figure for Exercises 15–22

15. $1\dfrac{1}{4}$ 16. 1.75   17. $-\dfrac{1}{2}$ 18. $-\dfrac{2}{3}$

19. $-2\dfrac{2}{3}$ 20. −1.3   21. 0.6   22. $\dfrac{9}{4}$

In Exercises 23–30, determine which symbol, > or <, should be used to make each statement true.

23. 0 ? −3   24. −2 ? −6   25. −5 ? 2

26. −7 ? −4   27. $-\dfrac{1}{2}$? $-\dfrac{1}{4}$ 28. $-2\dfrac{1}{3}$? $-2\dfrac{2}{3}$

29. −0.2 ? −0.7   30. $\dfrac{1}{4}$? −0.5

31. In 1985, Caesar's Head, South Carolina, had a record low temperature of −19°F. On the same day, Mt. Mitchell, North Carolina, had a record low temperature of −34°F. Which city was colder?

32. The average depth of the Pacific Ocean is −12,925 feet, and the average depth of the Atlantic Ocean is −11,730 feet. Which ocean is deeper?

6.2 Adding Signed Numbers

In this section we show how to add signed numbers. We can represent a signed number on the number line by an arrow that begins at the point representing 0 and ends at the point representing that particular number.

EXAMPLE 1   Represent 4 by an arrow.

SOLUTION

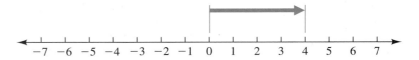

This arrow represents a movement of 4 units to the *right*. Any *positive* number is represented by an arrow directed to the *right*. The arrow need not start at zero so long as it has a length equal to the number it represents.

EXAMPLE 2   Represent −5 by an arrow.

SOLUTION

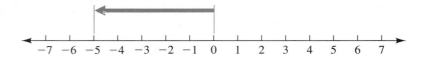

This arrow represents a movement of 5 units to the *left*. Any *negative* number is represented by an arrow directed to the *left*.

We can also represent the addition of signed numbers by means of arrows.

EXAMPLE 3   Add 3 to 2 by means of arrows.

SOLUTION

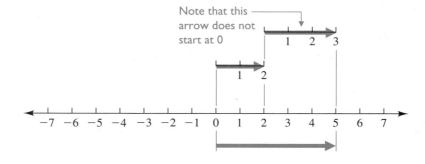

To add 3 to 2 on the number line, we begin by drawing the arrow representing 2. We then draw the arrow representing 3, starting at the arrowhead end of the arrow representing 2. These two movements represent a net movement to the right of 5 units. Therefore, $2 + 3 = 5$.

EXAMPLE 4   Add −7 to 5 by means of arrows.

SOLUTION

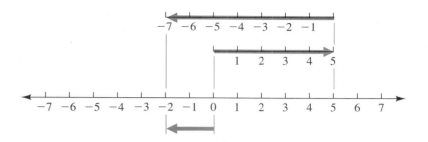

We begin by drawing the arrow representing 5. We then draw the arrow representing −7, starting at the arrowhead end of the arrow representing 5. These two movements represent a net movement to the left of 2 units. Therefore, $5 + (-7) = -2$.

EXAMPLE 5   Add −4 to −3 by means of arrows.

SOLUTION

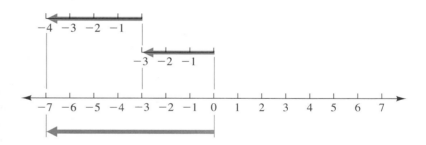

We begin by drawing the arrow representing −3. We then draw the arrow representing −4, starting at the arrowhead end of the arrow representing −3. These two movements represent a net movement to the left of 7 units. Therefore, $-3 + (-4) = -7$.

Absolute Value   The **absolute value** of a number is the distance between that number and 0 on the number line, *with no regard for direction*. See Figure 6. The absolute value of a real number x is written $|x|$.

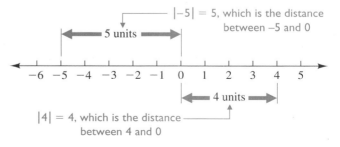

Figure 6

EXAMPLE 6

Examples of absolute value:

a. $|9| = 9$ A positive number
b. $|0| = 0$ Zero
c. $|-4| = 4$ A positive number

The absolute value of a number can never be negative

A signed number has two distinct parts: its absolute value and its sign.

Sign ⟶   Absolute value   Sign ⟶   Absolute value   Sign ⟶   Absolute value

$-\ 2$ $+\ 5$ 13

Sign understood to be $+$

 Note The absolute value of a signed number is the number written without its sign.

Adding signed numbers by means of arrows is easy to understand, but it's a very slow process. The following rules give an easier and faster method for adding signed numbers. The previous four examples can be used to show why the following rules are true.

Adding signed numbers

1. When the numbers have the *same sign*	First: Add their absolute values.
	Second: Give the sum the same sign as both numbers.
2. When the numbers have *different signs*	First: Subtract the smaller absolute value from the larger absolute value.
	Second: Give the result the sign of the number with the larger absolute value.

In the following examples we show how to add signed numbers by means of the rules.

EXAMPLE 7

Find $(-7) + (-11)$.

SOLUTION   Because -7 and -11 have the *same* sign, we *add* their absolute values: $7 + 11 = 18$. The sum has the same sign $(-)$. Therefore,

$$(-7) + (-11) = -18$$

EXAMPLE 8

Find $(-24) + (17)$.

SOLUTION   Because -24 and 17 have *different* signs, we *subtract* the smaller absolute value from the larger absolute value: $24 - 17 = 7$. The answer has the sign of the number with the larger absolute value. Since -24 has the larger absolute value, the sign is $-$.

$$(-24) + (17) = -7$$

EXAMPLE 9

Find $(-18) + (32)$.

SOLUTION Because -18 and 32 have *different* signs, we *subtract* the smaller absolute value from the larger absolute value: $32 - 18 = 14$. The answer has the sign of the number with the larger absolute value. Since 32 has the larger absolute value, the sign is $+$.

$$(-18) + (32) = +14, \quad \text{or} \quad 14$$

EXAMPLE 10

Find $(-29) + (-35)$.

SOLUTION

The numbers have the *same* sign

Add their absolute values: $29 + 35 = 64$

$$(-29) + (-35) = -64$$

The answer has the same sign $(-)$

EXAMPLE 11

Find $(-9) + (+23)$.

SOLUTION

The numbers have *different* signs

Subtract their absolute values: $23 - 9 = 14$

$$(-9) + (+23) = +14$$

The answer has the sign of the number with the larger absolute value $(+)$

EXAMPLE 12

Find $\left(-2\dfrac{1}{2}\right) + \left(-4\dfrac{1}{3}\right)$.

SOLUTION LCD $= 6$

$$\left(-2\dfrac{1}{2}\right) = \left(-2\dfrac{3}{6}\right)$$

Because the signs are the *same*, *add* their absolute values

$$+ \left(-4\dfrac{1}{3}\right) = \left(-4\dfrac{2}{6}\right)$$

$$-6\dfrac{5}{6}$$ The sum has the same sign $(-)$

Therefore,

$$\left(-2\dfrac{1}{2}\right) + \left(-4\dfrac{1}{3}\right) = -6\dfrac{5}{6}$$

EXAMPLE 13

Find the following, using a calculator. To enter a negative number, press the $\boxed{+/-}$ after the numeral.

SOLUTION

a. $(-6) + (3)$ *Key in* $6 \boxed{+/-} \boxed{+} 3 \boxed{=}$
                       Answer -3
b. $(-2) + (-5)$ *Key in* $2 \boxed{+/-} \boxed{+} 5 \boxed{+/-} \boxed{=}$
                       Answer -7

Additive Inverses If the sum of two numbers is zero, they are called **additive inverses** or **opposites** of each other. To find the additive inverse, or opposite, of a signed number, simply change the sign of the number.

Additive inverse property	If a represents any real number, then $$a + (-a) = 0$$ Opposites

EXAMPLE 14
a. The opposite of 5 is -5. $\quad (5) + (-5) = 0$
b. The additive inverse of -10 is 10. $\quad (-10) + (+10) = 0$
c. The opposite of $\dfrac{3}{4}$ is $-\dfrac{3}{4}$. $\quad \left(+\dfrac{3}{4}\right) + \left(-\dfrac{3}{4}\right) = 0$

Exercises 6.2

In Exercises 1–56, find the sums.

1. $(4) + (5)$
2. $(6) + (2)$
3. $(-3) + (-4)$
4. $(-7) + (-1)$
5. $(-6) + (5)$
6. $(-8) + (3)$
7. $(7) + (-3)$
8. $(9) + (-4)$
9. $(-2) + (-9)$
10. $(-4) + (-8)$
11. $(-5) + (8)$
12. $(-3) + (6)$
13. $(5) + (6)$
14. $(4) + (2)$
15. $(2) + (-5)$
16. $(6) + (-8)$
17. $(-1) + (-8)$
18. $(-5) + (-5)$
19. $(-9) + (2)$
20. $(-7) + (4)$
21. $(3) + (-3)$
22. $(-9) + (9)$
23. $(8) + (-1)$
24. $(6) + (-4)$
25. $(-3) + (-9)$
26. $(-5) + (-7)$
27. $(6) + (-15)$
28. $(4) + (-12)$
29. $(-8) + (2)$
30. $(-6) + (-5)$
31. $(4) + (-1)$
32. $(7) + (-5)$
33. $(-3) + (-6)$
34. $(-9) + (4)$
35. $(-7) + (9)$
36. $(-8) + (-5)$
37. $(-18) + (-3)$
38. $(-16) + (9)$
39. $(-8) + (15)$
40. $(-3) + (-17)$
41. $(15) + (-5)$
42. $(18) + (-6)$
43. $(-27) + (-13)$
44. $(-42) + (-12)$
45. $(-80) + (121)$
46. $(69) + (-134)$
47. $(-105) + (73)$
48. $(218) + (-113)$
49. $\left(-1\dfrac{1}{2}\right) + \left(-3\dfrac{2}{5}\right)$
50. $\left(-2\dfrac{1}{2}\right) + \left(-5\dfrac{1}{4}\right)$
51. $\left(4\dfrac{5}{6}\right) + \left(-1\dfrac{1}{3}\right)$
52. $\left(6\dfrac{3}{4}\right) + \left(-2\dfrac{1}{8}\right)$
53. $(-7.3) + (-5.48)$
54. $(-12.67) + (-4.092)$
55. $(1.03) + (-0.946)$
56. $(-43.2) + (9.85)$

57. At 6 A.M. the temperature in Hibbing, Minnesota, was $-35°F$. If the temperature had risen $53°F$ by 2 P.M., what was the temperature at that time?

58. At midnight in Billings, Montana, the temperature was $-50°F$. By noon the temperature had risen $67°F$. What was the temperature at noon?

59. Find the opposite of 7.

60. Find the additive inverse of 4.

61. Find the additive inverse of -6.

62. Find the opposite of -18.

63. Find the opposite of $\dfrac{2}{3}$.

64. Find the opposite of $-\dfrac{5}{7}$.

In Exercises 65–68, use a calculator to find the sums.

65. $(-8) + 3$
66. $(-9) + (-6)$
67. $(-12) + (-26)$
68. $(14) + (-36)$

6.3 Subtracting Signed Numbers

The opposite of a number is used in the definition of subtraction.

Definition of subtraction

$$a - b = a + (-b)$$

In words: To subtract b from a, add the opposite of b to a.

This definition leads to the following rule for subtracting signed numbers.

Subtracting one signed number from another

1. Change the subtraction symbol to an addition symbol, and change the sign of the number being subtracted.
2. Add the resulting signed numbers, as shown in the previous section. For example,

$$(8) - (\ +5\ )$$
$$= (8) + (\ -5\ ) \quad \text{Change the sign of the number being subtracted}$$
$$= 3 \quad \text{Change the subtraction symbol to an addition symbol}$$

EXAMPLE 1 Find $(-3) - (-7)$.

SOLUTION

Change subtraction to addition
Change the sign of the number being subtracted

$$(-3) - (\ -\ 7)$$
$$= (-3) + (\ +\ 7)$$
$$= 4$$

EXAMPLE 2 Find $(-5) - (2)$.

SOLUTION

Change subtraction to addition
Change the sign of the number being subtracted

$$(-5) - (\ +\ 2)$$
$$= (-5) + (\ -\ 2)$$
$$= -7$$

EXAMPLE 3   Find $(9) - (-8)$.

SOLUTION

$$(+9) - (-8)$$
$$= (+9) + (+8)$$
$$= 17$$

EXAMPLE 4   Subtract 87 from 25.

SOLUTION   To subtract 87 from 25 means to find $25 - 87$.

$$(+25) - (+87)$$
$$= (+25) + (-87)$$
$$= -62$$

EXAMPLE 5   Find $\left(-3\dfrac{1}{2}\right) - \left(-2\dfrac{1}{4}\right)$.

SOLUTION   LCD = 4

$$\left(-3\frac{1}{2}\right) - \left(-2\frac{1}{4}\right) \quad \text{or}$$

$$= \left(-3\frac{1}{2}\right) + \left(+2\frac{1}{4}\right)$$

$$= \left(-3\frac{2}{4}\right) + \left(+2\frac{1}{4}\right)$$

$$= -1\frac{1}{4}$$

$$\left(-3\frac{1}{2}\right) = \left(-3\frac{2}{4}\right)$$

$$-\left(-2\frac{1}{4}\right) = +\left(+2\frac{1}{4}\right)$$

$$ -1\frac{1}{4}$$

Because the signs are *different, subtract* their absolute values

The sum has the sign of the number with the larger absolute value $(-)$

EXAMPLE 6   Find $(-16.5) - (9.83)$.

SOLUTION

$$(-16.5) - (+9.83) \quad \text{or}$$

$$= (-16.5) + (-9.83)$$

$$= -26.33$$

$$(-16.5) = (-16.50)$$

$$-(+9.83) = +(-9.83)$$

$$ -26.33$$

Because the signs are the *same, add* their absolute values

The sum has the same sign $(-)$

Exercises 6.3

In Exercises 1–56, find the differences.

1. $(10) - (4)$	**2.** $(12) - (5)$	**3.** $(-3) - (-2)$
4. $(-4) - (-3)$	**5.** $(-6) - (2)$	**6.** $(-8) - (5)$
7. $(9) - (-5)$	**8.** $(7) - (-3)$	**9.** $(2) - (-7)$
10. $(3) - (-5)$	**11.** $(-6) - (-8)$	**12.** $(-3) - (-5)$

13. $(-9) - (-2)$	**14.** $(-7) - (-4)$	**15.** $(3) - (7)$
16. $(1) - (9)$	**17.** $(-6) - (-9)$	**18.** $(-5) - (-1)$
19. $(-7) - (5)$	**20.** $(-8) - (2)$	**21.** $(-8) - (-1)$
22. $(-4) - (-2)$	**23.** $(-9) - (-3)$	**24.** $(-5) - (-2)$
25. $(6) - (-8)$	**26.** $(-3) - (7)$	**27.** $(-2) - (-9)$

28. $(4) - (6)$ 29. $(-5) - (7)$ 30. $(-9) - (-4)$

31. $(5) - (8)$ 32. $(7) - (-7)$ 33. $(-10) - (-6)$

34. $(9) - (15)$ 35. $(-15) - (-20)$

36. $(8) - (-17)$ 37. $(-16) - (9)$

38. $(-26) - (8)$ 39. $(34) - (89)$

40. $(47) - (53)$ 41. $(-156) - (-97)$

42. $(-203) - (-168)$ 43. $(384) - (-279)$

44. $(136) - (-275)$ 45. $\left(-4\frac{2}{3}\right) - \left(2\frac{1}{6}\right)$

46. $\left(5\frac{7}{8}\right) - \left(1\frac{1}{4}\right)$ 47. $\left(-3\frac{3}{4}\right) - \left(-2\frac{1}{6}\right)$

48. $\left(9\frac{5}{6}\right) - \left(-6\frac{4}{9}\right)$ 49. Subtract (-7) from (-10).

50. Subtract (-6) from (8).   51. $(-7.3) - (2.06)$

52. $(9.48) - (-26.4)$ 53. $(-56.2) - (-8.53)$

54. $(0.375) - (0.972)$

55. Subtract $\left(3\frac{1}{4}\right)$ from $\left(-5\frac{3}{10}\right)$.

56. Subtract $\left(1\frac{1}{2}\right)$ from $\left(4\frac{5}{6}\right)$.

57. At 5 A.M. the temperature at Mammoth Mountain, California, was $-7°F$. At noon the temperature was $42°F$. What was the rise in temperature?

58. At 4 A.M. the temperature in Massena, New York, was $-5.6°F$. At 1 P.M. the temperature was $37.5°F$. What was the rise in temperature?

59. A jeep starting from the shore of the Dead Sea $(-1,299$ ft) was driven to the top of a nearby hill having an elevation of 723 ft. What was the change in the jeep's altitude?

60. A dune buggy starting from the floor of Death Valley $(-282$ ft) was driven to the top of a nearby mountain having an elevation of 5,782 ft. What was the change in the dune buggy's altitude?

61. A scuba diver descends to a depth of 141 ft below sea level. His buddy dives 68 ft deeper. What is his buddy's altitude at the deepest point of her dive?

62. Mt. Everest (the highest known point on earth) has an altitude of 29,028 ft. The Mariana Trench in the Pacific Ocean (the lowest known point on earth) has an altitude of $-36,198$ ft. Find the difference in altitude of these two places.

 In Exercise 63–66, use a calculator fo find the differences.

63. $(8) - (-2)$ 64. $(-4) - (3)$

65. $(-17) - (-6)$ 66. $(-12) - (-20)$

6.4 Multiplying Signed Numbers

Terms Used in Multiplication   The numbers being multiplied are called **factors**; the answer is called the **product**. Below, 6 and 2 are factors of 12; 12 is the product of 6 and 2.

$$6 \times 2 = 12$$

Factors ⎯⎯⎯⎯   ⎯⎯ Product

Likewise, 3 and 4 are factors of 12 because $3 \times 4 = 12$.

Multiplicative Identity   Because multiplying any real number by 1 gives the identical real number, 1 is called the **multiplicative identity element**.

Multiplicative identity property	If a represents any real number, $$1 \cdot a = a \cdot 1 = a$$

Symbols Used in Multiplication   Multiplication can be shown in several different ways:

$3 \times 2 = 6$

$3 \cdot 2 = 6$ The multiplication dot " $\cdot$ " is written a little higher than the decimal point

$ab = a \cdot b$ When two expressions are written next to each other with no symbol of operation, it is understood that they are to be multiplied (*Exception*: When two *numbers* are written next to each other, they are *not* to be multiplied; for example, 23 does *not* mean $2 \cdot 3 = 6$)

$3x = 3 \cdot x$

$3(2) = 6$ The symbols ( ) are called *parentheses*

$(3)(2) = 6$ In this kind of multiplication, the double parentheses are not necessary

Multiplying Two Signed Numbers

Multiplication (by a positive integer) is a short method for doing repeated addition of the same number.

EXAMPLE 1

$3 \cdot 5 = \text{three 5's} = 5 + 5 + 5 = 15$

EXAMPLE 2

$6 \cdot 2 = \text{six 2's} = 2 + 2 + 2 + 2 + 2 + 2 = 12$

Carrying this same idea over into multiplying signed numbers, we have the following.

EXAMPLE 3

$3 \cdot (-2) = \text{three } -2\text{'s} = (-2) + (-2) + (-2) = -6$

EXAMPLE 4

$2 \cdot (-6) = \text{two } -6\text{'s} = (-6) + (-6) = -12$

From Examples 3 and 4 we see that *when two numbers having opposite signs are multiplied, their product is negative.*

EXAMPLE 5

$(-1)(5) = -5$ The opposite of 5

EXAMPLE 6

$(-1)(8) = -8$ The opposite of 8

From Examples 5 and 6 we see that multiplying a number by -1 gives the opposite of that number.

$$(-1) \cdot a = a \cdot (-1) = -a$$

We can use the above rule to find the product of two negative numbers.

EXAMPLE 7

$(-5)(-4) = (-1)(5)(-4)^*$ Because $-5 = (-1)(5)$

$\qquad\quad = (-1)(-20)$ Because $(5)(-4) = -20$

$\qquad\quad = +20$ Because -1 times a number gives its opposite and the opposite of -20 is $+20$

From Example 7 we see that *when two negative numbers are multiplied, their product is positive.* The fact that the product of two negative numbers is positive can also be seen from the following pattern:

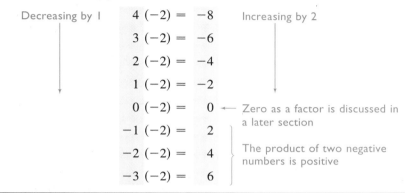

The rules for multiplying two signed numbers are summarized as follows:

Multiplying two signed numbers

1. Multiply their absolute values.
2. The product is *positive* when the signed numbers have the same sign.
   The product is *negative* when the signed numbers have different signs.

$$+ \cdot + = + \qquad\qquad + \cdot - = -$$
$$- \cdot - = + \qquad\qquad - \cdot + = -$$

EXAMPLE 8

Multiply $(-7)(4)$.

SOLUTION $(-7)(4) = -28$

— Product of their absolute values: $7 \cdot 4 = 28$

— The product is negative because the numbers have different signs

EXAMPLE 9

Multiply $(-14)(-10)$.

SOLUTION $(-14)(-10) = +140$

— $14 \cdot 10 = 140$

— The product is positive because the numbers have the same sign

*These arguments depend upon the commutative and associative properties, which are discussed in a later section.

EXAMPLE 10   Multiply $\left(4\dfrac{1}{2}\right)\left(-1\dfrac{1}{3}\right)$.

SOLUTION $\left(4\dfrac{1}{2}\right)\left(-1\dfrac{1}{3}\right) = -\left(\dfrac{\overset{3}{\cancel{9}}}{\underset{1}{\cancel{2}}} \cdot \dfrac{\overset{2}{\cancel{4}}}{\underset{1}{\cancel{3}}}\right) = -\dfrac{6}{1} = -6$

EXAMPLE 11   Multiply $(-2.7)(-4.6)$.

SOLUTION $(-2.7)(-4.6) = +(2.7 \times 4.6) = 12.42$

EXAMPLE 12   Multiply $(-2)(-5)(-3)$.

SOLUTION $(-2)(-5)(-3) = (+10)(-3) = -30$

 Note With an *odd* number of negative factors, the product is *negative*.

EXAMPLE 13   Multiply $(-2)(-5)(-3)(-4)$.

SOLUTION $(-2)(-5)(-3)(-4) = (+10)(-3)(-4) = (-30)(-4) = +120$

 Note With an *even* number of negative factors, the product is *positive*.

Multiplicative Inverses   If the product of two numbers is 1, they are called **multiplicative inverses** or **reciprocals** of each other. To find the multiplicative inverse, or reciprocal, of a signed number, write the signed number as a fraction, then invert the fraction. The sign of the fraction will not change.

Multiplicative inverse property

> If a represents any real number except zero, then
>
> $$a \cdot \frac{1}{a} = 1$$
>
> $\underset{\text{Reciprocals}}{\underset{\uparrow\quad\uparrow}{}}$

EXAMPLE 14

a. The reciprocal of 5 is $\dfrac{1}{5}$. $5 \cdot \dfrac{1}{5} = \dfrac{5}{1} \cdot \dfrac{1}{5} = \dfrac{5}{5} = 1$

b. The reciprocal of $-\dfrac{1}{3}$ is -3. $\left(-\dfrac{1}{3}\right)(-3) = +\left(\dfrac{1}{3} \cdot \dfrac{3}{1}\right) = \dfrac{3}{3} = 1$

c. The multiplicative inverse of $\dfrac{3}{4}$ is $\dfrac{4}{3}$. $\dfrac{3}{4} \cdot \dfrac{4}{3} = \dfrac{12}{12} = 1$

Exercises 6.4

In Exercises 1–44, find the products.

1. $3(-2)$	2. $4(-6)$	3. $(-5)(2)$	4. $(-7)(5)$
5. $(-8)(-2)$	6. $(-6)(-7)$	7. $8(-4)$	8. $9(-5)$
9. $(-6)(3)$	10. $(-8)(-5)$	11. $(9)(-6)$	12. $(-8)(8)$

13. $(-7)(-4)$	14. $9(-3)$	15. $(-5)(6)$	16. $(-3)(-9)$
17. $(-7)(9)$	18. $(-6)(8)$		19. $(-10)(-10)$
20. $(-9)(-9)$	21. $(8)(-7)$		22. $(12)(-6)$
23. $(-26)(10)$	24. $(-11)(12)$		25. $(-20)(-10)$

26. $(-30)(-20)$ **27.** $(75)(-15)$ **28.** $(86)(-13)$

29. $(-30)(5)$ **30.** $(-50)(6)$ **31.** $(-7)(-20)$

32. $(-9)(-40)$ **33.** $(-5)(-4)(-2)$ **34.** $(-3)(2)(-8)$

35. $(-4)(-2)(-1)(-7)$ **36.** $(3)(-2)(-6)(-5)$

37. $\left(2\dfrac{1}{4}\right)\left(-1\dfrac{1}{3}\right)$ **38.** $\left(-3\dfrac{3}{4}\right)\left(-2\dfrac{2}{5}\right)$

39. $\left(-1\dfrac{7}{8}\right)\left(-2\dfrac{4}{5}\right)$ **40.** $\left(-3\dfrac{1}{3}\right)\left(2\dfrac{1}{4}\right)$

41. $(2.74)(-100)$ **42.** $(-3.04)(-1{,}000)$

43. $(-4.6)(-8.3)$ **44.** $(-9.7)(0.52)$

45. Find the reciprocal of 7.

46. Find the multiplicative inverse of 4.

47. Find the multiplicative inverse of -6.

48. Find the reciprocal of -18.

49. Find the reciprocal of $\dfrac{2}{3}$.

50. Find the multiplicative inverse of $-\dfrac{5}{7}$.

In Exercises 51–54, use a calculator to find the products.

51. $(-9)(8)$ **52.** $(7)(-6)$

53. $(-15)(-25)$ **54.** $(-5)(-9)(-7)$

6.5   Dividing Signed Numbers

Division may be shown in several ways.

$$12 \div 4 = \frac{12}{4} = 4\overline{)12}$$

Terms Used in Division   The number we divide *into* is called the **dividend**; the number we divide *by* is called the **divisor**. The answer (which, in the case above, is 3) is known as the **quotient**.

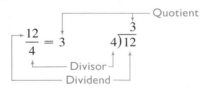

Division is the *inverse* operation of multiplication. Every division problem is related to a multiplication problem.

Because		it follows that	
Because	$4 \cdot 3 = 12,$	it follows that	$12 \div 3 = 4$
Because	$(4) \cdot (-3) = -12,$	it follows that	$(-12) \div (-3) = 4$
Because	$(-4) \cdot (-3) = 12,$	it follows that	$(12) \div (-3) = -4$
Because	$(-4) \cdot (3) = -12,$	it follows that	$(-12) \div (3) = -4$

Dividing Two Signed Numbers

From the examples above we see that we need the same rules for *dividing* signed numbers that we used for *multiplying* signed numbers.

Dividing one signed number by another

1. Divide their absolute values.
2. The quotient is *positive* when the signed numbers have the same sign.
   The quotient is *negative* when the signed numbers have different signs.

$$+ \div + = + \qquad\qquad + \div - = -$$

$$- \div - = + \qquad\qquad - \div + = -$$

EXAMPLE 1 Find $(-30) \div (5)$.

SOLUTION $(-30) \div (5) = -\ 6$

Quotient of their absolute values: $30 \div 5 = 6$

The quotient is negative because the numbers have different signs

---

EXAMPLE 2 Find $(-64) \div (-8)$.

SOLUTION $(-64) \div (-8) = +\ 8$

$64 \div 8 = 8$

The quotient is positive because the numbers have the same sign

---

EXAMPLE 3 Find $\dfrac{2\frac{2}{3}}{-1\frac{1}{9}}$.

SOLUTION $\dfrac{2\frac{2}{3}}{-1\frac{1}{9}} = \dfrac{\frac{8}{3}}{-\frac{10}{9}} = \dfrac{8}{3} \div \left(-\dfrac{10}{9}\right) = -\left(\dfrac{\overset{4}{\cancel{8}}}{\underset{1}{\cancel{3}}} \cdot \dfrac{\overset{3}{\cancel{9}}}{\underset{5}{\cancel{10}}}\right) = -\dfrac{12}{5} = -2\dfrac{2}{5}$

Exercises 6.5

In Exercises 1–36, find the quotients.

1. $(-10) \div (-5)$
2. $(-12) \div (-4)$
3. $(-8) \div (2)$

4. $(-6) \div (3)$
5. $\dfrac{+10}{-2}$
6. $\dfrac{+8}{-4}$

7. $\dfrac{-6}{-3}$
8. $\dfrac{-10}{-2}$
9. $(-40) \div (8)$

10. $(-60) \div (10)$
11. $16 \div (-4)$
12. $25 \div (-5)$

13. $(-15) \div (-5)$
14. $(-27) \div (-9)$
15. $12 \div (-4)$

16. $24 \div (-6)$
17. $\dfrac{-18}{-2}$
18. $\dfrac{-49}{-7}$

19. $\dfrac{-150}{10}$
20. $\dfrac{-250}{100}$
21. $36 \div (-12)$

22. $56 \div (-8)$
23. $(-45) \div 15$
24. $(-39) \div 13$

25. $\dfrac{-15}{6}$
26. $\dfrac{-27}{12}$
27. $\dfrac{7.5}{-0.5}$

28. $\dfrac{1.25}{-0.25}$
29. $\dfrac{-6.3}{-0.9}$
30. $\dfrac{-4.8}{-0.6}$

31. $\dfrac{-367}{100}$
32. $\dfrac{-4,860}{1,000}$
33. $\dfrac{2\frac{1}{2}}{-5}$

34. $\dfrac{3\frac{2}{5}}{-17}$
35. $\dfrac{-4\frac{1}{2}}{-1\frac{7}{8}}$
36. $\dfrac{-3\frac{3}{4}}{-2\frac{1}{10}}$

In Exercises 37–40, use a calculator to find the quotients.

37. $(-84) \div (7)$
38. $(-72) \div (-4)$
39. $(-132) \div (-12)$
40. $(75) \div (-1.5)$

6.6 Properties

Commutative Properties

Addition   If we change the *order* of the two numbers in an addition problem, we get the same sum.

$$(-6) + (2) = -4 \quad \text{and} \quad (2) + (-6) = -4$$

Therefore,
$$(-6) + (2) = (2) + (-6)$$

This property is called the **commutative property of addition**.

Commutative property of addition	If a and b represent any real numbers, then $$a + b = b + a$$

Subtraction   If we change the order of the two numbers in a subtraction problem, we *do not* get the same difference (except when the two numbers are equal).

$$3 - 2 = 1 \quad \text{but} \quad 2 - 3 = -1$$

Therefore,
$$3 - 2 \neq 2 - 3$$

This example shows that *subtraction is not commutative*.

Multiplication   If we change the order of the two numbers in a multiplication problem, we get the same product.

$$(-9)(3) = -27 \quad \text{and} \quad (3)(-9) = -27$$

Therefore,
$$(-9)(3) = (3)(-9)$$

This property is called the **commutative property of multiplication**.

Commutative property of multiplication	If a and b represent any real numbers, then $$a \cdot b = b \cdot a$$

Division   If we change the order of the two numbers in a division problem, we *do not* get the same quotient (except when the two numbers are equal).

$$10 \div 5 = 2 \quad \text{but} \quad 5 \div 10 = \frac{1}{2}$$

Therefore,
$$10 \div 5 \neq 5 \div 10$$

This example shows that *division is not commutative*.

Associative Properties

Addition   In an addition problem, the sum of three numbers is unchanged no matter how we *group* the numbers. Parentheses are used to show which two numbers are to be added first.

$$(2 + 3) + 4 = 5 + 4 = 9 \quad \text{and} \quad 2 + (3 + 4) = 2 + 7 = 9$$

Therefore,
$$(2 + 3) + 4 = 2 + (3 + 4)$$

This property is called the **associative property of addition**.

Associative property of addition

> If a, b, and c represent any real numbers, then
> $$(a + b) + c = a + (b + c)$$

Subtraction   A single example will show that *subtraction is not associative*.

$$(7 - 4) - 2 = 3 - 2 = 1 \quad \text{but} \quad 7 - (4 - 2) = 7 - 2 = 5$$

Therefore, $\qquad\qquad\qquad (7 - 4) - 2 \neq 7 - (4 - 2)$

Multiplication   In a multiplication problem, the product of three numbers is unchanged no matter how we group the numbers.

$$(3 \cdot 4) \cdot 2 = 12 \cdot 2 = 24 \quad \text{and} \quad 3 \cdot (4 \cdot 2) = 3 \cdot 8 = 24$$

Therefore, $\qquad\qquad\qquad (3 \cdot 4) \cdot 2 = 3 \cdot (4 \cdot 2)$

This property is called the **associative property of multiplication**.

Associative property of multiplication

> If a, b, and c represent any real numbers, then
> $$(a \cdot b) \cdot c = a \cdot (b \cdot c)$$

Division   A single example will show that *division is not associative*.

$$(16 \div 4) \div 2 = 4 \div 2 = 2 \quad \text{and} \quad 16 \div (4 \div 2) = 16 \div 2 = 8$$

Therefore, $\qquad\qquad\qquad (16 \div 4) \div 2 \neq 16 \div (4 \div 2)$

Summary

> 1. The *commutative property* says that *changing the order* of the numbers in an addition or multiplication problem gives the same answer.
> 2. The *associative property* says that *changing the grouping* of the numbers in an addition or multiplication problem gives the same answer.

EXAMPLE 1   State whether each of the following is true or false, and give the reason.

SOLUTION

a. $(-7) + 5 = 5 + (-7)$ *True*, because of the commutative property of addition (*order* of numbers changed)

b. $(+6)(-8) = (-8)(+6)$ *True*, because of the commutative property of multiplication (*order* of numbers changed)

c. $(9 + 4) + 5 = 9 + (4 + 5)$ *True*, because of the associative property of addition (*grouping* changed)

d. $(5 \cdot 2) \cdot 3 = 5 \cdot (2 \cdot 3)$ *True*, because of the associative property of multiplication (*grouping* changed)

e. $(+8) - (-7) = (-7) - (+8)$ *False*; subtraction is *not* commutative.

f. $a + (b + c) = (a + b) + c$ *True*, because of the associative property of addition (*grouping* changed)

g. $y \div z = z \div y$ *False*; division is *not* commutative.

h. $p \cdot (s \cdot r) = p \cdot (r \cdot s)$ *True*, because of the commutative property of multiplication (*order* changed)

Exercises 6.6

State whether each of the following is true or false, and give the reason.

1. $7 + 5 = 5 + 7$

2. $9 + 4 = 4 + 9$

3. $(2 + 6) + 3 = 2 + (6 + 3)$

4. $(1 + 8) + 7 = 1 + (8 + 7)$

5. $6 - 2 = 2 - 6$

6. $4 - 7 = 7 - 4$

7. $(a \cdot b) \cdot c = a \cdot (b \cdot c)$

8. $(p \cdot q) \cdot r = p \cdot (q \cdot r)$

9. $8 \div 4 = 4 \div 8$

10. $3 \div 6 = 6 \div 3$

11. $(p)(t) = (t)(p)$

12. $(m)(n) = (n)(m)$

13. $(4)(-5) = (-5) + (4)$

14. $(-7)(2) = (2) + (-7)$

15. $5 + (3 + 4) = 5 + (4 + 3)$

16. $6 + (8 + 2) = 6 + (2 + 8)$

17. $(3)(-2) = (-2)(3)$

18. $(-8)(-9) = (-9)(-8)$

19. $(12 \div 6) \div 2 = 12 \div (6 \div 2)$

20. $(30 \div 2) \div 5 = 30 \div (2 \div 5)$

21. $x - 4 = 4 - x$

22. $5 - y = y - 5$

23. $4 \cdot (6a) = (4 \cdot 6)a$

24. $(3 \cdot 5)x = 3 \cdot (5x)$

25. $H + 8 = 8 + H$

26. $4 + P = P + 4$

27. $(9 - 4) - 3 = 9 - (4 - 3)$

28. $(15 - 8) - 3 = 15 - (8 - 3)$

29. $a \cdot (b \cdot c) = a \cdot (c \cdot b)$

30. $2 \cdot (x \cdot y) = 2 \cdot (y \cdot x)$

6.7 Operations with Zero

Addition Involving Zero Because adding zero to a number gives the identical number for the sum, *zero* is called the **additive identity element**.

EXAMPLE 1

a. $9 + 0 = 9$

b. $0 + (-5) = -5$

Additive identity property	If a represents any real number, then $$a + 0 = 0 + a = a$$

Subtraction Involving Zero Because the subtraction $a - b$ has been defined as $a + (-b)$, the rules for subtractions involving zero are derived from the rules for addition.

Subtraction involving zero	If a represents any real number, then 1. $a - 0 = a$ 2. $0 - a = 0 + (-a) = -a$

Multiplication Involving Zero Because multiplication is a method for doing repeated addition of the same number, multiplying a number by zero gives a product of zero.

EXAMPLE 2 $4 \cdot 0 = 0 + 0 + 0 + 0 = 0$

Because of the commutative property of multiplication, it follows that

$$4 \cdot 0 = 0 \cdot 4 = 0$$

and

$$(-2) \cdot 0 = 0 \cdot (-2) = 0$$

Multiplication by zero

> If a represents any real number, then
> $$a \cdot 0 = 0 \cdot a = 0$$

Division Involving Zero   Every division problem is related to a multiplication problem.

$$\frac{6}{2} = 3 \quad \text{because} \quad 3 \cdot 2 = 6$$

Zero divided by a nonzero number is possible, and the quotient is zero.

EXAMPLE 3 $\dfrac{0}{2} = 0$ Because $0 \cdot 2 = 0$

A nonzero number divided by zero is impossible. Consider the problem

$$\frac{6}{0} = ?$$ ← Suppose the quotient is some number x

Then $\dfrac{6}{0} = x$ Means $x \cdot 0 = 6$, which is certainly false

$x \cdot 0 \neq 6$, because any number multiplied by zero $= 0$. Therefore, dividing any nonzero number by zero is undefined.

Zero divided by zero cannot be determined.

$$\frac{0}{0} = 0$$ Means $0 \cdot 0 = 0$, which is true

$$\frac{0}{0} = 1$$ Means $1 \cdot 0 = 0$, which is true

$$\frac{0}{0} = 5$$ Means $5 \cdot 0 = 0$, which is true

In other words, $\dfrac{0}{0} = 0$, 1, and also 5. In fact, it can be any number. Because there is *no* *unique* answer, we say that $\dfrac{0}{0}$ is undefined.

Division involving zero

> If a represents any real number, except 0, then
>
> 1. $\dfrac{0}{a} = 0$
>
> 2. $\dfrac{a}{0}$ is undefined ⎫
>
> 3. $\dfrac{0}{0}$ is undefined ⎬ Cannot divide by 0

Exercises   6.7

In Exercises 1–20, find the value of each expression, if it has one. If an expression does not have a value, give a reason.

1. $5 \cdot 0$

2. $0 \cdot 7$

3. $4 + 0$

4. $0 + 9$

5. $0 - 6$

6. $0 - 10$

7. $0 \div 12$

8. $0 \div (-15)$

9. $5 + (0 + 6)$

10. $(3 + 0) + 7$

11. $\dfrac{4}{0}$

12. $\dfrac{8}{0}$

13. $(0)(-15)$

14. $(-13)(0)$

15. $\dfrac{0}{0}$

16. $-\left(\dfrac{0}{0}\right)$

17. $0 + (-789)$

18. $(-546) + 0$

19. $\dfrac{-1}{0}$

20. $\dfrac{-156}{0}$

In Exercises 21–25, use a calculator to find the value.

21. $\dfrac{9}{0}$

22. $\dfrac{0}{-8}$

23. $\dfrac{0}{0}$

24. $0 - 74$

25. $(-6.5)(0)$

6.8   Powers of Signed Numbers

Earlier we considered products in which the same number is repeated as a factor. For example:

$$3 \cdot 3 \cdot 3 \cdot 3 = 3^4 = 81$$

In the symbol 3^4, the 3 is called the **base**; the 4 is called the **exponent** and is written above and to the right of the base 3. The entire symbol 3^4 is called the *fourth* **power** *of three* and is commonly read "three to the fourth power."

```
                    ┌── Exponent
              3⁴ = 81
      Base ──┘     └──── Fourth power of 3
```

If a base has an exponent that is exactly divisible by two, we say that it's an **even power** of the base. For example, 3^2, 5^4, and $(-2)^6$ are even powers. If a base has an exponent that is *not* exactly divisible by two, we say that it's an **odd power** of the base. For example, 3^1, 10^3, and $(-4)^5$ are odd powers.

Now that we have learned to multiply signed numbers, we consider powers of signed numbers.

EXAMPLE 1   Examples of powers of signed numbers:

a. $2^3 = 2 \cdot 2 \cdot 2 = 8$

b. $4^2 = 4 \cdot 4 = 16$

c. $(-3)^2 = (-3)(-3) = 9$

d. $(-1)^4 = (-1)(-1)(-1)(-1) = 1$

 Note An *even* power of a negative number is positive.

e. $(-2)^3 = (-2)(-2)(-2) = -8$

f. $(-1)^5 = (-1)(-1)(-1)(-1)(-1) = -1$

 Note An *odd* power of a negative number is negative.

g. $\left(\dfrac{2}{3}\right)^2 = \dfrac{2}{3} \cdot \dfrac{2}{3} = \dfrac{4}{9}$

A Word of Caution   A common error is to think that expressions such as $(-6)^2$ and -6^2 are the same. They are *not* the same. *The exponent applies only to the symbol immediately preceding it.*

$$(-6)^2 = (-6)(-6) = 36 \quad \text{The exponent applies to the ( )}$$

$$-6^2 = -(6 \cdot 6) = -36 \quad \text{The exponent applies only to the 6}$$

Therefore, $-6^2 \neq (-6)^2$.

EXAMPLE 2   Find the indicated powers.

SOLUTION

a. $(-5)^2$ $(-5)^2 = (-5)(-5) = 25$ Exponent applies to the ( )

b. -5^2 $-5^2 = -(5 \cdot 5) = -25$ Exponent applies only to 5

c. $-(-5)^2$ $-(-5)^2 = -(-5)(-5) = -(25) = -25$

d. $-(-2)^3$ $-(-2)^3 = -(-2)(-2)(-2) = -(-8) = 8$

Powers of Zero

> If a is any positive real number,
> $$0^a = 0$$

EXAMPLE 3 $0^5 = 0 \cdot 0 \cdot 0 \cdot 0 \cdot 0 = 0$

Zero as an Exponent   When any real number (other than 0) is raised to the 0 power, we get 1 (Example 4). The reason for this definition will be covered in a later section.

EXAMPLE 4   **a.** $2^0 = 1$ **b.** $10^0 = 1$ **c.** $(-5)^0 = 1$

In general,

> If a represents any real number except 0,
> $$a^0 = 1$$

 Note   The symbol 0^0 is not defined or used in this book.

EXAMPLE 5   Use a calculator to find the powers. (See Section 1.7 for a discussion on the power key $\boxed{y^x}$.)

SOLUTION

a. 2^9 *Key in* $2 \boxed{y^x} 9 \boxed{=}$

        Answer   512

b. $(-8)^5$ *Key in* $8 \boxed{+/-} \boxed{y^x} 5 \boxed{=}$

        Answer $-32{,}768$

Exercises 6.8

In Exercises 1–36, find the value of each expression.

1. 3^3 2. 2^4 3. $(-5)^2$ 4. $(-6)^3$ 5. 7^2

6. 3^4 7. 0^3 8. 0^4 9. $(-10)^1$ 10. $(-10)^2$

11. 10^3 12. 10^4 13. $(-10)^5$ 14. $(-10)^6$ 15. $(-2)^2$

16. $(-2)^3$ 17. -2^2 18. -3^2 19. 5^4 20. 25^2

21. 40^3 22. 400^2 23. -2^4 24. -9^2 25. $(-1)^5$

26. $(-1)^7$ 27. $(-1)^{98}$ 28. $(-1)^{99}$ 29. 3^0 30. $(-4)^0$

31. $-(-4)^2$ 32. $-(-3)^3$ 33. $\left(\dfrac{3}{4}\right)^2$

34. $\left(\dfrac{2}{3}\right)^3$ 35. $\left(-\dfrac{1}{10}\right)^3$ 36. $\left(-\dfrac{1}{2}\right)^4$

 In Exercises 37–42, use a calculator to find the value of each expression.

37. 3^{10} 38. 12^5 39. $(-4)^7$

40. $(-6)^9$ 41. $(-1.5)^4$ 42. $(-2.4)^0$

6.9 Square Roots of Signed Numbers

Finding the square root of a number is the *inverse* operation of squaring a number. Every square root problem has a related problem involving squaring a number.

Principal Square Root   Every positive number has both a positive and a negative square root. The positive square root is called the **principal square root**.

EXAMPLE 1   The number 9 has two square roots: $+3$ and -3.

$+3$ is the square root of 9 because $(+3)^2 = 9$.

-3 is the square root of 9 because $(-3)^2 = 9$.

3 is the principal square root of 9 because it is positive.

The radical sign, $\sqrt{}$, represents the principal square root of the number under it. Therefore, $\sqrt{9} = 3$.

EXAMPLE 2   Find the square roots.

SOLUTION

a. $\sqrt{16}$ $\sqrt{16} = 4$ Because $4^2 = 16$

b. $\sqrt{36}$ $\sqrt{36} = 6$ Because $6^2 = 36$

c. $\sqrt{1}$ $\sqrt{1} = 1$ Because $1^2 = 1$

d. $\sqrt{0}$ $\sqrt{0} = 0$ Because $0^2 = 0$

e. $\sqrt{\dfrac{4}{9}}$ $\sqrt{\dfrac{4}{9}} = \dfrac{2}{3}$ Because $\left(\dfrac{2}{3}\right)^2 = \dfrac{4}{9}$

f. $-\sqrt{4}$ $-\sqrt{4} = -2$ We know $\sqrt{4} = 2$; therefore, $-\sqrt{4} = -2$

 Note   Square roots of negative numbers, such as $\sqrt{-4}$, are imaginary numbers. Imaginary numbers are not real and cannot be graphed on the number line. (See Example 6B.)

Rational Numbers A **rational number** is a number that can be written in the form $\frac{a}{b}$, where a and b are integers ($b \neq 0$).

EXAMPLE 3

Examples of rational numbers:

a. $\frac{2}{3}$ All fractions are rational numbers

b. $-3 = \frac{-3}{1}$ All integers are rational numbers

c. $2\frac{1}{2} = \frac{5}{2}$ All mixed numbers are rational numbers

d. $0.25 = \frac{25}{100}$ All terminating decimals are rational numbers

e. $0.333\ldots = \frac{1}{3}$ All repeating decimals are rational numbers

f. $\sqrt{25} = 5 = \frac{5}{1}$ $\sqrt{25}$ simplifies to the integer 5, which is a rational number

The decimal representation of any rational number is either a terminating decimal or a repeating decimal. (See Section 3.10.)

Irrational Numbers An **irrational number** is a number whose decimal representation is a nonterminating, nonrepeating decimal.

EXAMPLE 4

Examples of irrational numbers:

a. $\sqrt{2} = 1.414213562\ldots$

b. $\sqrt{3} = 1.732050807\ldots$

c. $\pi = 3.1415926535\ldots$

d. $-\sqrt{5} = -2.236067977\ldots$

Real Numbers Together, the rational numbers and the irrational numbers form the set of **real numbers**. There is a one-to-one correspondence between the real numbers and the points on the number line (Figure 7).

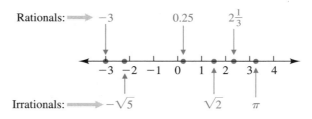

Figure 7

Finding Square Roots by Table or Calculator

Square roots can be approximated by using Table I (inside back cover) or by using a calculator with a square root key $\boxed{\sqrt{}}$.

EXAMPLE 5 Find $\sqrt{3}$ (a) using Table I and (b) using a calculator.

SOLUTION

N	$\sqrt{N}$
1	1.000
2	1.414
③	1.732
4	2.000
5	2.236

a. Locate 3 in the column headed N. Read the value of $\sqrt{3}$ in the column headed $\sqrt{N}$. Table I gives the square roots rounded off to three decimal places. Therefore, $\sqrt{3} \approx 1.732$.

b. *Key in* 3 $\boxed{\sqrt{}}$

Answer 1.7320508 ≈ 1.732

└── When rounded off to 3 decimal places, the answer is the same number obtained from the table

EXAMPLE 6 Find the value of each expression using a calculator.

SOLUTION

a. $-\sqrt{4}$ *Key in* 4 $\boxed{\sqrt{}}$ $\boxed{+/-}$

Answer -2

b. $\sqrt{-4}$ *Key in* 4 $\boxed{+/-}$ $\boxed{\sqrt{}}$

Answer E

The E in the display indicates that an error has been made. In this case, $\sqrt{-4}$ is not a real number.

Exercises 6.9

In Exercises 1–16, find the square roots.

1. $\sqrt{16}$ 2. $\sqrt{25}$ 3. $-\sqrt{4}$ 4. $-\sqrt{9}$

5. $\sqrt{81}$ 6. $\sqrt{36}$ 7. $\sqrt{100}$ 8. $\sqrt{144}$

9. $-\sqrt{49}$ 10. $-\sqrt{121}$ 11. $\sqrt{64}$ 12. $\sqrt{169}$

13. $-\sqrt{100}$ 14. $-\sqrt{144}$ 15. $\sqrt{\dfrac{16}{25}}$ 16. $\sqrt{\dfrac{49}{100}}$

In Exercises 17–24, find the square roots using Table I (inside back cover) or a calculator. Round off the answers to three decimal places.

17. $\sqrt{13}$ 18. $\sqrt{18}$ 19. $\sqrt{7}$ 20. $\sqrt{6}$

21. $\sqrt{50}$ 22. $\sqrt{75}$ 23. $\sqrt{184}$ 24. $\sqrt{155}$

Chapter 6   REVIEW

Natural Numbers
6.1

$\{1, 2, 3, \ldots\}$

Whole Numbers
6.1

$\{0, 1, 2, \ldots\}$

Integers
6.1

$\{\ldots, -3, -2, -1, 0, 1, 2, 3, \ldots\}$

Digits
6.1

$\{0, 1, 2, 3, 4, 5, 6, 7, 8, 9\}$

Rational Numbers
6.9

Rational numbers are numbers that can be written in the form $\dfrac{a}{b}$, where a and b are integers and $b \neq 0$. The decimal representation of a rational number is either a terminating or a repeating decimal.

Irrational Numbers
6.9

Irrational numbers are numbers whose decimal representation is a nonterminating, nonrepeating decimal.

Real Numbers
6.9

Real numbers are all the rational numbers and all the irrational numbers. There is a one-to-one correspondence between the real numbers and the points on the number line.

Operations Involving Zero
6.7, 6.8

Zero is a real number that is neither positive nor negative.

If a is any real number:

$$a + 0 = 0 + a = a$$

$$a - 0 = a$$

$$0 - a = 0 + (-a) = -a$$

$$a \cdot 0 = 0 \cdot a = 0$$

If a is any real number $\neq 0$:

$$\frac{0}{a} = 0$$

$$\frac{a}{0} \text{ is undefined}$$

$$\frac{0}{0} \text{ is undefined}$$

$$a^0 = 1$$
$$0^a = 0 \text{ if } a \text{ is positive}$$

Additive Identity Property
6.7

The **additive identity element** is 0.

$$0 + a = a + 0 = a$$

Additive Inverse Property
6.2

$$a + (-a) = 0$$

Opposites

Multiplicative Identity Property
6.4

The **multiplicative identity element** is 1.

$$1 \cdot a = a \cdot 1 = a$$

Multiplicative Inverse Property
6.4

$$a \cdot \frac{1}{a} = 1$$

Reciprocals

Commutative Property of Addition
6.6

$$a + b = b + a$$

Commutative Property of Multiplication
6.6

$$a \cdot b = b \cdot a$$

Associative Property of Addition
6.6

$$(a + b) + c = a + (b + c)$$

Associative Property of Multiplication
6.6

$$(a \cdot b) \cdot c = a \cdot (b \cdot c)$$

Absolute Value
6.2

The **absolute value** of a number is the distance between that number and 0 on the number line, with no regard for direction.
The absolute value of a number can never be negative.
The absolute value of a real number X is written $|X|$.

Adding Signed Numbers
6.2

1. When the numbers have the *same sign*

First: Add their absolute values.
Second: Give the sum the same sign as both numbers.

2. When the numbers have *different signs*

First: Subtract the smaller absolute value from the larger.
Second: Give the sum the sign of the number with the larger absolute value.

Subtracting Signed Numbers
6.3

$$a - b = a + (-b)$$

1. Change the subtraction symbol to an addition symbol, and change the sign of the number being subtracted.
2. Add the resulting signed numbers.

Multiplying Two Signed Numbers
6.4

1. Multiply their absolute values.
2. The product is *positive* when the signed numbers have the same sign. The product is *negative* when the signed numbers have different signs.

Dividing Signed Numbers
6.5

1. Divide their absolute values.

2. The quotient is *positive* when the signed numbers have the same sign. The quotient is *negative* when the signed numbers have different signs.

Powers of Signed Numbers
6.8

Raising a number to a power is repeated multiplication.

Exponent

$$3^4 = 3 \cdot 3 \cdot 3 \cdot 3 = 81$$

Base ——↑   ↑—— Fourth power of 3

Square Roots of Signed Numbers
6.9

Finding the square root of a number is the inverse of squaring a number.

$$\sqrt{9} = 3 \quad \text{because} \quad 3^2 = 9$$

Chapter 6   REVIEW EXERCISES

1. Write all the digits greater than 7.

2. Write all the whole numbers less than 3.

3. Write the largest digit.

4. Write the smallest natural number.

5. Write the smallest integer.

6. Write the largest integer less than zero.

7. Write the correct symbol, $<$ or $>$, to make each statement true.

   a. $2 \; ? \; 6$ **b.** $-3 \; ? \; 0$ **c.** $-3 \; ? \; -5$

   d. $2 \; ? \; -1$ **e.** $-1\frac{1}{2} \; ? \; -1$

8. State whether each of the following is true or false, and give the reason.

   a. $(6 \cdot 3) \cdot 4 = (6)(3 \cdot 4)$

   b. $5 + (-2) = (-2) + 5$

   c. $5 - (-2) = (-2) - 5$

   d. $a + (b + c) = (a + b) + c$

   e. $(c \cdot d) \cdot e = e \cdot (c \cdot d)$

   f. $5 + (x + 7) = (x + 7) + 5$

In Exercises 9–66, perform the indicated operation. If the indicated operation cannot be done, give a reason.

9. $(-6) \div (-2)$ 10. $(-5) - (-3)$ 11. $(5) + (-2)$

12. $(-3)(-4)$ 13. $(-7) + (-4)$ 14. $(8)(-9)$

15. $(-4) - (-8)$ 16. $18 \div (-3)$ 17. $(6) - (15)$

18. $(-9) + (3)$ 19. $(-7)(6)$ 20. $(-2) - (8)$

21. -3^2 22. $(-10) + (6)$ 23. $(-5) - (-8)$

24. $(-6) \div 2$ 25. 0^3 26. $(-5)^2$

27. $(-10) + (-2)$ 28. $(-4)(6)$ 29. $(-25) \div (-5)$

30. $0 \div (-4)$ 31. $\sqrt{36}$ 32. -2^4

33. $(4) - (-9)$ 34. $(1) + (-6)$ 35. $(-8)(-6)$

36. $(-5) + (14)$ 37. $(-56) \div 8$ 38. $(0)(-5)$

39. $(-9) + (-5)$ 40. $\sqrt{25}$ 41. $(-4) - (7)$

42. $(-3) + (-6)$ 43. 10^0 44. $(2) - (9)$

45. $(-9)(-7)(-2)$ 46. $(-3)^3$ 47. $(7) - (-5)$

48. $(-8) + (3)$ 49. $-\sqrt{9}$ 50. $(-1) - (-6)$

51. 4^1 52. $(-5)(-8)(-2)(-1)$

53. $(-9) + (-8)$ 54. 10^4

55. $(-3) - (8)$ 56. $(27) + (-43)$

57. $\dfrac{0}{0}$ 58. $\dfrac{-24}{18}$

59. $\left(4\frac{5}{6}\right) + \left(-2\frac{2}{3}\right)$ 60. $\left(-\frac{9}{10}\right) - \left(\frac{3}{4}\right)$

61. $\left(-3\frac{3}{4}\right)\left(3\frac{1}{3}\right)$ 62. $\dfrac{-3}{0}$

63. $\left(-2\frac{1}{4}\right) \div \left(-1\frac{5}{6}\right)$ 64. $\left(-3\frac{1}{6}\right) + \left(-2\frac{3}{4}\right)$

65. Subtract 9 from -7.   66. Subtract -2 from -10.

In Exercises 67–70, use Table I or a calculator. Round off the answers to three decimal places.

67. $\sqrt{53}$ 68. $\sqrt{92}$ 69. $\sqrt{153}$ 70. $\sqrt{185}$

Chapter 6 Critical Thinking and Writing Problems

Answer Problems 1–15 in your own words, using complete sentences.

1. Explain why -6 is less than -2.

2. Explain the meaning of the absolute value of a number.

3. Explain how to add -9 and 3.

4. Explain how to add -9 and -3.

5. Explain how to subtract 12 from 7.

6. Explain how to subtract -12 from 7.

7. If the product of two numbers is positive, what can you say about the numbers?

8. If the product of two numbers is negative, what can you say about the numbers?

9. Explain why $-4 \div 0$ is undefined.

10. Explain why $0 \div 0$ is undefined.

11. Explain the difference between $(-3)^2$ and -3^2.

12. Explain why $(-2)^3 \neq -6$.

13. Explain why $\sqrt{4}$ is a rational number, but $\sqrt{5}$ is an irrational number.

14. Explain how to determine whether $2 + (x + y) = 2 + (y + x)$ uses the associative property of addition or the commutative property of addition.

15. Explain how to determine whether $3 \cdot (7 \cdot x) = (3 \cdot 7) \cdot x$ uses the associative property of multiplication or the commutative property of multiplication.

16. Are $\frac{5}{6}$ and $-\frac{6}{5}$ multiplicative inverses? Why or why not?

17. Give an example to show that subtraction is not commutative.

18. Find the mystery $\boxed{?}$ integer.

   Clue 1: The integer is less than 0.

   Clue 2: It is an even number.

   Clue 3: Its absolute value is less than 4.

19. Find the mystery $\boxed{?}$ integer.

   Clue 1: It is a two-digit number less than 30.

   Clue 2: The square root of the number is a rational number.

   Clue 3: It is an odd number.

Chapter 6 · DIAGNOSTIC TEST

Allow yourself about 40 minutes to do these problems. Complete solutions for all problems, together with section references, are given in the answer section at the end of the book.

In Problems 1–3, write the symbol, $<$ or $>$, that should be used to make the statement true.

1. 3 ? 5   2. 0 ? -4 3. -5 ? -4

4. Write all the digits less than 3.

5. Write the smallest positive integer.

6. At 5 A.M. the temperature in Denver, Colorado, was $-20°$F. At noon the temperature was $38°$F. What was the rise in temperature?

7. Write -35 in words.

In Problems 8–11, state whether each is true or false, and give the reason.

8. $(-9) + (3) = (3) + (-9)$

9. $(3 + 5) + 6 = 3 + (5 + 6)$

10. $a \cdot (b \cdot c) = a \cdot (c \cdot b)$

11. $7 - 4 = 4 - 7$

In Problems 12–47, perform the indicated operation. If the indicated operation cannot be done, give a reason.

12. Add -6 and 2.   13. Subtract -10 from 20.

14. $(-10)(-5)$ 15. $(42) \div (-14)$

16. 6^2 17. $(8) + (-5)$

18. Multiply (-12) times 3.   19. Subtract 5 from -10.

20. $\dfrac{2}{0}$ 21. $(-5)(8)$

22. $\dfrac{15}{-18}$ 23. $(-4)^3$

24. $(12) + (-5)$ 25. $(-9) - (-7)$

26. $(7)(-6)$ 27. $(5) - (12)$

28. $(-2) - (-6)$ 29. $(-3) + (-8)$

30. 2^0 31. $(-9)(-3)(-2)$

32. $(-24) \div (-6)$ 33. $(6) - (10)$

34. $(-7) + (-2)$ 35. $(6)(-4)$

36. $(-4) + (9)$ 37. 0^2

38. $(-4) - (8)$ 39. $(-11) + (4)$

40. $\left(-1\dfrac{7}{8}\right)\left(2\dfrac{2}{5}\right)$ 41. $\left(-2\dfrac{3}{4}\right) + \left(-1\dfrac{5}{8}\right)$

42. $\dfrac{-33}{11}$ 43. $(-2)^4$

44. $0 \div (-12)$ 45. $(-5)(0)$

46. $(9) - (-5)$ 47. $(3) + (-8)$

In Problems 48–50, find the indicated root.

48. $\sqrt{49}$ 49. $\sqrt{81}$ 50. $-\sqrt{64}$

Evaluating Expressions

CHAPTER 7

One of the first places that a student uses algebra is in the evaluation of formulas in courses taken in science, business, and so on. In this chapter we apply the operations with signed numbers to evaluate algebraic expressions and formulas.

In the last chapter we performed all six arithmetic operations with signed numbers—addition, subtraction, multiplication, division, taking powers, and finding square roots—but each problem dealt with only one kind of operation:

$$2 + (-4) = -2 \qquad \text{Addition only}$$

$$(-3) \cdot (4 \cdot 5) = (-3)(20) = -60 \qquad \text{Multiplication only}$$

$$(-7) - 4 = -11 \qquad \text{Subtraction only}$$

In this chapter we will evaluate expressions in which more than one kind of operation is used. For example,

$$5 + 4 \cdot 6 = ? \qquad \text{Multiplication } and \text{ addition}$$

If the addition is done first, we get

$$5 + 4 \cdot 6$$
$$= \quad 9 \quad \cdot 6$$
$$= \quad\quad 54$$

If the multiplication is done first, we get

$$5 + 4 \cdot 6$$
$$= 5 + \quad 24$$
$$= \quad\quad 29$$

Obviously, both answers cannot be correct. In an earlier chapter we showed that multiplication is done before addition. Therefore, $5 + 4 \cdot 6 = 5 + 24 = 29$.

The next section will cover the correct order of operations with signed numbers.

7.1   Order of Operations

In evaluating expressions with more than one operation, we use the following order of operations:

Order of operations

> 1. Any expressions in parentheses are evaluated first.
> 2. Evaluations are done in this order.
>   First: Powers and roots
>   Second: Multiplication and division in order from *left to right*
>   Third: Addition and subtraction in order from *left to right*

EXAMPLE 1

$$-5 - 7 + 3 - (-6) \qquad \text{Addition and subtraction are done } left to right$$
$$= -5 + (-7) + 3 - (-6)$$
$$= \quad\quad -12 \quad + 3 - (-6)$$
$$= \quad\quad\quad -9 \quad\quad + \quad 6$$
$$= \quad\quad\quad\quad\quad -3$$

Note   If the expression in Example 1 is first rewritten as a sum, then the terms can be added in any order, because addition is commutative and associative.

$$-5 - \quad 7 \quad + 3 - (-6)$$
$$= -5 + (-7) + 3 + \quad 6$$
$$= \quad\quad -12 \quad + \quad 9$$
$$= \quad\quad\quad\quad -3$$

EXAMPLE 2 $-16 \div 2 \cdot 4$ Multiplication and division are done *left to right*

$= \quad -8 \quad \cdot 4$

$= \quad -32$

EXAMPLE 3 $-6 + 3 \cdot 5$ Multiplication is done before addition

$= -6 + \quad 15$

$= \quad\quad 9$

EXAMPLE 4 $-7^2 + (-5)^2$ Powers are done first

$= -49 + \quad 25$

$= \quad\quad -24$

A Word of Caution   A common error often made in Example 4 is to mistake $(-7)^2$ for -7^2.

$(-7)^2 = (-7)(-7) = 49$ The exponent 2 applies to (-7)

$-7^2 = -(7)(7) \quad = -49$ The exponent 2 applies only to 7

EXAMPLE 5 $3 + 5(-2) - 3^2$ Powers are done first

$= 3 + 5(-2) - 9$ Next multiply

$= 3 + (-10) - 9$ Addition and subtraction are done *left to right*

$= \quad -7 \quad + (-9)$

$= \quad\quad -16$

EXAMPLE 6 $-6(-8 + 5)$ Operations in parentheses are done first

$= -6 \quad (-3)$ The understood multiplication of -6 and -3 is done last

$= \quad 18$

> **A Word of Caution** After an expression in parentheses has been evaluated, the parentheses may be dropped *only* if all other operations are maintained. In Example 6, if the parentheses are dropped from the -3, the operation between -6 and -3 becomes subtraction, which is incorrect because it does not reflect the meaning of the original problem.
>
> $$-6(-8 + 5) = -6 - 3 = -9$$
> $$\uparrow$$
> Understood
> multiplication

EXAMPLE 7

$$2\sqrt{25} - 3(-2)^3 \qquad \text{Powers and roots are done first}$$
$$= 2 \cdot 5 - 3(-8) \qquad \text{Both multiplications are done next}$$
$$= 10 - (-24) \qquad \text{Subtraction is last}$$
$$= 10 + 24$$
$$= 34$$

EXAMPLE 8

$$7 - 2(4 + 3 \cdot 2) \qquad \text{Multiply inside parentheses}$$
$$= 7 - 2(4 + 6) \qquad \text{Add inside parentheses}$$
$$= 7 - 2(10) \qquad \text{Multiplication before subtraction}$$
$$= 7 - 20 \qquad \text{Subtraction is last}$$
$$= 7 + (-20)$$
$$= -13$$

EXAMPLE 9

Find the value of each expression using a calculator.

a. $3(-4) - \sqrt{225}$

SOLUTION

Key in   3 $\boxed{\times}$ 4 $\boxed{+/-}$ $\boxed{-}$ 225 $\boxed{\sqrt{}}$ $\boxed{=}$

Answer -27

b. $(-2)^4 - (9 + 8)$

SOLUTION

Key in   2 $\boxed{+/-}$ $\boxed{y^x}$ 4 $\boxed{-}$ $\boxed{(}$ 9 $\boxed{+}$ 8 $\boxed{)}$ $\boxed{=}$

Answer -1

Exercises 7.1

In Exercises 1–48, be sure to perform the operations in the correct order.

1. $12 - 8 - 6$

2. $15 - 9 - 4$

3. $-7 + 11 + 13 - 9$

4. $-2 - 8 + 14 - 6$

5. $7 + 2 \cdot 4$

6. $10 + 3 \cdot 6$

7. $9 + 3(-2)$

8. $-4 + 8(-3)$

9. $10 \div 2 \cdot 5$

10. $-24 \div 2 \cdot 3$

11. $12 \div 6 \div (-2)$

12. $-24 \div 12 \div (-2)$

13. $(-12) \div 2 \cdot (-3)$

14. $(-18) \div (-3) \cdot (-6)$

15. $6(8 - 2)$

16. $-5(10 - 6)$

17. $(-40)^2 \cdot 0 \cdot (-5)^2$

18. $(-500)^2 \cdot 0 \cdot (-3)^2$

19. $12 \cdot 4 + 16 \div 8$

20. $4(-3) + 15 \div 5$

21. $3 \cdot 2^2 + 8$

22. $2 \cdot 5^2 - 9$

23. $-3^2 - 4^2$

24. $-6^2 + (-5)^2$

25. $(-2)^2 + 5(-2)$

26. $(-4)3 - 3^2$

27. $2\sqrt{9} - 5(-6)$

28. $-3(4) - 4\sqrt{25}$

29. $-2(5 - 3) - 5(7 - 3)$

30. $-3(8 - 2) + 2(3 + 1)$

31. $5 + 3(4 + 1)$

32. $8 + 4(9 - 7)$

33. $-6 - 8 \div 2 \cdot 4$

34. $-6 - 8 \div (2 \cdot 4)$

35. $(-6 - 8) \div 2 \cdot 4$

36. $(-6 - 8 \div 2) \cdot 4$

37. $-4^2 - 2(-5) - 2\sqrt{9}$

38. $-8^2 + 7\sqrt{4} - 3(-4)$

39. $4 - (5 + 2 \cdot 3)$

40. $3 - (8 - 4 \div 2)$

41. $5 + 3(8 - 2 \cdot 3)$

42. $6 - 2(1 + 4 \cdot 2)$

43. $2 \cdot \dfrac{7}{16} + \dfrac{9}{20} \div \dfrac{3}{5}$

44. $\dfrac{15}{16} \div 3 - \dfrac{2}{3} \cdot \dfrac{3}{8}$

45. $\left(\dfrac{2}{3}\right)^2 + 3\dfrac{1}{3} \cdot \dfrac{1}{4}$

46. $\left(\dfrac{3}{4}\right)^2 - \dfrac{5}{8} \cdot 1\dfrac{1}{5}$

47. $2.3 + 5(3.7) \div 100$

48. $7.3 - 9(4.6) \div 10$

In Exercises 49–57, use a calculator to evaluate each expression.

49. $-18 - 16 \cdot 5$

50. $3^4 + 9(-12)$

51. $-2\sqrt{196} - 6$

52. $-50 + 2 \cdot 5^3$

53. $(-2)^6 - 7 \cdot 3$

54. $15(-6) - 18(-3)$

55. $(-23 + 17) \cdot 12$

56. $-6(34 - 4 \cdot 16)$

57. $(-19 + 8) \cdot (15 - 27)$

7.2 Grouping Symbols

Other symbols can be used with or without parentheses to indicate grouping.

( ) Parentheses

[ ] Brackets

{ } Braces

—— Bar

Fraction bar ⟶ $\dfrac{8 + 7}{9 - 4}$

Bar in a square root $\sqrt{6(4) - 8}$

All grouping symbols have the same meaning.

$$(8 + 4) - (9 - 7)$$
$$= [8 + 4] - [9 - 7]$$
$$= \{8 + 4\} - \{9 - 7\}$$
$$= \overline{8 + 4} - \overline{9 - 7}$$
$$= \quad 12 \quad - \quad 2$$
$$= \qquad 10$$

Different grouping symbols can be used in the same expression.

$$(8 + 4) - \{9 - 7\}$$
$$= [8 + 4] - (9 - 7)$$
$$= \{8 + 4\} - [9 - 7]$$
$$= \quad 12 \quad - \quad 2$$
$$= \qquad 10$$

When grouping symbols appear within other grouping symbols, *evaluate the inner grouping first.*

EXAMPLE 1

$10 - [3 - (2 - 7)]$ Evaluate the inner grouping first

$= 10 - [3 - (-5)\ ]$

$= 10 - [3 +\quad 5\quad]$ Evaluate inside the brackets [ ]

$= 10 -\quad\quad 8$

$=\quad\quad 2$

EXAMPLE 2

$\dfrac{(-4) + (-2)}{8 - 5}$ ⟵ This bar is a grouping symbol both for $(-4) + (-2)$
and for $8 - 5$. Notice that it can be used either above or below the numbers being grouped

$=\quad \dfrac{-6}{3}$

$=\quad\quad -2$

EXAMPLE 3

   This bar is a grouping symbol

$\sqrt{13^2 - 12^2}$ The expression under the bar is evaluated first

$= \sqrt{169 - 144}$

$=\quad \sqrt{25}$ Then the square root is taken

$=\quad\quad 5$

A Word of Caution   A common error in Example 3 is to write

$$\sqrt{13^2 - 12^2} = \sqrt{13^2} - \sqrt{12^2} = \sqrt{169} - \sqrt{144} = 13 - 12 = 1$$

Example 3 shows us that $\sqrt{13^2 - 12^2} = 5$, *not* 1. When taking the square root of a sum or a difference, remember that, in general, $\sqrt{a + b} \neq \sqrt{a} + \sqrt{b}$ and $\sqrt{a - b} \neq \sqrt{a} - \sqrt{b}$.

EXAMPLE 4

$5 - 2\{9 - [3 - (-2)]\}$

$= 5 - 2\{9 - [3 +\ 2\ ]\}$ Evaluate inside the brackets [ ]

$= 5 - 2\{9 -\quad\quad 5\}$ Evaluate inside the braces { }

$= 5 - 2\{\quad\quad 4\}$ Multiplication before subtraction

$= 5 -\quad\quad 8$

$=\quad\quad -3$

Exercises 7.2

Evaluate each expression.

1. $-3[4 - (-5)]$

2. $-2[-8 - (-3)]$

3. $24 - [(-6) + 18]$

4. $17 - [(-9) + 15]$

5. $4 - [7 - (-3)]$

6. $2 - [5 - (-4)]$

7. $20 - [5 - (7 - 10)]$

8. $16 - [8 - (2 - 7)]$

9. $9 + 2[3 - (-4)]$

10. $6 + 4[2 - (-1)]$

11. $\sqrt{3^2 + 4^2}$

12. $\sqrt{10^2 - 8^2}$

13. $\dfrac{7 + (-12)}{8 - 3}$

14. $\dfrac{(-14) + (-2)}{9 - 5}$

15. $\dfrac{10 + (-4)}{2 - 5} + 3$

16. $\dfrac{8 - 2}{5 - 7} - 1$

17. $\dfrac{6^2 - 4^2}{(6 - 4)^2}$

18. $\dfrac{3^2 - 9^2}{(3 - 9)^2}$

19. $[9 + (-3)][8 - (-2)]$

20. $[5 - (-4)][2 + (-7)]$

21. $-4 - 3[5 + (-7)]$

22. $-1 - 4[2 + (-5)]$

23. $-2^2 + [3 + (-6)]^2$

24. $-4^2 + [6 + (-1)]^2$

25. $5 - \{6 - [2 + (-7)]\}$

26. $7 - \{2 - [9 + (-1)]\}$

27. $6 - 3\{8 - [4 - (-1)]\}$

28. $4 - \{7 - 2[3 - (-4)]\}$

29. $4\{2 - [6 - 8 + (-1)]\}$

30. $8\{3 - [5 - (-2) + (-6)]\}$

31. $\dfrac{2 + 5(-4)}{3(4) - 6}$

32. $\dfrac{6 - 2(-3)}{4(-2) + 2}$

33. $\dfrac{2 \cdot 3^2 - 8}{3 - 2(4)}$

34. $\dfrac{5 \cdot 2^2 + (-2)}{9 + 3(-4)}$

35. $\dfrac{3^2 + 5}{2} - \dfrac{(-4)^2}{8}$

36. $\dfrac{2^3 - 2}{-3} + \dfrac{5^2 - 10}{5}$

37. $\sqrt{7^2 - 4(2)(3)}$

38. $\sqrt{5^2 - 4(4)(1)}$

39. $-4 + \sqrt{4^2 - 4(1)(3)}$

40. $-2 + \sqrt{2^2 - 4(1)(-8)}$

41. $15 - \{4 - [2 - 3(6 - 4)]\}$

42. $17 - \{6 - [9 - 2(2 - 7)]\}$

7.3 Finding the Value of Expressions That Have Variables and Numbers

In algebra we use letters to represent numbers. Such letters are called **variables**. The value of the variable may change in a particular problem or discussion.

A **constant** is an object or symbol that does not change its value in a particular problem or discussion. It is usually represented by a number symbol. In the expression $4x - 3y$, the constants are 4 and -3, and the variables are x and y.

An **algebraic expression** consists of numbers, variables, signs of operation, and signs of grouping (not *all* of those need be present).

In this section we make use of what we have already learned about signed numbers to help us find the value of algebraic expressions when the values of the variables are given.

Finding the value of an expression that has variables and numbers	1. Replace each variable by its number value in parentheses. 2. Carry out all arithmetic operations using the correct order of operations.

EXAMPLE 1 Find the value of $3x - 5y$ if $x = 10$ and $y = 4$.

SOLUTION $\quad 3\ x\ -\ 5\ y \qquad 3x = 3 \cdot x;\ \ 5y = 5 \cdot y$

$\qquad = 3\ (10) - 5\ (4)$

$\qquad = \quad 30 \quad - \quad 20$

$\qquad = \qquad\quad 10$

Notice that we simply replace each letter by its number value in parentheses, then carry out the arithmetic operations as we have done before.

> **A Word of Caution** When replacing a variable by a number, we enclose the number in *parentheses* to avoid the following common errors.
>
> Evaluate $3x$ when $x = -2$.
>
> | *Correct* | *Common error* |
> |---|---|
> | $3x = 3(-2) = -6$ | $3x = 3 - 2 = 1$ |
>
> Evaluate $4x^2$ when $x = -3$.
>
> | *Correct* | *Common errors* |
> |---|---|
> | $4x^2 = 4(-3)^2 = 4 \cdot 9 = 36$ | $4x^2 = 4 - 3^2 = 4 - 9 = -5$ |
> | | or $\quad 4x^2 = 4 - 3^2 = 4 + 9 = 13$ |

EXAMPLE 2

Find the value of $\dfrac{2a - b}{10c}$ if $a = -1$, $b = 3$, and $c = -2$.

SOLUTION

Remember: This bar is a grouping symbol

$$\frac{2a - b}{10c} = \frac{2(-1) - (3)}{10(-2)} = \frac{-2 - 3}{-20} = \frac{-5}{-20} = \frac{1}{4}, \text{ or } 0.25$$

EXAMPLE 3

Find the value of $2a - [b - (3x - 4y)]$ for $a = -3$, $b = 4$, $x = -5$, and $y = 2$.

SOLUTION

$$2a \quad - [b - (3x \quad - 4y \;)]$$
$$= 2(-3) - [4 - \{3(-5) - 4(2)\}] \quad \leftarrow \text{Notice that } \{ \; \} \text{ are used in place of ( ) to clarify the grouping}$$
$$= 2(-3) - [4 - \{ \; -15 \; - \; 8 \; \}]$$
$$= 2(-3) - [4 - \{ \quad\quad -23\}]$$
$$= 2(-3) - [4 + \quad\quad 23]$$
$$= \quad -6 \quad - \quad\quad [27]$$
$$= \quad\quad\quad -33$$

EXAMPLE 4

Evaluate $b - \sqrt{b^2 - 4ac}$ when $a = 3$, $b = -7$, and $c = 2$.

SOLUTION

This bar is a grouping symbol for $b^2 - 4ac$

$$b \quad - \sqrt{\; b^2 \quad - \quad 4ac}$$
$$= (-7) - \sqrt{(-7)^2 - 4(3)(2)}$$
$$= (-7) - \sqrt{\; 49 \quad - \quad 24}$$
$$= (-7) - \quad\quad \sqrt{25}$$
$$= (-7) - \quad\quad\quad 5$$
$$= -12$$

Exercises 7.3

In Exercises 1–36, evaluate the expression when $a = 3$, $b = -5$, $c = -1$, $x = 4$, and $y = -7$.

1. $4b$ 2. $5c$ 3. $2a - 10$ 4. $3b - 4$

5. b^2 6. y^2 7. $-y^2$ 8. $-b^2$

9. $3y^2$ 10. $6c^2$ 11. $2a - 3b$ 12. $3x - 2y$

13. $x - y - 2b$ 14. $a - b - 3y$ 15. $3b - ab + xy$

16. $4c + ax - by$ 17. $x^2 - y^2$ 18. $b^2 - c^2$

19. $4 + 3(x + y)$ 20. $5 - 2(a + c)$ 21. $2(a - b) - 3c$

22. $3(a - x) - 4b$ 23. $3x^2 - 10x + 5$ 24. $2y^2 - 7y + 9$

25. $a^2 - 2ab + b^2$ 26. $x^2 - 2xy + y^2$ 27. $(a - b)^2$

28. $(x - y)^2$ 29. $\dfrac{3x}{y + b}$ 30. $\dfrac{4a}{c - b}$

31. $\dfrac{2c - a}{2b}$ 32. $\dfrac{3y - c}{2ab}$

33. $2(a - 6) + 3(b + 7)$ 34. $4(x - 1) + 3(y - 2)$

35. $a - [b - (c + x)]$ 36. $y - [c - (a - b)]$

In Exercises 37–44, find the value of the expression when $E = -1$, $F = 3$, $G = -5$, $H = -4$, and $K = 0$.

37. $\dfrac{E + F}{EF}$ 38. $\dfrac{G + H}{GH}$

39. $\dfrac{(1 + G)^2 - 1}{H}$ 40. $\dfrac{1 - (1 + E)^2}{F}$

41. $2E - [F - (3K - H)]$ 42. $3H - [K - (4F - E)]$

43. $G - \sqrt{G^2 - 4EH}$ 44. $F - \sqrt{F^2 + 4HE}$

In Exercises 45–48, find the value of the expression when $x = \dfrac{1}{2}, y = -\dfrac{3}{4}, z = \dfrac{3}{8}$.

45. $\dfrac{3x^2}{z}$ 46. $\dfrac{y^2}{xz}$

47. $(x + y + z)^2$ 48. $x^2 + y^2 + z^2$

7.4 Evaluating Formulas

One reason for studying algebra is to prepare to use formulas. You will encounter formulas in many courses you take, as well as in real-life situations. In the examples and exercises we have listed the subject areas where the formulas are used.

Formulas are evaluated in the same way any expression having numbers and variables is evaluated.

EXAMPLE 1 The total amount in an account earning simple interest is given by the formula

$$A = P(1 + rt) \quad \text{(Business)}$$

where A is the amount in the account, P is the principal, r is the rate, and t is the time. Find A when $P = 1,000$, $r = 0.08$, and $t = 1.5$.

SOLUTION $A = P(1 + r \cdot t)$

$A = 1,000[1 + (0.08)(1.5)]$ Notice that [ ] were used in place of ( ) to clarify the grouping

$A = 1,000[1 + 0.12]$

$A = 1,000[1.12]$

$A = 1,120$

EXAMPLE 2 The formula to change Fahrenheit temperature to Celsius temperature is

$$C = \frac{5}{9}(F - 32) \quad \text{(Science)}$$

where C is degrees Celsius and F is degrees Fahrenheit. Find C when $F = -13$.

SOLUTION $C = \dfrac{5}{9}(F - 32)$

$C = \dfrac{5}{9}(-13 - 32)$

$C = \dfrac{5}{\underset{1}{9}}(\overset{-5}{-45})$

$C = -25$

EXAMPLE 3   The time it takes a pendulum to make one swing is given by the formula

$$T = \pi\sqrt{\dfrac{L}{g}} \quad \text{(Physics)}$$

where T is the time, $\pi \approx 3.14$, L is the length, and g is the force of gravity. Find T when $\pi \approx 3.14$, $L = 128$, and $g = 32$.

SOLUTION $T = \pi\sqrt{\dfrac{L}{g}}$

$T \approx (3.14)\sqrt{\dfrac{128}{32}}$

$T \approx (3.14)\sqrt{4}$

$T \approx (3.14)(2)$

$T \approx 6.28$

EXAMPLE 4   The sum of a geometric series is given by the formula

$$S = \dfrac{a(1 - r^n)}{1 - r} \quad \text{(Mathematics)}$$

where S is the sum, a is the first term, r is the ratio, and n is the number of terms. Find S when $a = -4$, $r = \dfrac{1}{2}$, and $n = 3$.

SOLUTION $S = \dfrac{a(1 - r^n)}{1 - r}$

$S = \dfrac{(-4)\left[1 - \left(\dfrac{1}{2}\right)^3\right]}{1 - \dfrac{1}{2}}$ $\left(\dfrac{1}{2}\right)^3 = \left(\dfrac{1}{2}\right)\left(\dfrac{1}{2}\right)\left(\dfrac{1}{2}\right) = \dfrac{1}{8}$

$S = \dfrac{(-4)\left[1 - \dfrac{1}{8}\right]}{1 - \dfrac{1}{2}}$

$S = \dfrac{(-4)\left[\dfrac{8}{8} - \dfrac{1}{8}\right]}{\dfrac{2}{2} - \dfrac{1}{2}}$

$$S = \frac{\dfrac{-1}{-4}\left[\dfrac{7}{\frac{8}{2}}\right]}{\dfrac{1}{2}}$$

$$S = \frac{-7}{2} \div \frac{1}{2}$$

$$S = \frac{-7}{\underset{1}{2}} \cdot \frac{\overset{1}{2}}{1}$$

$$S = -7$$

Exercises 7.4

Evaluate each formula, using the values of the variables given with the formula.

The area of a triangle is given by the formula

$$A = \frac{1}{2}bh \quad \text{(Geometry)}$$

where A is the area, b is the base, and h is the height.

1. Find A when $b = 15$ and $h = 14$.

2. Find A when $b = 27$ and $h = 36$.

Ohm's law states that

$$I = \frac{E}{R} \quad \text{(Electricity)}$$

where I is the current, E is the electromotive force, and R is the resistance.

3. Find I when $E = 110$ and $R = 22$.

4. Find I when $E = 220$ and $R = 33$.

The formula for simple interest is

$$I = Prt \quad \text{(Business)}$$

where I is the interest, P is the principal, r is the rate, and t is the time.

5. Find I when $P = 600$, $r = 0.09$, $t = 4.5$.

6. Find I when $P = 700$, $r = 0.08$, $t = 2.5$.

The formula to change Celsius to Fahrenheit is

$$F = \frac{9}{5}C + 32 \quad \text{(Chemistry)}$$

where F is degrees Fahrenheit and C is degrees Celsius.

7. Find F when $C = 25$.

8. Find F when $C = -25$.

The area of a circle is given by the formula

$$A = \pi r^2 \quad \text{(Geometry)}$$

where A is the area and r is the radius.

9. Find A when $\pi \approx 3.14$ and $r = 10$.

10. Find A when $\pi \approx 3.14$ and $r = 20$.

The distance a free-falling object travels is given by the formula

$$s = \frac{1}{2}gt^2 \quad \text{(Physics)}$$

where s is the distance, g is the force of gravity, and t is the time.

11. Find s when $g = 32$ and $t = 3$.

12. Find s when $g = 32$ and $t = 5$.

The value of an item that is depreciated each year is given by the formula

$$V = C - Crt \quad \text{(Business)}$$

where V is the present value, C is the original cost, r is the rate of depreciation, and t is the time.

13. Find V when $C = 500$, $r = 0.1$, $t = 2$.

14. Find V when $C = 1,000$, $r = 0.08$, $t = 5$.

The standard deviation of a binomial distribution is given by the formula

$$\sigma = \sqrt{npq} \quad \text{(Statistics)}$$

where σ is the standard deviation, n is the number of trials, p is the probability of a success, and q is the probability of a failure.

15. Find σ when $n = 100$, $p = 0.9$, $q = 0.1$.

16. Find σ when $n = 100$, $p = 0.8$, $q = 0.2$.

The formula to change Fahrenheit to Celsius is

$$C = \frac{5}{9}(F - 32)$$ *(Chemistry)*

where C is degrees Celsius and F is degrees Fahrenheit.

17. Find C when $F = -4$.

18. Find C when $F = 5$.

The formula to determine the dosage for a child is

$$C = \frac{a}{a + 12} \cdot A$$ *(Nursing)*

where C is the child's dosage, a is the age of the child, and A is the adult dosage.

19. Find C when $a = 6$ and $A = 30$.

20. Find C when $a = 4$ and $A = 48$.

The total amount in an account earning compound interest is given by the formula

$$A = P(1 + r)^t$$ *(Business)*

where A is the amount, P is the principal, r is the rate, and t is the time. Round off answers to two decimal places.

 21. Find A when $P = 1{,}000$, $r = 0.05$, $t = 10$.

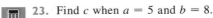 22. Find A when $P = 5{,}000$, $r = 0.12$, $t = 5$.

The Pythagorean theorem for right triangles is

$$c = \sqrt{a^2 + b^2}$$ *(Geometry)*

where c is the hypotenuse (longest side) and a and b are the other sides. Round off answers to one decimal place.

 23. Find c when $a = 5$ and $b = 8$.

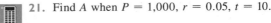 24. Find c when $a = 12$ and $b = 15$.

Chapter 7   R E V I E W

Grouping Symbols
7.2

( )   Parentheses

[ ]   Brackets

{ }   Braces

——   Bar

Fraction bar → $\dfrac{3x - 2}{5 + 7y}$

Bar in a square root ——┐

$\sqrt{5x - 12}$

Order of Operations
7.1

1. Any expressions in parentheses are evaluated first.

2. Evaluations are done in this order.
   First: Powers and roots
   Second: Multiplication and division in order from left to right
   Third: Addition and subtraction in order from left to right

Evaluating an Expression
or Formula
7.3, 7.4

1. Replace each variable by its number value in parentheses.

2. Carry out all operations using the correct order of operations.

Chapter 7   R E V I E W   E X E R C I S E S

In Exercises 1–20, evaluate each expression.

1. $26 - 14 + 8 - 11$

2. $11 - 7 \cdot 3$

3. $15 \div 5 \cdot 3$

4. $2 - [6 - (-8)]$

5. $-9(-7 + 5)$

6. $3 + 4(12 - 5)$

7. $-3^2 + (-4)^2$

8. $\sqrt{6^2 + 8^2}$

9. $6 - [8 - (3 - 4)]$

10. $7 - 3(5 - 4 \cdot 2)$

11. $2 \cdot 4 - 5^2 + 6 \cdot 2$

12. $4\sqrt{36} - 7(-3)$

13. $\dfrac{6 + (-14)}{3 - 7}$

14. $\dfrac{-5 - 3}{8} + 2(-8)$

15. $(-2)^3 + (-3)^2 - 4(-5)$

16. $-3(2 - 5) - 2(-4 + 9)$

17. $9 - \{6 - [4 - (7 - 8)]\}$

18. $9 - 3\{8 - [7 + (-1)]\}$ 19. $\dfrac{6 \cdot 2^2 - 8}{12 + 4(-5)}$

20. $\left(\dfrac{3}{4} - \dfrac{1}{3}\right) \div \left(\dfrac{4}{9} + \dfrac{2}{3}\right)$

In Exercises 21–32, find the value of each expression when $x = -2$, $y = 3$, and $z = -4$.

21. $4x^2$ 22. $x + yz$ 23. $x(y + z)$

24. $\sqrt{y^2 + z^2}$ 25. $3x - y + z$ 26. $6 - x(y - z)$

27. $5x^2 - 3x + 10$ 28. $y^2 - 2yz + z^2$

29. $x - 2(y - xz)$ 30. $\dfrac{(x + y)^2 - z^2}{x - 2y}$

31. $(x^2 + y^2)(x^2 - y^2)$ 32. $(x - y)(x^2 + xy + y^2)$

In Exercises 33–40, evaluate each formula, using the values of the variables given with the formula.

33. $C = \dfrac{a}{a + 12} \cdot A$; $a = 8, A = 35$

34. $I = Prt$; $P = 100, r = 0.07, t = 4.5$

35. $C = \dfrac{5}{9}(F - 32)$; $F = 15\dfrac{1}{2}$

36. $\sigma = \sqrt{npq}$; $n = 100, p = 0.5, q = 0.5$

37. $A = P(1 + rt)$; $P = 1{,}000, r = 0.06, t = 5$

38. $s = \dfrac{1}{2}gt^2$; $g = 32, t = 1\dfrac{1}{2}$

39. $F = \dfrac{9}{5}C + 32$; $C = -15$

40. $V = \pi r^2 h$; $\pi \approx 3.14, r = 5, h = 8$

Chapter 7 Critical Thinking and Writing Problems

1. In your own words, explain how to evaluate an expression.

2. If x is a negative number, will the value of x^2 be positive or negative?

3. A problem involves addition and multiplication. Which operation do you do first?

4. A problem involves multiplication and division. Which operation do you do first?

Problems 5–10 each have an error. Find the error, and in your own words, explain why it is wrong. Then work the problem correctly.

5. $\sqrt{16 + 9}$

   $= \sqrt{16} + \sqrt{9}$

   $= \ 4 \ + \ 3$

   $= 7$

6. $10 - (-3)^2$

   $= 10 + \ 3^2$

   $= 10 + \ 9$

   $= \ \ 19$

7. $24 \div (-6)(2)$

   $= 24 \div \ (-12)$

   $= \ \ \ -2$

8. $5(-4 + 1)$

   $= 5 \ \ - 3$

   $= \ \ 2$

9. $7 + 2[3 + (-5)]$

   $= 7 + 2[-2]$

   $= \ \ \ 9[-2]$

   $= \ \ \ -18$

10. $8 - [5 + (-2 + 4)]$

   $= 8 - 5 + \ \ \ 2$

   $= \ \ 3 \ + \ \ 2$

   $= \ \ \ \ \ \ 5$

Chapter 7   D I A G N O S T I C   T E S T

Allow yourself about 50 minutes to do these problems. Complete solutions for all problems, together with section references, are given in the answer section at the end of the book.

In Problems 1–12, evaluate each expression.

1. $17 - 9 - 6 + 11$ 2. $5 + 2 \cdot 3$ 3. $-12 \div 2 \cdot 3$

4. $-5^2 + (-4)^2$ 5. $2 \cdot 3^2 - 4$ 6. $3\sqrt{25} - 5(-4)$

7. $\dfrac{8 - 12}{-6 + 2}$ 8. $(2^3 - 7)(5^2 + 4^2)$

9. $10 - [6 - (5 - 7)]$ 10. $2 + (6 - 3 \cdot 4)$

11. $\sqrt{10^2 - 6^2}$ 12. $5 - 2[4 - (6 - 9)]$

In Problems 13–15, find the value of each expression when $a = -2$, $b = 4$, $c = -3$, $x = 5$, and $y = -6$.

13. $3a + bx - cy$ 14. $4x - [a - (3c - b)]$

15. $x^2 + 2xy - y^2$

In Problems 16–20, evaluate each formula, using the values of the variables given with the formula.

16. $C = \dfrac{5}{9}(F - 32)$; $F = -4$

17. $A = \pi r^2$; $\pi \approx 3.14$, $r = 20$

18. $C = \dfrac{a}{a + 12} \cdot A$; $a = 7$, $A = 38$

19. $A = P(1 + rt)$; $P = 600$, $r = 0.10$, $t = 2.5$

20. $V = C - Crt$; $C = 600$, $r = 0.04$, $t = 10$

Polynomials

CHAPTER

8

In this chapter we look in detail at a particular type of algebraic expression called a *polynomial*. Polynomials have the same importance in algebra that whole numbers have in arithmetic. Most of the work in arithmetic involves operations with whole numbers. In the same way, most of the work in algebra involves operations with polynomials.

8.1 Basic Definitions

The + and − signs in an algebraic expression break it into smaller pieces called **terms**. Each + and − sign is part of the term that follows it. *Exception*: An expression within grouping symbols is considered as a single piece even though it may contain + and − signs. See Examples 1b and 1c and Example 2.

EXAMPLE 1

a. $3x^2y - 5xy^3 + 7xy$

The − and + signs separate the algebraic expression into three terms

By rewriting the expression as a sum, we can see that the minus sign is part of the second term:

The minus sign is part of the second term

$$3x^2y \ + \ -5xy^3 \ + \ 7xy$$

First term   Second term   Third term

b. $3x^2 - 9x(2y + 5z)$

$$3x^2 \ + \ -9x(2y + 5z)$$

First term   Second term

c. $\dfrac{2 - x}{xy} + 5(2x^2 - y)$

$$\dfrac{2 - x}{xy} \ + \ 5(2x^2 - y)$$

First term   Second term

EXAMPLE 2

a. $3 + 2x - 1$ has three terms.
b. $3 + (2x - 1)$ has two terms.
c. $(3 + 2x - 1)$ has one term.

Recall that numbers that are multiplied to give a product are called the **factors** of that product.

EXAMPLE 3

a. $(3)(5) = 15$

Factors of 15

b. $(2)(x) = 2x$

Factors of $2x$

c. $(7)(a)(b)(c) = 7abc$

Product of factors

Factors of $7abc$

The **numerical coefficient** of a term is the factor that is a number. "The coefficient" of a term is understood to mean the *numerical* coefficient of that term.

EXAMPLE 4

Term	*Numerical coefficient*
a. $6w$	6
b. $-12xy^2$	-12
c. $\dfrac{3xy}{4} = \dfrac{3}{4}xy$	$\dfrac{3}{4}$
d. xy	1

Even though no number is written, the numerical coefficient can be considered to be 1 because $xy = 1(xy)$ (Also see part e)

| **e.** $-a^2$ | -1 |

-1 is the numerical coefficient of a^2 because $-a^2 = -1 \cdot a^2$

| **f.** $\dfrac{c}{5} = \dfrac{1}{5}c$ | $\dfrac{1}{5}$ |

Exercises 8.1

In Exercises 1–12, (a) determine the number of terms and (b) write the second term, if there is one.

1. $7xy$
2. $-6ab^2$
3. $x^2 - 5x + 3$
4. $3x^2 + 4x - 8$
5. $2(x - 5)$
6. $5z(3z^2 + 2z - 1)$
7. $x^2 - 5(x + 3)$
8. $3x^2 + 4(x - 8)$
9. $a(x + y) + b(x + y)$
10. $3x(x^2 - 9) - 5(x^2 - 9)$
11. $x^2 - \dfrac{x + y}{2} + y^2$
12. $\dfrac{a + b}{2} + \dfrac{a - b}{2}$

In Exercises 13–24, write the numerical coefficient of the second term.

13. $x^4 - 3x^2$
14. $4x^3 + 2x$
15. $x^2 + 2xy + y^2$
16. $a^3 - 5a^2 + a$
17. $5x^2 - x - 5$
18. $a^4 + a^2 - 1$
19. $x^2y + xy - xy^2$
20. $x^3y^2 - x^2y^2$
21. $\dfrac{2x}{3} + \dfrac{3x}{4}$
22. $\dfrac{4a}{5} - \dfrac{2b}{3}$
23. $\dfrac{a}{3} - \dfrac{b}{5} + \dfrac{c}{4}$
24. $\dfrac{5x}{6} + \dfrac{y}{4} - \dfrac{1}{2}$

8.2   Positive and Zero Exponents

In an earlier chapter we discussed bases, exponents, and powers of signed numbers. The same definitions are carried over to expressions with variables.

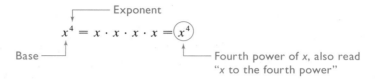

Consider the product $x^3 \cdot x^2$:

$$x^3 \cdot x^2 = (xxx)(xx) = xxxxx = x^5$$

3 factors ——————↑
+2 factors ——————↑
5 factors ——————↑

Therefore, $x^3 \cdot x^2 = x^{3+2} = x^5$.

This leads to the following rule of exponents:

RULE I Multiplying powers having the same base	$$x^a \cdot x^b = x^{a+b}$$ *In words:* When like bases are multiplied, the exponents are added.

Rule 1 of exponents can be extended to include more factors:

$$x^a x^b x^c \cdots = x^{a+b+c+\cdots}$$

EXAMPLE 1

a. $x^5 \cdot x^2 = x^{5+2} = x^7$

b. $x \cdot x^2 = x^{1+2} = x^3$

    └— $x = x^1$

 Note When no exponent is written, the exponent is understood to be 1.

c. $x^3 \cdot x^7 \cdot x^4 = x^{3+7+4} = x^{14}$

d. $10^7 \cdot 10^5 = 10^{7+5} = 10^{12} = 1,\!000,\!000,\!000,\!000$

                               12 zeros

e. $2 \cdot 2^3 \cdot 2^2 = 2^{1+3+2} = 2^6 = 64$

f. $2^a \cdot 2^b = 2^{a+b}$

g. $x^3 \cdot y^2$ *The rule does not apply* because the bases are different

| **A Word of Caution** | A common error in using Rule 1 is to multiply the bases. |

$$2^2 \cdot 2^3 = 4^5$$

as

$$2^2 \cdot 2^3 = 4 \cdot 8 = 32 \qquad \text{whereas} \qquad 4^5 = 1{,}024$$

When like bases are multiplied, add the exponents and keep the base the same; do *not* multiply the bases.

$$2^2 \cdot 2^3 = 2^{2+3} = 2^5 = 32$$

Consider the expression $(x^4)^2$:

$$(x^4)^2 = (x^4)(x^4) = x^4 \cdot x^4 = x^{4+4} = x^{2 \cdot 4} = x^{4 \cdot 2} = x^8$$

Note that $(x^4)^2 = x^{4 \cdot 2}$. In this case the exponents are multiplied. This leads to another rule:

RULE 2
Power of a power

$$(x^a)^b = x^{ab}$$

In words: When a power is raised to a power, the exponents are multiplied.

EXAMPLE 2

a. $(x^5)^4 = x^{5 \cdot 4} = x^{20}$

b. $(x)^4 = (x^1)^4 = x^{1 \cdot 4} = x^4$

c. $(y^4)^3 = y^{4 \cdot 3} = y^{12}$

d. $(10^3)^2 = 10^{3 \cdot 2} = 10^6 = 1{,}\underbrace{000{,}000}_{\text{6 zeros}}$

e. $(2^a)^b = 2^{a \cdot b} = 2^{ab}$

Consider the expression $\dfrac{x^5}{x^3}$:

$$\frac{x^5}{x^3} = \frac{xxxxx}{xxx} = \frac{xxx \cdot xx}{xxx \cdot 1} = \frac{xxx}{xxx} \cdot \frac{xx}{1} = 1 \cdot \frac{xx}{1} = xx = x^2$$

⎸— The value of this fraction is 1
(for $x \neq 0$)

Notice that $\dfrac{x^5}{x^3} = x^{5-3} = x^2$. In this case, three of the factors of the denominator cancel with three of the five factors of the numerator, leaving $5 - 3 = 2$ factors of x. This leads to a third rule:

RULE 3
Dividing powers having the same base

$$\frac{x^a}{x^b} = x^{a-b} \qquad (x \neq 0)$$

In words: When like bases are divided, the exponents are subtracted.

EXAMPLE 3

a. $\dfrac{x^6}{x^2} = x^{6-2} = x^4$ The exponent in the denominator is *subtracted from* the exponent in the numerator

b. $\dfrac{y^3}{y} = \dfrac{y^3}{y^1} = y^{3-1} = y^2$

c. $\dfrac{10^7}{10^3} = 10^{7-3} = 10^4 = 10{,}000$

d. $\dfrac{x^5}{y^2}$ Rule 3 does not apply when the bases are different

e. $\dfrac{8x^3}{2x} = \dfrac{\overset{4}{\cancel{8}}}{\underset{1}{\cancel{2}}} \cdot \dfrac{x^3}{x} = 4 \cdot x^2 = 4x^2$

f. $\dfrac{6a^3b^4}{9ab^2} = \dfrac{\overset{2}{\cancel{6}}}{\underset{3}{\cancel{9}}} \cdot \dfrac{a^3}{a} \cdot \dfrac{b^4}{b^2} = \dfrac{2}{3} \cdot \dfrac{a^2}{1} \cdot \dfrac{b^2}{1} = \dfrac{2}{3}a^2b^2, \text{ or } \dfrac{2a^2b^2}{3}$

g. $\dfrac{2^a}{2^b} = 2^{a-b}$

Why $x \neq 0$ in Rule 3   When writing Rule 3, we added the restriction $x \neq 0$. Consider the example $\dfrac{0^5}{0^2}$. By Rule 3, $\dfrac{0^5}{0^2} = 0^{5-2} = 0^3 = 0$. However, $\dfrac{0^5}{0^2} = \dfrac{0 \cdot 0 \cdot 0 \cdot 0 \cdot 0}{0 \cdot 0} = \dfrac{0}{0}$ is undefined (Section 6.7). For this reason, x cannot be zero in Rule 3.

In this book, unless otherwise noted, none of the variables has a value that makes a denominator zero.

> **A Word of Caution**   An expression like $\dfrac{x^5 - y^3}{x^2}$ is usually left unchanged.
>
> Rules 1 and 3 cannot be applied here because the numerator is not a product. This expression could be changed as follows:
>
> $$\dfrac{x^5 - y^3}{x^2} = \dfrac{x^5}{x^2} - \dfrac{y^3}{x^2} = x^3 - \dfrac{y^3}{x^2}$$
>
> A common mistake is to divide only x^5 by x^2, instead of dividing both x^5 and y^3 by x^2. Expressions of this form are discussed in a later section.

When Rule 3 was used, we were careful to choose the exponent in the numerator larger than the exponent in the denominator, so that the quotient always had a positive exponent. When the exponent in the denominator is larger than the exponent in the numerator, the quotient has a negative exponent. Negative exponents are discussed in a later section. Now we consider the case in which the exponents of the numerator and denominator are the same. For $x \neq 0$,

$$\dfrac{x^4}{x^4} = \dfrac{xxxx}{xxxx} = 1 \quad \text{Because a nonzero number divided by itself is 1}$$

and

$$\dfrac{x^4}{x^4} = x^{4-4} = x^0 \quad \text{Using Rule 3}$$

Therefore, we define $x^0 = 1$.

RULE 4 Zero exponent	$x^0 = 1 \qquad (x \neq 0)$

EXAMPLE 4

a. $a^0 = 1$ Provided $a \neq 0$

b. $10^0 = 1$

c. $6x^0 = 6 \cdot 1 = 6$ Provided $x \neq 0$

     └─── The 0 exponent applies only to x

Exercises   8.2

Use the rules of exponents to simplify each expression.

1. $x^5 \cdot x^8$
2. $H^3 \cdot H^4$
3. $(y^2)^5$
4. $(N^3)^4$

5. $\dfrac{x^7}{x^2}$
6. $\dfrac{y^8}{y^6}$
7. y^0
8. b^0

9. $a \cdot a^4$
10. $B^7 \cdot B$
11. $(x^4)^7$
12. $(v^3)^8$

13. $\dfrac{a^5}{a}$
14. $\dfrac{b^7}{b}$
15. $\dfrac{z^5}{z^4}$
16. $\dfrac{x^8}{x^7}$

17. $10^2 \cdot 10^5$
18. $2^3 \cdot 2^2$
19. $(10^2)^3$
20. $(10^7)^2$

21. $\dfrac{10^{11}}{10}$
22. $\dfrac{2^6}{2}$
23. $5x^0$
24. $3y^0$

25. x^2y^3
26. a^4b
27. $\dfrac{6x^2}{2x}$
28. $\dfrac{9y^3}{3y}$

29. $\dfrac{a^3}{b^2}$
30. $\dfrac{x^5}{y^3}$
31. $\dfrac{10x^4}{5x^3}$
32. $\dfrac{15y^5}{9y^2}$

33. $\dfrac{12h^4k^3}{8h^2k}$
34. $\dfrac{16a^5b^3}{12ab^2}$
35. $5^u \cdot 5^v$
36. $\dfrac{x^c}{x^d}$

37. $\dfrac{a^4 - b^3}{a^2}$
38. $\dfrac{x^6 + y^4}{y^2}$
39. $y^2 \cdot y^4$
40. $(x^3)^2$

41. $\dfrac{a^6}{a^4}$
42. $x^5 \cdot x$
43. $(z^2)^4$
44. $\dfrac{b^4}{b}$

45. $\dfrac{y^5}{y^5}$
46. $10^3 \cdot 10^2$
47. $(3^3)^2$
48. $\dfrac{5^3}{5^2}$

49. hk^2
50. $\dfrac{12x^3}{10x}$
51. $\dfrac{m^4}{n^2}$
52. $\dfrac{18z^6}{10z^4}$

53. $\dfrac{25a^3b^4}{15ab^3}$
54. $3^2 \cdot 5^3$
55. $2 \cdot 2^3 \cdot 2^2$

56. $3 \cdot 3^2 \cdot 3^3$
57. $x \cdot x^3 \cdot x^4$
58. $y \cdot y^5 \cdot y^3$

59. $x^2y^3x^5$
60. $z^3z^4w^2$
61. ab^3a^5

62. x^8yx^4
63. $(3^a)^b$
64. $2^a \cdot 2^b \cdot 2^c$

65. $(2^a)^0$

8.3   The Distributive Rule

When simplifying a **product of factors**, we use the commutative and associative properties of multiplication to change the order and grouping of the factors.

Simplifying a product of factors

> 1. *Multiply the numerical coefficients.* The sign of the product is negative if there are an odd number of negative factors. The sign is positive if there are an even number of negative factors (or no negative factors).
>
> 2. *Multiply the variables* (usually writing them in alphabetical order). Use Rule 1 of exponents to determine the exponent of each variable.

EXAMPLE 1

a. $(4x^4)(3x^3)$

$= (4)(3)(x^4 \cdot x^3)$

$= 12x^7$

— Rule 1 of exponents

— Product of the coefficients

b. $(-2a^2b)(5a^3b^3)$

$= (-2)(5)(a^2 \cdot a^3)(b \cdot b^3)$

$= -10a^5b^4$

c. $(4xy^2)(-3x^2y^4)(-2x^3y)$

$= (4)(-3)(-2)(x \cdot x^2 \cdot x^3)(y^2 \cdot y^4 \cdot y)$

$= 24x^6y^7$

d. $(5x^3)^2$

$= (5x^3)(5x^3)$

$= (5)(5)(x^3 \cdot x^3)$

$= 25x^6$

The Distributive Rule

One property of real numbers combines both addition and multiplication. Consider $3(10 + 1)$:

$$3(10 + 1) = 3(11) = 33$$

If we distribute the 3 and multiply, we get

$$3 \cdot 10 + 3 \cdot 1 = 30 + 3 = 33$$

Therefore, $3(10 + 1) = 3 \cdot 10 + 3 \cdot 1$.

This leads to a fundamental property of real numbers called the **distributive rule**.

The distributive rule

> $$a(b + c) = ab + ac$$
>
> The distributive rule can be extended to cover any number of terms within the parentheses.

Meaning of the Distributive Rule   When a factor is multiplied by an expression enclosed within grouping symbols, the factor must be multiplied by *each term* within the grouping symbols; then the products are added.

$$5(2 - 7) = (5)(2 + -7)$$

First product   Second product

Each term inside the parentheses is multiplied by the factor outside the parentheses; then these products are added

$$= (5)(2) + (5)(-7)$$
$$= \quad 10 \quad + \quad (-35)$$
$$= \qquad\quad -25$$

EXAMPLE 2

a. $2(x + y) = (2)(x) + (2)(y)$
$$= \quad 2x \quad + \quad 2y$$

b. $a(x - y) = (a)(x) + (a)(-y)$
$$= \quad ax \quad + \quad (-ay)$$
$$= \quad ax \quad - \quad ay$$

c. $x(x^2 + y) = (x)(x^2) + (x)(y)$
$$= \quad x^3 \quad + \quad xy$$

d. $3x(4x + x^3y) = (3x)(4x) + (3x)(x^3y)$
$$= \quad 12x^2 \quad + \quad 3x^4y$$

EXAMPLE 3

Example of applying the distributive rule when more than two terms appear within the parentheses:

$$-5a(4a^3 - 2a^2b + b^2) = (-5a)(4a^3 + -2a^2b + b^2)$$
$$= (-5a)(4a^3) + (-5a)(-2a^2b) + (-5a)(b^2)$$
$$= \quad -20a^4 \quad + \quad 10a^3b \quad - \quad 5ab^2$$

Further Extensions of the Distributive Rule   Because of the distributive rule and the commutative property of multiplication, it follows that

$$(b + c)a = a(b + c) = ab + ac = ba + ca$$

Therefore, $(b + c)a = ba + ca$. This result can be extended to include more terms:

$$(b + c + d + \cdots)a = ba + ca + da + \cdots$$

EXAMPLE 4

a. $(2x - 5)(4x) = (2x + -5)(4x)$
$$= (2x)(4x) + (-5)(4x)$$
$$= \quad 8x^2 \quad - \quad 20x$$

b. $(-2x^2 + xy - 5y^2)(-3xy)$

$$= (\boxed{-2x^2} + \boxed{xy} + \boxed{-5y^2})(\boxed{-3xy})$$

$$= (-2x^2)(-3xy) + (xy)(-3xy) + (-5y^2)(-3xy)$$

$$= \qquad 6x^3y \qquad - \qquad 3x^2y^2 \qquad + \qquad 15xy^3$$

Recall that multiplying a number by -1 gives the opposite of that number:

$$-1 \cdot a = -a$$

Therefore, to find the opposite of an algebraic expression, we multiply the expression by -1.

EXAMPLE 5

a. $- (2x + 3) = -1 (2x + 3)$

$$= (\ -1\ )(2x) + (\ -1\ )(3)$$

$$= \qquad -2x \qquad - \qquad 3$$

b. $-(a - b) = -1(a - b)$

$$= (-1)(a) + (-1)(-b)$$

$$= \qquad -a \qquad + \qquad b$$

 Note The opposite of an expression changes the sign of each term.

c. $-(3x^2 - 5x + 4) = -3x^2 + 5x - 4$

A Word of Caution A common error is to think that the distributive rule applies to expressions like $2(3 \cdot 4)$.

The distributive rule only applies when this is an addition or subtraction

$$2(3 \cdot 4) = (2 \cdot 3)(2 \cdot 4)$$

as

$$2(3 \cdot 4) = 2(12) = 24 \qquad \text{whereas} \qquad (2 \cdot 3)(2 \cdot 4) = (6)(8) = 48$$

Exercises 8.3

In Exercises 1–12, simplify each product of factors.

1. $(-2a)(4a^2)$ **2.** $(-3x)(5x^3)$ **3.** $(5x^2)(-7y)$

4. $(-3x^3)(4y)$ **5.** $(4x^2)^3$ **6.** $(3x^2)^2$

7. $(-5x^3)(2x^2)(-3x)$ **8.** $(4x^2)(-7x^2)(2x)$

9. $(2mn^2)(-4m^2n^2)(2m^3n)$ **10.** $(5x^2y)(-2xy^2)(-3xy)$

11. $(-5x^2y)(2yz^3)(-xz)$ **12.** $(3xyz)(-2x^2y)(-yz^2)$

In Exercises 13–50, find each product using the distributive rule.

13. $5(a + 6)$ **14.** $4(x + 10)$ **15.** $3(m - 4)$

16. $3(a - 5)$ **17.** $-2(x - 3)$ **18.** $-3(x - 5)$

19. $-3(2x^2 - 4x + 5)$ **20.** $-5(3x^2 - 2x - 7)$

21. $4x(3x^2 - 6)$ **22.** $3x(5x^2 - 10)$

23. $-2x(5x^2 + 3x - 4)$ **24.** $-4x(2x^2 - 5x + 3)$

25. $-1(x^2 - y^2)$ **26.** $-1(5x + 2y)$

27. $-(x + 3)$ **28.** $-(y - 7)$

29. $-(2x^2 + 4x - 7)$ **30.** $-(z^2 - 5z + 6)$

31. $(x - 4)6$ **32.** $(3 - 2x)(-5)$

33. $(y^2 - 4y + 3)7$ **34.** $(-9 + z - 2z^2)(-7)$

35. $(2x^2 - 3x + 5)4x$ **36.** $(3w^2 + 2w - 8)5w$

37. $x(xy - 3)$ **38.** $a(ab - 4)$

39. $3a(ab - 2a^2)$ **40.** $4x(3x - 2y^2)$

41. $(-2x + 4x^2y)(-3y)$ **42.** $(-3a + 2a^2z)(-2z)$

43. $-2xy(x^2y - y^2x - y - 5)$

44. $-3ab(8 - a^2 - b^2 + ab)$

45. $(3x^3 - 2x^2y + y^3)(-2xy)$

46. $(4z^3 - z^2y - y^3)(-2yz)$

47. $4ab^2(3a^4b^2 - 2ab - 5b)$

48. $6x^2y(2x^5y - 4x^3y^2 + xy^3)$

49. $(-2xy^2z^2)(6x^2y - 3yz - 4xz^2)$

50. $(-5a^2bc)(2b^2c - 5ac^3 - 3a^2b^3)$

$-8x^3 + 20x^2 - 12x$

8.4   Combining Like Terms

Like terms have the same variables with the same exponents. Only the numerical coefficients may differ.

EXAMPLE 1   Examples of like terms:

a. $2x$ and $4x$ are like terms. They are called "x-terms."
b. $-3x^2$ and $5x^2$ are like terms. They are called "x^2-terms."
c. $3xy$ and $6yx$ are like terms because $xy = yx$. They are called "xy-terms."
d. x^2y and $9x^2y$ are like terms. They are called "x^2y-terms."

 Note All constants are like terms.

e. $5, -3, \dfrac{1}{4}$, and 2.6 are like terms. They are called "constant terms."

EXAMPLE 2   Examples of unlike terms:

a. $2x$ and $3y$ are unlike terms.   The variables are different
b. $5x^2$ and $7x$ are unlike terms.   The variable x has different exponents
c. $4x^2y$ and $10xy^2$ are unlike terms.

Combining Like Terms   3 dollars + 5 dollars = (3 + 5) dollars = 8 dollars

3 cars   + 5 cars   = (3 + 5) cars   = 8 cars

$3x$ $+ 5x$ $= (3 + 5)x$ $= 8x$

This is an application of
the distributive rule

Combining like terms

1. Identify the like terms.
2. Find the sum of each group of like terms by adding their numerical coefficients, then multiply that sum by the common variables. (When no coefficient is written, it's understood to be 1.)

EXAMPLE 3

a. $2x + 4x = (2 + 4)x = 6x$

b. $e - 9e = (1 - 9)e = -8e$

When no coefficient is written,
it's understood to be 1

c. $5x^2y - 7x^2y = (5 - 7)x^2y = -2x^2y$

d. $4ab - 3ab + 6ab = (4 - 3 + 6)ab = 7ab$

e. $9x - 3x + 5y = (9 - 3)x + 5y = 6x + 5y$

These are the only like terms in the expression;
unlike terms cannot be combined

f. $3xy + 6yx = (3 + 6)xy = 9xy$

These are like terms because $xy = yx$

 Note When we combine like terms, we do not need to show the step that uses the distributive rule. We simply add the coefficients mentally and multiply this sum by the common variables. For example, in the expression $7x - 4x + 2x$,

Add these coefficients

$\boxed{7}x \;\boxed{-4}x\; \boxed{+2}x = 5x$

When we combine like terms, we are usually changing the grouping and the order in which the terms appear. The commutative and associative properties of addition guarantee that when we do this, the sum remains unchanged.

g. $12a - 7b - 9a + 4b = (12a - 9a) + (-7b + 4b)$

$= \quad 3a \quad + \quad (-3b)$

$= \quad 3a \quad - \quad 3b$

h. $3x^2 - 2x + 5 - 4x + 6 = 3x^2 + (-2x - 4x) + (5 + 6)$

$= 3x^2 + \quad (-6x) \quad + \quad 11$

$= 3x^2 - \quad 6x \quad + \quad 11$

Exercises 8.4

Combine like terms.

1. $15x - 3x$

2. $12a - 9a$

3. $5a - 12a$

4. $10x - 24x$

5. $-2x^2 + 6x^2$

6. $-4a^2 + 8a^2$

7. $x^3 + 7x^3$

8. $9y^4 + y^4$

9. $3xy - 10xy$

10. $5ab - 8ab$

11. $4x^2 + 3x$

12. $8a^3 - 2a^2$

13. $2a - 5a + 6a$

14. $3y - 4y + 5y$

15. $5x - 8x + x + 3x$

16. $3a - 5a + a - 2a$

17. $3x + 2y - 3x$

18. $4a - 2b + 2b$

19. $3mn - 5mn + 2mn$

20. $5cd - 8cd + 3cd$

21. $2xy - 5yx + xy$

22. $8mn - 7nm + 3nm$

23. $a^2b - 3a^2b$

24. $x^2y^2 - 5x^2y^2$

25. $5x + 7 - 2x + 3$

26. $3x - 7 + 6x + 2$

27. $x - 3 + 4x - 8$

28. $9x + 4 - x - 10$

29. $4x^2 - 3x - 8x^2 - 4x$

30. $x^2 + 2x + 4x^2 - 8x$

31. $5x^2 - 3x + 7 - 2x^2 + 8x - 9$

32. $7y^2 + 4y - 6 - 9y^2 - 2y + 7$

33. $x - 3x^3 + 2x^2 - 5x - 4x^2 + x^3$

34. $y - 2y^2 - 5y^3 - y + 3y^3 - y^2$

35. $x^3 - 2x^2 + 6x - 3x^2 - 7x + 3$

36. $a^2 - 4a - 6 + 7a^3 - 5a^2 - 3$

37. $5x^2 - xy + y^2 + 6xy - 7y^2$

38. $3a^2 + 6ab + b^2 - 2a^2 - 9ab$

39. $7x^2y - 2xy^2 - 4x^2y - xy^2$

40. $4xy^2 - 5x^2y - 2xy^2 + 3x^2y$

41. $\dfrac{2}{3}y^2 - \dfrac{1}{2}y^2 + \dfrac{5}{6}y^2$

42. $\dfrac{1}{2}x - \dfrac{3}{4}x + \dfrac{5}{8}x - \dfrac{3}{16}x$

43. $12.67 \text{ sec} + 9.08 \text{ sec} - 6.73 \text{ sec}$

44. $18.7 \text{ ft} + 69.5 \text{ ft} - 42.1 \text{ ft} - 26.3 \text{ ft}$

30) $5x^2 - 6x$

8.5   Addition and Subtraction of Polynomials

8.5A   Definition of a Polynomial

A **polynomial in x** is an algebraic expression having only terms of the form ax^n, where a is any real number and n is a whole number.

EXAMPLE 1   Examples of polynomials:

a. $3x$	A polynomial of one term is called a **monomial**
b. $4x^4 - 2x^2$	A polynomial of two unlike terms is called a **binomial**
c. $7x^2 - 5x + 2$	A polynomial of three unlike terms is called a **trinomial**
d. $x^3 + 3x^2 + 3x + 1$	We will not use special names for polynomials of more than three terms
e. 5	This is a polynomial of one term (monomial) because its only term has the form $5x^0 = 5 \cdot 1 = 5$
f. $6z^3 - 3z + 1$	Polynomials can be in any variable; this is a polynomial in z

EXAMPLE 2   Examples of algebraic expressions that are *not* polynomials:

a. $4x^{-2}$ This expression is *not* a polynomial because the exponent -2 is not a whole number

b. $\dfrac{2}{x-5}$ This is *not* a polynomial because the variable x is in the denominator; the term does not have the form ax^n

c. $\sqrt{2x+3}$ This is not a polynomial because the variable x is under the radical sign

Polynomials often have terms containing more than one variable.

A **polynomial in x and y** is an algebraic expression having only terms of the form $ax^n y^m$, where a is any real number and n and m are whole numbers.

EXAMPLE 3   Examples of polynomials in more than one variable:

a. $5x^2 y$ Monomial

b. $-3xy^3 + 2x^2 y$ Binomial

c. $4x^2 y^2 - 7xy + 6$ Trinomial

d. $x^3 y^2 \; - 2x \; + 3y^2 \; - 1$

$$-1 = -1x^0 y^0$$
$$3y^2 = 3x^0 y^2$$
$$-2x = -2x^1 y^0$$

e. $7uv^4 - 5u^2 v + 2u$ Polynomials can be in any variables; this is a polynomial in u and v

The **degree of a term** in a polynomial is the sum of the exponents of its variables.

EXAMPLE 4   **a.** $5x^3$ is third degree.

b. $6x^2 y$ is third degree because $6x^2 y = 6x^2 y^1$.

c. 14 is zero degree because $14 = 14x^0$.

d. $-2u^3 vw^2$ is sixth degree because $-2u^3 vw^2 = -2u^3 v^1 w^2$.

$$3 + 1 + 2 = 6$$

The **degree of a polynomial** is the same as that of its highest-degree term (provided like terms have been combined).

EXAMPLE 5

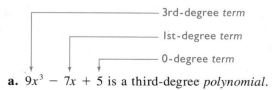

3rd-degree *term*

1st-degree *term*

0-degree *term*

a. $9x^3 - 7x + 5$ is a third-degree *polynomial*.

4th-degree *term*

6th-degree *term*

0-degree *term*

b. $14xy^3 - 11x^5y + 8$ is a sixth-degree *polynomial.*

c. $6a^2bc^3 + 12ab^6c^2$ is a ninth-degree *polynomial.*

Polynomials are usually written in **descending powers** of one of the variables. For example,

Exponents get smaller from left to right

$$8x^3 - 3x^2 + 5x^1 + 7$$

$7 = 7x^0$

EXAMPLE 6   Arrange $5 - 2x^2 + 4x$ in descending powers of x.

SOLUTION $-2x^2 + 4x + 5$

When a polynomial has more than one variable, it can be arranged in descending powers of any one of its variables.

EXAMPLE 7   Arrange $3x^3y - 5xy + 2x^2y^2 - 10$ (a) in descending powers of x, then (b) in descending powers of y.

a. $3x^3y + 2x^2y^2 - 5xy - 10$ Arranged in descending powers of x

b. $2x^2y^2 + 3x^3y - 5xy - 10$ Arranged in descending powers of y

Because y is the same power in both terms, the higher-degree term is written first

Exercises 8.5A

In Exercises 1–12, if the expression is a polynomial, write **(a)** the degree of the first term and **(b)** the degree of the polynomial. Otherwise, write "not a polynomial."

1. $2x^2 + 3x$ **2.** $5y^3 + 4y$ **3.** $3x^4 - x^5 + 5$

4. $5x^3 - x^6 + 2$ **5.** $7x + x^2$ **6.** $8 - x$

7. $x^2y - x^3y^2$ **8.** $2ab + 5ab^4$

9. $8m^3n - 12m^2n + 6mn^2 - n^3$

10. $x^3y^3 - 3x^2y + 3xy^2 - y^3$

11. $x^{-2} + 5x^{-1} + 4$ **12.** $\dfrac{1}{2x^2 - 5x}$

In Exercises 13–18, write each polynomial in descending powers of the indicated letter.

13. $x^3 + x - 3x^2 - 4$ Powers of x

14. $x^2 - x - 7 + 6x^3$ Powers of x

15. $7x^3 - 4x - 5 + 8x^5$ Powers of x

16. $10 - 3y^5 + 4y^2 - 2y^3$ Powers of y

17. $8xy^2 + xy^3 - 4x^2y$ Powers of y

18. $3x^3y + x^4y^2 - 3xy^3$ Powers of x

8.5B   Adding and Subtracting Polynomials

Addition of Polynomials

The commutative and associative properties allow us to add polynomials by combining like terms.

Adding polynomials

1. Remove grouping symbols.
2. Combine like terms.

It's helpful to underline like terms with the same kind of line before adding. Answers are usually written in descending powers of the variable.

EXAMPLE 8

a. $(3x^2 + 5x - 4) + (2x + 5) + (x^3 - 4x^2 + x)$

$= 3x^2 + 5x - 4 + 2x + 5 + x^3 - 4x^2 + x$

$= x^3 - x^2 + 8x + 1$

b. $(5x^3y^2 - 3x^2y^2 + 4xy^3) + (4x^2y^2 - 2xy^2) + (-7x^3y^2 + 6xy^2 - 3xy^3)$

$= 5x^3y^2 - 3x^2y^2 + 4xy^3 + 4x^2y^2 - 2xy^2 - 7x^3y^2 + 6xy^2 - 3xy^3$

$= -2x^3y^2 + x^2y^2 + xy^3 + 4xy^2$

We will do most addition of polynomials horizontally, as already shown. However, in a few cases it is convenient to use *vertical addition*. We use the following procedure to add polynomials vertically:

1. Arrange the polynomials under one another so that like terms are in the same vertical line.
2. Find the sum of the terms in each vertical line by adding their numerical coefficients.

EXAMPLE 9

Add $(3x^2 + 2x - 1)$, $(2x + 5)$, and $(4x^3 + 7x^2 - 6)$ vertically.

SOLUTION

$$
\begin{array}{r}
3x^2 + 2x - 1 \\
2x + 5 \\
4x^3 + 7x^2 - 6 \\
\hline
4x^3 + 10x^2 + 4x - 2
\end{array}
$$

Subtraction of Polynomials

We subtract polynomials the same way we subtract signed numbers. That is, we change the signs in the polynomial being subtracted, then add the resulting polynomials.

Subtracting polynomials

1. Change the sign of *each* term in the polynomial being subtracted.
2. Add the resulting polynomials.

EXAMPLE 10

a. Find $(-4x^3 + 8x^2 - 2x - 3) - (4x^3 - 7x^2 + 6x + 5)$.

SOLUTION ⎯⎯ This is the polynomial being subtracted

$(-4x^3 + 8x^2 - 2x - 3) - (4x^3 - 7x^2 + 6x + 5)$

$= -4x^3 + 8x^2 - 2x - 3 - 4x^3 + 7x^2 - 6x - 5$

$= -8x^3 + 15x^2 - 8x - 8$

b. Subtract $(-4x^2y + 10xy^2 + 9xy - 7)$ from $(11x^2y - 8xy^2 + 7xy + 2)$.

⎯⎯ This is the polynomial being subtracted

SOLUTION $(11x^2y - 8xy^2 + 7xy + 2) - (-4x^2y + 10xy^2 + 9xy - 7)$

$= 11x^2y - 8xy^2 + 7xy + 2 + 4x^2y - 10xy^2 - 9xy + 7$

$= 15x^2y - 18xy^2 - 2xy + 9$

c. Subtract $(2x^2 - 5x + 3)$ from the sum of $(8x^2 - 6x - 1)$ and $(4x^2 + 7x - 9)$.

SOLUTION $(8x^2 - 6x - 1) + (4x^2 + 7x - 9) - (2x^2 - 5x + 3)$

$= 8x^2 - 6x - 1 + 4x^2 + 7x - 9 - 2x^2 + 5x - 3$

$= 10x^2 + 6x - 13$

d. Find $(x^2 + 5) - [(x^2 - 3) + (2x^2 - 1)]$.

SOLUTION

$(x^2 + 5) - [(x^2 - 3) + (2x^2 - 1)]$

$= (x^2 + 5) - [x^2 - 3 + 2x^2 - 1]$ Removing the innermost ( )

$= (x^2 + 5) - [3x^2 - 4]$ Combining like terms inside the [ ]

$= x^2 + 5 - 3x^2 + 4$ Removing the grouping symbols

$= -2x^2 + 9$ Combining like terms

We will do most subtraction of polynomials horizontally, as already shown, but in a few cases it's convenient to use *vertical subtraction*. We use the following procedure to subtract polynomials vertically:

1. Write the polynomial being subtracted *under* the polynomial it's being subtracted from. Write like terms in the same vertical line.

2. Change the sign of *each* term in the polynomial being subtracted.

3. Find the sum of the resulting terms in each vertical line by adding their numerical coefficients.

EXAMPLE 11

Subtract $5x^2 + 3x - 6$
$\quad\quad\quad\;\; 3x^2 - 5x + 2$

MATH IS BORING + RIDICULOUSLY EASY !!

SOLUTION

$\begin{array}{r} 5x^2 + 3x - 6 \\ -(3x^2 - 5x + 2) \end{array} \longrightarrow \begin{array}{r} 5x^2 + 3x - 6 \\ -3x^2 + 5x - 2 \\ \hline 2x^2 + 8x - 8 \end{array}$ ⎯ Change the sign of *each* term in the polynomial being subtracted, then *add* the resulting terms

EXAMPLE 12   Subtract $(3x^2 - 2x - 7)$ from $(x^3 - 2x - 5)$ vertically.

SOLUTION

$$
\begin{array}{rl}
x^3 \quad\ - 2x - 5 & \quad x^3 \quad\ - 2x - 5 \\
-(\ + 3x^2 - 2x - 7) \longrightarrow & \quad - 3x^2 + 2x + 7 \quad \leftarrow \textit{Change signs and add} \\
\hline
& \quad x^3 - 3x^2 \quad\ + 2
\end{array}
$$

Exercises   8.5B

In Exercises 1–26, work the addition and subtraction problems.

1. $(3x - 5) + (2x + 7)$ 2. $(4y + 3) + (y - 9)$

3. $(3x - 5) - (2x + 7)$ 4. $(4y + 3) - (y - 9)$

5. $(x^2 + 12x) + (7x^2 - 4x)$ 6. $(4x^3 - 4x^2) + (x^3 - 6x^2)$

7. $(2x^2 + 8x) - (6x^2 - x)$ 8. $(4a^2 - a) - (-2a^2 + 5a)$

9. $(2m^2 - m + 4) + (3m^2 + m - 5)$

10. $(5n^2 + 8n - 7) + (6n^2 - 6n + 10)$

11. $(3x^2 + 4x - 10) - (5x^2 - 3x + 7)$

12. $(2a^2 - 3a + 9) - (3a^2 + 4a - 5)$

13. Subtract $(-5b^2 + 4b + 8)$ from $(8b^2 + 2b - 14)$.

14. Subtract $(-8c^2 - 9c + 6)$ from $(11c^2 - 4c + 7)$.

15. $(2x^3 - 4) + (4x^2 + 8x) + (-9x + 7)$

16. $(5 + 8z^2) + (4 - 7z) + (z^2 + 7z)$

17. $(6a - 5a^2 + 6) + (4a^2 + 6 - 3a)$

18. $(2b + 7b^2 - 5) + (4b^2 - 2b + 8)$

19. $(4a^2 + 6 - 3a) - (5a + 3a^2 - 4)$

20. $(8 + 3b^2 - 7b) - (2b + b^2 - 9)$

21. $(3x^2 - 4x + 8) - (9 - x^2 + 6x)$

22. $(8x^2 - 6 - x) - (7x + 3x^2 + 5)$

23. $(3x - 7) + (5 - x) - (4x - 2)$

24. $(7a + 2) - (3a - 8) + (a - 7)$

25. $(y^2 - 3y) - (4y - 6) - (2 - y^2)$

26. $(x^3 - 4x^2) - (5x^2 - x) - (7x - 9)$

In Exercises 27–30, add the polynomials vertically.

27.
$$
\begin{array}{r}
17a^3 \quad\ + 4a - 9 \\
8a^2 - 6a + 9 \\
\hline
\end{array}
$$

28.
$$
\begin{array}{r}
-\ b^3 + 5b^2 - 8 \\
-20b^4 + 2b^3 \quad\ + 7 \\
\hline
\end{array}
$$

29.
$$
\begin{array}{r}
14x^2y^3 - 11xy^2 + 8xy \\
-9x^2y^3 + 6xy^2 - 3xy \\
7x^2y^3 - 4xy^2 - 5xy \\
\hline
\end{array}
$$

30.
$$
\begin{array}{r}
12a^2b - 8ab^2 + 6ab \\
-7a^2b + 11ab^2 - 3ab \\
4a^2b - ab^2 - 13ab \\
\hline
\end{array}
$$

In Exercises 31–34, subtract the lower polynomial from the upper polynomial, vertically.

31.
$$
\begin{array}{r}
15x^3 - 4x^2 \quad\ + 12 \\
8x^3 \quad\ + 9x - 5 \\
\hline
\end{array}
$$

32.
$$
\begin{array}{r}
-14y^2 + 6y - 24 \\
7y^3 + 14y^2 - 13y \\
\hline
\end{array}
$$

33.
$$
\begin{array}{r}
8a^3 - 3a^2 + 2a + 9 \\
5a^3 - 7a^2 - a + 6 \\
\hline
\end{array}
$$

34.
$$
\begin{array}{r}
4x^2y^2 + x^2y - 6xy^2 - 2xy \\
3x^2y^2 + 5x^2y - 2xy^2 + 7xy \\
\hline
\end{array}
$$

35. $(7 - 8v^3 + 9v^2 + 4v) + (9v^3 - 8v^2 + 4v + 6)$

36. $(15 - 10w + w^2 - 3w^3) + (18 + 4w^3 + 7w^2 + 10w)$

37. $(7x^2y^2 - 3x^2y + xy + 7) - (3x^2y^2 + 7x^2y - 5xy + 4)$

38. $(4x^2y^2 + x^2y - 5xy - 4) - (5x^2y^2 - 3x^2y + 9)$

39. $(5x - 3) - [(x + 2) + (2x - 6)]$

40. $(8y + 1) - [(2y - 3) + (y + 7)]$

41. $(x^2 + 4) - [(x^2 - 5) - (3x^2 + 1)]$

42. $(3x^2 - 2) - [(4 - x^2) - (2x^2 - 1)]$

43. $(4x^2 - 5) - [(2 - x) - (x^2 + 3x)]$

44. $(9a^2 + 2) - [(7 + 3a) - (4a^2 - 5a)]$

45. Subtract $(2x^2 - 4x + 3)$ from the sum of $(5x^2 - 2x + 1)$ and $(-4x^2 + 6x - 8)$.

46. Subtract $(6y^2 + 3y - 4)$ from the sum of $(-2y^2 + y - 9)$ and $(8y^2 - 2y + 5)$.

47. Subtract $(10a^3 - 8a + 12)$ from the sum of $(11a^2 + 9a - 14)$ and $(-6a^3 + 17a)$.

48. Subtract $(15b^3 - 17b^2 + 6)$ from the sum of $(14b^2 - 8b - 11)$ and $(-9b^3 + 12b)$.

49. Subtract the sum of $(x^3y + 3xy^2 - 4)$ and $(2x^3y - xy^2 + 5)$ from the sum of $(5 + xy^2 + x^3y)$ and $(-6 - 3xy^2 + 4x^3y)$.

50. Subtract the sum of $(2m^2n - 4mn^2 + 6)$ and $(-3m^2n + 5mn^2 - 4)$ from the sum of $(5 + m^2n - mn^2)$ and $(3 + 4m^2n + 2mn^2)$.

51. Given the polynomials $(2x^2 - 5x - 7)$, $(-4x^2 + 8x - 3)$, and $(6x^2 - 2x + 1)$, subtract the sum of the first two from the sum of the last two.

52. Given the polynomials $(8y^2 + 10y - 9)$, $(13y^2 - 11y + 6)$, and $(-2y^2 - 16y + 4)$, subtract the sum of the first two from the sum of the last two.

53. $(7.2x^2 - 4.8x + 6.5) + (-2.8x^2 + 8.9x + 5.3)$

54. $(9.6x^2 + 5.7x - 9.8) + (7.9x^2 - 9.6x - 3.2)$

55. $(2.62x^2 - 8.95x - 6.08) - (7.4x^2 + 11.9x - 8.54)$

56. $(3.56x^2 + 17.4x - 5.84) - (14.7x^2 - 3.09x + 12.6)$

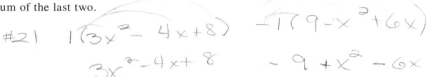

8.6 Multiplication of Polynomials

Multiplication of a Polynomial by a Monomial

The basis of multiplication of polynomials is the distributive rule and its extensions:

$$a(b + c + \cdots) = ab + ac + \cdots$$
$$(b + c + \cdots)a = ba + ca + \cdots$$

Multiplying a polynomial by a monomial	Multiply *each* term in the polynomial by the monomial, then add the results.

EXAMPLE 1

a. $3x^2(2x + 7) = (3x^2)(2x) + (3x^2)(7) = 6x^3 + 21x^2$

b. $-5x(3x^2 - 2x + 6) = (-5x)(3x^2) + (-5x)(-2x) + (-5x)(6)$
$$= -15x^3 + 10x^2 - 30x$$

Multiplication of a Polynomial by a Polynomial

Consider the product $(x + 2)(x + 5)$. To find this product, we will use the distributive rule twice.

First, we consider $(x + 2)$ as a single quantity and distribute it to each term in $(x + 5)$:

$$(x + 2)(x + 5) = (x + 2)(x) + (x + 2)(5)$$

Now we use the distributive rule again:

$$= (x)(x) + (2)(x) + (x)(5) + (2)(5)$$
$$= x^2 + 2x + 5x + 10$$
$$= x^2 + 7x + 10$$

The result is that *each term* of the first polynomial is multiplied by *each term* of the second polynomial, and then these products are added.

A convenient arrangement for multiplying two polynomials is the one used in arithmetic for multiplying two whole numbers:

$$
\begin{array}{r}
56 \\
\times \quad 23 \\
\hline
168 \\
112 \\
\hline
1288
\end{array}
$$

Product of 56 and 3

Product of 56 and 2

Notice that the second line in this arithmetic example is moved one place to the left—why?

We use this arrangement in multiplying the following binomials (Examples 2–5).

EXAMPLE 2 Find $(2x^2 - 5x)(3x + 7)$.

SOLUTION

$$
\begin{array}{r}
2x^2 - 5x \\
\times \quad 3x + 7 \\
\hline
14x^2 - 35x \\
6x^3 - 15x^2 \\
\hline
6x^3 - x^2 - 35x
\end{array}
$$

$\Leftarrow (2x^2 - 5x)7$

$\Leftarrow (2x^2 - 5x)3x$

Notice that the second line is moved one place to the left so that we have like terms in the same vertical line

EXAMPLE 3 Find $(3a^2b - 6ab^2)(2ab - 5)$.

SOLUTION

$$
\begin{array}{r}
3a^2b - 6ab^2 \\
\times \quad 2ab - 5 \\
\hline
- 15a^2b + 30ab^2 \\
6a^3b^2 - 12a^2b^3 \\
\hline
6a^3b^2 - 12a^2b^3 - 15a^2b + 30ab^2
\end{array}
$$

Notice that the second line is moved over far enough so that only like terms are in the same vertical line

EXAMPLE 4 Find $(3x^2 + 2x - 4)(5x + 2)$.

SOLUTION

$$
\begin{array}{r}
3x^2 + 2x - 4 \\
\times \quad 5x + 2 \\
\hline
6x^2 + 4x - 8 \\
15x^3 + 10x^2 - 20x \\
\hline
15x^3 + 16x^2 - 16x - 8
\end{array}
$$

The higher-degree polynomial is placed above the other, as in arithmetic

EXAMPLE 5 Find $(2m - 5 - 3m^2)(2 + m^2 - 3m)$.

SOLUTION

$$
\begin{array}{r}
-3m^2 + 2m - 5 \\
\times \quad m^2 - 3m + 2 \\
\hline
- 6m^2 + 4m - 10 \\
9m^3 - 6m^2 + 15m \\
-3m^4 + 2m^3 - 5m^2 \\
\hline
-3m^4 + 11m^3 - 17m^2 + 19m - 10
\end{array}
$$

Multiplication is simplified by first arranging the polynomials in descending powers of m

Powers of Polynomials Because raising to a power is repeated multiplication, we now have a method for finding powers of polynomials.

EXAMPLE 6 Find $(a - b)^3$.

SOLUTION

$$(a - b)^3 = \underbrace{(a - b)(a - b)}(a - b)$$

First find $(a - b)^2$ Then multiply by $(a - b)$

$$
\begin{array}{r}
a - b \\
\times \; a - b \\
\hline
- ab + b^2 \\
a^2 - ab \\
\hline
a^2 - 2ab + b^2
\end{array}
\qquad
\begin{array}{r}
a^2 - 2ab + b^2 \\
\times \; a - b \\
\hline
- a^2b + 2ab^2 - b^3 \\
a^3 - 2a^2b + ab^2 \\
\hline
a^3 - 3a^2b + 3ab^2 - b^3
\end{array}
$$

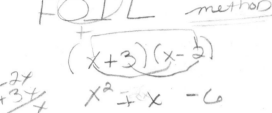

Exercises 8.6

Find the products.

1. $-3z^2(5z^3 - 4z^2 + 2z - 8)$

2. $-4m^3(2m^3 - 3m^2 + m - 5)$

3. $(-3x^2y + xy^2 - 4y^3)(-2xy)$

4. $(-4xy^2 - x^2y + 3x^3)(-3xy)$

5. $5ab^2(3a^2 - 6ab + 5b^2)$

6. $7x^2y(9x^2 + 3xy - 2y^2)$

7. $3xy^2z^3(x^4y^2z + 4x^3y^2z^3 - 2xyz^5)$

8. $4rs^4t^2(5r^6s^3t - r^4s^2t^2 + 8rst^5)$

9. $(x + 3)(x - 2)$

10. $(a - 4)(a + 3)$

11. $(2y + 5)(y - 4)$

12. $(3z - 2)(z + 5)$

13. $(x + 4)^2$

14. $(x - 5)^2$

15. $(2x^2 - 5)(x + 2)$

16. $(3y^2 - 2)(y - 2)$

17. $(x^2 - 4x + 2)(x + 3)$

18. $(x^2 + 5x - 8)(x - 2)$

19. $(z - 4)(z^2 + 4z + 16)$

20. $(a + 5)(a^2 - 5a + 25)$

21. $(2x + 5)(3x^2 + x - 4)$

22. $(3x - 1)(5x^2 - 6x - 2)$

23. $(7x + x^2 - 6)(4x - 3)$

24. $(2 - x^2 - 6x)(5x + 2)$

25. $(9 - x + 2x^2)(5 - x)$

26. $(4x - 1 + 3x^2)(3 - x)$

27. $(7 + 2x^3 + 3x)(x - 4)$

28. $(4 + a^3 - 2a^2)(a - 6)$

29. $(y^3 - 7y^2 + y - 4)(2y - 3)$

30. $(3x^3 + x^2 - 5x - 3)(3x - 1)$

31. $(x^2 - x + 6)(x^2 + 3x - 4)$

32. $(4y^2 + 3y - 5)(y^2 - 8y - 2)$

33. $(x^2 + 2x + 3)^2$

34. $(z^2 - 3z - 4)^2$

35. $(x + 2)^3$

36. $(x + 3)^3$

37. $(x + y)^2(x - y)^2$

38. $(x - 2)^2(x + 2)^2$

8.7   Product of Two Binomials

The product of two binomials occurs so often that is useful to learn a method that will allow us to obtain their product quickly.

When the first terms of each binomial are like terms and the last terms of each binomial are like terms, their product is a trinomial (Example 1).

EXAMPLE 1

Like terms

$$(x + 2)(x + 3) = x(x + 3) + 2(x + 3)$$ Distributive rule

Like terms $$= x^2 + 3x + 2x + 6$$

$$= x^2 + 5x + 6$$

Trinomial

Notice that we multiply each term in the first binomial by each term in the second binomial. The product $(x + 2)(x + 3)$ can also be found by the following method:

$(x + 2)(x + 3) = x^2$ Product of *First* terms

$(x + 2)(x + 3) = x^2 + 3x$ Product of *Outer* terms

$(x + 2)(x + 3) = x^2 + 3x + 2x$ Product of *Inner* terms

$(x + 2)(x + 3) = x^2 + 3x + 2x + 6$ Product of *Last* terms

$$= x^2 + 5x + 6$$

The method above is often referred to as **the FOIL method** because we are finding the sum of the products of the *First*, *Outer*, *Inner*, and *Last* terms.

EXAMPLE 2

Multiply $(x - 3)(x - 4)$.

SOLUTION

F First	O Outer	I Inner	L Last
$(x - 3)(x - 4)$	$(x - 3)(x - 4)$	$(x - 3)(x - 4)$	$(x - 3)(x - 4)$
$= x^2$	$-4x$	$-3x$	$+12$
$= x^2$		$-7x$	$+12$

We can shorten this procedure by adding the inner and outer terms mentally.

EXAMPLE 3

Multiply $(3x + 2)(4x - 5)$.

SOLUTION

$(3x + 2)(4x - 5)$ $(3x + 2)(4x - 5)$ $(3x + 2)(4x - 5)$

$8x$

$-15x$

$= 12x^2$ $-7x$ -10

Multiplying two binomials

1. *The first term of the product* is the *product* of the first terms of the binomials.
2. *The middle term of the product* is the *sum* of the inner and outer products.
3. *The last term of the product* is the *product* of the last terms of the binomials.

EXAMPLE 4

Multiply $(5x - 4y)(6x + 7y)$.

SOLUTION

$$(\; 5x \; - 4y)(\; 6x \; + 7y) \qquad (\; 5x \; - 4y \;)(\; 6x \; + 7y \;) \qquad (5x - 4y \;)(6x + 7y \;)$$

$$\underbrace{}_{-24xy} \qquad \underbrace{}_{+35xy}$$

$$= \qquad 30x^2 \qquad\qquad\qquad +11xy \qquad\qquad\qquad -28y^2$$

EXAMPLE 5

Multiply $(3x - 8y)(4x - 5y)$.

SOLUTION

$$(3x - 8y) \qquad (4x - 5y)$$

$$\underbrace{}_{-32xy}$$

$$\underbrace{}_{-15xy}$$

$$= 12x^2 \qquad -47xy \quad +40y^2$$

Practice this procedure until you can find the three terms of the product without having to write anything down.

The Square of a Binomial   We can square a binomial by multiplying it by itself.

EXAMPLE 6

a. $(a + b)^2 = (a + b) \quad (a + b)$

$$\underbrace{}_{ab}$$

$$\underline{ ab }$$

$$= \quad a^2 + 2ab + b^2$$

b. $(a - b)^2 = (a - b) \quad (a - b)$

$$\underbrace{}_{-ab}$$

$$\underline{ -ab }$$

$$= \quad a^2 - 2ab + b^2$$

The two special products shown in Example 6 occur so often that they are worth remembering.

Squaring a binomial

> 1. *The first term of the product* is the square of the first term of the binomial.
> 2. *The middle term of the product* is *twice* the product of the two terms of the binomial.
> 3. *The last term of the product* is the square of the last term of the binomial.
>
> $$(a + b)^2 = a^2 + 2ab + b^2$$
>
> $$(a - b)^2 = a^2 - 2ab + b^2$$

EXAMPLE 7

a. $(m + n)^2 = (m)^2 + 2(m)(n) + (n)^2 = m^2 + 2mn + n^2$
b. $(a - 3)^2 = (a)^2 + 2(a)(-3) + (-3)^2 = a^2 - 6a + 9$
c. $(2x - 5)^2 = (2x)^2 + 2(2x)(-5) + (-5)^2 = 4x^2 - 20x + 25$

A Word of Caution A common error in squaring a binomial is to forget the middle term. We cannot find $(a + b)^2$ simply by squaring a and b.

Incorrect method	*Correct method*
$(a + b)^2 = a^2 + b^2$	$(a + b)^2 = (a + b)(a + b) = a^2 + 2ab + b^2$

When squaring a binomial, do not forget this middle term——→

The Product of the Sum and Difference of Two Terms

$$\overset{\text{Sum}}{(a} \overset{}{+} \overset{\text{Difference}}{b)} \overset{}{(a} \overset{}{-} b) = \overset{F}{a^2} \overset{O}{- ab} \overset{I}{+ ab} \overset{L}{- b^2}$$

$$= a^2 - b^2$$

Notice that the inner and outer terms add to zero. The product is the difference of two squares.

The product of the sum and difference of two terms	$$(a + b)(a - b) = a^2 - b^2$$ *In words:* The product is equal to the square of the first term *minus* the square of the second term.

EXAMPLE 8

a. $(x + 4)(x - 4) = (x)^2 - (4)^2 = x^2 - 16$

b. $(2x + 3y)(2x - 3y) = (2x)^2 - (3y)^2 = 4x^2 - 9y^2$

c. $(m^2 - 5n^2)(m^2 + 5n^2) = (m^2)^2 - (5n^2)^2 = m^4 - 25n^4$

Exercises 8.7

Find the products.

1. $(x + 1)(x + 4)$

2. $(x + 3)(x + 1)$

3. $(a + 5)(a + 2)$

4. $(a + 7)(a + 1)$

5. $(m - 4)(m + 2)$

6. $(n - 3)(n + 7)$

7. $(8 + y)(9 - y)$

8. $(3 - z)(10 + z)$

9. $(x + 2)(x - 2)$

10. $(x + 3)(x - 3)$

11. $(h - 6)(h + 6)$

12. $(w - 7)(w + 7)$

13. $(a + 4)^2$

14. $(y - 5)^2$

15. $(x + 3)^2$

16. $(x + 5)^2$

17. $(4 - b)^2$

18. $(6 - b)^2$

19. $(3x + 4)(2x - 5)$

20. $(2y + 5)(4y - 3)$

21. $(2x - 3y)(2x + 3y)$

22. $(3x + 4y)(3x - 4y)$

23. $(2a + 5b)(a + b)$

24. $(3c + 2d)(c + d)$

25. $(4x - y)(2x + 7y)$

26. $(3x - 2y)(4x + 5y)$

27. $(7x - 10y)(7x - 10y)$

28. $(4u - 9v)(4u - 9v)$

29. $(3x + 4)^2$

30. $(2x + 5)^2$

31. $(5m - 2n)^2$

32. $(7a - 3b)^2$

33. $(x^2 + y^2)(2x^2 + 3y^2)$

34. $(4a^2 + b^2)(a^2 + 5b^2)$

35. $(m^2 - 3n^2)(4m^2 + 2n^2)$

36. $(8x^2 + y^2)(2x^2 - 3y^2)$

37. $(3x^2 + y^2)(3x^2 - y^2)$

38. $(a^2 - 4b^2)(a^2 + 4b^2)$

39. $(x^2 + y^2)^2$

40. $(2x^2 - 3y^2)^2$

41. $\left(\dfrac{2x}{3} + 1\right)\left(\dfrac{2x}{3} - 1\right)$

42. $\left(1 + \dfrac{3x}{5}\right)^2$

43. $\left(2 - \dfrac{x}{3}\right)^2$

44. $\left(\dfrac{3x}{4} - 3\right)\left(\dfrac{3x}{4} + 3\right)$

8.8 Division of Polynomials

8.8A Division of a Polynomial by a Monomial

We can add fractions if they have a common denominator: $\dfrac{a}{c} + \dfrac{b}{c} = \dfrac{a + b}{c}$. Reversing this process, we can write $\dfrac{4x^3 + 6x^2}{2x}$ as $\dfrac{4x^3}{2x} + \dfrac{6x^2}{2x}$ and then reduce each fraction to lowest terms:

$$\dfrac{4x^3 + 6x^2}{2x} = \dfrac{4x^3}{2x} + \dfrac{6x^2}{2x} \qquad \text{Dividing each term of the polynomial by the monomial}$$

$$= 2x^2 + 3x$$

Dividing a polynomial by a monomial

Divide *each* term in the polynomial by the monomial, then add the results.

EXAMPLE 1

a. $\dfrac{4x + 2}{2} = \dfrac{4x}{2} + \dfrac{2}{2} = 2x + 1$

b. $\dfrac{9x^3 - 6x^2 + 12x}{3x} = \dfrac{9x^3}{3x} + \dfrac{-6x^2}{3x} + \dfrac{12x}{3x} = 3x^2 - 2x + 4$

c. $\dfrac{4x^2 - 8x + 16}{-4x} = \dfrac{4x^2}{-4x} + \dfrac{-8x}{-4x} + \dfrac{16}{-4x} = -x + 2 - \dfrac{4}{x}$

d. $\dfrac{4a^2b^2c^2 - 6ab^2c^2 + 12abc^2}{-6abc} = \dfrac{4a^2b^2c^2}{-6abc} + \dfrac{-6ab^2c^2}{-6abc} + \dfrac{12abc^2}{-6abc}$

$$= -\dfrac{2}{3}abc + bc - 2c$$

Exercises 8.8A

Perform the divisions.

1. $\dfrac{3x + 6}{3}$

2. $\dfrac{10x + 15}{5}$

3. $\dfrac{4 + 8x}{4}$

4. $\dfrac{5 - 10x}{5}$

5. $\dfrac{6x - 8y}{-2}$

6. $\dfrac{5x - 10y}{-5}$

7. $\dfrac{2x^2 + 3x}{x}$

8. $\dfrac{4y^2 - 3y}{y}$

9. $\dfrac{3a^2b - ab}{ab}$

10. $\dfrac{5mn^2 - mn}{mn}$

11. $\dfrac{12z^3 - 16z^2 + 8z}{4z}$

12. $\dfrac{-15a^4 + 12a^3 - 18a^2}{3a^2}$ **13.** $\dfrac{5x^3 - 4x^2 + 10}{5x^2}$ **18.** $\dfrac{21m^2n^3 - 35m^3n^2 - 14m^2n^2}{7m^2n^2}$

14. $\dfrac{7y^3 - 5y^2 + 14}{7y^2}$ **15.** $\dfrac{-15x^2y^2z^2 - 30xyz}{-5xyz}$ **19.** $\dfrac{10x^7y^4 + 18x^4y^2 + 2x^3y}{2x^3y}$ **20.** $\dfrac{24a^6b^6 - 8a^3b^2 + 4ab^2}{4ab^2}$

16. $\dfrac{-24a^2b^2c^2 - 16abc}{-8abc}$ **17.** $\dfrac{13x^3y^2 - 26x^2y^3 + 39x^2y^2}{13x^2y^2}$

8.8B Division of a Polynomial by a Polynomial

When the divisor of a polynomial contains two or more terms, we use a method similar to long division of whole numbers.

EXAMPLE 2   Find $882 \div 21$.

SOLUTION

First term in the quotient $= \dfrac{\text{first term of dividend}}{\text{first term of divisor}} = \dfrac{8}{2} = 4$

Second term in the quotient $= \dfrac{4}{2} = 2$

$$
\begin{array}{r}
4\,2 \\
21\overline{)8\,8\,2} \\
\underline{8\,4} \quad \leftarrow \text{Multiply } 4 \cdot 21 = 84 \\
4\,2 \\
\underline{4\,2} \leftarrow \text{Multiply } 2 \cdot 21 = 42 \\
0
\end{array}
$$

EXAMPLE 3   Find $(x^2 - 3x - 10) \div (x + 2)$.

SOLUTION

Step 1.   First term in the quotient $= \dfrac{x^2}{x} = x$

$$
\begin{array}{r}
x \\
x + 2\overline{)\,x^2 - 3x - 10}
\end{array}
$$

Step 2. Multiply.

$$
\begin{array}{r}
x \\
x + 2\overline{)x^2 - 3x - 10} \\
x^2 + 2x \leftarrow \text{Multiply } x(x + 2) = x^2 + 2x
\end{array}
$$

Step 3.

$$
\begin{array}{r}
x \\
x + 2\overline{)x^2 - 3x - 10} \\
\ominus \quad \ominus \\
\underline{x^2 + 2x} \\
-5x - 10
\end{array}
$$

To subtract, *change* all signs and *add*

Bring down the next term

Step 4.   Second term in the quotient $= \dfrac{-5x}{x} = -5$

$$
\begin{array}{r}
x - 5 \\
x + 2\overline{)x^2 - 3x - 10} \\
\ominus \quad \ominus \\
\underline{x^2 + 2x} \\
-5x - 10
\end{array}
$$

Step 5. Multiply.

$$
\begin{array}{r}
x - 5 \\
x + 2\overline{)x^2 - 3x - 10} \\
\ominus \quad \ominus \\
x^2 + 2x \\
\hline
-5x - 10 \\
-5x - 10 \;\leftarrow \text{Multiply } -5(x + 2) = -5x - 10 \\
\end{array}
$$

Step 6.

$$
\begin{array}{r}
x - 5 \\
x + 2\overline{)x^2 - 3x - 10} \\
\ominus \quad \ominus \\
x^2 + 2x \\
\hline
-5x - 10 \\
\oplus \quad \oplus \\
-5x - 10 \quad \text{\textit{Change} signs and \textit{add}} \\
\hline
0
\end{array}
$$

EXAMPLE 4   Find $(6x^2 + x - 10) \div (2x + 3)$.

SOLUTION

$$
\begin{array}{r}
3x - 4 \quad \text{R 2,} \quad \text{or} \quad 3x - 4 + \dfrac{2}{2x + 3} \\
2x + 3\overline{)6x^2 + x - 10} \\
\ominus \quad \ominus \\
6x^2 + 9x \\
\hline
-8x - 10 \\
\oplus \quad \oplus \\
-8x - 12 \\
\hline
2 \quad \text{Remainder}
\end{array}
$$

EXAMPLE 5   Find $(27x - 19x^2 + 6x^3 + 10) \div (5 - 3x)$.

SOLUTION   We must arrange the terms of the dividend and divisor in *descending powers* of the variable *before* beginning the division.

$$
\begin{array}{r}
-2x^2 + 3x - 4 \quad \text{R 30} \\
-3x + 5\overline{)6x^3 - 19x^2 + 27x + 10} \\
\ominus \quad \oplus \\
6x^3 - 10x^2 \\
\hline
-9x^2 + 27x \\
\oplus \quad \ominus \\
-9x^2 + 15x \\
\hline
12x + 10 \\
\ominus \quad \oplus \\
12x - 20 \\
\hline
30
\end{array}
$$

Therefore, the answer is

$$
-2x^2 + 3x - 4 \quad \text{R 30,} \quad \text{or} \quad -2x^2 + 3x - 4 + \frac{30}{-3x + 5}
$$

EXAMPLE 6   Find $(17ab^2 + 12a^3 - 10b^3 - 11a^2b) \div (3a - 2b)$.

SOLUTION   We arrange the terms of the dividend and divisor in descending powers of a before we begin the division.

$$
\begin{array}{r}
4a^2 - ab + 5b^2 \\
3a - 2b\overline{)12a^3 - 11a^2b + 17ab^2 - 10b^3} \\
\ominus \quad \oplus \\
\underline{12a^3 - 8a^2b} \\
- 3a^2b + 17ab^2 \\
\oplus \quad \ominus \\
\underline{- 3a^2b + 2ab^2} \\
15ab^2 - 10b^3 \\
\ominus \quad \oplus \\
\underline{15ab^2 - 10b^3} \\
0
\end{array}
$$

EXAMPLE 7   Find $(x^3 - 1) \div (x - 1)$.

SOLUTION   If the dividend is missing any powers of the variable, it is important that we insert the missing terms so that like terms will be aligned. Thus, the dividend $x^3 - 1$ becomes $x^3 + 0x^2 + 0x - 1$.

$$
\begin{array}{r}
x^2 + x + 1 \\
x - 1\overline{)x^3 + 0x^2 + 0x - 1} \\
\ominus \quad \oplus \\
\underline{x^3 - x^2} \\
x^2 + 0x \\
\ominus \quad \oplus \\
\underline{x^2 - x} \\
x - 1 \\
\ominus \quad \oplus \\
\underline{x - 1} \\
0
\end{array}
$$

It is important to leave space for missing powers by using zeros in this way

EXAMPLE 8   Find $(2x^4 - x^3 - 7x^2 + 8) \div (x^2 - x - 2)$.

SOLUTION   When the divisor is a polynomial of more than two terms, we use exactly the same procedure as the one followed in Examples 3–7.

$$
\begin{array}{r}
2x^2 + x - 2 \quad \text{R } 4 \\
x^2 - x - 2\overline{)2x^4 - x^3 - 7x^2 + 0x + 8} \\
\ominus \quad \oplus \quad \oplus \\
\underline{2x^4 - 2x^3 - 4x^2} \\
x^3 - 3x^2 + 0x \\
\ominus \quad \oplus \quad \oplus \\
\underline{x^3 - x^2 - 2x} \\
- 2x^2 + 2x + 8 \\
\oplus \quad \ominus \quad \ominus \\
\underline{- 2x^2 + 2x + 4} \\
4
\end{array}
$$

Insert missing power, 0x

Exercises 8.8B

$$x + 2 \overline{)\, x^2 + 5x + 6}$$

Perform the divisions.

1. $(x^2 + 5x + 6) \div (x + 2)$ 2. $(x^2 + 5x + 6) \div (x + 3)$

3. $(x^2 - 9x + 20) \div (x - 4)$

4. $(x^2 - x - 12) \div (x + 3)$

5. $(6x^2 + 5x - 6) \div (3x - 2)$

6. $(20x^2 + 13x - 15) \div (5x - 3)$

7. $(15v^2 + 19v + 10) \div (5v - 7)$

8. $(15v^2 + 19v - 4) \div (3v + 8)$

9. $(8a^2 - 2ab - b^2) \div (2a - b)$

10. $(8a^2 - 2ab - b^2) \div (4a + b)$

11. $(6a^2 + 5ab + b^2) \div (2a + 3b)$

12. $(6a^2 + 5ab - b^2) \div (3a - 2b)$

13. $(x^3 - 10x + 24) \div (x + 4)$

This is hard!!!!

14. $(x^3 - 2x - 4) \div (x - 2)$

15. $(8x - 4x^3 + 10) \div (2 - x)$

16. $(12x - 15 - x^3) \div (3 - x)$

17. $(a^3 - 8) \div (a - 2)$ 18. $(c^3 - 27) \div (c - 3)$

19. $(4x^3 - 4x^2 + 5x + 4) \div (2x + 1)$

20. $(3x^3 - 5x^2 + 11x - 6) \div (3x - 2)$

21. $(5 + 3x^3 + 9x - 17x^2) \div (x - 5)$

22. $(12 - 8x + 2x^3 - 11x^2) \div (x - 6)$

23. $(2x^3 + x^2 + 28) \div (2x + 5)$

24. $(3x^3 - 8x^2 + 3) \div (3x + 1)$

25. $(1 + 64x^3) \div (4x + 1)$ 26. $(27 + 8x^3) \div (2x + 3)$

27. $(x^4 + 2x^3 - x^2 - 2x + 4) \div (x^2 + x - 1)$

28. $(x^4 - 2x^3 + 3x^2 - 2x + 7) \div (x^2 - x + 1)$

Chapter 8   REVIEW

Terms
8.1

The + and + signs in an algebraic expression break it up into smaller pieces called **terms**. A minus sign is part of the term that follows it. *Exception*: An expression within grouping symbols is considered as a single piece even though it may contain + and − signs.

Like Terms
8.4

Terms having the same variables with the same exponents are called **like terms**.

Factors
8.1

The numbers that are multiplied together to give a product are called the **factors** of that product.

Numerical Coefficients
8.1

The **numerical coefficient** of a term is the factor that is a number. *The coefficient of a term is understood to mean the *numerical* coefficient of that term.*

Distributive Rule
8.3

$$a(b + c) = ab + ac$$

When a factor is multiplied by an expression enclosed within grouping symbols, the factor must be multiplied by each term within the grouping symbols, then those products are added.

Combining Like Terms
8.4

1. Identify the like terms.

2. Find the sum of each group of like terms by adding their numerical coefficients, then multiply that sum by the common variables.

Rules for Exponents
8.2

1. $x^a x^b = x^{a+b}$

2. $(x^a)^b = x^{ab}$

3. $\dfrac{x^a}{x^b} = x^{a-b} \; (x \neq 0)$

4. $x^0 = 1 \; (x \neq 0)$

Polynomials
8.5A

A **polynomial** *in x* is an algebraic expression having only terms of the form ax^n, where a is any real number and n is a whole number. A *polynomial in x and y* is an algebraic expression having only terms of the form $ax^n y^m$, where a is any real number and n and m are whole numbers.

A **monomial** is a polynomial of one term.

A **binomial** is a polynomial of two unlike terms.

A **trinomial** is a polynomial of three unlike terms.

The Degree of a Term
8.5A

The **degree of a term** in a polynomial is the sum of the exponents of its variables. The **degree of a polynomial** is the same as that of its highest-degree term (provided like terms have been combined).

Adding Polynomials
8.5B

To add polynomials, remove grouping symbols, then combine like terms.

Subtracting Polynomials
8.5B

To subtract polynomials, change the sign of each term in the polynomial being subtracted, then add the resulting polynomials.

Multiplying a Polynomial by a Monomial
8.6

To multiply a polynomial by a monomial, multiply each term in the polynomial by the monomial, then add the results.

Multiplying a Polynomial by a Polynomial
8.6

To multiply a polynomial by a polynomial, multiply each term of the first polynomial by each term of the second polynomial, then add the results.

Multiplying Two Binomials
8.7

1. The first term of the product is the product of the first terms of the binomials.

2. The middle term of the product is the sum of the inner and outer products.

3. The last term of the product is the product of the last terms of the binomials.

Squaring a Binomial
8.7

$$(a + b)^2 = a^2 + 2ab + b^2$$
$$(a - b)^2 = a^2 - 2ab + b^2$$

Product of the Sum and Difference of Two Terms
8.7

$$(a + b)(a - b) = a^2 - b^2$$

Dividing a Polynomial by a Monomial
8.8A

To divide a polynomial by a monomial, divide each term in the polynomial by the monomial, then add the results.

Dividing a Polynomial by a Polynomial
8.8B

See Section 8.8B for the method.

Chapter 8   REVIEW EXERCISES

In Exercises 1–3, (a) determine the number of terms and (b) write the second term, if there is one.

1. $a^2 + 2ab + b^2$ 2. $6x - 2(x^2 + y^2)$ 3. $\dfrac{2x + y}{3} - 4x$

4. Write the polynomial $3x^4 - 6 + 7x^2 - x$ in descending powers of x.

5. In the polynomial $2xy + 3x^2y$, find:

   a. the degree of the first term

   b. the degree of the polynomial

In Exercises 6–17, use the rules of exponents to simplify each expression.

6. $x \cdot x^3$ 7. $m^6 \cdot m^3$ 8. $(m^6)^3$ 9. $\dfrac{m^6}{m^3}$

10. $2x^2 \cdot 3x^3$ 11. $\dfrac{8x^8}{2x^2}$ 12. $(x^4)^0$ 13. $y \cdot y^7 \cdot y^4$

14. $\dfrac{3^5}{3^3}$ 15. $2^4 \cdot 2^2$ 16. $(10^3)^2$ 17. 5^0

In Exercises 18–26, simplify the addition and subtraction problems.

18. $(5x^2y + 3xy^2 - 4y^3) + (2xy^2 + 4y^3 + 3x^2y)$

19. $(7ab^2 + 3a^3 - 5) + (6a^3 + 10 - 10ab^2)$

20. $(5x^2y + 3xy^2 - 4 + y^2) - (8 - 4x^2y + 2xy^2 - y^2)$

21. $(8a + 5a^2b - 3 + 6ab^2) - (5 + 4a - 3ab^2 + a^2b)$

22. $(4a^2 - 5a) - (6a + 3) + (2a^2 + 5)$

23. $(2x + 9) - [(3x - 5) - (2 - x)]$

24. Subtract $(3y^2 - 5y + 1)$ from $(5y^2 - 6y - 4)$.

25. Add
$$\begin{array}{l} 3a^3 + 4a^2 \qquad\quad - 3 \\ -6a^3 \qquad\quad - 7a + 5 \\ \underline{5a^3 - 9a^2 - \ 4a - 6} \end{array}$$

26. Subtract
$$\begin{array}{l} 5x^2y^2 - \ \ x^2y - 3xy \\ \underline{7x^2y^2 + 4x^2y - 8xy} \end{array}$$

In Exercises 27–41, find the products.

27. $(9x^8y^3)(-7x^4y^5)$ 28. $(-5ab^2)(-a^7b^3)(4a^2b)$

29. $2x(3x^2 - 4x)$ 30. $5x^2y(3xy^2 + 4x - 2z)$

31. $(2a - 4)(3a + 5)$ 32. $(3x + 4)^2$

33. $(x - 5y)(x + 5y)$ 34. $(m - 3)(4m - 6)$

35. $(3u + 1)(5u + 2)$ 36. $(7x + 4)(7x - 4)$

37. $(x^2 - 6)^2$ 38. $(5x - 4y)(7x + 2y)$

39. $(x^2 + y^2)(2x^2 - 3y^2)$ 40. $\left(\dfrac{2x}{3} + 4\right)^2$

41. $(x - 2)(x^2 + 2x + 4)$

In Exercises 42–48, perform the divisions.

42. $\dfrac{3x^2y - 6xy^2}{3xy}$ 43. $\dfrac{-15a^2b^3 + 4ab^2 - 10ab}{-5ab}$

44. $(4x^2 - 1) \div (2x + 1)$

45. $(6x^2 - 9x + 10) \div (2x - 5)$

46. $(10a^2 + 23ab - 5b^2) \div (5a - b)$

47. $(6x^3 + 7x^2 + 1) \div (3x - 1)$

48. $(2a^4 + a + a^2 - 3 - a^3) \div (2a^2 - a + 3)$

Chapter 8 Critical Thinking and Writing Problems

Answer Problems 1–6 in your own words, using complete sentences.

1. Explain why $4x^2y$ and $4xy^2$ are not like terms.

2. Explain how to add like terms.

3. Explain how to subtract polynomials.

4. Explain how to multiply polynomials.

5. Explain why $(a + b)^2 \neq a^2 + b^2$.

6. Explain why $3^2 \cdot 3^3 \neq 9^5$.

7. When do we use long division to divide polynomials?

8. In the following expressions, is $-2x$ a term or a factor?

    a. $(4x - 7) - 2x$ **b.** $(4x - 7)(-2x)$

9. True or false: $x^2 + x^3 = x^5$. Explain why.

Exercises 10–15 each have an error. Find the error, and in your own words, explain why it is wrong. Then work the problem correctly.

10. $2x(4x^2)(-3x^3) = 8x^3 - 6x^4$

11. $(7y - 5)(-4y) = 3y - 5$

12. $(x - 3)^2 = x^2 - 9$

13. $\quad (5x - 3) - [(2x + 7) - (x - 1)]$
$= 5x - 3 - 2x - 7 - x + 1$
$= 2x - 9$

14. Subtract $(x^2 - 6)$ from $(4x^2 - 2x)$.
$$\begin{array}{r} 4x^2 - 2x \\ -\ x^2\ \ \ \ -6 \\ \hline 3x^2 - 2x - 6 \end{array}$$

15. Divide $(x^3 - 7x + 12) \div (x - 3)$, using long division.
$$\begin{array}{r} x^2 \quad - 4 \\ x - 3\overline{)x^3 - 7x + 12} \\ \ominus \ \oplus \\ \underline{x^3 - 3x} \\ -4x + 12 \\ \oplus \quad \ominus \\ \underline{-4x + 12} \\ 0 \end{array}$$

Chapter 8   DIAGNOSTIC TEST

Allow yourself about 50 minutes to do these problems. Complete solutions for all problems, together with section references, are given in the answer section at the end of the book.

If the answer to a problem is a polynomial in one variable, write that answer in descending powers.

1. In the polynomial $x^2 - 4xy^2 + 5$, find:

   a. the degree of the first term

   b. the degree of the polynomial

   c. the numerical coefficient of the second term

2. Subtract $(10 - z + 2z^2)$ from $(-4z^2 - 5z + 10)$.

3. Add $-5x^3 + 2x^2 \qquad - 5$
   $7x^3 + 5x - 8$
   $3x^2 - 6x + 10$

4. Subtract $9a^2b - 2ab^2 - 3ab$
   $4a^2b + 6ab^2 - 7ab$

In Problems 5–8, simplify the addition and subtraction problems.

5. $(6x^2 - 5 - 11x) + (15x - 4x^2 + 7)$

6. $(3xy^2 - 4xy) - (6xy - 3xy^2) + (4xy + x^3)$

7. $(2m^2 - 5) - [(7 - m^2) - (4m^2 - 3)]$

8. Subtract $(5x^2 - 9x + 6)$ from the sum of $(2x - 8 - 7x^2)$ and $(12 - 2x^2 + 11x)$.

In Problems 9–13, use the rules of exponents to simplify.

9. $10^2 \cdot 10^3$

10. $(a^3)^5$

11. $\dfrac{12x^6}{2x^2}$

12. $x^3 \cdot x^5$

13. $\dfrac{2^8}{2^5}$

In Problems 14–22, find the products.

14. $(-3xy)(5x^3y)(-2xy^4)$

15. $4x(3x^2 - 2)$

16. $-2ab(5a^2 - 3ab^2 + 4b)$

17. $(2x - 5)(3x + 4)$

18. $(2y - 3)^2$

19. $(m + 3)(2m + 6)$

20. $(5a + 2)(5a - 2)$

21. $(4x - 2y)(3x - 5y)$

22. $(w - 3)(w^2 + 3w + 8)$

In Problems 23–25, perform the divisions.

23. $\dfrac{6x^3 - 4x^2 + 8x}{2x}$

24. $(10x^2 + x - 5) \div (2x - 1)$

25. $(2x^3 - 6 - 20x) \div (x + 3)$

Equations and Inequalities

CHAPTER

9

he main reason we study algebra is to equip ourselves with the tools necessary for solving problems. Most problems are solved by the use of equations. Some problems are solved by using inequalities.

In this chapter we show how to solve simple equations and inequalities. Methods for solving more difficult equations and inequalities will be given in later chapters.

9.1 Solving Equations Having Only One Number on the Same Side as the Variable

In algebra, an **equation** is a statement that two algebraic expressions are equal. An equation is made up of three parts:

1. The equal sign ($=$).
2. The expression to the left of the $=$ sign, called the left side (or left member) of the equation.
3. The expression to the right of the $=$ sign, called the right side (or right member) of the equation.

$$5x - 8 = 3x + 3$$

Left side ⟶ | Right side
Equal sign

The equal sign ($=$) in an equation means that the number represented by the left side *is the same* as the number represented by the right side.

A **solution** of an equation is a number that, when put in place of the variable, makes a true statement.

If we replace x by 3 in the equation

$$x + 2 = 5$$

we have

$$3 + 2 = 5$$

which is a true statement. Therefore, 3 is a solution of the equation.

Checking a solution of an equation	1. Replace the variable in the given equation by the solution in parentheses.
	2. Perform the indicated operations on both sides of the $=$ sign.
	3. If the resulting number on each side of the $=$ sign is the same, the solution is correct.

EXAMPLE 1 Is -3 a solution of the equation $5 - x = 2$?

SOLUTION
$$5 - x = 2$$
$$5 - (-3) \stackrel{?}{=} 2 \quad x \text{ was replaced by } (-3)$$
$$5 + 3 \stackrel{?}{=} 2$$
$$8 \neq 2 \quad \text{False}$$

Therefore, -3 is *not* a solution of $5 - x = 2$.

EXAMPLE 2   Is -4 a solution of the equation $2x + 3 = -5$?

SOLUTION $2x + 3 = -5$

$2(-4) + 3 \stackrel{?}{=} -5$ *x was replaced by (−4)*

$-8 + 3 \stackrel{?}{=} -5$

$-5 = -5$ True

Therefore, -4 *is* a solution of $2x + 3 = -5$.

Solving Equations

To *solve an equation* means to find the value of the variable that makes the statement true. An equation is solved when we succeed in getting the variable by itself on one side of the equal sign and only a single number on the other side.

We use four basic rules in solving equations. Each of these is illustrated in one of the following examples.

EXAMPLE 3   If we add the same signed number to both sides of an equation, the resulting sums must be equal:

$$\begin{array}{rr} 7 = & 7 \\ +\ 5 & +\ 5 \\ \hline 12 = & 12 \end{array}$$

We are adding $+5$ to both sides

The resulting sums are equal

EXAMPLE 4   If we subtract the same signed number from both sides of an equation, the resulting differences must be equal:

$$\begin{array}{rr} 7 = & 7 \\ -\ 5 & -\ 5 \\ \hline 2 = & 2 \end{array}$$

We are subtracting 5 from both sides

The resulting differences are equal

We could also work this example by *adding* -5 to both sides.

EXAMPLE 5   If we multiply both sides of an equation by the same signed number, the resulting products must be equal:

$$4 = 4$$
$$2\,(4) = 2\,(4)$$ Multiplying both sides by 2
$$8 = 8$$ The resulting products are equal

EXAMPLE 6   If we divide both sides of an equation by the same signed number (not zero), the resulting quotients must be equal:

$$6 = 6$$
$$\frac{6}{-2} = \frac{6}{-2}$$ We are dividing both sides by -2
$$-3 = -3$$ The resulting quotients are equal

We summarize these results in the following box:

<table>
<tr><td>**Four basic rules used in solving equations**</td><td>

1. *Addition rule:* The same number may be added to both sides.

2. *Subtraction rule:* The same number may be subtracted from both sides.

3. *Multiplication rule:* Both sides may be multiplied by the same nonzero number.

4. *Division rule:* Both sides may be divided by the same nonzero number.

</td></tr>
</table>

To know which rule to use in solving a particular equation, we make use of inverse operations.

Addition and subtraction are inverse operations.
Multiplication and division are inverse operations.

If we have $x - 3$, we have a subtraction of 3. The -3 can be removed by its inverse: the addition of 3 (Example 7).

If we have $x + 5$, we have an addition of 5. The $+5$ can be removed by its inverse: the subtraction of 5 (Example 8).

If we have $7 \cdot x$, we have a multiplication by 7. The $7 \cdot$ can be removed by its inverse: a division by 7 (Example 9).

If we have $\dfrac{x}{3}$, we have a division by 3. The division by 3 can be removed by its inverse: a multiplication by 3 (Example 10).

EXAMPLE 7   Solve $x - 3 = 6$.

SOLUTION

Remove by an addition of 3

$$
\begin{array}{rcl}
x - 3 & = & 6 \\
+\,3 & & +\,3 \\
\hline
x & = & 9
\end{array}
$$

Adding 3 to both sides gets x by itself on the left side

√ **Check for $x = 9$** $x - 3 = 6$

$(9) - 3 \overset{?}{=} 6$

$6 = 6$ True; therefore, the solution is correct

EXAMPLE 8   Solve $x + 5 = 3$.

SOLUTION

Remove by a subtraction of 5

$$
\begin{array}{rcl}
x + 5 & = & 3 \\
-\,5 & & -5 \\
\hline
x & = & -2
\end{array}
$$

Solution

√ **Check for $x = -2$** $x + 5 = 3$

$(-2) + 5 \overset{?}{=} 3$

$3 = 3$ True; the solution is correct

EXAMPLE 9   Solve $7x = 56$.

SOLUTION   ——Remove by a division by 7

$$7x = 56$$

$$\frac{\overset{1}{7}x}{\underset{1}{7}} = \frac{56}{7} \quad \text{Dividing both sides by 7}$$

$$x = 8 \quad \text{Solution}$$

✓ **Check for $x = 8$** $7x = 56$

$$7(8) \overset{?}{=} 56$$

$$56 = 56 \quad \text{True}$$

EXAMPLE 10   Solve $\frac{x}{3} = -5$.

SOLUTION   ——Remove by a multiplication of 3

$$\frac{x}{3} = -5$$

$$3\left(\frac{x}{3}\right) = 3(-5) \quad \text{Multiplying both sides by 3}$$

$$\overset{1}{3}\left(\frac{x}{\underset{1}{3}}\right) = 3(-5)$$

$$x = -15 \quad \text{Solution}$$

✓ **Check for $x = -15$** $\frac{x}{3} = -5$

$$\frac{-15}{3} \overset{?}{=} -5$$

$$-5 = -5 \quad \text{True}$$

In Examples 11 and 12, the variable x is on the right side of the equation.

EXAMPLE 11   Solve $-6 = 2 + x$.

SOLUTION   Because 2 is being added to x, we must subtract 2 from both sides to get x by itself on the right side.

$$\begin{array}{rl} -6 = & 2 + x \\ \underline{-2} & \underline{-2} \\ -8 = & x \quad \text{Solution} \end{array}$$

✓ **Check for $x = -8$** $-6 = 2 + x$

$$-6 \overset{?}{=} 2 + (-8)$$

$$-6 = -6 \quad \text{True}$$

A Word of Caution In Example 11, a common error is to add 6 to both sides of the equation, but this does *not* lead to a solution of the equation. An equation is solved when the variable is by itself on one side of the equal sign.

$$\begin{array}{r} -6 = \ \ 2 + x \\ +6 \quad +6 \\ \hline 0 = \ \ \ 8 + x \end{array}$$ ← Not solved for x because x is not by itself on the right side

EXAMPLE 12   Solve $-6 = 2x$.

SOLUTION Because 2 is being multiplied by x, we must divide both sides by 2 to get x by itself.

$$-6 = 2x$$

$$\frac{-6}{2} = \frac{\overset{1}{2}x}{\underset{1}{2}}$$

$$-3 = x \qquad \text{Solution}$$

✓ **Check for $x = -3$** $-6 = 2x$

$$-6 \overset{?}{=} 2(-3)$$

$$-6 = -6 \qquad \text{True}$$

 A Word of Caution In Example 12, a common error is to subtract 2 from both sides of the equation, but this does *not* lead to a solution of the equation. Unlike terms cannot be added or subtracted; therefore, only like terms should be written in a vertical line.

$$\left.\begin{array}{r} -6 = \ \ \ 2x \\ -2 \quad -2 \end{array}\right\}$$ Incorrect alignment because $2x$ and -2 are not like terms

$$-8 = \ \ \ ?$$ ← Cannot combine $2x$ and -2

Exercises 9.1

1. Is -3 a solution of the equation $x + 8 = 5$?

2. Is 2 a solution of the equation $x - 6 = -4$?

3. Is -4 a solution of the equation $3x = 12$?

4. Is -6 a solution of the equation $\dfrac{x}{2} = -3$?

5. Is 2 a solution of the equation $3x - 8 = -2$?

6. Is -5 a solution of the equation $2x + 4 = 6$?

In Exercises 7–50, solve and check the equations.

7. $x + 5 = 8$ 8. $x + 4 = 9$ 9. $x - 3 = 4$

10. $x - 7 = 2$ 11. $3 + x = -4$ 12. $2 + x = -5$

13. $2x = 8$ 14. $3x = 15$ 15. $21 = 7x$

16. $42 = 6x$ 17. $11x = -33$ 18. $12x = -48$

19. $x - 35 = 7$ 20. $x - 42 = 9$ 21. $9 = x + 5$

22. $11 = x + 8$ 23. $12 = x - 11$ 24. $14 = x - 15$

25. $\dfrac{x}{3} = 4$ 26. $\dfrac{x}{5} = 3$ 27. $\dfrac{x}{5} = -2$

28. $\dfrac{x}{6} = -4$ 29. $4 = \dfrac{x}{7}$ 30. $3 = \dfrac{x}{8}$

31. $-28 = -15 + x$ 32. $-47 = -18 + x$

33. $-17 + x = 28$ 34. $-14 + x = 33$

35. $5.6 + x = 2.8$ 36. $3.04 + x = 2.96$

37. $5x = 35$ 38. $-24 = 6x$

39. $12x = -36$

40. $54 = 18x$

45. $0.25x = 2$

46. $0.4x = 16$

41. $-13 = \dfrac{x}{9}$

42. $-15 = \dfrac{x}{8}$

47. $x - \dfrac{1}{2} = \dfrac{2}{5}$

48. $x - \dfrac{3}{5} = \dfrac{1}{3}$

43. $\dfrac{x}{10} = 3.14$

44. $\dfrac{x}{5} = 7.8$

49. $x + \dfrac{1}{3} = \dfrac{3}{4}$

50. $x + \dfrac{2}{9} = \dfrac{5}{6}$

9.2 Solving Equations Having Two Numbers on the Same Side as the Variable

Solving an equation when two numbers appear on the same side as the variable

All numbers that appear on the same side as the variable must be removed.

1. Remove those numbers being added or subtracted by using the inverse operation.
2. Remove the number multiplied by the variable by dividing both sides by that signed number.

EXAMPLE 1

Solve $2x - 3 = 11$.

SOLUTION The numbers 2 and -3 must be removed from the side containing the x.

Step 1. Because the 3 is subtracted, it is removed first.

$$\begin{array}{rcl} 2x - 3 &=& 11 \\ +3 & & +3 \\ \hline 2x &=& 14 \end{array}$$ Adding 3 to both sides

Step 2. Because the 2 is multiplied by the x, it is removed by dividing both sides by 2.

$$2x = 14$$
$$\frac{\overset{1}{2}x}{\underset{1}{2}} = \frac{14}{2}$$ Dividing both sides by 2
$$x = 7$$

✓ **Check for $x = 7$**
$$2x - 3 = 11$$
$$2(7) - 3 \overset{?}{=} 11$$
$$14 - 3 \overset{?}{=} 11$$
$$11 = 11$$

EXAMPLE 2

Solve $-12 = 3x + 15$.

SOLUTION 15 and 3 must be removed from the side with the x.

Step 1.
$$\begin{array}{rcl} -12 &=& 3x + 15 \\ -15 & & -15 \\ \hline -27 &=& 3x \end{array}$$ Subtracting 15 from both sides

Step 2.
$$\frac{-27}{3} = \frac{\overset{1}{3}x}{\underset{1}{3}}$$ Dividing both sides by 3
$$-9 = x$$

√ **Check for** $x = -9$ $-12 = 3x + 15$

$$-12 \overset{?}{=} 3(-9) + 15$$

$$-12 \overset{?}{=} -27 + 15$$

$$-12 = -12$$

EXAMPLE 3   Solve $2 - 4x = -10$.

SOLUTION   The numbers 2 and -4 must be removed from the side with x.

Step 1.
$$\begin{array}{rr} 2 - 4x = & -10 \\ -2 & -2 \\ \hline -4x = & -12 \end{array}$$
Subtracting 2 from both sides

Step 2.
$$\frac{\overset{1}{\cancel{-4}}x}{\underset{1}{\cancel{-4}}} = \frac{-12}{-4}$$
Dividing both sides by -4

$$x = 3$$

√ **Check for** $x = 3$ $2 - 4x = -10$

$$2 - 4(3) \overset{?}{=} -10$$

$$2 - 12 \overset{?}{=} -10$$

$$-10 = -10$$

Exercises 9.2

Solve and check the equations.

1. $4x + 1 = 9$
2. $5x + 2 = 12$
3. $6x - 2 = 10$
4. $7x - 3 = 4$
5. $2x - 15 = 11$
6. $3x - 4 = 14$
7. $4x + 2 = -14$
8. $5x + 5 = -10$
9. $14 = 9x - 13$
10. $25 = 8x - 15$
11. $-8 = 2x + 4$
12. $-1 = 3x + 8$
13. $4x + 14 = 6$
14. $5x + 23 = 3$
15. $-6 + 8x = 2$
16. $-9 + 2x = 1$
17. $-3x + 7 = 13$
18. $-8x + 1 = 9$
19. $-x - 6 = 4$
20. $-x - 14 = 3$

21. $2 - 6x = 38$
22. $8 - 7x = 29$
23. $9 - x = -6$
24. $7 - x = 4$
25. $20 = -5x - 10$
26. $-3 = -4x - 15$
27. $-1 = 7 - 2x$
28. $13 = 6 - x$
29. $9x + 2 = -1$
30. $10x - 4 = 2$
31. $3 + 12x = 11$
32. $5 - 8x = -5$
33. $5 = 6x - 3$
34. $4 = 15x + 16$
35. $9 = 7 - 4x$
36. $-7 = 18 + 10x$
37. $8x - 4.4 = 6$
38. $2x + 0.6 = 8$
39. $0.3x - 1.8 = -5.04$
40. $-1.6 = 0.5x + 2.05$

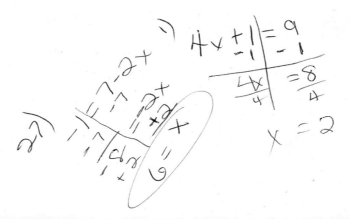

9.3

Solving Equations in Which the Variable Appears on Both Sides

All the equations we have discussed in this chapter so far have had the variable on only *one side* of the equation.

Solving an equation in which the variable appears on both sides	1. Combine like terms on each side of the equation (if there are any). 2. Remove the term containing the variable from one side by using the inverse operation. 3. Solve the resulting equation by the method given in the last section.

EXAMPLE 1   Solve $2x - 23 = -15 + 6x$.

SOLUTION 1   We can solve the equation by removing the x-term from the right side or from the left side. Here we will show how to remove the x-term from the *right* side.

To remove $6x$ from the right side, subtract $6x$ from both sides.

$$
\begin{array}{ll}
2x - 23 = -15 + 6x & \\
\underline{-6x \qquad\qquad\quad -6x} & \text{Subtracting } 6x \text{ from both sides} \\
-4x - 23 = -15 & \\
\underline{\quad +23 = +23} & \text{Adding 23 to both sides} \\
-4x \quad = \quad 8 & \\
\dfrac{\overset{1}{-4}x}{\underset{1}{-4}} = \dfrac{8}{-4} & \text{Dividing both sides by } -4 \\
x = -2 &
\end{array}
$$

✓ **Check for $x = -2$**

$$
\begin{aligned}
2x - 23 &= -15 + 6x \\
2(-2) - 23 &\overset{?}{=} -15 + 6(-2) \\
-4 - 23 &\overset{?}{=} -15 + (-12) \\
-27 &= -27
\end{aligned}
$$

SOLUTION 2   Now we will show how to remove the x-term from the *left* side. To remove $2x$ from the left side, subtract $2x$ from both sides.

$$
\begin{array}{ll}
2x - 23 = -15 + 6x & \\
\underline{-2x \qquad\qquad\quad -2x} & \text{Subtracting } 2x \text{ from both sides} \\
-23 = -15 + 4x & \\
\underline{+15 = +15} & \text{Adding 15 to both sides} \\
-8 = \qquad\quad 4x & \\
\dfrac{-8}{4} = \dfrac{\overset{1}{4}x}{\underset{1}{4}} & \text{Dividing both sides by 4} \\
-2 = x & \text{Same answer}
\end{array}
$$

Note   When we are solving equations, the answers $x = -2$ and $-2 = x$ have the same meaning. This fact is stated in the symmetric property of equality.

Symmetric property of equality	If $\quad a = b$
	then $\quad b = a$

Some students like to move all of the x-terms to the side with the largest amount of x-terms because this reduces the number of negative signs used in a problem. The choice is yours.

EXAMPLE 2 Solve $4x - 5 - x = 13 - 2x - 3$.

SOLUTION $4x - 5 - x = 13 - 2x - 3$ Combine like terms on each side

$$
\begin{array}{rcl}
3x - 5 =& 10 - 2x \\
+\ 2x & \underline{+\ 2x} \\
5x - 5 =& 10 \\
+\ 5 & \underline{+\ 5} \\
5x =& 15
\end{array}
$$

$\left\{\begin{array}{l}\text{Adding } 2x \text{ to both sides so that an} \\ x\text{-term remains on only one side}\end{array}\right.$

Adding 5 to both sides

$$\frac{\overset{1}{\cancel{5}}x}{\underset{1}{\cancel{5}}} = \frac{15}{5}$$

Dividing both sides by 5

$$x = 3$$

✓ **Check for** $x = 3$ $4x - 5 - x = 13 - 2x - 3$

$$4(3) - 5 - (3) \overset{?}{=} 13 - 2(3) - 3$$

$$12 - 5 - 3 \overset{?}{=} 13 - 6 - 3$$

$$4 = 4$$

Exercises   9.3

Solve the equations.

1. $9x = 5x + 12$
2. $4x = x + 15$
3. $x = 9 - 2x$
4. $3x = 8 - 5x$
5. $5x = 3x - 4$
6. $7x = 4x - 9$
7. $2x - 7 = x$
8. $5x - 8 = x$
9. $2x + 3 = x + 8$
10. $3x - 4 = 2x + 5$
11. $6x + 5 = 4x + 9$
12. $5x - 6 = 3x + 6$
13. $8x + 6 = 5x - 3$
14. $4x + 28 = 7 + x$
15. $3x - 6 = 9 - 2x$
16. $5x - 4 = 8 - x$
17. $x - 7 = 3x - 1$
18. $2x - 8 = 5x + 10$
19. $3 - x = 3x + 7$
20. $7 - 4x = 4x - 9$

21. $3x - 7 - x = 15 - 2x - 6$
22. $5x - 2 - x = 4 - 3x - 27$
23. $8x - 13 + 3x = 12 + 5x - 7$
24. $9x - 16 + 6x = 11 + 4x - 5$
25. $7 - 9x - 12 = 3x + 5 - 8x$
26. $13 - 11x - 17 = 5x + 4 - 10x$
27. $6x - 2 - x = 21 - 3x - 7$
28. $4x + 14 + 2x = 12 - 3x - 8$
29. $16 - 7x - 4 = 5x + 6 - 4x$
30. $3x + 15 - 6x - 10 = 9x - 6 - 3x + 5$

$9x = 5x + 12$
$-5x \quad -5x$
$4x = 12$
$\frac{4x}{4} = \frac{12}{4}$
$x = 3$

9.4 Solving Equations Containing Grouping Symbols

If an equation contains grouping symbols, first we remove them using the distributive property, then we solve the resulting equation using the methods discussed in the previous sections. The following box summarizes how to solve an equation.

Solving an equation

1. Remove grouping symbols using the distributive property.
2. Combine like terms on each side of the equation.
3. If the variable appears on both sides of the equation, remove the term containing the variable from one side by using the inverse operation.
4. Remove all numbers that appear on the same side as the variable.
   First: Remove those numbers that are added or subtracted (by using the inverse operation).
   Second: Remove the number that is multiplied by the variable (by dividing both sides by that signed number).

Check the solution in the *original* equation.

EXAMPLE 1 Solve $10x - 2(3 + 4x) = 7 - (x - 2)$.

SOLUTION

$$10x - 2(3 + 4x) = 7 - (x - 2)$$

$$10x - 6 - 8x = 7 - x + 2 \qquad \text{Removing grouping symbols}$$

$$2x - 6 = 9 - x \qquad \text{Combining like terms on each side}$$
$$\underline{+ x \qquad\qquad + x} \qquad \text{Adding } x \text{ to both sides}$$
$$3x - 6 = 9$$
$$\underline{+ 6 \qquad + 6} \qquad \text{Adding 6 to both sides}$$
$$3x = 15$$

$$\frac{\overset{1}{\cancel{3}}x}{\underset{1}{\cancel{3}}} = \frac{15}{3} \qquad \text{Dividing both sides by 3}$$

$$x = 5$$

✓ **Check for** $x = 5$

$$10x - 2(3 + 4x) = 7 - (x - 2)$$
$$10(5) - 2(3 + 4 \cdot 5) \overset{?}{=} 7 - (5 - 2)$$
$$10(5) - 2(3 + 20) \overset{?}{=} 7 - (3)$$
$$10(5) - 2(23) \overset{?}{=} 7 - 3$$
$$50 - 46 \overset{?}{=} 4$$
$$4 = 4$$

EXAMPLE 2   Solve $5(2 - 3x) - 4 = 5x + [-(2x - 10) + 8]$.

SOLUTION

$$5(2 - 3x) - 4 = 5x + [-(2x - 10) + 8]$$

$$\left.\begin{array}{l} 10 - 15x - 4 = 5x + [-2x + 10 + 8] \\ 10 - 15x - 4 = 5x + [-2x + 18] \end{array}\right\} \text{Removing grouping symbols}$$

$$10 - 15x - 4 = 5x - 2x + 18$$

$$\begin{array}{rcl} -15x + 6 & = & 3x + 18 \\ +15x & & +15x \end{array}$$ Combining like terms on both sides
Adding 15x to both sides

$$\begin{array}{rcl} 6 & = & 18x + 18 \\ -18 & & -18 \end{array}$$ Subtracting 18 from both sides

$$-12 = 18x$$

$$\frac{-12}{18} = \frac{\overset{1}{\cancel{18}}x}{\underset{1}{\cancel{18}}}$$ Dividing both sides by 18

$$-\frac{2}{3} = x$$

You should check to see that $x = -\dfrac{2}{3}$ makes the two sides of the equation equal.

Exercises 9.4

Solve the equations.

1. $3(3x + 5) = 7x + 5$
2. $2(7x + 9) = 8x + 6$
3. $4(2x - 1) = 5(x + 4)$
4. $3(3x - 2) = 6(x - 4)$
5. $3x + 2(x - 3) = 9$
6. $5x + 3(x + 1) = -5$
7. $5x - 3(x + 4) = 4$
8. $6x - 2(x - 3) = -2$
9. $3x - 4(2x - 3) = 2$
10. $9x - 5(3x + 3) = -3$
11. $x - (7 - x) = -1$
12. $x - (8 - 2x) = 4$
13. $9 - 4x = 5(9 - 8x)$
14. $10 - 7x = 4(11 - 6x)$
15. $3y - 2(2y - 7) = 2(3 + y) - 4$
16. $4a - 3(5a - 14) = 5(7 + a) - 9$
17. $6(3 - 4x) + 12 = 10x - 2(5 - 3x)$
18. $7(2 - 5x) + 27 = 18x - 3(8 - 4x)$
19. $2(3x - 6) - 3(5x + 4) = 5(7x - 8)$
20. $4(7z - 9) - 7(4z + 3) = 6(9z - 10)$
21. $6(5 - 2h) = 3(4h - 2) - 7(2 + 2h)$
22. $5(4 + 2k) = 8(3k - 2) - 4(1 + k)$

23. $2y - 3(4y - 8) = 2(5 + y) - 10$
24. $5(6 - 3a) + 18 = -9a - 3(4 - 2a)$
25. $3(2z - 6) - 2(6z + 4) = 5(z + 8)$
26. $7(2 - 5h) = 4(3h - 2) - 6(8 + 9h)$
27. $2[5 - 5(x - 4)] = 10 - 5x$
28. $3[2 - 4(x - 7)] = 26 - 8x$
29. $3[2h - 6] = 2\{2(3 - h) - 5\}$
30. $6(3h - 5) = 3\{4(1 - h) - 7\}$
31. $5(3 - 2x) - 10 = 4x + [-(2x - 5) + 15]$
32. $4(2 - 6x) - 6 = 8x + [-(3x - 11) + 20]$
33. $9 - 3(2x - 7) - 9x = 5x - 2[6x - (4 - x) - 20]$
34. $14 - 2(7 - 4x) - 4x = 8x - 3[2x - (5 - x) - 30]$
35. $10 - 6\left(\dfrac{1}{2}x - \dfrac{2}{3}\right) = 8\left(\dfrac{3}{4}x + \dfrac{5}{8}\right)$
36. $3\left[2 - 4\left(\dfrac{1}{4}x - \dfrac{1}{2}\right)\right] = 12\left(\dfrac{2}{3}x - \dfrac{5}{6}\right)$

9.5 Conditional Equations, Identities, and Equations with No Solution

There are different kinds of equations. In this section we discuss three types of equations: conditional equations, identical equations, and equations with no solution.

Conditional Equations   A **conditional equation** is an equation whose two sides are equal only when certain numbers (called *solutions*) are substituted for the variable. All equations given in this chapter so far have been conditional equations.

Consider the equation $\qquad\qquad x + 1 = 3$

We'll try $x = 5$. Then $\qquad\qquad x + 1 = 3$
becomes $\qquad\qquad\qquad\qquad\; 5 + 1 = 3$???
But we know that $\qquad\qquad\qquad 6 \neq 3$

We'll try $x = -1$. Then $\qquad\; x + 1 = 3$
becomes $\qquad\qquad\qquad\quad -1 + 1 = 3$???
But we know that $\qquad\qquad\qquad 0 \neq 3$

However, if $x = 2$, then $\qquad x + 1 = 3$
becomes $\qquad\qquad\qquad\qquad\; 2 + 1 = 3$
and $\qquad\qquad\qquad\qquad\qquad\; 3 = 3$

The two sides of $x + 1 = 3$ are not always equal. The *condition* $x = 2$ must exist for the two sides of $x + 1 = 3$ to be equal.

Identities   An **identity** (or *identical equation*) is an equation whose two sides are equal no matter what permissible number is substituted for the variable. Therefore, an identity has an endless number of solutions, because any real number will make its two sides equal.

EXAMPLE 1   The equation $2(5x - 7) = 10x - 14$ is an identity because

$$10x - 14 = 10x - 14 \quad \text{Both sides identical}$$

EXAMPLE 2   The equation $4(3x - 5) - 2x = 2(5x + 1) - 22$ is an identity because

$$12x - 20 - 2x = 10x + 2 - 22$$

$$10x - 20 = 10x - 20 \quad \text{Both sides identical}$$

Equations with No Solution   A solution of an equation is a number that, when put in place of the variable, makes the two sides of the equation equal. For some equations no such number can be found. In these cases we say that *the equation has* **no solution**.

EXAMPLE 3   Solve $x + 1 = x + 2$.

SOLUTION
$$
\begin{array}{rcl}
x + 1 &=& x + 2 \\
\underline{-x} & & \underline{-x} \qquad \text{Trying to get the } x\text{-term on only one side} \\
1 &\neq& 2 \qquad \text{The two sides are unequal}
\end{array}
$$

In this case we say that *the equation has no solution*, because no number substituted for x will make the two sides of the equation equal.

Solving an equation that has only one variable

1. Attempt to solve the equation (by the method given in the last section).
2. There are three possible results:
   a. *Conditional equation:* If a *single solution* is obtained, the equation is a conditional equation.
   b. *Identity:* If the two sides of the equation simplify *to the same expression*, the equation is an identity.
   c. *No solution:* If the two sides of the equation simplify to *unequal expressions*, the equation has no solution.

EXAMPLE 4   Solve $4x - 2(3 - x) = 12$.

SOLUTION
$$4x - 2(3 - x) = 12$$
$$4x - 6 + 2x = 12$$
$$6x - 6 = 12$$
$$6x = 18$$
$$x = 3 \quad \text{Conditional equation (single solution)}$$

EXAMPLE 5   Solve $4x - 2(3 + 2x) = -6$.

SOLUTION
$$4x - 2(3 + 2x) = -6$$
$$4x - 6 - 4x = -6$$
$$-6 = -6 \quad \text{Identity (both sides the same)}$$

EXAMPLE 6   Solve $4x - 2(3 + 2x) = 8$.

SOLUTION
$$4x - 2(3 + 2x) = 8$$
$$4x - 6 - 4x = 8$$
$$-6 \neq 8 \quad \text{No solution (sides unequal)}$$

Exercises 9.5

Find the solution of each conditional equation. If the equation is an identity, write "identity." If the equation has no solution, write "no solution."

1. $x + 3 = 8$

2. $4 - x = 6$

3. $2x + 5 = 7 + 2x$

4. $10 - 5y = 8 - 5y$

5. $6 + 4x = 4x + 6$

6. $7x + 12 = 12 + 7x$

7. $5x - 2(4 - x) = 6$

8. $8x - 3(5 - x) = 7$

9. $6x - 3(5 + 2x) = -15$

10. $4x - 2(6 + 2x) = -12$

11. $4x - 2(6 + 2x) = -15$

12. $6x - 3(5 + 2x) = -12$

13. $7(2 - 5x) - 32 = 10x - 3(6 + 15x)$

14. $6(3 - 4x) + 10 = 8x - 3(2 - 3x)$

15. $2(2x - 5) - 3(4 - x) = 7x - 20$

16. $3(x - 4) - 5(6 - x) = 2(4x - 21)$

17. $2[3 - 4(5 - x)] = 2(3x - 11)$

18. $3[5 - 2(7 - x)] = 6(x - 7)$

19. $5x - 3(2x - 1) = 7 - (x + 4)$

20. $7x - 4(x - 3) = 6x - (8 - x)$

Aaran

9.6 Inequalities

An *equation* is a statement that two expressions are *equal*. An **inequality** is a statement that two expressions are *not equal*. An inequality has three parts:

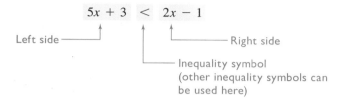

$$5x + 3 \; < \; 2x - 1$$

Left side ⟶

Right side

Inequality symbol
(other inequality symbols can
be used here)

In this text we will discuss only inequalities that have the symbols $<$, $>$, $\leq$, and $\geq$.

Types of Inequalities

Greater than ($>$) $a > b$ is read "*a* is greater than *b*."

EXAMPLE 1
a. Write: $11 > 2$.
 Read: "11 is greater than 2."
b. Write: $3x - 4 > 7$.
 Read: "3*x* minus 4 is greater than 7."

Less than ($<$) $a < b$ is read "*a* is less than *b*."

EXAMPLE 2
a. Write: $3 < 6$.
 Read: "3 is less than 6."
b. Write: $2x < 5 + x$.
 Read: "2*x* is less than 5 plus *x*."

Less than or equal to ($\leq$) $a \leq b$ is read "*a* is less than *or* equal to *b*." This means that if $\begin{cases} \text{either} & a < b \\ \text{or} & a = b \end{cases}$ is true, then $a \leq b$ is true.

EXAMPLE 3
a. $2 \leq 3$ is true, because $2 < 3$ is true.
b. $4 \leq 4$ is true, because $4 = 4$ is true.
c. Write: $7 \leq 5x - 2$.
 Read: "7 is less than or equal to 5*x* minus 2."

Greater than or equal to ($\geq$) $a \geq b$ is read "*a* is greater than *or* equal to *b*." This means that if $\begin{cases} \text{either} & a > b \\ \text{or} & a = b \end{cases}$ is true, then $a \geq b$ is true.

EXAMPLE 4
a. $5 \geq 1$ is true, because $5 > 1$ is true.
b. $7 \geq 7$ is true, because $7 = 7$ is true.
c. Write: $x + 6 \geq 10$.
 Read: "*x* plus 6 is greater than or equal to 10."

The **sense** of an inequality symbol refers to the direction in which the symbol points:

$$\left.\begin{array}{l} a > b \\ c > d \end{array}\right\} \text{Same sense} \qquad \left.\begin{array}{l} a < b \\ c > d \end{array}\right\} \text{Opposite sense}$$

$$\left.\begin{array}{l} e \leq f \\ g \leq h \end{array}\right\} \text{Same sense} \qquad \left.\begin{array}{l} e \geq f \\ g \leq h \end{array}\right\} \text{Opposite sense}$$

Facts About Inequalities

The basic rules used to solve inequalities are the same as those used to solve equations, with the exception that *the sense must be changed when multiplying or dividing both sides by a negative number.*

EXAMPLE 5

Examples of the rules used with inequalities:

a.
$$\begin{array}{r} 10 > \quad 5 \\ + 6 \quad + 6 \\ \hline 16 > \quad 11 \end{array}$$
Adding the same number, 6, to both sides

Sense is *not* changed

b.
$$\begin{array}{r} 7 < \quad 12 \\ - 2 \quad - 2 \\ \hline 5 < \quad 10 \end{array}$$
Subtracting the same number, 2, from both sides

Sense is *not* changed

c.
$$3 < 4$$
$$2(3) < 2(4)$$
$$6 < 8$$
Multiplying both sides by the same *positive* number, 2

Sense is *not* changed

d.
$$3 < 4$$
$$(-2)(3) \ ? \ (-2)(4)$$
$$-6 > -8$$
Multiplying both sides by the same *negative* number, -2

Sense is changed

e.
$$9 > 6$$
$$\frac{9}{3} > \frac{6}{3}$$
$$3 > 2$$
Dividing both sides by the same *positive* number, 3

Sense is *not* changed

f.
$$9 > 6$$
$$\frac{9}{-3} \ ? \ \frac{6}{-3}$$
$$-3 < -2$$
Dividing both sides by the same *negative* number, -3

Sense is changed

Solving Inequalities

An inequality is solved when we have nothing but the variable on one side of the inequality symbol and everything else on the other side.

The method of solving inequalities is very much like the method used for solving equations. We show how the methods are alike and different in the following summary.

	Solving Equations	*Solving Inequalities*
Addition rule	The same number may be added to both sides.	The same number may be added to both sides.
Subtraction rule	The same number may be subtracted from both sides.	The same number may be subtracted from both sides.
Multiplication rule	Both sides may be multiplied by the same nonzero number.	1. Both sides may be multiplied by the same *positive* number. 2. When both sides are multiplied by the same *negative* number, *the sense must be changed.*
Division rule	Both sides may be divided by the same nonzero number.	1. Both sides may be divided by the same *positive* number. 2. When both sides are divided by the same *negative* number, *the sense must be changed.*

EXAMPLE 6   Solve $x + 3 < 7$, and graph the solution on the number line.

SOLUTION

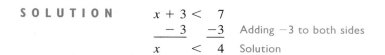

$$
\begin{array}{rl}
x + 3 <\ & 7 \\
-3\ & -3 \quad \text{Adding } -3 \text{ to both sides} \\
\hline
x \quad < \ & 4 \quad \text{Solution}
\end{array}
$$

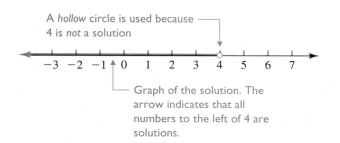

A *hollow* circle is used because 4 is *not* a solution

Graph of the solution. The arrow indicates that all numbers to the left of 4 are solutions.

The solution $x < 4$ means that if we replace x by any number less than 4 in the given inequality, we get a true statement. For example, if we replace x by 3 (which is less than 4),

$$x + 3 < 7$$
$$3 + 3 < 7$$
$$6 < 7 \quad \text{True}$$

EXAMPLE 7   Solve $2x - 1 \geq -7$, and graph the solution on the number line.

SOLUTION

$$2x - 1 \geq -7$$
$$\underline{+ 1 \quad\quad +1} \quad \text{Adding 1 to both sides}$$
$$2x \quad\quad \geq -6$$

$$\frac{2x}{2} \geq \frac{-6}{2} \quad \begin{array}{l}\text{Dividing both sides by 2}\\ \text{Sense is } not \text{ changed}\end{array}$$

$$x \geq -3 \quad \text{Solution}$$

A *solid* circle is used because
-3 *is* a solution

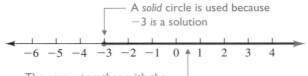

The arrow together with the
solid circle indicates that -3
and all numbers to the right of
-3 are solutions

EXAMPLE 8   Solve $x + 8 > 6x - 2$, and graph the solution on the number line.

SOLUTION 1

$$x + 8 > 6x - 2$$
$$\underline{-6x \quad\quad\quad -6x} \quad \text{Getting } x\text{-terms on the left side}$$
$$-5x + 8 > \quad\quad - 2$$
$$\underline{\quad - 8 \quad\quad\quad - 8}$$
$$-5x \quad > \quad\quad - 10$$

$$\frac{-5x}{-5} < \frac{-10}{-5} \quad \begin{array}{l}\text{Dividing both sides by } -5\\ \textit{Sense is changed}\end{array}$$

$$x < 2 \quad \text{Solution}$$

SOLUTION 2

$$x + 8 > 6x - 2$$
$$\underline{-x \quad\quad\quad - x} \quad \text{Getting } x\text{-terms on the right side}$$
$$8 > 5x - 2$$
$$\underline{+ 2 \quad\quad\quad + 2}$$
$$10 > 5x$$

$$\frac{10}{5} > \frac{5x}{5} \quad \begin{array}{l}\text{Dividing both sides by 5}\\ \text{Sense is } not \text{ changed}\end{array}$$

$$2 > x \quad \text{Solution}$$

Note that the two inequalities $x < 2$ and $2 > x$ have the same meaning.
Here is the graph of the solution.

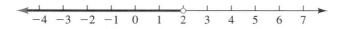

 A Word of Caution   When we are solving equations, the answers $x = a$ and $a = x$ have the same meaning. This is because of the symmetric property of equality. Example 8 shows that inequalities are *not* symmetric. When the sides of an inequality are interchanged, the sense of the inequality must be changed if the new inequality is to have the same meaning as the original one.

$$\text{If} \quad a < b$$
$$\text{then} \quad b > a$$

Although we may solve inequalities with the variable on the left side or the right side, it is easier to graph the solution if the variable is on the left side. (When the variable is on the left side, the inequality symbol points the same way as the arrow on the graph.) Therefore, it is good practice to always write the final answer with the variable on the left and change the sense of the inequality if you interchange the sides.

EXAMPLE 9 Solve $3x - 2(2x - 7) \le 2(3 + x) - 4$.

SOLUTION
$$3x - 2(2x - 7) \le 2(3 + x) - 4$$
$$3x - 4x + 14 \le 6 + 2x - 4$$
$$-x + 14 \le 2 + 2x$$
$$\underline{-2x} \qquad \underline{-2x}$$
$$-3x + 14 \le 2$$
$$\underline{-14} \quad \underline{-14}$$
$$-3x \le -12$$
$$\frac{-3x}{-3} \ge \frac{-12}{-3} \qquad \text{Dividing both sides by } -3$$
$$\qquad\qquad\qquad\qquad \textit{Sense is changed}$$
$$x \ge 4 \qquad\qquad \text{Solution}$$

The method used to solve inequalities may be summarized as follows.

Solving an inequality

Proceed in the same way used to solve an equation, except the sense must be changed when

1. Both sides are multiplied by a *negative* number.
2. Both sides are divided by a *negative* number.
3. The sides are interchanged.

Exercises 9.6

Solve the inequalities, and graph the solution on the number line.

1. $x - 5 < 2$
2. $x + 3 \ge -4$
3. $-3x < 12$
4. $-2x \ge -6$
5. $-x \ge 5$
6. $-x \ge -8$
7. $5x + 4 \le 19$
8. $7x - 3 < 18$
9. $6 - x > 2$
10. $2 - x \ge -3$
11. $6x + 2 \ge x - 8$
12. $4x + 28 > 7 + x$
13. $5x + 7 > 13 + 11x$
14. $6x + 7 \ge 3 + 8x$
15. $2x - 9 > 3(x - 2)$
16. $3x - 11 > 5(x - 1)$
17. $4(6 - 2x) \le 5x - 2$
18. $3x - 6 \le 3(2 - x)$
19. $2x - 3(x + 4) \le -7$
20. $4x - 2(3 - x) \le 12$
21. $6(3 - 4x) + 12 \ge 10x - 2(5 - 3x)$
22. $6(2x - 5) + 29 > 3x - 7(11 - 4x)$
23. $5(3 - 2x) + 25 > 4x - 6(10 - 3x)$
24. $7(2 - 5x) + 27 > 18x - 3(8 - 4x)$
25. $2[3 - 5(x - 4)] < 10 - 5x$
26. $3[2 - 4(x - 7)] < 26 - 8x$
27. $4x - \{6 - [2x - (x + 3)]\} \le 6$
28. $2x - \{4 - [5x - (8 - x)]\} \ge -20$

Chapter 9 REVIEW

Equation
9.1

An **equation** is a statement that two expressions are *equal*.

Solution of an Equation
9.1

A **solution** of an equation is a number that, when put in place of the variable, makes a true statement.

Conditional Equations
9.5

A **conditional equation** is an equation whose two sides are equal only when certain numbers (called *solutions*) are substituted for the variable.

Identities
9.5

An **identity** (or *identical equation*) is an equation whose two sides are equal no matter what permissible number is substituted for the variable.

Equations with No Solution
9.5

An equation has **no solution** if there is no number that, when put in place of the variable, makes the two sides of the equation equal.

Solving an Equation
9.1–9.4

1. Remove grouping symbols.
2. Combine like terms on each side of the equation.
3. Attempt to solve the resulting equation.
   There are three possible results.
   a. **Conditional equation:** If a *single solution* is obtained, the equation is a conditional equation.
   b. **Identity:** If the two sides of the equation simplify *to the same expression*, the equation is an identity.
   c. **No solution:** If the two sides of the equation simplify to *unequal expressions*, the equation has no solution.

Checking the Solution of an Equation
9.1

1. Replace the variable in the original equation by the solution in parentheses.
2. Perform the indicated operations on both sides of the equal sign.
3. If the resulting number on each side of the equal sign is the same, the solution is correct.

Inequality
9.6

An **inequality** is a statement that two expressions are *not equal*.

Solution of an Inequality
9.6

A solution of an inequality is a number that, when put in place of the variable, makes the inequality a true statement.

Solving an Inequality
9.6

To solve an inequality, proceed in the same way used to solve an equation, except the sense must be changed when

1. Both sides are multiplied by a *negative* number.
2. Both sides are divided by a *negative* number.
3. The sides are interchanged.

Graphing an Inequality
9.6

Graph first-degree inequalities on the number line as shown.

To graph the solution $x > c$:

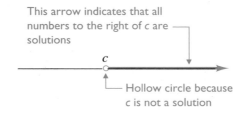

This arrow indicates that all numbers to the right of c are solutions

Hollow circle because c is not a solution

To graph the solution $x \leq b$:

This arrow indicates that all numbers to the left of b are solutions

b

Solid circle because b is a solution

Chapter 9   REVIEW EXERCISES

1. Is -4 a solution of the equation $2x - 3 = -11$?

2. Is -5 a solution of the equation $6 - 2x = -4$?

In Exercises 3–20, find the solution of each conditional equation, and check your answers. Identify any equations that are identities and any that have no solution.

3. $3x - 5 = 4$ 4. $17 - 5x = 2$ 5. $2 = 20 - 9x$

6. $2x + 1 = 2x + 7$ 7. $7.5 = \dfrac{A}{10}$ 8. $\dfrac{C}{7} - 15 = 13$

9. $7 - 2(M - 4) = 5$ 10. $10 - 4(2 - 3x) = 2 + 12x$

11. $6R - 8 = 6(2 - 3R)$ 12. $7x - (2x - 4) = 3(x + 1)$

13. $4(2x - 5) - (3 - x) = -17$

14. $6 - \{2x - [3 + x]\} = -5$

15. $15(4 - 5V) = 16(4 - 6V) + 10$

16. $56T - 18 = 7(8T - 4)$

17. $5x - 7(4 - 2x) + 8 = 10 - 9(11 - x)$

18. $9 - 3(x - 2) = 3(5 - x)$

19. $4[-24 - 6(3x - 5) + 22x] = 0$

20. $2[-7y - 3(5 - 4y) + 10] = 10y - 12$

In Exercises 21–26, solve each inequality, and graph the solution on the number line.

21. $3x - 5 < 4$ 22. $x + 4 \geq -2$

23. $9x - 4(2 + 2x) \geq -15$ 24. $2x + 7 > 11 + 4x$

25. $3[2 - 5(x - 1)] > 3 - 6x$

26. $2(x + 4) - 13 \leq 7x - 5(2x - 3)$

Chapter 9 Critical Thinking and Writing Problems

Answer Problems 1–8 in your own words, using complete sentences.

1. Explain how to check a solution of an equation.

2. Explain what your *first* step would be in solving the equation $12 = 3x + 6$.

3. Explain what your *first* step would be in solving the equation $4x - 2 + x = 3 + 6x - 8$.

4. Explain how to determine whether an equation is an identity.

5. Explain how to determine whether an equation has no solution.

6. Explain the difference between solving an equation and solving an inequality.

7. Explain why 6 is not a solution for $4x < 24$.

8. Explain how to determine, when graphing an inequality, whether the dot is hollow or solid.

9. Is the following statement true or false? Explain why.
   If $-x > 4$
   then $x > -4$

10. Which is the correct graph of $-2 \leq x$? Explain why.

   a.

   b.

   c.
   $$\begin{array}{ccccccc} & | & | & | & | & | & \\ & -4 & -3 & -2 & -1 & 0 & \end{array}$$

   d.
   $$\begin{array}{ccccccc} & | & | & \circ & | & | & \\ & -4 & -3 & -2 & -1 & 0 & \end{array}$$

Exercises 11–13 each have an error. Find the error, and in your own words, explain why it is wrong. Then work the problem correctly.

11.
$$3x + 6 = -9$$
$$\underline{\ -6\quad -6}$$
$$3x\quad = -15$$

$$3x = -15$$
$$\underline{-3\quad -3}$$
$$x = -18$$

12.
$$-2 = x + 5$$
$$\underline{+2 = \quad +2}$$
$$x = 7$$

13.
$$3x - 2(x - 3) = -8$$
$$3x - 2x - 6 = -8$$
$$x - 6 = -8$$
$$x = -2$$

Chapter 9   DIAGNOSTIC TEST

Allow yourself about an hour to do these problems. Complete solutions for all problems, together with section references, are given in the answer section at the end of the book.

1. Is -3 a solution of the equation $4x - 5 = -7$?

In Problems 2–16, solve each equation. If the equation is an identity, write "identity." If the equation has no solution, write "no solution."

2. $x - 3 = 7$

3. $10 = x + 4$

4. $3x + 2 = 14$

5. $15 - 2x = 7$

6. $-5 = \dfrac{x}{7}$

7. $4x + 5 = 17 - 2x$

8. $6x - 3 = 5x + 2$

9. $9 + 7x = 5x - 1$

10. $2x - 5 = 8x - 3$

11. $3z - 21 + 5z = 4 - 6z + 17$

12. $6k - 3(4 - 5k) = 9$

13. $6x - 2(3x - 5) = 10$

14. $2m - 4(3m - 2) = 5(6 + m) - 7$

15. $2(3x + 5) = 14 + 3(2x - 1)$

16. $3[7 - 6(y - 2)] = -3 + 2y$

In Problems 17–20, solve each inequality, and graph the solution on the number line.

17. $5x - 2 > 10 - x$

18. $4x + 5 < x - 4$

19. $2(x - 4) - 5 \geq 6 + 3(2x - 1)$

20. $7x - 2(5 + 4x) \leq -7$

Word Problems

CHAPTER 10

The main reason for studying algebra is to equip oneself with the tools necessary to solve problems. Most problems are expressed in words. In this chapter we show how to change the words of a written problem into an equation. That equation can then be solved by the methods learned in the last chapter.

10.1 Changing Word Expressions into Algebraic Expressions

In solving word problems, it is helpful to break them up into smaller expressions. In this section we show how to change these small *word* expressions into *algebraic* expressions.

10.1A Representing Word Expressions

A list of key word expressions and their corresponding algebraic operations follows:

+	−	×	÷
add	subtract	multiply	divided by
sum	difference	times	quotient
plus	minus	product	per
increased by	decreased by	of	
more than	diminished by		
	less		
	less than		
	subtracted from		

Note Because subtraction and division are not commutative, the order in which we write the terms is very important. In most cases we write the terms in the order stated. The exceptions are

less than

subtracted from   Means the terms are written in *reverse*
order (see Examples 1d and 1i)
subtract . . . from

EXAMPLE 1   Change each word expression into an algebraic expression.

SOLUTION

a. The sum of A and B $A + B$

b. The product of L and W LW

c. Two decreased by C $2 - C$

d. Two subtracted from C $C - 2$

e. The quotient of x squared and y increased by six $\dfrac{x^2}{y} + 6$

f. Twice x plus three $2x + 3$

g. Twice the sum of x plus three $2(x + 3)$

h. Twice x less four $2x - 4$

i. Twice x less than four $4 - 2x$

Exercises 10.1A

Change each word expression into an algebraic expression.

1. The sum of x and ten
2. A added to B
3. Five less than A
4. B diminished by C
5. The product of six and z
6. A multiplied by B
7. C decreased by D
8. 10 less than A
9. Ten added to x
10. x diminished by 4
11. Subtract the product of U and V from x.
12. Subtract x from the product of P and Q.
13. The product of five and the square of x
14. The product of ten and the cube of x
15. The square of the sum of A and B
16. The square of the quotient of A divided by B
17. The sum of x and seven, divided by y
18. T divided by the sum of x and nine
19. The quotient of A divided by the sum of C and ten
20. The quotient of the sum of A and C divided by B
21. Fifteen more than twice F
22. Twice the sum of x and y
23. Five times y subtracted from three times x
24. Ten times x decreased by ten times y
25. The product of x and the difference, six less than y
26. The product of seven and the difference, x less than y
27. Three times the sum of A and five
28. Five times the difference of two and C
29. Twice the difference of six and B
30. The sum of x and four subtracted from ten
31. The sum of six times x and 7
32. The difference of four times x and three

10.1.B Representing Word Expressions Containing Unknown Numbers

Word expressions may contain unknown numbers. To change such a word expression into an algebraic expression, we must first represent each unknown number in terms of the same variable.

Changing a word expression into an algebraic expression	1. Identify which number or numbers are unknown.
	2. Represent one of the unknown numbers by a variable. Express any other unknown number in terms of the same variable.
	3. Change the word expression into an algebraic expression, using the variable representations in place of the unknown numbers.

EXAMPLE 2 Change the word expression *the cost of five stamps* into an algebraic expression.

SOLUTION

Step 1. The cost of one stamp is the unknown number.
Step 2. Let c represent the cost of one stamp.
Step 3. Then $5c$ is the algebraic expression for *the cost of five stamps*. Because one stamp costs c cents, 5 stamps will cost 5 times c cents, or $5c$.

Many words that end in *er*, such as *older, younger, faster, slower, longer, shorter,* and *later,* imply either addition or subtraction. If the smaller unknown is represented by x, then the larger unknown must be represented by an addition (see Solution 1 in Example 3); but if the larger unknown in represented by x, then the smaller unknown is represented by a subtraction (see Solution 2).

EXAMPLE 3

Represent both unknown numbers in the word expression *Flora is ten years older than José* in terms of the same variable.

SOLUTION 1

Step 1. There are two unknown numbers in this expression: Flora's age and José's age.
Step 2. Let x = José's age
        Then $x + 10$ = Flora's age   Because Flora is older than José

SOLUTION 2

Step 1. There are two unknown numbers in this expression: Flora's age and José's age.
Step 2. Let x = Flora's age
        Then $x - 10$ = José's age   Because José is younger than Flora

---

EXAMPLE 4

Represent both unknown numbers in the word expression *the sum of two numbers is ten* in terms of the same variable.

SOLUTION   When the *total* of two unknown quantities is given, we may represent one quantity as x and the other quantity as "total − x" (total minus x).

Step 1. There are two unknown numbers.
Step 2. Let x = one unknown number
        Then $10 - x$ = the other unknown number

Notice that the sum of the two numbers is 10:

$$x + (10 - x) = 10$$

EXAMPLE 5

Represent the length and the width in the word expression *the length of a rectangle is three feet more than twice the width* in terms of the same variable.

SOLUTION   Because the word expression describes the length *in terms of the width*, we will represent the width by the variable w and the length in terms of the same variable.

Step 1. There are two unknown numbers: the length and the width.
Step 2. Let w = width
        Then $2w + 3$ = length

Consecutive Integers   Consecutive integers follow one another in order, such as 5, 6, and 7. **Consecutive integers** are integers that differ by 1.

If x = first consecutive integer

then $x + 1$ = second consecutive integer

and $x + 2$ = third consecutive integer

Consecutive Even Integers   Even integers are integers that are divisible by 2. **Consecutive even integers** are even integers that differ by 2, such as 4, 6, and 8.

If x = first consecutive even

then $x + 2$ = second consecutive even

and $x + 4$ = third consecutive even

Consecutive Odd Integers   Odd integers are not divisible by 2. **Consecutive odd integers** are odd integers that differ by 2, such as 5, 7, and 9.

$$If \quad x = \text{first consecutive odd}$$

$$then \ x + 2 = \text{second consecutive odd}$$

$$and \ x + 4 = \text{third consecutive odd}$$

EXAMPLE 6   Change the word expression *the sum of three consecutive integers* into an algebraic expression.

SOLUTION

Step 1. There are three unknown numbers.
Step 2. Let $\quad x =$ first consecutive integer
Then $x + 1 =$ second consecutive integer
and $\quad x + 2 =$ third consecutive integer
Step 3. Then the sum is $x + (x + 1) + (x + 2)$.

EXAMPLE 7   Change the word expression *the product of two consecutive odd integers* into an algebraic expression.

SOLUTION

Step 1. There are two unknown numbers.
Step 2. Let $\quad x =$ first consecutive odd
Then $x + 2 =$ second consecutive odd
Step 3. Then their product is $x(x + 2)$.

Exercises 10.1B

In the following exercises, (a) *If there is only one unknown number*, represent it by a variable and then change the word expression into an algebraic expression; (b) *if there is more than one unknown number*, represent each number in terms of the same variable.

1. Goh's salary plus seventy-five dollars

2. Jaime's salary less forty-two dollars

3. Two less than the number of children in Mr. Moore's family

4. Two more players were added to Omar's team.

5. Four times Joyce's age

6. One-fourth of Rene's age

7. Twenty times the cost of a record increased by eighty-nine cents

8. Seventeen cents less than five times the cost of a ball-point pen

9. One-fifth the cost of a hamburger

10. The length of the building divided by eight

11. Five times the speed of a car, plus one hundred miles per hour

12. Twice the speed of a car, diminished by forty miles per hour

13. Ten less than five times the square of an unknown number

14. Eight more than four times the cube of an unknown number

15. The sum of eight and an unknown number is divided by the square of that unknown number

16. The sum of the square of an unknown number and eleven is divided by the unknown number.

17. The sum of thirty-two and nine-fifths the Celsius temperature

18. Five-ninths times the result of subtracting thirty-two from the Fahrenheit temperature

19. Twice the result of subtracting an unknown number from five

20. The square of the result of subtracting twelve from an unknown number

21. The length of a rectangle is twelve centimeters more than its width.

22. The altitude of a triangle is seven centimeters less than its base.

23. The sum of two numbers is sixty.

24. The sum of two numbers is −22.

25. Sanjay is five years younger than his brother.

26. Mrs. Lopez is twenty-one years older than her daughter.

27. Tom has $139 more than Miri.

28. Pete has $53 less than Anna.

29. The product of two consecutive integers

30. The product of two consecutive even integers

31. The sum of three consecutive odd integers

32. The sum of four consecutive integers

33. The length of a rectangle is four inches less than twice the width.

34. The width of a rectangle is ten inches less than three times the length.

35. Nine more than three times an unknown number

36. Eight less than six times an unknown number

37. Twice the first consecutive even integer minus the second even integer

38. The first consecutive odd integer plus three times the second odd integer

10.2   Solving Word Problems

In this section we show how we change the words of a written problem into an equation. Then we solve the equation using the method discussed earlier.

Solving a word problem

1. Represent the unknown number by a variable. Express any other unknown numbers in terms of the same variable.
2. Break the word problem into small pieces.
3. Represent each piece by an algebraic expression.
4. Arrange the algebraic expressions into an equation.
5. Solve the equation.
6. Check the solution in the word statement.

Number Problems

EXAMPLE 1   Seven increased by three times an unknown number is thirteen. What is the unknown number?

SOLUTION   First we break the word problem into small pieces, then we represent each piece by an algebraic expression.

Let x stand for the unknown number.

Seven	increased by	three times	an unknown number	is	thirteen.
7	+	3 ·	x	=	13

Now we solve the equation:

$$7 + 3x = 13$$

$$3x = 6 \quad \text{Subtracting 7 from both sides}$$

$$x = 2 \quad \text{Dividing both sides by 3}$$

Therefore, the unknown number is 2.

Last, we check $x = 2$ in the word statement.

Seven	increased by	three times	an unknown number	is	thirteen.
7	+	3 ·	(2)	=	13

$$7 + 3(2) \stackrel{?}{=} 13 \qquad \text{The unknown number is replaced by 2}$$
$$7 + 6 \stackrel{?}{=} 13$$
$$13 = 13$$

 Note To check a word problem, we must check the solution in the word statement. The reason we check in the word statement is that we may have made an error in writing the equation, which would not be discovered if we substituted the solution in the equation.

EXAMPLE 2 Four times an unknown number is equal to twice the sum of five and that unknown number. Find the unknown number.

SOLUTION Let x = the unknown number

Four times	an unknown number	is equal to	twice the sum of five and that unknown number.
4 ·	x	=	2 · (5 + x)

$$4x = 2(5 + x)$$
$$4x = 10 + 2x \qquad \text{Using the distributive rule}$$
$$2x = 10 \qquad \text{Subtracting 2x from both sides}$$
$$x = 5 \qquad \text{Dividing both sides by 2}$$

Therefore, the unknown number is 5.

√ **Check** Checking the solution in the word statement is left to the student.

EXAMPLE 3 When seven is subtracted from six times an unknown number, the result is eleven. What is the unknown number?

SOLUTION Let x = the unknown number

When seven	is subtracted from	six times an unknown number	the result is	eleven.
	6 · x −	7	=	11

$$6x - 7 = 11$$
$$6x = 18 \qquad \text{Adding 7 to both sides}$$
$$x = 3 \qquad \text{Dividing both sides by 6}$$

Therefore, the unknown number is 3. The check is left to the student.

Consecutive Integer Problems

EXAMPLE 4   The sum of three consecutive even integers is 42. Find the integers.

SOLUTION   Let x = first consecutive even
Then $x + 2$ = second consecutive even
and $x + 4$ = third consecutive even

The sum of three consecutive even integers	is	42.
$x + (x + 2) + (x + 4)$	=	42

$$x + (x + 2) + (x + 4) = 42$$

$$3x + 6 = 42 \quad \text{Combining like terms}$$

$$3x = 36 \quad \text{Subtracting 6 from both sides}$$

$$x = 12 \quad \text{Dividing both sides by 3}$$

Therefore,

$$x = 12 \quad \text{First integer}$$

$$x + 2 = 14 \quad \text{Second integer}$$

$$x + 4 = 16 \quad \text{Third integer}$$

√   *Check*   The sum of the even integers is $12 + 14 + 16 = 42$.

Geometry Problems

A list of geometric formulas is given on the inside front cover of this book. A complete discussion of geometric formulas is provided in a later chapter.

EXAMPLE 5   The length of a rectangle is four inches more than the width. If the perimeter is forty inches, find the length and the width.

SOLUTION   There are two unknown numbers, the width and the length.

Let w = width   Because the length is 4 more
Then $w + 4$ = length   than the width

$$P = 2l + 2w$$

$$40 = 2(w + 4) + 2w$$

$$40 = 2w + 8 + 2w \quad \text{Using the distributive rule}$$

$$40 = 4w + 8 \quad \text{Combining like terms}$$

$$32 = 4w \quad \text{Subtracting 8 from both sides}$$

$$8 = w \quad \text{Dividing both sides by 4}$$

Therefore, $w = 8$ inches   For the width

and $w + 4 = 12$ inches   For the length

√   *Check*   The perimeter is $2(12 \text{ in.}) + 2(8 \text{ in.}) = 24 \text{ in.} + 16 \text{ in.} = 40 \text{ in.}$

Exercises 10.2

In the following word problems, (a) represent the unknown numbers by a variable, (b) set up an equation and solve, and (c) answer the question.

1. When twice an unknown number is added to thirteen, the sum is twenty-five. What is the unknown number?

2. When twenty-five is added to three times an unknown number, the sum is thirty-four. What is the unknown number?

3. Five times an unknown number, decreased by eight, is twenty-two. What is the unknown number?

4. Four times an unknown number, decreased by five, is fifteen. What is the unknown number?

5. An unknown number divided by twelve equals six. What is the unknown number?

6. The quotient of an unknown number and five is ten. What is the unknown number?

7. Seven minus an unknown number is equal to the unknown number plus one. What is the unknown number?

8. Six plus an unknown number is equal to twelve decreased by the unknown number. What is the unknown number?

9. When seven is added to an unknown number, the result is twice that unknown number. What is the unknown number?

10. When three times an unknown number is subtracted from twenty, the result is the unknown number. What is the unknown number?

11. What is the unknown number if twice the sum of five and the unknown number is equal to twenty-six?

12. What is the unknown number if four times the sum of nine and the unknown number is equal to twenty?

13. Twice the sum of four and an unknown number is equal to ten less than the unknown number. What is the unknown number?

14. Six subtracted from an unknown number is equal to three times the difference of the unknown number and eight. What is the unknown number?

15. What are the three consecutive integers whose sum is 63?

16. What are the three consecutive odd integers whose sum is 21?

17. There are three consecutive even integers such that the sum of the first two integers minus the third integer is four. What are the integers?

18. There are two consecutive integers such that twice the first integer minus the second integer is eight. What are the integers?

19. There are three consecutive odd integers such that the sum of the first two integers is equal to four times the third. What are the integers?

20. There are three consecutive even integers such that the sum of the first two integers is equal to three times the third. What are the integers?

21. The length of a rectangle is 12 inches. What is the width if the perimeter is 36 inches?

22. The perimeter of a square is 64 centimeters. What is the length of a side of the square?

23. The length of a rectangle is three feet more than the width. If the perimeter is 50 feet, what are the length and the width?

24. The width of a rectangle is six inches less than the length. What are the length and the width if the perimeter is 40 inches?

25. The area of a triangle is 24 square meters, and the base is 8 meters. What is the height of the triangle?

26. The length of a rectangle is three inches less than twice the width. If the perimeter is 24 inches, what are the length and the width?

10.3 Mixture Problems

Mixture problems involve mixing two or more dry ingredients.

Two important facts necessary to solve mixture problems

1. $\left(\begin{array}{c} Amount \text{ of} \\ \text{ingredient A} \end{array} \right) + \left(\begin{array}{c} amount \text{ of} \\ \text{ingredient B} \end{array} \right) = \left(\begin{array}{c} amount \text{ of} \\ \text{mixture} \end{array} \right)$

2. $\left(\begin{array}{c} Total \; cost \text{ of} \\ \text{ingredient A} \end{array} \right) + \left(\begin{array}{c} total \; cost \text{ of} \\ \text{ingredient B} \end{array} \right) = \left(\begin{array}{c} total \; cost \text{ of} \\ \text{mixture} \end{array} \right)$

EXAMPLE 1   A wholesaler makes up a 60-pound mixture of two kinds of coffee. Brand A costs $3 per pound, and Brand B costs $6 per pound. How many pounds of each kind must be used if the mixture is to cost $5 per pound?

SOLUTION Because the *total* number of pounds is 60, we will represent the number of pounds of Brand A as x and the number of pounds of Brand B as the "total $-x$," or $60 - x$.

Let $\quad x =$ number of pounds of Brand A
Then $60 - x =$ number of pounds of Brand B

Some students find it helpful to use a chart to keep track of the useful information.

	Cost per Pound ·	Number of Pounds =	Total Cost
Brand A	3	x	$3x$
Brand B	6	$60 - x$	$6(60 - x)$
Mixture	5	60	$5(60)$

$$\boxed{\text{Cost of Brand A}} + \boxed{\text{cost of Brand B}} = \boxed{\text{cost of mixture}}$$

$$3x + 6(60 - x) = 5(60)$$

$$3x + 360 - 6x = 300 \qquad \text{Using the distributive rule}$$

$$360 - 3x = 300 \qquad \text{Combining like terms}$$

$$-3x = -60 \qquad \text{Subtracting 360 from both sides}$$

$$x = 20 \qquad \text{Dividing both sides by } -3$$

Therefore, $\qquad\qquad\qquad\qquad x = 20$ lb of Brand A

$$60 - x = 40 \text{ lb of Brand B}$$

√ **Check** Total cost of Brand A $= 3(20) = \quad\$ 60$
Total cost of Brand B $= 6(40) = \underline{+\ 240}$
Total cost of mixture $= 5(60) = \quad\$300$

EXAMPLE 2   A farmer wants to mix 10 bushels of soybeans costing $7.50 per bushel with corn costing $6.00 per bushel. How many bushels of corn must he use to make a mixture costing $7.25 per bushel?

SOLUTION   Let $x =$ bushels of corn.

	Cost per Bushel ·	Number of Bushels =	Total Cost
Soybeans	7.50	10	$7.50(10)$
Corn	6.00	x	$6.00x$
Mixture	7.25	$x + 10$	$7.25(x + 10)$

$$\boxed{\text{Cost of soybeans}} + \boxed{\text{cost of corn}} = \boxed{\text{cost of mixture}}$$

$$7.50(10) + 6.00x = 7.25(x + 10)$$

$$750(10) + 600x = 725(x + 10) \quad \text{Multiply both sides by 100}$$

$$7500 + 600x = 725x + 7250$$

$$7500 = 125x + 7250$$
$$250 = 125x$$
$$2 = x$$

Therefore, $x = 2$ bushels of corn.

√ **Check**   Total cost of soybeans $= \$7.50(10) =$ $\$75.00$
                Total cost of corn $= \$6.00(2)$ $= + \ \ 12.00$
                Total cost of mixture $= \$7.25(12) =$ $\$87.00$

The next example involves mixing coins.

EXAMPLE 3

Tonya has $3.20 in nickels, dimes, and quarters. If there are 7 more dimes than quarters and 3 times as many nickels as quarters, how many of each kind of coin does she have?

SOLUTION   Let $Q =$ number of quarters
                 Then $Q + 7 =$ number of dimes   Because there are 7 more
                                                dimes than quarters
                 and $3Q =$ number of nickels   Because there are 3 times as
                                                  many nickels as quarters

	Value of One Coin ·	Number of Coins =	Total Value
Quarters	25	Q	$25Q$
Dimes	10	$Q + 7$	$10(Q + 7)$
Nickels	5	$3Q$	$5(3Q)$
Mixture			320

← This time we are given the *total* value of the *mixture*

Value of quarters $+$ value of dimes $+$ value of nickels $= 320¢$ ← $\$3.20 = 320¢$

     $25Q$ $+$ $10(Q + 7)$ $+$ $5(3Q)$ $= \ \ 320$

$$25Q + 10(Q + 7) + 5(3Q) = 320$$
$$25Q + 10Q + 70 + 15Q = 320$$
$$50Q = 250$$
$$Q = 5$$

Therefore, she has

$$Q = 5 \text{ quarters}$$
$$Q + 7 = 12 \text{ dimes}$$
$$3Q = 15 \text{ nickels}$$

√ **Check**   5 quarters $=$ $5(\$0.25) =$ $\$1.25$
              12 dimes $= 12(\$0.10) =$ 1.20
              15 nickels $= 15(\$0.05) = + \ 0.75$
                                     $\$3.20$

Exercises 10.3

In the following mixture problems, (a) represent the un-known number(s) by a variable, (b) set up an equation and solve, and (c) answer the question.

1. René wants to make a 12-pound mixture of peanuts and cashews. Peanuts cost $4 per pound, and cashews cost $8 per pound. How many pounds of each kind of nut must be used if the mixture is to cost $5 per pound?

2. Randy mixes dried figs with dried apricots to make an 8-pound mixture costing $2.70 per pound. If dried figs costs $1.80 per pound and dried apricots cost $4.20 per pound, how many pounds of each are used?

3. Mrs. Martinez mixed 15 pounds of English toffee candy costing $1.25 per pound with caramels costing $1.50 per pound. How many pounds of caramels must she use to make a mixture costing $1.35 per pound?

4. 20 pounds of kidney beans costing 45¢ per pound is mixed with green beans costing 70¢ per pound. How many pounds of green beans should be used to make a mixture costing 60¢ per pound?

5. A 10-pound mixture of nuts and raisins costs $25. If raisins cost $1.90 per pound and nuts $3.40 per pound, how many pounds of each are used?

6. A 50-pound mixture of Delicious and Jonathan apples costs $14.50. If the Delicious apples cost 30¢ per pound and the Jonathan apples cost 20¢ per pound, how many pounds of each kind are there?

7. Mr. Wong wants to mix 30 bushels of soybeans with corn to make a 100-bushel mixture costing $4.85 per bushel. How much can he afford to pay for each bushel of corn if soybeans cost $8.00 per bushel?

8. Mrs. Lavalle wants to mix 6 pounds of Brand A with Brand B to make a 10-pound mixture costing $11.50. How much can she afford to pay per pound for Brand B if Brand A costs $1.23 per pound?

9. Bill has 13 coins in his pocket that have a total value of 95¢. If these coins consist of nickels and dimes, how many of each kind are there?

10. Miko has 11 coins that have a total value of 85¢. If the coins are only nickels and dimes, how many of each kind are there?

11. Jennifer has 12 coins that have a total value of $2.20. The coins are nickels and quarters. How many of each kind of coin are there?

12. Angelo has 18 coins consisting of nickels and quarters. If the total value of the coins is $2.50, how many of each kind of coin does he have?

13. Theo has $4.00 in nickels, dimes, and quarters. If there are 4 more quarters than nickels and 3 times as many dimes as nickels, how many of each kind of coin does he have?

14. Yoko has $5.50 in nickels, dimes, and quarters. If there are 7 more dimes than nickels and twice as many quarters as dimes, how many of each kind of coin does she have?

10.4   Solution Problems

Solution problems involve mixing two or more liquid ingredients. They can be solved by the same method we used for mixture problems.

EXAMPLE 1   How many liters (L) of a 20% alcohol solution must be added to 3 liters of a 90% alcohol solution to make an 80% solution?

SOLUTION   Let x = number of liters of 20% solution.

	Percent of Alcohol	· Number of Liters =	Total Amount of Alcohol
20% Solution	0.20	x	$0.20x$
90% Solution	0.90	3	$0.90(3)$
Mixture	0.80	$x + 3$	$0.80(x + 3)$

Amount of alcohol in 20% solution	+	amount of alcohol in 90% solution	=	amount of alcohol in 80% solution
$0.20x$	+	$0.90(3)$	=	$0.80(x + 3)$

$$0.20x + 0.90(3) = 0.80(x + 3)$$

$$2x + 9(3) = 8(x + 3) \quad \text{Multiply by 10}$$

$$2x + 27 = 8x + 24$$

$$27 = 6x + 24$$

$$3 = 6x$$

$$\frac{1}{2} = x$$

Therefore, we need $\frac{1}{2}$ liter of the 20% solution.

✓ **Check** $\frac{1}{2}$ L of 20% solution $= \frac{1}{2}(0.20) = \quad 0.10$

$\qquad\qquad$ 3 L of 90% solution $= 3(0.90) \ = + 2.70$

$\qquad\qquad$ $3\frac{1}{2}$ L of 80% solution $= 3.5(0.80) = \quad 2.80$

EXAMPLE 2 How many milliliters (mL) of water must be added to a 25% solution of glycerin to make 10 milliliters of a 5% solution? (*Note:* Water is 0% glycerin.)

SOLUTION Let $\quad x =$ number of milliliters of water

$\qquad\qquad$ Then $10 - x =$ number of milliliters of 25% solution

	Percent of Glycerin	·	Number of Milliliters	=	Total Amount of Glycerin
Water	0		x		0
25% Solution	0.25		$10 - x$		$0.25(10 - x)$
Mixture	0.05		10		$0.05(10)$

Amount of glycerin in water	+	amount of glycerin in 25% solution	=	amount of glycerin in 5% solution
0	+	$0.25(10 - x)$	=	$0.05(10)$

$$0 + 25(10 - x) = 5(10) \quad \text{Multiply by 100}$$

$$250 - 25x = 50$$

$$-25x = -200$$

$$x = 8$$

Therefore, we need 8 milliliters of water.

✓ **Check** 8 mL of water (0%) $= \quad 8(0) = \quad 0$

$\qquad\qquad$ 2 mL of 25% solution $= \quad 2(0.25) = + 0.50$

$\qquad\qquad$ 10 mL of 5% solution $= 10(0.05) = \quad 0.50$

Exercises 10.4

In the following solution problems, (a) represent the unknown number(s) by a variable, (b) set up an equation and solve, and (c) answer the question.

1. How many cubic centimeters (cc) of a 20% solution of sulfuric acid must be mixed with 100 cubic centimeters of a 50% solution to make a 25% solution of sulfuric acid?

2. How many pints of a 2% solution of disinfectant must be mixed with 5 pints of a 12% solution to make a 4% solution of disinfectant?

3. If 100 gallons of 75% glycerin solution is made up by combining a 30% glycerin solution with a 90% glycerin solution, how much of each solution must be used?

4. If 1,600 cubic centimeters of 10% dextrose solution is made up by combining a 20% dextrose solution with a 4% dextrose solution, how much of each solution must be used?

5. How many milliliters of water must be added to 500 milliliters of a 40% solution of sodium bromide to reduce it to a 25% solution? (*Hint:* Water is a 0% solution.)

6. How many liters of pure alcohol must be added to 10 liters of a 20% solution of alcohol to make a 50% solution? (*Hint:* Pure alcohol is a 100% solution.)

7. How many ounces of a 10% salt solution should be mixed with 20 ounces of a 25% solution to make a 16% solution?

8. How many milliliters of a 4% acid solution should be mixed with 100 milliliters of a 20% solution to make an 8% solution?

9. A chemist needs 50 milliliters of a 30% iodine solution. If she has only a 50% solution and a 25% solution in stock, how much of each will be needed?

10. How many quarts of pure antifreeze must be added to a 20% antifreeze solution to make 8 quarts of a 50% solution? (*Hint:* Pure antifreeze is a 100% solution.)

10.5   Distance Problems

A physical law relating *distance* traveled d, *rate* of travel r, and *time* of travel t is

$$r \cdot t = d \quad \text{Distance formula}$$

For example, you know that if you are driving your car at an average speed of 50 mph, then

$$r \cdot t = d$$

you travel a distance of 100 miles in 2 hr: $50(2) = 100$

you travel a distance of 150 miles in 3 hr: $50(3) = 150$

and so on.

Method for solving distance-rate-time problems

1. Draw the blank chart:

	r	$\cdot$	t	$=$	d

2. Fill in two columns of the chart using a single variable and the given information.

3. Use the formula $r \cdot t = d$ to fill in the remaining column.

4. Write the equation by using information in the chart with an unused fact given in the problem.

5. Solve the resulting equation.

EXAMPLE 1 Mr. Maxwell takes 2 hours to drive to the airport in the morning, but he takes 3 hours to return home over the same route during the evening rush hour. If his average morning speed is 20 mph faster than his average evening speed, how far is it from his home to the airport?

SOLUTION Let x = speed returning home
Then $x + 20$ = speed to airport

	r $\cdot$ t $=$ d		
Going to Airport	$x + 20$	2	
Returning Home	x	3	

Use the formula $r \cdot t = d$ to find what goes here

Use the given information to fill these columns

	r $\cdot$ t $=$ d		
Going to Airport	$x + 20$	2	$2(x + 20)$
Returning Home	x	3	$3x$

We also know that the two distances are equal.

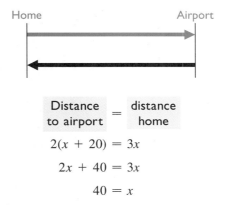

$$\frac{\text{Distance to airport}}{} = \frac{\text{distance}}{\text{home}}$$

$$2(x + 20) = 3x$$
$$2x + 40 = 3x$$
$$40 = x$$

The problem asks for the *distance* from his home to the airport.

Distance to airport = $2(x + 20) = 2(40 + 20) = 2(60) = 120$ mi
Distance home = $3x = 3(40) = 120$ mi ⟵ Distances are equal

EXAMPLE 2 Two trains are 600 miles apart and traveling toward each other. Train A's speed is 50 mph, and train B's speed is 70 mph. How long will it take for the trains to meet?

SOLUTION Let x = time for each train.

	r $\cdot$ t $=$ d		
Train A	50	x	$50x$
Train B	70	x	$70x$

⟵ $d = r \cdot t = 50 \cdot x$

⟵ $d = r \cdot t = 70 \cdot x$

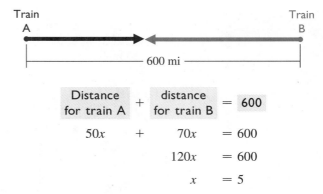

$$\text{Distance for train A} + \text{distance for train B} = 600$$

$$50x + 70x = 600$$
$$120x = 600$$
$$x = 5$$

Therefore, they will meet in 5 hours.

✓ **Check**   Distance for train A = 50(5) =   250 mi
Distance for train B = 70(5) = + 350 mi
                                600 mi

EXAMPLE 3   A boat cruises downstream for 4 hours. Because of the stream's current, it takes the boat 5 hours to go back upstream. If the speed of the stream is 3 mph, find the speed of the boat in still water.

SOLUTION   Let x = speed of boat in still water
Then $x + 3$ = speed of boat downstream
and $x - 3$ = speed of boat upstream

	r	$\cdot$	t	$=$	d
Downstream	$x + 3$		4		$4(x + 3)$
Upstream	$x - 3$		5		$5(x - 3)$

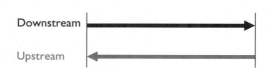

$$\text{Distance downstream} = \text{distance upstream}$$

$$4(x + 3) = 5(x - 3)$$
$$4x + 12 = 5x - 15$$
$$12 = x - 15$$
$$27 = x$$

Therefore, the speed of the boat is 27 mph.

✓ **Check**   Distance downstream = $4(x + 3)$ = $4(27 + 3)$ = $4(30)$ = 120 mi
Distance upstream   = $5(x - 3)$ = $5(27 - 3)$ = $5(24)$ = 120 mi

Exercises 10.5

In the following word problems, (a) represent the unknown number(s) using a variable, (b) set up an equation and solve, and (c) answer the question(s).

1. The Malone family left San Diego by car at 7 A.M., bound for San Francisco. Their neighbors the King family left in their car at 8 A.M., also bound for San Francisco. By traveling 9 mph faster, the Kings overtook the Malones at 1 P.M.

   a. What was the average speed of each car?

   b. What was the total distance traveled by each car before they met?

2. The Duran family left Ames, Iowa, by car at 6 A.M., bound for Yellowstone National Park. Their neighbors the Silva family left in their car at 8 A.M., also bound for Yellowstone. By traveling 10 mph faster, the Silvas overtook the Durans at 4 P.M.

   a. What was the average speed of each car?

   b. What was the total distance traveled before they met?

3. Tran hiked from his camp to a lake in the mountains and returned to camp later in the day. He walked at a rate of 2 mph going to the lake and 5 mph coming back. If the trip to the lake took 3 hours longer than the trip back:

   a. How long did it take Tran to hike to the lake?

   b. How far is it from his camp to the lake?

4. Lee hiked from her camp up to an observation tower in the mountains and returned to camp later in the day. She walked up at the rate of 2 mph and jogged back at the rate of 6 mph. The trip to the tower took 2 hours longer than the return trip.

   a. How long did it take her to hike to the tower?

   b. How far is it from her camp to the tower?

5. Anita and Felix live 54 miles apart. Both leave their homes at 7 A.M. by bicycle, riding toward each other. They meet at 10 A.M. If Felix's average speed is 2 mph faster than Anita's speed, how fast does each cycle?

6. Danny and Yolanda live 60 miles apart. Both leave their homes at 10 A.M. by bicycle, riding toward each other. They meet at 2 P.M. If Yolanda's average speed is 3 mph slower than Danny's speed, how fast does each cycle?

7. Abdul paddled a kayak downstream for 3 hours. After having lunch, he took 5 hours to paddle back upstream. If the speed of the stream is 2 mph, how fast does Abdul row in still water? How far downstream did he travel?

8. The Chang family sailed their houseboat upstream for 4 hours, but it took them only 2 hours to sail back downstream. If the speed of the houseboat in still water is 15 mph, what was the speed of the stream? How far upstream did the Changs travel?

9. Mr. Zaleva flew his private plane from his office to his company's storage facility, bucking a 20-mph head wind all the way. He flew home the same day with the same wind at his back. The *round trip* took 10 hours of flying time. If the plane makes 100 mph in still air, how far is the storage facility from his office?

10. A motor boat cruised from the marina to an island and then back to the marina. The speed of the boat in still water was 30 mph, and the speed of the current was 6 mph. If the *total* trip took 5 hours, how far is it from the marina to the island?

10.6 Variation Problems

A *variation* is an equation that relates one variable to one or more other variables by means of multiplication, division, or both. Variation problems are used in physics and chemistry.

Direct Variation **Direct variation** is a type of variation relating one variable to another by the formula

$$y = kx$$

⎣— k is called the variation constant

EXAMPLE 1

The circumference C of a circle varies directly with the diameter d, according to the formula

$$C = \pi d$$

└──── Variation constant

If the diameter is 2 inches, the circumference $C = 3.14(2) = 6.28$ inches. If the diameter is 3 inches, the circumference $C = 3.14(3) = 9.42$ inches.

In a direct variation problem where k is positive, when one variable increases, the other variable will also increase.

Solving a variation problem

1. Choose the correct variation formula for the given problem.
2. Substitute the first set of given numbers into the formula and solve for k.
3. Substitute the value of k and the third given number into the formula and solve for the unknown variable.

EXAMPLE 2

y varies directly with x. If $y = 10$ when $x = 2$, find y when $x = 6$.

SOLUTION y **varies directly with** x.

Step 1. $y \quad = k \qquad \cdot x$

Step 2. Find k:

$$y = kx$$
$$10 = k(2) \quad \text{Substitute } y = 10 \text{ and } x = 2$$
$$\frac{10}{2} = \frac{2k}{2}$$
$$5 = k$$

Step 3. Find y when $x = 6$:

$$y = kx$$
$$y = 5(6) \quad \text{Substitute } k = 5 \text{ and } x = 6$$
$$y = 30$$

EXAMPLE 3

The distance s a spring stretches varies directly with the force F applied. If a 4-pound force stretches a spring 2 inches, how far will a 10-pound force stretch it?

SOLUTION s **varies directly with** F.

Step 1. $s \quad = k \qquad \cdot F$

Step 2. Find k:

$$s = kF$$
$$2 = k(4) \qquad \text{Substitute } s = 2 \text{ and } F = 4$$
$$\frac{2}{4} = \frac{4k}{4}$$
$$\frac{1}{2} = k$$

Step 3. Find s when $F = 10$:

$$s = kF$$

$$s = \frac{1}{2}(10) \quad \text{Substitute } k = \frac{1}{2} \text{ and } F = 10$$

$$s = 5 \text{ in.}$$

A 10-pound force will stretch the spring 5 inches.

If the formula of the variation is $y = kx$, we say that "y varies directly with x." If the formula of the variation is $y = kx^2$, we say that "y varies directly with x^2."

EXAMPLE 4 y varies directly with x^2. If $y = 20$ when $x = 2$, find y when $x = 6$.

SOLUTION y varies directly with x^2.

Step 1. $y = k \qquad \cdot x^2$

Step 2. Find k:

$$y = kx^2$$

$$20 = k(2)^2 \quad \text{Substitute } y = 20 \text{ and } x = 2$$

$$\frac{20}{4} = \frac{4k}{4}$$

$$5 = k$$

Step 3. Find y when $x = 6$:

$$y = kx^2$$

$$y = 5(6)^2 \quad \text{Substitute } k = 5 \text{ and } x = 6$$

$$y = 5(36)$$

$$y = 180$$

Inverse Variation   **Inverse variation** is a type of variation relating one variable to another by the formula

$$y = \frac{k}{x} \quad \longleftarrow \text{Variation constant}$$

EXAMPLE 5   For a 300-mile trip the formula relating the car's rate r and the time t is

$$r \cdot t = 300$$

or $\qquad\qquad t = \dfrac{300}{r} \quad \longleftarrow \text{Variation constant}$

If the car averages 50 mph, the trip will take $t = \dfrac{300}{50} = 6$ hours. If the car averages 60 mph, the trip will take $t = \dfrac{300}{60} = 5$ hours.

In an inverse variation problem where k is positive, when one variable increases, the other variable will decrease.

EXAMPLE 6 y varies inversely with x. If $y = 6$ when $x = 2$, find y when $x = 3$.

SOLUTION $\boxed{y}$ $\boxed{\text{varies}}$ $\boxed{\text{inversely with } x.}$

Step 1. $y = \dfrac{k}{x}$

Step 2. Find k:

$$y = \frac{k}{x}$$

$$6 = \frac{k}{2} \qquad \text{Substitute } y = 6 \text{ and } x = 2$$

$$2 \cdot 6 = \frac{k}{\underset{1}{2}} \cdot \overset{1}{2}$$

$$12 = k$$

Step 3. Find y when $x = 3$:

$$y = \frac{k}{x}$$

$$y = \frac{12}{3} \qquad \text{Substitute } k = 12 \text{ and } x = 3$$

$$y = 4$$

EXAMPLE 7   The pressure P varies inversely with the volume V. If $P = 30$ when $V = 500$, find P when $V = 200$.

SOLUTION $\boxed{P}$ $\boxed{\text{varies}}$ $\boxed{\text{inversely with } V.}$

Step 1. $P = \dfrac{k}{V}$

Step 2. Find k:

$$P = \frac{k}{V}$$

$$30 = \frac{k}{500} \qquad \text{Substitute } P = 30 \text{ and } V = 500$$

$$500 \cdot 30 = \frac{k}{\underset{1}{500}} \cdot \overset{1}{500}$$

$$15,000 = k$$

Step 3. Find P when $V = 200$:

$$P = \frac{k}{V}$$

$$P = \frac{15,000}{200} \qquad \text{Substitute } k = 15,000 \text{ and } V = 200$$

$$P = 75$$

If the formula of the variation is $y = \dfrac{k}{x^2}$, we say that "y varies inversely with x^2."

EXAMPLE 8 y varies inversely with x^2. If $y = -3$ when $x = 4$, find y when $x = -6$.

SOLUTION y varies inversely with x^2.

Step 1. $y = \dfrac{k}{x^2}$

Step 2. Find k:

$$y = \frac{k}{x^2}$$

$$-3 = \frac{k}{4^2} \qquad \text{Substitute } y = -3 \text{ and } x = 4$$

$$16(-3) = \frac{k}{\overset{1}{\cancel{16}}} \cdot \overset{1}{\cancel{16}}$$

$$-48 = k$$

Step 3. Find y when $x = -6$:

$$y = \frac{k}{x^2}$$

$$y = \frac{-48}{(-6)^2} \qquad \text{Substitute } k = -48 \text{ and } x = -6$$

$$y = \frac{-48}{36} = -\frac{4}{3}$$

The formulas for variation can be summarized as follows:

Direct variation		*Inverse variation*	
y varies directly with x	$y = kx$	y varies inversely with x	$y = \dfrac{k}{x}$
y varies directly with x^2	$y = kx^2$	y varies inversely with x^2	$y = \dfrac{k}{x^2}$

Exercises 10.6

1. y varies directly with x. If $y = 14$ when $x = 2$, find y when $x = 4$.

2. y varies directly with x. If $y = 6$ when $x = -2$, find y when $x = 10$.

3. y varies directly with x. If $y = -9$ when $x = -3$, find y when $x = 4$.

4. y varies directly with x. If $y = -6$ when $x = 3$, find y when $x = 5$.

5. y varies inversely with x. If $y = 6$ when $x = 5$, find y when $x = 10$.

6. y varies inversely with x. If $y = 2$ when $x = -3$, find y when $x = \dfrac{1}{2}$.

7. y varies inversely with x. If $y = 3$ when $x = -2$, find y when $x = 3$.

8. y varies inversely with x. If $y = -\dfrac{1}{2}$ when $x = 16$, find y when $x = -2$.

9. y varies directly with x^2. If $y = -20$ when $x = -2$, find y when $x = 5$.

10. y varies directly with x^2. If $y = -12$ when $x = -2$, find y when $x = 5$.

11. y varies directly with x^2. If $y = 90$ when $x = -3$, find y when $x = 10$.

12. M varies directly with P^2. If $M = 2$ when $P = 2$, find M when $P = 8$.

13. F varies inversely with d^2. If $F = 3$ when $d = -4$, find F when $d = 8$.

14. L varies inversely with r^2. If $L = 16$ when $r = -3$, find L when $r = 4$.

15. C varies inversely with v^2. If $C = 6$ when $v = -3$, find C when $v = 6$.

16. y varies inversely with x^2. If $y = \dfrac{1}{2}$ when $x = 4$, find y when $x = 8$.

17. The distance s an object falls (in a vacuum) varies directly with the square of the time t it takes to fall. If an object falls 64 feet in 2 seconds, how far will it fall in 3 seconds?

18. The air resistance R on a car varies directly with the square of the car's velocity v. If the air resistance is 400 pounds at 60 mph, find the air resistance at 90 mph.

19. The intensity I of light received from a light source varies inversely with the square of the distance d from the source. If the light intensity is 15 candelas at a distance of 10 feet from the light source, what is the light intensity at a distance of 15 feet?

20. The pressure P of a gas (at constant temperature) varies inversely with its volume V. If the pressure is 15 pounds per square inch when the volume is 350 cubic inches, find the pressure when the volume is 70 cubic inches.

21. The amount s a spring is stretched varies directly with the force F applied. If a 5-pound force stretches a spring 3 inches, how far will a 2-pound force stretch it?

22. The resistance R of a boat moving through water varies directly with the square of its speed S. If its resistance is 50 pounds at a speed of 10 knots, what is the resistance at 20 knots?

23. The volume V of gas (at constant temperature) varies inversely with its pressure P (Boyle's law). If the volume is 1,600 cubic centimeters at a pressure of 250 millimeters of mercury, find the volume at a pressure of 400 millimeters of mercury.

24. The sound intensity I (loudness) varies inversely with the square of the distance d from the source. If a pneumatic drill has a sound intensity of 75 decibels at 50 feet, find its sound intensity at 150 feet.

25. The weight W of an object varies inversely with the square of the distance d separating that object from the center of the earth. If a man at the surface of the earth weighs 160 pounds, find his weight at an altitude of 8,000 miles. (*Hint:* At the surface of the earth, the man is 4,000 miles from the earth's center. Therefore, at an altitude of 8,000 miles, he is 12,000 miles from the earth's center.)

26. The pressure p in water varies directly with the depth d. If the pressure at a depth of 100 feet is 43.3 pounds per square inch (neglecting atmospheric pressure), find the pressure at a depth of 60 feet.

Chapter 10 REVIEW

Solving Word Problems
10.2–10.5

1. Read the problem, and determine what is unknown.
2. Represent each unknown in terms of the same variable.
3. Break up the word statement into small pieces, and represent each piece by an algebraic expression.
4. After each of the pieces has been written as an algebraic expression, fit them together into an equation.
5. Solve the equation for the variable.
6. Check the solution in the *original word statement*.

Solving Mixture Problems
10.3

1. $\left(\begin{array}{c}\text{Amount of} \\ \text{ingredient A}\end{array}\right) + \left(\begin{array}{c}\text{amount of} \\ \text{ingredient B}\end{array}\right) = \left(\begin{array}{c}\text{amount of} \\ \text{mixture}\end{array}\right)$

2. $\left(\begin{array}{c}\text{Total cost of} \\ \text{ingredient A}\end{array}\right) + \left(\begin{array}{c}\text{total cost of} \\ \text{ingredient B}\end{array}\right) = \left(\begin{array}{c}\text{total cost of} \\ \text{mixture}\end{array}\right)$

Solving Distance Problems
10.5

1. Draw the blank chart:

	r	$\cdot$	t	$=$	d

2. Fill in two columns using a single variable and the given information.
3. Use the formula $r \cdot t = d$ to fill in the remaining column.
4. Write the equation by using information in the chart along with an unused fact given in the problem.
5. Solve the resulting equation.

Solving a Variation Problem
10.6

1. Choose the correct variation formula for the given problem.

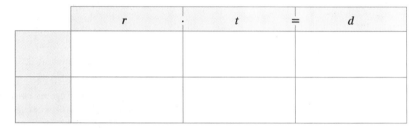

Direct variation		*Inverse variation*	
y varies directly with x	$y = kx$	y varies inversely with x	$y = \dfrac{k}{x}$
y varies directly with x^2	$y = kx^2$	y varies inversely with x^2	$y = \dfrac{k}{x^2}$

2. Substitute the first set of given numbers into the formula and solve for k.
3. Substitute the value of k and the third given number into the formula and solve for the unknown variable.

Chapter 10   REVIEW EXERCISES

1. Eight minus an unknown number is equal to the unknown number plus four. Find the unknown number.

2. When three times an unknown number is added to fifteen, the sum is forty-two. Find the unknown number.

3. Find three consecutive integers whose sum is -21.

4. Find two consecutive odd integers such that twice the first minus the second is 9.

5. The length of a rectangle is five feet more than the width. If the perimeter is seventy feet, find the length and the width.

6. The area of a triangle is 90 square centimeters, and the base is 18 centimeters. What is the height of the triangle?

7. Tokie has 27 coins that have a total value of $2.15. If all the coins are nickels or dimes, how many of each does she have?

8. Staci has $2.40 in nickels, dimes, and quarters. If there are 4 more quarters than nickels and 4 times as many dimes as nickels, how many of each kind of coin does she have?

9. A 10-pound mixture of almonds and walnuts costs $7.20. If walnuts cost 69¢ per pound and almonds 79¢ per pound, how many pounds of each kind are there?

10. A dealer makes up a 100-pound mixture of Colombian coffee costing $1.70 per pound and Brazilian coffee costing $1.50 per pound. How many pounds of each kind must he use in order for the mixture to cost $1.56 per pound?

11. How many cubic centimeters of a 50% phenol solution must be added to 400 cubic centimeters of a 5% solution to make it a 10% solution?

12. How many milliliters of water must be added to a 40% solution of sodium bromide to make 40 milliliters of a 25% solution?

13. A carload of campers leaves Los Angeles for Lake Havasu at 8:00 A.M. A second carload of campers leaves

Los Angeles at 9:00 A.M. and travels 10 mph faster over the same road. If the second car overtakes the first at 1:00 P.M., what is the average speed of each car?

14. Two airplanes take off from the same airport, traveling in opposite directions. Plane A is traveling 50 mph faster than plane B, and at the end of 4 hours they are 1,400 miles apart. Find the speed of each plane.

15. y varies inversely with x. If $y = -6$ when $x = 2$, find x when $y = -3$.

16. y varies directly with x. If $y = -6$ when $x = 2$, find y when $x = 4$.

17. y varies directly with x^2. If $y = 16$ when $x = 2$, find y when $x = 3$.

18. y varies inversely with x^2. If $y = 2$ when $x = -3$, find y when $x = 2$.

19. The electrical resistance R of a wire varies inversely with the square of its diameter d. If the resistance of a wire having a diameter of 0.02 inch is 9 ohms, find the resistance of a wire of the same length and material having a diameter of 0.03 inch.

20. In a business, the revenue R varies directly with the number of items sold n, if the price is fixed. If the revenue is $12,000 when 800 items are sold, how many items must be sold for the revenue to be $15,000?

21. Don has 10 coins consisting of dimes and quarters. After paying for a 35-cent phone call, he had $1.10 left. How many dimes did he have after the phone call?

22. It takes a boat 6 hours to go upstream to the marina but only 4 hours to go back to the dock downstream. If the speed of the boat in still water is 20 mph, what is the speed of the current?

Chapter 10 Critical Thinking and Writing Problems

Answer Problems 1–3 in your own words, using complete sentences.

1. Explain how to check a solution to a word problem.

2. Explain the difference in the meaning between the following expressions:

   a. x less 6

   b. x less than 6

3. Explain the difference in the meaning between the following expressions:

   a. 5 subtract y

   b. 5 subtracted from y

In Exercises 4–7, each of the word problems is set up incorrectly. Find the error, and in your own words, explain why it is wrong. Then set up the problem correctly.

4. Find three consecutive odd integers whose sum is 87.

$$x + (x + 1) + (x + 2) = 87$$

5. How many ounces of a 5% salt solution should be added to 10 ounces of a 30% solution to get a 15% solution?
Let x = ounces of 5% solution.

$$0.05x + 0.30(10) = 0.15(10)$$

6. How many ounces of a 5% salt solution should be added to a 30% solution to get 10 ounces of a 15% solution?
Let x = ounces of 5% solution.

$$0.05x + 0.30x = 0.15(10)$$

7. Sun Ai leaves San Francisco, heading for Mexico, at 50 mph. One hour later Ho leaves San Francisco at 60 mph, heading for Mexico. How long must Ho drive before he catches up to Sun Ai?
Let x = time for Ho.

	r	$\cdot$	t	$=$	d
Sun Ai	50		$x - 1$		$50(x - 1)$
Ho	60		x		$60x$

$$50(x - 1) = 60x$$

Chapter 10   DIAGNOSTIC TEST

Allow yourself about an hour to do these problems. Complete solutions for all problems, together with section references, are given in the answer section at the end of the book.

1. When 16 is added to three times an unknown number, the sum is 37. Find the unknown number.

2. Find three consecutive even integers such that their sum is 54.

3. Mario has 15 coins consisting of dimes and quarters. If the total value is $3.00, how many of each kind of coin are there?

4. The length of a rectangle is ten inches more than the width. If the perimeter is sixty inches, find the length and the width.

5. When 4 is subtracted from five times an unknown number, the result is the same as when 6 is added to three times the unknown number. Find the unknown number.

6. A wholesaler wants to make a 10-pound mixture of two kinds of tea. Brand A costs $2.50 per pound, and Brand

B costs $3.75 per pound. How many pounds of each kind of tea must be used if the mixture is to cost $3.00 per pound?

7. y varies inversely with x. If $y = 3$ when $x = -4$, find y when $x = \dfrac{4}{5}$.

8. How many pints of a 2% solution of disinfectant must be mixed with 4 pints of a 12% solution to make a 4% solution of disinfectant?

9. It took John 4 hours to drive to the mountains but 5 hours to drive home. If his average speed driving to the mountains is 12 mph faster than his average speed driving home, how far is it from his home to the mountains?

10. The pressure p in water varies directly with the depth d. If the pressure at a depth of 40 feet is 17.32 pounds per square inch (neglecting atmospheric pressure), find the pressure at a depth of 70 feet.

Chapters 1–10   CUMULATIVE REVIEW EXERCISES

In Exercises 1–10, perform the indicated operations. Reduce all fractions to lowest terms, and change any improper fractions to mixed numbers.

1. $7\dfrac{4}{15} + 9\dfrac{5}{6}$

2. $3\dfrac{3}{4} \div 1\dfrac{1}{8}$

3. $4\dfrac{1}{8} + 12.6$

4. $51 \div 0.06$

5. $-4^2 - 6 \cdot (-2)$

6. $9 - 2[6 - (3 - 5)]$

7. $(3x^2 - 6x - 2) - (x^2 - 4x + 5)$

8. $(2x - 5)^2$

9. $(y^2 - 3y + 4)(y - 2)$

10. $(2x^2 - 10x + 12) \div (x - 3)$

In Exercises 11–14, solve each equation or inequality.

11. $3x + 6 + 2x = 9 - x - 15$

12. $8 + 4(x - 3) = 2(8 + 3x)$

13. $3[2m - 2(6 - m)] = -12$

14. $2x - 3(4 + x) \geq -7$

15. Change $8\frac{1}{3}\%$ to a fraction in lowest terms.

16. 150 is what percent of 120?

17. Find the cost of $4\frac{3}{4}$ yards of molding at $0.69 per yard. Round off the answer to the nearest cent.

18. A car burns $2\frac{1}{2}$ quarts of oil on an 800-mile trip. How many quarts of oil can the owner expect to use on a 4,800-mile trip?

19. Seven subtracted from twice an unknown number is equal to nine. Find the number.

20. y varies directly with x^2. If $y = 12$ when $x = 2$, find y when $x = 5$.

21. How many ounces of a 10% acid solution should be mixed with 12 ounces of a 30% solution to make a 25% acid solution?

22. Two boats are 400 miles apart and traveling toward each other. Boat A is traveling 20 mph faster than Boat B. If they meet in 5 hours, find the speed of each boat.

23. Evaluate $xy^2 - 4xy$ if $x = 2$ and $y = -3$.

In Exercises 24 and 25, use the following heights:
{62 in., 56 in., 59 in., 68 in., 72 in., 79 in.}.

24. Find the mean height.   25. Find the median height.

Factoring

CHAPTER

11

n this chapter we discuss factoring. Factoring is essential for working with fractions and solving certain kinds of equations.

II.I Prime Factorization and Greatest Common Factor (GCF)

Products and Positive Factors   We know that $2 \cdot 3 = 6$. There are two ways of looking at this fact. We can start with the $2 \cdot 3$ and *find the product* 6. Or we can start with the 6 and ask ourselves what positive integers multiplied together will give us 6, then think of $2 \cdot 3$. When we do this we are *finding the factors of* 6, or, more simply, *factoring* 6.

We find the product of 2 and 3 as follows:

$$\overrightarrow{2 \cdot 3 = 6} \qquad \text{Starting with } 2 \cdot 3 \text{ and finding the } \textit{product } 6$$

Product

We factor 6 as follows:

$$\overrightarrow{6 = 2 \cdot 3} \qquad \text{Starting with 6 and finding the } \textit{factors } 2 \text{ and } 3$$

Factors

In the latter case, we call 2 and 3 **factors** (or **divisors**) of 6. But because $6 = 1 \cdot 6$, 1 and 6 are also factors of 6. Therefore, 6 has four positive factors: 1, 2, 3, and 6.

Prime Factorization of Positive Integers

Just as prime factoring was useful when we worked with arithmetic fractions, prime factoring will be helpful later when we work with algebraic fractions.

A **prime number** is a positive integer greater than 1 that can be divided evenly only by itself and 1. A prime number has no factors other than itself and 1.

A **composite number** is a positive integer greater than 1 that can be divided evenly by some integer other than itself and 1. A composite number has factors other than itself and 1.

Note that 1 is neither prime nor composite.

EXAMPLE 1

a. 9 is a composite number because it has a factor other than itself and 1, since $3 \cdot 3 = 9$.
b. 17 is a prime number because 1 and 17 are the only integral factors of 17.
c. 45 is a composite number because it has factors 3, 5, 9, and 15, in addition to 1 and 45.
d. 31 is a prime number because 1 and 31 are the only integral factors of 31.

A partial list of prime numbers is 2, 3, 5, 7, 11, 13, 17, 19, 23, 29, ...

The **prime factorization** of a positive integer is the indicated product of all its factors that are themselves prime numbers.

$$\left.\begin{array}{l} 18 = 2 \cdot 9 \\ 18 = 3 \cdot 6 \end{array}\right\} \quad \begin{array}{l} \text{These } \textit{are not prime} \\ \text{factorizations because 9 and 6} \\ \text{are not prime numbers} \end{array}$$

$$\left.\begin{array}{l} 18 = 2 \cdot 9 = 2 \cdot 3 \cdot 3 = 2 \cdot 3^2 \\ 18 = 3 \cdot 6 = 3 \cdot 2 \cdot 3 = 2 \cdot 3^2 \end{array}\right\} \quad \begin{array}{l} \text{These } \textit{are prime} \text{ factorizations} \\ \text{because all the factors are} \\ \text{prime numbers} \end{array}$$

Note that the two ways we factored 18 led to the *same* prime factorization, $2 \cdot 3^2$. The prime factorization of *any* positive integer (greater than 1) is unique.

Method for Finding the Prime Factorization   The smallest prime is 2; the next smallest is 3; the next smallest is 5; and so on. To find the prime factorization of 24, we first try to divide 24 by the smallest prime, 2. Two does divide 24 and gives a quotient of 12. Next we try to divide 12 by the smallest prime, 2. Two does divide 12 and gives a quotient of 6. Next we try to divide 6 by the smallest prime, 2. Two does divide 6 and gives a quotient of 3. This process ends here because the final quotient 3 is itself a prime. The work of finding the prime factorization of a number can be conveniently arranged as follows:

$$
\begin{array}{r|r}
2 & 24 \\ \hline
2 & 12 \\ \hline
2 & 6 \\ \hline
  & 3
\end{array}
$$

The prime factorization is the product of these numbers

Therefore, $24 = 2 \cdot 2 \cdot 2 \cdot 3 = 2^3 \cdot 3$, where $2^3 \cdot 3$ is the prime factorization of 24.

EXAMPLE 2   Find the prime factorization of the following numbers.

a. 30

SOLUTION

$$
\begin{array}{r|r}
2 & 30 \\ \hline
3 & 15 \\ \hline
  & 5
\end{array}
$$

Prime factorization of $30 = 2 \cdot 3 \cdot 5$

b. 36

SOLUTION

$$
\begin{array}{r|r}
2 & 36 \\ \hline
2 & 18 \\ \hline
3 & 9 \\ \hline
  & 3
\end{array}
$$

Prime factorization of $36 = 2 \cdot 2 \cdot 3 \cdot 3$
$\qquad\qquad\qquad\qquad = 2^2 \cdot 3^2$

When we are trying to find a prime factor of a number, we need try no prime whose square is greater than that number (Example 3).

EXAMPLE 3   Find the prime factorization of 97.

SOLUTION   Primes in order of size

2   does not divide 97.
3   does not divide 97.
5   does not divide 97.
7   does not divide 97.
11   is not possible because $11^2 = 121$, which is greater than 97.

Therefore, 97 is prime.

Greatest Common Factor (GCF)

The **greatest common factor (GCF)** of two or more integers is the greatest integer that is a factor of all of the integers.

<table>
<tr><td>**Finding the greatest common factor (GCF)**</td><td>1. Prime factor each number. Repeated factors should be expressed as powers.
2. Write each different factor that is common to all of the numbers.
3. Raise each factor to the *lowest* power to which it occurs in any of the numbers.
4. The GCF is the product of all the powers found in Step 3.</td></tr>
</table>

EXAMPLE 4   Find the GCF of 24 and 60.

SOLUTION

Step 1. Prime factor each number.

$$
\begin{array}{r|l} 2 & 24 \\ 2 & 12 \\ 2 & 6 \\ & 3 \end{array}
\qquad
\begin{array}{r|l} 2 & 60 \\ 2 & 30 \\ 3 & 15 \\ & 5 \end{array}
$$

$$24 = 2^3 \cdot 3 \qquad 60 = 2^2 \cdot 3 \cdot 5$$

Step 2. Write each different factor that is common to both numbers.

$$2, \quad 3$$

Step 3. Raise each factor to the lowest power to which it occurs in any of the numbers.

$$2^2, \quad 3^1$$

Step 4. Multiply:

$$\text{GCF} = 2^2 \cdot 3^1 = 12$$

We can also find the GCF of algebraic expressions.

EXAMPLE 5   Find the GCF of $6x^3$ and $15x^2$.

SOLUTION

Step 1. $6x^3 = 2 \cdot 3 \cdot x^3$ and $15x^2 = 3 \cdot 5 \cdot x^2$ Each expression in prime-factored form

Step 2. $3, \quad x$ Factors common to both expressions
Step 3. $3^1, \quad x^2$ Lowest powers of the factors
Step 4. $\text{GCF} = 3^1 \cdot x^2 = 3x^2$

EXAMPLE 6   Find the GCF of $5x^3y^2$, $10x^2y^3$, and $25xy^4$.

SOLUTION

Step 1. $5x^3y^2 = 5 \cdot x^3 \cdot y^2$; $10x^2y^3 = 2 \cdot 5 \cdot x^2 \cdot y^3$; $25xy^4 = 5^2 \cdot x \cdot y^4$
Step 2. $5, \quad x, \quad y$ Factors common to all expressions
Step 3. $5^1, \quad x^1, \quad y^2$ Lowest powers of the factors
Step 4. $\text{GCF} = 5^1 \cdot x^1 \cdot y^2 = 5xy^2$

Factoring a Polynomial That Has a Common Factor

Each type of factoring depends on a particular special product. Factoring a polynomial that has a common factor depends on products found by using the distributive rule. Consider the product $2(x + 5)$:

$$2(x + 5) = 2x + 10 \quad \text{By the distributive rule}$$

Therefore, $2x + 10$ factors into $2(x + 5)$.

$$2x + 10 = \boxed{2} \; (x + 5)$$

GCF ————↑   ↑———— Polynomial factor

Factoring a polynomial that has a greatest common factor

1. Find the GCF.
2. To find the polynomial factor, divide each term of the polynomial being factored by the GCF.
3. Check by multiplying the GCF and the polynomial factor, using the distributive rule.

EXAMPLE 7

Factor $6x^3 + 15x^2$.

SOLUTION The GCF is $3x^2$ (see Example 5). To find the polynomial factor, divide each term by $3x^2$.

$$6x^3 + 15x^2$$
$$= 3x^2(\quad + \quad) \qquad \text{The polynomial factor has as many terms as the original expression}$$
$$= 3x^2(\; 2x \; + \; 5 \;)$$

This term is $\dfrac{15x^2}{3x^2} = 5$

This term is $\dfrac{6x^3}{3x^2} = 2x$

Therefore, $6x^3 + 15x^2$ factors into $3x^2(2x + 5)$.

√ **Check** $\quad 3x^2(2x + 5) = 3x^2 \cdot 2x + 3x^2 \cdot 5 = 6x^3 + 15x^2$

EXAMPLE 8

Factor $5x^3y^2 - 10x^2y^3 + 25xy^4$.

SOLUTION The GCF is $5xy^2$ (see Example 6). To find the polynomial factor, divide each term by $5xy^2$.

$$5x^3y^2 - 10x^2y^3 + 25xy^4$$
$$= 5xy^2(x^2 - 2xy + 5y^2)$$

This term is $\dfrac{25xy^4}{5xy^2} = 5y^2$

This term is $-\dfrac{10x^2y^3}{5xy^2} = -2xy$

This term is $\dfrac{5x^3y^2}{5xy^2} = x^2$

Therefore, $5x^3y^2 - 10x^2y^3 + 25xy^4$ factors into $5xy^2(x^2 - 2xy + 5y^2)$.

✓ **Check** $5xy^2(x^2 - 2xy + 5y^2) = 5x^3y^2 - 10x^2y^3 + 25xy^4$

EXAMPLE 9

Factor $-3xy + 6y^2 - 3y$.

SOLUTION   Prime factor each term to find the GCF.

$$-3xy + 6y^2 - 3y = -3 \cdot x \cdot y + 2 \cdot 3 \cdot y^2 - 3 \cdot y$$

$$GCF = 3 \cdot y = 3y$$

The polynomial factor $= \dfrac{-3xy}{3y} + \dfrac{6y^2}{3y} + \dfrac{-3y}{3y} = -x + 2y - 1$

Therefore, $-3xy + 6y^2 - 3y = 3y(-x + 2y - 1)$.

ALTERNATIVE SOLUTION   Because the first term, $-3xy$, is negative, we can factor out $-3y$. Using GCF $= -3y$, we have

The polynomial factor $= \dfrac{-3xy}{-3y} + \dfrac{6y^2}{-3y} + \dfrac{-3y}{-3y} = x - 2y + 1$

Therefore, $-3xy + 6y^2 - 3y = -3y(x - 2y + 1)$.

 Note   Although both answers are correct, the second answer is preferred.

✓ **Check** $3y(-x + 2y - 1) = -3xy + 6y^2 - 3y$

$-3y(x - 2y + 1) = -3xy + 6y^2 - 3y$

Sometimes a common factor is not a monomial.

EXAMPLE 10

Factor $a(x + y) + b(x + y)$.

SOLUTION   This expression has two terms. $(x + y)$ is a common factor of both terms.

$$\overbrace{a\ (x + y)}^{\text{First term}} + \overbrace{b\ (x + y)}^{\text{Second term}}$$

$(x + y)$ is a common binomial factor of both terms

$= (x + y)\ (a + b)$

This term is $\dfrac{b(x + y)}{x + y} = b$

This term is $\dfrac{a(x + y)}{x + y} = a$

Therefore, $a(x + y) + b(x + y)$ factors into $(x + y)(a + b)$.

Exercises   *II.I*

In Exercises 1–8, state whether each of the numbers is prime or composite. To justify your answer, give the set of all positive integral factors for each number.

1. 21	**2.** 31
3. 41	**4.** 51
5. 18	**6.** 20
7. 16	**8.** 101

In Exercises 9–16, prime factor each number.

9. 12   10. 63   11. 27   12. 32

13. 75   14. 84   15. 144   16. 180

In Exercises 17–32, find the greatest common factor (GCF) for each set of numbers or expressions.

17. 6, 9   18. 8, 20   19. 18, 60

20. 15, 75   21. 4, 20, 12   22. 12, 18, 24

23. $10x^4$, $15x^3$ 24. $36x^5$, $30x^2$ 25. $15x^2y$, $6xy^2$

26. $16a^2b^3$, $40a^2b$ 27. $6x^2$, $4x$, 12 28. $8y^5$, $16y^3$, 20

29. $4x^5y^4$, $6x^4y^2$, $9x^3y^2$ 30. $5x^5y^3$, $10x^3y^4$, $6xy^3$

31. $12a^2b^2$, $9ab$, $18a^4$ 32. $20x^3y^5$, $50x^2y^2$, $100y^4$

In Exercises 33–74, factor each expression, if possible.

33. $2x - 8$ 34. $3x - 9$ 35. $5a + 10$

36. $7b + 14$ 37. $6y - 3$ 38. $15z - 5$

39. $-15x - 10$ 40. $-6a - 15$ 41. $8x + 12y$

42. $18a + 24b$ 43. $9x^2 + 3x$ 44. $8y^2 - 4y$

45. $-4y^2 + 6y$ 46. $-18x^2 + 6x$ 47. $10a^3 - 25a^2$

48. $27b^2 - 18b^4$ 49. $21w^2 - 20z^2$ 50. $15x^3 - 16y^3$

51. $15xy + 20y$ 52. $30xy + 27x$ 53. $2a^2b + 4ab^2$

54. $3mn^2 + 6m^2n^2$ 55. $12c^3d^2 - 18c^2d^3$

56. $15ab^3 - 45a^2b^4$ 57. $4x^3 - 12x - 24x^2$

58. $18y - 6y^2 - 30y^3$ 59. $24a^4 + 8a^2 - 40$

60. $45b^3 - 15b^4 - 30$ 61. $18x^3y^5 + 24x^4y^4 - 12x^5y^2$

62. $30a^3b^4 - 15a^6b^3 + 45a^8b^2$

63. $15h^2k - 8hk^2 + 9st$ 64. $10uv^3 + 5u^2v - 4wz$

65. $-14x^8y^9 + 42x^5y^4 - 28xy^3$

66. $-21u^7v^8 - 63uv^5 + 35u^2v^5$

67. $32m^5n^7 - 24m^8n^9 - 40m^3n^6$

68. $18x^3y^4 - 12x^2y^3 - 48x^4y^3$

69. $m(a + b) + n(a + b)$ 70. $x(m - n) - y(m - n)$

71. $x(y + 1) - 1(y + 1)$ 72. $a(c - d) + b(c - d)$

73. $3a(a - 2b) + 2(a - 2b)$

74. $2e(3e - f) - 3(3e - f)$

II.2   Factoring by Grouping

Sometimes we can factor an expression by rearranging its terms into smaller groups and then finding the GCF of *each group*. The GCF does not have to be a monomial.

> **Factoring an expression of four terms by grouping**
>
> 1. Arrange the four terms into two groups of two terms each. Each group of two terms must have a GCF.
> 2. Factor each group by using its GCF.
> 3. Factor the two-term expression resulting from Step 2, if the two terms have a GCF.

EXAMPLE 1   Factor $ax + ay + bx + by$.

SOLUTION   GCF = a ⎯⎯⎯⎯⎯⎯⎯⎯↓   ↓ ⎯⎯⎯ GCF = b

$$ax + ay + bx + by$$

$$= a(x + y) + b(x + y) \quad (x + y) \text{ is the GCF of these two terms}$$

$$= (x + y)(a + b)$$

This term is $\dfrac{b(x + y)}{(x + y)} = b$

This term is $\dfrac{a(x + y)}{(x + y)} = a$

Therefore, $ax + ay + bx + by = (x + y)(a + b)$.

EXAMPLE 2

Factor $2x^2 + 6xy - 3x - 9y$.

SOLUTION   GCF $= 2x$ ——————————————— GCF $= -3$

$$2x^2 + 6xy - 3x - 9y$$

$$= 2x(x + 3y) - 3(x + 3y) \qquad (x + 3y) \text{ is the GCF}$$

$$= (x + 3y)(2x - 3)$$

Therefore, $2x^2 + 6xy - 3x - 9y = (x + 3y)(2x - 3)$.

It's sometimes possible to group terms differently and still factor the expression. We obtain the same factors no matter what grouping is used.

EXAMPLE 3

Factor $ab - b + a - 1$.

SOLUTION

One grouping	*A different grouping*
$ab - b + a - 1$	$ab + a - b - 1$
$= b(a - 1) + 1(a - 1)$	$= a(b + 1) - 1(b + 1)$
$= (a - 1)(b + 1)$	$= (b + 1)(a - 1)$

——————————— Same factors ———————————

Therefore, $ab - b + a - 1 = (a - 1)(b + 1) = (b + 1)(a - 1)$.

A Word of Caution   An expression is *not* factored until it has been written as a *single* term that is a product of factors. Consider Example 1 again:

$$ax + ay + bx + by$$

$$= \underbrace{a(x + y)}_{\text{First term}} + \underbrace{b(x + y)}_{\text{Second term}}$$

The expression is *not* in factored form because it has *two terms*

$$= \underbrace{(x + y)(a + b)}_{\text{Single term}}$$

Factored form of $ax + ay + bx + by$

Expressions with more than four terms may also be factored by grouping. However, in this book we only consider factoring expressions of four terms by grouping.

Exercises 11.2

Factor each expression, if possible.

1. $am + bm + an + bn$

2. $cu + cv + du + dv$

3. $mx - nx - my + ny$

4. $ah - ak - bh + bk$

5. $xy + x - y - 1$

6. $ad - d + a - 1$

7. $3a^2 - 6ab + 2a - 4b$

8. $2h^2 - 6hk + 5h - 15k$

9. $6e^2 - 2ef - 9e + 3f$

10. $8m^2 - 4mn - 6m - 3n$

11. $ax + ay + 2bx + 2by$

12. $hw - kw - hz + kz$

13. $ef + f - e + 1$

14. $2s^2 - 6st + 5s - 15t$

15. $10xy - 15y + 8x - 12$

16. $35 - 42m - 18mn + 15n$

17. $x^3 + 3x^2 - 2x - 6$

18. $2a^3 - 8a^2 + 3a - 12$

11.3   Factoring the Difference of Two Squares

Each type of factoring depends on a particular *special product*. GCF factoring is based on products found by using the distributive rule. The kind of factoring discussed in this section depends on the special product $(a + b)(a - b)$:

$$(a + b)(a - b) = a^2 - ab + ab - b^2$$
$$= a^2 - b^2$$

Thus, the product of the sum and difference of two terms equals the difference of two squares. To factor the difference of two squares, we need to reverse this process.

Finding the product
$$(a + b)(a - b) = a^2 - b^2$$
Finding the factors

Therefore, $a^2 - b^2$ *factors into* $(a + b)(a - b)$.

Factoring the difference of two squares ($a^2 - b^2$)

The factors of the difference of two squares are the sum and the difference of the two terms that were squared:

$$a^2 - b^2 = (a + b)(a - b)$$

One factor has $+$; the other has $-$

EXAMPLE 1   Factor $x^2 - 4$.

SOLUTION
$$x^2 - 4 = x^2 - 2^2$$
$$= (x + 2)(x - 2)$$

EXAMPLE 2   Factor $25x^2 - 9y^2$.

SOLUTION
$$25x^2 - 9y^2 = (5x)^2 - (3y)^2$$
$$= (5x + 3y)(5x - 3y)$$

EXAMPLE 3   Factor $49a^6 - 81b^4$.

S O L U T I O N $49a^6 - 81b^4 = (7a^3)^2 - (9b^2)^2$

$$= (7a^3 + 9b^2)(7a^3 - 9b^2)$$

EXAMPLE 4   Factor $50x - 2x^3$.

S O L U T I O N $50x - 2x^3 = 2x(25 - x^2)$ *Remove the common factor first*

$$= 2x(5^2 - x^2)$$

$$= 2x(5 + x)(5 - x)$$

Exercises   11.3

Factor each expression, if possible.

1. $x^2 - 9$
2. $a^2 - 36$
3. $b^2 - 1$
4. $1 - x^2$
5. $m^2 - n^2$
6. $u^2 - v^2$
7. $4c^2 - 25$
8. $16d^2 - 1$
9. $25a^2 - 4b^2$
10. $9x^2 - 100y^2$
11. $64a^2 - 49b^2$
12. $36x^2 - 121y^2$
13. $2x^2 - 8$
14. $3a^2 - 3$
15. $5xy^2 - 5xa^2$
16. $4ab^2 - 4ac^2$
17. $3x^3 - 75x$
18. $2y^3 - 18y$
19. $9h^2 - 10k^2$
20. $16e^2 - 15f^2$
21. $x^6 - a^4$
22. $b^2 - y^6$
23. $49u^4 - 36v^4$
24. $81m^6 - 100n^4$
25. $a^2b^2 - c^2d^2$
26. $m^2n^2 - r^2s^2$
27. $49 - 25w^2z^2$
28. $36 - 25u^2v^2$
29. $3x^4 - 27x^2$
30. $6x^2 - 6x^4$
31. $4a^3b - 9ab^3$
32. $16x^4y - 25x^2y^3$

11.4   Factoring a Trinomial Whose Leading Coefficient Is 1

The **leading coefficient** of a polynomial is the numerical coefficient of its *highest-degree* term.

EXAMPLE 1
a. The leading coefficient of $x^2 - 2x + 8$ is 1.
b. The leading coefficient of $2x^2 - 3x + 5$ is 2.
c. The leading coefficient of $2y - 5y^2 - 7$ is -5.

The easiest type of trinomial to factor is one having a leading coefficient of 1. Consider the product $(x + 2)(x + 5)$:

$$(x + 2)(x + 5) = x^2 + 7x + 10$$

Therefore, $x^2 + 7x + 10$ factors into $(x + 2)(x + 5)$:

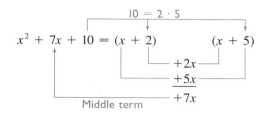

This shows that the last terms of the binomials must be factors of the last term of the trinomial. Also notice that the sum of the inner and outer products must equal the middle term of the trinomial.

EXAMPLE 2

Factor $x^2 + 5x + 6$.

SOLUTION The first term of each binomial factor must be x in order to give the x^2 in the trinomial:

$$(x \quad)(x \quad)$$

The product of the last terms of the binomials must be 6. Because $6 = 1 \cdot 6$ or $6 = 2 \cdot 3$, we could have either

$$(x + 1)(x + 6) \quad \text{or} \quad (x + 2)(x + 3)$$

The sum of the inner and outer products must equal the middle term of the trinomial, $5x$:

$$(x + 1) \qquad (x + 6)$$
$$+1x$$
$$+6x$$
$$+7x \longleftarrow \text{Incorrect middle term}$$

$$(x + 2) \qquad (x + 3)$$
$$+2x$$
$$+3x$$
$$+5x \longleftarrow \text{Correct middle term}$$

Therefore, $x^2 + 5x + 6$ factors into $(x + 2)(x + 3)$.

EXAMPLE 3

Factor $m^2 - 9m + 8$.

SOLUTION The first terms of the binomial factors must be m in order to give m^2 in the trinomial:

$$(m \quad)(m \quad)$$

The product of the last terms of the binomials must be 8, and their sum must be -9. Therefore, we have either

$$(m - 1) \qquad (m - 8) \quad \text{or} \quad (m - 2) \qquad (m - 4) \qquad 8 = 1 \cdot 8$$
$$-1m \qquad\qquad -2m \qquad\qquad\qquad = 2 \cdot 4$$
$$-8m \qquad\qquad -4m$$
$$-9m \longleftarrow \text{Correct middle term} \qquad -6m \longleftarrow \text{Incorrect middle term}$$

Therefore, $m^2 - 9m + 8 = (m - 1)(m - 8)$.

EXAMPLE 4   Factor $a^2 + 4a - 12$.

SOLUTION   The first terms of the binomial factors must be a in order to give a^2 in the trinomial:

$$(a \quad)(a \quad)$$

The product of the last terms of the binomials must be -12, and their sum must be $+4$. We find the correct pair by trial:

$$(a + 6) \qquad (a - 2) \qquad 12 = 1 \cdot 12$$

$$+6a$$
$$-2a$$
$$+4a$$

$$= 2 \cdot 6$$
$$= 3 \cdot 4$$

Therefore, $a^2 + 4a - 12 = (a + 6)(a - 2)$.

Rule of Signs   Look at the signs in Examples 2, 3, and 4.

1. If the last term of the trinomial is $+$, the signs of the binomials are the same. And if the middle term of the trinomial is $+$, then both signs are $+$.

$$x^2 + 5x + 6 = (x + 2)(x + 3)$$

Signs are the *same*

Both are $+$

2. If the last term of the trinomial is $+$, the signs of the binomials are the same. And if the middle term of the trinomial is $-$, then both signs are $-$.

$$m^2 - 9m + 8 = (m - 1)(m - 8)$$

Signs are the *same*

Both are $-$

3. If the last term of the trinomial is $-$, then the signs of the binomials are different.

$$a^2 + 4a - 12 = (a + 6)(a - 2)$$

Signs are *different*

The method of factoring a trinomial having a leading coefficient of 1 is summarized as follows.

Factoring a trinomial whose leading coefficient is 1

Arrange the trinomial in descending powers of one of the variables.

1. The product of the first term of each binomial factor equals the first term of the trinomial.
2. List all pairs of factors of the coefficient of the last term of the trinomial.
3. Select the pair of factors so that the sum of the inner and outer products equals the middle term of the trinomial.

EXAMPLE 5   Factor $x^2 + 9x + 20$.

SOLUTION   When the last term of the trinomial is $+$, the signs of the binomials are the same. Because the middle term is $+$, both signs are $+$:

$$(x + \quad)(x + \quad)$$

List all pairs of factors for the coefficient of the last term of the trinomial, $+20$; then select the pair that gives the correct middle term, $9x$:

$$(x + 4) \qquad (x + 5) \qquad 20 = 1 \cdot 20$$
$$+4x \qquad\qquad = 2 \cdot 10$$
$$+5x \qquad\qquad = 4 \cdot 5$$
$$+9x$$

Therefore, $x^2 + 9x + 20 = (x + 4)(x + 5)$.

EXAMPLE 6   Factor $x^2 - 8x - 20$.

SOLUTION   Because the last term of the trinomial is $-$, the signs of the binomials are different:

$$(x + \quad)(x - \quad)$$

From the list of factors for 20, pick the pair that gives the correct middle term, $-8x$:

$$(x + 2) \qquad (x - 10) \qquad 20 = 1 \cdot 20$$
$$+ 2x \qquad\qquad = 2 \cdot 10$$
$$-10x \qquad\qquad = 4 \cdot 5$$
$$- 8x$$

Therefore, $x^2 - 8x - 20 = (x + 2)(x - 10)$.

EXAMPLE 7   Factor $30 - 13y + y^2$.

SOLUTION $\qquad\qquad y^2 - 13y + 30$ Arranged in descending powers

When the last term of the trinomial is $+$, the signs of the binomials are the same. Since the middle term is $-$, both are $-$:

$$(y - \quad)(y - \quad)$$

List all pairs of factors of 30, then select the correct pair that gives the middle term, $-13y$:

$$(y - 3) \qquad (y - 10) \qquad 30 = 1 \cdot 30$$
$$- 3y \qquad\qquad = 2 \cdot 15$$
$$-10y \qquad\qquad = 3 \cdot 10$$
$$-13y \qquad\qquad = 5 \cdot 6$$

Therefore, $y^2 - 13y + 30 = (y - 3)(y - 10)$.

Exercises   *II.4*

Factor each expression, if possible.

1. $x^2 + 6x + 8$ 2. $x^2 + 9x + 8$ 3. $x^2 + 5x + 4$

4. $x^2 + 4x + 4$ 5. $k^2 + 7k + 6$ 6. $k^2 + 5k + 6$

7. $7u + u^2 + 10$ 8. $11u + u^2 + 10$ 9. $y^2 - 2y + 8$

10. $y^2 - 7y + 8$ 11. $b^2 - 9b + 14$ 12. $b^2 - 15b + 14$

13. $z^2 - 9z + 20$ 14. $z^2 - 12z + 20$ 15. $x^2 - 11x + 18$

16. $x^2 - 9x + 18$ 17. $x^2 + 9x - 10$ 18. $y^2 - 3y - 10$

19. $z^2 - z - 6$ 20. $m^2 + 5m - 6$ 21. $t^2 + 11t - 30$ 30. $s^2 + 22st - 48t^2$ 31. $w^2 + 2w - 24$

22. $m^2 - 17m - 30$ 23. $u^4 - 16u^2 + 64$ 32. $r^2 - 9r - 20$ 33. $n^2 - 10n - 24$

24. $v^4 - 30v^2 - 64$ 25. $16 - 8v + v^2$ 34. $36 - 15y + y^2$ 35. $w^2 - 8wz - 48z^2$

26. $16 - 10v + v^2$ 27. $b^2 - 11bd - 60d^2$ 36. $p^2 - 9pt - 52t^2$

28. $c^2 + 17cx - 60x^2$ 29. $r^2 - 13rs - 48s^2$

II.5   Factoring a Trinomial Whose Leading Coefficient Is Greater Than 1

When the leading coefficient equals 1, we need to consider only the factors of the last term. When the leading coefficient is not equal to 1, however, we must consider the factors of the first term as well as the factors of the last term.

Trial Method

Factoring a trinomial whose leading coefficient is greater than 1

Arrange the trinomial in descending powers of one of the variables.

1. Make a blank outline, and fill in all obvious information.
2. List all pairs of factors for the coefficients of the first term *and* the last term of the trinomial.
3. Select the correct pairs of factors so that the sum of the inner and outer products equals the middle term of the trinomial.

Check your factoring by multiplying the binomial factors to see whether their product is the given trinomial.

EXAMPLE 1   Factor $2x^2 + 7x + 5$.

SOLUTION   The signs for both binomial factors are $+$:

$$(\quad + \quad)(\quad + \quad)$$

The first term of each binomial factor must contain an x in order to give the x^2 in the first term of the trinomial $2x^2$:

$$(x + \quad)(x + \quad)$$

The product of the first terms of the binomial factors must be $2x^2$. Because the coefficient 2 has only two factors, 1 and 2, we will use $1x$ and $2x$:

$$(1x + \quad)(2x + \quad)$$

The product of the last terms of the binomial factors must be 5, which has only two factors, 1 and 5. Therefore, we could have either

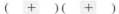

$$(1x + 5) \qquad (2x + 1) \qquad \text{or} \qquad (1x + 1) \qquad (2x + 5)$$

$+10x$

$+ 1x$

$+11x$ ← Incorrect middle term

$+2x$

$+5x$

$+7x$ ← Correct middle term

The sum of the inner and outer products must equal the middle term of the trinomial. We found the correct pair by trial. Therefore, $2x^2 + 7x + 5$ factors into $(x + 1)(2x + 5)$.

EXAMPLE 2   Factor $5x^2 + 13x + 6$.

SOLUTION   Make a blank outline, and fill in all the obvious information:

$$(x + \quad)(x + \quad)$$ The letter in each binomial must be x so that the first term of the trinomial contains x^2; the sign in each binomial is $+$

Next, list all pairs of factors for the first coefficient and for the last term of the trinomial:

Factors of the first coefficient	*Factors of the last term*
$5 = 1 \cdot 5$	$6 = 1 \cdot 6$
	$= 2 \cdot 3$

$$(1x + \quad)(5x + \quad)$$ Because I and 5 are the only factors of 5, we can fill them in next

Now we must select the correct pair of factors for the last term, 6, so that the sum of the inner and outer products is $13x$. We find the correct pair by trial:

$$(1x + 2) \qquad (5x + 3) \qquad 6 = 1 \cdot 6$$
$$+10x$$
$$+ 3x \qquad\qquad = 2 \cdot 3$$
$$\overline{+13x}$$

Therefore, $5x^2 + 13x + 6$ factors into $(x + 2)(5x + 3)$.

EXAMPLE 3   Factor $6x^2 - 19x + 15$.

SOLUTION   Make a blank outline:

$$(x - \quad)(x - \quad)$$ Because the last term is $+$ and the middle term is $-$, the signs in both binomials must be $-$

Select the correct pairs of factors by trial:

$$6 = 1 \cdot 6 \qquad (2x - 3) \qquad\qquad (3x - 5) \qquad 15 = 1 \cdot 15$$
$$2 \cdot 3 \qquad\qquad\quad - 9x \qquad\qquad\qquad\qquad 3 \cdot 5$$
$$-10x$$
$$\overline{-19x}$$

Therefore, $6x^2 - 19x + 15 = (2x - 3)(3x - 5)$.

EXAMPLE 4   Factor $12a^2 + 7ab - 10b^2$.

SOLUTION $(a + b)(a - b)$ Because the last term is $-$, the signs in the binomials must be different

$$12 = 1 \cdot 12 \qquad (4a + 5b) \qquad (3a - 2b) \qquad 10 = 1 \cdot 10$$
$$= 2 \cdot 6 \qquad \qquad {+}15ab \qquad \qquad = 2 \cdot 5$$
$$= 3 \cdot 4 \qquad \qquad -\,8ab$$
$$\qquad\qquad\qquad 7ab$$

Therefore, $12a^2 + 7ab - 10b^2 = (4a + 5b)(3a - 2b)$.

Master Product Method (Optional)

The master product method for factoring trinomials makes use of factoring by grouping.

Factoring a trinomial by the master product method

Arrange the trinomial in descending powers of one of the variables:

$$ax^2 + bx + c$$

1. Find the master product (MP) by multiplying the first and last coefficients of the trinomial being factored (MP $= a \cdot c$).
2. Write the pairs of factors of the master product (MP).
3. Choose the pair of factors whose sum is the coefficient of the middle term (b).
4. Rewrite the given trinomial, replacing the middle term by the sum of two terms whose coefficients are the pair of factors found in Step 3.
5. Factor the Step 4 expression by grouping.

Check your factoring by multiplying the binomial factors to see whether their product is the given trinomial.

EXAMPLE 5   Factor $5x + 2x^2 + 3$.

SOLUTION $2x^2 + 5x + 3$ Arranged in descending powers

Master product $= (2)(+3) = +6$

List the factors of 6:

$$1 \cdot 6$$
$$2 \cdot 3 \quad (+2) + (+3) = 5 \quad \text{Choose factors whose sum is the middle coefficient, 5}$$

Rewrite the middle term $5x$ as the sum $2x + 3x$. Therefore,

$$2x^2 + 5x + 3 = 2x^2 + 2x + 3x + 3$$
$$= 2x(x + 1) + 3(x + 1) \quad \text{Factor by grouping}$$
$$= (x + 1)(2x + 3)$$

EXAMPLE 6

Factor $3m^2 - 2m - 8$.

SOLUTION

$$3m^2 - 2m - 8$$

Master product $= (3)(-8) = -24$

$$\left.\begin{array}{l} 1 \cdot 24 \\ 2 \cdot 12 \\ 3 \cdot 8 \\ 4 \cdot 6 \end{array}\right\}$$ Factors of 24

$(+4) + (-6) = -2$ The middle coefficient

Therefore, $3m^2 - 2m - 8 = 3m^2 + 4m - 6m - 8$

$ = m(3m + 4) - 2(3m + 4)$ Factor by grouping

$ = (3m + 4)(m - 2)$

EXAMPLE 7

Factor $12a^2 + 7ab - 10b^2$.

SOLUTION

$$12a^2 + 7ab - 10b^2$$

Master product $= (12)(-10) = -120$

$$\left.\begin{array}{l} 1 \cdot 120 \\ 2 \cdot 60 \\ 3 \cdot 40 \\ 4 \cdot 30 \\ 5 \cdot 24 \\ 6 \cdot 20 \\ 8 \cdot 15 \end{array}\right\}$$ Factors of 120

$(-8) + (+15) = +7$ The middle coefficient

Therefore, $12a^2 + 7ab - 10b^2 = 12a^2 - 8ab + 15ab - 10b^2$

$ = 4a(3a - 2b) + 5b(3a - 2b)$ Factor by grouping

$ = (3a - 2b)(4a + 5b)$

  Note The master product method of factoring trinomials can also be used with trinomials whose leading coefficient is 1. However, we think the method presented in the previous section is shorter and simpler for trinomials of that type.

Exercises *11.5*

Factor each expression by the trial method or the master product method, if possible.

1. $3x^2 + 7x + 2$

2. $3x^2 + 5x + 2$

3. $5x^2 + 7x + 2$

4. $5x^2 + 11x + 2$

5. $7x + 4x^2 + 3$

6. $13x + 4x^2 + 3$

7. $5x^2 + 20x + 4$

8. $5x^2 + 11x + 4$

9. $5a^2 - 16a + 3$

10. $5m^2 - 8m + 3$

11. $3b^2 - 22b + 7$

12. $3u^2 - 10u + 7$

13. $5z^2 - 36z + 7$

14. $5z^2 - 12z + 7$

15. $3n^2 + 14n - 5$

16. $3n^2 - 2n - 5$

17. $5k^2 - 34k - 7$

18. $5k^2 + 2k - 7$

19. $7x^2 + 23xy + 6y^2$

20. $7a^2 + 43ab + 6b^2$

29. $35k^2 - 12k + 1$

30. $3e^2 - 20e - 7$

21. $7h^2 - 11hk + 4k^2$

22. $7h^2 - 16hk + 4k^2$

31. $7x^2 + 2x - 5$

32. $3a^2 + 8ab + 5b^2$

23. $3t^2 + 19tz - 6z^2$

24. $3w^2 - 11wx - 6x^2$

33. $4m^2 - 16mn + 7n^2$

34. $6s^2 - 17st - 5t^2$

25. $6 - 17v + 5v^2$

26. $6 - 11v + 5v^2$

35. $17u - 12 + 5u^2$

36. $8v^4 - 14v^2 - 15$

27. $6e^4 - 7e^2 - 20$

28. $10f^4 - 29f^2 - 21$

II.6 Factoring Completely

Consider the following example:

EXAMPLE 1   Factor $27x^2 - 12y^2$.

SOLUTION $27x^2 - 12y^2$ 3 is the GCF

This factor can be factored again:

$= 3(\ 9x^2 - 4y^2\ )$ $(9x^2 - 4y^2) = (3x + 2y)(3x - 2y)$

$= 3(3x + 2y)(3x - 2y)$

We say that $27x^2 - 12y^2 = 3(3x + 2y)(3x - 2y)$ has been *completely factored*. We will consider an expression to be completely factored if no more factoring can be done (by *any* method we have discussed).

Example 1 illustrates the importance of the following statement: *Always remove a greatest common factor first when possible.*

Factoring an expression completely

> **1.** Factor out any common factors *first*, if possible.
> **2.** If the expression has two terms, look for the difference of squares.
> **3.** If the expression has three terms, look for a factorable trinomial.
> **4.** If the expression has four terms, see whether it can be factored by grouping.
> **5.** *Check to see whether any factor already obtained can be factored again.*

EXAMPLE 2   Factor $3x^3 - 27x + 5x^2 - 45$.

SOLUTION $\underline{3x^3 - 27x} + \underline{5x^2 - 45}$ Factor by grouping

$= 3x(x^2 - 9) + 5(x^2 - 9)$

This factor can be factored again:

$= (\ x^2 - 9\ )(3x + 5)$ $x^2 - 9 = (x + 3)(x - 3)$

$= (x + 3)(x - 3)(3x + 5)$

EXAMPLE 3   Factor $5x^2 + 10x - 40$.

SOLUTION $5x^2 + 10x - 40$ 5 is the GCF

This factor can be factored again:

$= 5(\ x^2 + 2x - 8\ )$ $(x^2 + 2x - 8) = (x - 2)(x + 4)$

$= 5(x - 2)(x + 4)$

EXAMPLE 4 Factor $a^4 - b^4$.

SOLUTION $a^4 - b^4$

This factor can be factored again:
$(a^2 - b^2) = (a + b)(a - b)$

$= (a^2 + b^2)(\ a^2 - b^2\ )$

$= (a^2 + b^2)(a + b)(a - b)$

A Word of Caution   Although the difference of squares does factor, the sum of squares does not factor.

Try to factor $a^2 + b^2$:

$a^2 + b^2 \neq (a + b)(a + b)$ because $(a + b)(a + b) = a^2 + 2ab + b^2$

$a^2 + b^2 \neq (a - b)(a - b)$ because $(a - b)(a - b) = a^2 - 2ab + b^2$

$a^2 + b^2 \neq (a + b)(a - b)$ because $(a + b)(a - b) = a^2 - b^2$

Therefore, the *sum* of squares *cannot* be factored.

Exercises 11.6

Factor each expression *completely*, if possible.

1. $2x^2 - 8y^2$
2. $3x^2 - 27y^2$
3. $5a^4 - 20b^2$
4. $6m^2 - 54n^4$
5. $x^4 - y^4$
6. $a^4 - 16$
7. $4v^2 + 14v - 8$
8. $6v^2 - 27v - 15$
9. $8z^2 - 12z - 8$
10. $18z^2 - 21z - 9$
11. $12x^2 + 10x - 8$
12. $45x^2 - 6x - 24$
13. $ab^2 - 2ab + a$
14. $au^2 - 2au + a$
15. $3a^2 - 75b^2$
16. $4h^4 - 36b^2$
17. $m^4 - 1$
18. $10x^2 + 25x - 15$
19. $10y^2 + 14y - 5$
20. $30w^2 + 27w - 21$
21. $x^2 + 4$
22. $81c^4 + 16$
23. $4m^3n^3 - mn^5$
24. $2t^2r^4 - 18t^4$
25. $5wz^2 + 5w^2z - 10w^3$
26. $12x^2y - 42xy^2 + 36y^3$
27. $x^4 - 81$
28. $16y^8 - z^4$

29. $a^5b^2 - 4a^3b^4$
30. $x^2y^4 - 100x^4y^2$
31. $2ax^2 - 8a^3y^2$
32. $3b^2x^4 - 12b^2y^2$
33. $2u^3 + 2u^2v - 12uv^2$
34. $3m^3 - 3m^2n - 36mn^2$
35. $8h^3 - 20h^2k + 12hk^2$
36. $15h^2k - 35hk^2 + 10k^3$
37. $12 + 4x - 3x^2 - x^3$
38. $45 - 9z - 5z^2 + z^3$
39. $6my - 4nz + 15mz - 5zn$
40. $10xy + 5mn - 6xy - nm$
41. $6ac - 6bd + 6bc - 6ad$
42. $10cy - 6cz + 5dy - 3dz$
43. $24x^2 + 30x - 9$
44. $18x^2 - 24x - 10$
45. $45a^3 - 65a^2 + 20a$
46. $24y^4 - 28y^3 - 40y^2$
47. $x^3 - 4x^2 - 4x + 16$
48. $x^3 + x^2 - 9x - 9$
49. $27a^2 + 36ab + 12b^2$
50. $2x^4 - 16x^3 + 32x^2$
51. $x^4 - 2x^2 + 1$
52. $y^4 - 8y^2 + 16$
53. $8x^3y^2 + 4x^2y^3 - 12xy^4$
54. $3a^4b^2 - 9a^3b^3 - 12a^2b^4$

11.7 Solving Equations by Factoring

Factoring has many applications. In this section we use factoring to solve equations.

Equations with a first-degree term as the highest-degree term are called **first-degree**, or **linear**, **equations**. All of the equations solved so far have been first-degree equations. Equations with a second-degree term as the highest-degree term are called **second-degree**, or **quadratic**, **equations**.

EXAMPLE 1 Examples of linear and quadratic equations:

a. $5x - 3 = 0$ is a linear (or first-degree) equation in one variable.
b. $2x^2 - 4x + 7 = 0$ is a quadratic (or second-degree) equation in one variable.

Before we can solve quadratic equations, we need the zero factor property, which states that if the product of two factors is zero, one or both of the factors must be zero.

Zero factor property

If	$a \cdot b = 0$
Then	$a = 0$ or $b = 0$

EXAMPLE 2 Solve $(x - 1)(x - 2) = 0$.

SOLUTION Because $(x - 1)(x - 2) = 0$,

$$(x - 1) = 0 \quad \text{or} \quad (x - 2) = 0$$

$$\text{If} \quad \begin{aligned} x - 1 &= 0 \\ +1 \;\; &\;\; +1 \\ \hline \end{aligned} \qquad \text{If} \quad \begin{aligned} x - 2 &= 0 \\ +2 \;\; &\;\; +2 \\ \hline \end{aligned}$$

$$\text{then} \quad x = 1 \qquad \qquad \text{then} \quad x = 2$$

Therefore, 1 and 2 are solutions for the equation $(x - 1)(x - 2) = 0$.

✓ **Check for $x = 1$**

$(x - 1)(x - 2) = 0$

$(1 - 1)(1 - 2) \overset{?}{=} 0$

$(0)(-1) \overset{?}{=} 0$

$0 = 0$

Check for $x = 2$

$(x - 1)(x - 2) = 0$

$(2 - 1)(2 - 2) \overset{?}{=} 0$

$(1)(0) \overset{?}{=} 0$

$0 = 0$

The same method can be used when a product of more than two factors is equal to zero.

EXAMPLE 3 Solve $2x(x - 3)(x + 4) = 0$.

SOLUTION $2x(x - 3)(x + 4) = 0$

$$\begin{aligned} 2x &= 0 \\ \frac{2x}{2} &= \frac{0}{2} \\ x &= 0 \end{aligned} \qquad \begin{aligned} x - 3 &= 0 \\ +3 \;\; &\;\; +3 \\ \hline x &= 3 \end{aligned} \qquad \begin{aligned} x + 4 &= 0 \\ -4 \;\; &\;\; -4 \\ \hline x &= -4 \end{aligned}$$

We will use the zero factor property to solve quadratic and higher-degree equations.

Solving an equation by factoring	**1.** Write all nonzero terms on one side of the equation by adding the same expression to both sides. *Only zero* must remain on the other side. Then arrange the polynomial in descending powers.
	2. Factor the polynomial.
	3. Set each factor equal to zero, and solve for the variable.
	4. Check apparent solutions in the *original* equation.

EXAMPLE 4   Solve $4x^2 - 16x = 0$.

SOLUTION   The polynomial must be factored first.

$$4x^2 - 16x = 0 \qquad \text{A quadratic equation}$$

$$4x(x - 4) = 0 \qquad \text{Factor}$$

$$
\begin{array}{c|c}
4x = 0 & x - 4 = 0 \quad \text{Set each factor equal to 0} \\
\dfrac{4x}{4} = \dfrac{0}{4} & \underline{+4 \quad +4} \\
x = 0 & x \qquad = 4
\end{array}
$$

EXAMPLE 5   Solve $x^2 - x - 6 = 0$.

SOLUTION $x^2 - x - 6 = 0$

$$(x + 2)(x - 3) = 0 \qquad \text{Factor}$$

$$
\begin{array}{c|c}
x + 2 = 0 & x - 3 = 0 \quad \text{Set each factor equal to 0} \\
\underline{-2 \quad -2} & \underline{+3 \quad +3} \\
x \quad = -2 & x \quad = 3
\end{array}
$$

EXAMPLE 6   Solve $3x^2 + 12 = 12x$.

SOLUTION   Subtract $12x$ from both sides to make the right side zero, and arrange the polynomial in descending powers.

$$3x^2 + 12 = 12x$$

$$3x^2 - 12x + 12 = 0 \qquad \text{Subtract } 12x \text{ from both sides}$$

$$3(x^2 - 4x + 4) = 0 \qquad \text{Factor out the GCF}$$

$$3(x - 2)(x - 2) = 0 \qquad \text{Factor the trinomial}$$

$$
\begin{array}{c|c|c}
3 \neq 0 & x - 2 = 0 & x - 2 = 0 \\
 & x = 2 & x = 2
\end{array}
$$

 Note   In Example 6, the constant factor 3 has no effect on the solution. The solution $x = 2$ is called a double root.

Sometimes it is more convenient to get the polynomal on the *right* side of the equal sign and the zero on the *left* side.

EXAMPLE 7   Solve $4 - x = 3x^2$.

SOLUTION $0 = 3x^2 + x - 4$

$0 = (x - 1)(3x + 4)$

In this case it's convenient to arrange all nonzero terms on the *right* side in descending order

$x - 1 = 0$	$3x + 4 = 0$
$x = 1$	$3x = -4$
	$x = -\dfrac{4}{3}$

EXAMPLE 8   Solve $\dfrac{x - 1}{x - 3} = \dfrac{12}{x + 1}$.

SOLUTION
$$\frac{x - 1}{x - 3} = \frac{12}{x + 1}$$

This is a proportion

$$(x - 1)(x + 1) = 12(x - 3)$$

Product of means = product of extremes

$$x^2 - 1 = 12x - 36$$

$$x^2 - 12x + 35 = 0$$

$$(x - 7)(x - 5) = 0$$

$x - 7 = 0$	$x - 5 = 0$
$\underline{+\ 7 \quad + 7}$	$\underline{+\ 5 \quad + 5}$
$x = 7$	$x = 5$

✖ **A Word of Caution**   The product must equal *zero*, or else no conclusions can be drawn about the factors. Suppose

$$(x - 1)(x - 3) = \boxed{8}$$

⌐— No conclusion can be drawn because the product $\neq 0$

A common mistake is to think that

if $\qquad\qquad\qquad\qquad (x - 1)(x - 3) = 8$

then $\qquad\qquad x - 1 = 8 \qquad\qquad x - 3 = 8$

$\qquad\qquad\qquad\qquad\quad x = 9 \qquad\qquad\quad x = 11$

Both of these assumptions are incorrect.

✓ **Check** $x = 9$	**Check** $x = 11$
$(x - 1)(x - 3) = 8$	$(x - 1)(x - 3) = 8$
$(9 - 1)(9 - 3) \overset{?}{=} 8$	$(11 - 1)(11 - 3) \overset{?}{=} 8$
$8 \ \cdot \ 6 \ \overset{?}{=} 8$	$10 \ \cdot \ 8 \ \overset{?}{=} 8$
$48 \neq 8$	$80 \neq 8$

The correct solution is

$$(x - 1)(x - 3) = 8$$
$$x^2 - 4x + 3 = 8 \qquad \text{Remove parentheses}$$
$$x^2 - 4x - 5 = 0 \qquad \text{Subtract 8 from both sides}$$
$$(x - 5)(x + 1) = 0 \qquad \text{Factor}$$

$x - 5 = 0$	$x + 1 = 0$	Set each factor
$x = 5$	$x = -1$	equal to 0

Exercises 11.7

Solve each equation.

1. $(x - 5)(x + 4) = 0$
2. $(x + 7)(x - 2) = 0$
3. $3x(x - 4) = 0$
4. $5x(x + 6) = 0$
5. $(x + 10)(2x - 3) = 0$
6. $(x - 8)(3x + 2) = 0$
7. $3(x + 2)(x - 1) = 0$
8. $5(x + 6)(x + 3) = 0$
9. $2x(3x - 2)(5x + 9) = 0$
10. $4x(2x - 1)(4x - 3) = 0$
11. $x^2 + 9x + 8 = 0$
12. $x^2 + 6x + 8 = 0$
13. $x^2 - x - 12 = 0$
14. $x^2 + x - 12 = 0$
15. $x^2 - 18 = 7x$
16. $x^2 - 20 = 8x$
17. $6x^2 - 10x = 0$
18. $6y^2 - 21y = 0$
19. $4w^2 = 24w$
20. $5m^2 = 20m$
21. $x^2 - 36 = 0$
22. $x^2 - 49 = 0$
23. $2x^2 = 50$
24. $5x^2 = 45$
25. $5a^2 = 16a - 3$
26. $3z^2 = 22z - 7$
27. $3u^2 = 2u + 5$
28. $5k^2 = 34k + 7$

29. $18 = m^2 + 3m$
30. $5w = w^2 - 24$
31. $7n + 6 = 3n^2$
32. $13x - 3 = 4x^2$
33. $9x = 3x^2$
34. $12t = 6t^2$
35. $2x^2 = 12x - 18$
36. $5x^2 + 80 = 40x$
37. $6 = 6x^2 + 9x$
38. $16x - 10 = 6x^2$
39. $(y - 3)(y - 6) = -2$
40. $(x - 3)(x - 5) = 3$
41. $(x - 2)(x - 3) = 2$
42. $u(u - 9) = -14$
43. $x(x - 4) = 12$
44. $x(x - 2) = 15$
45. $2x^3 + x^2 = 3x$
46. $4x^3 = 10x - 18x^2$
47. $2a^3 - 10a^2 = 0$
48. $4b^3 - 24b^2 = 0$
49. $\dfrac{x - 1}{x + 4} = \dfrac{3}{x}$
50. $\dfrac{2x + 5}{3x} = \dfrac{x}{2x - 5}$
51. $\dfrac{x + 2}{x + 1} = \dfrac{10}{x + 5}$

11.8 Word Problems Solved by Factoring

Number Problems

EXAMPLE 1 The difference of two numbers is 3. Their product is 10. What are the two numbers?

SOLUTION Let $\quad x = $ smaller number

Then $\quad x + 3 = $ larger number $\qquad$ Since their difference is 3

Their product is 10

$$x(x + 3) \quad = \quad 10 \qquad \textit{Remember: It is incorrect to say}$$
$$\textit{that } x = 10 \textit{ or } x + 3 = 10$$

$$x^2 + 3x = 10$$

$$x^2 + 3x - 10 = 0$$

$$(x - 2)(x + 5) = 0$$

$x - 2 = 0$	$x + 5 = 0$	
$x = 2$	$x = -5$	Smaller number
$x + 3 = 5$	$x + 3 = -2$	Larger number

Therefore, the numbers 2 and 5 are one solution, and the numbers -5 and -2 are another solution.

 A Word of Caution   In Example 1, a mistake students commonly make after solving for x is to say that the answer is 2 and -5. This is incorrect, because the difference between 2 and -5 is 7 (not 3) and the product of 2 and -5 is -10 (not 10).

Two possible values for the *smaller number* are $x = 2$ and $x = -5$. To find the larger corresponding numbers, we must add 3 to each of these values. There are two pairs of numbers that are solutions; the first is 2 and 5, and the second solution is -5 and -2.

Consecutive Integer Problems

EXAMPLE 2   Find two consecutive odd integers such that the sum of their squares is 34.

SOLUTION   Let $x = $ first consecutive odd
Then $x + 2 = $ second consecutive odd

The sum of their squares	is	34

$$x^2 + (x + 2)^2 = 34$$

$$x^2 + x^2 + 4x + 4 = 34$$

$$2x^2 + 4x - 30 = 0$$

$$2(x^2 + 2x - 15) = 0$$

$$2(x - 3)(x + 5) = 0$$

$2 \neq 0$	$x - 3 = 0$	$x + 5 = 0$	
	$x = 3$	$x = -5$	First odd integer
	$x + 2 = 5$	$x + 2 = -3$	Second odd integer

There are two solutions. One solution is 3 and 5, and the other is -5 and -3.

Geometry Problems

EXAMPLE 3   The length of a rectangle is 2 inches more than its width. The area of the rectangle is 24 square inches. Find the length and the width.

S O L U T I O N There are two unknown numbers, the width and the length.

Let w = width
Then $w + 2$ = length

$$\boxed{\text{Area} = 24 \quad w}$$
$$w + 2$$

$$\text{Area} = \text{width} \times \text{length}$$
$$24 = w(w + 2)$$
$$24 = w^2 + 2w$$
$$0 = w^2 + 2w - 24$$
$$0 = (w - 4)(w + 6)$$

$$w - 4 = 0 \qquad \bigm| \qquad w + 6 = 0$$

Width $\qquad w = 4 \qquad \bigm| \qquad w = -6 \longleftarrow$ −6 is not a solution because length cannot be negative

Length $\quad w + 2 = 6$

Therefore, the rectangle has a width of 4 inches and a length of 6 inches.

Exercises *11.8*

In the following exercises, (a) represent the unknown numbers by variables, (b) set up an equation and solve it, and (c) solve the problem.

1. The difference of two numbers is 5. Their product is 14. Find the numbers.

2. The difference of two numbers is 6. Their product is 27. Find the numbers.

3. The sum of two numbers is 12. Their product is 35. Find the numbers.

4. The sum of two numbers is −4. Their product is −12. Find the numbers.

5. Find two consecutive odd integers whose product is 63.

6. Find two consecutive even integers whose product is 48.

7. Find two consecutive integers whose product is 11 more than their sum.

8. Find two consecutive integers whose product is 5 more than their sum.

9. Find three consecutive integers such that the product of the first two is 7 more than the third integer.

10. Find three consecutive integers such that the product of the first two is 23 more than the third integer.

11. Find two consecutive integers such that the sum of their squares is 25.

12. Find two consecutive even integers such that the sum of their squares is 20.

13. One number is three times another number. The sum of their squares is 40. Find the numbers.

14. One number is three more than another number. The sum of their squares is 89. Find the numbers.

15. The length of a rectangle is 5 feet more than its width. Its area is 84 square feet. Find the length and the width.

16. The width of a rectangle is 3 feet less than its length. Its area is 28 square feet. Find the length and the width.

17. The base of a triangle is 3 inches more than its altitude. Its area is 20 square inches. Find the base and the altitude.

18. The sum of the base and altitude of a triangle is 15 centimeters. The area of the triangle is 27 square centimeters. Find the base and the altitude.

19. The width of a rectangle is 4 yards less than its length. The area is 17 more (numerically) than its perimeter. Find the length and the width.

20. The area of a square is twice its perimeter (numerically). What is the length of its side?

21. One square has a side 3 centimeters shorter than the side of a second square. The area of the larger square is 4 times as great as the area of the smaller square. Find the length of the side of each square.

22. One square has a side 2 meters longer than the side of a second square. The area of the larger square is 16 times as great as the area of the smaller square. Find the length of the side of each square.

23. The fourth term of a proportion is 7 more than its first term. Find the proportion if its second term is 3 and its third term is 6.

24. The third term of a proportion is 2 more than its second term. Find the proportion if its first term is 3 and its fourth term is 8.

Chapter 11   R E V I E W

Prime Number
11.1
A **prime number** is a positive integer greater than 1 that can be divided evenly only by itself and 1. A prime number has no factors other than itself and 1.

Composite Number
11.1
A **composite number** is a positive integer greater than 1 that can be exactly divided by some integer other than itself and 1. A composite number has factors other than itself and 1.

Prime Factorization
11.1
The **prime factorization** of a positive integer is the indicated product of all its factors that are themselves prime numbers.

Methods of Factoring
11.1   1. Greatest common factor (GCF)

11.3   2. Difference of two squares: $a^2 - b^2 = (a + b)(a - b)$

    3. Trinomial

11.4   a. Leading coefficient equal to 1

11.5   b. Leading coefficient greater than 1

11.2   4. Grouping

Factoring Completely
11.6
1. Factor out any common factors first, if possible.
2. If the expression has two terms, look for the difference of squares.
3. If the expression has three terms, look for a factorable trinomial.
4. If the expression has four terms, see whether it can be factored by grouping.
5. Check to see whether any factor already obtained can be factored again.

Solving an Equation by Factoring
11.7
1. Write *all* nonzero terms on one side of the equation by adding the same expression to both sides. *Only zero must remain on the other side.* Then arrange the polynomial in descending powers.
2. Factor the polynomial.
3. Set each factor equal to zero, and solve for the variable.
4. Check apparent solutions in the original equation.

Chapter 11   R E V I E W   E X E R C I S E S

In Exercises 1–5, find the prime factorization of each number.

1. 12
2. 36
3. 31
4. 42
5. 210

In Exercises 6–23, factor each expression completely, if possible.

6. $8x - 4$
7. $m^2 - 4$
8. $x^2 + 10x + 21$
9. $-2x^3 + 18x$
10. $z^2 - 7z - 18$
11. $4x^2 - 25x + 6$
12. $9k^2 - 144$
13. $8 - 2a^2$
14. $ab + 2b - a - 2$
15. $xy - 4x + 3y - 12$
16. $3x^2 + 4x - 4$
17. $15u^2v - 3uv$
18. $a^4 - b^4$
19. $x^8 - 1$
20. $x^4 - 5x^2 + 4$
21. $4x^2y - 8xy^2 + 4xy$
22. $15a^2 + 15ab - 30b^2$
23. $6u^3v^2 - 9uv^3 - 12uv$

In Exercises 24–32, solve each of the equations.

24. $(x - 5)(x + 3) = 0$
25. $x^2 - 5x - 14 = 0$
26. $m^2 = 18 + 3m$
27. $4x^2 = 36$
28. $4x^2 = 36x$
29. $6e^2 = 13e + 5$
30. $(x + 5)(x - 1) = 7$
31. $6x^3 = 9x^2 + 6x$

32. $\dfrac{x - 2}{9} = \dfrac{4}{x + 7}$

33. The difference of two numbers is 3. Their product is 28. Find the numbers.

34. Find three consecutive integers such that the product of the first two minus the third is 7.

35. The width of a rectangle is 3 less than its length. Its area is 40. Find the length and the width of the rectangle.

36. One side of a square is 6 feet longer than the side of a second square. The area of the larger square is 16 times as great as the area of the smaller square. Find the length of the side of each square.

37. The area of a square is 3 times its perimeter. What is the length of its side?

38. The length of a rectangle is 6 yards more than its width. The area is 12 more (numerically) than its perimeter. Find the length and the width.

 ## Chapter II Critical Thinking and Writing Problems

Answer Problems 1–7 in your own words, using complete sentences.

1. Explain the meaning of the GCF.

2. Explain why $a^2 + b^2$ cannot be factored.

3. Explain why $x^2 + x + 1$ cannot be factored.

4. Explain why $x^2 - 5x + 6 \neq (x - 6)(x + 1)$.

5. Explain why $x^2 - 5x - 6 \neq (x - 2)(x - 3)$.

6. Explain how to determine whether an equation is first-degree or second-degree.

7. Explain how to solve a second-degree equation.

Are the following expressions factored *completely*? Explain your answer.

8. $3x^2 - 6x - 24 = (3x + 6)(x - 4)$

9. $x^2 - 16 = (x^2 + 4)(x + 2)(x - 2)$

10. $x^3 - 2x^2 + 5x - 10 = x^2(x - 2) + 5(x - 2)$

11. $36x^2 - 4 = (6x + 2)(6x - 2)$

Each of the following problems has an error. Find the error, and in your own words, explain why it is wrong. Then work the problem correctly.

12. Factor $12x^4 - 16x^3 + 2x^2$.
$$12x^4 - 16x^3 + 2x^2 = 2x^2(6x^2 - 8x)$$

13. Factor $3x^2 - 75$.
$$3x^2 - 75 = 3(x^2 - 25)$$
$$= 3(x + 5)(x - 5)$$

$3 \neq 0$ $\quad$ | $\quad$ $x + 5 = 0$ $\quad$ | $\quad$ $x - 5 = 0$
$\quad$ | $\quad$ $x = -5$ $\quad$ | $\quad$ $x = 5$

14. Solve $x^2 - 3x + 2 = 6$.
$$x^2 - 3x + 2 = 6$$
$$(x - 1)(x - 2) = 6$$
$x - 1 = 6$ $\quad$ | $\quad$ $x - 2 = 6$
$\quad x = 7$ $\quad$ | $\quad$ $\quad x = 8$

15. Solve $2x^2 = 12x$.
$$2x^2 = 12x$$
$$\frac{2x^2}{2x} = \frac{12x}{2x}$$
$$x = 6$$

Chapter II   DIAGNOSTIC TEST

Allow yourself about an hour to do these problems. Complete solutions for all problems, together with section references, are given in the answer section at the end of the book.

1. State whether each of the following numbers is prime or composite.

    a. 18   b. 21   c. 31

In Problems 2 and 3, find the prime factorization of each number.

2. 45   3. 160

In Problems 4–18, factor each expression completely, if possible.

4. $5x + 10$ 5. $3x^3 - 6x^2$

6. $16x^2 - 49y^2$ 7. $2z^2 - 8$

8. $z^2 + 6z + 8$ 9. $m^2 + m - 6$

10. $5w^2 - 12w + 7$ 11. $3v^2 + 14v - 5$

12. $x^2 + 25$ 13. $4x^2 - 11x + 6$

14. $x^2 + 6x - 8$ 15. $5n - mn - 5 + m$

16. $6h^2k - 8hk^2 + 2k^3$ 17. $x^4 - 8x^2 - 9$

18. $24a^2 + 2a - 12$

In Problems 19–23, solve each equation by factoring.

19. $x^2 - 12x + 20 = 0$ 20. $3x^2 = 8 - 2x$ 21. $3x^2 = 12x$

22. $36y = 18y^2 + 18$ 23. $(x + 1)(x + 3) = 24$

24. The length of a rectangle is 3 feet more than its width. Its area is 28 square feet. Find the length and the width.

25. Find two consecutive odd integers such that their product is 7 more than their sum.

Algebraic Fractions

CHAPTER
12

In this chapter we define algebraic fractions. We then discuss how to perform necessary operations with them and how to solve equations and word problems involving them. A knowledge of the different methods of factoring is essential in work with algebraic fractions.

12.1　Simplifying Algebraic Fractions

A **simple algebraic fraction** (also called a **rational expression**) is an algebraic expression of the form

$$\text{Terms of fraction} \quad \begin{cases} \dfrac{P}{Q} \end{cases} \quad \begin{array}{l} \text{Numerator} \\ \text{Fraction bar} \\ \text{Denominator (cannot be zero)} \end{array}$$

where P and Q are polynomials and $Q \neq 0$.

Excluded Values

Any value of the variable (or variables) that makes the denominator Q equal to zero must be excluded.

Finding excluded values

> If the denominator contains a variable, set the denominator equal to zero and solve the equation. The solutions are the values of the variable that must be excluded.

EXAMPLE 1　Find the excluded values.

a. $\dfrac{x}{3}$

SOLUTION　There is no variable in the denominator; therefore, there are *no* excluded values.

b. $\dfrac{x - 5}{x - 2}$

SOLUTION　Set the denominator equal to zero and solve.

$$x - 2 = 0$$
$$x = 2$$

Therefore, the excluded value is 2; that is, $x \neq 2$.

c. $\dfrac{x^2 + 2}{x^2 - 3x - 4}$

SOLUTION　Set the denominator equal to zero and solve.

$$x^2 - 3x - 4 = 0$$
$$(x - 4)(x + 1) = 0$$

$$x - 4 = 0 \qquad \qquad x + 1 = 0$$
$$x = 4 \qquad \qquad x = -1$$

Therefore, the excluded values are 4 and -1; that is, $x \neq 4$ and $x \neq -1$.

Excluded values will be discussed again when we solve equations having fractions.

The Three Signs of a Fraction

Every fraction has three signs associated with it.

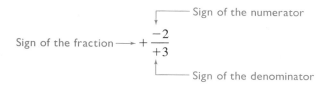

Rule for signs of a fraction

If any two of the three signs of a fraction are changed, the value of the fraction is unchanged.

We illustrate the rule for signs of a fraction by applying it to the fraction $+\dfrac{+8}{+4}$:

$$+\frac{-8}{-4} = +\left(\frac{-8}{-4}\right) = +(+2) = 2$$

Here, we changed the signs of the numerator and the denominator

$$-\frac{-8}{+4} = -\left(\frac{-8}{+4}\right) = -(-2) = 2$$

Here, we changed the signs of the fraction and the numerator

$$-\frac{+8}{-4} = -\left(\frac{+8}{-4}\right) = -(-2) = 2$$

Here, we changed the signs of the fraction and the denominator

This rule of signs is helpful in simplifying some expressions with fractions.

EXAMPLE 2 Examples of changing two signs of a fraction to obtain an equivalent fraction:

a. $-\dfrac{-5}{xy} = +\dfrac{+5}{xy} = \dfrac{5}{xy}$

Here, we changed the signs of the fraction and the numerator

b. $-\dfrac{1}{2-x} = +\dfrac{1}{-(2-x)} = \dfrac{1}{-2+x} = \dfrac{1}{x-2}$

Here, we changed the signs of the fraction and the denominator

The simplification in Example 2b can be explained as follows:

$$-(2 - x) = (-1)(2 - x) = -2 + x = x - 2 \quad \text{Distributive rule}$$

This fact is summarized below.

The negative of a binomial difference interchanges the terms:

Negative

$$-(\ b - a\ ) = a - b$$

Terms interchanged

EXAMPLE 3

a. $-(x - y) = y - x$
b. $-(d - 5) = 5 - d$

Reducing Fractions to Lowest Terms

We reduce fractions to lowest terms in algebra for the same reason we do in arithmetic: It makes them simpler and easier to work with. After this section, it is understood that *all fractions are to be reduced to lowest terms* (unless otherwise indicated).

A fraction can be reduced only when the numerator and the denominator have a common factor. In arithmetic, we can reduce a fraction as follows:

2 is a factor of the numerator

$$\frac{6}{8} = \frac{2 \cdot 3}{2 \cdot 4} = \frac{2 \cdot 3}{2 \cdot 4} = \frac{3}{4} \quad \begin{array}{l}\text{Both numerator and}\\ \text{denominator were divided by 2}\end{array}$$

2 is a factor of the denominator

This work is usually shown in a shortened form:

$$\frac{\overset{3}{6}}{\underset{4}{8}} = \frac{3}{4}$$

The procedure for reducing fractions in algebra is the same as the method used in arithmetic.

Reducing a fraction to lowest terms

1. Factor the numerator and denominator completely.
2. Divide the numerator and denominator by all factors common to both.

EXAMPLE 4

Examples of reducing algebraic fractions to lowest terms:

a. $\dfrac{4x^2y}{2xy} = \dfrac{\overset{2}{4}\overset{x}{x^2}\overset{1}{y}}{\underset{1}{2}\underset{1}{x}\underset{1}{y}} = \dfrac{2x}{1} = 2x \quad \begin{array}{l}\text{Here, the numerator and denominator}\\ \text{are already factored}\end{array}$

b. $\dfrac{x - 3}{x^2 - 9} = \dfrac{\overset{1}{(x - 3)}}{(x + 3)\underset{1}{(x - 3)}} = \dfrac{1}{x + 3}$

 Note A factor of 1 will always remain in the numerator.

c. $\dfrac{x^2 - 4x - 5}{x^2 + 5x + 4} = \dfrac{(\cancel{x+1})(x - 5)}{(\cancel{x+1})(x + 4)} = \dfrac{x - 5}{x + 4}$

d. $\dfrac{3x^2 - 5xy - 2y^2}{6x^3y + 2x^2y^2} = \dfrac{(x - 2y)(\cancel{3x+y})}{2x^2y(\cancel{3x+y})} = \dfrac{x - 2y}{2x^2y}$

e. $\dfrac{y - x}{x - y} = \dfrac{-(\cancel{x-y})}{\cancel{x-y}} = \dfrac{-1}{1} = -1$

 Note   Since $y - x$ is the negative of $x - y$, they may be canceled directly if a factor of -1 is used.

$$\dfrac{y - x}{x - y} = \dfrac{\overset{-1}{\cancel{y-x}}}{\underset{1}{\cancel{x-y}}} = \dfrac{-1}{1} = -1$$

f. $\dfrac{a^2 - 3ab + 2b^2}{b^2 - a^2} = \dfrac{(\overset{-1}{\cancel{a-b}})(a - 2b)}{(\underset{1}{\cancel{b-a}})(b + a)} = \dfrac{-1(a - 2b)}{b + a} = \dfrac{2b - a}{b + a}$

g. $\dfrac{x + 3}{x + 6}$ cannot be reduced.   Neither x nor 3 is a factor of the numerator or denominator

h. $\dfrac{x + y}{x}$ cannot be reduced.   See the following Word of Caution

(arrow) — x is *not a factor* of the numerator

✗   *A Word of Caution*   A common error made in reducing fractions is to forget that the number the numerator and denominator are divided by *must be a factor of both* (see Examples 4g and 4h).

3 is *not* a factor of the numerator

$$\dfrac{\cancel{3} + 2}{\cancel{3}} = 2 \quad \textit{Incorrect} \text{ reduction}$$

The above reduction is incorrect because

$$\dfrac{3 + 2}{3} = \dfrac{5}{3} \neq 2$$

Exercises   *12.1*

In Exercises 1–12, what value(s) of the variable (if any) must be excluded?

1. $\dfrac{3x + 4}{x - 2}$ **2.** $\dfrac{5 - 4x}{x + 3}$ **3.** $\dfrac{x}{10}$

4. $\dfrac{y}{20}$ **5.** $\dfrac{3 + x}{(x - 1)(x + 2)}$ **6.** $\dfrac{x - 4}{3x(x - 2)}$

7. $\dfrac{z}{4}$ **8.** $\dfrac{x - 1}{3}$ **9.** $\dfrac{x^2 + 4}{x^2 - x - 2}$

10. $\dfrac{x^2 - 2}{x^2 - 4}$ **11.** $\dfrac{x - 6}{2x^2 + 6x}$ **12.** $\dfrac{x + 2}{5x^2 - 5x}$

In Exercises 13–20, use the rule about the three signs of a fraction to find the missing term.

13. $-\dfrac{5}{6} = \dfrac{?}{6}$ **14.** $\dfrac{2}{-x} = \dfrac{?}{x}$

15. $\dfrac{5}{-y} = \dfrac{-5}{?}$ **16.** $\dfrac{2 - x}{-9} = \dfrac{?}{9}$

17. $\dfrac{7}{5 - x} = \dfrac{?}{x - 5}$ **18.** $\dfrac{1 - x}{-8} = \dfrac{?}{8}$

19. $\dfrac{6 - y}{5} = \dfrac{y - 6}{?}$ **20.** $-\dfrac{3}{x - 4} = \dfrac{3}{?}$

In Exercises 21–58, reduce each fraction to lowest terms.

21. $\dfrac{9}{12}$

22. $\dfrac{8}{14}$

23. $\dfrac{6ab^2}{3ab}$

24. $\dfrac{10m^2n}{5mn}$

25. $\dfrac{4x^2y}{2xy}$

26. $\dfrac{12x^3y}{4xy}$

27. $\dfrac{5x - 10}{x - 2}$

28. $\dfrac{3x + 12}{x + 4}$

29. $\dfrac{5x - 6}{6 - 5x}$

30. $\dfrac{4 - 3z}{3z - 4}$

31. $\dfrac{5x^2 + 30x}{10x^2 - 40x}$

32. $\dfrac{4x^3 - 4x^2}{12x^2 - 12x}$

33. $\dfrac{2 + 4}{4}$

34. $\dfrac{3 + 9}{3}$

35. $\dfrac{5 + x}{5}$

36. $\dfrac{x + 8}{8}$

37. $\dfrac{3xy}{6x^2 + 3xy}$

38. $\dfrac{5ab}{5ab + 5a^2}$

39. $\dfrac{x^2 - 1}{x + 1}$

40. $\dfrac{x^2 - 4}{x - 2}$

41. $\dfrac{3x - 12}{4 - x}$

42. $\dfrac{2x - 4}{2 - x}$

43. $\dfrac{x^2 - 11x + 30}{x^2 - 9x + 20}$

44. $\dfrac{x^2 - 9}{x^2 + 5x + 6}$

45. $\dfrac{x^2 - 16}{x^2 - x - 12}$

46. $\dfrac{x^2 + x - 20}{x^2 + 2x - 15}$

47. $\dfrac{6x^2 - x - 2}{10x^2 + 3x - 1}$

48. $\dfrac{8x^2 - 10x - 3}{12x^2 + 11x + 2}$

49. $\dfrac{x^2 - y^2}{(x + y)^2}$

50. $\dfrac{a^2 - 9b^2}{(a - 3b)^2}$

51. $\dfrac{(a - 2b)(b - a)}{(2a + b)(a - b)}$

52. $\dfrac{8(n - 2m)}{(3n + m)(2m - n)}$

53. $\dfrac{2y^2 + xy - 6x^2}{3x^2 + xy - 2y^2}$

54. $\dfrac{10y^2 + 11xy - 6x^2}{4x^2 - 4xy - 15y^2}$

55. $\dfrac{8x^2 - 2y^2}{2ax - ay + 2bx - by}$

56. $\dfrac{3x^2 - 12y^2}{ax + 2by + 2ay + bx}$

57. $\dfrac{xy + 2x + 3y + 6}{xy + x + 3y + 3}$

58. $\dfrac{ax - 2x + 3ay - 6y}{3ax - 6x + ay - 2y}$

12.2   Multiplying and Dividing Fractions

Multiplying Fractions

In arithmetic, we can multiply two fractions as follows:

$$\frac{4}{9} \cdot \frac{3}{8} = \frac{\overset{1}{4} \cdot \overset{1}{3}}{\underset{3}{9} \cdot \underset{2}{8}} = \frac{1}{6}$$

We multiply fractions in algebra the same way we multiply fractions in arithmetic.

Multiplying fractions

1. Factor the numerator and denominator of the fractions.
2. Divide the numerator and denominator by all factors common to both.
3. The answer is the product of the factors remaining in the numerator divided by the product of the factors remaining in the denominator:

$$\frac{P}{Q} \cdot \frac{R}{S} = \frac{P \cdot R}{Q \cdot S}$$

where $Q \neq 0$ and $S \neq 0$.

EXAMPLE 1   Examples of multiplying algebraic fractions:

a. $\dfrac{1}{m^2} \cdot \dfrac{m}{5} = \dfrac{1 \cdot \overset{1}{m}}{\underset{m}{m^2} \cdot 5} = \dfrac{1}{5m}$ A factor of 1 will always remain in the numerator

b. $\dfrac{2y^3}{3x^2} \cdot \dfrac{12x}{5y^2} = \dfrac{2\overset{y}{y^3} \cdot \overset{4}{12}\overset{1}{x}}{\underset{1}{3}\underset{x}{x^2} \cdot 5\underset{1}{y^2}} = \dfrac{8y}{5x}$

c. $\dfrac{x}{2x - 6} \cdot \dfrac{4x - 12}{x^2} = \dfrac{x}{2(x - 3)} \cdot \dfrac{4(x - 3)}{x^2} = \dfrac{\overset{1}{x} \cdot \overset{2}{4}(\overset{1}{x - 3})}{\underset{1}{2}(\underset{1}{x - 3}) \cdot \underset{x}{x^2}} = \dfrac{2}{x}$

d. $\dfrac{10xy^3}{x^2 - y^2} \cdot \dfrac{2x^2 + xy - y^2}{15x^2y} = \dfrac{\overset{2 \; 1 \; y^2}{\cancel{10xy^3}}}{\cancel{(x+y)}(x - y)} \cdot \dfrac{(\overset{1}{\cancel{x+y}})(2x - y)}{\underset{3 \; x \; 1}{\cancel{15x^2y}}}$

$$= \dfrac{2y^2(2x - y)}{3x(x - y)}$$

Dividing Fractions

In arithmetic, we can divide two fractions as follows:

$$\dfrac{3}{5} \div \dfrac{4}{7} = \dfrac{3}{5} \cdot \boxed{\dfrac{7}{4}} = \dfrac{3 \cdot 7}{5 \cdot 4} = \dfrac{21}{20}$$

└── Invert the second fraction and multiply

This method works because

$$\dfrac{3}{5} \div \dfrac{4}{7} = \dfrac{\dfrac{3}{5}}{\dfrac{4}{7}} = \dfrac{\dfrac{3}{5}}{\dfrac{4}{7}} \cdot \dfrac{\boxed{\dfrac{7}{4}}}{\dfrac{7}{4}} = \dfrac{\dfrac{3}{5} \cdot \dfrac{7}{4}}{\dfrac{4}{7} \cdot \dfrac{7}{4}} = \dfrac{\dfrac{3}{5} \cdot \dfrac{7}{4}}{1} = \dfrac{3}{5} \cdot \dfrac{7}{4}$$

└── The value of this fraction is 1

Therefore, $\dfrac{3}{5} \div \dfrac{4}{7} = \dfrac{3}{5} \cdot \dfrac{7}{4}$.

We divide algebraic fractions the same way we divide fractions in arithmetic.

Dividing fractions

Invert the second fraction and multiply:

$$\dfrac{P}{Q} \div \dfrac{R}{S} = \dfrac{P}{Q} \cdot \dfrac{S}{R}$$

First fraction ──┘ └── Second fraction

where $Q \neq 0$, $R \neq 0$, and $S \neq 0$.

EXAMPLE 2

Examples of dividing algebraic fractions:

a. $\dfrac{4}{3x} \div \dfrac{12}{x^3} = \dfrac{\overset{1}{\cancel{4}}}{3x} \cdot \dfrac{\overset{x^2}{\cancel{x^3}}}{\underset{1}{\cancel{12}}} = \dfrac{x^2}{9}$

b. $\dfrac{4r^3}{9s^2} \div \dfrac{8r^2s^4}{15rs} = \dfrac{\overset{1 \; r}{\cancel{4r^3}}}{\underset{3 \; s}{\cancel{9s^2}}} \cdot \dfrac{\overset{5 \; 1}{\cancel{15rs}}}{\underset{2 \; 1}{\cancel{8r^2s^4}}} = \dfrac{5r^2}{6s^5}$

c. $\dfrac{3y^3 - 3y^2}{16y^5 + 8y^4} \div \dfrac{y^2 + 2y - 3}{4y + 12} = \dfrac{3\overset{1}{\cancel{y^2}}(\cancel{y-1})}{\underset{2 \; y^2}{\cancel{8y^4}}(2y + 1)} \cdot \dfrac{4(y+3)}{(\cancel{y-1})(\cancel{y+3})}$

$$= \dfrac{3}{2y^2(2y + 1)}$$

d. $\dfrac{y^2 - x^2}{4xy - 2y^2} \div \dfrac{2x - 2y}{2x^2 + xy - y^2} = \dfrac{(y + x)(\overset{-1}{\cancel{y - x}})}{2y(\underset{1}{\cancel{2x - y}})} \cdot \dfrac{(x + y)(\overset{1}{\cancel{2x - y}})}{2(\underset{1}{\cancel{x - y}})}$

$$= \dfrac{-(x + y)^2}{4y} \text{ or } -\dfrac{(x + y)^2}{4y}$$

Exercises 12.2

Perform the indicated operations.

1. $\dfrac{5}{6} \div \dfrac{5}{3}$

2. $\dfrac{3}{8} \div \dfrac{21}{12}$

3. $\dfrac{4a^3}{5b^2} \cdot \dfrac{10b}{8a^2}$

4. $\dfrac{6d}{8c} \cdot \dfrac{12c^2}{9d}$

5. $\dfrac{3x^2}{16} \div \dfrac{x}{8}$

6. $\dfrac{4y^3}{7} \div \dfrac{4y^2}{21}$

7. $\dfrac{3x^4y^2z}{18xy} \cdot \dfrac{15z}{x^3yz^2}$

8. $\dfrac{21a^2b^5}{4b^2} \cdot \dfrac{6c}{7b^2c^3}$

9. $\dfrac{5x^3}{4y^2} \cdot \dfrac{8y}{10x^2}$

10. $\dfrac{a^3}{2a+6} \div \dfrac{a^2}{a+3}$

11. $\dfrac{x}{x+2} \cdot \dfrac{5x+10}{x^3}$

12. $\dfrac{y-2}{y} \cdot \dfrac{6}{3y-6}$

13. $\dfrac{b^3}{a+3} \div \dfrac{4b^2}{2a+6}$

14. $\dfrac{3n-6}{15n} \div \dfrac{n-2}{20n^2}$

15. $\dfrac{a+2}{a+1} \cdot \dfrac{a^2+a}{a^2-4}$

16. $\dfrac{2x-1}{y} \div \dfrac{2x^2+x-1}{y^2}$

17. $\dfrac{a+4}{a-4} \div \dfrac{a^2+8a+16}{a^2-16}$

18. $\dfrac{x^3-5x}{9x} \div \dfrac{4x^3-20x}{12x^2}$

19. $\dfrac{5}{z+4} \cdot \dfrac{z^2-16}{(z-4)^2}$

20. $\dfrac{u+3}{u^2-9} \cdot \dfrac{2u-6}{6}$

21. $\dfrac{3a-3b}{4c+4d} \cdot \dfrac{2c+2d}{b-a}$

22. $\dfrac{2y-2x}{6} \cdot \dfrac{x+y}{x^2-y^2}$

23. $\dfrac{4x-8}{4} \cdot \dfrac{x+2}{x^2-4}$

24. $\dfrac{2a+2b}{3} \cdot \dfrac{a-b}{a^2-b^2}$

25. $\dfrac{4a+4b}{ab^2} \div \dfrac{3a+3b}{a^2b}$

26. $\dfrac{x^2-y^2}{x^2-2xy+y^2} \div \dfrac{x+y}{x-y}$

27. $\dfrac{2u^2-14u}{5u^2} \cdot \dfrac{15u^3}{4u-28}$

28. $\dfrac{-3v}{18v+90} \cdot \dfrac{6v^3+30v^2}{v^3}$

29. $\dfrac{3e^2}{28e-42} \div \dfrac{6e^3}{14e^2-21e}$

30. $\dfrac{12f+16}{15f} \div \dfrac{6f^3+8f^2}{20f^4}$

31. $\dfrac{x^2+x-2}{x-1} \div \dfrac{x^2+5x+6}{x^2}$

32. $\dfrac{z^3}{z+4} \div \dfrac{z-1}{z^2+3z-4}$

33. $\dfrac{x^2+10x+25}{x^2-25} \cdot \dfrac{5-x}{x+5}$

34. $\dfrac{3-y}{y+1} \cdot \dfrac{4+5y+y^2}{y^2+y-12}$

35. $\dfrac{x-y}{9x+9y} \div \dfrac{x^2-y^2}{3x^2+6xy+3y^2}$

36. $\dfrac{x^2-y^2}{8x^2-16xy+8y^2} \div \dfrac{x+y}{4x-4y}$

37. $\dfrac{2x^3y+2x^2y^2}{6x} \div \dfrac{x^2y^2-xy^3}{y-x}$

38. $\dfrac{2b^2c-2bc^2}{b+c} \div \dfrac{4bc^2-4b^2c}{4b+4c}$

39. $\dfrac{x^2-8x-9}{x^3-3x^2-x+3} \cdot \dfrac{x^2-9}{x^2-9x}$

40. $\dfrac{x^2-2xy+y^2}{6x^2-9xy} \div \dfrac{x^2-xy-6y^2}{2x^2-5xy-3y^2} \cdot \dfrac{3x+6y}{2x^2-xy-y^2}$

41. $\dfrac{a^2-3ab+2b^2}{2a+2b} \cdot \dfrac{2a^2+3ab+b^2}{a^2-ab-2b^2} \div \dfrac{4a^2+10ab+4b^2}{4a+8b}$

12.3   Adding and Subtracting Like Fractions

Like fractions are fractions that have the same denominator.

EXAMPLE 1
a. $\dfrac{5}{8}$ and $\dfrac{7}{8}$ are like fractions.

└───── Same denominator

b. $\dfrac{3}{2x}$ and $\dfrac{5}{2x}$ are like fractions.

└───── Same denominator

Unlike fractions are fractions that have different denominators.

EXAMPLE 2
a. $\dfrac{5}{8}$ and $\dfrac{7}{6}$ are unlike fractions.

└── Different denominators

b. $\dfrac{3}{2x}$ and $\dfrac{5}{x+2}$ are unlike fractions.

└── Different denominators

Adding Like Fractions

In arithmetic, we can add like fractions as follows:

$$\frac{5}{8} + \frac{7}{8} = \frac{5+7}{8} = \frac{12}{8} = \frac{\overset{3}{\cancel{12}}}{\underset{2}{\cancel{8}}} = \frac{3}{2} \quad \text{Reduced form}$$

This method is also used for adding like fractions in algebra.

Adding like fractions	1. Add the numerators. 2. Write the sum of the numerators over the denominator of the like fractions: $$\frac{P}{Q} + \frac{R}{Q} = \frac{P+R}{Q}$$ where $Q \neq 0$. 3. Reduce the resulting fraction to lowest terms.

EXAMPLE 3

Examples of adding like algebraic fractions:

a. $\dfrac{3}{2x} + \dfrac{5}{2x} = \dfrac{3+5}{2x} = \dfrac{8}{2x} = \dfrac{\overset{4}{\cancel{8}}}{\underset{1}{\cancel{2x}}} = \dfrac{4}{x}$ Reduced form

b. $\dfrac{15}{d-5} + \dfrac{-3d}{d-5} = \dfrac{15-3d}{d-5} = \dfrac{3(\overset{-1}{\cancel{5-d}})}{\underset{1}{\cancel{d-5}}} = \dfrac{-3}{1} = -3$

When the denominators are negatives of each other, we can make the fractions like fractions by changing *two* signs of the fraction. (See Example 4.)

EXAMPLE 4

Add $\dfrac{9}{x-2} + \dfrac{5}{2-x}$.

SOLUTION

Changing the sign of the numerator and denominator

$$\frac{9}{x-2} + \frac{5}{2-x} = \frac{9}{x-2} + \frac{-5}{-(2-x)} = \frac{9}{x-2} + \frac{-5}{x-2} = \frac{9-5}{x-2} = \frac{4}{x-2}$$

This denominator can be changed to $x-2$ by a sign change

Subtracting Like Fractions

Subtracting like fractions	1. Subtract the numerators. 2. Write the difference of the numerators over the denominator of the like fractions: $$\frac{P}{Q} - \frac{R}{Q} = \frac{P-R}{Q}$$ where $Q \neq 0$. 3. Reduce the resulting fraction to lowest terms.

EXAMPLE 5   Examples of subtracting like fractions:

a. $\dfrac{3}{4a} - \dfrac{5}{4a} = \dfrac{3-5}{4a} = \dfrac{-2}{4a} = -\dfrac{\overset{1}{2}}{\underset{2}{4a}} = -\dfrac{1}{2a}$

b. $\dfrac{4x}{2x-y} - \dfrac{2y}{2x-y} = \dfrac{4x-2y}{2x-y} = \dfrac{2(2x-y)}{(2x-y)} = \dfrac{2(2\cancel{x}\overset{1}{-}y)}{(2\cancel{x}\underset{1}{-}y)} = \dfrac{2}{1} = 2$

 A Word of Caution   If the fraction being subtracted has more than one term in the numerator, it is important to put parentheses around that numerator so that the entire numerator will be subtracted. (See Example 6.)

EXAMPLE 6   Subtract $\dfrac{2x+3}{x+3} - \dfrac{x-3}{x+3}$.

SOLUTION

Note: These parentheses are important

$$\dfrac{2x+3}{x+3} - \dfrac{x-3}{x+3} = \dfrac{2x+3-(x-3)}{x+3} = \dfrac{2x+3-x+3}{x+3} = \dfrac{x+6}{x+3}$$

Exercises 12.3

Perform the indicated operations.

1. $\dfrac{7}{a} + \dfrac{2}{a}$ **2.** $\dfrac{5}{b} + \dfrac{2}{b}$ **3.** $\dfrac{2}{3a} + \dfrac{4}{3a}$

4. $\dfrac{8}{5z} + \dfrac{2}{5z}$ **5.** $\dfrac{6}{x} - \dfrac{2}{x}$ **6.** $\dfrac{x+1}{a+b} + \dfrac{x-1}{a+b}$

7. $\dfrac{6}{x-y} - \dfrac{2}{x-y}$ **8.** $\dfrac{7}{m+n} - \dfrac{1}{m+n}$

9. $\dfrac{2y}{y+1} + \dfrac{2}{y+1}$ **10.** $\dfrac{3x}{x-4} - \dfrac{12}{x-4}$

11. $\dfrac{3}{x+3} + \dfrac{x}{x+3}$ **12.** $\dfrac{m}{m-4} - \dfrac{4}{m-4}$

13. $\dfrac{2a}{2a+3} + \dfrac{3}{2a+3}$ **14.** $\dfrac{2x+1}{x+4} - \dfrac{x-3}{x+4}$

15. $\dfrac{9x}{3x-y} - \dfrac{3y}{3x-y}$ **16.** $\dfrac{5a}{a-2b} - \dfrac{10b}{a-2b}$

17. $\dfrac{8}{z-4} - \dfrac{2z}{z-4}$ **18.** $\dfrac{15}{m-5} - \dfrac{3m}{m-5}$

19. $\dfrac{4x+y}{2x+y} + \dfrac{2x+2y}{2x+y}$ **20.** $\dfrac{7x-1}{3x-1} - \dfrac{x+1}{3x-1}$

21. $\dfrac{x-3}{y-2} - \dfrac{x+5}{y-2}$ **22.** $\dfrac{z+4}{a-b} - \dfrac{z+3}{a-b}$

23. $\dfrac{a+2}{2a+1} - \dfrac{1-a}{2a+1}$ **24.** $\dfrac{6x-1}{3x-2} - \dfrac{3x+1}{3x-2}$

25. $\dfrac{-x}{x-2} - \dfrac{2}{2-x}$ **26.** $\dfrac{-2a}{b-2a} + \dfrac{-b}{2a-b}$

27. $\dfrac{-15w}{1-5w} - \dfrac{3}{5w-1}$ **28.** $\dfrac{-35}{6w-7} - \dfrac{30w}{7-6w}$

29. $\dfrac{7z}{8z-4} + \dfrac{6-5z}{4-8z}$ **30.** $\dfrac{13-30w}{15-10w} - \dfrac{10w+17}{10w-15}$

31. $\dfrac{x^2+x}{x+2} - \dfrac{x+4}{x+2}$ **32.** $\dfrac{2x^2-3}{2x-3} - \dfrac{6-2x^2}{2x-3}$

12.4   Lowest Common Denominator (LCD)

We ordinarily use the lowest common denominator when adding or subtracting unlike fractions.

In arithmetic, the **lowest common denominator (LCD)** is the smallest number that is exactly divisible by each of the denominators. Consider the expression

$$\frac{1}{2} + \frac{3}{4}$$

It's possible to determine the LCD for this expression by inspection: LCD = 4. Because $\frac{1}{2} = \frac{2}{4}$,

$$\frac{1}{2} + \frac{3}{4} = \frac{2}{4} + \frac{3}{4} = \frac{2 + 3}{4} = \frac{5}{4}$$

Now, however, consider the expression

$$\frac{7}{12} + \frac{4}{15}$$

Here the LCD is not easily found by inspection. We find the LCD in this case as follows.

1. Find the prime factorization of each denominator:

$$
\begin{array}{c|c}
2 & 12 \\
2 & 6 \\
& 3
\end{array}
\qquad\qquad
\begin{array}{c|c}
3 & 15 \\
& 5
\end{array}
$$

$$15 = 3 \cdot 5$$

$$12 = 2^2 \cdot 3$$

2. Write down each different factor that appears in the prime factorizations:

$$2, \quad 3, \quad 5$$

3. Raise each factor to the highest power to which it occurs in any denominator:

$$2^2, \quad 3, \quad 5$$

4. Multiply:

$$\text{LCD} = 2^2 \cdot 3 \cdot 5 = 60$$

This same method is used for finding the LCD for algebraic fractions.

Finding the LCD

1. Factor each denominator completely. Repeated factors should be expressed as powers.
2. Write each different factor that appears.
3. Raise each factor to the *highest* power to which it occurs in *any* denominator.
4. The LCD is the product of all the powers found in Step 3.

EXAMPLE 1   Find the LCD for $\dfrac{2}{x} + \dfrac{5}{x^3}$.

SOLUTION

Step 1. The denominators are already factored.
Step 2. x is the only different factor.

Step 3. x^3 is the highest power in any denominator.
Step 4. Therefore, LCD $= x^3$.

EXAMPLE 2

Find the LCD for $\dfrac{7}{18x^2y} + \dfrac{5}{8xy^4}$.

SOLUTION

Step 1. $2 \cdot 3^2 \cdot x^2 \cdot y; \quad 2^3 \cdot x \cdot y^4$ Denominators in factored form
Step 2. $2, \quad 3, \quad x, \quad y$ All the different factors
Step 3. $2^3, \quad 3^2, \quad x^2, \quad y^4$ Highest powers of the factors
Step 4. LCD $= 2^3 \cdot 3^2 \cdot x^2 \cdot y^4 = 72x^2y^4$

EXAMPLE 3

Find the LCD for $\dfrac{2}{x} + \dfrac{x}{x + 2}$.

SOLUTION

Step 1. The denominators are already factored.
Step 2. $x, \quad (x + 2)$
Step 3. $x^1, \quad (x + 2)^1$
Step 4. LCD $= x(x + 2)$

EXAMPLE 4

Find the LCD for $\dfrac{8}{3x + 3} - \dfrac{5}{x^2 + 2x + 1}$.

SOLUTION

Step 1. $3x + 3 = 3(x + 1); \quad x^2 + 2x + 1 = (x + 1)^2$
Step 2. $3, \quad (x + 1)$
Step 3. $3^1, \quad (x + 1)^2$
Step 4. LCD $= 3(x + 1)^2$

Exercises   12.4

Find the LCD in each exercise. Do *not* add the fractions.

1. $\dfrac{x}{2} - \dfrac{x}{5}$ **2.** $\dfrac{3y}{4} + \dfrac{y}{12}$ **3.** $\dfrac{3}{2} + \dfrac{4}{y}$

4. $\dfrac{5}{x^2} - \dfrac{7}{3}$ **5.** $\dfrac{x + 1}{4} - \dfrac{x + 3}{2}$ **6.** $\dfrac{y - 2}{3} + \dfrac{y + 5}{9}$

7. $\dfrac{3}{2x} + \dfrac{5}{3x}$ **8.** $\dfrac{2}{5z} - \dfrac{4}{7z}$ **9.** $\dfrac{4}{x} + \dfrac{6}{x^3}$

10. $\dfrac{3}{a^2} + \dfrac{5}{a^4}$ **11.** $\dfrac{5}{4y^2} + \dfrac{9}{6y}$ **12.** $\dfrac{5}{9x^2y} - \dfrac{7}{6xy^2}$

13. $\dfrac{7}{12u^3v^2} - \dfrac{11}{18uv^3}$ **14.** $\dfrac{13}{50x^3y^4} - \dfrac{17}{20x^2y^5}$

15. $\dfrac{4}{a} + \dfrac{a}{a + 3}$ **16.** $\dfrac{5}{b} + \dfrac{b}{b - 5}$

17. $\dfrac{x}{2x + 4} - \dfrac{5}{4x}$ **18.** $\dfrac{4}{3x} + \dfrac{2x}{3x + 6}$

19. $\dfrac{x}{x^2 + 4x + 4} + \dfrac{1}{x + 2}$ **20.** $\dfrac{2x}{x^2 - 2x + 1} - \dfrac{5}{x - 1}$

21. $\dfrac{x - 4}{x^2 + 3x + 2} + \dfrac{3x + 1}{x^2 + 2x + 1}$

22. $\dfrac{2x + 3}{x^2 - x - 12} + \dfrac{x - 4}{x^2 + 6x + 9}$

23. $\dfrac{9y}{x^2 - y^2} + \dfrac{6x}{(x + y)^2}$

24. $\dfrac{5}{8z^3} - \dfrac{8z}{z^2 - 4} + \dfrac{5z}{9z + 18}$

25. $\dfrac{x^2 + 1}{12x^3 + 24x^2} - \dfrac{4x + 3}{x^2 - 4x + 4} + \dfrac{1}{x^2 - 4}$

26. $\dfrac{2y + 5}{y^2 + 6y + 9} - \dfrac{7y}{y^2 - 9} - \dfrac{11}{8y^2 - 24y}$

12.5 Adding and Subtracting Unlike Fractions

Equivalent fractions are fractions that have the same value. If a fraction is multiplied by 1, its value is unchanged.

$$\frac{2}{2} = 1 \qquad \frac{x}{x} = 1 \qquad \frac{x+2}{x+2} = 1$$

Multiplying a fraction by expressions like these will produce equivalent fractions

For example,

$$\frac{5}{6} = \frac{5}{6} \cdot \frac{2}{2} = \frac{10}{12}$$

A fraction equivalent to $\dfrac{5}{6}$

$$\frac{x}{x+2} = \frac{x}{x+2} \cdot \frac{x}{x} = \frac{x^2}{x(x+2)}$$

A fraction equivalent to $\dfrac{x}{x+2}$

$$\frac{2}{x} = \frac{2}{x} \cdot \frac{x+2}{x+2} = \frac{2x+4}{x(x+2)}$$

A fraction equivalent to $\dfrac{2}{x}$

Adding unlike fractions

1. Find the LCD.
2. Convert each fraction to an equivalent fraction having the LCD as its denominator.
3. Add the resulting like fractions.
4. Reduce the resulting fraction to lowest terms.

EXAMPLE 1

Add $\dfrac{5}{4x^2y} + \dfrac{3}{2xy^3}$.

SOLUTION

Step 1. LCD $= 4x^2y^3$

Multiply the numerator and denominator by y^2 to obtain an equivalent fraction whose denominator is the LCD, $4x^2y^3$

Step 2. $\dfrac{5}{4x^2y} \cdot \dfrac{y^2}{y^2} + \dfrac{3}{2xy^3} \cdot \dfrac{2x}{2x}$

Multiply the numerator and denominator by $2x$ to obtain an equivalent fraction whose denominator is the LCD, $4x^2y^3$

$$= \frac{5y^2}{4x^2y^3} + \frac{6x}{4x^2y^3}$$

Step 3. $= \dfrac{5y^2 + 6x}{4x^2y^3}$

EXAMPLE 2

Add $\dfrac{2}{x} + \dfrac{x}{x+2}$.

SOLUTION

Step 1. LCD $= x(x+2)$

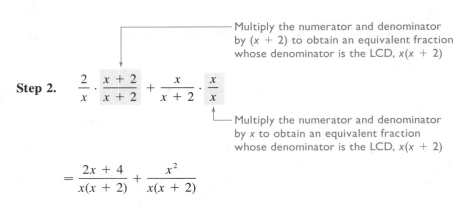

Multiply the numerator and denominator
by $(x + 2)$ to obtain an equivalent fraction
whose denominator is the LCD, $x(x + 2)$

Step 2. $\dfrac{2}{x} \cdot \dfrac{x + 2}{x + 2} + \dfrac{x}{x + 2} \cdot \dfrac{x}{x}$

Multiply the numerator and denominator
by x to obtain an equivalent fraction
whose denominator is the LCD, $x(x + 2)$

$$= \dfrac{2x + 4}{x(x + 2)} + \dfrac{x^2}{x(x + 2)}$$

Step 3. $= \dfrac{x^2 + 2x + 4}{x(x + 2)}$

EXAMPLE 3   Add $3 - \dfrac{2a}{a + 4}$.

SOLUTION

Step 1. LCD $= (a + 4)$

Step 2. $\dfrac{3}{1} \cdot \dfrac{a + 4}{a + 4} - \dfrac{2a}{a + 4}$

$$= \dfrac{3a + 12}{a + 4} - \dfrac{2a}{a + 4}$$

Step 3. $= \dfrac{3a + 12 - 2a}{a + 4}$

$$= \dfrac{a + 12}{a + 4}$$

 Note   The fraction obtained in Step 3 cannot be reduced because neither a nor 4 is a
factor in the numerator and denominator.

EXAMPLE 4   Add $\dfrac{20}{x^2 - 25} + \dfrac{2}{x + 5}$.

SOLUTION

Step 1. Factor the denominators to find the LCD:

$$\dfrac{20}{x^2 - 25} + \dfrac{2}{x + 5}$$

$$= \dfrac{20}{(x + 5)(x - 5)} + \dfrac{2}{x + 5}$$

The LCD is $(x + 5)(x - 5)$.

Step 2. $\dfrac{20}{(x+5)(x-5)} + \dfrac{2}{x+5} \cdot \dfrac{x-5}{x-5}$

$= \dfrac{20}{(x+5)(x-5)} + \dfrac{2x-10}{(x+5)(x-5)}$

Step 3. $= \dfrac{20+2x-10}{(x+5)(x-5)}$

$= \dfrac{2x+10}{(x+5)(x-5)}$

Step 4. Factor the numerator and reduce:

$\dfrac{2(\cancel{x+5})}{(\cancel{x+5})(x-5)}$

$= \dfrac{2}{x-5}$

EXAMPLE 5

Add $\dfrac{6}{x^2-9} - \dfrac{2}{x^2-4x+3}$.

SOLUTION

Step 1. Factor the denominators to find the LCD:

$\dfrac{6}{x^2-9} - \dfrac{2}{x^2-4x+3}$

$= \dfrac{6}{(x+3)(x-3)} - \dfrac{2}{(x-3)(x-1)}$

The LCD is $(x+3)(x-3)(x-1)$.

Step 2. $\dfrac{6}{(x+3)(x-3)} \cdot \dfrac{x-1}{x-1} - \dfrac{2}{(x-3)(x-1)} \cdot \dfrac{x+3}{x+3}$

$= \dfrac{6x-6}{(x+3)(x-3)(x-1)} - \dfrac{2x+6}{(x-3)(x-1)(x+3)}$

Step 3. $= \dfrac{6x-6-(2x+6)}{(x+3)(x-3)(x-1)}$

Note: The parentheses in the numerator are important. When subtracting fractions, be sure to subtract the *entire* numerator of the second fraction. Thus, $2x+6$ becomes $-2x-6$.

$= \dfrac{6x-6-2x-6}{(x+3)(x-3)(x-1)}$

$= \dfrac{4x-12}{(x+3)(x-3)(x-1)}$

Step 4. Factor the numerator and reduce:

$= \dfrac{4(\cancel{x-3})}{(x+3)(\cancel{x-3})(x-1)}$

$= \dfrac{4}{(x+3)(x-1)}$

A Word of Caution   Many errors are made in reducing, or "canceling." Because of the definition of multiplication of fractions, it is correct to divide out common factors when the fractions are being multiplied:

We may reduce across the multiplication sign

$$\frac{x}{3x-9} \cdot \frac{3}{x^2-3x} = \frac{\overset{1}{x}}{\underset{1}{3}(x-3)} \cdot \frac{\overset{1}{3}}{\underset{1}{x}(x-3)} = \frac{1}{(x-3)^2}$$

But it is incorrect to divide out common factors when the fractions are being added or subtracted:

Cannot reduce across the subtraction sign

$$\frac{x}{3x-9} - \frac{3}{x^2-3x} = \frac{\overset{1}{x}}{\underset{1}{3}(x-3)} \diagdown \frac{\overset{1}{3}}{\underset{1}{x}(x-3)}$$

When fractions are added or subtracted, the final answer is reduced to lowest terms by dividing out the common factors.

$$\frac{x}{3x-9} - \frac{3}{x^2-3x}$$

$$= \frac{x}{3(x-3)} - \frac{3}{x(x-3)} \qquad \text{LCD} = 3x(x-3)$$

$$= \frac{x}{3(x-3)} \cdot \frac{x}{x} - \frac{3}{x(x-3)} \cdot \frac{3}{3}$$

$$= \frac{x^2}{3x(x-3)} - \frac{9}{3x(x-3)}$$

$$= \frac{x^2-9}{3x(x-3)} \qquad \text{Factor the numerator to reduce the final answer}$$

$$= \frac{(x+3)(\overset{1}{x-3})}{x(\underset{1}{x-3})}$$

$$= \frac{x+3}{x}$$

Exercises   12.5

In Exercises 1–48, perform the indicated additions and subtractions.

1. $\dfrac{3}{a^2} + \dfrac{2}{a^3}$ 2. $\dfrac{5}{u} + \dfrac{4}{u^3}$ 3. $\dfrac{3}{x} + \dfrac{4}{x^2}$

4. $\dfrac{1}{3} - \dfrac{1}{a} + \dfrac{2}{a^2}$ 5. $\dfrac{1}{2} + \dfrac{3}{x} - \dfrac{5}{x^2}$ 6. $\dfrac{2}{3} - \dfrac{1}{y} + \dfrac{4}{y^2}$

7. $\dfrac{2}{xy} - \dfrac{3}{y}$ 8. $\dfrac{5}{ab} - \dfrac{4}{a}$ 9. $3 + \dfrac{4}{y}$

10. $5 + \dfrac{2}{x}$ 11. $3x - \dfrac{3}{x}$ 12. $4y - \dfrac{6}{y}$

13. $\dfrac{5}{6x} + \dfrac{1}{3x^2}$ 14. $\dfrac{3}{4y^2} + \dfrac{1}{6y}$ 15. $\dfrac{5}{9x^2y} - \dfrac{7}{6xy^2}$

16. $\dfrac{3}{10ab^3} - \dfrac{1}{4a^2b}$ 17. $\dfrac{3}{a} + \dfrac{a}{a+3}$ 18. $\dfrac{5}{b} + \dfrac{b}{b-5}$

19. $\dfrac{x}{x-2} - \dfrac{1}{x}$ 20. $\dfrac{4}{x+4} - \dfrac{1}{4}$ 21. $\dfrac{3}{2x+4} - \dfrac{1}{2x}$

45. $\dfrac{x}{x^2+4x+4} + \dfrac{1}{x+2}$ 46. $\dfrac{2x}{x^2-2x+1} - \dfrac{5}{x-1}$

22. $\dfrac{5}{3x-9} - \dfrac{2}{3x}$ 23. $x + \dfrac{2}{x} - \dfrac{3}{x-2}$

47. $\dfrac{4x}{x^2-2x-3} - \dfrac{3}{x-3}$ 48. $\dfrac{5x}{x^2-x-6} - \dfrac{2}{x+2}$

In Exercises 49–64, perform the indicated operations.

24. $m - \dfrac{3}{m} + \dfrac{2}{m+4}$ 25. $\dfrac{a+b}{b} + \dfrac{b}{a-b}$

49. $\dfrac{2x}{x^2+3x-4} \cdot \dfrac{x^2+5x-6}{2x+12}$

50. $\dfrac{x^2+5x+6}{x^2-2x-15} \cdot \dfrac{3x-15}{15}$

26. $\dfrac{x-y}{x} - \dfrac{x}{x+y}$ 27. $\dfrac{3}{2e-2} - \dfrac{2}{3e-3}$

51. $\dfrac{2}{x^2-1} + \dfrac{1}{x^2+3x+2}$

28. $\dfrac{2}{3f+6} - \dfrac{1}{5f+10}$ 29. $\dfrac{3}{m-2} - \dfrac{5}{2-m}$

52. $\dfrac{1}{x^2-7x+12} + \dfrac{6}{x^2-9}$

53. $\dfrac{a^2}{5a+10} \div \dfrac{a^2-5a}{a^2-3a-10}$

30. $\dfrac{7}{n-5} - \dfrac{2}{5-n}$ 31. $\dfrac{3}{x-2} - \dfrac{2}{x+2}$

54. $\dfrac{2b^3}{b^2+8b} \div \dfrac{8b+8}{b^2+9b+8}$

32. $\dfrac{5}{x+5} - \dfrac{3}{x-5}$ 33. $\dfrac{x-1}{x+2} - \dfrac{x-2}{x+1}$

55. $\dfrac{3x}{x^2-x-20} - \dfrac{5}{x^2-7x+10}$

34. $\dfrac{x+2}{x-3} - \dfrac{x+3}{x-2}$ 35. $\dfrac{x+1}{x-1} - \dfrac{x-1}{x+1}$

56. $\dfrac{x}{x^2+x-6} - \dfrac{3}{x^2+11x+24}$

57. $\dfrac{x-4}{x^2+3x+2} + \dfrac{3x+1}{x^2+2x+1}$

36. $\dfrac{x-5}{x+5} - \dfrac{x+5}{x-5}$ 37. $\dfrac{2x+4}{x^2+4x} + \dfrac{x}{4x+16}$

58. $\dfrac{2x+3}{x^2-x-12} + \dfrac{x-4}{x^2+6x+9}$

59. $\dfrac{4x^2-y^2}{4x^2+y^2} \cdot \dfrac{4x^2+2xy}{4x^2+4xy+y^2}$

38. $\dfrac{x+5}{2x+6} + \dfrac{3}{x^2+3x}$ 39. $\dfrac{3x+1}{6x-6} - \dfrac{x+1}{3x-3}$

60. $\dfrac{a^2-6ab+9b^2}{3ab-9b^2} \cdot \dfrac{a^2+9b^2}{a^2-9b^2}$

40. $\dfrac{x+1}{3x-6} - \dfrac{x+2}{4x-8}$ 41. $\dfrac{y}{x^2-xy} + \dfrac{x}{y^2-xy}$

61. $\dfrac{2x}{x+2} + \dfrac{x}{x-2} - \dfrac{3x+2}{x^2-4}$

62. $\dfrac{3x}{x+5} - \dfrac{x}{x-5} + \dfrac{7x+15}{x^2-25}$

42. $\dfrac{b}{ab-a^2} - \dfrac{a}{b^2-ab}$ 43. $\dfrac{2x}{x-3} - \dfrac{2x}{x+3} + \dfrac{36}{x^2-9}$

63. $\dfrac{2x}{x^2-9x+14} \cdot \dfrac{x^2-6x-7}{x^2+4x} \div \dfrac{x^2-1}{x^2+2x-8}$

44. $\dfrac{x}{x+4} - \dfrac{x}{x-4} - \dfrac{32}{x^2-16}$

64. $\dfrac{x+1}{x^2-9} \div \dfrac{3x}{x^2+8x+15} \cdot \dfrac{x^2-3x}{x^2+6x+5}$

12.6   Complex Fractions

A **simple fraction** is a fraction that has only one fraction bar, such as

$$\frac{2}{x}, \quad \frac{3+y}{12}, \quad \frac{7a-7b}{ab^2}, \quad \frac{5}{x+y}$$

A **complex fraction** is a fraction that has more than one fraction bar, such as

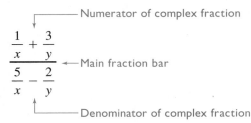

$$\frac{\frac{2}{x}}{3}, \quad \frac{a}{\frac{1}{c}}, \quad \frac{\frac{3}{z}}{\frac{5}{z}}, \quad \frac{\frac{3}{x} - \frac{2}{y}}{\frac{5}{x} + \frac{3}{y}}$$

The simple fractions that make up the numerator and denominator of a complex fraction are known as *secondary fractions*:

┌────── Numerator of complex fraction   ┌──┬── Secondary fractions

$$\frac{\frac{1}{x} + \frac{3}{y}}{\frac{5}{x} - \frac{2}{y}}$$ ←──Main fraction bar $$\frac{\frac{1}{x} + \frac{3}{y}}{\frac{5}{x} - \frac{2}{y}}$$

└────── Denominator of complex fraction   └──┴── Secondary fractions

Simplifying complex fractions

Method 1. Multiply both numerator and denominator of the complex fraction by the LCD of the secondary fractions; then simplify the results.

Method 2. First, simplify the numerator and denominator of the complex fraction; then divide the simplified numerator by the simplified denominator.

☞ **Note**   In some of the following examples, simplifying a complex fraction is easier by method 1 than by method 2. In others, the opposite is true.

EXAMPLE 1   Simplify $\dfrac{\frac{1}{2} + \frac{3}{4}}{\frac{5}{6} - \frac{2}{3}}$.

SOLUTION 1   The LCD of the secondary denominators, 2, 4, 6, and 3, is 12.

$$\frac{12}{12} \cdot \frac{\frac{1}{2} + \frac{3}{4}}{\frac{5}{6} - \frac{2}{3}} = \frac{12\left(\frac{1}{2} + \frac{3}{4}\right)}{12\left(\frac{5}{6} - \frac{2}{3}\right)} = \frac{\frac{12}{1}\left(\frac{1}{2}\right) + \frac{12}{1}\left(\frac{3}{4}\right)}{\frac{12}{1}\left(\frac{5}{6}\right) + \frac{12}{1}\left(\frac{2}{3}\right)} = \frac{6 + 9}{10 - 8} = \frac{15}{2} = 7\frac{1}{2}$$

└── This is 1

SOLUTION 2

$$\frac{\frac{1}{2} + \frac{3}{4}}{\frac{5}{6} - \frac{2}{3}} = \frac{\frac{2}{4} + \frac{3}{4}}{\frac{5}{6} - \frac{4}{6}} = \frac{\frac{5}{4}}{\frac{1}{6}} = \frac{5}{4} \div \frac{1}{6} = \frac{5}{\underset{2}{4}} \cdot \frac{\overset{3}{6}}{1} = \frac{15}{2} = 7\frac{1}{2}$$

EXAMPLE 2   Simplify $\dfrac{\frac{4b^2}{9a^2}}{\frac{8b}{3a^3}}$. ←── Main fraction line

SOLUTION 1 The LCD of the secondary denominators, $9a^2$ and $3a^3$, is $9a^3$.

$$\frac{9a^3}{9a^3}\left(\frac{\dfrac{4b^2}{9a^2}}{\dfrac{8b}{3a^3}}\right) = \frac{\dfrac{9a^3}{1}\left(\dfrac{4b^2}{9a^2}\right)}{\dfrac{9a^3}{1}\left(\dfrac{8b}{3a^3}\right)} = \frac{4ab^2}{24b} = \frac{ab}{6}$$

⎯⎯ The value of this fraction is 1

SOLUTION 2

$$\frac{\dfrac{4b^2}{9a^2}}{\dfrac{8b}{3a^3}} = \frac{4b^2}{9a^2} \div \frac{8b}{3a^3} = \frac{\overset{1}{\cancel{4b^2}}\overset{b}{}}{\underset{3}{\cancel{9a^2}}\underset{1}{}} \cdot \frac{\overset{1}{\cancel{3a^3}}\overset{a}{}}{\underset{2}{\cancel{8b}}\underset{1}{}} = \frac{ab}{6}$$

EXAMPLE 3 Simplify $\dfrac{1 + \dfrac{2}{x}}{1 - \dfrac{4}{x^2}}$.

SOLUTION 1 The LCD of the secondary denominators, x and x^2, is x^2.

$$\frac{x^2}{x^2}\left[\frac{1 + \dfrac{2}{x}}{1 - \dfrac{4}{x^2}}\right] = \frac{\dfrac{x^2}{1}\cdot\dfrac{1}{1} + \dfrac{x^2}{1}\cdot\dfrac{2}{x}}{\dfrac{x^2}{1}\cdot\dfrac{1}{1} - \dfrac{x^2}{1}\cdot\dfrac{4}{x^2}} = \frac{x^2 + 2x}{x^2 - 4} = \frac{x(\overset{1}{\cancel{x+2}})}{(\underset{1}{\cancel{x+2}})(x-2)} = \frac{x}{x-2}$$

SOLUTION 2

$$\frac{1 + \dfrac{2}{x}}{1 - \dfrac{4}{x^2}} = \frac{\dfrac{x}{x} + \dfrac{2}{x}}{\dfrac{x^2}{x^2} - \dfrac{4}{x^2}} = \frac{\dfrac{x+2}{x}}{\dfrac{x^2-4}{x^2}} = \frac{x+2}{x} \div \frac{x^2-4}{x^2}$$

$$= \frac{\overset{1}{\cancel{x+2}}}{\underset{1}{\cancel{x}}} \cdot \frac{\overset{x}{\cancel{x^2}}}{(\underset{1}{\cancel{x+2}})(x-2)}$$

$$= \frac{x}{x-2}$$

Exercises 12.6

Simplify each of the complex fractions.

1. $\dfrac{\dfrac{3}{4} - \dfrac{1}{2}}{\dfrac{5}{8} + \dfrac{1}{4}}$

2. $\dfrac{\dfrac{5}{6} - \dfrac{1}{3}}{\dfrac{2}{9} + \dfrac{1}{6}}$

3. $\dfrac{\dfrac{3}{5} + 2}{2 - \dfrac{3}{4}}$

4. $\dfrac{\dfrac{3}{16} + 5}{6 - \dfrac{7}{8}}$

5. $\dfrac{\dfrac{5x^3}{3y^4}}{\dfrac{10x}{9y}}$

6. $\dfrac{\dfrac{8a^4}{5b}}{\dfrac{4a^3}{15b^2}}$

7. $\dfrac{\dfrac{18cd^2}{5a^3b}}{\dfrac{12cd^2}{15ab^2}}$

8. $\dfrac{\dfrac{8x^2y}{7z^3}}{\dfrac{12xy^2}{21z^5}}$

9. $\dfrac{\dfrac{3a^3}{5b^2}}{\dfrac{6a^2}{10b^3}}$

10. $\dfrac{\dfrac{x-2}{4}}{\dfrac{3x-6}{12}}$

11. $\dfrac{\dfrac{x+3}{5}}{\dfrac{2x+6}{10}}$

12. $\dfrac{\dfrac{a-4}{3}}{\dfrac{2a-8}{9}}$

13. $\dfrac{\dfrac{a}{b} + 1}{\dfrac{a}{b} - 1}$

14. $\dfrac{2 + \dfrac{x}{y}}{2 - \dfrac{x}{y}}$

15. $\dfrac{\dfrac{c}{d} + 2}{\dfrac{c^2}{d^2} - 4}$

28. $\dfrac{\dfrac{1}{y^2} - 9}{\dfrac{1}{y} + 3}$

29. $\dfrac{\dfrac{1}{a^2} - \dfrac{1}{b^2}}{\dfrac{1}{a} + \dfrac{1}{b}}$

30. $\dfrac{\dfrac{1}{c^2} - \dfrac{1}{4}}{\dfrac{1}{c} - \dfrac{1}{2}}$

16. $\dfrac{\dfrac{x^2}{y^2} - 1}{\dfrac{x}{y} - 1}$

17. $\dfrac{1}{1 - \dfrac{1}{x}}$

18. $\dfrac{1}{1 + \dfrac{1}{y}}$

31. $\dfrac{2 + \dfrac{6}{x + 3}}{x - \dfrac{18}{x + 3}}$

32. $\dfrac{x - \dfrac{2}{x - 1}}{5 + \dfrac{10}{x - 1}}$

19. $\dfrac{x + \dfrac{1}{y}}{x}$

20. $\dfrac{\dfrac{1}{a} - b}{b}$

21. $\dfrac{\dfrac{1}{x} + x}{\dfrac{1}{x} - x}$

33. $\dfrac{\dfrac{1}{x} + \dfrac{1}{x + 2}}{\dfrac{1}{x} - \dfrac{1}{x + 2}}$

34. $\dfrac{\dfrac{3}{2} - \dfrac{12}{x + 4}}{\dfrac{x}{x + 4} - \dfrac{1}{2}}$

22. $\dfrac{a - \dfrac{4}{a}}{a + \dfrac{4}{a}}$

23. $\dfrac{1 - \dfrac{1}{b}}{3 - \dfrac{3}{b}}$

24. $\dfrac{x + \dfrac{x}{y}}{1 + \dfrac{1}{y}}$

35. $\dfrac{\dfrac{2}{x - 2} - \dfrac{1}{x + 2}}{\dfrac{8}{x^2 - 4} + \dfrac{2}{x - 2}}$

36. $\dfrac{\dfrac{3}{x + 5} + \dfrac{2}{x - 5}}{\dfrac{6}{x^2 - 25} - \dfrac{1}{x - 5}}$

25. $\dfrac{1 - \dfrac{1}{a^2}}{\dfrac{1}{a} - \dfrac{1}{a^2}}$

26. $\dfrac{\dfrac{1}{y} + \dfrac{x}{y^2}}{1 - \dfrac{x^2}{y^2}}$

27. $\dfrac{\dfrac{x^2}{y^2} - 4}{\dfrac{x}{y} + 2}$

12.7   Solving Equations Having Fractions

The multiplication rule for equations says that we may multiply both sides of an equation by any nonzero number. If an equation contains fractions, multiplying both sides by its LCD will clear the equation of all fractions. The resulting equation can then be solved using the methods learned in earlier chapters.

Excluded Values   Any value of the variable that makes the denominator equal to zero must be excluded. (See Section 12.1.) Therefore, in solving equations with fractions, it is important to note any excluded values, because they cannot be solutions to the original equation.

Equations Having Fractions That Simplify to First-Degree Equations

Solving an equation having fractions that simplifies to a first-degree equation

1. Find the LCD, and note any excluded values.
2. Remove fractions by multiplying each term on both sides of the equation by the LCD.
3. Remove grouping symbols.
4. Combine like terms.
5. Get all terms with the variable on one side, and get all other terms on the other side.
6. Divide both sides by the coefficient of the variable.
7. Check the apparent solution in the original equation. Any excluded value must be rejected as a solution.

EXAMPLE 1

Solve $\dfrac{x}{2} + \dfrac{x}{3} = 5$.

SOLUTION The LCD is 6. There are no variables in the denominators; therefore, there are no excluded values.

$$6\left(\dfrac{x}{2}\right) + 6\left(\dfrac{x}{3}\right) = 6\,(5)$$

Multiplying each term on both sides by 6

$$\dfrac{\overset{3}{6}}{1}\left(\dfrac{x}{\underset{1}{2}}\right) + \dfrac{\overset{2}{6}}{1}\left(\dfrac{x}{\underset{1}{3}}\right) = \dfrac{6}{1}\left(\dfrac{5}{1}\right)$$

$$3x + 2x = 30$$

$$5x = 30$$

$$x = 6$$

✓ **Check for** $x = 6$ $\dfrac{x}{2} + \dfrac{x}{3} = 5$

$$\dfrac{6}{2} + \dfrac{6}{3} \overset{?}{=} 5$$

$$3 + 2 = 5$$

EXAMPLE 2

Solve $\dfrac{x-4}{2} - \dfrac{x}{5} = \dfrac{1}{10}$.

SOLUTION The LCD is 10. There are no excluded values.

$$\dfrac{\overset{5}{10}}{1}\left(\dfrac{x-4}{\underset{1}{2}}\right) - \dfrac{\overset{2}{10}}{1}\left(\dfrac{x}{\underset{1}{5}}\right) = \dfrac{\overset{1}{10}}{1}\left(\dfrac{1}{\underset{1}{10}}\right)$$

Multiplying each term on both sides by 10

$$5(x-4) - 2x = 1$$

$$5x - 20 - 2x = 1$$

$$3x = 21$$

$$x = 7$$

✓ **Check for** $x = 7$ $\dfrac{x-4}{2} - \dfrac{x}{5} = \dfrac{1}{10}$

$$\dfrac{7-4}{2} - \dfrac{7}{5} \overset{?}{=} \dfrac{1}{10}$$

$$\dfrac{3}{2} - \dfrac{7}{5} \overset{?}{=} \dfrac{1}{10}$$

$$\dfrac{15}{10} - \dfrac{14}{10} = \dfrac{1}{10}$$

When an equation has only one fraction on each side of the equal sign, it is a proportion. Proportions can be solved by setting the product of the means equal to the product of the extremes. (See Example 3.) Proportions can also be solved by multiplying both sides by the LCD.

EXAMPLE 3   Solve $\dfrac{9x}{x-3} = 6$.

SOLUTION   The excluded value is 3; that is, $x \neq 3$.

$$\dfrac{9x}{x-3} = \dfrac{6}{1} \cdot \left(\dfrac{x-3}{1}\right) \qquad \text{Writing } \dfrac{6}{1} \text{ makes the equation a proportion}$$

$$9x \cdot 1 = 6(x-3) \qquad \text{Product of means = product of extremes}$$

$$9x = 6x - 18 \qquad \text{Remove the parentheses}$$

$$\underline{-6x \qquad -6x}$$

$$3x = -18 \qquad \text{Get } x\text{-terms on one side}$$

$$x = -6 \qquad \text{Divide both sides by 3}$$

✓ **Check for $x = -6$** $\dfrac{9x}{x-3} = 6$

$$\dfrac{9(-6)}{-6-3} \overset{?}{=} 6$$

$$\dfrac{-54}{-9} \overset{?}{=} 6$$

$$6 = 6$$

EXAMPLE 4   Solve $\dfrac{x}{x-3} = \dfrac{3}{x-3} + 4$.

SOLUTION   The LCD is $x - 3$. The excluded value is 3; that is, $x \neq 3$.

$$\dfrac{\overset{1}{(\cancel{x-3})}}{1} \cdot \dfrac{x}{(\underset{1}{\cancel{x-3}})} = \dfrac{\overset{1}{(\cancel{x-3})}}{1} \cdot \dfrac{3}{(\underset{1}{\cancel{x-3}})} + \dfrac{(x-3)}{1} \cdot \dfrac{4}{1}$$

$$x = 3 + 4(x-3)$$

$$x = 3 + 4x - 12$$

$$x = 4x - 9$$

$$9 = 3x$$

$$3 = x \longleftarrow \text{Because 3 is an excluded value, this equation has } \textit{no solution}$$

If we try to check the value 3 in the equation, we come up with a division by zero, which is undefined.

✓ **Check for $x = 3$** $\dfrac{x}{x-3} = \dfrac{3}{x-3} + 4$

$$\dfrac{3}{3-3} \overset{?}{=} \dfrac{3}{3-3} + 4$$

$$\dfrac{3}{0} \overset{?}{=} \dfrac{3}{0} + 4$$

$$\uparrow \qquad \uparrow$$

Undefined

Equations Having Fractions That Simplify to Second-Degree Equations

After fractions and grouping symbols have been removed, there may be second-degree terms. When this is the case, the equation can sometimes be solved by factoring.

Solving an equation having fractions that simplifies to a second-degree equation

1. Find the LCD, and note any excluded values.
2. Remove fractions by multiplying each term on both sides of the equation by the LCD.
3. Remove grouping symbols.
4. Combine like terms.
5. Get *all* nonzero terms to one side by adding the same expression to both sides. *Only zero must remain on the other side.* Then arrange the terms in descending powers.
6. Factor the polynomial.
7. Set each factor equal to zero, and then solve for the variable.
8. Check the apparent solutions in the original equation. Any excluded values must be rejected as solutions.

EXAMPLE 5 Solve $\dfrac{2}{x} + \dfrac{3}{x^2} = 1$.

SOLUTION The LCD is x^2. The excluded value is 0; that is, $x \neq 0$.

$$\frac{x^2}{1}\left(\frac{2}{x}\right) + \frac{x^2}{1}\left(\frac{3}{x^2}\right) = \frac{x^2}{1}\left(\frac{1}{1}\right)$$

Second-degree term

$$2x + 3 = x^2$$
$$0 = x^2 - 2x - 3$$
$$0 = (x - 3)(x + 1)$$

$$x - 3 = 0 \qquad \bigg| \qquad x + 1 = 0$$
$$x = 3 \qquad \bigg| \qquad x = -1$$

✓ *Check for $x = 3$*   *Check for $x = -1$*

$$\frac{2}{x} + \frac{3}{x^2} = 1 \qquad\qquad \frac{2}{x} + \frac{3}{x^2} = 1$$

$$\frac{2}{3} + \frac{3}{3^2} \stackrel{?}{=} 1 \qquad\qquad \frac{2}{-1} + \frac{3}{(-1)^2} \stackrel{?}{=} 1$$

$$\frac{2}{3} + \frac{1}{3} = 1 \qquad\qquad -2 + 3 = 1$$

EXAMPLE 6 Solve $\dfrac{8}{x} = \dfrac{3}{x + 1} + 3$.

SOLUTION The LCD is $x(x + 1)$. The excluded values are 0 and -1; that is, $x \neq 0$ and $x \neq -1$.

$$\frac{\overset{1}{x(x+1)}}{1}\cdot\frac{8}{\underset{1}{x}} = \frac{\overset{1}{x(x+1)}}{1}\cdot\frac{3}{\underset{1}{(x+1)}} + \frac{x(x+1)}{1}\cdot\frac{3}{1}$$

$$8(x+1) = 3x + 3x(x+1)$$

Second-degree term

$$8x + 8 = 3x + 3x^2 + 3x$$

$$0 = 3x^2 - 2x - 8$$

$$0 = (3x+4)(x-2)$$

$$3x + 4 = 0 \qquad\qquad x - 2 = 0$$

$$3x = -4 \qquad\qquad\quad x = 2$$

$$x = -\frac{4}{3}$$

You should check to see that both $x = 2$ and $x = -\dfrac{4}{3}$ make the two sides of the equation equal.

A Word of Caution We have used the LCD to add and subtract fractions, to simplify complex fractions, and to solve equations with fractions. Notice how the LCD is used in each case to avoid some common errors.

To solve an equation, multiply each term on both sides by the LCD to remove fractions.

$$\frac{1}{2} + \frac{1}{x} = 1 \qquad \text{LCD} = 2x$$

$$\frac{2x}{1}\cdot\frac{1}{2} + \frac{2x}{1}\cdot\frac{1}{x} = 2x \cdot 1$$

$$x + 2 = 2x$$

$$2 = x \quad\longleftarrow\text{The answer is a solution}$$

To add fractions, convert each fraction to an equivalent fraction having the LCD as its denominator.

$$\frac{1}{2} + \frac{1}{x} = \frac{1}{2}\cdot\frac{x}{x} + \frac{1}{x}\cdot\frac{2}{2} \qquad \text{LCD} = 2x$$

$$= \frac{x}{2x} + \frac{2}{2x}$$

$$= \frac{x+2}{2x} \qquad\longleftarrow\;\text{The answer is a simplified fraction}$$

Remember: It is *incorrect* to remove fractions from an addition problem.

$$\frac{1}{2} + \frac{1}{x} = \frac{1}{2}\cdot\frac{2x}{1} + \frac{1}{x}\cdot\frac{2x}{1} = x + 2$$

> *To simplify a complex fraction*, multiply both numerator and denominator by the LCD.

$$\dfrac{\dfrac{1}{2}}{\dfrac{1}{x}} = \dfrac{\dfrac{2x}{1} \cdot \dfrac{1}{2}}{\dfrac{2x}{1} \cdot \dfrac{1}{x}} \qquad \text{LCD} = 2x$$

$$= \dfrac{x}{2} \longleftarrow \quad \begin{array}{l}\text{The answer is a}\\ \text{simplified fraction}\end{array}$$

Exercises 12.7

Solve the equations.

1. $\dfrac{x}{3} + \dfrac{x}{4} = 7$

2. $\dfrac{x}{5} + \dfrac{x}{3} = 8$

3. $\dfrac{a}{2} - \dfrac{a}{5} = 6$

4. $\dfrac{b}{3} - \dfrac{b}{7} = 12$

5. $\dfrac{x}{5} + \dfrac{x}{2} = 7$

6. $\dfrac{8}{z} = 4$

7. $\dfrac{9}{2x} = 3$

8. $\dfrac{14}{3x} = 7$

9. $z + \dfrac{1}{z} = \dfrac{17}{z}$

10. $y + \dfrac{3}{y} = \dfrac{12}{y}$

11. $\dfrac{2}{x} - \dfrac{2}{x^2} = \dfrac{1}{2}$

12. $\dfrac{3}{x} - \dfrac{4}{x^2} = \dfrac{1}{2}$

13. $\dfrac{M-2}{5} + \dfrac{M}{3} = \dfrac{1}{5}$

14. $\dfrac{y+2}{4} + \dfrac{y}{5} = \dfrac{1}{4}$

15. $\dfrac{7}{x+4} = \dfrac{3}{x}$

16. $\dfrac{5}{x+6} = \dfrac{2}{x}$

17. $\dfrac{3x}{x-2} = 5$

18. $\dfrac{4}{x+3} = \dfrac{2}{x}$

19. $\dfrac{2z-4}{3} + \dfrac{3z}{2} = \dfrac{5}{6}$

20. $\dfrac{y-1}{2} + \dfrac{y}{5} = \dfrac{3}{10}$

21. $\dfrac{1}{x} + \dfrac{12}{x^2} = 1$

22. $\dfrac{2}{x} + \dfrac{1}{x^2} = 3$

23. $\dfrac{x}{x+1} = \dfrac{4x}{3x+2}$

24. $\dfrac{x}{3x-4} = \dfrac{3x}{2x+2}$

25. $\dfrac{2x}{3x+1} = \dfrac{4x}{5x+1}$

26. $\dfrac{4y-3}{2} = \dfrac{5y}{y+2}$

27. $\dfrac{6}{y-2} = \dfrac{3}{y}$

28. $\dfrac{5}{x-3} = \dfrac{2}{x}$

29. $\dfrac{x}{x^2+1} = \dfrac{2}{1+2x}$

30. $\dfrac{3x}{3x^2+2} = \dfrac{1}{x+1}$

31. $\dfrac{2x-1}{3} + \dfrac{3x}{4} = \dfrac{5}{6}$

32. $\dfrac{3z-2}{4} + \dfrac{3z}{8} = \dfrac{3}{4}$

33. $\dfrac{x}{x-2} = \dfrac{2}{x-2} + 5$

34. $\dfrac{x}{x+5} = 4 - \dfrac{5}{x+5}$

35. $\dfrac{1}{x-2} - \dfrac{4}{x+2} = \dfrac{1}{5}$

36. $\dfrac{4}{x+1} = \dfrac{3}{x} + \dfrac{1}{15}$

37. $\dfrac{2}{x-3} + 2 = \dfrac{6}{x^2-3x}$

38. $1 + \dfrac{2}{y-1} = \dfrac{2}{y^2-y}$

39. $\dfrac{3}{2x+5} + \dfrac{x}{4} = \dfrac{3}{4}$

40. $\dfrac{5}{2x-1} - \dfrac{x}{6} = \dfrac{4}{3}$

12.8 Literal Equations

Literal equations are equations that have more than one variable. When we solve a literal equation for one of its variables, the solution will contain the other variables as well as numbers.

EXAMPLE 1

a. Solve $x + y = 5$ for x.

> **SOLUTION** $x + y = 5$ Subtract y from both sides
>
> $x = 5 - y$ Solution for x

b. Solve $x + y = 5$ for y.

SOLUTION $x + y = 5$ Subtract x from both sides

$y = 5 - x$ Solution for y

Solving a literal equation

1. Remove fractions (if there are any) by multiplying both sides by the LCD.
2. Remove grouping symbols (if there are any).
3. Get all terms containing the variable you are solving for on one side, and get all other terms on the other side.
4. Factor out the variable you are solving for (if it appears in more than one term).
5. Divide both sides by the coefficient of the variable you are solving for.

EXAMPLE 2   Solve $3x + 4y = 12$ for x.

SOLUTION $3x + 4y = 12$

$3x = 12 - 4y$ Subtracting $4y$ from both sides

$\dfrac{3x}{3} = \dfrac{12 - 4y}{3}$ Dividing both sides by 3

$x = \dfrac{12 - 4y}{3}$ Solution for x

EXAMPLE 3   Solve $\dfrac{4ab}{d} = 15$ for a.

Sometimes a literal equation can be solved as a proportion

SOLUTION $\dfrac{4ab}{d} = \dfrac{15}{1}$ This is a proportion

$4ab \cdot 1 = 15 \cdot d$ Product of means = product of extremes

$\dfrac{\overset{1}{4}a\overset{1}{b}}{\underset{1}{4b}} = \dfrac{15d}{4b}$ Dividing both sides by $4b$

$a = \dfrac{15d}{4b}$ Solution

EXAMPLE 4   Solve $A = P(1 + rt)$ for t.

SOLUTION $A = P(1 + rt)$

$A = P + Prt$ Removing parentheses by using the distributive rule

$A - P = Prt$ Getting the P-term to the other side by subtraction

$\dfrac{A - P}{Pr} = \dfrac{\overset{1}{P}r\overset{}{t}}{\underset{1}{Pr}}$ Dividing both sides by Pr

$\dfrac{A - P}{Pr} = t$ Solution

EXAMPLE 5 Solve $I = \dfrac{nE}{R + nr}$ for n.

SOLUTION

$$\frac{I}{1} = \frac{nE}{R + nr}$$ This is a proportion

$$1(nE) = I(R + nr)$$ Product of means = product of extremes

$$nE = IR + Inr$$ Removing parentheses using distributive rule

$$nE - Inr = IR$$ Getting all terms containing n on one side of the equation

$$n(E - Ir) = IR$$ Factoring out the common factor n

$$\frac{n(\overset{1}{\cancel{E - Ir}})}{(\underset{1}{\cancel{E - Ir}})} = \frac{IR}{(E - Ir)}$$ Dividing both sides by $(E - Ir)$

$$n = \frac{IR}{E - Ir}$$ Solution

Exercises 12.8

Solve for the letter specified after each equation.

1. $2x + y = 4$; x

2. $x + 3y = 6$; y

3. $y - z = -8$; z

4. $m - n = -5$; n

5. $2x - y = -4$; y

6. $3y - z = -5$; z

7. $2x - 3y = 6$; x

8. $3x - 2y = 6$; x

9. $x + 2y = 5$; x

10. $x - y = -4$; y

11. $2x - y = -4$; x

12. $3x - 4y = 12$; y

13. $2(x - 3y) = x + 4$; x

14. $3x - 14 = 2(y - 2x)$; x

15. $PV = k$; V

16. $I = Prt$; P

17. $\dfrac{3xy}{z} = 10$; x

18. $\dfrac{PV}{T} = 100$; V

19. $y = mx + b$; x

20. $P = 2l + 2w$; w

21. $A = P(1 + rt)$; r

22. $b = c(1 + xy)$; x

23. $S = \dfrac{a}{1 - r}$; r

24. $I = \dfrac{E}{R + r}$; R

25. $ax + bx = c$; x

26. $az - 2z = 4$; z

27. $z = \dfrac{Rr}{R + r}$; R

28. $c = \dfrac{ax}{x + 3}$; x

29. $\dfrac{1}{F} = \dfrac{1}{u} + \dfrac{1}{v}$; u

30. $\dfrac{1}{a} = \dfrac{1}{b} - \dfrac{1}{c}$; b

31. $C = \dfrac{5}{9}(F - 32)$; F

32. $A = \dfrac{h}{2}(B + b)$; B

33. $\dfrac{m + n}{x} - a = \dfrac{m - n}{x} + c$; x

34. $\dfrac{a - b}{x} + c = \dfrac{a + b}{x} - h$; x

12.9 Word Problems Involving Fractions

Number Problems

EXAMPLE 1 The sum of a number and its reciprocal is $\dfrac{13}{6}$. Find the number.

SOLUTION Let x = the number

Then $\dfrac{1}{x}$ = its reciprocal

$$\boxed{\text{Sum of a number and its reciprocal}} \;\; \text{is} \;\; \frac{13}{6}$$

$$x + \frac{1}{x} = \frac{13}{6} \qquad \text{LCD} = 6x$$

$$6x(x) + \overset{1}{6x}\left(\frac{1}{\underset{1}{\cancel{x}}}\right) = \overset{1}{6x}\left(\frac{13}{\underset{1}{\cancel{6}}}\right)$$

$$6x^2 + 6 = 13x$$

$$6x^2 - 13x + 6 = 0$$

$$(2x - 3)(3x - 2) = 0$$

$$2x - 3 = 0 \qquad \bigg| \qquad 3x - 2 = 0$$

$$2x = 3 \qquad \bigg| \qquad 3x = 2$$

$$x = \frac{3}{2} \qquad \bigg| \qquad x = \frac{2}{3}$$

Therefore, there are two answers: $\dfrac{3}{2}$ and $\dfrac{2}{3}$.

✓ **Check**

$$\boxed{\text{Number}} + \boxed{\text{reciprocal}}$$

$$\frac{3}{2} + \frac{2}{3} = \frac{9}{6} + \frac{4}{6} = \frac{13}{6}$$

$$\boxed{\text{Number}} + \boxed{\text{reciprocal}}$$

$$\frac{2}{3} + \frac{3}{2} = \frac{4}{6} + \frac{9}{6} = \frac{13}{6}$$

EXAMPLE 2   The denominator of a fraction exceeds the numerator by 3. If 4 is added to the numerator and 2 is subtracted from the denominator, the resulting fraction is $\dfrac{3}{2}$. Find the original fraction.

SOLUTION   Let x = numerator
Then $x + 3$ = denominator

$$\boxed{\begin{array}{c}\text{If 4 is added to the}\\ \text{numerator and 2}\\ \text{is subtracted from}\\ \text{the denominator}\end{array}} \quad \boxed{\begin{array}{c}\text{the resulting}\\ \text{fraction is}\end{array}} \quad \boxed{\dfrac{3}{2}}$$

$$\frac{x + 4}{x + 3 - 2} = \frac{3}{2}$$

$$\frac{x + 4}{x + 3 - 2} = \frac{3}{2}$$

$$\frac{x + 4}{x + 1} = \frac{3}{2} \qquad \text{A proportion}$$

$$3(x + 1) = 2(x + 4)$$

$$3x + 3 = 2x + 8$$

$$x + 3 = 8$$
$$x = 5$$

Therefore, the original fraction is $\dfrac{x}{x + 3} = \dfrac{5}{8}$.

√ **Check**

$$\frac{5 + 4}{8 - 2} \stackrel{?}{=} \frac{3}{2}$$

$$\frac{9}{6} \stackrel{?}{=} \frac{3}{2}$$

$$\frac{3}{2} = \frac{3}{2}$$

Work Problems

The basic principle used to solve work problems is

$$\text{Rate} \times \text{time} = \left(\begin{array}{c}\text{amount}\\\text{of work}\end{array}\right)$$

Suppose John can mow a lawn in 2 hours. Then

$$\text{John's } rate = \frac{1 \text{ lawn}}{2 \text{ hr}} = \frac{1}{2} \text{ lawn per hr}$$

If John works for 8 hours,

$$\text{Rate} \times \text{time} = \left(\begin{array}{c}\text{amount}\\\text{of work}\end{array}\right)$$

$$\frac{1}{2} \quad \cdot \quad 8 \quad = \quad 4 \text{ lawns}$$

If John works for x hours,

$$\text{Rate} \times \text{time} = \left(\begin{array}{c}\text{amount}\\\text{of work}\end{array}\right)$$

$$\frac{1}{2} \quad \cdot \quad x \quad = \quad \frac{x}{2} \text{ lawns}$$

EXAMPLE 3 Spyro can mow a lawn in 2 hours, and Assim can mow the same lawn in 4 hours. How long will it take them to mow the lawn working together?

SOLUTION Let x = time each works

$$\text{Spyro's rate} = \frac{1 \text{ lawn}}{2 \text{ hr}} = \frac{1}{2} \text{ lawn per hr}$$

$$\text{Assim's rate} = \frac{1 \text{ lawn}}{4 \text{ hr}} = \frac{1}{4} \text{ lawn per hr}$$

	Rate	· Time	= Amount of work
Spyro	$\dfrac{1}{2}$	x	$\dfrac{x}{2}$
Assim	$\dfrac{1}{4}$	x	$\dfrac{x}{4}$

Amount Spyro mows	+	amount Assim mows	=	1 lawn

$$\frac{x}{2} \quad + \quad \frac{x}{4} \quad = \quad 1 \qquad \text{LCD} = 4$$

$$\overset{2}{4}\left(\frac{x}{\underset{1}{2}}\right) + \overset{1}{4}\left(\frac{x}{\underset{1}{4}}\right) = 4(1)$$

$$2x + x = 4$$

$$3x = 4$$

$$x = \frac{4}{3} = 1\frac{1}{3}$$

Therefore, it will take them $1\dfrac{1}{3}$ hours working together.

√ **Check**

$$\text{Spyro's work} = \frac{1 \text{ lawn}}{2 \text{ hr}} \cdot \frac{4}{3} \text{ hr} = \frac{2}{3} \text{ lawn}$$

$$\text{Assim's work} = \frac{1 \text{ lawn}}{4 \text{ hr}} \cdot \frac{4}{3} \text{ hr} = + \frac{1}{3} \text{ lawn}$$

$$\overline{ 1 \text{ lawn}}$$

EXAMPLE 4

Mary can wash a car in 45 minutes, and Yuki can wash a car in 30 minutes. How long will it take them to wash 10 cars working together?

SOLUTION Let x = time each works

$$\text{Mary's rate} = \frac{1 \text{ car}}{45 \text{ min}} = \frac{1}{45} \text{ car per min}$$

$$\text{Yuki's rate} = \frac{1 \text{ car}}{30 \text{ min}} = \frac{1}{30} \text{ car per min}$$

	Rate	· Time	= Amount of work
Mary	$\dfrac{1}{45}$	x	$\dfrac{x}{45}$
Yuki	$\dfrac{1}{30}$	x	$\dfrac{x}{30}$

Amount Mary washes	+	amount Yuki washes	=	10 cars

$$\frac{x}{45} \quad + \quad \frac{x}{30} \quad = \quad 10 \qquad \text{LCD} = 90$$

$$\overset{2}{\cancel{90}}\left(\frac{x}{\underset{1}{45}}\right) + \overset{3}{\cancel{90}}\left(\frac{x}{\underset{1}{30}}\right) = 90(10)$$

$$2x + 3x = 900$$

$$5x = 900$$

$$x = 180$$

Therefore, it will take them 180 minutes, or 3 hours.

√ **Check**

$$\text{Mary's work} = \frac{1 \text{ car}}{45 \text{ min}} \cdot 180 \text{ min} = \quad 4 \text{ cars}$$

$$\text{Yuki's work} = \frac{1 \text{ car}}{30 \text{ min}} \cdot 180 \text{ min} = + \ 6 \text{ cars}$$

$$\underline{\phantom{= + 6 \text{ cars}}}$$

$$10 \text{ cars}$$

EXAMPLE 5

The hot water faucet can fill the bath tub in 10 minutes, and the cold water faucet can fill the tub in 5 minutes. The drain can empty the tub in 15 minutes. How long will it take to fill the tub if both faucets are turned on, but the drain is accidentally left open for 3 minutes?

SOLUTION   Let $x = $ time both faucets are on

$$\text{Hot water's rate} = \frac{1 \text{ tub}}{10 \text{ min}} = \frac{1}{10} \text{ tub per min}$$

$$\text{Cold water's rate} = \frac{1 \text{ tub}}{5 \text{ min}} = \frac{1}{5} \text{ tub per min}$$

$$\text{Drain's rate} = \frac{1 \text{ tub}}{15 \text{ min}} = \frac{1}{15} \text{ tub per min}$$

	Rate	· Time	= Amount of work
Hot water	$\frac{1}{10}$	x	$\frac{x}{10}$
Cold water	$\frac{1}{5}$	x	$\frac{x}{5}$
Drain	$\frac{1}{15}$	3	$\frac{3}{15} = \frac{1}{5}$

Amount hot water fills	+	amount cold water fills	−	amount drain empties	=	1 full tub

└─ Because drain *empties*

$$\frac{x}{10} \quad + \quad \frac{x}{5} \quad - \quad \frac{1}{5} \quad = \quad 1 \qquad \text{LCD} = 10$$

$$\overset{1}{\cancel{10}}\left(\frac{x}{\underset{1}{10}}\right) + \overset{2}{\cancel{10}}\left(\frac{x}{\underset{1}{5}}\right) - \overset{2}{\cancel{10}}\left(\frac{1}{\underset{1}{5}}\right) = 10(1)$$

$$x + 2x - 2 = 10$$

$$3x - 2 = 10$$

$$3x = 12$$
$$x = 4$$

Therefore, it will take 4 minutes to fill the tub.

√   *Check*

$$\text{Hot water} = \frac{1 \text{ tub}}{10 \text{ min}} \cdot 4 \text{ min} = \frac{4}{10} = \frac{2}{5} \text{ tub}$$

$$\text{Cold water} = \frac{1 \text{ tub}}{5 \text{ min}} \cdot 4 \text{ min} = \frac{4}{5} \text{ tub}$$

$$\text{Drain} = \frac{1 \text{ tub}}{15 \text{ min}} \cdot 3 \text{ min} = \frac{3}{15} = \frac{1}{5} \text{ tub}$$

$$\text{And} \quad \frac{2}{5} + \frac{4}{5} - \frac{1}{5} = \frac{5}{5} = 1 \text{ tub}$$

Exercises   12.9

In the following word problems, (a) represent the unknown numbers using a variable, (b) set up an equation and solve it, and (c) solve the problem.

1. The sum of a number and its reciprocal is $\frac{25}{12}$. Find the number.

2. The sum of a number and its reciprocal is $\frac{29}{10}$. Find the number.

3. The difference between a number and its reciprocal is $\frac{8}{3}$. Find the number.

4. The difference between a number and its reciprocal is $\frac{3}{2}$. Find the number.

5. The denominator of a fraction exceeds the numerator by 2. If 1 is subtracted from the numerator and 5 is added to the denominator, the resulting fraction is $\frac{1}{5}$. Find the original fraction.

6. The denominator of a fraction exceeds the numerator by 4. If 3 is added to the numerator and the denominator, the resulting fraction is $\frac{2}{3}$. Find the original fraction.

7. The denominator of a fraction is twice the numerator. If 2 is added to the numerator and 8 is added to the denominator, the resulting fraction is $\frac{2}{5}$. Find the original fraction.

8. The denominator of a fraction is three times the numerator. If 1 is subtracted from the numerator and 3 is added to the denominator, the resulting fraction is $\frac{2}{9}$. Find the original fraction.

9. Lee can do a job in 2 hours, and Dwight can do the same job in 6 hours. How long will it take them to do the job working together?

10. Ken can do a job in 4 hours, and Betty can do the same job in 6 hours. How long will it take them to do the job working together?

11. Pipe A can fill a tank in 3 hours, pipe B can fill the tank in 2 hours, and pipe C can fill the tank in 6 hours. How long will it take to fill the tank if all three pipes are turned on?

12. Pipe A can fill a tank in 6 hours, pipe B can fill the tank in 3 hours, and pipe C can drain the tank in 8 hours. How long will it take to fill the tank if all three pipes are turned on?

13. Carlo can paint a wall in 30 minutes, and Marcia can paint a wall in 50 minutes. How long will it take them to paint 4 walls working together?

14. Juan can type a page in 10 minutes, and Monica can type a page in 15 minutes. How long will it take them to type a 20-page report?

15. The old machine can sort the mail in 4 hours, and the new machine can sort the mail in 3 hours. How long will it take to sort the mail with both machines working if the old machine breaks down after 1 hour?

16. Pipe A can fill a water tank in 6 hours, and pipe B can fill the tank in 3 hours. Pipe C can drain the tank in 8 hours. How long will it take pipes A and B to fill the tank if pipe C is accidentally left open for 1 hour?

17. Tonya and Irma live 54 miles apart. Both leave their homes at 7 A.M. by bicycle, riding toward each other. They meet at 10 A.M. If Irma's average speed is four-fifths of Tonya's, how fast does each cycle?

18. Hahn and Cathy live 60 miles apart. Both leave their homes at 10 A.M. by bicycle, riding toward each other. They meet at 2 P.M. If Cathy's average speed is two-thirds of Hahn's, how fast does each cycle?

19. Shakil scored 70, 85, and 83 on three exams. If she must average at least 80 to get a B in the class, what must she score on the fourth exam to receive a B?

20. Ernest scored 95, 88, and 91 on three exams. If he must average at least 90 to get an A in the class, what must he score on the final exam to receive an A if the final exam is counted as two tests?

Chapter 12   R E V I E W

Simple Fractions
12.1, 12.6

A **simple fraction** is a fraction having only one fraction bar.

$$\frac{P}{Q}\begin{matrix}\longleftarrow \text{Numerator}\\ \longleftarrow \text{Denominator}\end{matrix}$$

where $Q \neq 0$.

Excluded Values
12.1

Any value of the variable that makes the denominator equal to zero must be excluded.

Finding Excluded Values
12.1

If the denominator contains a variable, set the denominator equal to zero and solve the equation. The solutions are the values of the variable that must be excluded.

The Three Signs of a Fraction
12.1

Every fraction has three signs associated with it: the sign of the entire fraction, the sign of the numerator, and the sign of the denominator. *If any two of the three signs of a fraction are changed, the value of the fraction is unchanged.*

Reducing a Fraction to Lowest Terms
12.1

1. Factor the numerator and denominator completely.

2. Divide the numerator and denominator by all factors common to both.

Multiplying Fractions
12.2

1. Factor the numerator and denominator of the fractions.

2. Divide the numerator and denominator by all factors common to both.

3. The answer is the product of factors remaining in the numerator divided by the product of factors remaining in the denominator:

$$\frac{P}{Q} \cdot \frac{R}{S} = \frac{P \cdot R}{Q \cdot S}$$

where $Q \neq 0$ and $S \neq 0$.

Dividing Fractions
12.2

To divide fractions, invert the second fraction and multiply:

$$\frac{P}{Q} \div \frac{R}{S} = \frac{P}{Q} \cdot \frac{S}{R}$$

where $Q \neq 0$, $R \neq 0$, and $S \neq 0$.

Adding or Subtracting Like Fractions
12.3

1. Write the sum or difference of the numerators over the denominator of the like fractions:

$$\frac{P}{Q} + \frac{R}{Q} = \frac{P + R}{Q} \quad \text{or} \quad \frac{P}{Q} - \frac{R}{Q} = \frac{P - R}{Q}$$

where $Q \neq 0$.

2. Reduce the resulting fraction to lowest terms.

Finding the LCD
12.4

1. Factor each denominator completely. Repeated factors should be expressed as powers.

2. Write down each different factor that appears.

3. Raise each factor to the *highest* power to which it occurs in *any* denominator.

4. The LCD is the product of all the powers found in Step 3.

Adding Unlike Fractions
12.5

1. Find the LCD.

2. Convert each fraction to an equivalent fraction having the LCD as its denominator.

3. Add the resulting like fractions.

4. Reduce the fraction found in Step 3 to lowest terms.

Complex Fractions
12.6

A **complex fraction** is a fraction having more than one fraction bar.

Simplifying Complex Fractions
12.6

Method 1. Multiply both the numerator and the denominator of the complex fraction by the LCD of the secondary fractions; then simplify the results.

Method 2. First, simplify the numerator and denominator of the complex fraction; then divide the simplified numerator by the simplified denominator.

Solving an Equation That Has Fractions
12.7

1. Find the LCD, and note any excluded values.

2. Remove fractions by multiplying each term on both sides of the equation by the LCD.

3. Remove grouping symbols.

4. Combine like terms.

If a *first-degree equation* is obtained in Step 3:

5. Get all terms with the variable on one side, and get the remaining terms on the other side.

6. Divide both sides by the coefficient of the variable.

If a *second-degree* (quadratic) *equation* is obtained in Step 3:

5. Get all nonzero terms on one side, and arrange them in descending powers. *Only zero must remain on the other side.*

6. Factor the polynomial.

7. Set each factor equal to zero, and then solve for the variable.

Check the apparent solutions in the original equation. Any excluded values must be rejected as solutions.

Literal Equations
12.8

Literal equations are equations that have more than one variable.

Solving a Literal Equation
12.8

To solve a literal equation, proceed in the same way you would to solve an equation with a single variable. The solution will be expressed in terms of the other variables given in the literal equation, as well as numbers.

Chapter 12   REVIEW EXERCISES

In Exercises 1–3, what value(s) of the variable (if any) must be excluded?

1. $\dfrac{2x - 1}{x + 4}$

2. $7x + \dfrac{2}{3x}$

3. $\dfrac{x - 1}{x^2 - 3x - 10}$

In Exercises 4–6, use the rule about the three signs of a fraction to find the missing term.

4. $-\dfrac{2}{5} = \dfrac{-2}{?}$

5. $\dfrac{5}{2 - x} = \dfrac{-5}{?}$

6. $\dfrac{2}{a - b} = -\dfrac{?}{b - a}$

In Exercises 7–12, reduce each fraction to lowest terms.

7. $\dfrac{4x^3 y}{2xy^2}$

8. $\dfrac{2 + 4m}{2}$

9. $\dfrac{a^2 - 4}{a + 2}$

10. $\dfrac{x + 3}{x^2 - x - 12}$

11. $\dfrac{x^2 - 2xy + y^2}{y^2 - x^2}$

12. $\dfrac{a - b}{ax + ay - bx - by}$

In Exercises 13–23, perform the indicated operations.

13. $\dfrac{7}{z} - \dfrac{2}{z}$

14. $5 - \dfrac{3}{2x}$

15. $\dfrac{3x}{x + 1} - \dfrac{2x - 1}{x + 1}$

16. $\dfrac{-5a^2}{3b} \div \dfrac{10a}{9b^2}$

17. $\dfrac{3x + 6}{6} \cdot \dfrac{2x^2}{4x + 8}$

18. $\dfrac{x + 4}{5} - \dfrac{x - 2}{3}$

19. $\dfrac{a-2}{a-1} + \dfrac{a+1}{a+2}$

20. $\dfrac{x-y}{xy^2} - \dfrac{y-x}{x^2y}$

21. $\dfrac{3x}{x-3} + \dfrac{9}{3-x}$

22. $\dfrac{x^2}{9-x^2} \div \dfrac{x^2+x}{x^2-2x-3}$

23. $\dfrac{3}{x^2+5x+4} - \dfrac{2}{x^2+6x+8}$

In Exercises 24–26, simplify the complex fractions.

24. $\dfrac{\dfrac{5k^2}{3m^2}}{\dfrac{10k}{9m}}$

25. $\dfrac{\dfrac{x}{y}+3}{\dfrac{x}{y}-3}$

26. $\dfrac{\dfrac{y}{x}-\dfrac{y^2}{x^2}}{1-\dfrac{y}{x}}$

In Exercises 27–32, solve the equations.

27. $\dfrac{6}{m} = 5$

28. $\dfrac{2m}{3} - m = 1$

29. $\dfrac{z}{5} - \dfrac{z}{8} = 3$

30. $\dfrac{2x+1}{3} = \dfrac{5x-4}{2}$

31. $\dfrac{4}{2z} + \dfrac{2}{z} = 1$

32. $\dfrac{3}{x} - \dfrac{8}{x^2} = \dfrac{1}{4}$

In Exercises 33–38, solve for the variable specified after each equation.

33. $3x - 4y = 12$; x

34. $\dfrac{2m}{n} = P$; n

35. $V = LWH$; H

36. $E = \dfrac{mv^2}{gr}$; m

37. $\dfrac{F-32}{C} = \dfrac{9}{5}$; C

38. $\dfrac{ax+b}{x} = c$; x

39. The denominator of a fraction exceeds the numerator by twenty. If seven is added to the numerator and subtracted from the denominator, the resulting fraction equals $\dfrac{6}{7}$. Find the fraction.

40. Twice a number plus three times its reciprocal is 7. Find the number.

41. Eduardo can wash a window in 10 minutes, and Cindy can wash a window in 8 minutes. How long will it take them to wash the 45 office windows, working together?

42. Yelena and Chun live 21 miles apart. Both leave their homes, riding their bicycles toward each other. They meet in $1\dfrac{1}{2}$ hours. If Chun's average speed is 2 mph less than Yelena's, how fast does each cycle?

Chapter 12 Critical Thinking and Writing Problems

Answer Problems 1–6 in your own words, using complete sentences.

1. Explain how to find excluded values for a fraction.

2. Explain how to reduce a fraction.

3. Explain how to find the LCD.

4. Explain how to add fractions.

5. Explain how to divide fractions.

6. Explain how to solve an equation having fractions.

7. True or false: $\dfrac{x+8}{y+8} = \dfrac{x+8}{y+8} = \dfrac{x}{y}$. Explain why.

8. True or false: $\dfrac{2a+6}{2} = \dfrac{2a+6}{2} = a + 6$. Explain why.

Each of the following problems has an error. Find the error, and in your own words, explain why it is wrong. Then work the problem correctly.

9. Subtract $\dfrac{3x}{x-5} - \dfrac{2x-5}{x-5}$.

$\dfrac{3x}{x-5} - \dfrac{2x-5}{x-5}$

$= \dfrac{x-5}{x-5}$

$= 1$

10. Add $\dfrac{x}{x+1} + \dfrac{1}{x+4}$.

$\dfrac{x}{x+1} + \dfrac{1}{x+4}$

$= (\cancel{x+1})(x+4) \cdot \dfrac{x}{\cancel{x+1}} + (x+1)(\cancel{x+4}) \cdot \dfrac{1}{\cancel{x+4}}$

$= x(x+4) + x + 1$

$= x^2 + 4x + x + 1$

$= x^2 + 5x + 1$

11. Add $\dfrac{3}{x} + \dfrac{3}{x+2}$.

$$\dfrac{3}{x} + \dfrac{3}{x+2}$$

$$= \dfrac{3 \cdot 2}{x+2} + \dfrac{3}{x+2}$$

$$= \dfrac{5}{x+2} + \dfrac{3}{x+2}$$

$$= \dfrac{8}{x+2}$$

12. Simplify $\dfrac{3 - \dfrac{9}{x}}{1 - \dfrac{9}{x^2}}$.

$$\dfrac{3 - \dfrac{9}{x}}{1 - \dfrac{9}{x^2}}$$

$$= \dfrac{\dfrac{x}{1} \cdot \dfrac{3}{1} - \dfrac{x}{1} \cdot \dfrac{9}{x}}{\dfrac{x^2}{1} \cdot \dfrac{1}{1} - \dfrac{x^2}{1} \cdot \dfrac{9}{x^2}}$$

$$= \dfrac{3x - 9}{x^2 - 9}$$

$$= \dfrac{3(\overset{1}{\cancel{x-3}})}{(x+3)(\underset{1}{\cancel{x-3}})}$$

$$= \dfrac{3}{x+3}$$

Chapter 12   D I A G N O S T I C   T E S T

Allow yourself about an hour to do these problems. Complete solutions for all problems, together with section references, are given in the answer section at the end of the book.

1. What value(s) of the variable must be excluded, if any?

   a. $\dfrac{3x}{x-4}$ **b.** $\dfrac{5x+4}{x^2+2x}$

2. Use the rule about the three signs of a fraction to find the missing term.

   a. $-\dfrac{-4}{5} = \dfrac{4}{?}$ **b.** $\dfrac{-3}{x-y} = \dfrac{?}{y-x}$

In Problems 3–5, reduce each fraction to lowest terms.

3. $\dfrac{6x^3y}{9x^2y^2}$ **4.** $\dfrac{x^2+8x+16}{x^2-16}$

5. $\dfrac{6a^2+11ab-10b^2}{6a^2b-4ab^2}$

In Problems 6–11, perform the indicated operations. (Be sure to reduce fractions to lowest terms.)

6. $\dfrac{2}{x^2-4} \cdot \dfrac{x^2-2x-8}{2x-8}$ **7.** $\dfrac{3y}{y-5} + \dfrac{15}{5-y}$

8. $\dfrac{1}{6x} + \dfrac{3}{4x^2}$ **9.** $\dfrac{b}{b-1} - \dfrac{b+1}{b}$

10. $\dfrac{x^2}{x^2-3x} \div \dfrac{3x-15}{x^2-8x+15}$

11. $\dfrac{2}{x^2+4x+3} - \dfrac{1}{x^2+5x+6}$

In Problems 12 and 13, simplify each complex fraction.

12. $\dfrac{\dfrac{9x^5}{10y}}{\dfrac{3x^2}{20y^3}}$ **13.** $\dfrac{1 + \dfrac{2}{a}}{1 - \dfrac{4}{a^2}}$

In Problems 14–19, solve each equation.

14. $\dfrac{y}{3} - \dfrac{y}{4} = 1$ **15.** $\dfrac{x-2}{5} = \dfrac{x+1}{2} + \dfrac{3}{5}$

16. $\dfrac{3}{a+4} = \dfrac{5}{a}$ **17.** $\dfrac{1}{x} + \dfrac{6}{x^2} = 1$

18. Solve for x: $3x - 4y = 9$.

19. Solve for P: $PM = Q + PN$.

20. Sid can unload a truck full of dirt in 20 minutes, and Jorgé can unload the same truck in 30 minutes. How long would it take them to unload the truck working together?

Graphing

CHAPTER

13

any algebraic relationships are easier to understand if a picture called a *graph* is drawn. In this chapter we discuss how to draw such graphs.

13.1 The Rectangular Coordinate System

Earlier we discussed how any real number can be represented by a point on a *number line* (Figure 1).

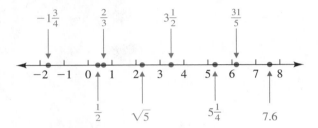

Figure 1   Horizontal number line

Now we will draw a second number line vertically, with its zero point at the zero point of the horizontal number line; these two lines form the **axes** of a **rectangular coordinate system**. The rectangular coordinate system consists of a vertical number line called the **vertical axis**, or **y-axis**, and a horizontal line called the **horizontal axis**, or **x-axis**, that meet at a point called the **origin**. The vertical and horizontal axes determine the **plane** of the rectangular coordinate system; they also divide this plane into four **quadrants** (Figure 2).

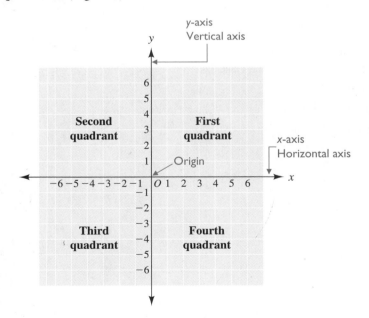

Figure 2   Rectangular coordinate system

With a single number line, we needed only a *single* real number to represent a point on that line. With two lines forming a rectangular coordinate system, we need a *pair* of real numbers to represent a point in the plane.

Graphing Points

A point is represented by an **ordered pair** of numbers enclosed in parentheses. The point (3, 2) is shown in Figure 3. We call 3 and 2 the **coordinates** of the point (3, 2). The first number, 3, is called the **x-coordinate** (or **horizontal coordinate**). The second number, 2, is called the **y-coordinate** (or **vertical coordinate**). To graph the point (3, 2), start at the origin and move 3 units *right*, then move 2 units *up*.

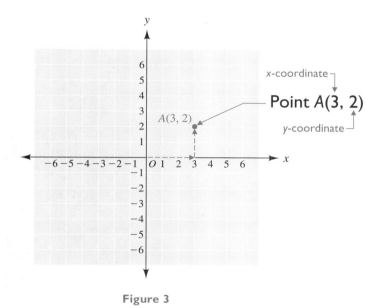

Figure 3

The *x*-coordinate of an ordered pair tells us the horizontal movement; to the right is positive, and to the left is negative. The *y*-coordinate tells us the vertical movement; up is positive, and down is negative.

 Note   When the order is changed in an ordered pair, we get a different point. For example, (1, 4) and (4, 1) are two different points (see Figure 4).

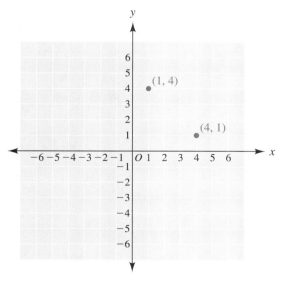

Figure 4

EXAMPLE 1   Graph the following points on Figure 5.

SOLUTION

a. (3, 5)   Start at the origin and move *right* 3 units, then move *up* 5 units (point A).

b. (−5, 2)   Start at the origin and move *left* 5 units, then move *up* 2 units (point B).

c. (−5, −4)   Start at the origin and move *left* 5 units, then move *down* 4 units (point C).

d. (0, −3)   Start at the origin, but because the first number is zero, do not move either right or left; just move *down* 3 units (point D).

e. (4, −6)   Start at the origin and move *right* 4 units, then move *down* 6 units (point E).

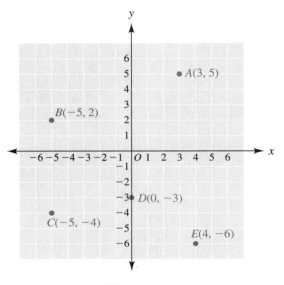

Figure 5

The phrase *plot the points* means the same as *graph the points*.

Exercises   13.1

1. Graph each of the following points.

    a. (3, 1)   **b.** (−4, −2)   **c.** (0, 3)

    d. (5, −4)   **e.** (4, 0)   **f.** (−2, 4)

2. Graph each of the following points.

    a. (2, 4)   **b.** (2, −4)   **c.** (3, 0)

    d. (−3, −2)   **e.** (0, 0)   **f.** (0, −4)

In Exercises 3 and 4, use the figure on the right.

3. Give the coordinates of each of the following points.

    a. R   **b.** N   **c.** U   **d.** S

4. Give the coordinates of each of the following points.

    a. M   **b.** P   **c.** Q   **d.** T

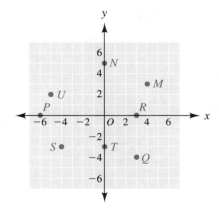

Figure for Exercises 3 and 4

In Exercises 5 and 6, use the figure below.

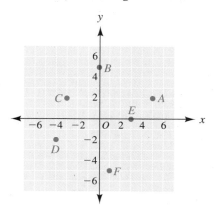

Figure for Exercises 5 and 6

5. Write the *x*-coordinate of each of the following points.

   a. *A*   **b.** *C*   **c.** *E*   **d.** *F*

6. Write the *y*-coordinate of each of the following points.

   a. *B*   **b.** *D*   **c.** *E*   **d.** *F*

7. What name is given to the point (0, 0)?

8. What is the *x*-coordinate of the origin?

9. Draw the triangle whose vertices (corners) have the following coordinates:

   $$A(0, 0) \qquad B(3, 2) \qquad C(-4, 5)$$

10. Draw the triangle whose vertices (corners) have the following coordinates:

   $$A(-2, -3) \qquad B(-2, 4) \qquad C(3, 5)$$

13.2 Graphing Lines

In the preceding section we showed how to graph points. In this section we show how to graph straight lines.

Table of Values

The equation $y = x + 1$ has two variables. A solution to this equation is a pair of numbers, one for x and one for y, that satisfies the equation. For example, if $x = 0$, then

$$y = x + 1$$
$$y = 0 + 1 = 1$$

The values $x = 0$ and $y = 1$ form an ordered pair $(0, 1)$, which represents a solution to the equation $y = x + 1$. We can find other solutions to the equation by choosing several values for x and finding the corresponding values for y. Four of the solutions are listed in the following **table of values**.

Table of values

Equation: $y = x + 1$

When $x = \boxed{0}$, $y = 0 + 1 = \boxed{1}$

When $x = \boxed{2}$, $y = 2 + 1 = \boxed{3}$

When $x = \boxed{5}$, $y = 5 + 1 = \boxed{6}$

When $x = \boxed{-3}$, $y = -3 + 1 = \boxed{-2}$

x	*y*
0	1
2	3
5	6
-3	-2

Each of these ordered pairs represents a solution to the equation $y = x + 1$

When we plot, in Figure 6, the four ordered pairs contained in the table of values, the points appear to lie in a straight line. This suggests that the graph of $y = x + 1$ is a straight line. In fact, any first-degree equation (in no more than two variables) has a graph that is a straight line. Such equations are called **linear** (line ar) **equations**.

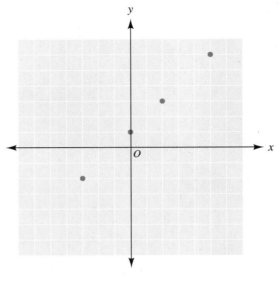

Figure 6

In Figure 7 we have drawn the line through the points listed in the table of values. This line represents the graph of all the solutions to the equation $y = x + 1$; therefore, any point on the line should satisfy the equation. We put arrowheads on both ends of the line to indicate that the line can be extended beyond the graph.

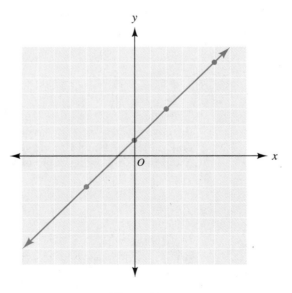

Figure 7

We can draw a straight line if we know two points that lie on that line. Although only two points are *necessary* to draw the line, it is *advisable* to plot a third point as a **checkpoint**. If the three points do not lie in a straight line, we know that a mistake has been made in some calculation of the coordinates of the points.

EXAMPLE 1

Graph the equation $3x - 5y = 15$.

SOLUTION Substitute any three values for x, and find the corresponding values for y. We will let x be 0, 2, and 5.

Equation: $3x - 5y = 15$

If $x = \boxed{0}$, $3(0) - 5y = 15$
$-5y = 15$
$y = \boxed{-3}$

If $x = \boxed{2}$, $3(2) - 5y = 15$
$6 - 5y = 15$
$-5y = 9$
$y = -\dfrac{9}{5} = \boxed{-1\dfrac{4}{5}}$

If $x = \boxed{5}$, $3(5) - 5y = 15$
$15 - 5y = 15$
$-5y = 0$
$y = \boxed{0}$

x	y
0	-3
2	$-1\dfrac{4}{5}$
5	0

Plot the three points from the table of values, and draw a straight line through them, as shown in Figure 8.

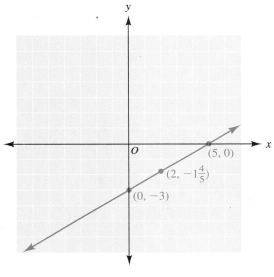

Figure 8

Intercepts

Students often ask which points to use when plotting the graph of a straight line. We usually start by choosing zeros.

x	y
	0
0	

For example, suppose the equation of the line is $3x - 2y = -6$.

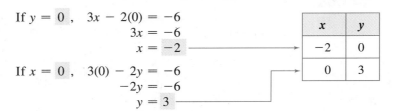

If $y = \boxed{0}$, $3x - 2(0) = -6$
$$3x = -6$$
$$x = \boxed{-2}$$

If $x = \boxed{0}$, $3(0) - 2y = -6$
$$-2y = -6$$
$$y = \boxed{3}$$

x	y
-2	0
0	3

The points we find by this method are called the *x*-intercept and the *y*-intercept. The **x-intercept** of an equation is the point where its graph meets the *x*-axis. The *y*-value at this point is zero (Figure 9). The **y-intercept** of an equation is the point where its graph meets the *y*-axis. The *x*-value at this point is zero (Figure 9).

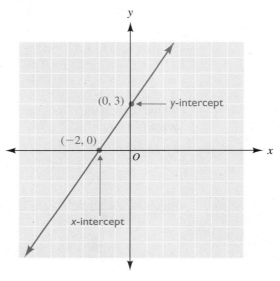

Figure 9

EXAMPLE 2   Graph the equation $4x + 3y = 12$.

SOLUTION   The *intercepts method* of graphing a straight line is often easier to use than the method shown in Example 1.

x-intercept   Set $y = 0$. Then $4x + 3y = 12$ becomes

$$4x + 3(0) = 12$$
$$4x = 12$$
$$x = 3$$

x	y
3	0
0	

Therefore, the *x*-intercept is (3, 0).

y-intercept   Set $x = 0$. Then $4x + 3y = 12$ becomes

$$4(0) + 3y = 12$$
$$3y = 12$$
$$y = 4$$

x	y
3	0
0	4

Therefore, the *y*-intercept is (0, 4).

Checkpoint Set $x = 6$. Then $4x + 3y = 12$ becomes

$$4(6) + 3y = 12$$
$$24 + 3y = 12$$
$$3y = -12$$
$$y = -4$$

x	y
3	0
0	4
6	−4

Therefore, the checkpoint is $(6, -4)$.

  Plot the x-intercept $(3, 0)$, the y-intercept $(0, 4)$, and the checkpoint $(6, -4)$; then draw the straight line through them, as shown in Figure 10.

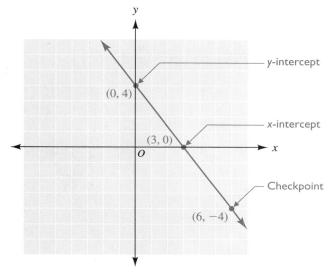

Figure 10

EXAMPLE 3 Graph the equation $3x - 4y = 0$.

SOLUTION

x-intercept Set $y = 0$. Then $3x - 4y = 0$ becomes

$$3x - 4(0) = 0$$
$$3x = 0$$
$$x = 0$$

Therefore, the x-intercept is $(0, 0)$ (the origin). Because the line goes through the origin, the y-intercept is also $(0, 0)$.

  We have found only one point on the line: $(0, 0)$. Therefore, we must find another point on the line. To find another point, we must set one variable equal to a number and then solve the equation for the other variable. Let's set $y = 3$. Then $3x - 4y = 0$ becomes

$$3x - 4(3) = 0$$
$$3x = 12$$
$$x = 4$$

This gives the point $(4, 3)$ on the line.

Checkpoint   Set $x = -4$. Then $3x - 4y = 0$ becomes

$$3(-4) - 4y = 0$$
$$-12 - 4y = 0$$
$$-4y = 12$$
$$y = -3$$

x	y
0	0
4	3
-4	-3

Therefore, the checkpoint is $(-4, -3)$.

Plot the points $(0, 0)$, $(4, 3)$, and $(-4, -3)$; then draw the straight line through them as shown in Figure 11.

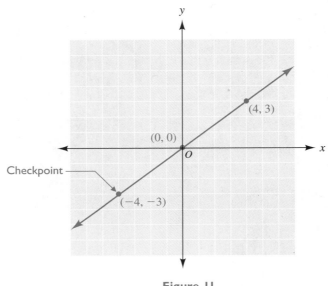

Figure 11

Some equations of a line have only one variable. Such equations have graphs that are either vertical or horizontal lines (Examples 4 and 5).

EXAMPLE 4   Graph the equation $x = 3$.

SOLUTION   The equation $x = 3$ is equivalent to $0y + x = 3$.

If $y = 5$, $0(5) + x = 3$
$$0 + x = 3$$
$$x = 3$$

If $y = -2$, $0(-2) + x = 3$
$$0 + x = 3$$
$$x = 3$$

x	y
3	5
3	-2

We can see that no matter what value y has in this equation, x is always 3. Therefore, all the points having an x-value of 3 lie in a vertical line whose x-intercept is (3, 0) (Figure 12).

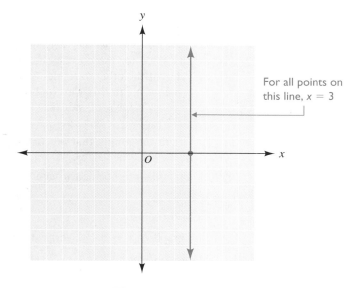

For all points on this line, $x = 3$

Figure 12

EXAMPLE 5   Graph the equation $y + 4 = 0$.

SOLUTION

$$y + 4 = 0$$
$$y = -4$$

In the equation $y + 4 = 0$, no matter what value x has, y is always -4. Therefore, all the points having a y-value of -4 lie in a horizontal line whose y-intercept is (0, -4) (Figure 13).

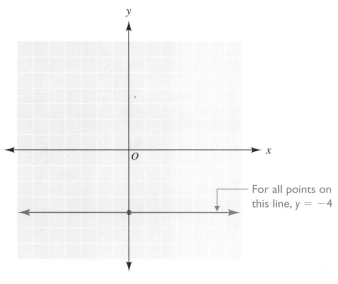

For all points on this line, $y = -4$

Figure 13

The methods used in these examples are summarized as follows:

Graphing a straight line

Table of values

1. Substitute a number in the equation for one of the variables, and solve for the corresponding value of the other variable. Find three points in this way, and list them in a table of values. (The third point is a checkpoint.)

2. Graph the points that are listed in the table of values.

3. Draw a straight line through the points.

Special cases:

The graph of $x = a$ is a vertical line through $(a, 0)$.

The graph of $y = b$ is a horizontal line through $(0, b)$.

Intercepts method

1. Find the x-intercept: Set $y = 0$, then solve for x.

2. Find the y-intercept: Set $x = 0$, then solve for y.

3. Draw a straight line through the x- and y-intercepts.

4. If both intercepts are $(0, 0)$, an additional point must be found before the line can be drawn (Example 3).

 Note Sometimes the x- and y-intercepts are very close together. To draw the line through them accurately would be very difficult. In this case, find another point on the line far enough away from the intercepts so that it is easy to draw an accurate line. To find the other point, set either variable equal to a number, and then solve the equation for the other variable.

Exercises 13.2

In Exercises 1–28, graph the equation.

1. $x + y = 3$
2. $x - y = 4$
3. $2x - 3y = 6$
4. $3x - 4y = 12$
5. $y = 8$
6. $x = 9$
7. $x + 5 = 0$
8. $y + 2 = 0$
9. $3x - 5y = 15$
10. $2x - 5y = 10$
11. $y = -\frac{1}{2}x$
12. $y = -2x$
13. $4 - x = y$
14. $6 - x = y$
15. $y = x$
16. $x + y = 0$
17. $x - y = 0$
18. $x = 0$
19. $y = \frac{2}{3}x$
20. $y = \frac{7}{3}x$
21. $4x - 3y = 12$

22. $6x - 3y = 18$
23. $3x - 2y = 12$
24. $3x - 4y = -4$
25. $3x - 4y = 24$
26. $7x + 2y = 14$
27. $y = -\frac{3}{2}x + 4$

28. $y = -\frac{3}{4}x + 1$

In Exercises 29 and 30, (a) graph the two equations on the same set of axes. (b) What are the coordinates of the point where the two lines cross?

29. $x - y = 5$
    $x + y = 1$

30. $3x - 4y = -12$
    $3x + y = 18$

13.3   Slope of a Line

If we imagine a line as representing a hill, then the **slope** of the line is a measure of the steepness of the hill. To measure the slope of a line, we choose any two points on the line, $P_1(x_1, y_1)$ and $P_2(x_2, y_2)$.

☞ **Note** A small number written lower than and to the right of a variable is called a **subscript**. The subscripted expression x_1 is read "x sub 1," and x_2 is read "x sub 2." Subscripts are used to indicate different values of the variable.

Consider the two points $P_1(x_1, y_1)$ and $P_2(x_2, y_2)$ shown in Figure 14. From this figure it can be seen that

The change in x from P_1 to P_2 is $x_2 - x_1$. This horizontal movement is called the **run**.

The change in y from P_1 to P_2 is $y_2 - y_1$. This vertical movement is called the **rise**.

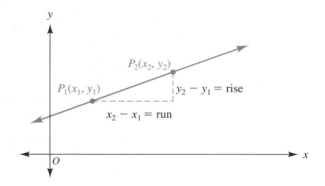

Figure 14

The letter m is used to represent the slope of a line. The slope is defined as the ratio of the rise to the run.

Slope of a line

$$\text{Slope} = \frac{\text{rise}}{\text{run}}$$

$$m = \frac{y_2 - y_1}{x_2 - x_1}$$

where $P_1(x_1, y_1)$ and $P_2(x_2, y_2)$ are any two points on the line.

EXAMPLE 1 Find the slope of the line through the points $(-3, 5)$ and $(6, -1)$ in Figure 15.

SOLUTION

Let $\qquad P_1 = (-3, 5)$

and $\qquad P_2 = (6, -1)$

Then $\qquad m = \dfrac{y_2 - y_1}{x_2 - x_1}$

$\qquad\qquad = \dfrac{(-1) - (5)}{(6) - (-3)}$

$\qquad\qquad = \dfrac{-6}{9} = -\dfrac{2}{3}$

The slope is not changed if the points P_1 and P_2 are interchanged.

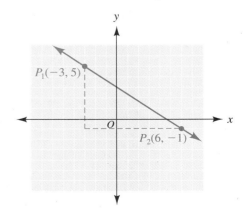

Figure 15

Let $\qquad$ $P_1 = (6, -1)$

and $\qquad$ $P_2 = (-3, 5)$

Then $\qquad$ $m = \dfrac{y_2 - y_1}{x_2 - x_1} = \dfrac{(5) - (-1)}{(-3) - (6)} = \dfrac{6}{-9} = -\dfrac{2}{3}$

EXAMPLE 2 Find the slope of the line through the points $A(-2, -4)$ and $B(5, 1)$ in Figure 16.

SOLUTION

$$m = \dfrac{y_2 - y_1}{x_2 - x_1}$$

$$= \dfrac{(1) - (-4)}{(5) - (-2)}$$

$$= \dfrac{5}{7}$$

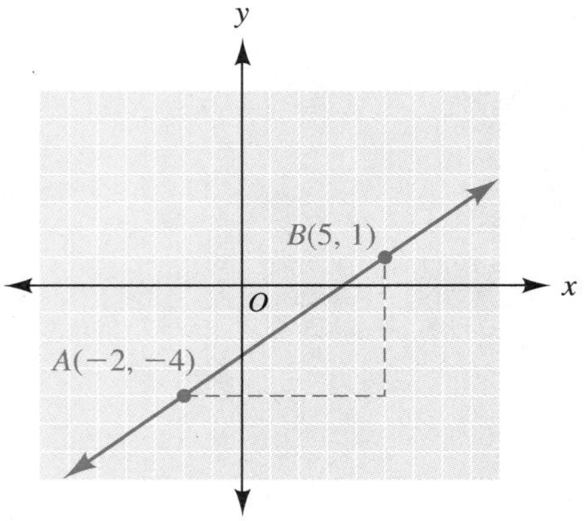

Figure 16

EXAMPLE 3 Find the slope of the horizontal line through the points $E(-4, -3)$ and $F(2, -3)$ in Figure 17.

SOLUTION

$$m = \dfrac{y_2 - y_1}{x_2 - x_1}$$

$$= \dfrac{(-3) - (-3)}{(2) - (-4)}$$

$$= \dfrac{0}{6} = 0$$

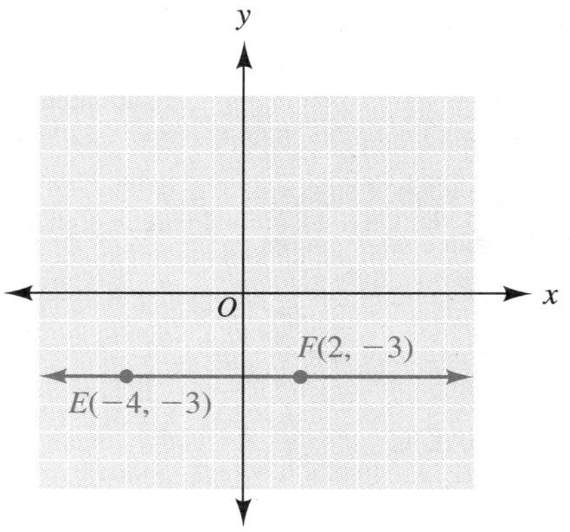

Figure 17

Note The slope of any horizontal line is zero.

EXAMPLE 4   Find the slope of the vertical line through the points $R(4, 5)$ and $S(4, -2)$ in Figure 18.

SOLUTION

$$m = \frac{y_2 - y_1}{x_2 - x_1}$$

$$= \frac{(-2) - (5)}{(4) - (4)}$$

$$= \frac{-7}{0} \quad \text{Undefined}$$

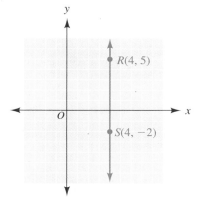

Figure 18

 Note Because we cannot divide by zero, the slope of a vertical line is not defined.

Meaning of the Sign of the Slope

The slope of a line is positive if a point moving along the line in the positive *x*-direction (to the right) rises (Figure 16).

The slope of a line is negative if a point moving along the line in the positive *x*-direction falls (Figure 15).

The slope is zero if the line is horizontal (Figure 17).

The slope is undefined if the line is vertical (Figure 18).

Slope-Intercept Form of the Equation of a Line

If we know the slope and the *y*-intercept of a line, we can find the equation of the line in **slope-intercept form**. Similarly, if we have an equation in slope-intercept form, we can find the slope and the *y*-intercept.

For a line whose slope is *m*, let $(0, b)$ be the *y*-intercept and let $P(x, y)$ represent any other point on the line (Figure 19). Then

$$m = \frac{y - b}{x - 0}$$

$$m = \frac{y - b}{x}$$

$$m \cdot x = \frac{y - b}{x} \cdot x$$

$$mx = y - b$$

$$mx + b = y$$

or $y = m\,x + b$ Slope-intercept form

Slope ⎯⎯↑   ⎿⎯⎯ *y*-intercept

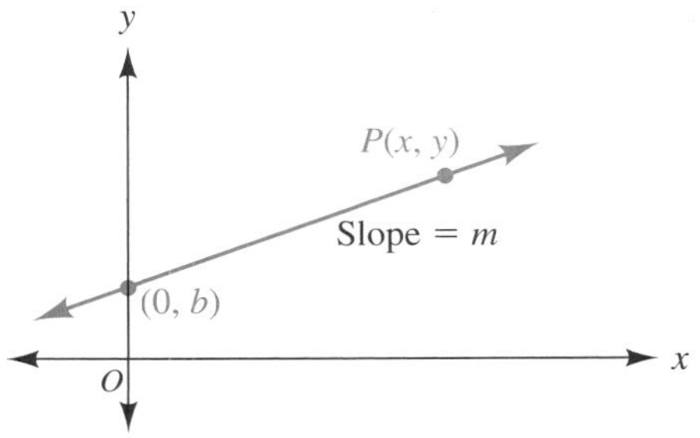

Figure 19

Slope-intercept form of the equation of a line

$$y = mx + b$$

where m = the slope of the line and b = the y-intercept of the line.

EXAMPLE 5

Examples of identifying the slope and the y-intercept:

Equation of line	Slope	y-intercept
a. $y = 2x + 3$	2	3
b. $y = -\dfrac{1}{2}x + 5$	$-\dfrac{1}{2}$	5
c. $y = x - 6$	1	−6
d. $y = 4x$ $(y = 4x + 0)$	4	0
e. $y = 4$ $(y = 0 \cdot x + 4)$	0	4

In **a.**, the arrows point to: Slope, y-intercept.

Graphing Lines Using the Slope and the y-Intercept

For the equation $y = \dfrac{2}{3}x + 1$, we know that the slope is $\dfrac{2}{3}$ and the y-intercept is 1.

First we graph the y-intercept $(0, 1)$ (Figure 20). Now we can locate a second point on the line by using the slope.

$$\text{Slope} = \frac{\text{rise}}{\text{run}} = \frac{2}{3}$$

Since the rise is 2 and the run is 3, starting from the y-intercept we will move *up* 2 units and then to the *right* 3 units. Now we have the second point $(3, 3)$, and we can draw the line (Figure 21).

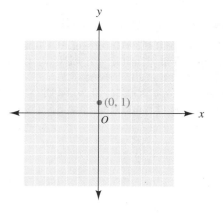

Figure 20

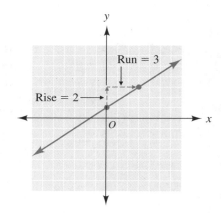

Figure 21

Graphing lines using the slope-intercept method

1. Graph the y-intercept.
2. To locate a second point, start from the y-intercept and move *up* the number of units indicated by the numerator of the slope (move *down* if the numerator is negative), then move *right* the number of units indicated by the denominator of the slope.
3. Draw a straight line through the points.

EXAMPLE 6

Graph $y = -\dfrac{1}{2}x + 5$, using the slope-intercept method.

SOLUTION

Step 1. We graph the y-intercept $(0, 5)$.

Step 2. The slope is $-\dfrac{1}{2}$. Thus,

$$\frac{\text{rise}}{\text{run}} = \frac{-1}{2}$$

From the y-intercept, we move *down* 1 unit and then *right* 2 units, as shown in Figure 22.

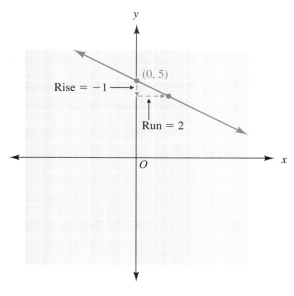

Figure 22

EXAMPLE 7

Graph $3x - y = 4$, using the slope-intercept method.

SOLUTION To identify the slope and the y-intercept, we must put the equation in slope-intercept form; that is, solve for y:

$$3x - y = 4$$
$$-y = -3x + 4$$
$$\frac{-y}{-1} = \frac{-3x + 4}{-1}$$
$$y = 3x - 4$$

Step 1. We graph the y-intercept $(0, -4)$.

Step 2. The slope is 3. Thus,

$$\frac{\text{rise}}{\text{run}} = \frac{3}{1}$$

From the y-intercept, we move *up* 3 units and then *right* 1 unit, as shown in Figure 23.

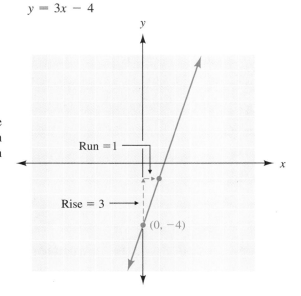

Figure 23

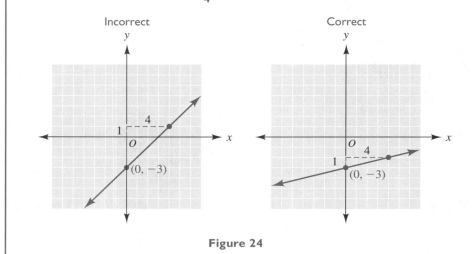

 A Word of Caution When graphing the slope, be sure to *start from the y-intercept*, NOT the origin. Figure 24 shows correct and incorrect ways of graphing $y = \frac{1}{4}x - 3$.

Figure 24

Exercises 13.3

In Exercises 1–18, find the slope of the line through the given pair of points.

1. $(-4, 8)$ and $(7, -9)$ 2. $(-1, -3)$ and $(5, 2)$

3. $(-1, -1)$ and $(5, -3)$ 4. $(16, -14)$ and $(8, -11)$

5. $(-5, -3)$ and $(4, -3)$ 6. $(-2, -4)$ and $(3, -4)$

7. $(-7, 5)$ and $(8, -3)$ 8. $(12, -9)$ and $(-5, -4)$

9. $(-6, -15)$ and $(4, -5)$ 10. $(-7, 8)$ and $(-4, 5)$

11. $(-2, 5)$ and $(-2, 8)$ 12. $(6, -4)$ and $(6, -11)$

13. $(-4, 3)$ and $(-1, 1)$ 14. $(6, 3)$ and $(10, -2)$

15. $(-3, 2)$ and $(6, 4)$ 16. $(-2, -2)$ and $(4, -1)$

17. $(-10, 2)$ and $(-4, -8)$ 18. $(-3, 8)$ and $(9, 12)$

In Exercises 19–30, identify the slope and the y-intercept.

19. $y = \frac{2}{3}x + 4$ 20. $y = -\frac{3}{4}x + 6$

21. $y = -\frac{1}{2}x - \frac{2}{3}$ 22. $y = \frac{1}{6}x - \frac{3}{4}$

23. $y = -x$ 24. $y = -3$

25. $y = 0$ 26. $x = 5$

27. $3x + 4y = 12$ 28. $x + 4y = 16$

29. $2x - 3y = 6$ 30. $5x - 4y = -20$

In Exercises 31–42, graph the line using the slope-intercept method.

31. $y = \frac{3}{4}x - 2$ 32. $y = \frac{2}{5}x - 3$

33. $y = -\frac{1}{2}x + 5$ 34. $y = -\frac{2}{3}x + 4$

35. $y = 3x - 6$ 36. $y = 2x + 3$

37. $y = -x + 2$ 38. $y = -3x$

39. $3x + 2y = 12$ 40. $2x + 5y = -10$

41. $3x - 5y = -15$ 42. $5x - 4y = 20$

13.4   Equations of Lines

In the last two sections we discussed the *graph* of a straight line. In this section we show how to write the *equation* of a line when certain facts about the line are known.

Standard Form of the Equation of a Line

Standard form of the equation of a line

$$Ax + By = C$$

where A, B, and C are real numbers and A and B are not both 0.

Whenever possible, write the standard form with A *positive* and A, B, and C *integers*.

EXAMPLE 1   Write $-\dfrac{2}{3}x + \dfrac{1}{2}y = 1$ in standard form.

SOLUTION   The LCD is 6.

$$\frac{6}{1}\left(-\frac{2}{3}x\right) + \frac{6}{1}\left(\frac{1}{2}y\right) = \frac{6}{1}\left(\frac{1}{1}\right)$$

$$-4x \quad + \quad 3y \quad = \quad 6$$

$$4x \quad - \quad 3y \quad = \quad -6 \qquad \text{Standard form}$$

Point-Slope Form of the Equation of a Line

Let $P_1(x_1, y_1)$ be a known point on a line whose slope is m. Let $P(x, y)$ represent any other point on the line (Figure 25).

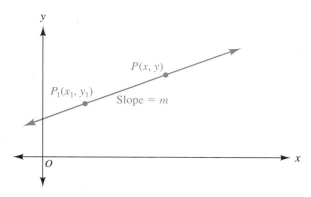

Figure 25

Then

$$m = \frac{y - y_1}{x - x_1}$$

$$\frac{m}{1} \cdot \frac{x - x_1}{1} = \frac{y - y_1}{x - x_1} \cdot \frac{x - x_1}{1}$$

$$m(x - x_1) = y - y_1$$

or

$$y - y_1 = m(x - x_1) \qquad \text{Point-slope form}$$

Point-slope form of the equation of a line

$$y - y_1 = m(x - x_1)$$

where m = the slope of the line and $P_1(x_1, y_1)$ is a known point on the line.

In the following examples we choose the particular form of the equation of the line that makes the best use of the given information.

EXAMPLE 2   Write the standard form of the equation of the line that passes through $(2, -3)$ and has a slope of 4.

SOLUTION   Because we are given a point and the slope, we will write the equation in point-slope form and substitute $x_1 = 2$, $y_1 = -3$, and $m = 4$.

$$y - y_1 = m(x - x_1) \qquad \text{Point-slope form}$$

$$y - (-3) = 4(x - 2) \qquad \text{Substituting}$$

$$y + 3 = 4x - 8$$

$$-4x + y = -11$$

$$4x - y = 11 \qquad \text{Standard form}$$

EXAMPLE 3   Write the standard form of the equation of the line that passes through $(-1, 4)$ and has a slope of $-\dfrac{2}{3}$.

SOLUTION   Again, we are given a point and the slope, so we will write the equation in point-slope form and substitute $x_1 = -1$, $y_1 = 4$, and $m = -\dfrac{2}{3}$.

$$y - y_1 = m(x - x_1) \qquad \text{Point-slope form}$$

$$y - (4) = -\frac{2}{3}[x - (-1)] \qquad \text{Substituting}$$

$$3(y - 4) = \frac{\overset{1}{3}}{1} \cdot \frac{-2}{\underset{1}{3}}(x + 1) \qquad \text{Multiplying by 3}$$

$$3y - 12 = -2x - 2$$

$$2x + 3y = 10 \qquad \text{Standard form}$$

EXAMPLE 4   Find the equation of the line that passes through the points $(-5, 3)$ and $(-15, -9)$.

SOLUTION   Find the slope of the two given points:

$$m = \frac{(-9) - (3)}{(-15) - (-5)} = \frac{-12}{-10} = \frac{6}{5}$$

Use this slope with *either* given point to find the equation of the line.

Using the point $(-5, 3)$	*Using the point $(-15, -9)$*	
$y - y_1 = m(x - x_1)$	$y - y_1 = m(x - x_1)$	Point-slope form
$y - (3) = \dfrac{6}{5}[x - (-5)]$	$y - (-9) = \dfrac{6}{5}[x - (-15)]$	
$5(y - 3) = \dfrac{\overset{1}{5}}{1} \cdot \dfrac{6}{\underset{1}{5}}(x + 5)$	$5(y + 9) = \dfrac{\overset{1}{5}}{1} \cdot \dfrac{6}{\underset{1}{5}}(x + 15)$	
$5y - 15 = 6x + 30$	$5y + 45 = 6x + 90$	
$-6x + 5y = 45$	$-6x + 5y = 45$	
$6x - 5y = -45$	$6x - 5y = -45$	Standard form

This shows that the same equation is obtained no matter which of the given points is used.

EXAMPLE 5 Write the standard form of the e

y-intercept of −2.

SOLUTION We are give

intercept form of the equatio

$$y =$$

$$y$$

$$4y$$

$$3x + 4y = -8$$

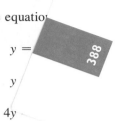

Exercises 13.4

In Exercises 1–8, write e

1. $3x = 2y - 4$

3. $y = -\frac{3}{4}x - 2$

5. $y = \frac{2}{3}x -$

7. $y -$

Equations for Horizontal and Vertical Lines

> The equation of a horizontal line is $y = b$. The slope of a horizontal line is 0.
>
> The equation of a vertical line is $x = a$. The slope of a vertical line is undefined.

EXAMPLE 6 Find the equation of the horizontal line through $(-5, 3)$.

SOLUTION A horizontal line has an equation of the form $y = b$. Because the y-coordinate of $(-5, 3)$ is 3, the equation of the line is

$$y = 3$$

EXAMPLE 7 Find the equation of the vertical line through $(-5, 3)$.

SOLUTION A vertical line has an equation of the form $x = a$. Because the x-coordinate of $(-5, 3)$ is -5, the equation of the line is

$$x = -5$$

The different forms for the equation of a line are summarized in the following box:

Standard form	$Ax + By = C$	where A and B are not both 0
Point-slope form	$y - y_1 = m(x - x_1)$	where m is the slope and (x_1, y_1) is a known point
Slope-intercept form	$y = mx + b$	where m is the slope and b is the y-intercept
A vertical line	$x = a$	where the line passes through $(a, 0)$ and the slope is undefined
A horizontal line	$y = b$	where the line passes through $(0, b)$ and the slope is 0

ch equation in standard form.

2. $2x = 3y + 7$

4. $y = -\dfrac{3}{5}x - 4$

$\dfrac{1}{6}$

6. $y = \dfrac{1}{2}x + \dfrac{3}{4}$

$= \dfrac{3}{4}(x + 1)$

8. $y + 1 = -\dfrac{1}{3}(x - 2)$

**rcises 9–14, write the standard form of the equa-
of the line that passes through the given point and
s the indicated slope.**

9. $(3, 4)$, $m = \dfrac{1}{2}$

10. $(5, 6)$, $m = \dfrac{1}{3}$

11. $(-1, -2)$, $m = -\dfrac{2}{3}$

12. $(-2, -3)$, $m = -\dfrac{5}{4}$

13. $(-6, 3)$, $m = -\dfrac{1}{2}$

14. $(5, -7)$, $m = -\dfrac{3}{4}$

**In Exercises 15–20, write the standard form of the equa-
tion of the line having the indicated slope and y-intercept.**

15. $m = \dfrac{3}{4}$, y-intercept $= -3$

16. $m = -\dfrac{2}{3}$, y-intercept $= -4$

17. $m = \dfrac{2}{7}$, y-intercept $= -2$

18. $m = \dfrac{5}{4}$, y-intercept $= -3$

19. $m = -\dfrac{2}{5}$, y-intercept $= \dfrac{1}{2}$

20. $m = -\dfrac{5}{3}$, y-intercept $= \dfrac{3}{4}$

**In Exercises 21–28, find the standard form of the equa-
tion of the line that passes through the given points.**

21. $(4, -1)$ and $(2, 4)$ **22.** $(5, -2)$ and $(3, 1)$

23. $(0, 0)$ and $(3, 4)$ **24.** $(0, 0)$ and $(-2, -5)$

25. $(4, 3)$ and $(4, -2)$ **26.** $(-5, 2)$ and $(5, 2)$

27. $(-3, 4)$ and $(5, -2)$ **28.** $(5, -3)$ and $(-2, -4)$

29. Write the equation of the horizontal line through $(3, 5)$.

30. Write the equation of the vertical line through $(3, 5)$.

31. Write the equation of the vertical line through $(-6, -2)$.

32. Write the equation of the horizontal line through $(-6, -2)$.

13.5   Graphing Curves

Earlier in the chapter we showed how to graph straight lines. In this section we show how to graph *curves*.

Two points are all that we need to draw a straight line. To draw a curved line, however, we must find more than two points.

Graphing a curve
1. Use the equation to make a table of values. **2.** Plot the points from the table of values. **3.** Draw a smooth curve through the points, joining them in order from left to right.

EXAMPLE 1   Graph the equation $y = x^2 - x - 2$.

SOLUTION   Make a table of values by substituting values of x in the equation and finding the corresponding values for y.

$$y = x^2 - x - 2$$

If $x = -2$, then $y = (-2)^2 - (-2) - 2 = 4 + 2 - 2 = 4$

If $x = -1$, then $y = (-1)^2 - (-1) - 2 = 1 + 1 - 2 = 0$

If $x = 0$, then $y = (0)^2 - (0) - 2 = -2$

If $x = 1$, then $y = (1)^2 - (1) - 2 = 1 - 1 - 2 = -2$

If $x = 2$, then $y = (2)^2 - (2) - 2 = 4 - 2 - 2 = 0$

If $x = 3$, then $y = (3)^2 - (3) - 2 = 9 - 3 - 2 = 4$

x	y
-2	4
-1	0
0	-2
1	-2
2	0
3	4

In Figure 26 we graph these points and draw a smooth curve through them. In drawing the smooth curve, we start with the point in the table of values having the smallest x-value. We draw to the point having the next larger x-value, continuing in this way through all the points. The graph of the equation $y = x^2 - x - 2$ is called a **parabola**.

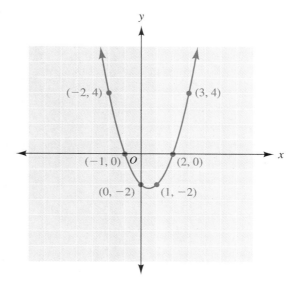

Figure 26

EXAMPLE 2   Graph the equation $y = x^3 - 4x$.

SOLUTION   First, make a table of values.

$$y = x^3 - 4x$$

If $x = -3$, then $y = (-3)^3 - 4(-3) = -27 + 12 = -15$

If $x = -2$, then $y = (-2)^3 - 4(-2) = -8 + 8 = 0$

If $x = -1$, then $y = (-1)^3 - 4(-1) = -1 + 4 = 3$

If $x = 0$, then $y = (0)^3 - 4(0) = 0$

If $x = 1$, then $y = (1)^3 - 4(1) = 1 - 4 = -3$

If $x = 2$, then $y = (2)^3 - 4(2) = 8 - 8 = 0$

If $x = 3$, then $y = (3)^3 - 4(3) = 27 - 12 = 15$

x	y
-3	-15
-2	0
-1	3
0	0
1	-3
2	0
3	15

In Figure 27 we graph these points and draw a smooth curve through them, joining the points in order from left to right.

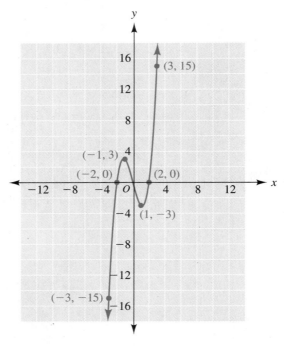

Figure 27

Exercises   *13.5*

In Exercises 1–8, complete the table of values for each equation, and then draw its graph.

1. $y = x^2$

x	y
−3	
−2	
−1	
0	
1	
2	
3	

2. $y = \dfrac{x^2}{4}$

x	y
−3	
−2	
−1	
0	
1	
2	
3	

3. $y = \dfrac{x^2}{2}$

x	y
−3	
−2	
−1	
0	
1	
2	
3	

4. $y = x^2 + 4x$

x	y
−5	
−4	
−3	
−2	
−1	
0	
1	

5. $y = x^2 - 2x$

x	y
−2	
−1	
0	
1	
2	
3	
4	

6. $y = 3x - x^2$

x	y
−2	
−1	
0	
1	
2	
3	
4	

9. Use integer values of x from −2 to values for the equation $y = x^3$. Graph draw a smooth curve through them.

10. Use integer values of x from −2 to +2 to make a table values for the equation $y = -x^3$. Graph the points, and draw a smooth curve through them.

11. Use integer values of x from −2 to +2 to make a table of values for the equation $y = x^3 - 3x + 4$. Graph the points, and draw a smooth curve through them.

12. Use integer values of x from −2 to +2 to make a table of values for the equation $y = 1 - 2x - x^3$. Graph the points, and draw a smooth curve through them.

7. $y = 2x - x^2$

x	y
−2	
−1	
0	
1	
2	
3	
4	

8. $y = 2x + x^2$

x	y
−4	
−3	
−2	
−1	
0	
1	
2	

13.6 Graphing Inequalities in the Plane

Any line in a plane divides that plane into two **half-planes**. For example, in Figure 28 the line *AB* divides the plane into the two half-planes shown.

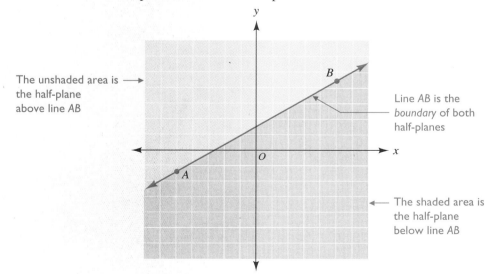

The unshaded area is the half-plane above line *AB*

Line *AB* is the *boundary* of both half-planes

The shaded area is the half-plane below line *AB*

Figure 28

all the solutions to a first-degree *inequality* (in no more than two vari-
.-plane. The equation of the **boundary line** of the half-plane is obtained
, the inequality sign by an equal sign.

391

ɔw to Determine When the Boundary
ıs a Dashed or a Solid Line

1. *If equality is included*, the boundary line is a *solid* line.
2. *If equality is not included*, the boundary line is a *dashed* line.

See the examples in Figure 29.

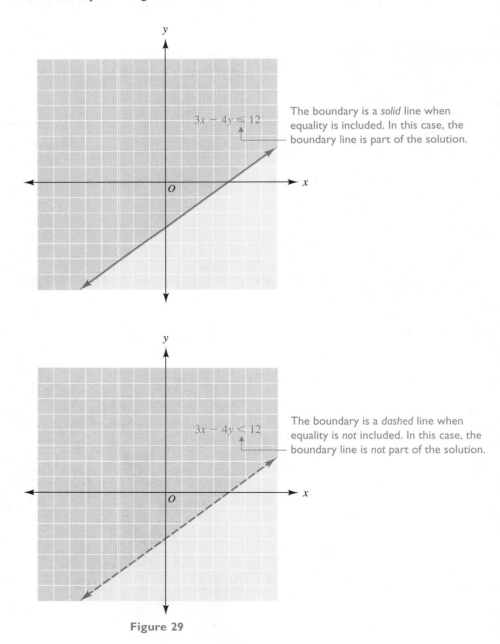

$3x - 4y \leq 12$

The boundary is a *solid* line when
equality is included. In this case, the
boundary line is part of the solution.

$3x - 4y < 12$

The boundary is a *dashed* line when
equality is *not* included. In this case, the
boundary line is *not* part of the solution.

Figure 29

How to Determine the Correct Half-Plane

1. *If the boundary does not go through the origin*, substitute the coordinates of the origin (0, 0) into the inequality.

   a. If the resulting inequality is *true*, the solution is the half-plane containing (0, 0).

   b. If the resulting inequality is *false*, the solution is the half-plane *not* containing (0, 0).

2. *If the boundary goes through the origin*, select a point *not* on the boundary and substitute the coordinates of this point into the inequality.

   a. If the resulting inequality is *true*, the solution is the half-plane containing the point selected.

   b. If the resulting inequality is *false*, the solution is the half-plane *not* containing the point selected.

Graphing a first-degree inequality in a plane	1. *Graph the boundary line.* The equation of the boundary line is obtained by replacing the inequality sign by an equal sign. a. The boundary line is *solid* if the equality is included ($\leq$, $\geq$). b. The boundary line is *dashed* if the equality is not included ($<$, $>$). 2. *Select and shade the correct half-plane.* Choose a test point [usually (0, 0)] that is not on the boundary line, and substitute its coordinates into the inequality. a. If the resulting inequality is *true*, the solution is the half-plane containing the test point. b. If the resulting inequality is *false*, the solution is the half-plane *not* containing the test point.

EXAMPLE 1   Graph the inequality $2x - 3y < 6$.

SOLUTION

Step 1. To graph the boundary line, replace the inequality sign by an equal sign:

$$2x - 3y < 6$$
$$\downarrow$$
$$2x - 3y = 6 \quad \text{Boundary line}$$

x-intercept   Set $y = 0$. Then $2x - 3y = 6$ becomes

$$2x - 3(0) = 6$$
$$2x = 6$$
$$x = 3$$

y-intercept   Set $x = 0$. Then $2x - 3y = 6$ becomes

$$2(0) - 3y = 6$$
$$-3y = 6$$
$$y = -2$$

x	y
3	0
0	-2

The boundary is a *dashed* line because the inequality is *not* included.

$$2x - 3y < 6$$

Step 2. The solution of the inequality is only one of the two half-planes determined by the boundary line. Substitute the coordinates of the origin (0, 0) into the inequality:

$$2x - 3y < 6$$

$$2(0) - 3(0) < 6$$

$$0 < 6 \quad \text{True}$$

Therefore, the half-plane containing the origin is the solution. The solution is the shaded area in Figure 30.

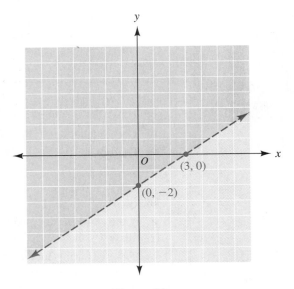

Figure 30

E X A M P L E 2 Graph the inequality $3x + 4y \le -12$.

S O L U T I O N

Step 1. Graph the boundary line, $3x + 4y = -12$.

  x-intercept Set $y = 0$. Then $3x + 4y = -12$ becomes

$$3x + 4(0) = -12$$

$$3x = -12$$

$$x = -4$$

  y-intercept Set $x = 0$. Then $3x + 4y = -12$ becomes

$$3(0) + 4y = -12$$

$$4y = -12$$

$$y = -3$$

x	y
−4	0
0	−3

The boundary is a *solid* line because the equality is included.

$$3x + 4y \le -12$$

Step 2. Select the correct half-plane by substituting the coordinates of the origin $(0, 0)$:

$$3x + 4y \leq -12$$
$$3(0) + 4(0) \leq -12$$
$$0 \leq -12 \quad \text{False}$$

Therefore, the solution is the (shaded) half-plane *not* containing $(0, 0)$. (See Figure 31.)

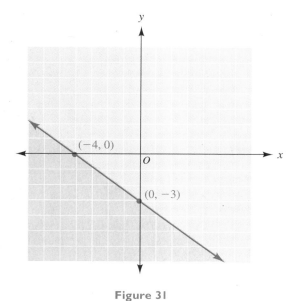

Figure 31

Some inequalities have equations with *only one variable*. Such inequalities have graphs whose boundaries are either *vertical* or *horizontal* lines.

E X A M P L E 3 Graph the inequality $x + 4 < 0$.

S O L U T I O N

Step 1. Graph the boundary line:

$$x + 4 = 0$$
$$x = -4$$

The boundary is a *dashed* line because the equality is *not* included.

$$x + 4 < 0$$

Step 2. Select the correct half-plane by substituting the coordinates of the origin $(0, 0)$:

$$x + 4 < 0$$
$$(0) + 4 < 0$$
$$4 < 0 \quad \text{False}$$

Therefore, the solution is the half-plane that does *not* contain $(0, 0)$. (See Figure 32.)

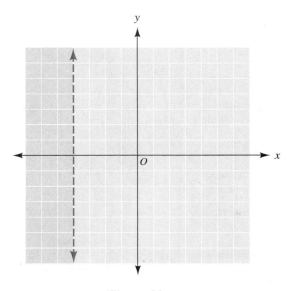

Figure 32

Earlier we discussed how to graph an inequality such as $x + 4 < 0$ on a *single number line*. Because $x + 4 < 0$, $x < -4$. The solution is graphed on the number line in Figure 33.

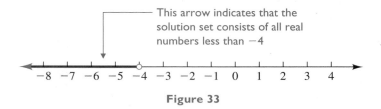

Figure 33

Figure 32 shows that the solution set of this same inequality, $x + 4 < 0$, represents an entire half-plane when it's plotted in the rectangular coordinate system.

EXAMPLE 4   Graph the inequality $2y - 5x \geq 0$.

SOLUTION

Step 1. Graph the boundary line, $2y - 5x = 0$.

      x-intercept   Set $y = 0$.

$$2(0) - 5x = 0$$
$$-5x = 0$$
$$x = 0$$

      y-intercept   Set $x = 0$.

$$2y - 5(0) = 0$$
$$2y = 0$$
$$y = 0$$

Therefore, the boundary lines pass through the origin (0, 0).

To find a second point on the line, set $x = 2$:

$$2y - 5(2) = 0$$
$$2y - 10 = 0$$
$$2y = 10$$
$$y = 5$$

x	y
0	0
2	5

This gives the point $(2, 5)$ on the line. The boundary is a *solid* line because the equality is included.

$$2y - 5x \geq 0$$

Step 2. Since the boundary goes through the origin $(0, 0)$, select a point *not* on the boundary, such as $(1, 0)$. Substitute the coordinate of $(1, 0)$ into $2y - 5x \geq 0$.

$$2(0) - 5(1) \geq 0$$
$$-5 \geq 0 \quad \text{False}$$

Therefore, the solution is the half-plane *not* containing $(1, 0)$. (See Figure 34.)

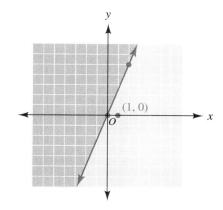

Figure 34

Exercises 13.6

Graph each of the following inequalities in the plane.

1. $x + 2y < 4$
2. $3x + y < 6$
3. $2x - 3y > 6$
4. $5x - 2y > 10$
5. $x \geq -2$
6. $y \geq -3$
7. $3y - 4x \geq 12$
8. $3x + 2y \geq -6$
9. $x + y > 0$
10. $x - y > 0$
11. $3x - 4y \geq 10$
12. $2x - 3y \leq 11$
13. $4x + 2y < 8$
14. $2x - 7y \geq 14$
15. $5x - 3y \geq 15$
16. $4x - y \leq 0$
17. $y > \dfrac{2}{3}x + 3$
18. $y < \dfrac{1}{2}x - 4$
19. $y \leq -\dfrac{1}{4}x + 2$
20. $y \geq -\dfrac{4}{3}x - 1$
21. $y - 4 \geq 0$
22. $x + 1 \leq 0$
23. $\dfrac{x}{5} - \dfrac{y}{3} > 1$
24. $\dfrac{y}{4} - \dfrac{x}{5} \geq 1$

Chapter 13   REVIEW

Ordered Pairs
13.1

An **ordered pair** of numbers is used to represent a point in the plane.

$$(a, b)$$

x-coordinate ⎯⎯⏌ ⎿⎯ y-coordinate

Graphing a Straight Line
13.2, 13.3

Method 1 (Table of Values)

1. Substitute a number in the equation for one of the variables, and solve for the corresponding value of the other variable. Find three points in this way, and list them in a table of values. (The third point is a checkpoint.)

2. Graph the points that are listed in the table of values.

3. Draw a straight line through the points.

Method 2 (Intercepts Method)

1. Find the x-intercept: Set $y = 0$, then solve for x.

2. Find the y-intercept: Set $x = 0$, then solve for y.

3. Draw a straight line through the x- and y-intercepts.

4. If both intercepts are $(0, 0)$, an additional point must be found before the line can be drawn.

Method 3 (Slope-Intercept Method)

1. Graph the y-intercept.

2. To locate a second point, start from the y-intercept and move up the number of units indicated by the numerator of the slope (move down if the numerator is negative), then move to the right the number of units indicated by the denominator of the slope.

3. Draw a straight line through the points.

Graphing a Curve
13.5

1. Use the equation to make a table of values.

2. Plot the points from the table of values.

3. Draw a smooth curve through the points, joining them in order from left to right.

Graphing a First-Degree Inequality in a Plane
13.6

1. *Graph the boundary line.* The equation of the boundary line is obtained by replacing the inequality sign by an equal sign.

   a. The boundary line is *solid* if the equality is included ($\leq$, $\geq$).

   b. The boundary line is *dashed* if the equality is not included ($<$, $>$).

2. *Select and shade the correct half-plane.* Choose a test point [usually $(0, 0)$] not on the boundary line, and substitute its coordinates into the inequality.

   a. If the resulting inequality is true, the solution is the half-plane containing the test point.

   b. If the resulting inequality is false, the solution is the half-plane *not* containing the test point.

Slope of a Line
13.3

The slope m of the line through points $P_1(x_1, y_1)$ and $P_2(x_2, y_2)$ is found using the formula

$$m = \frac{y_2 - y_1}{x_2 - x_1}$$

Equations of a Line
13.3, 13.4

1. *Standard form*: $Ax + By = C$, where A and B are not both 0

2. *Point-slope form*: $y - y_1 = m(x - x_1)$, where (x_1, y_1) is a known point on the line and m is the slope

3. *Slope-intercept form*: $y = mx + b$, where m is the slope and b is the y-intercept of the line

4. $y = b$ is the equation of a *horizontal* line. The slope of a horizontal line is 0.

5. $x = a$ is the equation of a *vertical* line. The slope of a vertical line is undefined.

Chapter 13   REVIEW EXERCISES

1. Draw a rectangle whose vertices (corners) have the following coordinates:

$$A(2, -5), \quad B(2, 1), \quad C(-3, 1), \quad D(-3, -5)$$

In Exercises 2–10, graph each equation by any convenient method.

2. $3x + 2y = 12$ 3. $x - y = 5$ 4. $x = y$

5. $x = -2$ 6. $y + 3 = 0$ 7. $4y - 5x = 20$

8. $x + 2y = 0$ 9. $y = \frac{1}{3}x - 4$ 10. $y = -\frac{3}{2}x + 5$

In Exercises 11 and 12, complete the table of values and then draw the graph.

11. $y = \frac{x^2}{2}$

12. $y = x^3 - 3x$

x	y
-4	
-2	
-1	
0	
1	
2	
4	

x	y
-3	
-2	
-1	
0	
1	
2	
3	

13. Find the slope of the line through $(2, -6)$ and $(-3, 5)$.

14. Find the slope and y-intercept of the line $4x - 15y = 30$.

15. Write the standard form of the equation of the line having a slope of $-\frac{1}{2}$ and a y-intercept of 6.

16. Write the standard form of the equation of the line passing through $(3, -5)$ and having a slope of $-\frac{2}{3}$.

17. Find the standard form of the line through $(-3, 4)$ and $(1, -2)$.

18. Find the equation of the vertical line through $(-7, 3)$.

In Exercises 19–22, graph the inequalities in the plane.

19. $2x - 5y > 10$ 20. $x + 4y \leq 0$

21. $y > 3$ 22. $\frac{x}{2} - \frac{y}{6} \geq 1$

Chapter 13 Critical Thinking and Writing Problems

Answer Problems 1–4 in your own words, using complete sentences.

1. Explain the difference between $(2, -3)$ and $(-3, 2)$.

2. Although only two points are needed to graph a line, explain why it is a good idea to find a third point.

3. Explain how to recognize the equation of a vertical line.

4. Explain how to recognize the equation of a horizontal line.

5. Find y if the line through $(2, y)$ and $(5, 3)$ has a slope of 4.

6. Find x if the line through $(x, 3)$ and $(1, 9)$ has a slope of -2.

7. Graph the line with x-intercept 3 and y-intercept 2. Find the slope of the line. What is the equation of the line, in standard form?

8. Starting with the standard form of a line, $Ax + By = C$, solve for y. What can you say about the slope of the line?

9. In the following graphs, is the slope positive, negative, zero, or undefined?

a.

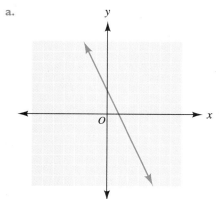

b.

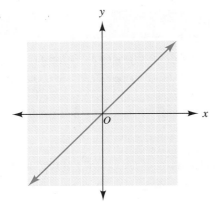

c.

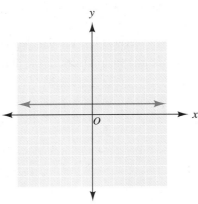

d.

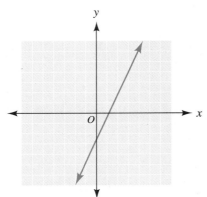

e.

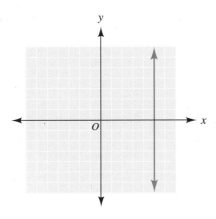

f.

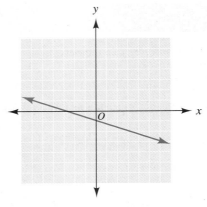

10. Explain how to determine, when graphing a first-degree inequality in the plane, whether the boundary line is dashed or solid.

Two lines in a plane are parallel if they do not intersect even when they are extended. Parallel lines have the same slope.

11. True or false: $2x - 3y = 12$ and $y = \dfrac{2}{3}x - 6$ are parallel lines. Explain your answer.

12. Find the standard form of the equation of the line through $(1, -4)$ and parallel to $y = \dfrac{5}{2}x + 6$.

13. Find the standard form of the equation of the line through $(-2, 7)$ and parallel to $3x + y = 5$.

14. Find the standard form of the equation of the line through $(-3, -5)$ and parallel to $x = 2$.

Chapter 13 • DIAGNOSTIC TEST

Allow yourself about 50 minutes to do these problems. Complete solutions for all problems, together with section references, are given in the answer section at the end of the book.

1. Graph the following points:

$$A(0, -3), \quad B(4, 0), \quad C(-3, 2), \quad D(-1, -4)$$

2. Find the slope and y-intercept of the line $5x + 2y = 8$.

In Problems 3–5, graph each equation by any convenient method.

3. $2x - 3y = -12$ 4. $y = -\dfrac{2}{5}x + 4$ 5. $x - 3 = 0$

6. Complete the table of values and draw the graph for $y = x^2 + x - 6$.

x	y
-4	
-3	
-2	
-1	
0	
1	
2	
3	

7. Graph the inequality $3x - 2y \leq 6$ in the plane.

8. Find the slope of the line through $(-5, 2)$ and $(1, -1)$.

9. Find the equation of the horizontal line through $(2, 5)$.

10. Find the standard form of the equation of the line with a slope of $-\dfrac{2}{3}$ and a y-intercept of -4.

11. Find the standard form of the equation of the line through $(-2, 5)$ with a slope of $\dfrac{1}{4}$.

12. Find the standard form of the equation of the line through $(-4, 2)$ and $(1, -3)$.

Systems of Equations

CHAPTER

14

n previous chapters we showed how to solve a single equation for a single variable. In this chapter we show how to solve systems of two linear equations in two variables.

14.1 Graphical Method

When two equations in the same two variables are considered together, they are called a **system of two equations in two variables.** A **solution** of a system of two equations in two variables is an ordered pair that, when substituted into both equations, makes *both* equations true statements.

EXAMPLE 1 Is $(2, 3)$ a solution of the system $\left\{\begin{array}{l} 2x + y = 7 \\ 3x - y = 3 \end{array}\right\}$?

SOLUTION Substitute $x = 2$ and $y = 3$ into Equation 1:

$$2x + y = 7$$
$$2(2) + 3 \overset{?}{=} 7$$
$$4 + 3 = 7 \quad \text{True}$$

Then substitute $x = 2$ and $y = 3$ into Equation 2:

$$3x - y = 3$$
$$3(2) - 3 \overset{?}{=} 3$$
$$6 - 3 = 3 \quad \text{True}$$

Therefore, $(2, 3)$ *is* a solution of the system.

EXAMPLE 2 Is $(-4, 2)$ a solution of the system $\left\{\begin{array}{l} 3x + 2y = -8 \\ x - 3y = -2 \end{array}\right\}$?

SOLUTION Substitute $x = -4$ and $y = 2$ into Equation 1:

$$3x + 2y = -8$$
$$3(-4) + 2(2) \overset{?}{=} -8$$
$$-12 + 4 \overset{?}{=} -8$$
$$-8 = -8 \quad \text{True}$$

Then substitute $x = -4$ and $y = 2$ into Equation 2:

$$x - 3y = -2$$
$$-4 - 3(2) \overset{?}{=} -2$$
$$-4 - 6 \overset{?}{=} -2$$
$$-10 = -2 \quad \text{False}$$

Therefore, $(-4, 2)$ is *not* a solution of the system.

In an earlier chapter we showed how to graph a straight line. The graph of *each* equation in a system of linear equations in two variables is a straight line.

EXAMPLE 3 Solve the system $\left\{\begin{array}{l} x + y = 6 \\ x - y = 2 \end{array}\right\}$ graphically.

SOLUTION

Line 1: $x + y = 6$

x-intercept If $y = 0$, then $x = 6$

y-intercept If $x = 0$, then $y = 6$

x	y
6	0
0	6

Therefore, line 1 goes through $(6, 0)$ and $(0, 6)$.

Line 2: $x - y = 2$

x-intercept If $y = 0$, then $x = 2$

y-intercept If $x = 0$, then $y = -2$

x	y
2	0
0	-2

Therefore, line 2 goes through $(2, 0)$ and $(0, -2)$.

Next, we graph both equations on the same set of axes (Figure 1).

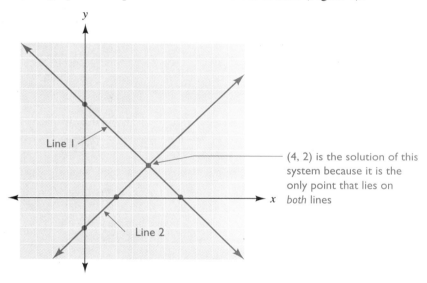

$(4, 2)$ is the solution of this system because it is the only point that lies on *both* lines

Figure 1 Intersecting lines

The coordinates of any point on line 1 satisfy the equation of line 1. The coordinates of any point on line 2 satisfy the equation of line 2. The only point that lies on *both* lines is $(4, 2)$. Therefore, it is the only point whose coordinates satisfy *both* equations.

√ **Check for (4, 2)** $(4, 2)$ must satisfy *both* equations to be a solution.

Line 1	Line 2
$x + y = 6$	$x - y = 2$
$4 + 2 = 6$ True	$4 - 2 = 2$ True

When the system has a solution, it's called a **consistent system**. When each equation in the system has a different graph, as in Example 3, the system is called **independent**. The system in Example 3 is a consistent, independent system.

EXAMPLE 4 Solve the system $\begin{Bmatrix} 2x - 3y = 6 \\ 6x - 9y = 36 \end{Bmatrix}$ graphically.

SOLUTION

Line 1 $2x - 3y = 6$ has intercepts $(3, 0)$ and $(0, -2)$.

Line 2 $6x - 9y = 36$ has intercepts $(6, 0)$ and $(0, -4)$.

Graph both equations on the same set of axes (Figure 2). There is no solution because these lines never meet. Lines that never meet, such as these, are called **parallel lines**.

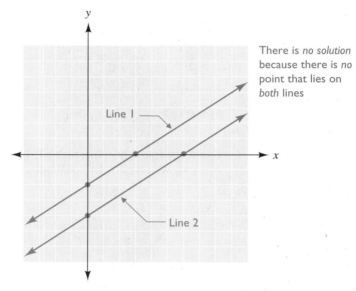

There is *no solution* because there is *no* point that lies on *both* lines

Figure 2   Parallel lines

When the system has no solution, as in Example 4, it's called an **inconsistent system**.

EXAMPLE 5 Solve the system $\begin{Bmatrix} 3x + 5y = 15 \\ 6x + 10y = 30 \end{Bmatrix}$ graphically.

SOLUTION

Line 1 $3x + 5y = 15$ has intercepts $(5, 0)$ and $(0, 3)$.

Line 2 $6x + 10y = 30$ has intercepts $(5, 0)$ and $(0, 3)$.

Graph both equations on the same set of axes (Figure 3). Because both lines go through the same two points, they must be the same line.

To find one of the many solutions, pick a value for one of the variables and substitute it into either equation. For example, let $x = 1$ in Equation 1.

$$3x + 5y = 15$$
$$3(1) + 5y = 15$$
$$3 + 5y = 15$$
$$5y = 12$$
$$y = \frac{12}{5} = 2\frac{2}{5}$$

Therefore, $\left(1, 2\frac{2}{5} \right)$ is a solution for this system. The intercepts $(5, 0)$ and $(0, 3)$ are also solutions.

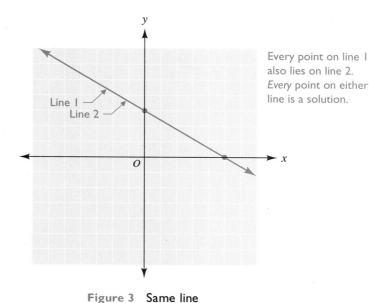

Every point on line I also lies on line 2. *Every* point on either line is a solution.

Line I
Line 2

Figure 3 Same line

When each equation in the system has the same graph, as in Example 5, the system is called **dependent**.

Solving a system of equations by the graphical method	1. Graph each equation of the system on the same set of axes.

1. Graph each equation of the system on the same set of axes.
2. There are three possibilities:
   a. The lines intersect in exactly *one point*. The solution is the *ordered pair* representing the point of intersection. (See Figure 1.)
   b. The lines are *parallel*. There is *no solution* for the system. The system is called "inconsistent." (See Figure 2.)
   c. The lines are the *same*. There are infinitely *many solutions*; any ordered pair that represents a point on the line is a solution. The system is called "dependent." (See Figure 3.)

Check your solution in *both* of the equations.

Exercises 14.1

In Exercises 1–4, determine whether the given ordered pair is a solution of the system.

1. $2x + 3y = -5$
   $5x - 2y = 4$
   $(2, -3)$

2. $x - 4y = 8$
   $2x - 3y = 5$
   $(4, -1)$

3. $3x - y = -5$
   $4x + y = -2$
   $(-1, 2)$

4. $x + 2y = -10$
   $6x - 3y = 0$
   $(-2, -4)$

In Exercises 5–20, find the solution of each system graphically. Check your solution in both equations. Write "inconsistent" if no solution exists (parallel lines). Write "dependent" if many solutions exist (same line).

5. $2x + y = 6$
   $2x - y = -2$

6. $2x - y = -4$
   $x + y = 1$

7. $x - 2y = -6$
   $4x + 3y = 20$

8. $x - 3y = 6$
   $4x + 3y = 9$

17. $2x + y = 4$
    $y = \dfrac{1}{3}x - 3$

18. $3x + 2y = 12$
    $y = 2x - 1$

9. $x + 2y = 0$
   $x - 2y = -2$

10. $2x + y = 0$
    $2x - y = -6$

19. $x - 2y = -4$
    $y = -\dfrac{1}{2}x + 6$

20. $x + 2y = -4$
    $y = \dfrac{3}{2}x + 2$

11. $4y - 2x = 8$
    $3x - 6y = -18$

12. $4x - 2y = 8$
    $-6x + 3y = 6$

13. $10x - 4y = 20$
    $6y - 15x = -30$

14. $12x - 9y = 36$
    $6y - 8x = -24$

15. $2x + y = -4$
    $x - y = -5$

16. $3x + 2y = 0$
    $3x - 2y = 12$

14.2   Addition Method

The graphical method for solving a system of equations has two disadvantages: (1) it is slow, and (2) it is not an exact method of solution. The method we discuss in this section, an algebraic method, has neither of these disadvantages.

In the *addition method* for solving a system, the equations are added to eliminate one variable.

EXAMPLE 1   Solve the system $\left\{\begin{array}{l} x + y = 6 \\ x - y = 2 \end{array}\right\}$.

SOLUTION   Equation 1: $x + y = 6$
                Equation 2: $\underline{x - y = 2}$
                              $2x = 8$ Adding the equations vertically

                              $x = 4$

Then, substituting $x = 4$ into Equation 1, we have

$$x + y = 6$$
$$4 + y = 6$$
$$y = 2$$

Therefore, the solution for the system is $x = 4$ and $y = 2$, written as the ordered pair $(4, 2)$.

The system in Example 1 is simple because one variable can be *eliminated* by adding the equations directly. Sometimes we must multiply one or both equations by a number to make the coefficients of one of the variables add to zero.

EXAMPLE 2   Solve the system $\left\{\begin{array}{l} x - 2y = -6 \\ 4x + 3y = 20 \end{array}\right\}$.

SOLUTION   If both sides of Equation 1 are multiplied by -4, the coefficients of x will add to zero.

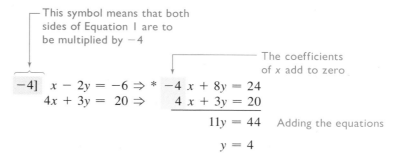

This symbol means that both sides of Equation 1 are to be multiplied by -4

The coefficients of x add to zero

$$-4]\ \ x - 2y = -6 \Rightarrow *\ \ -4\,x + 8y = 24$$
$$4x + 3y = 20 \Rightarrow \ \ \underline{4\,x + 3y = 20}$$
$$11y = 44 \quad \text{Adding the equations}$$
$$y = 4$$

When we have found the value of one variable, that value may be substituted into *either* Equation 1 or Equation 2 to find the value of the other variable. Usually one equation is easier to work with than the other.

Substituting $y = 4$ into Equation 1, we have

$$x - 2y = -6$$
$$x - 2(4) = -6$$
$$x - 8 = -6$$
$$x = 2$$

Therefore, the solution of the system is $(2, 4)$.

✓ **Check for (2, 4)** $(2, 4)$ must satisfy *both* equations to be a solution.

Equation 1	*Equation 2*
$x - 2y\ \ \ = -6$	$4x + 3y\ \ \ = 20$
$2 - 2(4) \overset{?}{=} -6$	$4(2) + 3(4) \overset{?}{=} 20$
$2 -\ \ 8\ \ = -6$ True	$8\ +\ 12\ = 20$ True

EXAMPLE 3 Solve the system $\begin{cases} 3x + 4y = 6 \\ 2x + 3y = 5 \end{cases}$.

SOLUTION If the coefficients of x in the equations are interchanged and Equation 1 is multiplied by 2 and Equation 2 by -3, the resulting coefficients of x will add to zero. One of the numbers is made negative so that the resulting coefficients of x will have opposite signs.

This pair of numbers is found by interchanging the coefficients of x and making one number negative

The coefficients of x add to zero

$$2\ ]\ 3\,x + 4y = 6 \Rightarrow\ \ \ 6\,x + 8y =\ \ \ \ 12$$
$$-3\ ]\ 2\,x + 3y = 5 \Rightarrow\ \ \underline{-6\,x - 9y = -15}$$
$$-y = -\ 3 \quad \text{Adding the equations}$$
$$y = 3$$

---

*The symbol $\Rightarrow$, read "*implies*," means that the second statement is true if the first statement is true. For example,

$$x - 2 = 0 \Rightarrow x = 2$$

is read "$x - 2 = 0$ implies $x = 2$" and means that $x = 2$ is true if $x - 2 = 0$ is true.

The value $y = 3$ may be substituted into either Equation 1 or Equation 2 to find the value of x. It's usually easier to substitute into the equation having the smaller coefficients. We substitute $y = 3$ into Equation 2:

$$2x + 3y = 5$$
$$2x + 3(3) = 5$$
$$2x + 9 = 5$$
$$2x = -4$$
$$x = -2$$

Therefore, the solution of the system is $(-2, 3)$.

✓ *Check for* $(-2, 3)$

Equation 1	*Equation 2*
$3x + 4y = 6$	$2x + 3y = 5$
$3(-2) + 4(3) \stackrel{?}{=} 6$	$2(-2) + 3(3) \stackrel{?}{=} 5$
$-6 + 12 \stackrel{?}{=} 6$	$-4 + 9 \stackrel{?}{=} 5$
$6 = 6$ True	$5 = 5$ True.

How to Choose the Number to Multiply by Each Equation   In Example 3 we multiplied Equation 1 by 2 and Equation 2 by -3 in order to make the coefficients of x add to zero. In this same system we could make the coefficients of y add to zero as follows:

This pair of numbers is found by interchanging the coefficients of y; one of these numbers is made negative so that the resulting coefficients of y will have opposite signs

The coefficients of y add to zero

$$3 \,] \; 3x + 4\,y = 6 \Rightarrow \quad 9x + 12\,y = \quad 18$$
$$-4 \,] \; 2x + 3\,y = 5 \Rightarrow -8x - 12\,y = -20$$

EXAMPLE 4   Determine the number to multiply by each equation in the system $\begin{cases} 10x - 9y = 5 \\ 15x + 6y = 4 \end{cases}$.

SOLUTION   To make the coefficients of x add to zero, we multiply by 3 and -2:

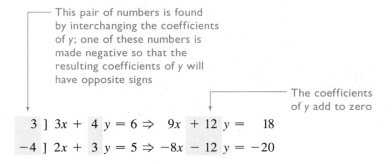

This pair is found by interchanging the coefficients of x

$$15 \,] \quad 3 \,] \; 10\,x - 9y = 5 \Rightarrow \quad 30\,x - 27y = \quad 15$$
$$-10 \,] \; -2 \,] \; 15\,x + 6y = 4 \Rightarrow -30\,x - 12y = -8$$

This pair is found by reducing the ratio $\dfrac{15}{10} = \dfrac{3}{2}$

To make the coefficients of y add to zero, we multiply by 2 and 3:

This pair is found by interchanging the coefficients of y

$$6 \;] \quad 2 \;] \quad 10x - 9y = 5 \Rightarrow 20x - 18y = 10$$

$$9 \;] \quad 3 \;] \quad 15x + 6y = 4 \Rightarrow 45x + 18y = 12$$

This pair is found by reducing the ratio $\dfrac{6}{9} = \dfrac{2}{3}$

EXAMPLE 5

Solve the system $\begin{cases} 2x + 3y = 4 \\ 5x + 6y = 11 \end{cases}$.

SOLUTION

$$5 \;] \quad 2x + 3y = 4 \Rightarrow \quad 10x + 15y = \quad 20$$
$$-2 \;] \quad 5x + 6y = 11 \Rightarrow \quad \underline{-10x - 12y = -22}$$
$$3y = \quad -2$$
$$y = -\frac{2}{3}$$

Substitute $y = -\dfrac{2}{3}$ into $2x + 3y = 4$:

$$2x + 3y = 4$$

$$2x + \frac{\overset{1}{3}}{1}\left(\frac{-2}{\underset{1}{3}}\right) = 4$$

$$2x - 2 = 4$$

$$2x = 6$$

$$x = 3$$

Therefore, the solution of the system is $\left(3, -\dfrac{2}{3}\right)$. The check is left for the student.

Systems That Have No Solution (Inconsistent Systems)

Earlier we found that a system of equations whose graphs are parallel lines has no solution. Here, we show how to identify systems that have no solutions by using the addition method.

EXAMPLE 6

Solve the system $\begin{cases} 2x - 3y = 6 \\ 6x - 9y = 36 \end{cases}$.

SOLUTION

$$6 \;] \quad 3 \;] \quad 2x - 3y = \quad 6 \Rightarrow \quad 6x - 9y = \quad 18$$
$$-2 \;] \quad -1 \;] \quad 6x - 9y = 36 \Rightarrow \quad \underline{-6x + 9y = -36}$$
$$0 = -18 \quad \text{False}$$

No values for x and y can make $0 = -18$.

When both variables are eliminated and a *false* statement occurs, there is *no solution* for the system. The system is inconsistent.

Systems That Have More Than One Solution (Dependent Systems)

Earlier we found that a system whose equations have the same line for a graph has an infinite number of solutions. Here, we show how to identify such systems using the addition method.

EXAMPLE 7 Solve the system $\begin{cases} 4x + 6y = 4 \\ 6x + 9y = 6 \end{cases}$.

SOLUTION

$$9] \quad 3] \ 4x + 6\,y = 4 \Rightarrow \quad 12x + 18y = \quad 12$$
$$-6] \ -2] \ 6x + 9\,y = 6 \Rightarrow -12x - 18y = -12$$
$$\underline{\hspace{6cm}}$$
$$0 = \quad 0 \quad \text{True}$$

When both variables are eliminated and a *true* statement occurs, there are *many solutions* for the system. Any ordered pair that satisfies Equation 1 or Equation 2 is a solution of the system. The system is dependent.

The addition method is summarized in the following box:

Solving a system of equations by the addition method

1. Multiply the equations by numbers that make the coefficients of one of the variables add to zero.
2. Add the equations.
3. There are three possibilities:
   a. *The resulting equation can be solved for one of the variables.* Substitute this value into either of the equations to find the value of the other variable.
   b. *Both variables are eliminated, and a false statement occurs.* There is no solution (inconsistent system).
   c. *Both variables are eliminated, and a true statement occurs.* There are many solutions (dependent system).

Check your solutions in *both* of the original equations.

Exercises 14.2

Find the solution of each system by the addition method. Check your solutions. Write "inconsistent" if no solution exists. Write "dependent" if many solutions exist.

1. $2x - y = -4$
   $x + y = -2$

2. $2x + y = 6$
   $x - y = 0$

3. $x - 2y = 10$
   $x + y = 4$

4. $x + 4y = 4$
   $x - 2y = -2$

5. $x - 3y = 6$
   $4x + 3y = 9$

6. $2x + 5y = 2$
   $3x - 5y = 3$

7. $x + y = 2$
   $3x - 2y = -9$

8. $x + 2y = -4$
   $2x - y = -3$

9. $x + 2y = 0$
   $2x - y = 0$

10. $2x + y = 0$
    $x - 3y = 0$

11. $4x + 3y = 2$
    $3x + 5y = -4$

12. $5x + 7y = 1$
    $3x + 4y = 1$

13. $6x - 10y = 6$
    $9x - 15y = -4$

14. $7x - 2y = 7$
    $21x - 6y = 6$

15. $3x - 5y = -2$
    $10y - 6x = 4$

16. $15x - 9y = -3$
    $6y - 10x = 2$

17. $x - y = -5$
    $3x + y = -3$

18. $x - 2y = 3$
    $3x + 7y = -4$

19. $x - 2y = 6$
    $3x - 2y = 12$

20. $4x - 3y = 7$
    $2x + 7y = 12$

21. $2x + 6y = 2$
    $3x + 9y = 3$

22. $2x - 5y = 4$
    $5y - 2x = 6$

23. $2x - 5y = 4$
    $3x - 4y = -1$

24. $5x - 2y = -9$
    $7x + 3y = -1$

14.3 Substitution Method

Another algebraic method of solving a system of equations is called the substitution method. We use the substitution method when we can easily solve for one of the variables in terms of the other variable.

Solving a system of equations by the substitution method

1. Solve one equation for one of the variables in terms of the other variable.
2. Substitute the expression obtained in Step 1 into the *other* equation (in place of the variable solved for in Step 1).
3. There are three possibilities:
   a. *The resulting equation can be solved for one of the variables.* Substitute this value into either of the system's equations to find the value of the other variable.
   b. *Both variables are eliminated, and a false statement occurs.* There is no solution (inconsistent system).
   c. *Both variables are eliminated, and a true statement occurs.* There are many solutions (dependent system).

Check your solutions in *both* of the original equations.

EXAMPLE 1 Solve the system $\begin{cases} y = x + 4 \\ x + 2y = 5 \end{cases}$.

SOLUTION

Step 1. Equation 1, $y = x + 4$, is already solved for y in terms of x. If we substitute this expression for y into the second equation, we will have an equation in one variable, x.

Step 2. Substitute $x + 4$ for y in Equation 2:

$$x + 2y = 5$$
$$x + 2(\,x + 4\,) = 5$$
$$x + 2x + 8 = 5$$
$$3x + 8 = 5$$
$$3x = -3$$
$$x = -1$$

Step 3. Then substitute -1 for x in Equation 1:

$$y = x + 4$$
$$y = -1 + 4$$
$$y = 3$$

✓ **Check** $(-1, 3)$

Equation 1	*Equation 2*
$y = x + 4$	$x + 2y = 5$
$3 = -1 + 4$ True	$-1 + 2(3) \stackrel{?}{=} 5$
	$-1 + 6 = 5$ True

Therefore, the solution is $(-1, 3)$.

EXAMPLE 2   Solve the system $\begin{Bmatrix} x - 2y = & 11 \\ 3x + 5y = & -11 \end{Bmatrix}$.

SOLUTION

Step 1. Because the coefficient of x in Equation 1 is 1, we can easily solve Equation 1 for x in terms of y.

$$x - 2y = 11$$

x has a coefficient of 1

$$x = 11 + 2y$$

Step 2. Substitute $11 + 2y$ for x in Equation 2:

$$3x + 5y = -11$$
$$3(\;11 + 2y\;) + 5y = -11$$
$$33 + 6y + 5y = -11$$
$$11y = -44$$
$$y = -4$$

Step 3. Substitute -4 for y in the equation found in Step 1:

$$x = 11 + 2y$$
$$x = 11 + 2(-4)$$
$$x = 11 - 8$$
$$x = 3$$

✓   **Check (3, −4)**

Equation 1	*Equation 2*
$x - 2y = 11$	$3x + 5y = -11$
$3 - 2(-4) \stackrel{?}{=} 11$	$3(3) + 5(-4) \stackrel{?}{=} -11$
$3 + 8 = 11$ True	$9 - 20 = -11$ True

Therefore, the solution is $(3, -4)$.

EXAMPLE 3   Solve the system $\begin{Bmatrix} \dfrac{x}{2} + \dfrac{y}{3} = 2 \\ \dfrac{x}{4} + \dfrac{y}{2} = 2 \end{Bmatrix}$.

SOLUTION   First, remove fractions by multiplying each equation by its LCD.

$$6\left(\frac{x}{2}\right) + 6\left(\frac{y}{3}\right) = 6\,(2) \qquad \text{The LCD for Equation 1 is 6}$$

$$3x \quad + \quad 2y \quad = \quad 12$$

$$4\left(\frac{x}{4}\right) + 4\left(\frac{y}{2}\right) = 4\,(2) \qquad \text{The LCD for Equation 2 is 4}$$

$$x \quad + \quad 2y \quad = \quad 8$$

Coefficient is 1

Step 1. Solve Equation 2 for x:

$$x + 2y = 8$$
$$x = 8 - 2y$$

Step 2. Substitute $8 - 2y$ for x in Equation 1:

$$3x + 2y = 12$$
$$3(\,8 - 2y\,) + 2y = 12$$
$$24 - 6y + 2y = 12$$
$$24 - 4y = 12$$
$$-4y = -12$$
$$y = 3$$

Step 3. Substitute 3 for y in the equation found in Step 1:

$$x = 8 - 2y$$
$$x = 8 - 2(3)$$
$$x = 8 - 6$$
$$x = 2$$

✓ *Check* $(2, 3)$

Equation 1	*Equation 2*
$\dfrac{x}{2} + \dfrac{y}{3} = 2$	$\dfrac{x}{4} + \dfrac{y}{2} = 2$
$\dfrac{2}{2} + \dfrac{3}{3} \stackrel{?}{=} 2$	$\dfrac{2}{4} + \dfrac{3}{2} \stackrel{?}{=} 2$
$1 + 1 = 2$ True	$\dfrac{1}{2} + \dfrac{3}{2} \stackrel{?}{=} 2$
	$\dfrac{4}{2} = 2$ True

Therefore, the solution is $(2, 3)$.

Systems That Have No Solution (Inconsistent Systems)

We now show how to identify systems that have no solution by using the substitution method.

EXAMPLE 4 Solve the system $\begin{cases} 6x - 2y = 3 \\ 3x - 1 = y \end{cases}$.

SOLUTION

Step 1. Equation 2, $3x - 1 = y$, is already solved for y.
Step 2. Substitute $3x - 1$ for y in Equation 1:

$$6x - 2y = 3$$
$$6x - 2(\,3x - 1\,) = 3$$
$$6x - 6x + 2 = 3$$
$$2 = 3 \quad \text{False}$$

No values of x and y can make $2 = 3$.

When both variables are eliminated and a *false* statement occurs, there is *no solution* for the system. The system is inconsistent.

Systems That Have More Than One Solution (Dependent Systems)

We now show how to identify systems that have more than one solution by using the substitution method.

EXAMPLE 5   Solve the system $\begin{Bmatrix} 3x - 12y = 6 \\ x - 4y = 2 \end{Bmatrix}$.

└── Coefficient is 1

SOLUTION

Step 1. Solve Equation 2 for x:

$$x - 4y = 2$$

$$x = 4y + 2$$

Step 2. Substitute $4y + 2$ for x in Equation 1:

$$3x - 12y = 6$$

$$3(\,4y + 2\,) - 12y = 6$$

$$12y + 6 - 12y = 6$$

$$6 = 6 \quad \text{True}$$

When both variables are eliminated and a *true* statement occurs, there are *many solutions* for the system. The system is dependent.

Exercises   14.3

Find the solution of each system using the substitution method. Write "inconsistent" if no solution exists. Write "dependent" if many solutions exist.

1. $2x - 3y = 1$
   $x = y + 2$

2. $y = 2x + 3$
   $3x + 2y = 20$

3. $3x + 4y = 2$
   $y = x - 37$

4. $2x + 3y = 11$
   $x = y - 2$

5. $4x + y = 2$
   $7x + 3y = 1$

6. $5x + 7y = 1$
   $x + 4y = -5$

7. $4x + y = 3$
   $8x + 2y = 6$

8. $x + 2y = 3$
   $4x + 8y = 12$

9. $x - y = 1$
   $y = 2x - 3$

10. $x + 2y = -1$
    $x = 5 + y$

11. $x - 4y = 9$
    $3x + 8y = 7$

12. $4x + y = 11$
    $3x - 4y = -6$

13. $x - 3y = 8$
    $3y - x = 8$

14. $5x - y = 4$
    $y - 5x = 2$

15. $3x = y$
    $2x - y = 2$

16. $5y = x$
    $3y - x = 2$

17. $2x + y = 2$
    $4x - y = 1$

18. $6x + y = 7$
    $3x - y = -1$

19. $2x - 3y = -1$
    $3x = y + 2$

20. $5x - 3y = -6$
    $2y = x + 4$

21. $\dfrac{x}{2} + \dfrac{y}{8} = 1$
    $\dfrac{x}{2} + \dfrac{y}{5} = 4$

22. $\dfrac{x}{3} - \dfrac{y}{2} = 6$
    $\dfrac{x}{2} + \dfrac{y}{8} = 2$

23. $\dfrac{x}{4} + \dfrac{y + 1}{2} = 2$
    $\dfrac{x}{2} + \dfrac{y - 2}{3} = 1$

24. $\dfrac{x}{3} + \dfrac{y - 6}{9} = 2$
    $\dfrac{x}{3} + \dfrac{y - 5}{4} = 1$

14.4   Using Systems of Equations to Solve Word Problems

In solving word problems that involve more than one unknown, it is sometimes difficult to represent each unknown in terms of a single variable. In this section we eliminate that difficulty by using a different variable for each unknown.

| **Solving a word problem using a system of equations** | **1.** Read the problem completely, and determine *how many* unknown numbers there are.
2. Draw a diagram showing the relationships in the problem whenever possible.
3. Represent *each* unknown number by a *different* variable.
4. Use the word statements to write a system of equations. *There must be as many equations as variables.*
5. Solve the system of equations using one of the following:
   a. Addition method
   b. Substitution method
   c. Graphical method |

In Example 1 the word problem is first solved by using a single variable and a single equation; then it is solved by using two variables and a system of equations.

EXAMPLE 1   The sum of two numbers is 20. Their difference is 6. What are the numbers?

SOLUTION 1   Let x = larger number
Then $20 - x$ = smaller number

$$\boxed{\text{Their difference}}\ \ \boxed{\text{is}}\ \ \boxed{6}$$

$$x - (20 - x) = 6$$
$$x - 20 + x = 6$$
$$2x = 26$$
$$x = 13 \quad \text{Larger number}$$
$$20 - x = 7 \quad \text{Smaller number}$$

The difficulty in using the one-variable method to solve this problem is that some students have difficulty deciding whether to represent the second unknown number by $x - 20$ or by $20 - x$. ($x - 20$ is incorrect.)

SOLUTION 2   Let x = larger number
and y = smaller number

$$\boxed{\text{The sum of two numbers}}\ \ \boxed{\text{is}}\ \ \boxed{20}$$

(1) $$x + y = 20$$

$$\boxed{\text{Their difference}}\ \ \boxed{\text{is}}\ \ \boxed{6}$$

(2) $$x - y = 6$$

Using the addition method, we have

$$
\begin{array}{r}
x + y = 20 \\
x - y = 6 \\
\hline
2x = 26
\end{array}
$$

$$x = 13 \quad \text{Larger number}$$

Substitute $x = 13$ into Equation 1 to find the smaller number:

$$x + y = 20$$
$$13 + y = 20$$
$$y = 7 \quad \text{Smaller number}$$

EXAMPLE 2

Yuki has 17 coins in her purse that have a total value of $1.15. If she has only nickels and dimes, how many of each are there?

SOLUTION   Let D = number of dimes
Then N = number of nickels

Number of nickels	+	number of dimes	=	17 coins

(1) $N + D = 17$

The coins in her purse have a total value of $1.15.

Amount of money in nickels	+	amount of money in dimes	=	total amount of money

(2) $0.05N + 0.10D = 1.15$

Multiply Equation 1 by -5 and Equation 2 by 100:

$$-5]\quad N + D = 17 \Rightarrow -5N - 5D = -85$$
$$100]\quad 0.05N + 0.10D = 1.15 \Rightarrow \underline{5N + 10D = 115}$$
$$5D = 30$$
$$D = 6 \text{ dimes}$$

Substitute $D = 6$ into Equation 1:

$$N + D = 17$$
$$N + 6 = 17$$
$$N = 11 \text{ nickels}$$

Therefore, Yuki has 11 nickels and 6 dimes.

✓ *Check*

$$11 \text{ nickels} = 11(\$0.05) = \$0.55$$
$$\underline{+\ 6 \text{ dimes} = 6(\$0.10) = +\ 0.60}$$
$$17 \text{ coins} \qquad\qquad = \$1.15$$

EXAMPLE 3

A 50-pound mixture of two different grades of coffee costs $40.50. If grade A costs $0.95 a pound and grade B costs $0.75 a pound, how many pounds of each grade were used?

SOLUTION   Let a = number of pounds of grade A
and b = number of pounds of grade B

Because the sum of the weights of the two ingredients must equal the weight of the mixture,

Pounds of A	+	pounds of B	=	pounds of mixture

(1) $a + b = 50 \qquad \Rightarrow b = 50 - a$

Because the sum of the costs of the two ingredients must equal the cost of the mixture,

$$\begin{array}{c} \text{Cost} \\ \text{of A} \end{array} + \begin{array}{c} \text{cost} \\ \text{of B} \end{array} = \begin{array}{c} \text{cost of} \\ \text{mixture} \end{array}$$

(2) $0.95a + 0.75b = 40.50$

$95a + 75b = 4050$ Multiplying by 100

$95a + 75(\,50 - a\,) = 4050$ Substituting $50 - a$ for b

$95a + 3750 - 75a = 4050$

$20a = 300$

$a = 15$ lb of grade A

and $b = 50 - a = 50 - 15$

$b = 35$ lb of grade B

✓ **Check**

15 lb of grade A
+ 35 lb of grade B

50 lb of mixture

Cost of grade A $= \$0.95(15) = \quad \$14.25$
Cost of grade B $= \$0.75(35) = + \;\; 26.25$
Cost of mixture $\qquad\qquad = \quad \$40.50$

Exercises   14.4

Solve the following word problems by using a system of equations. (a) Represent the unknown numbers using two variables. (b) Set up two equations and solve them. (c) Solve the problem.

1. The sum of two numbers is 30. Their difference is 12. What are the numbers?

2. Half the sum of two numbers is 15. Half their difference is 8. Find the numbers.

3. The sum of two angles is 90°. Their difference is 40°. Find the angles.

4. The sum of two angles is 180°. Their difference is 70°. Find the angles.

5. Find two numbers such that twice the smaller plus three times the larger is 34 and five times the smaller minus twice the larger is 9.

6. Find two numbers such that five times the larger plus three times the smaller is 47 and four times the larger minus twice the smaller is 20.

7. A 20-pound mixture of almonds and hazelnuts costs $19.75. If almonds cost $0.85 a pound and hazelnuts cost $1.40 a pound, find the number of pounds of each.

8. A 100-pound mixture of two different grades of coffee costs $145.00. If grade A costs $1.80 a pound and grade B costs $1.30 a pound, how many pounds of each grade were used?

9. A rectangle is 1 foot 6 inches longer than it is wide. Its perimeter is 19 feet. Find the length and the width.

10. A rectangle is 2 feet 6 inches longer than it is wide. Its perimeter is 25 feet. Find the length and the width.

11. Don spent $3.40 for 22 stamps. If he bought only 10-cent and 25-cent stamps, how many of each kind did he buy?

12. Amina spent $10.25 for 50 stamps. If she bought only 25-cent and 10-cent stamps, how many of each kind did she buy?

13. A fraction has the value $\frac{2}{3}$. If 4 is added to the numerator and the denominator is decreased by 2, the resulting fraction has the value $\frac{6}{7}$. What is the original fraction?

14. A fraction has the value $\frac{3}{4}$. If 4 is added to its numerator and 8 is subtracted from its denominator, the value of the resulting fraction is one. What is the original fraction?

15. Several families went to a movie together. They spent $10.60 for 8 tickets. If adult tickets cost $1.95 and children's tickets cost $0.95, how many of each kind of ticket were bought?

16. A class received $233 for selling 200 tickets to the school play. If student tickets cost $1 each and nonstudent tickets cost $2 each, how many nonstudents attended the play?

17. A mail-order office paid $725 for a total of 40 rolls of stamps in two denominations. If one kind costs $15 per roll and the other kind costs $20 per roll, how many rolls of each kind were bought?

18. An office manager paid $176 for 20 boxes of legal-size and letter-size file folders. If legal-size folders cost $10 a box and letter-size folders cost $8 a box, how many boxes of each kind were bought?

19. Luis worked at two jobs during the week for a total of 30 hours. For this he received a total of $140.00. If he was paid $4.00 an hour as a tutor and $5.00 an hour as a cashier, how many hours did he work at each job?

20. Linda worked at two jobs during the week for a total of 26 hours. For this she received a total of $90.00. If she was paid $3.00 an hour as a lab assistant and $4.00 an hour as a clerk-typist, how many hours did she work at each job?

21. A tie and a pin cost $1.10. The tie cost $1 more than the pin. What is the cost of each?

22. A number of birds are resting on two limbs of a tree. One limb is above the other. A bird on the lower limb says to the birds on the upper limb, "If one of you will come down here, we will have an equal number on each limb." A bird from above replies, "If one of you will come up here, we will have twice as many up here as you will have down there." How many birds were sitting on each limb?

Chapter 14   R E V I E W

Solution of a System
14.1

A **solution of a system of two equations in two variables** is an ordered pair that, when substituted into both equations, makes them both true.

Solving a System
of Equations
14.1

In solving a system of equations, there are three possibilities:

14.2

1. *There is only one solution.*

14.3

    a. Graphical method: The lines intersect at one point.

    b. Addition method:

    c. Substitution method: } The equations can be solved for a single ordered pair.

2. *There is no solution.*

    a. Graphical method: The lines are parallel.

    b. Addition method:

    c. Substitution method: } Both variables are eliminated, and a *false* statement occurs.

3. *There are many solutions.*

    a. Graphical method: Both equations have the same line for a graph.

    b. Addition method:

    c. Substitution method: } Both variables are eliminated, and a *true* statement occurs.

Word Problems Using a
System of Equations
14.4

To solve a word problem using a system of equations:

1. Read the problem completely, and determine how many unknown numbers there are.

2. Draw a diagram showing the relationships in the problem whenever possible.

3. Represent each unknown number by a different variable.

4. Use the word statements to write a system of equations. *There must be as many equations as variables.*

5. Solve the system of equations using one of the following:

    a. Addition method

    b. Substitution method

    c. Graphical method

Chapter 14   R E V I E W   E X E R C I S E S

In Exercises 1–4, find the solution of each system graphically. Write "inconsistent" if no solution exists. Write "dependent" if many solutions exist.

1. $x + y = 6$
$x - y = 4$

2. $4x + 5y = 22$
$4x - 3y = 6$

3. $2x - 3y = 3$
$3y - 2x = 6$

4. $3x - 4y = 12$
$y = \dfrac{3}{4}x - 3$

In Exercises 5–8, find the solution of each system using the addition method. Write "inconsistent" if no solution exists. Write "dependent" if many solutions exist.

5. $x + 5y = 11$
$3x + 4y = 11$

6. $4x - 8y = 4$
$3x - 6y = 3$

7. $3x - 5y = 15$
$5y - 3x = 8$

8. $\dfrac{x}{2} - \dfrac{y}{4} = 4$
$\dfrac{x}{3} + \dfrac{y}{4} = 1$

In Exercises 9–12, solve each system using the substitution method. Write "inconsistent" if no solution exists. Write "dependent" if many solutions exist.

9. $x = y + 2$
$4x - 5y = 3$

10. $8x - 3y = 1$
$y = 2x - 3$

11. $x + 2y = 4$
$4x - y = -2$

12. $\dfrac{x}{3} + \dfrac{y}{6} = -1$
$\dfrac{x}{4} + \dfrac{y}{12} = -1$

In Exercises 13–16, solve each system by any convenient method. Write "inconsistent" if no solution exists. Write "dependent" if many solutions exist.

13. $4x + 3y = 8$
$8x + 7y = 12$

14. $6x - 2y = -2$
$3x = y + 1$

15. $5x - 4y = -7$
$-6x + 8y = 2$

16. $\dfrac{x - 1}{6} + \dfrac{y + 5}{9} = 2$
$\dfrac{x + 1}{4} - \dfrac{y + 2}{6} = 1$

17. The sum of two numbers is 84. Their difference is 22. What are the numbers?

18. The sum of two numbers is 13. If three times the larger number minus four times the smaller number is 4, what are the numbers?

19. A rectangle is 4 feet longer than it is wide. Its perimeter is 36 feet. Find the length and the width.

20. Goh worked at two jobs during the week for a total of 32 hours. For this he received a total of $216. If he was paid $8 an hour as a tutor and $6 an hour as a waiter, how many hours did he work at each job?

21. A mail-order office paid $730 for a total of 80 rolls of stamps in two denominations. If one kind costs $10 per roll and the other kind costs $8 per roll, how many rolls of each kind were bought?

22. An office manager paid $148 for 20 boxes of legal-size and letter-size file folders. If legal-size folders cost $9 a box and letter-size folders cost $7 a box, how many boxes of each kind were bought?

Chapter 14 Critical Thinking and Writing Problems

Answer Problems 1–5 in your own words, using complete sentences.

1. Explain how to check a solution to a system of two equations in two variables.

2. Explain how to solve a system by graphing.

3. For the following systems, would you choose the addition method or the substitution method as the more convenient? Explain why.

a. $\begin{cases} 2x - 7y = 1 \\ y = 9x - 13 \end{cases}$

b. $\begin{cases} 5x + 3y = 6 \\ x - 2y = 7 \end{cases}$

c. $\begin{cases} 4x - 3y = 5 \\ 5x + 2y = 1 \end{cases}$

4. Explain how to recognize that a system has no solution (is inconsistent) when solving by

a. Graphing

b. Addition

c. Substitution

5. Explain how to recognize that a system has many solutions (is dependent) when solving by

a. Graphing

b. Addition

c. Substitution

6. The solution to the system $\begin{cases} Ax - 3y = 9 \\ 2x + By = 2 \end{cases}$ is $(3, 2)$.

   a. Find A, the coefficient of x in the first equation.

   b. Find B, the coefficient of y in the second equation.

Each of the following problems has an error. Find the error, and in your own words, explain why it is wrong. Then work the problem correctly.

7. Solve by substitution: $\begin{cases} x + 2y = 3 \\ 2x + 3y = 8 \end{cases}$.

   Solve for x in terms of y:
   $x + 2y = 3$
   $\quad x = 3 - 2y$
   Substitute $3 - 2y$ for x:
   $\quad x + 2y = 3$
   $3 - 2y + 2y = 3$
   $\qquad\qquad 3 = 3 \quad$ True

   Answer: Dependent

8. Solve by addition: $\begin{cases} 2x + y = 1 \\ 5x + 3y = 4 \end{cases}$.

   $\begin{aligned} -3] \; 2x + \; y = 1 &\Rightarrow -6x - 3y = -3 \\ 5x + 3y = 4 &\Rightarrow \underline{5x + 3y = 4} \\ &\qquad -x \quad = 1 \\ &\qquad x = -1 \end{aligned}$

   Answer: $x = -1$

Chapter 14   DIAGNOSTIC TEST

Allow yourself fifty minutes to do these problems. Complete solutions for all problems, together with section references, are given at the end of the book.

In Problems 1–7, write "inconsistent" if no solution exists. Write "dependent" if many solutions exist.

1. Solve graphically: $\begin{cases} 3x + 2y = 2 \\ 2x - 3y = 10 \end{cases}$.

2. Solve by addition: $\begin{cases} 3x - 4y = 1 \\ 5x - 3y = 9 \end{cases}$.

3. Solve by addition: $\begin{cases} 3x - 4y = 1 \\ 5x - 6y = 5 \end{cases}$.

4. Solve by substitution: $\begin{cases} 3x - 5y = 14 \\ x = y + 2 \end{cases}$.

5. Solve by substitution: $\begin{cases} 4x + y = 2 \\ 2x - 3y = 8 \end{cases}$.

6. Solve by any convenient method: $\begin{cases} 4x - 3y = 7 \\ x - 2y = -2 \end{cases}$.

7. Solve by any convenient method: $\begin{cases} 6x - 9y = 2 \\ 15y - 10x = -5 \end{cases}$.

8. The sum of two numbers is 18. Their difference is 42. What are the numbers?

9. Rosa paid $19.53 for 15 records at a special sale. If classical records sold for $2.99 per disc and pop records for $0.88 per disc, how many of each kind did she buy?

10. The length of a rectangle is three centimeters more than its width. Its perimeter is 102 centimeters. Find the length and the width.

Exponents and Radicals

CHAPTER

15

n this chapter we complete the discussion of exponents. Then we discuss the simplification of and the operations with square roots.

15.1   Negative Exponents

Earlier we discussed positive and zero exponents. At that time the following four rules were given.

RULE 1. $x^a \cdot x^b = x^{a+b}$

When powers of the same base are *multiplied*, their exponents are *added*. For example,

$$x^2 \cdot x^5 = x^{2+5} = x^7$$

RULE 2. $(x^a)^b = x^{a \cdot b}$

When a power is raised to a power, the exponents are *multiplied*. For example,

$$(x^3)^2 = x^{3 \cdot 2} = x^6$$

RULE 3. $\dfrac{x^a}{x^b} = x^{a-b}$

When powers of the same base are *divided*, the exponent in the denominator is *subtracted from* the exponent in the numerator. For example,

$$\frac{x^5}{x^2} = x^{5-2} = x^3$$

RULE 4. $x^0 = 1$

Any nonzero number raised to the zero power equals 1. For example,

$$3^0 = 1$$

Negative Exponents   When we used Rule 3 earlier, the exponent in the numerator was always larger than the exponent in the denominator. Now we consider the case in which the exponent of the numerator is smaller than the exponent of the denominator. Consider the expression $\dfrac{x^3}{x^5}$:

$$\frac{x^3}{x^5} = \frac{xxx}{xxxxx} = \frac{xxx \cdot 1}{xxx \cdot xx} = \frac{xxx}{xxx} \cdot \frac{1}{xx} = 1 \cdot \frac{1}{xx} = \frac{1}{x^2}$$

└── The value of this fraction is 1

However, if we use Rule 3,

$$\frac{x^3}{x^5} = x^{3-5} = x^{-2}$$

Therefore, we define $x^{-2} = \dfrac{1}{x^2}$.

RULE 5a
Negative exponent

$$x^{-n} = \frac{1}{x^n} \qquad (x \neq 0)$$

EXAMPLE 1

Examples of the use of Rule 5a:

a. $x^{-5} = \dfrac{1}{x^5}$ **b.** $10^{-4} = \dfrac{1}{10^4}$

Suppose we have a negative exponent in the denominator. Consider $\dfrac{1}{x^{-2}}$:

$$\frac{1}{x^{-2}} = \frac{1}{\dfrac{1}{x^2}} = 1 \div \frac{1}{x^2} = 1 \cdot \frac{x^2}{1} = x^2$$

RULE 5b
Reciprocal of negative exponent

$$\frac{1}{x^{-n}} = x^n \qquad (x \neq 0)$$

EXAMPLE 2

Examples of the use of Rule 5b:

a. $\dfrac{1}{x^{-4}} = x^4$ **b.** $\dfrac{1}{y^{-3}} = y^3$

 Note Why do we write $x \neq 0$ in Rules 5a and 5b? If the base were 0, it would lead to division by 0, which is undefined. For example,

$$0^{-2} = \frac{1}{0^2} = \frac{1}{0 \cdot 0} = \frac{1}{0}$$

which is undefined.

In this book, unless otherwise noted, none of the variables has a value that makes the denominator zero.

Rules 5a and 5b lead to the following statement:

> A *factor* can be moved either from the numerator to the denominator or from the denominator to the numerator simply by changing the sign of its exponent.

 Note The above procedure does not change the sign of the *expression*.

EXAMPLE 3

Examples of writing expressions with only positive exponents:

a. $a^{-3}\, b^4 = \dfrac{a^{-3}}{1} \cdot \dfrac{b^4}{1} = \dfrac{1}{a^3} \cdot \dfrac{b^4}{1} = \dfrac{b^4}{a^3}$ The factor a^{-3} was moved from the numerator to the denominator by changing the sign of its exponent

b. $\dfrac{h^5}{k^{-4}} = \dfrac{h^5}{1} \cdot \dfrac{1}{k^{-4}} = \dfrac{h^5}{1} \cdot \dfrac{k^4}{1} = h^5\, k^4$ k^{-4} was moved from the denominator to the numerator by changing the sign of its exponent

These steps need not be written; we include them here to show why this method works .

The exponent -3 applies *only* to x

c. $2x^{-3} = \dfrac{2}{1} \cdot \dfrac{x^{-3}}{1} = \dfrac{2}{1} \cdot \dfrac{1}{x^3} = \dfrac{2}{x^3}$

d. $7^{-1}x^4 = \dfrac{7^{-1}}{1} \cdot \dfrac{x^4}{1} = \dfrac{1}{7} \cdot \dfrac{x^4}{1} = \dfrac{x^4}{7}$

e. $\dfrac{a^{-2}b^4}{c^5 d^{-3}} = \dfrac{a^{-2}}{1} \cdot \dfrac{b^4}{1} \cdot \dfrac{1}{c^5} \cdot \dfrac{1}{d^{-3}} = \dfrac{1}{a^2} \cdot \dfrac{b^4}{1} \cdot \dfrac{1}{c^5} \cdot \dfrac{d^3}{1} = \dfrac{b^4 d^3}{a^2 c^5}$

E X A M P L E   4   Examples of writing expressions without fractions, using negative exponents if necessary:

a. $\dfrac{m^5}{n^2} = \dfrac{m^5}{1} \cdot \dfrac{1}{n^2} = \dfrac{m^5}{1} \cdot \dfrac{n^{-2}}{1} = m^5 n^{-2}$

b. $\dfrac{a}{bc^2} = \dfrac{a^1}{1} \cdot \dfrac{1}{b^1} \cdot \dfrac{1}{c^2} = \dfrac{a^1}{1} \cdot \dfrac{b^{-1}}{1} \cdot \dfrac{c^{-2}}{1} = ab^{-1}c^{-2}$

A Word of Caution   An expression that is *not* a factor *cannot* be moved from the numerator to the denominator of a fraction simply by changing the sign of its exponent.

The $+$ sign indicates that a^{-1} is a *term* rather than a factor of the numerator

$$\frac{a^{-1} + b}{c^2} = \frac{b}{ac^2} \qquad \text{Incorrect}$$

The correct method is

$$\frac{a^{-1} + b}{c^2} = \frac{\dfrac{1}{a} + b}{c^2} \qquad \begin{array}{l}\text{Expressions of this kind were}\\ \text{simplified earlier}\end{array}$$

Using the Rules of Exponents with Positive, Negative, and Zero Exponents

All the rules for positive and zero exponents can also be used with negative exponents.

E X A M P L E   5   Examples of applying the rules of exponents:

a. $a^4 \cdot a^{-3} = a^{4+(-3)} = a^1 = a$ ⠀⠀⠀⠀⠀ Rule 1

b. $x^{-5} \cdot x^2 = x^{-5+2} = x^{-3} = \dfrac{1}{x^3}$ ⠀⠀⠀ Rules 1 and 5

c. $\left(y^{-2}\right)^{-1} = y^{(-2)(-1)} = y^2$ ⠀⠀⠀⠀ Rule 2

d. $\left(x^2\right)^{-4} = x^{2(-4)} = x^{-8} = \dfrac{1}{x^8}$ ⠀⠀ Rules 2 and 5

e. $\dfrac{y^{-2}}{y^{-6}} = y^{(-2)-(-6)} = y^{-2+6} = y^4$ ⠀ Rule 3

f. $\dfrac{z^{-4}}{z^{-2}} = z^{(-4)-(-2)} = z^{-4+2} = z^{-2} = \dfrac{1}{z^2}$ ⠀ Rules 3 and 5

g. $h^3 h^{-1} h^{-2} = h^{3+(-1)+(-2)} = h^0 = 1$ Rules 1 and 4

h. $x^{4n} \cdot x^{-n} = x^{4n+(-n)} = x^{3n}$ Rule 1

EXAMPLE 6

Examples of simplifying fractions using the rules of exponents and writing the results using only positive exponents:

a. $\dfrac{12x^{-2}}{4x^{-3}} = \dfrac{\overset{3}{\cancel{12}}}{\underset{1}{\cancel{4}}} \cdot \dfrac{x^{-2}}{x^{-3}} = \dfrac{3}{1} \cdot \dfrac{x^{-2-(-3)}}{1} = 3x$

b. $\dfrac{5a^4 b^{-3}}{10a^{-2}b^{-4}} = \dfrac{\overset{1}{\cancel{5}}}{\underset{2}{\cancel{10}}} \cdot \dfrac{a^4}{a^{-2}} \cdot \dfrac{b^{-3}}{b^{-4}} = \dfrac{1}{2} \cdot a^{4-(-2)}b^{-3-(-4)} = \dfrac{a^6 b}{2}$

c. $\dfrac{9xy^{-3}}{15x^3 y^{-4}} = \dfrac{\overset{3}{\cancel{9}}}{\underset{5}{\cancel{15}}} \cdot \dfrac{x}{x^3} \cdot \dfrac{y^{-3}}{y^{-4}} = \dfrac{3}{5} \cdot \dfrac{x^{1-3}}{1} \cdot \dfrac{y^{-3-(-4)}}{1}$

$= \dfrac{3}{5} \cdot \dfrac{x^{-2}}{1} \cdot \dfrac{y}{1} = \dfrac{3}{5} \cdot \dfrac{1}{x^2} \cdot \dfrac{y}{1} = \dfrac{3y}{5x^2}$

Evaluating Expressions That Have Numerical Bases

EXAMPLE 7

a. $4^{-2} = \dfrac{1}{4^2} = \dfrac{1}{4 \cdot 4} = \dfrac{1}{16}$

b. $10^6 \cdot 10^{-2} = 10^{6+(-2)} = 10^4 = \underbrace{10{,}000}_{\text{4 zeros}}$

c. $(2^3)^{-1} = 2^{-3} = \dfrac{1}{2^3} = \dfrac{1}{2 \cdot 2 \cdot 2} = \dfrac{1}{8}$

d. $\dfrac{3^{-1}}{3^{-4}} = 3^{-1-(-4)} = 3^3 = 3 \cdot 3 \cdot 3 = 27$

This exponent applies only to the 5

e. $\dfrac{-5^2}{(-5)^2} = \dfrac{-(5 \cdot 5)}{(-5)(-5)} = \dfrac{-25}{25} = -1$

This exponent applies to the (-5)

A Word of Caution   A common error is to think that a negative exponent will give a negative answer.

$$3^{-2} = -9 \quad \text{Incorrect}$$

The correct method is

$$3^{-2} = \dfrac{1}{3^2} = \dfrac{1}{9}$$

Simplifying Exponential Expressions

An expression with exponents is considered **simplified** when each different base appears only once and its exponent is a single integer.

EXAMPLE 8   Simplify $\dfrac{x^{-4}y^3}{y^2}$. Write the answer using only positive exponents.

SOLUTION

$$\frac{x^{-4}y^3}{y^2} = \frac{y^3}{x^4y^2} \quad \longleftarrow \begin{array}{l}\text{Not simplified because the}\\ \text{base } y \text{ appears twice}\end{array}$$

$$= \frac{y}{x^4} \quad \longleftarrow \text{Simplified form}$$

Exercises 15.1

In Exercises 1–24, rewrite each expression using only positive exponents.

1. x^{-4}
2. y^{-7}
3. $4a^{-3}$
4. $3x^{-2}$

5. $\dfrac{1}{a^{-4}}$
6. $\dfrac{3}{b^{-5}}$
7. $x^{-2}y^3$
8. x^3y^{-2}

9. $r^{-2}st^{-4}$
10. $h^{-3}k^5n^{-1}$
11. $\dfrac{x^{-3}}{y^2}$
12. $\dfrac{a^{-2}}{b}$

13. $\dfrac{h^2}{k^{-4}}$
14. $\dfrac{m^3}{n^{-2}}$
15. $\dfrac{3x^{-4}}{y}$
16. $\dfrac{5a^{-5}}{b}$

17. $ab^{-2}c^0$
18. $x^{-3}y^0z$
19. $\dfrac{a^3b^0}{c^{-2}}$
20. $\dfrac{d^0e^2}{f^{-3}}$

21. $\dfrac{p^4r^{-1}}{t^{-2}}$
22. $\dfrac{u^5v^{-2}}{w^{-3}}$
23. $\dfrac{5x^{-1}}{y^{-2}}$
24. $\dfrac{2x^{-3}}{y^{-7}}$

In Exercises 25–30, write each expression without fractions, using negative exponents if necessary.

25. $\dfrac{1}{x^2}$
26. $\dfrac{1}{y^3}$
27. $\dfrac{h}{k}$
28. $\dfrac{m}{n}$

29. $\dfrac{x^2}{yz^5}$
30. $\dfrac{a^3}{b^2c}$

In Exercises 31–66, simplify each expression. Write the answer using only positive exponents.

31. $x^{-3} \cdot x^4$
32. $y^6 \cdot y^{-2}$
33. $x^{-7} \cdot x^4$

34. $a^2 \cdot a^{-3}$
35. $(x^2)^{-4}$
36. $(z^3)^{-2}$

37. $(a^{-2})^{-3}$
38. $(b^{-5})^{-2}$
39. $\dfrac{x^2}{x^{-4}}$

40. $\dfrac{y^5}{y^{-1}}$
41. $\dfrac{y^{-2}}{y^5}$
42. $\dfrac{z^{-2}}{z^2}$

43. $x^4x^{-1}x^{-3}$
44. $y^{-2}y^{-3}y^5$
45. ab^2a^{-8}

46. $xy^{-5}y^2$
47. $\dfrac{x^{-3} \cdot x^2}{x^{-8}}$
48. $\dfrac{n^3 \cdot n^{-5}}{n^{-7}}$

49. $\dfrac{8x^{-3}}{12x}$
50. $\dfrac{15y^{-2}}{10y}$
51. $\dfrac{20h^{-2}}{35h^{-4}}$

52. $\dfrac{35k^{-1}}{28k^{-4}}$
53. $\dfrac{7x^{-3}y}{14y^{-2}}$
54. $\dfrac{24m^{-4}p}{16m^{-2}}$

55. $\dfrac{15m^0n^{-2}}{5m^{-3}n^4}$
56. $\dfrac{14x^0y^{-3}}{12x^{-2}y^{-4}}$
57. $(x^2)^0$

58. $(x^{5n})^0$
59. $x^{3m} \cdot x^{-m}$
60. $y^{-2n} \cdot y^{5n}$

61. $(x^{3b})^{-2}$
62. $(y^{2a})^{-3}$
63. $\dfrac{x^{2a}}{x^{-5a}}$

64. $\dfrac{a^{3x}}{a^{-5x}}$
65. $\dfrac{x+y^{-1}}{y}$
66. $\dfrac{a^{-1}-b}{b}$

In Exercises 67–90, evaluate each expression.

67. 2^{-3}
68. 6^{-1}
69. 5^{-2}
70. 8^{-2}

71. $10^4 \cdot 10^{-2}$
72. $3^{-2} \cdot 3^3$
73. $\dfrac{4^{-1}}{4^{-3}}$
74. $\dfrac{5^{-2}}{5^{-5}}$

75. $(10^4)^{-2}$
76. $(2^{-3})^2$
77. $2^{-3} \cdot 2^{-1}$
78. $3^{-1} \cdot 3^{-2}$

79. $(10^{-1})^{-4}$
80. $(4^{-3})^{-1}$
81. $\dfrac{2^{-2}}{2^3}$
82. $\dfrac{10^{-1}}{10^2}$

83. $3^0 \cdot 5^2$
84. $7^2 \cdot 5^0$
85. $\dfrac{6^{-2}}{6^0}$
86. $\dfrac{10^0}{10^{-2}}$

87. $\dfrac{10^{-3} \cdot 10^2}{10^5}$
88. $\dfrac{2^3 \cdot 2^{-4}}{2^2}$
89. $\dfrac{-4^2}{(-4)^2}$
90. $\dfrac{(-3)^4}{-3^4}$

15.2   General Rule of Exponents

Power of a Product   Consider the expression $(xy)^3$:

$$(xy)^3 = (xy) \cdot (xy) \cdot (xy) = x \cdot x \cdot x \cdot y \cdot y \cdot y = x^3y^3$$

This leads to the following rule:

| **RULE 6** Power of a product | $$(xy)^n = x^n y^n$$ To raise a product to a power, distribute the exponent to each factor. |

EXAMPLE 1 Examples of the use of Rule 6:

a. $(ab)^4 = a^4 b^4$

b. $(2x)^3 = 2^3 x^3 = 8x^3$

> Be sure to distribute the exponent to the numerical base as well as to the literal base

c. $(4x)^{-1} = 4^{-1} x^{-1} = \dfrac{1}{4} \cdot \dfrac{1}{x} = \dfrac{1}{4x}$

d. $(x^2 y^3)^5 = (x^2)^5 (y^3)^5 = x^{2 \cdot 5} y^{3 \cdot 5} = x^{10} y^{15}$ Rules 6 and 2

Power of a Quotient   Consider the expression $\left(\dfrac{x}{y}\right)^3$:

$$\left(\frac{x}{y}\right)^3 = \left(\frac{x}{y}\right) \cdot \left(\frac{x}{y}\right) \cdot \left(\frac{x}{y}\right) = \frac{x \cdot x \cdot x}{y \cdot y \cdot y} = \frac{x^3}{y^3}$$

This leads to the following rule:

| **RULE 7** Power of a quotient | $$\left(\frac{x}{y}\right)^n = \frac{x^n}{y^n}$$ To raise a quotient to a power, distribute the exponent of each factor. |

EXAMPLE 2 Examples of the use of Rule 7:

a. $\left(\dfrac{a}{b}\right)^7 = \dfrac{a^7}{b^7}$

b. $\left(\dfrac{3}{x}\right)^{-2} = \dfrac{3^{-2}}{x^{-2}} = \dfrac{x^2}{3^2} = \dfrac{x^2}{9}$

c. $\left(\dfrac{x^4}{y^2}\right)^3 = \dfrac{(x^4)^3}{(y^2)^3} = \dfrac{x^{4 \cdot 3}}{y^{2 \cdot 3}} = \dfrac{x^{12}}{y^6}$ Rules 7 and 2

Rules 2, 6, and 7 of exponents can be combined into the following general rule.

| **RULE 8** General rule of exponents | $$\left(\frac{x^a y^b}{z^c}\right)^n = \frac{x^{an} y^{bn}}{z^{cn}}$$ |

When using Rule 8, notice the following:

1. x, y, and z are *factors* of the expression within the parentheses. They are *not* separated by + or − signs. See Example 3g.

2. The exponent of each factor within the parentheses is multiplied by the exponent outside the parentheses.

EXAMPLE 3   Examples of simplifying expressions using the rules of exponents and writing the answers using only positive exponents:

a. $(x^3y^{-1})^5 = x^{3 \cdot 5}y^{(-1)5} = x^{15}y^{-5} = \dfrac{x^{15}}{y^5}$

b. $\left(\dfrac{x^{-3}}{y^6}\right)^{-2} = \dfrac{x^{(-3)(-2)}}{y^{6(-2)}} = \dfrac{x^6}{y^{-12}} = x^6y^{12}$

The rules of exponents apply to *numerical* bases as well as to literal bases

c. $\left(\dfrac{2a^{-3}b^2}{c^5}\right)^3 = \dfrac{2^{1 \cdot 3}\, a^{(-3)3}b^{2 \cdot 3}}{c^{5 \cdot 3}} = \dfrac{2^3 a^{-9}b^6}{c^{15}} = \dfrac{8b^6}{a^9 c^{15}}$

d. $\left(\dfrac{3^{-7}x^{10}}{y^{-4}}\right)^0 = 1$ The zero power of any nonzero expression is 1

e. $\left(\dfrac{x^5y^4}{x^3y^7}\right)^2 = (x^2y^{-3})^2 = x^4y^{-6} = \dfrac{x^4}{y^6}$

Simplify the expression within the parentheses first whenever possible

f. $\left(\dfrac{10^{-2} \cdot 10^5}{10^4}\right)^3 = \left(\dfrac{10^3}{10^4}\right)^3 = (10^{-1})^3 = 10^{-3} = \dfrac{1}{10^3} = \dfrac{1}{1,000}$

g. $(x^2 + y^3)^4$ Rules 2 and 8 *cannot* be used here because the + sign means that x^2 and y^3 are *not* factors; they are *terms* of the expression being raised to the fourth power

Exercises   15.2

Simplify each expression. Write the answer using only positive exponents.

1. $(xy)^5$

2. $(mn)^3$

3. $(4x)^2$

4. $(3x)^3$

5. $(5x)^{-1}$

6. $(2x)^{-3}$

7. $\left(\dfrac{5}{x}\right)^{-2}$

8. $\left(\dfrac{2}{m}\right)^{-4}$

9. $(a^2b^3)^2$

10. $(x^4y^5)^3$

11. $(2z^3)^2$

12. $(3w^2)^3$

13. $(m^{-2}n)^4$

14. $(p^{-3}r)^5$

15. $(x^{-2}y^3)^{-4}$

16. $(w^{-3}z^4)^{-2}$

17. $(3a^{-2})^{-3}$

18. $(9x^{-4})^{-1}$

19. $\left(\dfrac{rs^3}{t^4}\right)^3$

20. $\left(\dfrac{x^3y}{z^2}\right)^2$

21. $\left(\dfrac{3x^3}{y}\right)^2$

22. $\left(\dfrac{6x}{y^6}\right)^2$

23. $\left(\dfrac{M^{-2}}{N^3}\right)^4$

24. $\left(\dfrac{R^5}{S^{-4}}\right)^3$

25. $\left(\dfrac{a^2b^{-4}}{b^{-5}}\right)^2$

26. $\left(\dfrac{x^{-2}y^2}{x^{-3}}\right)^3$

27. $\left(\dfrac{10^2 \cdot 10^{-1}}{10^{-2}}\right)^2$

28. $\left(\dfrac{3^{-3} \cdot 3^2}{3^{-2}}\right)^3$

29. $\left(\dfrac{mn^{-1}}{m^3}\right)^{-2}$

30. $\left(\dfrac{ab^{-2}}{a^2}\right)^{-3}$

31. $\left(\dfrac{x^4}{x^{-1}y^{-2}}\right)^{-1}$

32. $\left(\dfrac{x^3}{x^{-2}y^{-4}}\right)^{-1}$

33. $\left(\dfrac{8s^{-3}}{4st^2}\right)^{-2}$

34. $\left(\dfrac{10u^{-4}}{5uv^3}\right)^{-2}$

35. $(10^0k^{-4})^{-2}$

36. $(6^0z^{-5})^{-2}$

37. $(x^3 + y^4)^5$

38. $(a^5 - b^2)^6$

39. $\left(\dfrac{r^7s^8}{r^9s^6}\right)^0$

40. $\left(\dfrac{t^5u^6}{t^8u^7}\right)^0$

41. $\left(\dfrac{6m^{-4}p}{m^{-2}}\right)^{-1}$

42. $\left(\dfrac{10x^{-6}y}{x^{-4}}\right)^{-1}$

43. $\left(\dfrac{10 \cdot 10^{-2}}{10^{-3}}\right)^3$

44. $\left(\dfrac{m^{-1}n^3}{m}\right)^{-2}$

45. $\left(\dfrac{x^2}{x^{-3}y^{-2}}\right)^{-1}$

46. $\left(\dfrac{a^{-3}b^2}{a^{-4}}\right)^3$

47. $\left(\dfrac{x^2y^{-3}}{x^4y^2}\right)^{-2}$

48. $\left(\dfrac{t^{-1}u^2}{t^4u^{-3}}\right)^{-3}$

49. $\left(\dfrac{8yz^{-3}}{z^{-2}}\right)^{-1}$

50. $\left(\dfrac{9u^{-3}}{uv^2}\right)^{-2}$

15.3 Scientific Notation

In science we often work with very large and very small numbers. For example, a light-year, the distance light travels in one year, is approximately 5,880,000,000,000 miles. Hydrogen, the smallest atom, has a diameter of about 0.0000001 millimeter. Writing and calculating with such numbers is difficult because there are so many zeros. Scientific notation gives us a more convenient way to work with these very large and very small numbers.

Before we can write numbers in scientific notation, we must understand positive and negative powers of 10. Earlier we discussed positive powers of 10. We now extend this system of powers of 10 to include the negative powers of 10. See Figure 1.

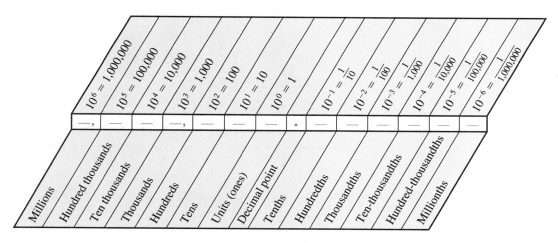

Figure 1 Powers of 10

Multiplying by a Positive Power of 10

To multiply a decimal by a *positive* power of 10, move the decimal point to the *right* the same number of places as the power of 10.

EXAMPLE 1

a. $7.5 \times 10^2 = 7.50_\wedge = 750$

b. $0.048 \times 10^3 = 0.048_\wedge = 48$

c. $272 \times 10^5 = 272.00000_\wedge = 27,200,000$

Multiplying by a Negative Power of 10

Recall that when we divided a decimal by a positive power of 10, the decimal point moved to the left. But multiplying by a negative power of 10 is the same as dividing by a positive power of 10. Therefore, to multiply by a *negative* power of 10, move the decimal point to the *left* the same number of places as the power of 10.

EXAMPLE 2

a. $7.5 \times 10^{-2} = \dfrac{7.5}{10^2} = {}_{\wedge}07.5 = 0.075$

b. $6.04 \times 10^{-3} = \dfrac{6.04}{10^3} = {}_{\wedge}006.04 = 0.00604$

c. $97 \times 10^{-1} = 9_{\wedge}7. = 9.7$

d. $3.2 \times 10^{-4} = {}_{\wedge}0003.2 = 0.00032$

Scientific Notation

A number is written in **scientific notation** if it is a number between 1 and 10 multiplied by a power of 10. To write a number in scientific notation, place the decimal point after the first nonzero digit and then multiply by the appropriate power of 10.

EXAMPLE 3

	Decimal notation	*Scientific notation*
a.	$245 =$	$2_{\wedge}4\ 5. = 2.45 \times 10^2$
b.	$24.5 =$	$2_{\wedge}4.5 = 2.45 \times 10^1$
c.	$2.45 =$	$2.4\ 5 = 2.45 \times 10^0$
d.	$0.245 =$	$0.2_{\wedge}4\ 5 = 2.45 \times 10^{-1}$
e.	$0.0245 =$	$0.0\ 2_{\wedge}4\ 5 = 2.45 \times 10^{-2}$
f.	$0.00245 =$	$0.0\ 0\ 2_{\wedge}4\ 5 = 2.45 \times 10^{-3}$

The preceding examples lead us to the following rule:

Writing a number in scientific notation

1. Place a caret ($\wedge$) to the right of the first nonzero digit.
2. Draw an arrow *from* the caret to the actual decimal point.
3. The exponent of 10 is $+$ if the arrow points right ($\rightarrow$); it is $-$ if the arrow points left ($\leftarrow$).
4. The number part of the exponent of 10 is equal to the number of places between the caret and the actual decimal point.
5. Write the number with decimal point after the first nonzero digit and multiply by the power of 10 found in steps 3 and 4.

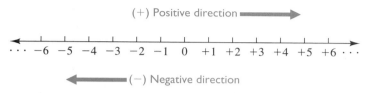

Note If the arrow is drawn toward the decimal point when the number is written in decimal notation, the sign used for the exponent of 10 is the same as that used on the number line. (See Figure 2.)

(+) Positive direction ➡

··· −6 −5 −4 −3 −2 −1 0 +1 +2 +3 +4 +5 +6 ···

⬅ (−) Negative direction

Figure 2

EXAMPLE 4

Examples of changing numbers from decimal to scientific notation:

a. $92{,}900{,}000. = 9{\wedge}2\,9\,0\,0\,0\,0\,0. = 9.29 \times 10^7$

b. $0.0056 = 0.0\,0\,5{\wedge}6 = 5.6 \times 10^{-3}$

c. $0.01745 = 0.0\,1{\wedge}7\,4\,5 = 1.745 \times 10^{-2}$

d. $684.5 = 6{\wedge}8\,4.5 = 6.845 \times 10^2$

EXAMPLE 5

Examples of changing numbers from scientific notation to decimal notation:

a. $2.54 \times 10^{-2} = 0.0\,2{\wedge}5\,4 = 0.0254$

b. $1.609 \times 10^3 = 1{\wedge}6\,0\,9. = 1{,}609$

c. $4.67 \times 10^{-5} = 0.0\,0\,0\,0\,4{\wedge}6\,7 = 0.0000467$

d. $3.57 \times 10^5 = 3{\wedge}5\,7\,0\,0\,0. = 357{,}000$

Calculations Using Scientific Notation

EXAMPLE 6

Find $2.8 \times 10^3 \times 1.6 \times 10^2$.

SOLUTION $2.8 \times 10^3 \times 1.6 \times 10^2 = (2.8 \times 1.6) \times (10^3 \times 10^2)$

$$= 4.48 \times 10^5$$

EXAMPLE 7

Find $\dfrac{7.5 \times 10^{-2}}{1.5 \times 10^4}$.

SOLUTION

$$\frac{7.5 \times 10^{-2}}{1.5 \times 10^4} = \frac{7.5}{1.5} \times \frac{10^{-2}}{10^4}$$

$$= 5 \times 10^{-2-4}$$

$$= 5 \times 10^{-6}$$

EXAMPLE 8   Find $\dfrac{0.065 \times 24{,}000}{0.0006}$.

SOLUTION   Write each number in scientific notation, then perform the indicated operations:

$$\frac{0.065 \times 24{,}000}{0.0006} = \frac{6.5 \times 10^{-2} \times 2.4 \times 10^4}{6 \times 10^{-4}}$$

$$= \frac{15.6 \times 10^2}{6 \times 10^{-4}} = 2.6 \times 10^{2-(-4)} = 2.6 \times 10^6$$

When the result of a calculation in decimal notation exceeds the calculator's display, the calculator may automatically convert to scientific notation.

EXAMPLE 9   Use a calculator to find the following:

a. $6{,}000{,}000 \times 20{,}000$

SOLUTION

Key in   6000000 $\boxed{\times}$ 20000 $\boxed{=}$

Answer   1.2   11
or   1.2 ^{11} } Either display means 1.2×10^{11}

b. $\dfrac{0.0000438}{300{,}000}$

SOLUTION

Key in   .0000438 $\boxed{\div}$ 300000 $\boxed{=}$

Answer   1.46   −10
or   1.46 $^{-10}$ } $= 1.46 \times 10^{-10}$

Exercises 15.3

For Exercises 1–20, complete the following table.

	Decimal notation	Scientific notation
1.	748	
2.	25,000	
3.	0.063	
4.		6.7×10^3
5.	0.001732	
6.		2.81×10^{-2}
7.		3.47×10^6
8.	86.48	
9.		1.91×10^{-6}
10.	588,000	

	Decimal notation	Scientific notation
11.	0.0000563	
12.	27,800	
13.	0.000058	
14.	1,761,000	
15.		6.547×10^{-9}
16.	0.00000078	
17.		4.77×10^{-10}
18.	5,780,000,000,000	
19.		8.36×10^7
20.		1.05×10^{-5}

In Exercises 21–30, perform the indicated operations, and write the answers in scientific notation.

 In Exercises 31–36, use a calculator to perform the indicated operations. Write the answers in scientific notation.

21. $3.4 \times 10^2 \times 2.5 \times 10^{-5}$

22. $2.3 \times 10^{-1} \times 4.06 \times 10^5$

31. $7,200,000 \times 5,000$

32. 0.0000009×0.00087

23. $\dfrac{8.1 \times 10^3}{2.7 \times 10^{-2}}$

24. $\dfrac{7.8 \times 10^{-1}}{1.3 \times 10^5}$

33. $\dfrac{0.000024}{6,400,000}$

34. $\dfrac{66,000,000}{0.00096}$

25. $203,000 \times 0.0004$

26. $0.00003 \times 2,600$

35. $\dfrac{380 \times 21,000}{0.0000035}$

36. $\dfrac{0.0057 \times 0.0003}{7,600,000}$

27. $\dfrac{0.00036}{0.024}$

28. $\dfrac{450,000}{0.0018}$

29. $\dfrac{0.0055 \times 1200}{0.00011}$

30. $\dfrac{0.072}{0.12 \times 50,000}$

15.4 Square Roots

Finding the square root of a number is the inverse operation of squaring a number. Every positive number has both a positive and a negative square root. The positive square root is called the **principal square root**. The principal square root of N is written $\sqrt{N}$. The negative square root of N is written $-\sqrt{N}$. The number under the radical sign is called the **radicand**.

$$\sqrt{N} \qquad \text{Read ``the square root of } N\text{''}$$

Radical sign ⟶   ⟵ Radicand

EXAMPLE 1

a. $\sqrt{25} = 5$ Because $5^2 = 25$

b. $\sqrt{100} = 10$ Because $10^2 = 100$

c. $\sqrt{\dfrac{4}{9}} = \dfrac{2}{3}$ Because $\left(\dfrac{2}{3}\right)^2 = \dfrac{4}{9}$

d. $-\sqrt{64} = -8$ Because $\sqrt{64} = 8$

$$-\sqrt{64} = -8$$

Rational Numbers   A **rational number** is a number that can be written in the form $\dfrac{a}{b}$, where a and b are integers ($b \neq 0$). The decimal representation of any rational number is either a terminating decimal or a repeating decimal.

EXAMPLE 2   Examples of rational numbers:

a. $\dfrac{2}{3}$ All fractions are rational numbers

b. $4 = \dfrac{4}{1}$ All integers are rational numbers

c. $0.25 = \dfrac{25}{100}$ All terminating decimals are rational numbers

d. $0.333\ldots = \dfrac{1}{3}$ All repeating decimals are rational numbers

e. $2\dfrac{1}{2} = \dfrac{5}{2}$ All mixed numbers are rational numbers

f. $\sqrt{25} = 5 = \dfrac{5}{1}$ $\sqrt{25}$ simplifies to the integer 5, which is a rational number

---

Irrational Numbers   An **irrational number** is a number whose decimal representation is a nonterminating, nonrepeating decimal.

E X A M P L E   3   Examples of irrational numbers:

a. $\sqrt{2} = 1.414213562\ldots$

b. $\sqrt{5} = 2.236067977\ldots$

c. $\pi = 3.1415926535\ldots$

Real Numbers   Together, the rational numbers and the irrational numbers form the set of **real numbers** and, therefore, can be represented by points on the number line.

☞ (Note)   Square roots of negative numbers, such as $\sqrt{-4}$, are imaginary numbers. Imaginary numbers are not real and cannot be graphed on the number line.

Exercises   15.4

1. Which of the following are rational numbers?

$$-5, \quad \frac{3}{4}, \quad 2\frac{1}{2}, \quad \sqrt{2}, \quad 3.5, \quad \sqrt{4}, \quad 3$$

2. Which of the following are irrational numbers?

$$\frac{5}{6}, \quad -4, \quad \sqrt{3}, \quad 3\frac{1}{4}, \quad \sqrt{9}, \quad 0.7, \quad \pi$$

In Exercises 3 and 4, identify the radicand in each expression.

3. **a.** $\sqrt{17}$ **b.** $\sqrt{x+1}$ 4. **a.** $2\sqrt{15}$ **b.** $3\sqrt{5x}$

In Exercises 5–14, find the square roots.

5. $\sqrt{36}$ 6. $\sqrt{64}$ 7. $\sqrt{1}$ 8. $\sqrt{0}$

9. $\sqrt{81}$ 10. $\sqrt{144}$ 11. $-\sqrt{4}$ 12. $-\sqrt{25}$

13. $\sqrt{\dfrac{16}{49}}$ 14. $\sqrt{\dfrac{9}{100}}$

15.5   Simplifying Square Roots

We consider square roots of two kinds:

   square roots *without* fractions in the radicand

   square roots *with* fractions in the radicand

15.5A   Square Roots Without Fractions in the Radicand

We know that the number 9 has two square roots, $+3$ and -3. The symbol $\sqrt{9}$ represents the principal square root; thus, $\sqrt{9} = 3$. The expression x^2 also has two square roots, $+x$ and $-x$. Because x could be either a positive or a negative number, we do not know which is the principal square root. However, we may obtain a positive answer if we use the absolute value symbol. Thus, $\sqrt{x^2} = |x|$.

In this chapter we assume that all variables represent positive numbers unless otherwise indicated. For this reason we do not use the absolute value symbol to indicate a principal square root.

We will write $\sqrt{x^2} = x$

Similarly, $\sqrt{x^4} = x^2$ Because $(x^2)^2 = x^4$

$\sqrt{x^6} = x^3$ Because $(x^3)^2 = x^6$

Notice that to find the square root of a number with an even exponent, we divide the exponent by 2. To find the square root of a number with an odd exponent, we will need Rule 9, below.

EXAMPLE 1 Does $\sqrt{4 \cdot 9} = \sqrt{4} \cdot \sqrt{9}$?

SOLUTION $\sqrt{4 \cdot 9} = \sqrt{36} = 6$

$\sqrt{4} \cdot \sqrt{9} = 2 \cdot 3 = 6$

Therefore, $\sqrt{4 \cdot 9} = \sqrt{4} \cdot \sqrt{9}$.

This leads to the first rule of radicals.

RULE 9
Square root of a product

$$\sqrt{a \cdot b} = \sqrt{a} \cdot \sqrt{b}$$

We will use Rule 9 to help us find the principal square root of a product.

Finding the principal square root

1. Find the square root of each factor. (Numerical and literal factors are done the same way.)

 a. *If the exponent of the factor is an even number*, divide the exponent by 2:

 Even exponent
 $$\sqrt{x^6} = x^{6 \div 2} = x^3$$

 b. *If the exponent of the factor is an odd number*, write the factor as the product of two factors—one factor having an even exponent and the other factor having an exponent of 1:

 Odd exponent
 $$\sqrt{x^9} = \sqrt{x^8 x^1} = \sqrt{x^8}\sqrt{x^1}$$
 $$= x^4\sqrt{x}$$

2. Multiply the square roots of all the factors found in Step 1.

EXAMPLE 2 Examples of finding the square root of a factor whose exponent is an *even* number:

 a. $\sqrt{5^2} = 5^{2/2} = 5^1$ Dividing the exponent by 2

 b. $\sqrt{x^2} = x^{2/2} = x^1$

 c. $\sqrt{y^4} = y^{4/2} = y^2$

EXAMPLE 3   Examples of finding the square root of a factor whose exponent is an *odd* number:

a. $\sqrt{3^5} = \sqrt{3^4 \cdot 3^1}$ ← $3^5 = 3^4 \cdot 3^1$

$\quad\quad = \sqrt{3^4}\sqrt{3}$ —— The factor 3^4 is the highest power of 3 whose exponent, 4, is exactly divisible by 2

$\quad\quad = 3^2\sqrt{3}$

$\quad\quad = 9\sqrt{3}$

b. $\sqrt{x^7} = \sqrt{x^6 \cdot x^1}$ ← $x^7 = x^6 \cdot x^1$

$\quad\quad = \sqrt{x^6}\sqrt{x}$ —— The factor x^6 is the highest power of x whose exponent, 6, is exactly divisible by 2

$\quad\quad = x^3\sqrt{x}$

When finding the square root of a number, first express it in prime-factored form.

EXAMPLE 4   Examples of finding the square root of a number:

a. $\sqrt{48} = \sqrt{2^4 \cdot 3}$

$\quad\quad = \sqrt{2^4}\sqrt{3}$ —— Prime-factored form of 48

$\quad\quad = 2^2\sqrt{3}$

$\quad\quad = 4\sqrt{3}$

$$\begin{array}{r|l} 2 & 48 \\ 2 & 24 \\ 2 & 12 \\ 2 & 6 \\ \hline & 3 \end{array} \quad 48 = 2^4 \cdot 3$$

b. $\sqrt{75} = \sqrt{3 \cdot 5^2}$

$\quad\quad = \sqrt{3}\sqrt{5^2}$ —— Prime-factored form of 75

$\quad\quad = \sqrt{3} \cdot 5$

$\quad\quad = 5\sqrt{3}$

$$\begin{array}{r|l} 5 & 75 \\ 5 & 15 \\ \hline & 3 \end{array} \quad 75 = 3 \cdot 5^2$$

A convenient arrangement of the work for finding the principal square root of a product is shown in Example 5.

EXAMPLE 5   Examples of finding the square root of a product:

a. $\sqrt{360} = \sqrt{2^3 \cdot 3^2 \cdot 5}$

$\quad\quad = \sqrt{2^2 \cdot 2 \cdot 3^2 \cdot 5}$

$\quad\quad = \quad 2 \quad \cdot \quad 3 \cdot \sqrt{2 \cdot 5}$

$\quad\quad = 6\sqrt{10}$

$$\begin{array}{r|l} 2 & 360 \\ 2 & 180 \\ 2 & 90 \\ 3 & 45 \\ 3 & 15 \\ \hline & 5 \end{array} \quad 360 = 2^3 \cdot 3^2 \cdot 5$$

b. $\sqrt{12x^4y^3} = \sqrt{2^2 \cdot 3 \cdot x^4 \cdot y^2 \cdot y}$

$\quad\quad = \quad 2 \quad \cdot \quad x^2 \cdot y\sqrt{3y}$

$\quad\quad = 2x^2y\sqrt{3y}$

$$\begin{array}{r|l} 2 & 12 \\ 2 & 6 \\ \hline & 3 \end{array} \quad 12 = 2^2 \cdot 3$$

c. $\sqrt{24a^5b^7} = \sqrt{2^3 \cdot 3 \cdot a^5 \cdot b^7}$

$\qquad = \sqrt{2^2 \cdot 2 \cdot 3 \cdot a^4 \cdot a \cdot b^6 \cdot b}$

$\qquad\qquad\quad | \qquad\qquad | \qquad\quad |$

$\qquad = \quad 2 \quad \cdot \quad a^2 \quad \cdot \quad b^3 \cdot \sqrt{2 \cdot 3 \cdot a \cdot b}$

$\qquad = 2a^2b^3\sqrt{6ab}$

$$\begin{array}{r|l} 2 & 24 \\ 2 & 12 \\ 2 & 6 \\ & 3 \end{array} \quad 24 = 2^3 \cdot 3$$

Sometimes the square root of a number can be found by inspection, if you can see that it has a factor that is a perfect square (Example 6).

EXAMPLE 6

Examples of finding the square root by inspection:

a. $\sqrt{12} = \sqrt{4 \cdot 3}$ 4 is a factor of 12 and is a perfect square

$\qquad = \sqrt{4}\sqrt{3}$

$\qquad = 2\sqrt{3}$

b. $\sqrt{50} = \sqrt{25 \cdot 2}$ 25 is a factor of 50 and is a perfect square

$\qquad = \sqrt{25} \cdot \sqrt{2}$

$\qquad = 5\sqrt{2}$

c. $5\sqrt{18} = 5\sqrt{9 \cdot 2}$ 9 is a factor of 18 and is a perfect square

$\qquad = 5 \cdot \sqrt{9} \cdot \sqrt{2}$

$\qquad = 5 \cdot 3 \cdot \sqrt{2}$

$\qquad = 15\sqrt{2}$

It's true for a product that, by Rule 9, $\sqrt{9 \cdot 16} = \sqrt{9} \cdot \sqrt{16}$, but what about a sum? See Example 7.

EXAMPLE 7

Does $\sqrt{9 + 16} = \sqrt{9} + \sqrt{16}$?

SOLUTION $\qquad\qquad \sqrt{9 + 16} = \sqrt{25} = 5$

$\qquad\qquad\qquad\qquad\quad \sqrt{9} + \sqrt{16} = 3 + 4 = 7$

Because $5 \neq 7$, the square root of a sum does not equal the sum of the square roots.

Multiplying a Square Root by Itself

By Rule 9, $\qquad\qquad\qquad\qquad \sqrt{a}\sqrt{b} = \sqrt{ab}$

If $b = a$, then $\qquad\qquad\qquad \sqrt{a}\sqrt{a} = \sqrt{aa}$

$\qquad\qquad\qquad\qquad\qquad\quad \sqrt{a}\sqrt{a} = \sqrt{a^2}$

$\qquad\qquad\qquad\qquad\qquad\quad \sqrt{a}\sqrt{a} = a$

RULE 10
Multiplying a square root by itself

$\sqrt{a}\sqrt{a} = a$

or $\qquad\qquad (\sqrt{a})^2 = a$

EXAMPLE 8

a. $\sqrt{5}\sqrt{5} = 5$

b. $\sqrt{7}\sqrt{7} = 7$

c. $\sqrt{2x}\sqrt{2x} = 2x$

d. $(\sqrt{x})^2 = x$

e. $(\sqrt{x-1})^2 = x - 1$

Simplifying a Square Root

A square root is considered **simplified** if no prime factor of the radicand has an exponent greater than 1.

Exercises 15.5A

In Exercises 1–40, simplify each square root.

1. $\sqrt{z^2}$ 2. $\sqrt{b^6}$ 3. $\sqrt{x^4}$ 4. $\sqrt{a^8}$

5. $\sqrt{16x^6}$ 6. $\sqrt{64w^4}$ 7. $\sqrt{12}$ 8. $\sqrt{20}$

9. $\sqrt{18}$ 10. $\sqrt{45}$ 11. $\sqrt{8}$ 12. $\sqrt{32}$

13. $\sqrt{24}$ 14. $\sqrt{54}$ 15. $\sqrt{x^3}$ 16. $\sqrt{y^5}$

17. $\sqrt{m^7}$ 18. $\sqrt{n^9}$ 19. $\sqrt{h^3k^2}$

20. $\sqrt{a^2b^5}$ 21. $3\sqrt{28}$ 22. $2\sqrt{27}$

23. $4\sqrt{50}$ 24. $5\sqrt{40}$ 25. $\sqrt{75x^3}$

26. $\sqrt{63a^5}$ 27. $\sqrt{20x^4y^6}$ 28. $\sqrt{8a^8b^2}$

29. $\sqrt{36x^{36}}$ 30. $\sqrt{64y^{64}}$ 31. $\sqrt{20x^2y^3}$

32. $\sqrt{8a^4b^7}$ 33. $\sqrt{16x^5y^3}$ 34. $\sqrt{4m^7n^5}$

35. $\sqrt{27a^3b^4}$ 36. $\sqrt{18c^5d^2}$ 37. $\sqrt{250x^4y^3}$

38. $\sqrt{72h^6k^5}$ 39. $\sqrt{25 + 144}$ 40. $\sqrt{25} + \sqrt{144}$

In Exercises 41–50, find the product.

41. $\sqrt{3}\sqrt{3}$ 42. $\sqrt{2}\sqrt{2}$ 43. $\sqrt{x}\sqrt{x}$

44. $\sqrt{y}\sqrt{y}$ 45. $\sqrt{5x}\sqrt{5x}$ 46. $\sqrt{7a}\sqrt{7a}$

47. $(\sqrt{5})^2$ 48. $(\sqrt{x})^2$ 49. $(\sqrt{6x})^2$

50. $(\sqrt{x+2})^2$

15.5B Square Roots with Fractions in the Radicand

EXAMPLE 9   Does $\sqrt{\dfrac{4}{9}} = \dfrac{\sqrt{4}}{\sqrt{9}}$?

SOLUTION $\sqrt{\dfrac{4}{9}} = \dfrac{2}{3}$ Because $\left(\dfrac{2}{3}\right)^2 = \dfrac{4}{9}$

and $\dfrac{\sqrt{4}}{\sqrt{9}} = \dfrac{2}{3}$

Therefore, $\sqrt{\dfrac{4}{9}} = \dfrac{\sqrt{4}}{\sqrt{9}}$.

This leads to another rule of radicals:

RULE 11
Square root of
a quotient

$$\sqrt{\dfrac{a}{b}} = \dfrac{\sqrt{a}}{\sqrt{b}} \qquad (b \neq 0)$$

A square root is not considered simplified if the radicand contains a fraction. Rule 11 can be used to simplify square roots of fractions.

EXAMPLE 10

Examples of finding the square root of a fraction:

a. $\sqrt{\dfrac{25}{36}} = \dfrac{\sqrt{25}}{\sqrt{36}} = \dfrac{5}{6}$

b. $\sqrt{\dfrac{x^4}{y^6}} = \dfrac{\sqrt{x^4}}{\sqrt{y^6}} = \dfrac{x^2}{y^3}$

c. $\sqrt{\dfrac{50h^3}{2h}} = \sqrt{\dfrac{\overset{25}{\cancel{50}}h^3}{\underset{1}{\cancel{2}}h}} = \sqrt{25h^2} = 5h$

     ⎣———————————————— Simplify the fraction first

d. $\sqrt{\dfrac{8}{9}} = \dfrac{\sqrt{4 \cdot 2}}{\sqrt{9}} = \dfrac{2\sqrt{2}}{3}$

A third condition that must be satisfied in simplifying square roots is that no denominator contains a square root.

Consider $\dfrac{1}{\sqrt{5}}$. If we multiply this fraction by 1 written as $\dfrac{\sqrt{5}}{\sqrt{5}}$, the denominator will become $\sqrt{5} \cdot \sqrt{5} = 5$:

$$\dfrac{1}{\sqrt{5}} = \dfrac{1}{\sqrt{5}} \cdot \dfrac{\sqrt{5}}{\sqrt{5}} = \dfrac{\sqrt{5}}{5} \longleftarrow \text{Simplified form}$$

     ⎣—— Multiply the numerator and the denominator by $\sqrt{5}$

Changing an irrational denominator to a rational number is called *rationalizing the denominator*.

EXAMPLE 11

Examples of rationalizing the denominator:

a. $\sqrt{\dfrac{1}{x}} = \dfrac{\sqrt{1}}{\sqrt{x}} \cdot \dfrac{\sqrt{x}}{\sqrt{x}} = \dfrac{\sqrt{x}}{x}$

b. $\sqrt{\dfrac{9}{7}} = \dfrac{\sqrt{9}}{\sqrt{7}} = \dfrac{3}{\sqrt{7}} \cdot \dfrac{\sqrt{7}}{\sqrt{7}} = \dfrac{3\sqrt{7}}{7}$

c. $\dfrac{6}{\sqrt{3}} = \dfrac{6}{\sqrt{3}} \cdot \dfrac{\sqrt{3}}{\sqrt{3}} = \dfrac{\overset{2}{\cancel{6}}\sqrt{3}}{\underset{1}{\cancel{3}}} = 2\sqrt{3}$

Requirements for the simplified form of an expression having square roots

1. No prime factor of a radicand has an exponent greater than 1.
2. No radicand contains a fraction.
3. No denominator contains a square root.

Exercises 15.5B

Simplify each of the following expressions.

1. $\sqrt{\dfrac{4}{81}}$ 2. $\sqrt{\dfrac{16}{25}}$ 3. $\sqrt{\dfrac{9}{49}}$ 4. $\sqrt{\dfrac{9}{100}}$

5. $\sqrt{\dfrac{a^2}{b^6}}$ 6. $\sqrt{\dfrac{y^4}{x^2}}$ 7. $\sqrt{\dfrac{x^6}{v^8}}$ 8. $\sqrt{\dfrac{x^4}{y^6}}$

9. $\sqrt{\dfrac{4x^2}{9}}$ 10. $\sqrt{\dfrac{36a^4}{49}}$ 11. $\sqrt{\dfrac{2m^2}{18}}$ 12. $\sqrt{\dfrac{20}{5x^2}}$

13. $\sqrt{\dfrac{2x^3}{50x}}$ 14. $\sqrt{\dfrac{3a}{27a^5}}$ 15. $\sqrt{\dfrac{12}{25}}$ 16. $\sqrt{\dfrac{27}{16}}$

17. $\sqrt{\dfrac{50}{81}}$ 18. $\sqrt{\dfrac{18}{49}}$ 19. $\sqrt{\dfrac{x^3}{y^4}}$ 20. $\sqrt{\dfrac{a^5}{b^2}}$

21. $\sqrt{\dfrac{1}{2}}$ 22. $\sqrt{\dfrac{1}{3}}$ 23. $\sqrt{\dfrac{1}{a}}$ 24. $\sqrt{\dfrac{1}{y}}$

25. $\sqrt{\dfrac{4}{5}}$ 26. $\sqrt{\dfrac{9}{10}}$ 27. $\dfrac{8}{\sqrt{2}}$ 28. $\dfrac{10}{\sqrt{5}}$

29. $\dfrac{2x}{\sqrt{x}}$ 30. $\dfrac{5mn}{\sqrt{m}}$

15.6   Adding Square Roots

Like square roots are square roots having the same radicand.

EXAMPLE 1   Examples of like square roots:

a. $3\sqrt{5},\ 2\sqrt{5},\ -7\sqrt{5}$

b. $2\sqrt{x},\ -9\sqrt{x},\ 11\sqrt{x}$

Unlike square roots are square roots having different radicands.

EXAMPLE 2   Examples of unlike square roots:

a. $2\sqrt{15},\ -6\sqrt{11},\ 8\sqrt{24}$

b. $5\sqrt{y},\ 3\sqrt{x},\ -4\sqrt{13}$

Adding Like Square Roots

Like square roots are added in the same way as like terms: We add their coefficients and then multiply that sum by the like square root.

$$\left. \begin{array}{l} 3c\ +\ 2c\ =\ (3+2)c\ =\ 5c \\ 3\sqrt{7}\ +\ 2\sqrt{7} = (3+2)\sqrt{7} = 5\sqrt{7} \end{array} \right\} \quad \text{Applications of the distributive rule}$$

EXAMPLE 3   **a.** $5\sqrt{2} + 3\sqrt{2} = (5+3)\sqrt{2} = 8\sqrt{2}$

b. $7\sqrt{x} - 3\sqrt{x} = (7-3)\sqrt{x} = 4\sqrt{x}$

Adding Unlike Square Roots

Simplifying *unlike* square roots *sometimes* results in *like* square roots, which can then be added.

Adding unlike square roots	1. Simplify each square root.
	2. Combine like square roots by adding their coefficients and then multiplying that sum by the like square root.

EXAMPLE 4

$$\sqrt{8} + \sqrt{18} = \sqrt{4 \cdot 2} + \sqrt{9 \cdot 2}$$
$$= 2\sqrt{2} + 3\sqrt{2}$$
$$= 5\sqrt{2}$$

EXAMPLE 5

$$\sqrt{12} - \sqrt{27} + \sqrt{75} = \sqrt{4 \cdot 3} - \sqrt{9 \cdot 3} + \sqrt{25 \cdot 3}$$
$$= 2\sqrt{3} - 3\sqrt{3} + 5\sqrt{3}$$
$$= 4\sqrt{3}$$

EXAMPLE 6

$$3\sqrt{20} - \sqrt{45} - \sqrt{50} = 3 \cdot \sqrt{4 \cdot 5} - \sqrt{9 \cdot 5} - \sqrt{25 \cdot 2}$$
$$= 3 \cdot 2\sqrt{5} - 3\sqrt{5} - 5\sqrt{2}$$
$$= 6\sqrt{5} - 3\sqrt{5} - 5\sqrt{2}$$
$$= 3\sqrt{5} - 5\sqrt{2}$$

Exercises 15.6

Find the sums, and simplify.

1. $2\sqrt{3} + 5\sqrt{3}$
2. $4\sqrt{2} + 3\sqrt{2}$
3. $5\sqrt{5} - \sqrt{5}$
4. $8\sqrt{6} - \sqrt{6}$
5. $7\sqrt{x} + 2\sqrt{x}$
6. $3\sqrt{a} - 2\sqrt{a}$
7. $5\sqrt{7} + 2\sqrt{7} - \sqrt{7}$
8. $6\sqrt{3} - \sqrt{3} + 2\sqrt{3}$
9. $6\sqrt{2} + 2\sqrt{2} - 4\sqrt{3}$
10. $4\sqrt{6} + 3\sqrt{5} - 2\sqrt{6}$
11. $5\sqrt{x} - 2\sqrt{y} + 3\sqrt{x}$
12. $\sqrt{a} + 7\sqrt{a} - 6\sqrt{b}$
13. $\sqrt{50} + \sqrt{8}$
14. $\sqrt{20} + \sqrt{125}$
15. $\sqrt{32} - \sqrt{18}$
16. $\sqrt{63} - \sqrt{28}$

17. $5\sqrt{12} + \sqrt{75}$
18. $3\sqrt{24} + \sqrt{54}$
19. $2\sqrt{48} - 4\sqrt{27}$
20. $2\sqrt{250} - 4\sqrt{90}$
21. $\sqrt{16x} + \sqrt{9x}$
22. $\sqrt{64x} + \sqrt{4x}$
23. $\sqrt{25} - \sqrt{20}$
24. $\sqrt{49} - \sqrt{48}$
25. $\sqrt{75} + 2\sqrt{27} - \sqrt{48}$
26. $\sqrt{18} + 5\sqrt{8} - \sqrt{32}$
27. $\sqrt{45} - 3\sqrt{12} + \sqrt{20}$
28. $\sqrt{54} - 5\sqrt{28} + \sqrt{24}$
29. $\sqrt{64} + \sqrt{50} - 2\sqrt{72}$
30. $\sqrt{36} - \sqrt{150} + 3\sqrt{24}$

15.7 Multiplying Square Roots

Earlier in the chapter we used Rule 9 in the following direction:

$$\overrightarrow{\sqrt{ab} = \sqrt{a}\sqrt{b}}$$ Using Rule 9 to find the square root of a product
$$\sqrt{4 \cdot 3} = \sqrt{4}\sqrt{3}$$
$$= 2\sqrt{3}$$

In this section we use Rule 9 in the opposite way:

$$\overrightarrow{\sqrt{a}\sqrt{b}} = \sqrt{ab}$$

Using Rule 9 to find the product of square roots

$$\sqrt{2}\sqrt{8} = \sqrt{2 \cdot 8}$$

$$= \sqrt{16} = 4$$

This means:

$$\left(\begin{array}{c}\text{Product of}\\\text{square roots}\end{array}\right) = \left(\begin{array}{c}\text{square root}\\\text{of product}\end{array}\right)$$

$$\sqrt{a}\sqrt{b} = \sqrt{ab}$$

EXAMPLE 1

Product of   Square root
square roots   of product
↓   ↓

a. $\sqrt{2}\sqrt{32} = \sqrt{2 \cdot 32} = \sqrt{64} = 8$

b. $\sqrt{4x}\sqrt{x} = \sqrt{4x \cdot x} = \sqrt{4x^2} = 2x$

c. $\sqrt{3y}\sqrt{6y^2} = \sqrt{3y \cdot 6y^2} = \sqrt{18y^3}$ Multiplying square roots

$$= \sqrt{9 \cdot 2 \cdot y^2 \cdot y}$$

$$= 3y\sqrt{2y}$$

$\Big\}$ Simplifying

d. $2\sqrt{3} \cdot 3\sqrt{8} = (2 \cdot 3)\sqrt{3 \cdot 8} = 6\sqrt{24}$ Multiplying square roots

$$= 6\sqrt{4 \cdot 6}$$

$$= 6 \cdot 2\sqrt{6}$$

$$= 12\sqrt{6}$$

$\Big\}$ Simplifying

EXAMPLE 2

Find $\sqrt{2}(3\sqrt{2} + 5)$.

SOLUTION $\sqrt{2}(3\sqrt{2} + 5) = \sqrt{2} \cdot 3\sqrt{2} + \sqrt{2} \cdot 5$ Distribute $\sqrt{2}$

$$= 3\sqrt{4} + 5\sqrt{2}$$

$$= 3 \cdot 2 + 5\sqrt{2}$$

$$= 6 + 5\sqrt{2}$$

EXAMPLE 3

Find $(2\sqrt{3} - 5)(4\sqrt{3} - 6)$.

SOLUTION $(2\sqrt{3} - 5)$ $(4\sqrt{3} - 6)$ Using FOIL

$2\sqrt{3} \cdot 4\sqrt{3} = 8 \cdot 3$ $-20\sqrt{3}$ $(-5)(-6)$

$$-12\sqrt{3}$$

$$= 24 - 32\sqrt{3} + 30$$

$$= 54 - 32\sqrt{3}$$

EXAMPLE 4 Find $(3 + \sqrt{5})^2$.

SOLUTION We use the formula for squaring a binomial.

$$(a + b)^2 = a^2 + 2ab + b^2$$
$$(3 + \sqrt{5})^2 = (3)^2 + 2(3)(\sqrt{5}) + (\sqrt{5})^2$$
$$= 9 + 6\sqrt{5} + 5$$
$$= 14 + 6\sqrt{5}$$

EXAMPLE 5 Find $(4 + \sqrt{3})(4 - \sqrt{3})$.

SOLUTION We use the formula for the product of the sum and difference of two terms.

$$(a + b)(a - b) = a^2 - b^2$$
$$(4 + \sqrt{3})(4 - \sqrt{3}) = (4)^2 - (\sqrt{3})^2$$
$$= 16 - 3$$
$$= 13 \longleftarrow \text{A rational number}$$

Exercises 15.7

Find the products, and simplify.

1. $\sqrt{3}\sqrt{12}$
2. $\sqrt{2}\sqrt{8}$
3. $\sqrt{9x}\sqrt{x}$

4. $\sqrt{25y}\sqrt{y}$
5. $\sqrt{2a}\sqrt{18a^3}$
6. $\sqrt{5x^5}\sqrt{20x}$

7. $\sqrt{6}\sqrt{2}$
8. $\sqrt{2}\sqrt{10}$
9. $\sqrt{3x}\sqrt{6x}$

10. $\sqrt{8y}\sqrt{3y^3}$
11. $\sqrt{5a}\sqrt{10a^2}$
12. $\sqrt{3x^2}\sqrt{15x^3}$

13. $3\sqrt{5} \cdot \sqrt{5}$
14. $5\sqrt{2} \cdot \sqrt{2}$
15. $2\sqrt{3} \cdot 4\sqrt{3}$

16. $3\sqrt{5} \cdot 2\sqrt{5}$
17. $\sqrt{12} \cdot 5\sqrt{2}$
18. $\sqrt{2} \cdot 3\sqrt{14}$

19. $3\sqrt{8} \cdot 2\sqrt{5}$
20. $2\sqrt{18} \cdot 5\sqrt{3}$
21. $\sqrt{2}(\sqrt{2} + 1)$

22. $\sqrt{3}(\sqrt{3} - 1)$
23. $\sqrt{5}(2\sqrt{5} + 3)$
24. $\sqrt{7}(3\sqrt{7} + 2)$

25. $\sqrt{x}(\sqrt{x} - 2)$
26. $\sqrt{y}(4 - \sqrt{y})$
27. $\sqrt{6}(\sqrt{6} - \sqrt{2})$

28. $\sqrt{2}(\sqrt{10} + \sqrt{2})$
29. $(\sqrt{7} + 2)(\sqrt{7} + 3)$

30. $(\sqrt{3} + 2)(\sqrt{3} + 4)$
31. $(3 - 2\sqrt{5})(4 - \sqrt{5})$

32. $(7 - 3\sqrt{2})(1 - \sqrt{2})$
33. $(5\sqrt{3} - 2)(2\sqrt{3} + 4)$

34. $(2\sqrt{5} + 4)(3\sqrt{5} - 3)$
35. $(3 - \sqrt{2})(3 + \sqrt{2})$

36. $(2 + \sqrt{7})(2 - \sqrt{7})$
37. $(\sqrt{5} + \sqrt{3})(\sqrt{5} - \sqrt{3})$

38. $(\sqrt{6} - \sqrt{2})(\sqrt{6} + \sqrt{2})$
39. $(\sqrt{3} + 4)^2$

40. $(2 - \sqrt{5})^2$
41. $(3 - \sqrt{x})^2$

42. $(\sqrt{y} + 5)^2$
43. $(\sqrt{5} + \sqrt{3})^2$

44. $(\sqrt{6} + \sqrt{2})^2$

15.8 Dividing Square Roots

Earlier in the chapter we used Rule 11 in the following direction:

$$\overrightarrow{\sqrt{\frac{a}{b}} = \frac{\sqrt{a}}{\sqrt{b}}}$$

Using Rule 11 to find the square root of a quotient

$$\sqrt{\frac{4}{9}} = \frac{\sqrt{4}}{\sqrt{9}} = \frac{2}{3}$$

In this section we use Rule 11 in the opposite direction:

$$\frac{\sqrt{a}}{\sqrt{b}} = \sqrt{\frac{a}{b}} \qquad \text{Using Rule 11 to find the} \atop \text{quotient of square roots}$$

$$\frac{\sqrt{8}}{\sqrt{2}} = \sqrt{\frac{8}{2}} = \sqrt{4} = 2$$

This means:

$$\left(\begin{matrix} \text{Quotient of} \\ \text{square roots} \end{matrix}\right) = \left(\begin{matrix} \text{square root} \\ \text{of quotient} \end{matrix}\right)$$

$$\frac{\sqrt{a}}{\sqrt{b}} = \sqrt{\frac{a}{b}} \qquad (b \neq 0)$$

EXAMPLE 1

$$\underset{\substack{\uparrow \\ \text{Quotient of} \\ \text{square roots}}}{} \qquad \underset{\substack{\uparrow \\ \text{Square root} \\ \text{of quotient}}}{}$$

a. $\dfrac{\sqrt{32}}{\sqrt{2}} = \sqrt{\dfrac{32}{2}} = \sqrt{16} = 4$

b. $\dfrac{\sqrt{x^5}}{\sqrt{x}} = \sqrt{\dfrac{x^5}{x}} = \sqrt{x^4} = x^2$

c. $\dfrac{\sqrt{150}}{\sqrt{3}} = \sqrt{\dfrac{150}{3}} = \sqrt{50} = \sqrt{25 \cdot 2} = 5\sqrt{2}$

d. $\dfrac{\sqrt{10}}{\sqrt{15}} = \sqrt{\dfrac{10}{15}} = \sqrt{\dfrac{2}{3}} = \dfrac{\sqrt{2}}{\sqrt{3}} \cdot \dfrac{\sqrt{3}}{\sqrt{3}} = \dfrac{\sqrt{6}}{3}$

         └── Multiply by $\dfrac{\sqrt{3}}{\sqrt{3}}$ to rationalize the denominator

e. $\dfrac{2}{\sqrt{20}} = \dfrac{2}{\sqrt{20}} \cdot \dfrac{\sqrt{5}}{\sqrt{5}} = \dfrac{2\sqrt{5}}{\sqrt{100}} = \dfrac{\overset{1}{\cancel{2}}\sqrt{5}}{\underset{5}{\cancel{10}}} = \dfrac{\sqrt{5}}{5}$

         └── Multiply by $\dfrac{\sqrt{5}}{\sqrt{5}}$ to make the denominator a perfect square

---

 Note In Example 1e we could have simplified the denominator first:

$$\frac{2}{\sqrt{20}} = \frac{2}{\sqrt{4 \cdot 5}} = \frac{\overset{1}{\cancel{2}}}{\underset{1}{\cancel{2}}\sqrt{5}} = \frac{1}{\sqrt{5}} \cdot \frac{\sqrt{5}}{\sqrt{5}} = \frac{\sqrt{5}}{5}$$

Rationalizing a Binomial Denominator That Contains Square Roots

The **conjugate** of a binomial is a binomial that has the same two terms but with the sign of the second term changed. The conjugate of $a + b$ is $a - b$.

EXAMPLE 2

a. The conjugate of $1 + \sqrt{3}$ is $1 - \sqrt{3}$.
b. The conjugate of $\sqrt{5} - \sqrt{3}$ is $\sqrt{5} + \sqrt{3}$.
c. The conjugate of $\sqrt{x} + 2$ is $\sqrt{x} - 2$.

The product of a binomial containing square roots and its conjugate is a rational number. For example,

$$(1 - \sqrt{2})(1 + \sqrt{2}) = (1)^2 - (\sqrt{2})^2 = 1 - 2 = -1 \quad \text{A rational number}$$

Because of this fact, the following procedure should be used when a binomial denominator contains a square root.

Rationalizing a binomial denominator that contains square roots	Multiply the numerator and the denominator by the conjugate of the denominator: $$\frac{a}{b + \sqrt{c}} \cdot \frac{b - \sqrt{c}}{b - \sqrt{c}} = \frac{a(b - \sqrt{c})}{b^2 - c}$$

EXAMPLE 3

a. $\dfrac{2}{1 + \sqrt{3}} = \dfrac{2}{1 + \sqrt{3}} \cdot \dfrac{1 - \sqrt{3}}{1 - \sqrt{3}} = \dfrac{2(1 - \sqrt{3})}{1 - 3} = \dfrac{\overset{1}{2}(1 - \sqrt{3})}{\underset{-1}{-2}} = \sqrt{3} - 1$

Multiply the numerator and denominator by $1 - \sqrt{3}$, the conjugate of the denominator; the value of this fraction is 1, and multiplying $\dfrac{2}{1 + \sqrt{3}}$ by 1 does not change its value

b. $\dfrac{6}{\sqrt{5} - \sqrt{3}} = \dfrac{6}{\sqrt{5} - \sqrt{3}} \cdot \dfrac{\sqrt{5} + \sqrt{3}}{\sqrt{5} + \sqrt{3}} = \dfrac{6(\sqrt{5} + \sqrt{3})}{5 - 3} = \dfrac{\overset{3}{6}(\sqrt{5} + \sqrt{3})}{\underset{1}{2}}$

$$= 3\sqrt{5} + 3\sqrt{3}$$

Multiply the numerator and denominator by $\sqrt{5} + \sqrt{3}$, the conjugate of the denominator

Exercises 15.8

In Exercises 1–24, find the quotients, and simplify.

1. $\dfrac{\sqrt{20}}{\sqrt{5}}$ 2. $\dfrac{\sqrt{27}}{\sqrt{3}}$ 3. $\dfrac{\sqrt{7}}{\sqrt{28}}$ 4. $\dfrac{\sqrt{2}}{\sqrt{50}}$

5. $\dfrac{\sqrt{x^3}}{\sqrt{x}}$ 6. $\dfrac{\sqrt{y^7}}{\sqrt{y^3}}$ 7. $\dfrac{\sqrt{a}}{\sqrt{a^5}}$ 8. $\dfrac{\sqrt{x^3}}{\sqrt{x^9}}$

9. $\dfrac{\sqrt{18x^5}}{\sqrt{2x}}$ 10. $\dfrac{\sqrt{72a^3}}{\sqrt{2a}}$ 11. $\dfrac{\sqrt{40}}{\sqrt{5}}$ 12. $\dfrac{\sqrt{60}}{\sqrt{3}}$

13. $\dfrac{\sqrt{24}}{\sqrt{2}}$ 14. $\dfrac{\sqrt{48}}{\sqrt{6}}$ 15. $\dfrac{\sqrt{2}}{\sqrt{6}}$ 16. $\dfrac{\sqrt{3}}{\sqrt{15}}$

17. $\dfrac{\sqrt{6}}{\sqrt{10}}$ 18. $\dfrac{\sqrt{20}}{\sqrt{50}}$ 19. $\dfrac{\sqrt{14}}{\sqrt{21}}$ 20. $\dfrac{\sqrt{6}}{\sqrt{21}}$

21. $\dfrac{2}{\sqrt{8}}$ 22. $\dfrac{5}{\sqrt{20}}$ 23. $\dfrac{3}{\sqrt{12}}$ 24. $\dfrac{2}{\sqrt{32}}$

In Exercises 25–28, write the conjugate for each expression.

25. $\sqrt{2} - 1$ 26. $3 - \sqrt{3}$ 27. $\sqrt{5} + \sqrt{2}$ 28. $\sqrt{x} + 2$

In Exercises 29–38, rationalize the denominators, and simplify.

29. $\dfrac{3}{\sqrt{2} - 1}$ 30. $\dfrac{5}{\sqrt{2} + 1}$

31. $\dfrac{6}{\sqrt{5} + \sqrt{2}}$

32. $\dfrac{12}{\sqrt{6} - \sqrt{2}}$

35. $\dfrac{8}{\sqrt{6} - 2}$

36. $\dfrac{9}{3 + \sqrt{5}}$

33. $\dfrac{4}{2 + \sqrt{2}}$

34. $\dfrac{6}{3 - \sqrt{3}}$

37. $\dfrac{x - 4}{\sqrt{x} + 2}$

38. $\dfrac{y - 9}{\sqrt{y} - 3}$

15.9   Radical Equations

A **radical equation** is an equation in which the variable appears in a radicand. In this text we will consider only radical equations with square roots.

EXAMPLE 1   Examples of radical equations:

a. $\sqrt{x} = 7$

b. $\sqrt{x + 2} = 3$

c. $\sqrt{2x - 3} = \sqrt{x} + 5$

> If two numbers are equal, then their squares are equal.
>
> $$\text{If} \qquad a = b$$
> $$\text{then} \quad a^2 = b^2$$

Because squaring a number is the inverse operation of taking the square root of a number, we can use this last property to remove the square root from an equation.

A Word of Caution   The squaring process may introduce an extra answer, called an **extraneous root**, that does not satisfy the original equation. This is because the squares of two numbers may be equal, but the numbers themselves may not be equal. It is true that $4^2 = (-4)^2$, but $4 \neq -4$. Therefore, all apparent solutions *must* be checked in the original equation.

Solving a radical equation with square roots

1. Isolate the radical term.
2. Square both sides of the equation.
3. Solve the resulting equation for the variable.
4. *Check* apparent solutions in the original equation. (Extra answers may occur because of the squaring process. See Example 4.)

EXAMPLE 2   Solve $\sqrt{x} = 7$.

SOLUTION

$$\sqrt{x} = 7$$
$$(\sqrt{x})^2 = (7)^2$$
$$x = 49$$

✓ **Check** $\sqrt{x} = 7$

$\sqrt{49} \overset{?}{=} 7$

$7 = 7$

EXAMPLE 3   Solve $\sqrt{x + 2} = 3$.

SOLUTION

$$\sqrt{x + 2} = 3$$
$$(\sqrt{x + 2})^2 = (3)^2$$
$$x + 2 = 9$$
$$x = 7$$

✓ **Check** $\sqrt{x + 2} = 3$

$\sqrt{7 + 2} \overset{?}{=} 3$

$\sqrt{9} \overset{?}{=} 3$

$3 = 3$

EXAMPLE 4   Solve $\sqrt{2x + 1} + 1 = x$.

SOLUTION $\sqrt{2x + 1} = x - 1$ Isolate the radical term

$(\sqrt{2x + 1})^2 = (x - 1)^2$

(handwritten: (x-1)(x-1), $x^2 - x - x + 1$, $x^2 - 2x + 1$)

When squaring $(x - 1)$, do not forget this middle term

$$2x + 1 = x^2 - 2x + 1$$
$$0 = x^2 - 4x$$
$$0 = x(x - 4)$$

$x = 0$ | $x - 4 = 0$

$x = 4$

✓ **Check for** $x = 0$ $\sqrt{2x + 1} + 1 = x$

$\sqrt{2(0) + 1} + 1 \overset{?}{=} 0$

$\sqrt{1} + 1 \overset{?}{=} 0$ The symbol $\sqrt{1}$ *always* stands

$1 + 1 \overset{?}{=} 0$ for the *principal* square root of I, which is I (*not* −I)

$2 \neq 0$

Therefore, *0 is not a solution* of $\sqrt{2x + 1} + 1 = x$ because it does not satisfy the equation (it's an extraneous root).

✓ **Check for** $x = 4$ $\sqrt{2x + 1} + 1 = x$

$\sqrt{2(4) + 1} + 1 \overset{?}{=} 4$

$\sqrt{9} + 1 \overset{?}{=} 4$

$3 + 1 \overset{?}{=} 4$

$4 = 4$

Therefore, 4 *is the only solution* that satisfies the equation.

Exercises 15.9

Solve each equation.

1. $\sqrt{x} = 5$ 2. $\sqrt{x} = 10$ 3. $\sqrt{x} = 8$

4. $\sqrt{5x} = 10$ 5. $\sqrt{2x} = 4$ 6. $\sqrt{3x} = 6$

7. $\sqrt{x-3} = 2$ 8. $\sqrt{x+4} = 6$ 9. $\sqrt{x-6} = 3$

10. $\sqrt{6x+1} = 5$ 11. $\sqrt{2x+1} = 9$ 12. $\sqrt{5x-4} = 4$

13. $\sqrt{3x+1} = 5$ 14. $\sqrt{7x+8} = 6$ 15. $\sqrt{9x-5} = 7$

16. $\sqrt{9-2x} = \sqrt{5x-12}$ 17. $\sqrt{x+1} = \sqrt{2x-7}$

18. $\sqrt{3x-2} = \sqrt{x+4}$ 19. $\sqrt{3x-2} = x$

20. $\sqrt{5x-6} = x$ 21. $x = \sqrt{3x+10}$

22. $\sqrt{3x+2} = 3x$ 23. $\sqrt{4x-1} = 2x$

24. $\sqrt{6x-1} = 3x$ 25. $\sqrt{x-3} + 5 = x$

26. $\sqrt{4x+5} + 5 = 2x$ 27. $2x = \sqrt{2x+3} + 3$

28. $\sqrt{x-6} + 8 = x$ 29. $x = \sqrt{2x+4} + 2$

30. $\sqrt{7-x} + x = 1$

Chapter 15 REVIEW

The Rules of Exponents
15.1, 15.2

$$x^a x^b = x^{a+b} \qquad\qquad \text{Rule 1}$$

$$(x^a)^b = x^{ab} \qquad\qquad \text{Rule 2}$$

$$\frac{x^a}{x^b} = x^{a-b} \quad (x \neq 0) \qquad \text{Rule 3}$$

$$x^0 = 1 \quad (x \neq 0) \qquad \text{Rule 4}$$

$$x^{-n} = \frac{1}{x^n} \quad (x \neq 0) \qquad \text{Rule 5a}$$

$$\frac{1}{x^{-n}} = x^n \quad (x \neq 0) \qquad \text{Rule 5b}$$

$$(xy)^n = x^n y^n \qquad\qquad \text{Rule 6}$$

$$\left(\frac{x}{y}\right)^n = \frac{x^n}{y^n} \quad (y \neq 0) \qquad \text{Rule 7}$$

$$\left(\frac{x^a y^b}{z^c}\right)^n = \frac{x^{an} y^{bn}}{z^{cn}} \quad (z \neq 0) \quad \text{Rule 8}$$

Simplifying Expressions Having Exponents
15.1

An expression having exponents is considered simplified when each different base appears only once and its exponent is a single integer.

Scientific Notation
15.3

To write a number in scientific notation, place the decimal point after the first nonzero digit, then multiply by the appropriate power of 10. The exponent in the power of 10 tells how many places (and the direction) to move the decimal point to get it back to its original position.

Square Roots
15.4, 15.5

The square root of a number N is a number that, when squared, gives N. Every positive number has both a positive and a negative square root. The positive square root is called the **principal square root**. The principal square root of N is written $\sqrt{N}$. The negative square root of N is written $-\sqrt{N}$.

$$\sqrt{a \cdot b} = \sqrt{a} \cdot \sqrt{b} \quad \text{Rule 9}$$

$$\sqrt{a} \cdot \sqrt{a} = a \qquad\qquad \text{Rule 10}$$

$$\sqrt{\frac{a}{b}} = \frac{\sqrt{a}}{\sqrt{b}} \qquad\qquad \text{Rule 11}$$

Simplifying Expressions Having Square Roots
15.5

1. No prime factor of a radicand has an exponent greater than 1.

2. No radicand contains a fraction.

3. No denominator contains a square root.

Adding Square Roots
15.6

1. Simplify each square root.

2. Combine like square roots by adding their coefficients and then multiplying that sum by the like square root.

Multiplying Square Roots
15.7

Use Rule 9 in the opposite direction: $\sqrt{a}\sqrt{b} = \sqrt{ab}$. Simplify the results.

Dividing Square Roots
15.8

Use Rule 11 in the opposite direction: $\dfrac{\sqrt{a}}{\sqrt{b}} = \sqrt{\dfrac{a}{b}}$. Simplify the results.

Rationalizing a Monomial Denominator
15.5B

$$\frac{1}{\sqrt{a}} = \frac{1}{\sqrt{a}} \cdot \frac{\sqrt{a}}{\sqrt{a}} = \frac{\sqrt{a}}{a}$$

Rationalizing a Binomial Denominator
15.8

$$\frac{a}{b + \sqrt{c}} = \frac{a}{b + \sqrt{c}} \cdot \frac{b - \sqrt{c}}{b - \sqrt{c}} = \frac{a(b - \sqrt{c})}{b^2 - c}$$

Solving a Radical Equation
15.9

1. Isolate the radical term.

2. Square both sides of the equation.

3. Solve the resulting equation for the variable.

4. *Check* apparent solutions in the original equation. (Extra answers—*extraneous roots*—may occur because of the squaring process.)

Chapter 15   R E V I E W   E X E R C I S E S

In Exercises 1–24, simplify each expression. Write the answer using only positive exponents.

1. $x^{-4} \cdot x^{-7}$

2. $a^5 \cdot a^{-3}$

3. $c^{-5} \cdot d^0$

4. $\dfrac{p^5}{p^{-2}}$

5. $\dfrac{x^{-4}}{x^5}$

6. $\dfrac{m^0}{m^{-3}}$

7. $(p^{-3})^5$

8. $(m^{-4})^{-2}$

9. $(h^0)^{-4}$

10. $(x^2y^3)^4$

11. $(p^{-1}r^3)^{-2}$

12. $2a^{-3}$

13. $(3x^4)^2$

14. $(4b^3)^{-2}$

15. $\left(\dfrac{x^2y^3}{z^4}\right)^5$

16. $\left(\dfrac{x^{-3}}{y^0z^2}\right)^{-4}$

17. $\left(\dfrac{u^{-5}}{v^2w^{-4}}\right)^3$

18. $(5a^3b^{-4})^{-2}$

19. $\left(\dfrac{4h^2}{ij^{-2}}\right)^{-3}$

20. $\left(\dfrac{x^{10}y^5}{x^5y}\right)^3$

21. $\left(\dfrac{6x^{-5}y^8}{3x^2y^{-4}}\right)^0$

22. $x^{3d} \cdot x^d$

23. $(x^{4a})^{-2}$

24. $\dfrac{6^{2x}}{6^{-x}}$

In Exercises 25–27, write each expression without fractions. Use negative exponents if necessary.

25. $\dfrac{x^2}{y^3}$

26. $\dfrac{m^2}{n^{-3}}$

27. $\dfrac{a^3}{b^2c^5}$

In Exercises 28–33, evaluate each expression.

28. 4^{-2}

29. $(3^{-1})^3$

30. $5^6 \cdot 5^{-3}$

31. $\dfrac{2^{-1}}{2^{-5}}$

32. $\dfrac{(-8)^2}{-8^2}$

33. $\left(\dfrac{10^{-4} \cdot 10}{10^{-2}}\right)^5$

In Exercises 34–36, write the numbers in scientific notation.

34. 0.000225

35. $960,000$

36. $\dfrac{1}{200}$

In Exercises 37–39, write the numbers in decimal notation.

37. 7.8×10^3

38. 4.06×10^{-5}

39. 1.207×10^{-2}

In Exercises 40–42, write each number in scientific notation, then perform the indicated operations. Leave the answer in scientific notation.

40. $1,600 \times 0.00006$

41. $\dfrac{78,000}{0.026}$

42. $\dfrac{0.0035 \times 540}{0.00027}$

43. Which of the numbers $\sqrt{3}$, $2\frac{1}{2}$, $2\sqrt{5}$, 3.6, $\sqrt{5}$, $\frac{5}{2}$

    a. are irrational numbers?   **b.** have like square roots?

44. What is the radicand in each expression?

    a. $5\sqrt{9x}$ **b.** $\sqrt{\dfrac{1}{2}}$ **c.** $\sqrt{x-5}$

In Exercises 45–65, perform the indicated operations, and simplify.

45. $\sqrt{81}$ **46.** $\sqrt{48}$ **47.** $\sqrt{2}\sqrt{32}$

48. $\sqrt{a^6}$ **49.** $\sqrt{x^3}$ **50.** $\sqrt{16x^2y^4}$

51. $\sqrt{a^3b^5}$ **52.** $3\sqrt{5}\cdot 2\sqrt{10}$ **53.** $3\sqrt{5}+\sqrt{5}$

54. $\sqrt{\dfrac{18x}{2x^3}}$ **55.** $\sqrt{18}-\sqrt{8}$ **56.** $\dfrac{\sqrt{6}}{\sqrt{15}}$

57. $\dfrac{6}{\sqrt{3}}$ **58.** $3\sqrt{12}-\sqrt{75}$ **59.** $\sqrt{2}(5\sqrt{2}+3)$

60. $(\sqrt{5}+3)(\sqrt{5}-3)$ **61.** $(\sqrt{3}+2)^2$

62. $(3\sqrt{2}+1)(2\sqrt{2}-1)$ **63.** $\dfrac{8}{\sqrt{3}-2}$

64. $\dfrac{10}{\sqrt{6}-2}$ **65.** $\sqrt{45}-2\sqrt{27}+\sqrt{20}$

In Exercises 66–71, solve each equation.

66. $\sqrt{x}=4$ **67.** $\sqrt{3a}=6$

68. $\sqrt{2x-1}=5$ **69.** $\sqrt{5a-4}=\sqrt{3a+2}$

70. $\sqrt{7x-6}=x$ **71.** $\sqrt{2x-1}+2=x$

Chapter 15 Critical Thinking and Writing Problems

Answer Problems 1–10 in your own words, using complete sentences.

1. Explain how to write a number in scientific notation.

2. Explain why 15×10^4 is not in scientific notation. Rewrite 15×10^4 in scientific notation.

3. Explain how to add like square roots.

4. True or false: $2 + 4\sqrt{3} = 6\sqrt{3}$. Explain why.

5. True or false: $2 \cdot 4\sqrt{3} = 8\sqrt{3}$. Explain why.

6. True or false: $\sqrt{36 \cdot 64} = \sqrt{36} \cdot \sqrt{64}$. Explain why.

7. True or false: $\sqrt{36 + 64} = \sqrt{36} + \sqrt{64}$. Explain why.

8. Explain how to rationalize the denominator if the denominator is a monomial (one term) with a square root.

9. Explain how to rationalize the denominator if the denominator is a binomial (two terms) with a square root.

10. Explain why the following expressions are *not* considered simplified.

    a. $3\sqrt{12}$ **b.** $x^2\sqrt{y^3}$

    c. $\sqrt{\dfrac{3}{5}}$ **d.** $\dfrac{1}{\sqrt{7}}$

Each of the following problems has an error. Find the error, and in your own words, explain why it is wrong. Then work the problem correctly.

11. $5x^{-1} = \dfrac{1}{5x}$

12. $\left(\dfrac{4x^4}{y^2}\right)^2 = \dfrac{4x^8}{y^4}$

13. $(2x^{-2})^{-3} = 2^{-3}x^6 = -8x^6$

14. $(\sqrt{6}+2)^2 = 6 + 4 = 10$

15. $\sqrt{75} + 5\sqrt{12} = \sqrt{25 \cdot 3} + 5\sqrt{4 \cdot 3}$

$$= 5\sqrt{3} + 5 + 2\sqrt{3}$$
$$= 5\sqrt{3} + 7\sqrt{3}$$
$$= 12\sqrt{3}$$

16. $\sqrt{x+15} - 3 = x$

$$\sqrt{x+15} = x + 3$$
$$(\sqrt{x+15})^2 = x^2 + 3^2$$
$$x + 15 = x^2 + 9$$
$$0 = x^2 - x - 6$$
$$0 = (x-3)(x+2)$$

$x - 3 = 0$ | $x + 2 = 0$

$x = 3$ | $x = -2$

Chapter 15   DIAGNOSTIC TEST

Allow yourself about one hour to do these problems. Complete solutions for all problems, together with section references, are given in the answer section at the end of the book.

In Problems 1–6, simplify each expression. Write the answer using only positive exponents.

1. $a^{-6} \cdot a^2$ 2. $(a^{-3}b)^2$ 3. $x^{3a} \cdot x^{2a}$

4. $\dfrac{x^5}{x^{-2}}$ 5. $\left(\dfrac{3x^{-4}}{y^2}\right)^2$ 6. $\left(\dfrac{x^5y}{x^2y^3}\right)^3$

In Problems 7–9, evaluate each expression.

7. $(10^{-3})^2$ 8. $\dfrac{2^{-2} \cdot 2^3}{2^{-4}}$ 9. $\dfrac{-4^2}{(-4)^2}$

10. Write $\dfrac{x^2}{yz^3}$ without fractions, using negative exponents if necessary.

11. Write each of the following in scientific notation.

     a. 723,000

     b. 0.0048

12. Write each number in scientific notation, then perform the indicated operation. Leave your answer in scientific notation.

$$\frac{4{,}200}{0.00014}$$

In Problems 13–23, perform the indicated operations, and simplify.

13. $\sqrt{2}\sqrt{18x^2}$ 14. $\sqrt{3}(2\sqrt{3} - 5)$ 15. $\sqrt{72}$

16. $\sqrt{\dfrac{18}{2m^2}}$ 17. $\dfrac{\sqrt{16}}{\sqrt{50}}$ 18. $\dfrac{4}{\sqrt{3} + 1}$

19. $\sqrt{12x^4y^3}$ 20. $2\sqrt{20} + \sqrt{45}$

21. $(3\sqrt{2} + 5)(2\sqrt{2} + 1)$ 22. $\sqrt{\dfrac{4}{5}}$

23. $\sqrt{28} + \sqrt{75} - \sqrt{27}$

In Problems 24 and 25, solve each equation.

24. $\sqrt{4x + 5} = 5$ 25. $\sqrt{5x - 6} = x$

Quadratic Equations

CHAPTER

16

n this chapter we discuss methods for solving quadratic equations. We have already solved quadratic equations by factoring in previous chapters.

16.1   General Form of a Quadratic Equation

A **quadratic equation** is an equation in one variable whose highest-degree term is a second-degree term.

EXAMPLE 1

Examples of quadratic equations:

a. $3x^2 + 7x + 2 = 0$

b. $\dfrac{x^2}{2} = \dfrac{2x}{3} - 1$

c. $x^2 - 4 = 0$

d. $3x = 12x^2$

Any quadratic equation can be arranged as follows:

The general form of a quadratic equation

$$ax^2 + bx + c = 0$$

where a, b, and c are integers and $a > 0$.

Notice that a is the coefficient of the x^2-term and is positive, b is the coefficient of the x-term, and c is the constant term.

Changing a quadratic equation into general form

1. Remove fractions by multiplying each term on both sides by the LCD.
2. Remove grouping symbols.
3. Combine like terms.
4. Arrange all nonzero terms in descending powers on one side, leaving only zero on the other side.

EXAMPLE 2

Change the quadratic equation into general form, and identify a, b, and c.

a. $2x^2 + 7 = 5x$

SOLUTION

$$2x^2 + 7 = 5x$$
$$2x^2 - 5x + 7 = 0 \quad \text{General form}$$

$a = 2$, $b = -5$, and $c = 7$.

b. $4x^2 = 9$

SOLUTION

$$4x^2 = 9$$
$$4x^2 - 9 = 0 \quad \text{General form}$$

$a = 4$ and $c = -9$. b, the coefficient of the x-term, is understood to be 0.

c. $3x = 12x^2$

SOLUTION Subtract $3x$ from both sides to make a, the coefficient of x^2, positive:

$$3x = 12x^2$$

$$0 = 12x^2 - 3x \quad \text{General form}$$

$a = 12$ and $b = -3$. c, the constant term, is understood to be 0.

d. $\dfrac{x^2}{2} = \dfrac{2x}{3} - 1$

SOLUTION Multiply each term on both sides by the LCD, 6:

$$\frac{x^2}{2} = \frac{2x}{3} - 1$$

$$\frac{\overset{3}{6}}{1} \cdot \frac{x^2}{\underset{1}{2}} = \frac{\overset{2}{6}}{1} \cdot \frac{2x}{\underset{1}{3}} - 6 \cdot 1$$

$$3x^2 = 4x - 6$$

$$3x^2 - 4x + 6 = 0 \qquad \text{General form}$$

$a = 3$, $b = -4$, and $c = 6$.

e. $(x + 2)(2x - 3) = 2x - 3$

SOLUTION Use FOIL to remove parentheses.

$$(x + 2)(2x - 3) = 2x - 3$$

$$2x^2 + x - 6 = 2x - 3$$

$$2x^2 - x - 3 = 0 \qquad \text{General form}$$

$a = 2$, $b = -1$, and $c = -3$.

Exercises 16.1

Write each quadratic equation in general form, and identify a, b, and c.

1. $2x^2 = 5x + 3$

2. $3x^2 = 4 - 2x$

3. $6x^2 = x$

4. $x^2 = 9x$

5. $16 = 9x^2$

6. $25 = 4x^2$

7. $x^2 - \dfrac{5x}{4} = \dfrac{2}{3}$

8. $2x^2 = \dfrac{x}{5} - \dfrac{1}{2}$

9. $\dfrac{3x}{2} + 5 = x^2$

10. $4 - x^2 = \dfrac{2x}{3}$

11. $x(x - 3) = 4$

12. $2x(x + 1) = 12$

13. $(x + 3)(x - 1) = 8$

14. $(x - 4)(x - 3) = x - 5$

15. $(x - 2)(x + 1) = 3x(x + 2)$

16.2 Solving Quadratic Equations by Factoring

We will use the same method for solving quadratic equations by factoring that we used in previous chapters.

Solving a quadratic equation by factoring

1. Arrange the equation in general form:

$$ax^2 + bx + c = 0$$

2. Factor the polynomial.
3. Set each factor equal to zero, and solve for the variable.
4. Check apparent solutions in the original equation.

EXAMPLE 1 Solve $x^2 - 2x = 8$.

SOLUTION

$$x^2 - 2x = 8$$
$$x^2 - 2x - 8 = 0 \qquad \text{Arrange in general form}$$
$$(x - 4)(x + 2) = 0 \qquad \text{Factor}$$

$$x - 4 = 0 \quad | \quad x + 2 = 0 \qquad \text{Set each factor equal to 0}$$
$$x = 4 \quad | \quad x = -2$$

√ **Check for $x = 4$**   **Check for $x = -2$**

$$x^2 - 2x = 8 \qquad\qquad x^2 - 2x = 8$$
$$(4)^2 - 2(4) \overset{?}{=} 8 \qquad (-2)^2 - 2(-2) \overset{?}{=} 8$$
$$16 - 8 = 8 \qquad\qquad 4 + 4 = 8$$

Therefore, 4 and -2 are solutions.

EXAMPLE 2 Solve $5 - 2x^2 = 3x$.

SOLUTION This time we will arrange all the terms on the right side so that a will be positive.

$$5 - 2x^2 = 3x$$
$$0 = 2x^2 + 3x - 5 \qquad \text{General form}$$
$$0 = (2x + 5)(x - 1) \qquad \text{Factor}$$

$$2x + 5 = 0 \quad | \quad x - 1 = 0 \qquad \text{Set each factor equal to 0}$$
$$2x = -5 \quad | \quad x = 1$$
$$x = -\frac{5}{2} \quad |$$

EXAMPLE 3 Solve $\dfrac{x^2}{8} = x - \dfrac{3}{2}$.

SOLUTION Multiply each term on both sides by the LCD, 8:

$$\frac{x^2}{8} = x - \frac{3}{2}$$
$$\frac{\overset{1}{8}}{1} \cdot \frac{x^2}{\underset{1}{8}} = 8 \cdot x - \frac{\overset{4}{8}}{1} \cdot \frac{3}{\underset{1}{2}}$$
$$x^2 = 8x - 12$$

$$x^2 - 8x + 12 = 0 \qquad \text{General form}$$
$$(x - 6)(x - 2) = 0 \qquad \text{Factor}$$
$$x - 6 = 0 \quad | \quad x - 2 = 0 \qquad \text{Set each factor}$$
$$x = 6 \quad | \quad x = 2 \qquad \text{equal to 0}$$

EXAMPLE 4 Solve $(6x + 2)(x - 4) = 2 - 11x$.

SOLUTION
$$(6x + 2)(x - 4) = 2 - 11x$$
$$6x^2 - 22x - 8 = 2 - 11x$$
$$6x^2 - 11x - 10 = 0 \qquad \text{General form}$$
$$(3x + 2)(2x - 5) = 0 \qquad \text{Factor}$$
$$3x + 2 = 0 \quad | \quad 2x - 5 = 0 \qquad \text{Set each factor equal to 0}$$
$$3x = -2 \quad | \quad 2x = 5$$
$$x = -\frac{2}{3} \quad | \quad x = \frac{5}{2}$$

Exercises 16.2

Solve by factoring.

1. $x^2 + x - 6 = 0$
2. $x^2 + 4x - 5 = 0$
3. $x^2 - 2x = 15$
4. $x^2 - x = 12$
5. $2x^2 = 1 + x$
6. $3x^2 = 2 - 5x$
7. $2 - x^2 = x$
8. $9x - x^2 = 8$
9. $4x + 9 = 5x^2$
10. $10x - 3 = 8x^2$
11. $\frac{x^2}{2} = x - \frac{3}{8}$
12. $\frac{x^2}{3} = 1 - \frac{x}{6}$
13. $\frac{x}{2} + \frac{2}{x} = \frac{5}{2}$
14. $\frac{x}{10} = \frac{4}{5} + \frac{2}{x}$
15. $x(x + 4) = 12$
16. $x(x - 4) = 18 - x$
17. $(x + 2)(x + 3) = x + 3$
18. $(x - 2)(x + 1) = 6x - 12$

16.3 Incomplete Quadratic Equations

An **incomplete quadratic equation** is one in which b or c (or both) is zero; a, the coefficient of x^2, *cannot* be zero. If a were zero, the equation would not be quadratic.

EXAMPLE 1 Examples of incomplete quadratic equations:

a. $12x^2 + 5 = 0$ $\quad b = 0$
b. $7x^2 - 2x = 0$ $\quad c = 0$
c. $3x^2 = 0$ $\qquad b \text{ and } c = 0$

Solving Quadratic Equations When $c = 0$

An incomplete quadratic equation with $c = 0$ is solved by factoring out the GCF.

Solving a quadratic equation when c = 0

1. Arrange the equation in general form.
2. Factor out the GCF.
3. Set each factor equal to zero, and solve for the variable.
4. Check apparent solutions in the original equation.

EXAMPLE 2 Solve $12x^2 = 3x$.

SOLUTION

$$12x^2 - 3x = 0 \quad \text{General form}$$

$$3x(4x - 1) = 0 \quad \text{GCF} = 3x$$

$3x = 0$	$4x - 1 = 0$
$x = 0$	$4x = 1$
	$x = \dfrac{1}{4}$

✓ **Check for $x = 0$**   **Check for $x = \dfrac{1}{4}$**

$$12x^2 = 3x \qquad\qquad 12x^2 = 3x$$

$$12(0)^2 \overset{?}{=} 3(0) \qquad 12\left(\frac{1}{4}\right)^2 \overset{?}{=} 3\left(\frac{1}{4}\right)$$

$$12 \cdot 0 \overset{?}{=} 3(0) \qquad \frac{\overset{3}{\cancel{12}}}{1} \cdot \frac{1}{\underset{4}{\cancel{16}}} \overset{?}{=} \frac{3}{1}\left(\frac{1}{4}\right)$$

$$0 = 0 \qquad\qquad\qquad \frac{3}{4} = \frac{3}{4}$$

Therefore, 0 and $\dfrac{1}{4}$ are solutions.

 A Word of Caution A common error is to divide both sides of the equation by $3x$:

$$12x^2 = 3x$$

$$\frac{12x^2}{3x} = \frac{3x}{3x} \quad \text{Dividing both sides by } 3x$$

$$4x = 1$$

$$x = \frac{1}{4}$$

Using this method, we found the solution $x = \dfrac{1}{4}$, but not $x = 0$. By dividing both sides of the equation by the variable x, we lost the solution $x = 0$. Never divide both sides by an expression containing the variable you are solving for, because you will lose solutions.

EXAMPLE 3 Solve $\dfrac{2}{5}x = 3x^2$.

SOLUTION

$$\frac{\overset{1}{5}}{1} \cdot \frac{2}{\underset{1}{5}}x = 5 \cdot 3x^2 \quad \text{LCD} = 5$$

$$2x = 15x^2$$

$$15x^2 - 2x = 0 \qquad \text{General form}$$

$$x(15x - 2) = 0 \qquad \text{GCF} = x$$

$$x = 0 \qquad\qquad 15x - 2 = 0$$

$$15x = 2$$

$$x = \frac{2}{15}$$

Solving Quadratic Equations When $b = 0$

Consider the incomplete quadratic equation $x^2 = 4$. This equation can be solved by factoring.

$$x^2 = 4$$

$$x^2 - 4 = 0$$

$$(x + 2)(x - 2) = 0$$

$$x + 2 = 0 \qquad\qquad x - 2 = 0$$

$$x = -2 \qquad\qquad x = 2$$

The solutions are $x = -2$ and $x = 2$. We can write these answers together using the $\pm$ sign:

$$x = \pm 2 \quad \text{Read "x equals plus or minus 2"}$$

Some incomplete quadratic equations with $b = 0$ cannot be factored over the integers, so an alternative method is needed. Because taking the square root of a number is the inverse operation of squaring a number, we can remove the square on the variable and solve the equation. Note the use of the $\pm$ sign.

$$x^2 = 4$$

$$\sqrt{x^2} = \pm\sqrt{4} \qquad \text{Taking the square root of both sides}$$

$$x = \pm 2 \qquad\qquad \text{Use the } \pm \text{ sign to obtain both solutions}$$

Solving a quadratic equation when $b = 0$

1. Isolate the squared term.
2. Take the square root of both sides. Use the $\pm$ sign to obtain both solutions.
3. Simplify the square roots.
4. Check apparent solutions in the original equation.

EXAMPLE 4 Solve $x^2 - 12 = 0$.

SOLUTION $x^2 = 12 \qquad \text{Isolate } x^2$

$$\sqrt{x^2} = \pm\sqrt{12} \qquad \text{Take the square root of both sides}$$

$$x = \pm 2\sqrt{3} \qquad \text{Simplify the square roots}$$

There are two solutions, $x = 2\sqrt{3}$ and $x = -2\sqrt{3}$.

✓ **Check for** $x = 2\sqrt{3}$ **Check for** $x = -2\sqrt{3}$

$$x^2 - 12 = 0 \qquad\qquad x^2 - 12 = 0$$

$$(2\sqrt{3})^2 - 12 \overset{?}{=} 0 \qquad (-2\sqrt{3})^2 - 12 \overset{?}{=} 0$$

$$4 \cdot 3 - 12 \overset{?}{=} 0 \qquad\qquad 4 \cdot 3 - 12 \overset{?}{=} 0$$

$$12 - 12 = 0 \qquad\qquad 12 - 12 = 0$$

EXAMPLE 5   Solve $3x^2 - 5 = 0$.

SOLUTION $3x^2 - 5 = 0$

$$3x^2 = 5$$

$$x^2 = \frac{5}{3} \qquad\qquad \text{Isolate } x^2$$

$$\sqrt{x^2} = \pm\sqrt{\frac{5}{3}} \qquad\qquad \text{Take the square root of both sides}$$

$$x = \pm\frac{\sqrt{5}}{\sqrt{3}} \cdot \frac{\sqrt{3}}{\sqrt{3}} \qquad \text{Rationalize the denominator}$$

$$x = \pm\frac{\sqrt{15}}{3}$$

We can use Table I or a calculator to evaluate $\sqrt{15}$ and express the answer as approximate decimals:

$$x = \pm\frac{\sqrt{15}}{3} \approx \pm\frac{3.873}{3} = \pm 1.291 \approx \pm 1.29 \qquad \text{Rounded off to two decimal places}$$

EXAMPLE 6   Solve $x^2 + 25 = 0$.

SOLUTION $x^2 + 25 = 0$

$$x^2 = -25 \qquad\qquad \text{Isolate } x^2$$

$$\sqrt{x^2} = \pm\sqrt{-25} \qquad \text{Take the square root of both sides}$$

$$x = \pm\sqrt{-25}$$

The solution is not a real number because the radicand is negative.

Equations such as the one in Example 6, where the radicand of a square root is negative, have roots that are *complex numbers*. Complex numbers are not discussed in this text.

Exercises 16.3

Solve each equation.

1. $x^2 - 9x = 0$

2. $x^2 - 16x = 0$

3. $x^2 - 9 = 0$

4. $x^2 - 16 = 0$

5. $x^2 = 8$

6. $x^2 = 20$

7. $5x^2 - 24 = 0$

8. $2x^2 - 27 = 0$

9. $8x^2 = 18x$

10. $6x^2 = 4x$

11. $x^2 + 9 = 0$

12. $x^2 + 4 = 0$

13. $\dfrac{2}{3}x = 4x^2$

14. $\dfrac{1}{2}x^2 = 5x$

15. $\dfrac{2x^2}{5} = 10$

16. $3 = \dfrac{x^2}{12}$

17. $(x + 4)(x - 4) = 2(x - 8)$

18. $(x - 3)(x + 3) = 3(2x - 3)$

19. $3x(x - 2) = 2(1 - 3x)$

20. $2x(x + 5) = 5(9 + 2x)$

16.4 The Quadratic Formula

The methods shown in previous sections can only be used to solve *some* quadratic equations. The method we show in this section can be used to solve *all* quadratic equations.

In Example 1 we use a method called *completing the square* to solve a quadratic equation.

EXAMPLE 1   Solve $x^2 - 4x + 1 = 0$.

SOLUTION

$$x^2 - 4x = -1$$

Take $\dfrac{1}{2}(-4) = -2$

Then $(-2)^2 = 4$

$$x^2 - 4x + 4 = -1 + 4 \qquad \text{Add 4 to both sides; this makes the left side a trinomial square}$$

$$(x - 2)(x - 2) = 3 \qquad \text{Factor the left side}$$

$$(x - 2)^2 = 3$$

$$\sqrt{(x - 2)^2} = \pm\sqrt{3} \qquad \text{Take the square root of both sides}$$

$$x - 2 = \pm\sqrt{3}$$

$$x = 2 \pm \sqrt{3} \qquad \text{Add 2 to both sides}$$

There are two solutions: $x = 2 + \sqrt{3}$ and $x = 2 - \sqrt{3}$.

✓ **Check for** $x = 2 + \sqrt{3}$

$$x^2 \quad - \quad 4x \quad + 1 = 0$$

$$(2 + \sqrt{3})^2 - 4(2 + \sqrt{3}) + 1 \overset{?}{=} 0$$

$$4 + 4\sqrt{3} + 3 - 8 - 4\sqrt{3} + 1 \overset{?}{=} 0$$

$$0 = 0$$

We leave the check for $x = 2 - \sqrt{3}$ to you.

Deriving the Quadratic Formula

The method of completing the square can be used to solve *any* quadratic equation. We now use it to solve the general form of the quadratic equation and in this way derive the *quadratic formula*.

$$ax^2 + bx + c = 0 \qquad \text{General form}$$

$$ax^2 + bx = 0 - c \qquad \text{Subtract } c \text{ from both sides}$$

$$x^2 + \frac{b}{a}x = -\frac{c}{a} \qquad \text{Divide both sides by } a$$

$$\text{Take } \frac{1}{2}\left(\frac{b}{a}\right) = \frac{b}{2a}$$

$$\text{Then } \left(\frac{b}{2a}\right)^2 = \frac{b^2}{4a^2}$$

$$x^2 + \frac{b}{a}x + \frac{b^2}{4a^2} = \frac{b^2}{4a^2} - \frac{c}{a} \qquad \text{Add } \frac{b^2}{4a^2} \text{ to both sides; this makes the left side a trinomial square}$$

$$\left(x + \frac{b}{2a}\right)\left(x + \frac{b}{2a}\right) = \frac{b^2}{4a^2} - \frac{4ac}{4a^2} \qquad \text{Factor the left side, and add the fractions on the right side}$$

$$\left(x + \frac{b}{2a}\right)^2 = \frac{b^2 - 4ac}{4a^2}$$

$$\sqrt{\left(x + \frac{b}{2a}\right)^2} = \pm\sqrt{\frac{b^2 - 4ac}{4a^2}} \qquad \text{Take the square root of both sides}$$

$$x + \frac{b}{2a} = \pm\frac{\sqrt{b^2 - 4ac}}{\sqrt{4a^2}} \qquad \text{Simplify radicals}$$

$$x + \frac{b}{2a} = \pm\frac{\sqrt{b^2 - 4ac}}{2a}$$

$$x = -\frac{b}{2a} \pm \frac{\sqrt{b^2 - 4ac}}{2a} \qquad \text{Add } -\frac{b}{2a} \text{ to both sides}$$

Therefore,

$$x = \frac{-b \pm \sqrt{b^2 - 4ac}}{2a} \qquad \text{Quadratic formula}$$

The procedure for using the quadratic formula can be summarized as follows:

Solving a quadratic equation by formula

1. Arrange the equation in general form:
$$ax^2 + bx + c = 0$$

2. Substitute the values of a, b, and c into the *quadratic formula*:
$$x = \frac{-b \pm \sqrt{b^2 - 4ac}}{2a} \qquad (a \neq 0)$$

3. Simplify your answers.

4. Check apparent solutions in the original equation.

EXAMPLE 2 Solve $x^2 - 5x + 6 = 0$ by formula.

SOLUTION Substitute $a = 1$, $b = -5$, and $c = 6$ into the formula:

$$x = \frac{-b \pm \sqrt{b^2 - 4ac}}{2a}$$

$$x = \frac{-(-5) \pm \sqrt{(-5)^2 - 4(1)(6)}}{2(1)}$$

$$x = \frac{5 \pm \sqrt{25 - 24}}{2} = \frac{5 \pm \sqrt{1}}{2}$$

$$x = \frac{5 \pm 1}{2} = \begin{cases} \dfrac{5 + 1}{2} = \dfrac{6}{2} = 3 \\ \dfrac{5 - 1}{2} = \dfrac{4}{2} = 2 \end{cases}$$

This equation can be solved by factoring.

$$x^2 - 5x + 6 = 0$$

$$(x - 2)(x - 3) = 0$$

$$x - 2 = 0 \qquad \bigg| \qquad x - 3 = 0$$

$$x = 2 \qquad \bigg| \qquad x = 3$$

Solving a quadratic equation by factoring is ordinarily shorter than using the formula. Therefore, first check to see whether the equation can be solved by factoring. If it cannot, use the formula. (See Examples 3–6.)

EXAMPLE 3 Solve $x^2 - 6x - 3 = 0$.

SOLUTION Substitute $a = 1$, $b = -6$, and $c = -3$ into the formula:

$$x = \frac{-b \pm \sqrt{b^2 - 4ac}}{2a}$$

$$x = \frac{-(-6) \pm \sqrt{(-6)^2 - 4(1)(-3)}}{2(1)}$$

$$x = \frac{6 \pm \sqrt{36 + 12}}{2} = \frac{6 \pm \sqrt{48}}{2}$$

$$x = \frac{6 \pm 4\sqrt{3}}{2} = \frac{\overset{1}{2}(3 \pm 2\sqrt{3})}{\underset{1}{2}} = 3 \pm 2\sqrt{3}$$

EXAMPLE 4 Solve $\dfrac{1}{4}x^2 = 1 - x$.

SOLUTION

$$\frac{1}{4}x^2 = 1 - x$$

$$\frac{\overset{1}{4}}{1} \cdot \frac{1}{\underset{1}{4}}x^2 = 4 \cdot 1 - 4 \cdot x \qquad \text{Multiply each term by the LCD, 4}$$

$$x^2 = 4 - 4x$$

$$x^2 + 4x - 4 = 0 \qquad \text{Change the equation to general form}$$

Substitute $a = 1$, $b = 4$, and $c = -4$ into the formula:

$$x = \frac{-b \pm \sqrt{b^2 - 4ac}}{2a}$$

$$x = \frac{-(4) \pm \sqrt{(4)^2 - 4(1)(-4)}}{2(1)}$$

$$x = \frac{-4 \pm \sqrt{16 + 16}}{2} = \frac{-4 \pm \sqrt{32}}{2}$$

$$x = \frac{-4 \pm 4\sqrt{2}}{2} = \frac{\overset{1}{2}(-2 \pm 2\sqrt{2})}{\underset{1}{2}} = -2 \pm 2\sqrt{2}$$

EXAMPLE 5   Solve $4x^2 - 5x + 2 = 0$.

SOLUTION   Substitute $a = 4$, $b = -5$, and $c = 2$ into the formula:

$$x = \frac{-b \pm \sqrt{b^2 - 4ac}}{2a}$$

$$x = \frac{-(-5) \pm \sqrt{(-5)^2 - 4(4)(2)}}{2(4)}$$

$$x = \frac{5 \pm \sqrt{25 - 32}}{8} = \frac{5 \pm \sqrt{-7}}{8} \quad \longleftarrow \text{The solution is not a real number because the radicand is negative}$$

 We can use Table I or a calculator to evaluate the square root and express the answers as approximate decimals.

EXAMPLE 6   Solve $x^2 - 5x + 3 = 0$.

SOLUTION   Substitute $a = 1$, $b = -5$, and $c = 3$ into the formula, and express the answers as decimals correct to two decimal places:

$$x = \frac{-b \pm \sqrt{b^2 - 4ac}}{2a}$$

$$x = \frac{-(-5) \pm \sqrt{(-5)^2 - 4(1)(3)}}{2(1)}$$

$$x = \frac{5 \pm \sqrt{25 - 12}}{2} = \frac{5 \pm \sqrt{13}}{2} \qquad \sqrt{13} \approx 3.606 \text{ from Table I}$$

$$x \approx \frac{5 \pm 3.606}{2} = \begin{cases} \dfrac{5 + 3.606}{2} = \dfrac{8.606}{2} = 4.303 \approx 4.30 \\[2mm] \dfrac{5 - 3.606}{2} = \dfrac{1.394}{2} = 0.697 \approx 0.70 \end{cases}$$

We suggest that you use a calculator to check these answers in the original equation.

A Word of Caution A common error in reducing answers is to forget that the number the numerator and denominator are divided by must be a *factor* of both.

$\dfrac{2 \pm \sqrt{3}}{2}$ cannot be reduced because 2 is not a factor of the numerator:

2 is *not* a factor of the numerator

$$x = \frac{2 \pm \sqrt{3}}{2} = \frac{\overset{1}{\cancel{2}} \pm \sqrt{3}}{\underset{1}{\cancel{2}}} = 1 \pm \sqrt{3} \qquad \textit{Incorrect reducing}$$

$\dfrac{2 \pm 2\sqrt{3}}{2}$, however, can be reduced:

2 *is* a factor of the numerator

$$x = \frac{2 \pm 2\sqrt{3}}{2} = \frac{\overset{1}{\cancel{2}}(1 \pm \sqrt{3})}{\underset{1}{\cancel{2}}} = 1 \pm \sqrt{3} \qquad \textit{Correct reducing}$$

Exercises **16.4**

Use the quadratic formula to solve each of the following equations.

1. $3x^2 - x - 2 = 0$

2. $2x^2 + 3x - 2 = 0$

3. $x^2 - 4x + 1 = 0$

4. $x^2 - 4x - 1 = 0$

5. $x^2 - 4x + 2 = 0$

6. $4x^2 = 12x - 7$

7. $2x^2 = 8x - 5$

8. $3x^2 = 6x - 2$

9. $3x^2 + 2x + 1 = 0$

10. $4x^2 + 3x + 2 = 0$

11. $x(x - 2) = 3$

12. $(x + 1)(x + 2) = 12$

13. $\dfrac{x}{2} + \dfrac{2}{x} = \dfrac{5}{2}$

14. $\dfrac{x}{3} + \dfrac{2}{x} = \dfrac{7}{3}$

15. $x^2 = \dfrac{3 - 5x}{2}$

16. $\dfrac{x}{3} + \dfrac{1}{x} = \dfrac{7}{6}$

17. $\dfrac{x^2}{2} = x + 1$

18. $\dfrac{4}{x + 6} + x = 0$

16.5 Word Problems Involving Quadratic Equations

Geometry Problems

EXAMPLE 1   The length of a rectangle is twice the width. If its area is 40 square feet, find the length and the width.

SOLUTION   Let x = width
Then $2x$ = length

Area = 40

x

$2x$

$$\text{Area} = \text{length} \times \text{width}$$

$$40 = 2x \cdot x$$

$$40 = 2x^2$$

$$20 = x^2 \qquad \text{Isolate } x^2$$

$$\pm\sqrt{20} = \sqrt{x^2} \qquad \text{Take the square root of both sides}$$

$$\pm\, 2\sqrt{5} = x \qquad \text{Simplify}$$

$-2\sqrt{3}$ is not a solution because length cannot be negative

Therefore, $\qquad\qquad\qquad\qquad$ width $= x = 2\sqrt{5}$ ft

and $\qquad\qquad\qquad\qquad\qquad$ length $= 2x = 4\sqrt{5}$ ft

√ **Check**   Length $\times$ width $= (4\sqrt{5})(2\sqrt{5}) = 8 \cdot 5 = 40$

Number Problems

EXAMPLE 2

The difference between a number and its reciprocal is 2. Find the number.

SOLUTION   Let $x =$ the number

Then $\dfrac{1}{x} =$ its reciprocal

| Difference between number and reciprocal | is | 2 |

$$x - \frac{1}{x} \qquad\quad = 2 \qquad \text{LCD} = x$$

$$x\,(x) - x\left(\frac{1}{x}\right) = x\,(2)$$

$$x^2 - 1 = 2x$$

$$x^2 - 2x - 1 = 0 \qquad \text{General form}$$

Because $x^2 - 2x - 1 = 0$ does not factor, we will use the quadratic formula. Substitute $a = 1$, $b = -2$, and $c = -1$ into the formula:

$$x = \frac{-b \pm \sqrt{b^2 - 4ac}}{2a}$$

$$x = \frac{-(-2) \pm \sqrt{(-2)^2 - 4(1)(-1)}}{2(1)}$$

$$x = \frac{2 \pm \sqrt{4 + 4}}{2} = \frac{2 \pm \sqrt{8}}{2}$$

$$x = \frac{2 \pm 2\sqrt{2}}{2} = \frac{\overset{1}{2}(1 \pm \sqrt{2})}{\underset{1}{2}}$$

$$x = 1 \pm \sqrt{2}$$

There are two answers: $1 + \sqrt{2}$ and $1 - \sqrt{2}$.

Exercises   *16.5*

In the following exercises, (a) represent the unknown numbers by variables, (b) set up an equation and solve it, and (c) solve the problem.

1. The area of a square is 24 in.^{2}. Find the length of a side.

2. The area of a square is 75 ft^2. Find the length of a side.

3. The length of a rectangle is three times the width. If its area is 54 square feet, find the length and the width.

4. The length of a rectangle is four times the width. Find the length and the width if the area is 45 cm^2.

5. The length of a rectangle is 2 inches more than its width. If its area is 4 in.^{2}, find the length and the width.

6. The width of a rectangle is 6 meters less than its length. Find the length and the width if the area is 9 square meters.

7. If each side of a square is increased by 1, the resulting square has twice the area of the original square. Find the length of the side of the original square.

8. If each side of a square is increased by 2, the resulting square has three times the area of the original square. Find the length of the side of the original square.

9. Find two consecutive integers such that the sum of their squares is 25.

10. If the product of two consecutive even integers is increased by 4, the result is 84. Find the integers.

11. The difference between a number and its reciprocal is 6. Find the number.

12. A number is equal to its reciprocal plus 4. Find the number.

Chapter 16   R E V I E W

Quadratic Equation
16.1

A **quadratic equation** is an equation in one variable whose highest-degree term is a second-degree term.

The **general form** of a quadratic equation is

$$ax^2 + bx + c = 0$$

where a, b, and c are integers and $a > 0$.

Methods of Solving
Quadratic Equations
16.2

Factoring

1. Arrange in general form.

2. Factor the polynomial.

3. Set each factor equal to zero, and solve for the variable.

4. Check solutions in the original equation.

16.3

Incomplete quadratic when $c = 0$

1. Arrange in general form.

2. Factor out the GCF.

3. Set each factor equal to zero, and solve for the variable.

4. Check solutions in the original equation.

16.3

Incomplete quadratic when $b = 0$

1. Isolate the squared term.

2. Take the square root of both sides. Use the $\pm$ sign to obtain both solutions.

3. Simplify the square roots.

4. Check solutions in the original equation.

16.4   *Formula*

1. Arrange in general form.

2. Substitute the values of a, b, and c into the quadratic formula

$$x = \frac{-b \pm \sqrt{b^2 - 4ac}}{2a}$$

3. Simplify the answers.

4. Check solutions in the original equation.

Chapter 16   REVIEW EXERCISES

In Exercises 1–4, write each equation in general form, and identify a, b, and c.

1. $4x^2 = 7 - 3x$

2. $x^2 = 10$

3. $3 = \frac{x}{2} - \frac{1}{4x}$

4. $(x - 2)(x + 5) = 2$

In Exercises 5–24, solve by any convenient method.

5. $x^2 + x = 6$

6. $x^2 = 3x + 10$

7. $x^2 - 25x = 0$

8. $x^2 - 49 = 0$

9. $x^2 - 2x - 4 = 0$

10. $x^2 - 4x + 1 = 0$

11. $x^2 = 5x$

12. $x^2 = 7x$

13. $5x^2 = 18$

14. $6x^2 + 27 = 0$

15. $x = \frac{9}{x} - 3$

16. $\frac{x}{2} + 3 = \frac{1}{2x}$

17. $\frac{x + 2}{3} = \frac{1}{x - 2} + \frac{2}{3}$

18. $\frac{x + 2}{4} = \frac{2}{x + 2} + \frac{1}{2}$

19. $5(x + 9) = x(x + 5)$

20. $3(x + 4) = x(x + 3)$

21. $3x^2 + 2x + 1 = 0$

22. $2x^2 - 3x + 4 = 0$

23. $(x + 5)(x - 2) = x(3 - 2x) + 2$

24. $(2x - 1)(3x + 5) = x(x + 7) + 4$

25. The length of a rectangle is twice the width. Find the length and the width if the area is 80 cm².

26. The square of a number is equal to twice the number plus 4. Find the number.

27. A number plus its reciprocal equals $\frac{10}{3}$. Find the number.

Chapter 16 Critical Thinking and Writing Problems

Answer Problems 1–4 in your own words, using complete sentences.

1. Explain why $5x^2 = 3x$ is not in general form.

2. Explain why $-2x^2 + 3x - 5 = 0$ is not in general form.

3. Explain when you should use the quadratic formula to solve an equation.

4. Explain why $1 - \sqrt{3}$ cannot represent the length of a rectangle.

5. Write the quadratic formula from memory.

6. One of the solutions to the equation $x^2 - 2x - 4 = 0$ is $1 + \sqrt{5}$. Check this solution in the original equation.

Each of the following problems has an error. Find the error, and in your own words, explain why it is wrong. Then work the problem correctly.

7. Solve $x^2 = 36$:

$$x^2 = 36$$
$$\sqrt{x^2} = \sqrt{36}$$
$$x = 6$$

8. Solve $x^2 + 4x + 1 = 0$:

$$x = -4 \pm \frac{\sqrt{4^2 - 4(1)(1)}}{2(1)}$$

$$= -4 \pm \frac{\sqrt{16 - 4}}{2} = -4 \pm \frac{\sqrt{12}}{2}$$

$$= -4 \pm \frac{2\sqrt{3}}{2}$$

$$= -4 \pm \sqrt{3}$$

9. Solve $x^2 - 3x - 5 = 0$:

$$x = \frac{-(-3) \pm \sqrt{-3^2 - 4(1)(-5)}}{2(1)}$$

$$= \frac{3 \pm \sqrt{-9 + 20}}{2}$$

$$= \frac{3 \pm \sqrt{11}}{2}$$

Chapter 16 ₒ DIAGNOSTIC TEST

Allow yourself about 50 minutes to do these problems. Complete solutions for all problems, together with section references, are given in the answer section at the end of the book.

In Problems 1 and 2, write each equation in general form, and identify a, b, and c.

1. $3x^2 = 9 - 2x$

2. $5(x - 1) = x(x + 2)$

In Problems 3–9, solve the equations by any convenient method. If necessary, use the quadratic formula:

If
$$ax^2 + bx + c = 0$$

then
$$x = \frac{-b \pm \sqrt{b^2 - 4ac}}{2a}$$

3. $x^2 + 3x = 10$

4. $3x^2 - 15x = 0$

5. $2x^2 = 9$

6. $2x^2 - 6x + 1 = 0$

7. $6x^2 = 5 - 7x$

8. $\dfrac{x^2}{8} = 2x$

9. $2(x + 1) = x^2$

10. The area of a square is 48 in.^{2}. Find the length of a side.

Geometry

CHAPTER 17

I n this chapter we consider some of the applications of arithmetic and algebra to the more common geometric figures. We live in *rectangular* rooms, we use *circular* plates, our roofs have *triangular* shapes, much of our food comes in *cylindrical* cans, we play with *spherical* basketballs, and so on.

17.1   Angles, Lines, and Polygons

In geometry, the terms *point*, *line*, and *plane* are undefined. We will give meaning to these terms by means of descriptions.

A **point** indicates a position or location. A point has no size. It is usually represented by a dot and named by a capital letter.

Points: *A* and *B*

A **line** (or **straight line**) is a set of points that has no width but extends forever in opposite directions. Two points determine a line. A line is named by identifying any two points on the line or by labeling it with a lowercase letter.

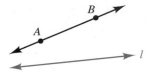

Lines: $\overleftrightarrow{AB}$ and *l*

A **line segment** is a part of a line that lies between two points. A line segment is named by giving its two endpoints.

Line segment: $\overline{AB}$

A **ray** is a portion of a line that has one endpoint and continues forever in one direction. A ray is named by giving its endpoint first, then any other point on the ray.

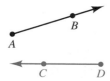

Rays: $\overrightarrow{AB}$ and $\overrightarrow{DC}$

An **angle** is formed by two rays with a common endpoint. The rays are the sides of the angle, and the common point is called the **vertex**. The symbol for angle is ∠. Angles may be named by a single letter (the vertex) or by three letters (the middle letter is always the vertex). We may also label the angle with a number.

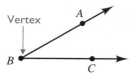

Angle: ∠*B*, ∠*ABC*, or ∠*CBA*

When two angles have the same vertex and a common side between them, they are called **adjacent angles**.

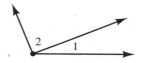

Adjacent angles: ∠1 and ∠2

Angles

Angles may be measured in degrees (°).

One revolution measures 360°.

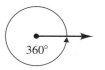

A **straight angle** $\left(\dfrac{1}{2}\text{ of a revolution}\right)$ measures 180°.

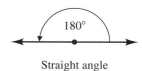

Straight angle

A **right angle** $\left(\dfrac{1}{4}\text{ of a revolution}\right)$ measures 90°. The square corner denotes a right angle.

Right angle

An angle that measures less than 90° is called an **acute angle**.

Acute angle

An angle that measures more than 90° and less than 180° is called an **obtuse angle**.

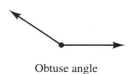

Obtuse angle

EXAMPLE 1 Classify each angle as acute, right, obtuse, or straight.

SOLUTION

a. $\angle A$ Acute
b. $\angle DBC$ Right
c. $\angle ABC$ Obtuse
d. $\angle C$ Acute
e. $\angle CDA$ Straight
f. $\angle BDA$ Obtuse

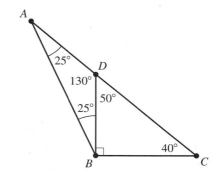

Two angles are **complementary** if their sum is 90°.

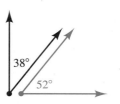

Complementary angles
38° + 52° = 90°

Two angles are **supplementary** if their sum is 180°.

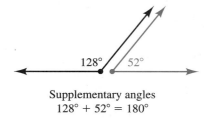

Supplementary angles
128° + 52° = 180°

| **E X A M P L E 2** | Find ∠1 if ∠2 is 67°. |

S O L U T I O N ∠1 and ∠2 form a straight angle. Therefore, their sum is 180°, and they are supplementary.

$$\angle 1 + \angle 2 = \quad 180$$
$$\angle 1 + 67 = \quad 180$$
$$\underline{\quad - 67 = -67}$$
$$\angle 1 = \quad 113°$$

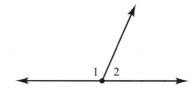

| **E X A M P L E 3** | Find two complementary angles if one angle is 20° more than the other angle. |

S O L U T I O N Let $x =$ one angle

$x + 20 =$ other angle

Because the two angles are complementary, their sum is 90°.

$$x + x + 20 = 90$$
$$2x + 20 = 90$$
$$2x = 70$$
$$x = 35$$
$$x + 20 = 55$$

Therefore, the angles are 35° and 55°.

Lines

Two lines **intersect** if they have only one point in common. The intersecting lines form four angles. When two lines intersect, the nonadjacent (or opposite) angles are called **vertical angles**. ∠1 and ∠3 are vertical angles; ∠2 and ∠4 also are vertical angles.

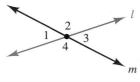

Intersecting lines

An important geometric theorem is stated in the following box.

When two lines intersect, the vertical angles are equal.

$\angle 1 = \angle 3$

$\angle 2 = \angle 4$

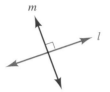

If two intersecting lines form a right angle, the lines are **perpendicular**. The symbol for perpendicular is $\perp$.

Perpendicular lines
$l \perp m$

If two lines in a plane do not intersect, even when they are extended, the lines are said to be **parallel**. The symbol for parallel is $\parallel$.

Parallel lines
$l \parallel m$

A line that crosses two or more lines is called a **transversal**. The eight angles formed in the figure on the right have special names:

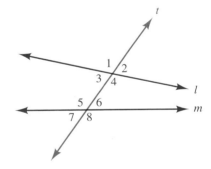

Line t is the transversal
of lines l and m

Alternate interior angles are nonadjacent angles on different sides of the transversal and between the two lines l and m. $\angle 3$ and $\angle 6$ are alternate interior angles. Another pair of alternate interior angles is $\angle 4$ and $\angle 5$.

Alternate exterior angles are nonadjacent angles on different sides of the transversal and outside of the two lines l and m. $\angle 1$ and $\angle 8$ are alternate exterior angles. Another pair of alternate exterior angles is $\angle 2$ and $\angle 7$.

Corresponding angles are nonadjacent angles on the same side of the transversal. One is an interior angle, and the other is an exterior angle. $\angle 2$ and $\angle 6$ are corresponding angles. Other pairs of corresponding angles are $\angle 1$ and $\angle 5$, $\angle 4$ and $\angle 8$, and $\angle 3$ and $\angle 7$.

The following box lists three properties of parallel lines.

If two parallel lines are cut by a transversal, then

1. the *alternate interior angles* are equal
   ($\angle 3 = \angle 6$, and $\angle 4 = \angle 5$).

2. the *alternate exterior angles* are equal
   ($\angle 1 = \angle 8$, and $\angle 2 = \angle 7$).

3. the *corresponding angles* are equal
   ($\angle 1 = \angle 5$, $\angle 2 = \angle 6$, $\angle 3 = \angle 7$, and
   $\angle 4 = \angle 8$).

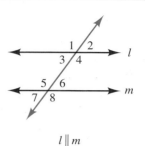

$l \parallel m$

 Note   The converse of the above properties is also true: If the alternate interior angles are equal, then the lines are parallel.

E X A M P L E   4   Given $l_1 \parallel l_2$, find

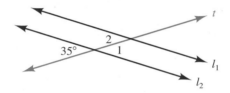

S O L U T I O N

a. $\angle 1$ Vertical angles are equal:

$$\angle 1 = 35°$$

b. $\angle 2$ Alternate interior angles are equal:

$$\angle 2 = \angle 1 = 35°$$

E X A M P L E   5   Given $\overleftrightarrow{AB} \parallel \overleftrightarrow{CD}$ and $\angle ABD = 140°$, find

S O L U T I O N

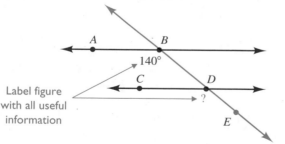

Label figure
with all useful
information

a. $\angle CDE$ Corresponding angles are equal:

$$\angle CDE = \angle ABD = 140°$$

b. $\angle CDB$ $\angle CDE$ and $\angle CDB$ form a straight angle, so their sum is 180°:

$$\angle CDE + \angle CDB = 180$$
$$140 + \angle CDB = 180$$
$$\angle CDB = 40°$$

 Note $\angle ABD + \angle CDB = 140° + 40° = 180°$. Example 5 illustrates another property: If two parallel lines are cut by a transversal, the interior angles on the same side of the transversal are supplementary.

Polygons

A **polygon** is a simple closed figure bounded by line segments called sides. Polygons are named according to the number of sides.

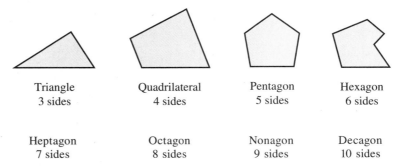

| Triangle | Quadrilateral | Pentagon | Hexagon |
| 3 sides | 4 sides | 5 sides | 6 sides |

| Heptagon | Octagon | Nonagon | Decagon |
| 7 sides | 8 sides | 9 sides | 10 sides |

A **regular polygon** is equilateral (all sides equal) and equiangular (all angles equal).

A **diagonal** of a polygon is a line segment joining two nonadjacent vertices.

Regular pentagon with
diagonal $\overline{AC}$

Exercises 17.1

1. Classify each angle as acute, right, obtuse, or straight.

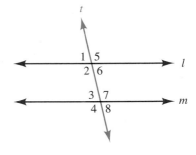

a. ∠ABD

b. ∠ABC

c. ∠ACD

d. ∠ADC

2. List pairs of

a. vertical angles

b. alternate interior angles

c. alternate exterior angles

d. corresponding angles

3. Find the complement of 25°.

4. Find the supplement of 25°.

5. Find the supplement of 73°.

6. Find the complement of 73°.

7. Find two supplementary angles if one angle is three times larger than the other angle.

8. Find two complementary angles if one angle is four times larger than the other angle.

9. Find two complementary angles if one angle is 30° more than the other angle.

10. Find two supplementary angles if one angle is 50° less than the other angle.

11. Find ∠1.

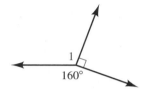

12. Find ∠1.

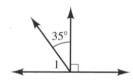

13. Given $l \perp m$, find ∠1 and ∠2.

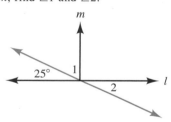

14. Given $l \perp m$, find ∠1 and ∠2.

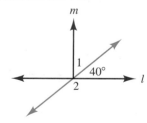

15. Given $\overleftrightarrow{DE} \parallel \overleftrightarrow{BC}$ and ∠ABC = 145°, find ∠CBE, ∠BED, and ∠DEF.

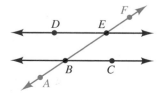

16. Given $\overleftrightarrow{AC} \parallel \overrightarrow{DE}$ and ∠DBC = 60°, find ∠FDE, ∠EDB, and ∠ABD.

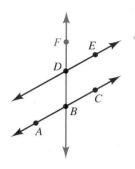

17. Given $l_1 \parallel l_2$, ∠2 = 75°, and ∠6 = 45°, find the remaining angles.

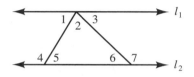

18. Given $l_1 \parallel l_2$, $m \perp l_1$, and ∠3 = 50°, find the remaining angles.

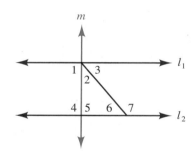

19. Name the polygons.

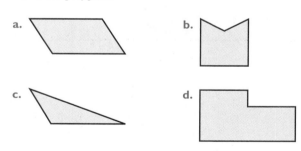

20. Draw the following polygons:

a. A regular hexagon

b. A triangle with one right angle

c. A quadrilateral with opposite sides parallel

d. A quadrilateral with all angles equal

17.2 Triangles

A **triangle** has three sides and three angles. The symbol for triangle is △.

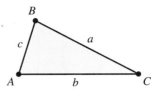

Triangle: △*ABC*

An **equilateral triangle** is a triangle with all sides equal. All three angles are also equal.

Equilateral triangle

An **isosceles triangle** has two sides equal. The angles opposite those sides are called base angles, and they are equal.

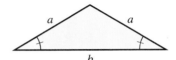

Isosceles triangle

A **scalene triangle** has no sides equal.

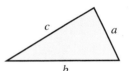

Scalene triangle

A **right triangle** has one right angle.

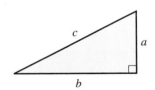

Right triangle

Sum of the Angles of a Triangle

Draw line *l* through *C* and parallel to $\overline{AB}$. Notice that

$$\angle 1 = \angle 4$$
$$\angle 3 = \angle 5$$

If two lines are parallel, the alternate interior angles are equal

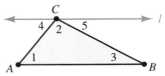

And ∠4 + ∠2 + ∠5 = 180°, a straight angle. Therefore,

$$\text{Sum of the angles of a triangle} = \angle 1 + \angle 2 + \angle 3$$
$$= \angle 4 + \angle 2 + \angle 5$$
$$= 180°$$

> The sum of the angles of a triangle equals 180°.

EXAMPLE 1 In right $\triangle ABC$, $\angle B = 26°$. Find $\angle A$.

SOLUTION

$$\text{Sum of angles} = 180°$$
$$\angle A + \angle B + \angle C = 180$$
$$\angle A + 26 + 90 = 180$$
$$\angle A + 116 = 180$$
$$\angle A = 64°$$

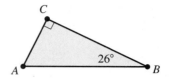

EXAMPLE 2 In $\triangle ABC$, $AC = BC$ and $\angle C = 40°$. Find $\angle A$ and $\angle B$.

SOLUTION Because $AC = BC$, $\triangle ABC$ is an isosceles triangle, where the base angles are equal ($\angle A = \angle B$).

Let $\angle A = \angle B = x$.

$$\angle A + \angle B + \angle C = 180$$
$$x + \quad x + \quad 40 = 180$$
$$2x + \quad 40 = 180$$
$$2x = 140$$
$$x = 70$$

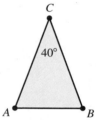

Therefore, $\angle A = \angle B = 70°$.

Sum of the Angles of a Quadrilateral

Diagonal $\overline{AC}$ divides quadrilateral $ABCD$ into two triangles.

$\angle 1 + \angle 2 + \angle 3 = 180$ The sum of the angles
$\angle 4 + \angle 5 + \angle 6 = 180$ of a triangle
equals 180°

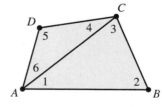

Since $\angle A = \angle 6 + \angle 1$ and $\angle C = \angle 3 + \angle 4$,

Sum of the angles of a quadrilateral $= \quad \angle A \quad + \angle B + \quad \angle C \quad + \angle D$

$$= \angle 6 + \angle 1 + \angle 2 + \angle 3 + \angle 4 + \angle 5$$
$$= 360°$$

> The sum of the angles of a quadrilateral equals 360°.

EXAMPLE 3 In quadrilateral $ABCD$, $\overline{AB} \perp \overline{AD}$, $\angle B = 130°$, and $\angle C = 75°$. Find $\angle D$.

SOLUTION

$$\angle A + \angle B + \angle C + \angle D = 360$$
$$90 + 130 + 75 + \angle D = 360$$
$$295 + \angle D = 360$$
$$\angle D = 65°$$

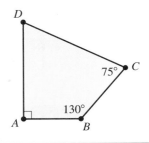

The Pythagorean Theorem

A triangle that has a right angle is called a **right triangle**. The side opposite the right angle is the **hypotenuse**, and the other two sides are called **legs**.

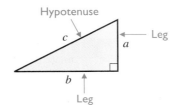

An important relationship between the sides of a right triangle was known to the Greeks about 500 B.C. and is named after the famous Greek mathematician Pythagoras.

The Pythagorean theorem

The square of the hypotenuse of a right triangle is equal to the sum of the squares of the other sides:

$$a^2 + b^2 = c^2$$

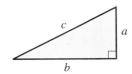

Note The Pythagorean theorem applies only to *right triangles*.

EXAMPLE 4 Examples of using the Pythagorean theorem to show whether or not a given triangle is a right triangle:

a. $3^2 + 4^2 \overset{?}{=} 5^2$

$9 + 16 \overset{?}{=} 25$

$25 = 25$

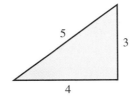

Therefore, the given triangle *is* a right triangle.

b. $2^2 + 3^2 \overset{?}{=} (\sqrt{11})^2$

$\quad\quad 4 + 9 \overset{?}{=} 11$

$\quad\quad\quad 13 \neq 11$

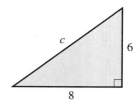

Therefore, the given triangle is *not* a right triangle.

EXAMPLE 5   Find the hypotenuse of a right triangle with legs 8 and 6.

SOLUTION $a^2 + b^2 = c^2$

$\quad\quad 6^2 + 8^2 = c^2$

$\quad\quad 36 + 64 = c^2$

$\quad\quad\quad\quad 100 = c^2$

$\quad\quad \pm\sqrt{100} = \sqrt{c^2}$ Take the square root of both sides

$\quad\quad\quad \pm 10 = c$

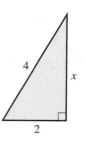

 Note $c = -10$ cannot be a solution of this geometric problem because we usually consider lengths as positive numbers. For this reason we will take only the positive (principal) square root when working with geometric figures.

Therefore, the hypotenuse is 10.

EXAMPLE 6   Find x using the Pythagorean theorem.

SOLUTION $a^2 + b^2 = c^2$

$\quad\quad x^2 + 2^2 = 4^2$

$\quad\quad x^2 + 4 = 16$

$\quad\quad\quad x^2 = 12$

$\quad\quad \sqrt{x^2} = \sqrt{12} = \sqrt{4 \cdot 3}$

$\quad\quad\quad x = 2\sqrt{3}$

EXAMPLE 7   In right $\triangle ABC$, $\angle A = \angle B = 45°$ and $AB = 8$. Find AC and BC.

SOLUTION   Because $\angle A = \angle B$, $\triangle ABC$ is an isosceles triangle, and $AC = BC$. Let $x = AC = BC$. Then

$\quad\quad a^2 + b^2 = c^2$

$\quad\quad x^2 + x^2 = 8^2$

$\quad\quad\quad 2x^2 = 64$

$\quad\quad\quad x^2 = 32$

$\quad\quad \sqrt{x^2} = \sqrt{32} = \sqrt{16 \cdot 2}$

$\quad\quad\quad x = 4\sqrt{2}$

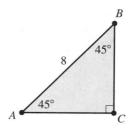

Therefore, $AC = BC = 4\sqrt{2}$.

Exercises 17.2

1. Classify each triangle as equilateral, isosceles, scalene, or right.

a.

b.

c.

d.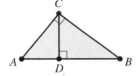

2. For right △ABC,   For right △BCD,

   a. name the hypotenuse   c. name the hypotenuse

   b. name the legs   d. name the legs

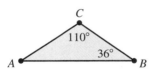

3. Find ∠A.

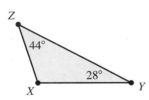

4. Find ∠X.

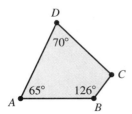

5. Find ∠C.

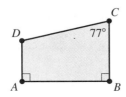

6. Find ∠D.

7. △ABC is an equilateral triangle. Find each angle.

8. △EFG is an isosceles triangle. Find the base angles if the other angle is 45°.

9. If one acute angle of a right triangle is 38°, find the other acute angle.

10. If one acute angle of a right triangle is 63°, find the other acute angle.

11. Find both acute angles of a right triangle if one angle is twice the other acute angle.

12. Find both acute angles of a right triangle if one acute angle is 40° more than the other acute angle.

13. In △ABC, ∠A is three times larger than ∠B, and ∠C is 10° less than ∠B. Find all three angles.

14. In △ABC, ∠B equals ∠A plus 20°, and ∠C equals the sum of ∠A and ∠B. Find all three angles.

15. Given $\overline{AC} \perp \overline{BC}$, $\overline{AB} \perp \overline{CD}$, and ∠B = 32°, find ∠A, ∠ACD, and ∠BCD.

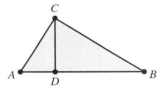

16. Give $\overline{DC} \perp \overline{BC}$, $\overline{EB} \perp \overline{BC}$, and ∠D = 53°, find ∠A, ∠BEA, and ∠BED.

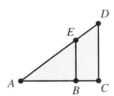

17. Given l ∥ m, ∠1 = 30°, and ∠5 = 80°, find the remaining angles.

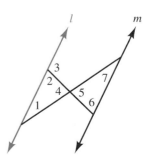

18. Given $l \parallel m$, $\angle 1 = 45°$, and $\angle 5 = 35°$, find the remaining angles.

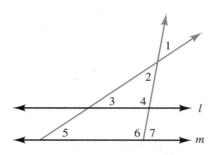

In Exercises 19–22, use the Pythagorean theorem to determine whether or not the given triangle is a right triangle.

19.

13   5   12

20.

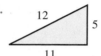

12   5   11

21.

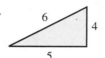

6   4   5

22.

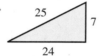

25   7   24

In Exercises 23–30, use the Pythagorean theorem to find x.

23.

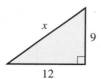

x 12   16

24.

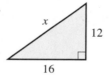

x 9   12

25.

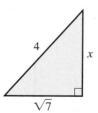

4   x $\sqrt{7}$

26.

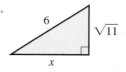

6    $\sqrt{11}$ x

27.

$3\sqrt{3}$ 3 x

28.

$2\sqrt{6}$ x 4

29.

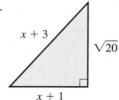

$x + 3$ $\sqrt{20}$ $x + 1$

30.

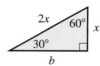

$x + 2$ $x + 1$ $\sqrt{17}$

31. One leg of a right triangle is 4 less than twice the other leg. If its hypotenuse is 10, how long are the two legs?

32. One leg of a right triangle is 2 less than twice the other leg. If the hypotenuse is 5, how long are the two legs?

33. In a 30°-60°-90° triangle, the hypotenuse is twice the length of the short leg (opposite the 30° angle).

Let x = short leg
Then $2x$ = hypotenuse

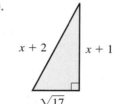

Find the length of the other leg in terms of x.

34. In a 45°-45°-90° triangle, the two legs are equal.

Let x = each leg

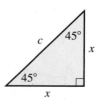

Find the hypotenuse in terms of x.

35. Use the results in Exercise 33 to find the missing sides of the 30°-60°-90° triangles.

a.

b.

c.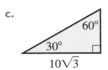

d.

36. Use the results in Exercise 34 to find the missing sides of the 45°-45°-90° triangles.

a.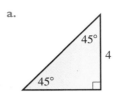

b.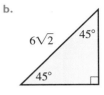

c.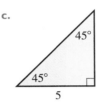

d.

17.3

Congruent Triangles

If two triangles have the same shape and the same size, the triangles are said to be **congruent**. In congruent triangles, the corresponding angles are equal, and the corresponding sides are equal. The symbol for congruent is ≅.

If two triangles are congruent, their corresponding angles are equal and their corresponding sides are equal.

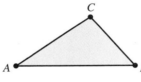

If	$\triangle ABC \cong \triangle DEF$		
then	$\angle A = \angle D,$	$\angle B = \angle E,$	$\angle C = \angle F$
and	$AB = DE,$	$BC = EF,$	$AC = DF$

EXAMPLE 1   **a.** Given $\triangle ABE \cong \triangle DCE$, name the corresponding parts that are equal.

SOLUTION

$\angle A = \angle D$

$\angle ABE = \angle DCE$

$\angle AEB = \angle DEC$

$AB = DC$

$BE = CE$

$AE = DE$

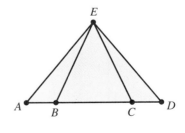

Note Corresponding sides lie opposite corresponding angles.

b. Given $\triangle ADB \cong \triangle BEA$, name the corresponding parts that are equal.

S O L U T I O N   It's helpful to label the corresponding parts that are equal.

$$\angle DAB = \angle EBA$$
$$\angle ADB = \angle BEA$$
$$\angle ABD = \angle BAE$$
$$AB = AB$$
$$AD = BE$$
$$DB = EA$$

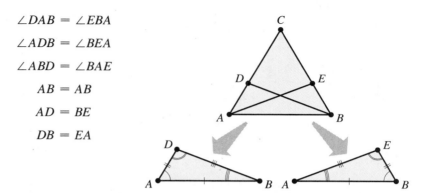

E X A M P L E 2   Given $\triangle ADC \cong \triangle BDC$, find BC and $\angle BCD$.

S O L U T I O N   Corresponding parts of congruent triangles are equal. Therefore, $BC = AC = 6$.

The sum of the angles of a triangle is 180°.

$$\angle A + \angle ADC + \angle ACD = 180$$
$$55 + \quad 90 \quad + \angle ACD = 180$$
$$145 \quad + \angle ACD = 180$$
$$\angle ACD = \quad 35$$

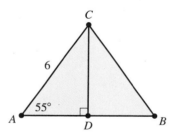

Therefore, $\angle BCD = \angle ACD = 35°$.

E X A M P L E 3   Given $\triangle ABE \cong \triangle CDE$, find x and y.

S O L U T I O N   Corresponding parts of congruent triangles are equal.

$$DE = BE$$
$$x = 10$$
$$DC = AB$$
$$2y = y + 8$$
$$y = 8$$

We can prove that two triangles are congruent by using one of the following properties:

SSS (side-side-side)
If three sides of one triangle are equal to three sides of another triangle, the triangles are congruent.

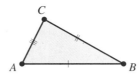

If $\qquad AB = DE, \qquad BC = EF, \qquad AC = DF$

then $\qquad\qquad \triangle ABC \cong \triangle DEF$

SAS (side-angle-side)
If two sides and the included angle of one triangle are equal to two sides and the included angle of another triangle, the triangles are congruent.

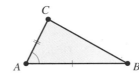

If $\qquad AB = DE, \qquad AC = DF, \qquad \angle A = \angle D$

then $\qquad\qquad \triangle ABC \cong \triangle DEF$

ASA (angle-side-angle)
If two angles and the included side of one triangle are equal to two angles and the included side of another triangle, the triangles are congruent.

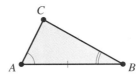

If $\qquad \angle A = \angle D, \qquad \angle B = \angle E, \qquad AB = DE$

then $\qquad\qquad \triangle ABC \cong \triangle DEF$

EXAMPLE 4 Identify the congruent triangles, and name the property used.

a. Given $AC = BC$ and $AD = BD$

SOLUTION It's helpful to label the corresponding parts that we know are equal.

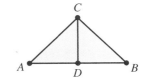

1. $AC = BC$ is given.

2. $AD = BD$ is given.

3. $DC = DC$ is a common side.

Therefore, $\triangle ADC \cong \triangle BDC$, by SSS.

b. Given $AE = BE$ and $CE = DE$

SOLUTION

1. $AE = BE$ is given.

2. $CE = DE$ is given.

3. $\angle AEC = \angle BED$ because vertical angles are equal.

Therefore, $\triangle AEC \cong \triangle BED$, by SAS.

c. Given $\overline{AB} \parallel \overline{CD}$ and $\overline{AC} \parallel \overline{BD}$

SOLUTION If two lines are parallel, the alternate interior angles are equal.

1. $\angle ABC = \angle DCB$

2. $\angle ACB = \angle DBC$

3. $CB = CB$ is a common side.

Therefore, $\triangle ABC \cong \triangle DCB$, by ASA.

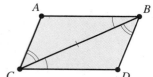

Exercises   17.3

In Exercises 1–6, for the given congruent triangles, name the corresponding parts.

1. $\triangle ABD \cong \triangle CDB$

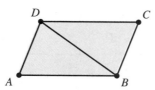

2. $\triangle AFD \cong \triangle BFE$

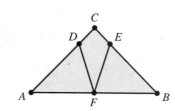

3. $\triangle AFD \cong \triangle BFE$

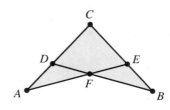

4. $\triangle AEC \cong \triangle BED$

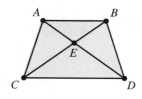

5. $\triangle ACD \cong \triangle CAB$

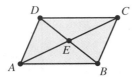

6. $\triangle ABE \cong \triangle CDE$

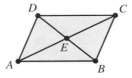

7. Given $\triangle ADC \cong \triangle BEF$, find $\angle F$ and $\angle FBE$.

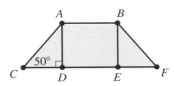

8. Given $\triangle ABC \cong \triangle CDA$, find $\angle DCA$ and $\angle DCB$.

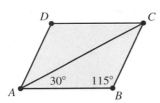

9. Given $\triangle ABE \cong \triangle DCE$, find $\angle D$ and $\angle ECB$.

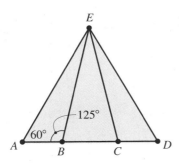

10. Given $\triangle AFD \cong \triangle BFE$, find $\angle B$ and $\angle DFE$.

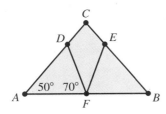

11. Given $\triangle ABE \cong \triangle CBD$, find x and y.

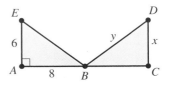

12. Given $\triangle ABD \cong \triangle FEC$, find x and y.

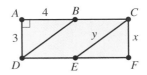

13. Given $\triangle AFD \cong \triangle BFE$, find x and y.

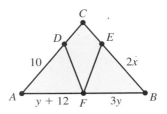

14. Given $\triangle ABE \cong \triangle CBD$, find x and y.

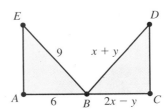

In Exercises 15–22, identify the congruent triangles, and name the property used.

15. Given $AB = AD$ and $BC = DC$

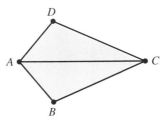

16. Given $AE = DE$ and $BE = CE$

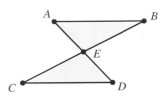

17. Given $\overline{AC} \parallel \overline{BD}$ and $\angle ABC = \angle DCB$

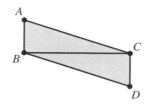

18. Given $\overline{AB} \perp \overline{BC}$, $\overline{DC} \perp \overline{BC}$, and $AB = DC$

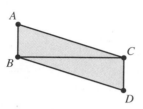

19. Given $\overline{AB} \perp \overline{CD}$ and $AD = BD$

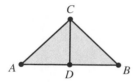

20. Given $\triangle ABC$ is an isosceles triangle and $AD = BD$

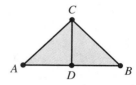

21. Given $\angle A = \angle B$ and $AF = BF$

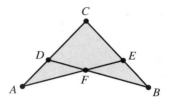

22. Given $\angle A = \angle B$ and $AC = BC$

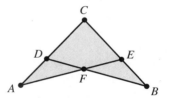

17.4   Similar Triangles

If two triangles have the same shape but not necessarily the same size, the triangles are said to be **similar**. In similar triangles, the corresponding angles are equal. The symbol for similar is $\sim$.

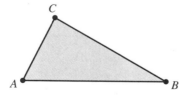

$$\triangle ABC \sim \triangle DEF$$
$$\angle A = \angle D$$
$$\angle B = \angle E$$
$$\angle C = \angle F$$

Because the sum of the angles of a triangle equals 180°, when two pairs of corresponding angles are equal, the third pair of corresponding angles must also be equal. Thus,

> If two angles of one triangle are equal to two angles of another triangle, then the triangles are similar.

EXAMPLE 1

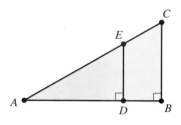

∠*A* is a common angle to both △*ADE* and △*ABC*. Also, ∠*ADE* = ∠*ABC* = 90°. Therefore, △*ADE* ∼ △*ABC*.

An important property of similar triangles is given in the following box:

> If two triangles are similar, their corresponding sides are proportional.
>
>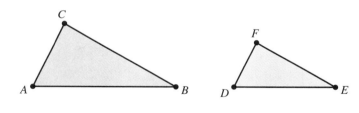
>
> If   △*ABC* ∼ △*DEF*
>
> then $\dfrac{AB}{DE} = \dfrac{BC}{EF} = \dfrac{AC}{DF}$

EXAMPLE 2   Given △*ABC* ∼ △*DEF*, find *DF*.

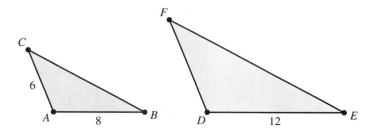

SOLUTION   Because the triangles are similar, the corresponding sides are proportional.

$$\frac{AB}{DE} = \frac{AC}{DF}$$

$$\frac{8}{12} = \frac{6}{DF}$$

$$8DF = 72 \qquad \text{Product of means = product of extremes}$$

$$DF = 9$$

EXAMPLE 3   Given $\overline{AB} \parallel \overline{DE}$, find DE.

SOLUTION $\angle A = \angle E$ and $\angle B = \angle D$. (If two lines are parallel, the alternate interior angles are equal.) Therefore, $\triangle ABC \sim \triangle EDC$.

$$\frac{AC}{EC} = \frac{AB}{DE}$$

$$\frac{4}{8} = \frac{5}{DE}$$

$$4DE = 40$$

$$DE = 10$$

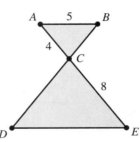

EXAMPLE 4   Find AD.

SOLUTION $\triangle ADE \sim \triangle ABC$ (Example 1).

Let $AD = x$
Then $AB = x + 8$

$$\frac{AD}{AB} = \frac{DE}{BC}$$

$$\frac{x}{x + 8} = \frac{6}{10}$$

$$10x = 6x + 48$$

$$4x = 48$$

$$x = 12$$

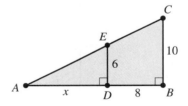

Exercises   17.4

In Exercises 1–4, find the missing sides.

1. $\triangle ABC \sim \triangle DEF$

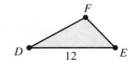

2. $\triangle ABC \sim \triangle DEF$

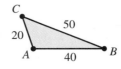
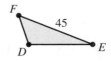

3. △ABC ~ △DEF

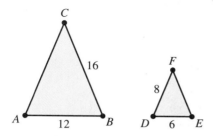

4. △ABC ~ △XYZ

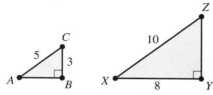

5. Given $\overline{AB} \parallel \overline{CD}$, find AE.

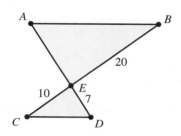

6. Given $\overline{AB} \parallel \overline{CD}$, find DE.

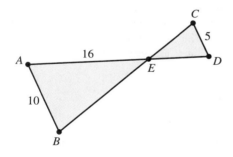

7. Given $\overline{BC} \parallel \overline{DE}$, find BC.

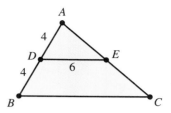

8. Given $\overline{BC} \parallel \overline{DE}$, find BD.

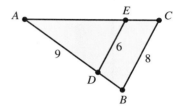

9. Given $\overline{AB} \perp \overline{DE}$ and $\overline{AB} \perp \overline{BC}$, find BD.

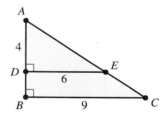

10. Given $\overline{AB} \perp \overline{BC}$ and $\overline{DE} \perp \overline{BC}$, find DC.

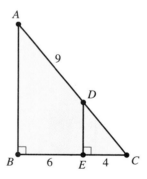

17.5   Perimeter

The word **perimeter** means *the distance around a figure*. To find the perimeter of a geometric figure, we sum the lengths of all its sides.

The perimeter of a **triangle** is the sum of its three sides, so

Perimeter of a triangle $= a + b + c$

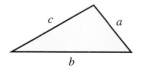

Triangle
$P = a + b + c$

A **parallelogram** is a quadrilateral whose opposite sides are parallel. It is also true that the opposite sides are equal. Therefore,

$$\text{Perimeter of a parallelogram} = a + b + a + b$$
$$= 2a + 2b$$

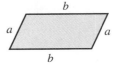

Parallelogram
$P = 2a + 2b$

A **rectangle** is a parallelogram with four right angles. If we label the sides length l and the width w, then

$$\text{Perimeter of a rectangle} = 2l + 2w$$

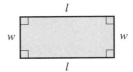

Rectangle
$P = 2l + 2w$

A **square** is a rectangle with all sides equal. If we let s represent the length of one side, then

$$\text{Perimeter of a square} = s + s + s + s = 4s$$

Square
$P = 4s$

An important geometric figure that is not a polygon is the **circle**. All points of a circle are the same distance from a point within it called the **center** O. The **radius** r of the circle is the distance from the center to any point on the circle. The **diameter** d of the circle is the greatest distance across the circle; it is twice the radius ($d = 2r$). The distance around a circle is called its **circumference** C. If we divide the circumference of any circle by its diameter, we always get the same number. That number has been named *pi* (a Greek letter) and is written π.

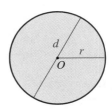

Circle
$d = 2r$
$C = \pi d$
$C = 2\pi r$

$$\frac{\text{circumference}}{\text{diameter}} = \frac{C}{d} = \pi$$

$$C = \pi d \qquad \text{Multiplying both sides by } d$$

$$C = 2\pi r \qquad \text{Since } d = 2r$$

Like $\sqrt{2}$ and $\sqrt{3}$, π is a irrational number. This means that π cannot be written exactly using decimals or fractions. When we want an approximate answer, we will use $\pi \approx 3.14$.

EXAMPLE 1

A circle has a radius of 5 in. Find the circumference.

SOLUTION

$$C = 2\pi r$$
$$C = 2\pi(5) = 10\pi \text{ in.}$$

or, using $\pi \approx 3.14$, $C \approx 10(3.14) = 31.4 \text{ in.}$

EXAMPLE 2   A rectangle has a length of 15 ft and a perimeter of 50 ft. Find the width.

SOLUTION   The formula for the perimeter of a rectangle is $P = 2l + 2w$. Substitute the given quantities into the formula, and solve for the unknown.

$$P = 2l + 2w$$
$$50 = 2(15) + 2w$$
$$50 = 30 + 2w$$
$$20 = 2w$$
$$10 = w$$
$$w = 10 \text{ ft}$$

EXAMPLE 3   Find the distance around each figure.

a.

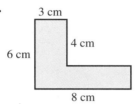

SOLUTION   Divide the figure into rectangles to determine the lengths of the missing sides.

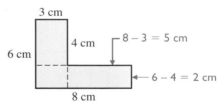

The perimeter is the sum of all the sides.

$$P = 6 + 3 + 4 + 5 + 2 + 8$$
$$P = 28 \text{ centimeters}$$

b.

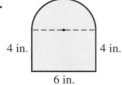

SOLUTION   The diameter of the semicircle (half circle) is 6. The distance along the curved portion of the figure is half the circumference of a circle.

$$\frac{1}{2}C = \frac{1}{2}\pi d = \frac{1}{2}\pi(6) = 3\pi$$

The total distance around the figure is

$$P = 3\pi + 4 + 6 + 4$$
$$P = (3\pi + 14) \text{ in.}$$

EXAMPLE 4   The perimeter of a square is 20 in. Find the length of its diagonal.

SOLUTION   First, find the length of a side:

$$P = 4s$$
$$20 = 4s$$
$$5 = s$$

To find the diagonal, use the Pythagorean theorem for right triangles:

$$a^2 + b^2 = c^2$$
$$5^2 + 5^2 = c^2$$
$$25 + 25 = c^2$$
$$50 = c^2$$
$$\sqrt{50} = \sqrt{c^2}$$
$$5\sqrt{2} = c$$

Therefore, the diagonal equals $5\sqrt{2}$ in.

Exercises   *17.5*

1. Find the perimeter of a triangle with sides 6 ft, 7 ft, and 9 ft.

2. Find the perimeter of a parallelogram with sides 8 m and 12 m.

3. Find the perimeter of a rectangle if the length is 16 in. and the width is 9 in.

4. Find the perimeter of a square if the length of a side is 6 cm.

5. Find the circumference of a circle with a diameter of 4 in.

6. Find the circumference of a circle with a radius of 4 in.

7. If the perimeter of a square is 64 in., find the length of a side.

8. If the perimeter of a rectangle is 32 ft and the width is 7 ft, find the length.

9. The circumference of a circle is 12π ft. Find the radius.

10. The circumference of a circle is 15π cm. Find the diameter.

In Exercises 11–16, find the distance around each figure.

11.

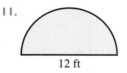

12 ft

12.

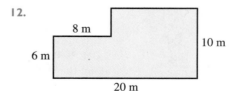

8 m, 6 m, 10 m, 20 m

13.

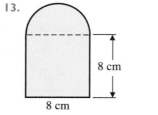

8 cm, 8 cm

14.

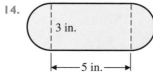

3 in., 5 in.

15.

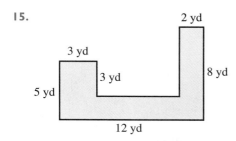

16.

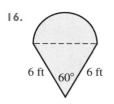

17. Find the perimeter of a right triangle with legs 5 cm and 12 cm.

18. Find the diagonal of a square if its perimeter is 32 ft.

19. The length of a rectangle is 15 in., and the diagonal is 17 in. Find the perimeter.

20. The width of a rectangle is 6 m, and the diagonal is 10 m. Find the perimeter.

21. The perimeter of a rectangle is 38 yd. If the length is 5 yd less than twice the width, find the length and width.

22. The perimeter of a parallelogram is 80 cm. If one side is 8 cm more than the other side, find the sides of the parallelogram.

17.6 Area

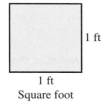

1 ft
1 ft
Square foot

If the side of a square has a length of 1 foot, the square is called a *square foot* (sq. ft). If the side of a square is 1 inch, the square is called a *square inch* (sq. in.), and if the side of a square is 1 meter, the square is called a *square meter* (sq. m). The **area** of a geometric figure is the space inside the lines. Area is measured in square units: square feet, square inches, square meters, and so on.

To measure the area of the rectangle, we see how many times a unit of area fits into it.

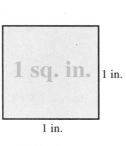

1 in.
1 in.
Unit of area
1 square inch
(1 sq. in.)

Length = 3 in.

Because the unit of area (1 sq. in.) fits into the space above six times, the area of the rectangle is 6 sq. in. We can find this area by multiplying the length times the width.

Area of a rectangle = *lw*

$$A = 3 \times 2 = 6 \text{ sq. in.}$$

— Area is 6 sq. in.

— Width is 2 in.

— Length is 3 in.

w

l

Rectangle
A = *lw*

A square is a rectangle in which the length and width are equal. If we let s represent the length of a side, then

$$\text{Area of a square} = s \cdot s = s^2$$

Square
$A = s^2$

The **height h of a parallelogram** is the perpendicular distance between a pair of parallel sides. To find the formula for the area of a parallelogram, slide the triangle on the left of the height to the other side of the parallelogram.

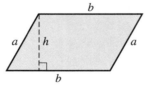

Now we have a rectangle whose area is length times width $= b \cdot h$. Because the areas of the parallelogram and the rectangle are equal,

$$\text{Area of a parallelogram} = bh$$

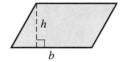

Parallelogram
$A = bh$

The **height h of a triangle** is the perpendicular distance from a vertex to the opposite side.

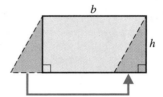

Extend line
when necessary

Area of a Triangle   To find the formula for the area of a triangle, draw parallelogram $ABCD$ with diagonal $\overline{BD}$.

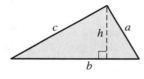

The diagonal $\overline{BD}$ divides the parallelogram into two triangles of equal area. Since the area of the parallelogram is bh,

$$\text{Area of a triangle} = \frac{1}{2}bh$$

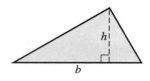

Triangle
$A = \frac{1}{2}bh$

Area of a Circle   The formula for the area of a circle with radius r is

$$\text{Area of a circle} = \pi r^2$$

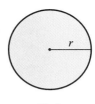

Circle
$A = \pi r^2$

 Note   A complete list of all the formulas given in this chapter is provided in the Chapter Review.

EXAMPLE 1   Find the area of each figure.

SOLUTION

a.

25 in.
13 in.
12 in.
15 in.

Notice that the height is 12 in. and not 13 in., because the height is measured perpendicularly to the base.

$$A = \frac{1}{2}bh$$

$$A = \frac{1}{2} \cdot 15 \cdot 12$$

$$A = 90 \text{ sq. in., or } 90 \text{ in.}^2 \quad \text{Area is measured in square uints}$$

b.

6 m

The radius of a circle is half the diameter. Since the diameter = 6 m, the radius = 3 m.

$$A = \pi r^2$$

$$A = \pi (3)^2$$

$$A = 9\pi \text{ sq. m, or } 9\pi \text{ m}^2$$

c.

4 ft
5 ft
3 ft
10 ft

Divide the figure into two rectangles, then add the areas of the rectangles to find the total area of the figure.

4 ft
10 − 4 = 6 ft
5 ft
3 ft
10 ft

$$A_1 = lw \qquad A_2 = lw$$

$$A_1 = 4 \cdot 5 \qquad A_2 = 6 \cdot 3$$

$$A_1 = 20 \qquad A_2 = 18$$

$$\text{Total area} = A_1 + A_2 = 20 + 18$$

$$= 38 \text{ sq. ft}$$

d.

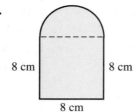

8 cm   8 cm

8 cm

Because $d = 8$ cm, $r = 4$ cm. The area of the semicircle is half the area of a circle.

$$\text{Half area of a circle} = A_1 = \frac{1}{2}\pi r^2$$

$$A_1 = \frac{1}{2}\pi(4)^2$$

$$A_1 = \frac{1}{2}\pi 16$$

$$A_1 = 8\pi \text{ cm}^2$$

$$\text{Area of a square} = A_2 = s^2$$

$$A_2 = (8)^2$$

$$A_2 = 64 \text{ cm}^2$$

$$\text{Total area} = A_1 + A_2 = (8\pi + 64) \text{ cm}^2$$

EXAMPLE 2

The perimeter of a rectangle is 28 in., and the length is 8 in. Find its area.

SOLUTION   Use the formula for perimeter of a rectangle to find the width.

$$P = 2l + 2w$$

$$28 = 2(8) + 2w$$

$$28 = 16 + 2w$$

$$12 = 2w$$

$$6 = w$$

w

8 in.

Then

$$A = lw$$

$$A = (8)(6)$$

$$A = 48 \text{ sq. in.}$$

EXAMPLE 3

Find the area of a right triangle with hypotenuse 5 m and leg 3 m.

SOLUTION   When we find the area of a triangle, the height and base must be perpendicular. Therefore, we must find AC first.

Because $\triangle ABC$ is a right triangle, use the Pythagorean theorem to find AC:

$$a^2 + b^2 = c^2$$

$$3^2 + b^2 = 5^2$$

$$9 + b^2 = 25$$

$$b^2 = 16$$

$$b = 4$$

$$AC = 4$$

A

5 m

C B

3 m

Then

$$A = \frac{1}{2}bh$$

$$A = \frac{1}{2}(3)(4) = 6 \text{ m}^2$$

Exercises 17.6

1. Find the area of a rectangle with length 12 ft and width 8 ft.

2. Find the area of a square with side 6 m.

3. Find the area of a triangle with height 3 in. and base 5 in.

4. Find the area of a parallelogram with height 10 cm and base 18 cm.

5. Find the area of a circle with radius 8 ft.

6. Find the area of a circle with diameter 8 in.

In Exercises 7–15, find the area of each figure.

7.
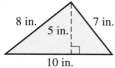
8 in. 7 in.
5 in.
10 in.

8.
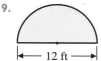
7 cm
5 cm
9 cm

9.
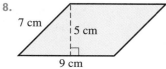
12 ft

10.
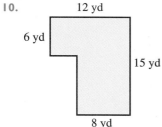
12 yd
6 yd
15 yd
8 yd

11.
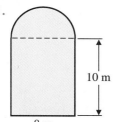
10 m
8 m

12.
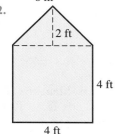
2 ft
4 ft
4 ft

13.
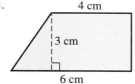
4 cm
3 cm
6 cm

14.
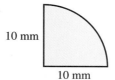
10 mm
10 mm

15.
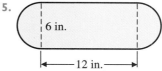
6 in.
12 in.

16. If the area of a rectangle is 60 cm² and the width is 5 cm, find the length.

17. If the area of a right triangle is 30 sq. in. and one leg is 12 in., find the other leg and the hypotenuse.

18. If the area of a circle is 36π sq. ft, find the radius.

19. If the area of a square is 64 m², find the length of a side.

20. Find the number of square inches in a square foot.

21. Find the number of square feet in a square yard.

22. Find the area of the circle if the side of the square is 4 in.

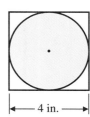
4 in.

23. Find the area of the square if the radius of the circle is 7 in.

7 in.

24. If the perimeter of a square is 20 cm, find its area.

25. If the circumference of a circle is 10π ft, find its area.

26. Given isosceles $\triangle ABC$ with $AC = BC = 10$ m and $CD = 8$ m, find the area.

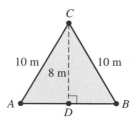

27. Given that the area of parallelogram $ABCD$ is 32 sq. in., $AD = 5$ in., and $DE = 4$ in., find the perimeter.

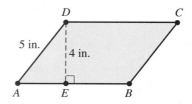

17.7   Volume

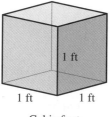

Cubic foot

A square box having equal length, width, and height is called a cube. If the length, width, and height are each 1 foot, the cube is called a *cubic foot* (cu. ft). If the length, width, and height are 1 inch, the cube is called a *cubic inch* (cu. in.). If the length, width, and height are each 1 yard, the cube is called a *cubic yard* (cu. yd). The **volume** of any container is a measure of the space inside that container. Volume is often measured in cubic units: cubic feet, cubic inches, cubic meters, and so on.

We often need to know the volume of a **rectangular box**. The top, bottom, and all sides of a rectangular box are rectangles. Examples of rectangular boxes are most classrooms, most rooms in houses and apartments, shipping boxes and crates, Kleenex boxes, laundry soap boxes, and so on.

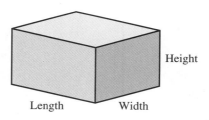

Rectangular box

A rectangular box is shown below. It has a length of 4 inches, a width of 3 inches, and a height of 2 inches.

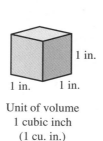

Unit of volume
1 cubic inch
(1 cu. in.)

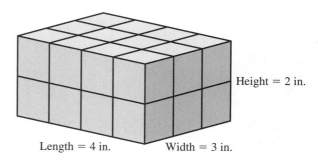

The unit of volume (1 cu. in.) fits into the top layer of the box 12 times. Since the box is made up of 2 layers each containing 12 cu. in., the volume of the box is $2 \cdot 12 = 24$ cu. in. If the height were 5 inches, there would be 5 layers each containing 12 cu. in., so the volume would be $5 \cdot 12 = 60$ cu. in.

Thus, we can find the volume of a rectangular box by multiplying the length $\times$ width $\times$ height:

Volume of a rectangular box $= lwh$

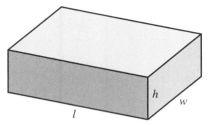

Rectangular box
$V = lwh$

A **cube** is a rectangular box in which the length, width, and height are all equal. If we represent the side of a cube by the letter s, then

Volume of a cube $= s \cdot s \cdot s = s^3$

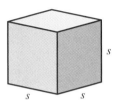

Cube
$V = s^3$

A right circular **cylinder** is shown. This cylinder is called *circular* because the top and bottom are circles. This cylinder is called *right* because the top and bottom form square corners with the sides. We can find the volume of the cylinder by multiplying the area of the bottom (πr^2) times the height h. Thus

Volume of a cylinder $= \pi r^2 h$

Cylinder
$V = \pi r^2 h$

A **sphere** is shown. If the radius (r) is the distance from the center to any point on the sphere, the formula for the volume is

Volume of a sphere $= \dfrac{4}{3} \pi r^3$

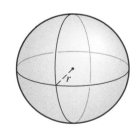

Sphere
$V = \frac{4}{3}\pi r^3$

EXAMPLE 1 Find the volume of a classroom having a length of 30 feet, a width of 25 feet, and a height of 8 feet.

SOLUTION $V = lwh$

$V = 30 \cdot 25 \cdot 8$

$V = 6{,}000$ cu. ft, or $6{,}000 \text{ ft}^3$ Volume is measured in cubic units

EXAMPLE **2**   An aquarium in the shape of a cube measures 10 in. on a side.

a. Find the volume of the aquarium.

SOLUTION $V = s^3$

$V = (10)^3$

$V = 1,000$ cu. in., or $1,000$ in.^{3}

b. Find the weight of the water in the aquarium if 1 cubic inch of water weighs 0.0361 pound.

SOLUTION   Weight of water $= 0.0361 \times 1,000 = 36.1$ lb

EXAMPLE **3**   Find the volume of a sphere with diameter 12 cm.

SOLUTION   Because the diameter $= 12$ cm, the radius $= 6$ cm.

$$V = \frac{4}{3}\pi r^3$$

$$V = \frac{4}{3}\pi(6)^3$$

$$V = \frac{4}{\underset{1}{3}} \cdot \pi \cdot \overset{2}{6} \cdot 6 \cdot 6$$

$$V = 288\pi \text{ cm}^3$$

EXAMPLE **4**   The figure shown is a cylinder capped with a hemisphere (half sphere). Find the total volume.

SOLUTION   The volume of the hemisphere is half the volume of a sphere.

Half volume of a sphere $= V_1 = \dfrac{1}{2} \cdot \dfrac{4}{3}\pi r^3$

$$V_1 = \frac{1}{2} \cdot \frac{4}{3}\pi(3)^3$$

$$V_1 = \frac{1}{\underset{1}{2}} \cdot \frac{\overset{2}{4}}{\underset{1}{3}} \cdot \pi \cdot \overset{9}{27}$$

$$V_1 = 18\pi \text{ m}^3$$

Volume of a cylinder $= V_2 = \pi r^2 h$

$$V_2 = \pi(3)^2(5)$$

$$V_2 = \pi \cdot 9 \cdot 5$$

$$V_2 = 45\pi \text{ m}^3$$

Total volume $= V_1 + V_2 = 18\pi + 45\pi = 63\pi \text{ m}^3$

Exercises   17.7

1. Find the volume of a rectangular box with length 15 ft, width 10 ft, and height 6 ft.

2. Find the volume of a cube with side 5 m.

3. Find the volume of a cylinder with radius 8 in. and height 10 in.

4. Find the volume of a cylinder with diameter 8 cm and height 12 cm.

5. Find the volume of a sphere with radius 6 in.

6. Find the volume of a sphere with diameter 6 ft.

In Exercises 7–9, find the total volume.

7.

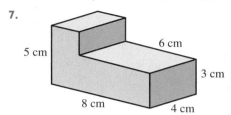

5 cm
6 cm
3 cm
8 cm
4 cm

8.

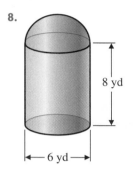

8 yd

6 yd

9.

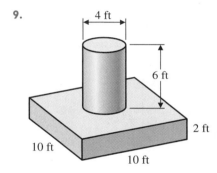

4 ft
6 ft
2 ft
10 ft
10 ft

10. Find the number of cubic feet in a cubic yard.

11. Find the number of cubic inches in a cubic foot.

12. The volume of a rectangular box is 90 cm³. If the length is 6 cm and the width is 5 cm, find the height.

13. The volume of a cylinder is 200π cu. in. If the radius is 5 in., find the height.

14. A classroom measures 10 yd long, 9 yd wide, and 3 yd high.

    a. Find the volume.

    b. If air weighs about 2 lb per cu. yd, find the weight of the air in this room.

15. An aquarium measures 24 in. long and 10 in. wide and is filled to a depth of 15 in.

    a. Find the volume.

    b. Find the weight of the water in the tank if 1 cu. in. of water weighs 0.0361 lb. (Round off to the nearest pound.)

16. A cylindrical cistern is 16 ft deep and 12 ft in diameter.

    a. Find its volume.

    b. If 1 cu. ft equals 7.48 gal, how many gallons of water will the cistern hold? (Use $\pi \approx 3.14$, and round off to the nearest gallon.)

17. A cylindrical water tank measures 20 in. in diameter and 50 in. high.

    a. Find its volume.

    b. If 1 gal equals 231 cu. in., how many gallons of water will the tank hold? (Use $\pi \approx 3.14$, and round off to the nearest gallon.)

Chapter 17   **REVIEW**

Angles
17.1
Two angles are *complementary* if their sum equals 90°. Two angles are *supplementary* if their sum equals 180°.

Lines
17.1
When two lines intersect, the *vertical angles* are equal.

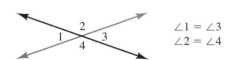

$\angle 1 = \angle 3$
$\angle 2 = \angle 4$

When two lines are *perpendicular*, they form a right angle (90°).

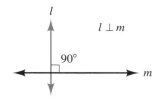

l

$l \perp m$

90°

m

When two *parallel* lines are cut by a transversal:

1. The *alternate interior angles* are equal ($\angle 3 = \angle 6$, and $\angle 4 = \angle 5$).

2. The *alternate exterior angles* are equal ($\angle 1 = \angle 8$, and $\angle 2 = \angle 7$).

3. The *corresponding angles* are equal ($\angle 1 = \angle 5$, $\angle 2 = \angle 6$, $\angle 3 = \angle 7$, and $\angle 4 = \angle 8$).

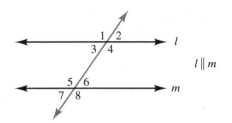

Triangles
17.2

The sum of the angles of a triangle equals 180°.

An *equilateral triangle* is a triangle with three equal sides and three equal angles.

An *isosceles triangle* has two equal sides and two equal angles.

A *scalene triangle* has no sides equal.

A *right triangle* has one right angle.

Pythagorean Theorem
17.2

In a right triangle, the square of the hypotenuse equals the sum of the squares of the other two sides.

$$a^2 + b^2 = c^2$$

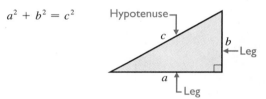

Congruent Triangles
17.3

If two triangles are congruent, their corresponding angles are equal and their corresponding sides are equal.

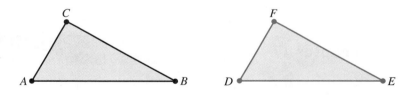

If $\triangle ABC \cong \triangle DEF$

then $\angle A = \angle D,$ $\angle B = \angle E,$ $\angle C = \angle F$

and $AB = DE,$ $BC = EF,$ $AC = DF$

SSS (side-side-side): If three sides of one triangle are equal to three sides of another triangle, the triangles are congruent.

SAS (side-angle-side): If two sides and the included angle of one triangle are equal to two sides and the included angle of another triangle, the triangles are congruent.

ASA (angle-side-angle): If two angles and the included side of one triangle are equal to two angles and the included side of another triangle, the triangles are congruent.

Similar Triangles
17.4

If two angles of one triangle are equal to two angles of another triangle, then the triangles are *similar*.

If two triangles are similar, their corresponding sides are proportional.

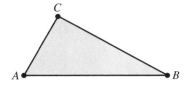

If $\triangle ABC \sim \triangle DEF$

then $\dfrac{AB}{DE} = \dfrac{BC}{EF} = \dfrac{AC}{DF}$

Perimeter and Area
17.5, 17.6

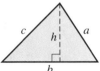

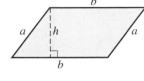

Triangle

$P = a + b + c$

$A = \frac{1}{2}bh$

Parallelogram

$P = 2a + 2b$

$A = bh$

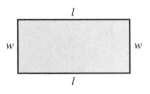

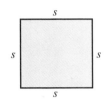

Rectangle

$P = 2l + 2w$

$A = lw$

Square

$P = 4s$

$A = s^2$

Circle

$d = 2r$

$C = \pi d$

$C = 2\pi r$

$A = \pi r^2$

Volume
17.7

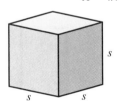

Rectangular box

$V = lwh$

Cube

$V = s^3$

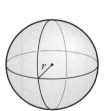

Cylinder

$V = \pi r^2 h$

Sphere

$V = \frac{4}{3}\pi r^3$

Chapter 17   R E V I E W   E X E R C I S E S

1. Find two complementary angles if one angle is five times larger than the other angle.

2. Find two supplementary angles if one angle is 40° less than the other angle.

3. Given $\overline{AB} \parallel \overline{DE}$, $\angle 6 = 30°$, and $\angle 7 = 65°$, find the remaining angles.

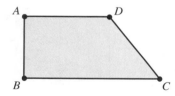

4. Given $l \parallel m$, $\angle 1 = 35°$, and $\angle 7 = 75°$, find the remaining angles.

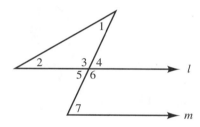

5. In quadrilateral $ABCD$, $\overline{AB} \perp \overline{BC}$, $\overline{AB} \perp \overline{AD}$, and $\angle C = 52°$. Find $\angle D$.

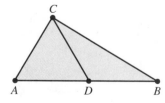

6. In $\triangle ABC$, $AC = AD = CD = DB$. Find $\angle B$.

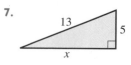

In Exercises 7–9, find the missing side of each right triangle.

7.

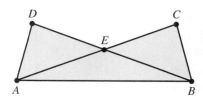

8.

9.

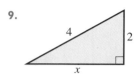

10. Find the diagonal of the rectangle with length 4 cm and width 2 cm.

11. Find the diagonal of a square if its area is 64 sq. in.

12. Given $\triangle ABD \cong \triangle CBE$, find $\angle C$ and $\angle DBE$.

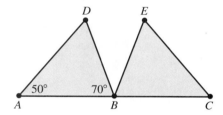

13. Given $\triangle AED \cong \triangle CEB$, find $\angle DAE$ and AE.

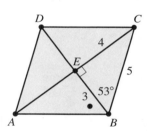

In Exercises 14 and 15, identify the congruent triangles, and name the property used.

14. Given $\angle D = \angle C$ and $DE = CE$

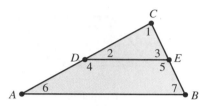

15. Given $\angle DAB = \angle CBA$ and $DA = CB$

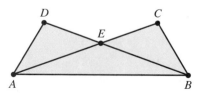

16. Given $\overline{AB} \parallel \overline{CD}$, find DE.

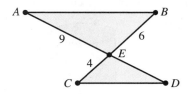

17. Given $\overline{AB} \perp \overline{BC}$, $\overline{DE} \perp \overline{BC}$, find EC.

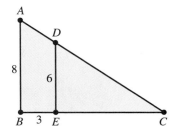

In Exercises 18–23, find the perimeter and the area of each figure.

18.

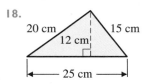

19.

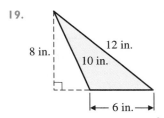

20.

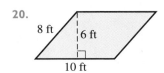

21.

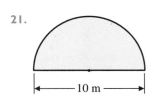

22.

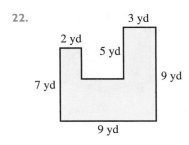

23.

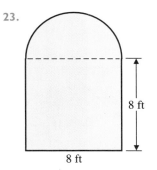

In Exercises 24–26, find the area of the shaded regions.

24.

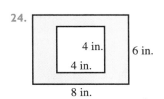

25.

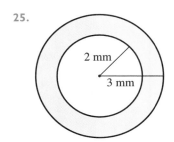

26.

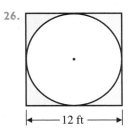

27. The perimeter of a rectangle is 22 m, and the length is 7 m. Find the area.

28. The area of a square is 36 cm². Find its perimeter.

29. The area of a circle is 64π sq. ft. Find its circumference.

30. The legs of a right triangle are 6 in. and 8 in. Find its perimeter.

31. The area of right $\triangle ABC$ is 10 sq. in., and $BC = 5$ in. Find AC.

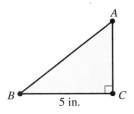

32. The area of parallelogram $ABCD$ is 32 m². If $AB = 6$ m and $AE = 4$ m, what is the perimeter?

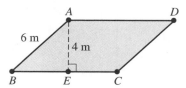

33. Find the volume of a cube with side 4 ft.

34. The volume of a rectangular box is 120 cm³. If the length is 8 cm and the width is 5 cm, find the height.

35. Find the volume of a hemisphere (half sphere) with diameter 12 m.

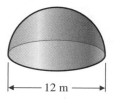

36. Find the volume of water in the cylindrical tank if one fourth of the water has been drained from the tank.

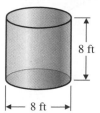

Chapter 17 Critical Thinking and Writing Problems

Answer Problems 1–3 in your own words, using complete sentences.

1. Explain the Pythagorean theorem.

2. Explain how to find the length of the diagonal of a rectangle when the length and the width of the rectangle are given.

3. Explain the difference between congruent triangles and similar triangles.

4. What is the common name used for a regular quadrilateral?

5. Let x = side of a square, then its area = x^2. If we double the length of the side, $s = 2x$, find how many times the area will increase.

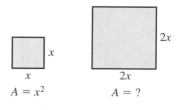

6. Let x = side of a square, then its area = x^2. If we triple the length of the side, $s = 3x$, find how many times the area will increase.

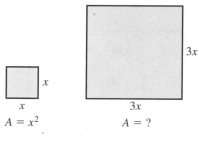

7. Let x = side of a cube; then its volume = x^3. If we double the length of the side, $s = 2x$, find how many times the volume will increase.

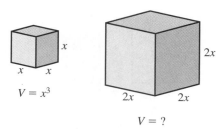

8. Let x = side of a cube, then its volume = x^3. If we triple the length of the side, $s = 3x$, find how many times the volume will increase.

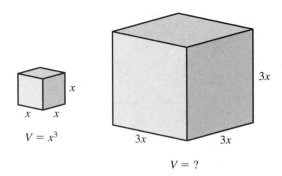

$V = x^3$

$V = ?$

Each of the following problems has an error. Find the error, and in your own words, explain why it is wrong. Then work the problem correctly.

9. Find the area of a rectangle with length 6 ft and width 4 ft:

$$A = lw = 6 \cdot 4 = 24 \text{ ft}$$

10. Find the circumference of the circle:

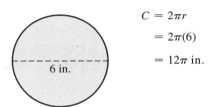

6 in.

$$C = 2\pi r$$
$$= 2\pi(6)$$
$$= 12\pi \text{ in.}$$

11. Find the distance around the figure:

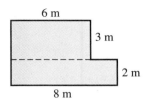

6 m

3 m

2 m

8 m

$P_1 = 2l + 2w$ $P_2 = 2l + 2w$

$P_1 = 2(6) + 2(3)$ $P_2 = 2(8) + 2(2)$

$P_1 = 12 + 6$ $P_2 = 16 + 4$

$P_1 = 18$ $P_2 = 20$

Total distance = $18 + 20 = 38$ m

Chapter 17	DIAGNOSTIC TEST

Allow yourself about 60 minutes to do these problems. Complete solutions for all problems, together with section references, are given in the answer section at the end of the book.

1. Given $l \parallel m$, $\angle 1 = 25°$, and $\angle 6 = 80°$, find the remaining angles.

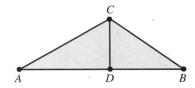

2. Given $\overline{AB} \perp \overline{CD}$, $\angle A = 30°$, and $\angle ACB = 115°$, find $\angle B$ and $\angle BCD$.

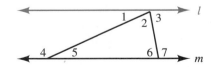

3. In right $\triangle ABC$, find BC.

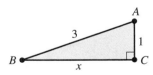

4. In right $\triangle ABC$, find AB.

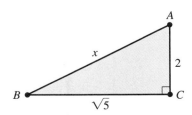

5. Find two supplementary angles if one angle is 30° more than the other angle.

6. Find the perimeter of a right triangle if the legs are 9 in. and 12 in.

7. Find the diagonal of a square if its perimeter is 20 cm.

8. Find the area of parallelogram *ABCD*.

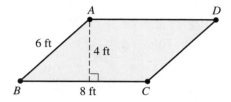

9. Find the perimeter.

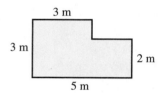

10. Find the area.

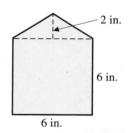

11. The area of a rectangle is 54 sq. ft, and the width is 6 ft. Find the perimeter of the rectangle.

12. The circumference of a circle is 12π ft. Find its area.

13. Find the volume of the rectangular box.

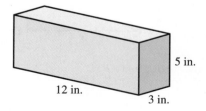

14. Find the volume of a sphere with diameter 6 m.

15. If $\overline{AB} \perp \overline{BC}$ and $\overline{DE} \perp \overline{BC}$, find *AB*.

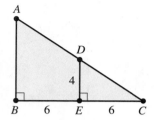

16. If $\overline{AB} \parallel \overline{DE}$, find *AD*.

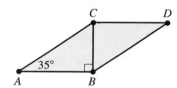

17. Given $\triangle ABC \cong \triangle DCB$, find $\angle DCB$ and $\angle CBD$.

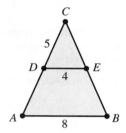

18. Given $\angle BAD = \angle CAD$ and $AB = AC$, identify the congruent triangles and name the property used.

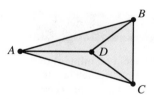

Chapters 1–17   C U M U L A T I V E   R E V I E W   E X E R C I S E S

In Exercises 1–13, perform the indicated operations. Reduce all fractions to lowest terms, and change any improper fractions to mixed numbers.

1. $27\frac{4}{9} - 12\frac{5}{6}$ 2. $5\frac{5}{6} \div 2\frac{1}{2}$ 3. $94.3 - 7.16$

4. 8.7×4.69 5. $8 - 2 \cdot 6 + (3 - 7)$

6. $(5x^2 - 3x - 9) - (2x^2 + x - 6)$

7. $(3x - 4y)(5x + 2y)$

8. $(4x - 3)^2$ 9. $\dfrac{x - 2}{x^2 + x - 6} \cdot \dfrac{x^2 + 3x}{3x^3}$

10. $\dfrac{12x}{x^2 - 8x + 12} \div \dfrac{2x + 4}{x^2 - 4}$ 11. $\dfrac{x^2}{x - 3} + \dfrac{9}{3 - x}$

12. $\dfrac{x - 3}{x + 4} - \dfrac{2}{x}$ 13. $\dfrac{x - \dfrac{25}{x}}{1 + \dfrac{5}{x}}$

In Exercises 14–17, solve each equation.

14. $5 - 3(x - 1) = 2x - (x + 4)$ 15. $3x^2 + x - 10 = 0$

16. $\dfrac{x + 7}{3} - \dfrac{x + 4}{6} = 2$ 17. $\sqrt{3x + 1} = 4$

18. Solve by addition. $2x + 3y = 4$
                         $5x + 4y = 3$

19. Solve by substitution. $x + 2y = 1$
                         $2x - 3y = 16$

20. Solve the inequality, and graph the solution on the number line.

$$4x - 2 > 6x + 8$$

21. Solve $P = 2L + 2W$ for L.

In Exercises 22–24, simplify, and write the answer using only positive exponents.

22. $(3x^3y^{-4})^2$ 23. $(a^{-2}b^5)(a^{-3}b^{-4})$ 24. $\dfrac{8xy^3}{4x^4y^{-2}}$

In Exercises 25–27, perform the indicated operations, and simplify.

25. $\sqrt{12x^4y^3}$ 26. $\sqrt{75} + 3\sqrt{20} - \sqrt{27}$

27. $(5\sqrt{2} - 3)(\sqrt{2} + 4)$

28. Divide, and round off to two decimal places: $2.4 \div 0.46$.

29. Divide, using long division:
$(2x^3 - 7x^2 + 10x - 8) \div (x - 2)$.

30. Change 0.076 to a fraction in lowest terms.

31. Evaluate $x^2 + xy - y^2$ if $x = -3$ and $y = 5$.

32. Perform the indicated operation, and write the answer in decimal form: $7.9 + \dfrac{5}{8}$.

33. Perform the indicated operation, and write the answer in scientific notation: $\dfrac{0.0072}{12,000}$.

34. Graph $5x - 2y = 10$.

35. Graph $y = -4$.

36. Find the equation of the line through the points $(3, 6)$ and $(5, -2)$.

37. Find the slope and the y-intercept of the line $3x + 2y = 8$.

38. Terry worked $6\frac{1}{2}$ hr on Monday, $4\frac{1}{4}$ hr on Tuesday, $3\frac{3}{4}$ hr on Wednesday, $2\frac{3}{4}$ hr on Thursday, and 4 hr on Friday. Find the average number of hours she worked.

39. If 3 pounds of fertilizer will cover 400 square feet of lawn, how many pounds of fertilizer must be used on a 1,000-square-foot lawn?

40. 34 is what percent of 40?

41. 65% of the students enrolled at a college are women. If there are 5,200 women students, how many students are enrolled at the college?

42. It took Wing 3 hours to drive to the mountains. Because of traffic, it took him 4 hours to drive home. If Wing's speed driving to the mountains was 15 mph faster than his speed coming home, find his speed coming home.

43. Felisa has 21 coins consisting of dimes and quarters. If the total value is $3.00, how many dimes does she have?

44. y varies directly with x. If $y = 9$ when $x = 6$, find y when $x = 10$.

45. A rectangle has a perimeter of 28 feet and a width of 5 feet. Find the area of the rectangle.

46. A circle has an area of 9π. Find its circumference.

47. In the right $\triangle ABC$, find AC.

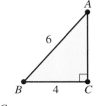

48. Given right $\triangle ABC$, $\overline{AB} \parallel \overline{DE}$, and $\angle A = 42°$, find $\angle CED$.

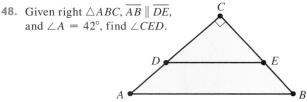

In Exercises 49 and 50, use the graph.

Store Sales

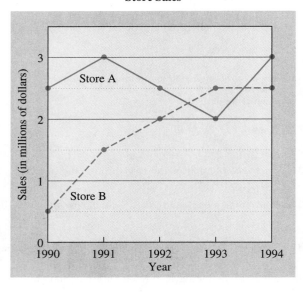

49. Between which two years did the sales increase the most for store A?

50. How much more did store A sell than store B in 1991?

Addition and Multiplication Facts Drill

APPENDIX A

his appendix contains addition and multiplication facts drill problems to improve your speed and accuracy.

A.I   Addition

Addition Facts   To show how to use the table of addition facts, we give an example. To find the sum $7 + 8 = 15$, we move right from 7 of the left column until we are directly below 8 of the top row (see the circled numbers in the table).

+	0	1	2	3	4	5	6	7	⑧	9
0	0	1	2	3	4	5	6	7	8	9
1	1	2	3	4	5	6	7	8	9	10
2	2	3	4	5	6	7	8	9	10	11
3	3	4	5	6	7	8	9	10	11	12
4	4	5	6	7	8	9	10	11	12	13
5	5	6	7	8	9	10	11	12	13	14
6	6	7	8	9	10	11	12	13	14	15
⑦	7	8	9	10	11	12	13	14	⑮	16
8	8	9	10	11	12	13	14	15	16	17
9	9	10	11	12	13	14	15	16	17	18

Table of basic addition facts

Addition Drill   The following addition combinations include most of the pairs of digits that can be added together. If you practice these daily, your addition skills will improve. A convenient way of practicing is to place a paper over the answers, then write or say the answers as quickly as possible. For variety in your practice, work from left to right across the page, then from right to left.

2	0	8	5	3	9	6	5	1	6
2	6	2	6	3	5	4	4	9	9
4	6	10	11	6	14	10	9	10	15

6	2	9	8	7	3	5	8	2	7
5	4	4	3	7	6	3	9	7	4
11	6	13	11	14	9	8	17	9	11

4	8	2	6	5	2	7	8	3	6
4	1	3	8	5	9	0	5	8	6
8	9	5	14	10	11	7	13	11	12

4	7	3	9	2	7	8	7	4	9
6	8	7	9	8	1	4	6	5	6
10	15	10	18	10	8	12	13	9	15

8	4	2	7	9	6	3	5	4	8
8	3	5	5	7	1	3	8	4	7
16	7	7	12	16	7	6	13	8	15

5	3	0	7	3	4	7	3	2	6
2	2	8	7	9	8	3	5	6	3
7	5	8	14	12	12	10	8	8	9

7	5	3	5	9	4	9	1	8	7
2	9	4	7	8	7	3	5	8	9
9	14	7	12	17	11	12	6	16	16

9	5	4	6	9	8	6	4	6	9
2	5	2	2	9	6	7	9	6	0
11	10	6	8	18	14	13	13	12	9

A.2 Multiplication

Multiplication Facts   To show how to use the table of multiplication facts, we give an example. To find the product, $7 \times 8 = 56$, we move right from 7 of the left column until we are directly below 8 of the top row (see the circled numbers in the table).

×	0	1	2	3	4	5	6	7	⑧	9
0	0	0	0	0	0	0	0	0	0	0
1	0	1	2	3	4	5	6	7	8	9
2	0	2	4	6	8	10	12	14	16	18
3	0	3	6	9	12	15	18	21	24	27
4	0	4	8	12	16	20	24	28	32	36
5	0	5	10	15	20	25	30	35	40	45
6	0	6	12	18	24	30	36	42	48	54
⑦	0	7	14	21	28	35	42	49	㊹56	63
8	0	8	16	24	32	40	48	56	64	72
9	0	9	18	27	36	45	54	63	72	81

Table of basic multiplication facts

Multiplication Drill   The following multiplication combinations include most of the pairs of digits that can be multiplied together. If you practice these daily, your multiplication skills will improve. A convenient way of practicing is to place a paper over the answers, then write or say the answers as quickly as possible. To get variety in your practice, work from left to right across the page, then from right to left.

1	5	9	3	7	6	4	5	7	9
9	3	2	4	7	3	2	6	2	4
9	15	18	12	49	18	8	30	14	36

8	4	6	5	3	9	3	7	4	8
7	6	6	7	2	5	7	8	5	6
56	24	36	35	6	45	21	56	20	48

6	7	3	0	8	4	5	3	9	6
7	4	8	6	8	9	2	3	8	5
42	28	24	0	64	36	10	9	72	30

9	7	3	8	4	6	0	4	7	5
9	0	9	5	4	9	8	3	6	5
81	0	27	40	16	54	0	12	42	25

7	3	6	3	7	2	1	8	4	3
9	3	4	5	7	8	7	9	8	6
63	9	24	15	49	16	7	72	32	18

5	4	9	2	5	9	6	8	2	6
9	7	7	2	4	0	6	4	3	8
45	28	63	4	20	0	36	32	6	48

2	8	6	0	8	4	7	2	9	5
5	8	2	5	3	4	5	6	9	8
10	64	12	0	24	16	35	12	81	40

9	2	7	5	2	9	8	2	8	1
6	7	3	5	4	3	1	9	2	6
54	14	21	25	8	27	8	18	16	6

Metric and English Measurement

APPENDIX B

The metric system is a decimal system of measurement that originated in France during the French Revolution. Today, most of the world uses the metric system* or will use it in the near future. The metric system is already used in our country in medicine, nursing, the pharmaceutical industry, the National Aeronautics and Space Administration, photography, the military, and sports. For example, you are probably familiar with the 100-meter race, 35-millimeter film and slides, the liter bottle, the 105-millimeter gun, and so on.

Astronauts on the moon use meters to describe their positions. In many countries—Mexico, for example—metric units are used for speeds and distances on road signs, weight and volume, clothing sizes, and so forth. The United States is committed to changing to the metric system in the near future, so learning the metric system is a necessary part of your education.

B.I Basic Metric Units of Measurement

We consider four of the original basic units used in the metric system:

The *meter* (m), used to measure *length*

The *liter* (L), used to measure *volume*

The *gram* (g), used to measure *weight*[†]

The *Celsius degree* (°C), used to measure *temperature*

The basic unit for measuring length in the metric system is the **meter** (m). It corresponds roughly to the yard in the English system of measurement. (See Figure 1.)

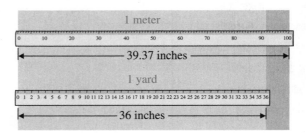

Figure I

The basic unit for measuring volume (capacity) in the metric system is the **liter** (L). It corresponds roughly to the quart in the English system of measurement. (See Figure 2.)

---

*The metric system used most widely is the *International System of Units* (SI System).

[†]To be exact, the gram measures *mass* and not weight. However, in everyday, nonscientific use, the gram is used for *weight*. For example, the *weight* limit on luggage for overseas flights is given as 20 kilo*grams* per person, and a bottle of tomato catsup is marked "NET *WEIGHT* 14 oz. 397 *GRAMS*."

1 liter = 1.06 qt

Figure 2

The original basic unit for measuring *weight* in the metric system is the **gram** (g). One gram is a small quantity of weight. A paper clip weighs about 1 gram. It takes approximately 454 grams to equal one pound. (See Figure 3.)

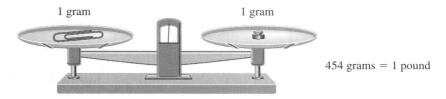

1 gram   1 gram

454 grams = 1 pound

Figure 3

The basic unit for measuring temperature in the metric system is the **Celsius degree**.

Some equivalent Celsius-Fahrenheit temperatures	
100°C = 212°F	Boiling point of water
37°C = 98.6°F	Normal body temperature
0°C = 32°F	Freezing point of water

B.2   Changing Units Within the Metric System

Basic units can be changed to larger or smaller units by means of *prefixes*. We consider the three most commonly used prefixes:

1. *Kilo* means 1,000. Therefore

$$1 \ kilo\text{meter (km)} = 1{,}000 \text{ meters (m)}$$

$$1 \ kilo\text{liter (kL)} = 1{,}000 \text{ liters (L)}$$

$$1 \ kilo\text{gram (kg)} = 1{,}000 \text{ grams (g)}$$

2. *Centi* means $\frac{1}{100}$. $\left(\text{Remember, 1 cent} = \frac{1}{100} \text{ dollar.}\right)$ Therefore,

$$1 \text{ } centi\text{meter (cm)} = \frac{1}{100} \text{ meter}$$

or $$100 \text{ cm} = 1 \text{ m}$$

$$1 \text{ } centi\text{liter (cL)} = \frac{1}{100} \text{ liter}$$

or $$100 \text{ cL} = 1 \text{ L}$$

$$1 \text{ } centi\text{gram (cg)} = \frac{1}{100} \text{ gram}$$

or $$100 \text{ cg} = 1 \text{ g}$$

3. *Milli* means $\frac{1}{1,000}$. Therefore,

$$1 \text{ } milli\text{meter (mm)} = \frac{1}{1,000} \text{ meter}$$

or $$1,000 \text{ mm} = 1 \text{ m}$$

$$1 \text{ } milli\text{liter (mL)} = \frac{1}{1,000} \text{ liter}$$

or $$1,000 \text{ mL} = 1 \text{ L}$$

$$1 \text{ } milli\text{gram (mg)} = \frac{1}{1,000} \text{ gram}$$

or $$1,000 \text{ mg} = 1 \text{ g}$$

All the prefixes involve either a multiplication or a division by a power of ten. Multiplying or dividing a number by a power of ten can be carried out just by moving the decimal point. Therefore, *to change to larger or smaller units in the metric system, it is only necessary to move the decimal point.* This is one of the main advantages to using the metric system.

We now study each prefix in more detail.

B.2A Kilo (Means 1,000)

Changing units involving the prefix *kilo*	
Large units to small units	

To change	kilometers to meters kiloliters to liters kilograms to grams	move the decimal point 3 places to the *right*. (See Example 1.)

Small units to large units

To change	meters to kilometers liters to kiloliters grams to kilograms	move the decimal point 3 places to the *left*. (See Example 2.)

EXAMPLE **1** Examples of changing from large units involving *kilo* to small units:

a. 0.42 km $= 0_{\wedge}420.$ m $= 420$ m
$\qquad\qquad\qquad\xrightarrow{}$
$\qquad\qquad\qquad\quad 3$

b. 6.039 kL $= 6_{\wedge}039.$ L $= 6,039$ L $\quad\longleftarrow$ Move the decimal point 3 places
$\qquad\qquad\qquad\xrightarrow{}$ to the *right*
$\qquad\qquad\qquad\quad 3$

c. 8.7 kg $= 8_{\wedge}700.$ g $= 8,700$ g
$\qquad\qquad\qquad\xrightarrow{}$
$\qquad\qquad\qquad\quad 3$

EXAMPLE **2** Examples of changing from small units to large units involving *kilo*:

a. $9,025$ m $= 9.025_{\wedge}$ km $= 9.025$ km
$\qquad\qquad\qquad\xleftarrow{}$
$\qquad\qquad\qquad\quad -3$

b. 640 L $= .640_{\wedge}$ kL $= 0.640$ kL $\quad\longleftarrow$ Move the decimal point 3 places
$\qquad\qquad\qquad\xleftarrow{}$ to the *left* (-3 means a movement
$\qquad\qquad\qquad\quad -3$ of 3 places to the left)

c. $62,300$ g $= 62.300_{\wedge}$ kg $= 62.3$ kg
$\qquad\qquad\qquad\xleftarrow{}$
$\qquad\qquad\qquad\quad -3$

Exercises B.2A

1. 1.8 km $= \underline{\ ?\ }$ m

2. 34 kL $= \underline{\ ?\ }$ L

3. 0.249 kg $= \underline{\ ?\ }$ g

4. 5.71 km $= \underline{\ ?\ }$ m

5. 60.5 L $= \underline{\ ?\ }$ kL

6. 322 g $= \underline{\ ?\ }$ kg

7. 275 g $= \underline{\ ?\ }$ kg

8. 56.4 L $= \underline{\ ?\ }$ kL

9. 0.78 km $= \underline{\ ?\ }$ m

10. 9.3 kg $= \underline{\ ?\ }$ g

11. $72,350$ g $= \underline{\ ?\ }$ kg

12. $2,365$ m $= \underline{\ ?\ }$ km

13. A pharmaceutical house's orders for hydrogen peroxide average 125 L per month. How many kiloliters is this per month?

14. A rectangular alfalfa field measures 0.90 km by 0.20 km. Find the area of this field in square meters (m^2).

B.2B Centi (Means 1/100)

Changing units involving the prefix *centi*

Small units to large units

To change $\begin{cases} \text{centimeters to meters} \\ \text{centiliters to liters} \\ \text{centigrams to grams} \end{cases}$ move the decimal point 2 places to the *left*. (See Example 3.)

Large units to small units

To change $\begin{cases} \text{meters to centimeters} \\ \text{liters to centiliters} \\ \text{grams to centigrams} \end{cases}$ move the decimal point 2 places to the *right*. (See Example 4.)

EXAMPLE 3   Examples of changing from small units involving *centi* to large units:

a. $155 \text{ cm} = 1.55_\wedge \text{ m} = 1.55 \text{ m}$

  -2

b. $76 \text{ cL} = .76_\wedge \text{ L} = 0.76 \text{ L}$

  -2

  Move the decimal point 2 places to the *left*

c. $4.9 \text{ cg} = .04_\wedge 9 \text{ g} = 0.049 \text{ g}$

  -2

EXAMPLE 4   Examples of changing from large units to small units involving *centi*:

a. $6.8 \text{ m} = 6_\wedge 80. \text{ cm} = 680 \text{ cm}$

  2

b. $0.47 \text{ L} = 0_\wedge 47. \text{ cL} = 47 \text{ cL}$

  2

  Move the decimal point 2 places to the *right*

c. $5.873 \text{ g} = 5_\wedge 87.3 \text{ cg} = 587.3 \text{ cg}$

  2

Exercises   B.2B

1. $279 \text{ cm} = \underline{?} \text{ m}$
2. $54 \text{ cL} = \underline{?} \text{ L}$
3. $8.3 \text{ cg} = \underline{?} \text{ g}$
4. $4{,}090 \text{ cm} = \underline{?} \text{ m}$
5. $2.5 \text{ m} = \underline{?} \text{ cm}$
6. $0.72 \text{ L} = \underline{?} \text{ cL}$
7. $3.906 \text{ g} = \underline{?} \text{ cg}$
8. $0.842 \text{ m} = \underline{?} \text{ cm}$
9. $632 \text{ cL} = \underline{?} \text{ L}$
10. $58.1 \text{ cg} = \underline{?} \text{ g}$
11. $0.0263 \text{ g} = \underline{?} \text{ cg}$
12. $0.092 \text{ L} = \underline{?} \text{ cL}$
13. Haruo's height is 1.82 m. Express his height in centimeters.
14. Hilda's gift weighed 1,430 g. Express the weight of the gift in centigrams.

B.2C   Milli (Means 1/1,000)

Changing units involving the prefix *milli*

Small units to large units

To change $\begin{cases} \text{millimeters to meters} \\ \text{milliliters to liters} \\ \text{milligrams to grams} \end{cases}$ move the decimal point 3 places to the *left*. (See Example 5.)

Large units to small units

To change $\begin{cases} \text{meters to millimeters} \\ \text{liters to milliliters} \\ \text{grams to milligrams} \end{cases}$ move the decimal point 3 places to the *right*. (See Example 6.)

EXAMPLE 5

Examples of changing from small units involving *milli* to large units:

a. $56 \text{ mm} = .056_\wedge \text{ m} = 0.056 \text{ m}$
$\qquad\qquad\qquad\quad -3$

b. $4{,}800 \text{ mL} = 4.800_\wedge \text{ L} = 4.8 \text{ L}$
$\qquad\qquad\qquad\qquad -3$

Move the decimal point 3 places to the *left*

c. $250 \text{ mg} = .250_\wedge \text{ g} = 0.25 \text{ g}$
$\qquad\qquad\qquad\quad -3$

EXAMPLE 6

Examples of changing from large units to small units involving *milli*:

a. $1.4 \text{ m} = 1_\wedge 400. \text{ mm} = 1{,}400 \text{ mm}$
$\qquad\qquad\qquad 3$

b. $0.68 \text{ L} = 0_\wedge 680. \text{ mL} = 680 \text{ mL}$
$\qquad\qquad\qquad\quad 3$

Move the decimal point 3 places to the *right*

c. $0.2050 \text{ g} = 0_\wedge 205.0 \text{ mg} = 205.0 \text{ mg}$
$\qquad\qquad\qquad\qquad 3$

Another commonly used metric unit is the *cubic centimeter*:

1 cubic centimeter (cc) = 1 milliliter (mL)

Exercises B.2C

1. $91 \text{ mm} = \underline{\ ?\ } \text{ m}$
2. $5{,}600 \text{ mL} = \underline{\ ?\ } \text{ L}$
3. $470 \text{ mg} = \underline{\ ?\ } \text{ g}$
4. $4{,}300 \text{ mm} = \underline{\ ?\ } \text{ m}$
5. $2.6 \text{ m} = \underline{\ ?\ } \text{ mm}$
6. $0.39 \text{ L} = \underline{\ ?\ } \text{ mL}$
7. $0.1080 \text{ g} = \underline{\ ?\ } \text{ mg}$
8. $0.0827 \text{ m} = \underline{\ ?\ } \text{ mm}$
9. $230 \text{ mL} = \underline{\ ?\ } \text{ L}$
10. $9{,}160 \text{ mg} = \underline{\ ?\ } \text{ g}$
11. $7.04 \text{ g} = \underline{\ ?\ } \text{ mg}$
12. $21.6 \text{ L} = \underline{\ ?\ } \text{ mL}$

13. $2 \text{ L} = \underline{\ ?\ } \text{ cc}$
14. $3.55 \text{ L} = \underline{\ ?\ } \text{ cc}$
15. $175 \text{ cc} = \underline{\ ?\ } \text{ L}$
16. $2{,}500 \text{ cc} = \underline{\ ?\ } \text{ L}$

17. A doctor recommends that Maria take 1.5 g of vitamin C a day. How many milligrams would she take in a day?

18. Find the volume in cubic centimeters of a container that holds 1.75 L of water (1 cc = 1 mL).

B.2D Additional Metric Prefixes

Figure 4 includes metric prefixes in addition to those we have previously discussed.

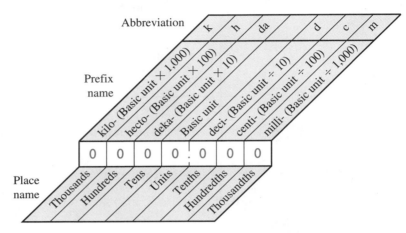

Figure 4 Relation between place names and metric prefixes

In the following box we show how Figure 4 can be used when changing from one of these metric units to another.

Changing from one metric unit to another	When we change from one prefix to another, count the steps moved in Figure 4 and note the direction. The decimal point moves the same number of places in the same direction.

To change from **kilo** gram to **centi** gram,

move the decimal point 5 places to right.

To change from **deci** liter to **hecto** liter,

move the decimal point 3 places to left.

EXAMPLE 7 **a.** Change 573 *milli*liters to *deci*liters.

SOLUTION   To change from milli to deci, we move 2 *steps to the left* in Figure 4. Therefore, we move the decimal point *2 places to the left*:

$$573 \text{ mL} = 5.73_{\wedge} \text{ dL} = 5.73 \text{ dL}$$

b. Change 0.094 *kilo*meters to *centi*meters.

S O L U T I O N   To change from kilo to centi, we move *5 steps to the right* in Figure 4. Therefore, we move the decimal point *5 places to the right*:

$$0.094 \text{ km} = 0{\underset{\scriptstyle 5}{\wedge}09400.} \text{ cm} = 9{,}400 \text{ cm}$$

Exercises   B.2D

Use Figure 4 to help you solve the following exercises.

1. 3.54 km = _?_ m	2. 275 cm = _?_ m
3. 47 kL = _?_ L	4. 144 cL = _?_ L
5. 2,546 g = _?_ kg	6. 386 g = _?_ kg
7. 3.4 dL = _?_ L	8. 784 mL = _?_ dL
9. 0.0516 km = _?_ dm	10. 0.074 dm = _?_ cm

11. 89.5 L = _?_ hL	12. 607 L = _?_ kL
13. 78.4 dam = _?_ km	14. 35.6 dm = _?_ hm
15. 456 hg = _?_ dg	16. 0.064 kg = _?_ cg
17. 3,402 mg = _?_ dag	18. 4,860 cg = _?_ kg
19. 5,614 mL = _?_ daL	20. 956 cL = _?_ hL
21. 3.75 m = _?_ mm	

B.2E   Temperature

When planning what clothing to wear or what activity to engage in, you usually check the temperature first. If you're traveling in countries that use the metric system, however, this can be confusing unless you can change Celsius to Fahrenheit. For example, if you hear that the temperature will be 30°C tomorrow, should you plan on going to the beach or going ice skating?

Changing Celsius to Fahrenheit   We can change degrees Celsius to degrees Fahrenheit by using the following formula:

Changing Celsius to Fahrenheit	$$F = \frac{9}{5}C + 32$$ where F = number of °F and C = number of °C.

EXAMPLE 8   30°C = _?_ °F

S O L U T I O N $F = \frac{9}{5}(\overset{6}{\cancel{30}}) + 32 = 54 + 32 = 86$

Therefore, 30°C = 86°F.

EXAMPLE 9   7°C = _?_ °F (Round the answer to the nearest degree.)

S O L U T I O N $F = \frac{9}{5}(7) + 32 = \frac{63}{5} + 32 = 12.6 + 32 = 44.6 \approx 45$

Therefore, 7°C ≈ 45°F.

Changing Fahrenheit to Celsius   We can change degrees Fahrenheit to degrees Celsius by using the following formula:

Changing Fahrenheit to Celsius	$$C = \frac{5}{9}(F - 32)$$ where C = number of °C and F = number of °F.

EXAMPLE 10   68°F = _?_ °C

SOLUTION $C = \frac{5}{9}(68 - 32) = \frac{5}{9}(\overset{4}{\underset{1}{36}}) = 20$

Therefore, 68°F = 20°C.

EXAMPLE 11   97°F = _?_ °C (Round the answer to the nearest degree.)

SOLUTION $C = \frac{5}{9}(97 - 32) = \frac{5}{9}(65) = 36.1 \approx 36$

Therefore, 97°F ≈ 36°C.

Exercises B.2E

Consider the given numbers to be exact. Round off the answers to the nearest degree.

1. 20°C = _?_ °F
2. 15°C = _?_ °F
3. 50°F = _?_ °C
4. 59°F = _?_ °C
5. 8°C = _?_ °F
6. 17°C = _?_ °F
7. 72°F = _?_ °C
8. 85°F = _?_ °C

9. A Frenchman traveling in the United States notes that a Fahrenheit thermometer reads 41°. What is the equivalent Celsius reading?

10. Is 10°C warmer or colder than 48°F? By how much?

B.3   The English System of Measurement

The English system that we use every day includes such units as inches, feet, yards, miles, pounds, and gallons. The relationships between some of these commonly used English system units are given in the following table:

Some common equivalent English units of measure	
Volume Units	1 tablespoon (tbsp) = 3 teaspoons (tsp) 1 cup = 16 tablespoons (tbsp) 1 cup = 8 ounces (oz) 1 pint (pt) = 2 cups 1 pint (pt) = 16 ounces (oz) 1 quart (qt) = 2 pints (pt) 1 gallon (gal) = 4 quarts (qt) 1 gallon (gal) = 231 cubic inches (cu in.)
Time Units	1 minute (min) = 60 seconds (sec) 1 hour (hr) = 60 minutes (min) 1 day (da) = 24 hours (hr) 1 week (wk) = 7 days (da) 1 year (yr) = 52 weeks (wk) 1 year (yr) = 365 days (da)
Length Units	1 foot (ft) = 12 inches (in.) 1 yard (yd) = 3 feet (ft) 1 mile (mi) = 5,280 feet (ft)
Weight Units	1 pound (lb) = 16 ounces (oz) 1 ton = 2,000 pounds (lb)

The cartoons in Figures 5 and 6 illustrate the derivation of two English units.

Figure 5 King Henry I decreed: "A yard should be the distance from the tip of my nose to the end of my thumb."

Figure 6 King Edward I decreed: "One inch should equal three barley corns laid end to end."

B.4 Changing Units Within the English System

It is often necessary to change from one unit of measure to another. This can be done by means of ratios called **unit fractions**. We now show the unit fraction method most commonly used in science.

Note We can change from one unit of measure to another without using unit fractions, but students are sometimes confused as to whether they should divide or multiply in making the conversion. The method of unit fractions minimizes this confusion.

EXAMPLE 1   Change 27 yards to feet.

SOLUTION   1 yd = 3 ft

If we form the ratio $\dfrac{3 \text{ ft}}{1 \text{ yd}}$ — Unit fraction, the ratio must equal 1 because the numerator and denominator are equal.

Therefore, $27 \text{ yd} = 27 \text{ yd} \times (1) = \dfrac{27 \text{ yd}}{1}\left(\dfrac{3 \text{ ft}}{1 \text{ yd}}\right) = (27 \times 3) \text{ ft} = 81 \text{ ft}$

This is a unit fraction because 1 yd = 3 ft

(Notice that units can be canceled as well as numbers.) Therefore, 27 yd = 81 ft.

We know that the value of the unit fraction must be 1. We now discuss how to select the correct unit fraction.

Selecting the correct unit fraction

1. The unit fraction must have two units:
   a. the unit we want in our answer
   b. the unit we want to get rid of
2. The unit fraction must be written so that the unit of measure we want to get rid of can be canceled.
   a. *If the unit we want to get rid of is in the numerator* of the given expression, that same unit must be in the denominator of the unit fraction chosen. (See Example 2.)
   b. *If the unit we want to get rid of is in the denominator* of the given expression, that same unit must be in the numerator of the unit fraction chosen. (See Example 7.)

EXAMPLE 2   Change 7 inches to feet.

SOLUTION

Feet is placed in the numerator so that we get feet in the answer

$$\dfrac{7 \text{ in.}}{1}\left(\dfrac{1 \text{ ft}}{12 \text{ in.}}\right)$$

Inches is placed in the denominator to cancel with *inches* in 7 inches

$$\dfrac{7 \text{ in.}}{1}\left(\dfrac{1 \text{ ft}}{12 \text{ in.}}\right) = \dfrac{7}{12} \text{ ft}$$

Therefore, $7 \text{ in.} = \dfrac{7}{12} \text{ ft}$.

If we try to use the ratio the other way, it won't work.

$$\frac{7 \text{ in.}}{1} \left(\frac{12 \text{ in.}}{1 \text{ ft}} \right)$$

Here we see that the inches will not cancel, so we will not be left with just feet in the answer

---

EXAMPLE 3 84 hr = _?_ da

SOLUTION

Because 1 day = 24 hr

$$\frac{84 \text{ hr}}{1} \left(\frac{1 \text{ day}}{24 \text{ hr}} \right) = \frac{84}{24} \text{ da} = \frac{7}{2} \text{ da} = 3\frac{1}{2} \text{ da}$$

Therefore, 84 hr = $3\frac{1}{2}$ da.

EXAMPLE 4 $7\frac{1}{2}$ gal = _?_ qt

SOLUTION

Because 4 qt = 1 gal

$$\frac{7\frac{1}{2} \text{ gal}}{1} \left(\frac{4 \text{ qt}}{1 \text{ gal}} \right) = \left(7\frac{1}{2} \times 4 \right) \text{ qt} = \left(\frac{15}{\underset{1}{2}} \times \frac{\overset{2}{4}}{1} \right) \text{ qt} = 30 \text{ qt}$$

Therefore, $7\frac{1}{2}$ gal = 30 qt.

 Note Sometimes it's necessary to use two or more unit fractions to change to the required units (see Example 5).

EXAMPLE 5 2 hr = _?_ sec

SOLUTION

Step 1. Change hours to minutes:

$$\frac{2 \text{ hr}}{1} \left(\frac{60 \text{ min}}{1 \text{ hr}} \right) = (2 \times 60) \text{ min} = 120 \text{ min}$$

Step 2. Change minutes to seconds:

$$\frac{120 \text{ min}}{1} \left(\frac{60 \text{ sec}}{1 \text{ min}} \right) = (120 \times 60) \text{ sec} = 7,200 \text{ sec}$$

Therefore, 2 hr = 7,200 sec.
This problem could be worked in one step, as follows:

$$\frac{2 \text{ hr}}{1} \left(\frac{60 \text{ min}}{1 \text{ hr}} \right) \left(\frac{60 \text{ sec}}{1 \text{ min}} \right) = (2 \times 60 \times 60) \text{ sec} = 7,200 \text{ sec}$$

 A Word of Caution The unit fraction must be written so that the unit of measure we want to get rid of can be canceled.

Sometimes the unit we want to get rid of is in the numerator (Example 6).

EXAMPLE 6   26 mi = ? ft

SOLUTION   *To get rid of miles* and be left with feet, we must put miles in the denominator:

Correct choice

$$\frac{26 \text{ mi}}{1} \left(\frac{5280 \text{ ft}}{1 \text{ mi}} \right)$$ Because miles cancel

Incorrect choice

$$\frac{26 \text{ mi}}{1} \left(\frac{1 \text{ mi}}{5280 \text{ ft}} \right)$$ Because miles do *not* cancel

Therefore, 26 mi = (26 × 5280) ft = 137,280 ft.

Sometimes the unit we want to get rid of is in the denominator (Example 7).

EXAMPLE 7 $2000 \text{ mi per hr} = 2000 \dfrac{\text{mi}}{\text{hr}} = \underline{?} \dfrac{\text{mi}}{\text{min}} = \underline{?} \text{ mi per min}$

SOLUTION   *To get rid of hours* and be left with minutes, we must put hours in the numerator:

Correct choice

$$\frac{2000 \text{ mi}}{\text{hr}} \left(\frac{1 \text{ hr}}{60 \text{ min}} \right)$$ Because hours cancel

Incorrect choice

$$\frac{2000 \text{ mi}}{\text{hr}} \left(\frac{60 \text{ min}}{1 \text{ hr}} \right)$$ Because hours do *not* cancel

Therefore, $2000 \text{ mi per hr} = \left(\dfrac{2000}{60} \right) \dfrac{\text{mi}}{\text{min}} \approx 33.33 \text{ mi per min.}$

EXAMPLE 8   Ernesto drives 10 miles in 20 minutes. What is his average speed in miles per hour (mph)?

SOLUTION $\dfrac{10 \text{ mi}}{\underset{1}{20 \text{ min}}} \left(\dfrac{\overset{3}{60 \text{ min}}}{1 \text{ hr}} \right) = 30 \dfrac{\text{mi}}{\text{hr}} = 30 \text{ mph}$

Exercises   B.4

Find the missing numbers.

1. 5 yd = ? ft

2. 3 ft = ? in.

3. $3\frac{1}{3}$ yd = ? ft

4. $5\frac{2}{3}$ ft = ? in.

5. 8 gal = ? qt

6. $2\frac{3}{4}$ gal = ? qt

7. 7 qt = ? pt

8. 2.5 qt = ? pt

9. $1\frac{1}{2}$ hr = ? min

10. 7.75 min = ? sec

11. $1\frac{3}{4}$ mi = ? ft

12. $3\frac{1}{4}$ da = ? hr

13. 10 yd = ? ft = ? in.

14. 8 gal = ? qt = ? pt

15. 24 in. = ? ft

16. 84 in. = ? ft

17. 90 sec = ? min

18. 150 min = ? hr

19. 5,280 ft = ? mi

20. 730 da = ? yr

21. 104 wk = ? yr

22. $\frac{1}{4}$ cup = ? oz

23. $\frac{3}{8}$ cup = ? tbsp

24. 1 gal = ? oz

25. $1\frac{1}{3}$ tbsp = ? tsp

26. 2 gal = ? cu. in.

27. $2\frac{1}{2}$ lb = ? oz

28. 2,200 lb = ? tons

29. $2\frac{1}{2}$ tons = ? lb

30. 48 oz = ? lb

31. Change 1.25 miles to feet.

32. Change 8,800 feet to miles.

33. Change 2 weeks to hours.

34. Change 5 miles to yards.

35. Change 2,160 minutes to days.

36. Change 60 ounces to pounds.

37. Louisa drove 8 mi in 12 min. What was her average speed in miles per hour?

38. Tran walks 2 mi in 50 min. What is his average walking speed in miles per hour?

39. Sound travels about 1,100 ft per sec. What is the speed of sound in miles per hour?

40. A certain glacier moves 100 yards per year. What is its speed in feet per month?

B.5 Simplifying Denominate Numbers

If you wish to describe how tall you are, how much you weigh, how old you are, or how much money you have in your pocket, you use certain *standard units of measure*. For example, you may be 69 *inches* tall, weigh 160 *pounds*, be 19 *years* old, and have 15 *dollars* in your pocket. Here the standard units of measure are inches, pounds, years, and dollars.

Numbers expressed in standard units of measure, such as inches, pounds, years, and dollars, are called **denominate numbers**. A de*nom*inate number is a number with a name (*nomen* means "name" in Latin). When we say 3 feet, the *3 feet* is a denominate number. When we say 5 hours, the *5 hours* is a denominate number. Numbers such as 5, 25, and $\frac{3}{4}$, which are not given with units, are called *abstract* numbers. When we say 35 + 10 = 45, we are using abstract numbers, but when we say 35 feet + 10 feet = 45 feet, we are using denominate numbers.

When we say *like numbers*, we mean denominate numbers expressed in the same units.

EXAMPLE 1

Examples of like numbers:

a. 7 feet and 20 feet
b. 15 pounds and 10 pounds

EXAMPLE 2

Examples of unlike numbers:

a. 5 feet and 10 inches
b. 2 hours and 15 minutes

To simplify a denominate number having two or more different units, we change a quantity in smaller units to larger units whenever possible.

EXAMPLE 3

Simplify 2 feet 18 in.

SOLUTION

$$2 \text{ ft} + 18 \text{ in.}$$
$$= 2 \text{ ft} + 1 \text{ ft} + 6 \text{ in.}$$
$$= 3 \text{ ft} + 6 \text{ in.}$$
$$= 3 \text{ ft } 6 \text{ in.}$$

ALTERNATIVE SOLUTION

Feet	Inches
2	~~18~~
1	6 ← 18 in. = 1 ft 6 in.
3 ft	6 in.

EXAMPLE 4

Simplify 14 hr 126 min.

SOLUTION

$$14 \text{ hr} + \overbrace{126 \text{ min}}$$
$$= \underbrace{14 \text{ hr} + 2 \text{ hr}} + 6 \text{ min}$$
$$= \quad 16 \text{ hr} \quad + 6 \text{ min}$$
$$= 16 \text{ hr } 6 \text{ min}$$

ALTERNATIVE SOLUTION

Hours	Minutes
14	~~126~~
2	6 ← 126 min = 2 hr 6 min
16 hr	6 min

Therefore, 14 hr 126 min = 16 hr 6 min.

EXAMPLE 5

Simplify 2 da 23 hr 150 min.

SOLUTION

Days	Hours	Minutes
2	23	~~150~~
	2	30 ← 150 min = 2 hr 30 min
	~~25~~	
1	1 ← 25 hr = 1 da 1 hr	
3 da	1 hr	30 min

Therefore, 2 da 23 hr 150 min = 3 da 1 hr 30 min.

EXAMPLE 6

Simplify 4 gal 11 qt 5 pt.

SOLUTION

Gallons	Quarts	Pints
4	11	~~5~~
	2	1 ← 5 pt = 2 qt 1 pt
	~~13~~	
3	1 ← 13 qt = 3 gal 1 qt	
7 gal	1 qt	1 pt

Exercises B.5

Simplify.

1. 4 ft 15 in.
2. 7 yd 5 ft
3. 2 wk 9 da
4. 2 da 36 hr
5. 3 gal 15 qt
6. 7 qt 11 pt
7. 5 yd 4 ft 27 in.
8. 10 yd 2 ft 34 in.
9. 2 hr 73 min 110 sec
10. 3 hr 82 min 125 sec
11. 3 gal 7 qt 5 pt
12. 4 gal 5 qt 6 pt
13. 2 mi 6,000 ft
14. 3 mi 6,400 ft
15. 2 yr 48 wk 75 da
16. 3 yr 51 wk 55 da
17. 2 tons 3,500 lb
18. 3 tons 2,250 lb
19. 4 lb 20 oz
20. 5 lb 56 oz
21. 2 da 23 hr 75 min

B.6 Adding Denominate Numbers

Only like numbers can be added.

$$5 \text{ dollars} + 3 \text{ dollars} = 8 \text{ dollars}$$
$$2 \text{ feet} + 4 \text{ feet} = 6 \text{ feet}$$
$$6 \text{ days} + 11 \text{ days} = 17 \text{ days}$$

Adding denominate numbers
1. Write the denominate numbers under one another, with like units in the same vertical line.
2. Add the numbers in each vertical line.
3. Simplify the denominate number found in Step 2.

EXAMPLE 1   Add 7 ft 3 in., 3 ft 5 in., and 2 ft 7 in.

SOLUTION

```
    7 ft   3 in.
    3      5
    2      7
   12     15
    1      3   ←——— 15 in. = 1 ft 3 in.
   13 ft   3 in.
```

EXAMPLE 2   Add 3 da 18 hr 42 min, 1 da 9 hr 29 min, and 1 da 14 hr 51 min.

SOLUTION

```
   3 da  18 hr   42 min
   1      9      29
   1     14      51
   5     41     122
                  2      2   ←——— 122 min = 2 hr 2 min
                 43
   1     19              ←——— 43 hr = 1 da 19 hr
   6 da  19 hr   2 min
```

Exercises B.6

Add and simplify.

1. 4 ft 5 in.
    3 ft 6 in.
    5 ft 2 in.

2. 2 yd 2 ft
    3 yd 2 ft
    4 yd 1 ft

3. 3 hr 15 min
    2 hr 50 min
    7 hr 24 min

4. 35 min 54 sec
    48 min 27 sec

5. 3 gal 2 qt 1 pt
    5 gal 3 qt 1 pt
    8 gal 1 qt 1 pt

6. 3 gal 2 qt 1 pt
    5 gal 3 qt 1 pt
    4 gal 1 qt 1 pt

7. 3 yd 2 ft 10 in.
    1 yd 1 ft 9 in.
    8 yd 2 ft 7 in.

8. 1 mi 4,000 ft
    2 mi 3,800 ft

9. 1 da 12 hr 15 min
    5 da 23 hr 54 min
    2 da 18 hr 47 min

10. 2 yr 41 wk 5 da
    1 yr 18 wk 4 da
    3 yr 27 wk 3 da

11. 3 tons 1,500 lb
    5 tons   450 lb
    7 tons 1,850 lb

12. 3 lb 5 oz
    8 lb 15 oz
    13 lb 9 oz

13. 7 lb 3 oz
    15 lb 9 oz
    22 lb 17 oz

14. 5 hr 17 min 35 sec
    3 hr 44 min 47 sec
    2 hr 53 min 24 sec

15. Add 2 yd 8 in., 3 yd 2 ft 5 in., and 4 yd 1 ft.

16. Add 3 hr 40 min, 2 hr 30 sec, and 5 hr 25 min 45 sec.

B.7 Subtracting Denominate Numbers

Only like numbers can be subtracted.

Subtracting denominate numbers

1. Write the number being subtracted under the number it's being subtracted from, writing like units in the same vertical line.
2. Subtract the numbers in each vertical line, borrowing when necessary from the first nonzero number to the left.
3. Simplify the answer.

EXAMPLE 1 Subtract 2 ft 4 in. from 10 ft 6 in.

SOLUTION
    10 ft 6 in.
  − 2 ft 4 in.
     8 ft 2 in.

EXAMPLE 2 Subtract 4 gal 3 qt from 7 gal 1 qt.

SOLUTION
     $\overset{6}{7}$ gal $\overset{5}{1}$ qt   — The 1 gallon *borrowed* makes
   − 4 gal 3 qt   4 quarts, which, when added
     2 gal 2 qt   to 1 quart already there, gives
                     5 quarts

EXAMPLE 3   Subtract 1 hr 50 min 40 sec from 3 hr 21 min.

SOLUTION

$$
\begin{array}{r}
\overset{2}{\cancel{3}}\ \text{hr}\ \overset{\overset{80}{\overset{20}{}}}{\cancel{21}}\ \text{min}\ \overset{60}{\cancel{0}}\ \text{sec} \\
-\ 1\ \text{hr}\ 50\ \text{min}\ 40\ \text{sec} \\
\hline
1\ \text{hr}\ 30\ \text{min}\ 20\ \text{sec}
\end{array}
$$

EXAMPLE 4   Subtract 1 yd 2 ft 10 in. from 3 yd 4 in.

SOLUTION

$$
\begin{array}{r}
\overset{2}{\cancel{3}}\ \text{yd}\ \overset{\overset{2}{3}}{\cancel{0}}\ \text{ft}\ \overset{16}{4}\ \text{in.} \\
-\ 1\ \text{yd}\ 2\ \text{ft}\ 10\ \text{in.} \\
\hline
1\ \text{yd}\ 0\ \text{ft}\ \ 6\ \text{in.}\ =\ 1\ \text{yd}\ 6\ \text{in.}
\end{array}
$$

Exercises   B.7

Subtract and simplify.

1. 8 ft 10 in.
    − 3 ft 4 in.

2. 9 ft 6 in.
    − 7 ft 2 in.

3. 13 yd 2 ft
    − 7 yd 1 ft

4. 7 yd 2 ft
    − 3 yd 1 ft

5. 8 gal 2 qt
    − 3 gal 3 qt 1 pt

6. 8 hr 15 min
    − 3 hr 50 min

7. 5 lb 3 oz
    − 2 lb 8 oz

8. 3 lb 7 oz
    − 1 lb 10 oz

9. 3 tons   700 lb
    − 1 ton 1,200 lb

10. 4 tons   500 lb
    − 2 tons 1,600 lb

11. 3 da 5 hr
    − 1 da 15 hr

12. 5 gal   1 pt
    − 3 gal 3 qt

13. 5 yd   6 in.
    − 2 yd 2 ft 9 in.

14. 3 yd   9 in.
    − 1 yd 2 ft 10 in.

15. 4 mi 3,000 ft
    − 1 mi 4,700 ft

16. 3 mi 2,000 ft
    − 2 mi 2,350 ft

17. 35 min 40 sec
    − 20 min 55 sec

18. 27 min 20 sec
    − 15 min 32 sec

19. 5 da 13 hr 22 min
    − 2 da 18 hr 45 min

20. 5 yr 43 wk 3 da
    − 2 yr 50 wk 5 da

21. Subtract 2 hr 4 min 29 sec from 5 hr 14 sec.

22. Subtract 4 yd 2 ft 8 in. from 6 yd 6 in.

B.8   Multiplying Denominate Numbers

Multiplying a denominate number by an abstract number

1. Multiply each part of the denominate number by the abstract number.
2. Simplify the product.

EXAMPLE 1   Multiply 4 × (2 ft 8 in.).

SOLUTION

$$
\begin{array}{r}
2\ \text{ft}\qquad 8\ \text{in.} \\
\times\ 4 \\
\hline
8\ \text{ft}\qquad \cancel{32}\ \text{in.} \\
\underline{2}\qquad\quad \underline{8} \\
10\ \text{ft}\qquad 8\ \text{in.}
\end{array}
$$

— Simplifying the product

EXAMPLE 2 Multiply $5 \times$ (4 gal 3 qt 1 pt).

SOLUTION

$$
\begin{array}{r}
4 \text{ gal} \quad 3 \text{ qt} \quad 1 \text{ pt} \\
\times\ 5 \\
\hline
20 \text{ gal}\ 15 \text{ qt}\ 5 \text{ pt} \\
2 \qquad 1 \\
\hline
17 \\
4 \qquad 1 \\
\hline
24 \text{ gal}\quad 1 \text{ qt}\ 1 \text{ pt}
\end{array}
$$

— Simplifying the product

Multiplying a Denominate Number by a Denominate Number Sometimes it is possible to multiply a denominate number by another denominate number.

EXAMPLE 3 Find the area of a rectangle that is 3 yd long and 2 yd wide.

SOLUTION

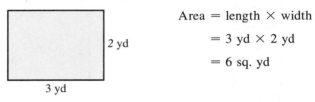

2 yd

3 yd

Area = length × width

= 3 yd × 2 yd

= 6 sq. yd

The reason we can multiply these two denominate numbers is that the following has meaning:

$$1 \text{ yd} \times 1 \text{ yd} = 1 \text{ sq. yd}$$

Therefore, the area is 6 sq. yd.

EXAMPLE 4 Find the number of worker-hours expended if a crew of 10 people work for 8 hours.

SOLUTION   10 workers × 8 hours = 80 worker-hours

The reason we can multiply these two denominate numbers is that the following has meaning:

1 worker × 1 hour = 1 worker-hour = work done by 1 person in 1 hour

EXAMPLE 5 Multiply 5 apples × 3 chairs.

SOLUTION   5 apples × 3 chairs = 15 apple-chairs

Because an apple-chair has no meaning that we can think of, multiplying these denominate numbers does not make sense.

Multiplying denominate numbers

1. A denominate number can always be multiplied by an abstract number. (See Examples 1 and 2.)
2. A denominate number can be multiplied by another denominate number only when the product of their units has meaning.

Exercises B.8

Multiply and simplify.

1. 4 × (3 wk 5 da)

2. 3 × (2 yr 225 da)

3. 6 × (5 mi 2,850 ft)

4. 5 × (3 mi 1,550 ft)

5. 5 × (2 yd 1 ft 3 in.)

6. 3 × (5 gal 2 qt 1 pt)

7. 4 × (1 hr 25 min 11 sec)

8. 7 × (2 yd 2 ft 6 in.)

9. 8 × (3 gal 3 qt 1 pt)

10. 6 × (2 hr 15 min 21 sec)

11. Find the area of a rectangle that is 176 ft long and 48 ft wide.

12. Find the area of a rectangle that is 28 in. by 17 in.

13. A baseball diamond is a square 90 ft on each side. Find the area enclosed in a baseball diamond.

14. A football field measures 100 yd between goal lines and measures 55 yd between side lines. Find the area enclosed by these lines.

15. A construction crew of 17 worked 12 eight-hour days to construct a building. Find the total number of worker-hours used to construct the building. Find the cost to construct the building if the cost per worker-hour was $16.50.

16. In designing a new vacuum cleaner, 956.4 worker-hours were used. Find the cost of designing this machine if the average cost of a worker-hour was $14.23.

B.9 Dividing Denominate Numbers

Dividing a Denominate Number by an Abstract Number

To divide a denominate number expressed in unlike units by an abstract number, we divide the largest unit first, then the next largest unit, and so on, until the division is completed. Our first example is simple and has no remainder.

EXAMPLE 1

Divide (10 yd 2 ft 8 in.) by 2.

SOLUTION

$$\frac{5 \text{ yd } 1 \text{ ft } 4 \text{ in.}}{2)\overline{10 \text{ yd } 2 \text{ ft } 8 \text{ in.}}}$$

Dividing a denominate number by an abstract number

1. Divide the largest unit by the abstract number.

2. If a remainder is left from Step 1, change it to the next smaller unit and add it to those units already there.

3. Divide the sum found in Step 2 by the abstract number.

4. Repeat Steps 2 and 3 until the division is complete.

EXAMPLE 2

Divide (11 yd 1 ft 6 in.) by 3.

SOLUTION

$$
\begin{array}{r}
3 \text{ yd} \quad 2 \text{ ft} \quad 6 \text{ in.} \\
3)\overline{11 \text{ yd} \quad 1 \text{ ft} \quad 6 \text{ in.}} \\
\underline{9} \\
2 \text{ yd} = \underline{6 \text{ ft}} \\
7 \text{ ft} \\
\underline{6} \\
1 \text{ ft} = \underline{12 \text{ in.}} \\
18 \text{ in.} \\
\underline{18} \\
0
\end{array}
$$

EXAMPLE 3   Divide (3 gal 2 qt 1 pt) by 4.

SOLUTION

$$
\begin{array}{r}
0 \text{ gal} \quad 3 \text{ qt} \quad 1\frac{1}{4} \text{ pt} \\
\overline{4)\,3 \text{ gal} \quad 2 \text{ qt} \quad 1 \quad \text{pt}} \\
\underline{0} \\
3 \text{ gal} = 12 \text{ qt} \\
14 \text{ qt} \\
\underline{12} \\
2 \text{ qt} = 4 \quad \text{pt} \\
5 \quad \text{pt} \\
\underline{4} \quad \text{pt} \\
1 \quad \text{pt} \longrightarrow \frac{1 \text{ pt}}{4} = \frac{1}{4} \text{ pt}
\end{array}
$$

This 1 pint has not been divided by 4; when it is, we get $\frac{1}{4}$ pint

Dividing a Denominate Number by a Denominate Number

Sometimes it is possible to divide a denominate number by another denominate number.

EXAMPLE 4   How many 3-foot shelves can be cut from a 12-foot board?

SOLUTION $\dfrac{12 \text{ ft}}{3 \text{ ft}} = \dfrac{\overset{4}{\cancel{12} \text{ ft}}}{\underset{1}{\cancel{3} \text{ ft}}} = 4$ Notice that the answer is an abstract number

EXAMPLE 5   If a sponsor buys enough TV time to permit a total of one hour for commercials, how many 3-minute commercials can the sponsor put on?

SOLUTION $\dfrac{1 \text{ hour}}{3 \text{ min}} = \dfrac{\overset{20}{\cancel{60} \text{ min}}}{\underset{1}{\cancel{3} \text{ min}}} = 20$

Exercises B.9

Divide and simplify.

1. (2 ft 6 in.) ÷ 2

2. (3 qt 1 pt) ÷ 3

3. (2 qt 1 pt) ÷ 4

4. (6 ft 8 in.) ÷ 5

5. (5 hr 30 min) ÷ 3

6. (6 hr 40 min) ÷ 4

7. (4 gal 3 qt 1 pt) ÷ 3

8. (5 yd 2 ft 6 in.) ÷ 3

9. (8 yd 2 ft 10 in.) ÷ 5

10. (5 gal 3 qt 1 pt) ÷ 6

11. (5 lb 8 oz) ÷ 7

12. (4 lb 10 oz) ÷ 8

13. (13 wk 5 da 15 hr) ÷ 3   14. (12 wk 4 da 10 hr) ÷ 5

15. (8 mi 4,500 ft) ÷ 6   16. (4 mi 4,000 ft) ÷ 12

17. How many 2-foot fence posts can be cut from a 16-foot board?

18. How many $1\frac{3}{4}$-foot stakes can be cut from a 14-foot board?

19. How many 15¢ postcards can $1.75 buy?

20. How many special-addressed envelopes costing 18 cents each can be bought for $17.50?

21. How many 2-minute radio commercials can be fitted into 1 hr and 30 min available just for commercials?

22. It takes 32 min to machine a special fitting. How many of these fittings can be made in an 8-hour work day?

B.10   Changing English System Units to Metric Units (and Vice Versa)

Have you ever had the experience of trying to use an American-made wrench on a bolt on a foreign-made car? The American-made wrench does not fit the metric-made bolt. Because over 90 percent of the people in the world today use the metric system, we must learn how to change the English system units that we use into metric units, and vice versa.

Here is a short table listing commonly used conversions between the metric and the English systems of measurement.

Common English-metric conversions

(These conversion factors have three-significant-digit accuracy.)

1 inch (in.) = 2.54 centimeters (cm)
39.4 inches (in.) ≈ 1 meter (m)
0.621 mile (mi) ≈ 1 kilometer (km)
1 mile (mi) ≈ 1.61 kilometers (km)
1 pound (lb) ≈ 454 grams (g)
2.20 pounds (lb) ≈ 1 kilogram (kg)
1.06 quarts (qt) ≈ 1 liter (L)
2.47 acres ≈ 1 hectare (ha)

Converting English units to metric units (and vice versa)

1. Select from the table the conversion factor that relates the units given in the problem.
2. If you cannot determine whether to multiply or divide by the conversion factor, use unit fractions.
3. Round off your answer to the allowable accuracy determined by the accuracy of the given numbers and the conversion factors used.

It is possible to get slightly different answers if conversion factors of different accuracy are used.

EXAMPLE 1

Mrs. Peralta weighs 62 kg. What is her weight in pounds?

SOLUTION
$$1 \text{ kg} = 2.20 \text{ lb}$$

$$62 \text{ kg} = \frac{62 \text{ kg}}{1}\left(\frac{2.20 \text{ lb}}{1 \text{ kg}}\right) = 62 \times 2.20 \text{ lb} \approx 140 \text{ lb}$$

62 kg = ? lb   1 lb   1 kg = 2.2 lb

Therefore, 62 kg ≅ 140 lb (rounded to the same accuracy as 62 kg).

EXAMPLE 2

A road sign in Mexico reads "85 kilometers to Ensenada." How far is this in miles?

SOLUTION
$$1 \text{ km} = 0.621 \text{ mi}$$

$$85 \text{ km} = \frac{85 \text{ km}}{1}\left(\frac{0.621 \text{ mi}}{1 \text{ km}}\right) = 85 \times 0.621 \text{ mi} \approx 53 \text{ mi}$$

Therefore, 85 km ≈ 53 mi (rounded to the same accuracy as 85 km).

EXAMPLE 3

Change 104 grams to ounces.

SOLUTION $104 \text{ g} = \dfrac{104 \text{ g}}{1}\left(\dfrac{1 \text{ lb}}{454 \text{ g}}\right)\left(\dfrac{16 \text{ oz}}{1 \text{ lb}}\right) = \dfrac{104 \times 16}{454} \text{ oz} \approx 3.67 \text{ oz}$

This is 1 because ——↑   ↑—— This is 1 because
1 lb = 454 g   1 lb = 16 oz

Therefore, 104 g ≈ 3.67 oz (rounded to the same accuracy as 104 g).

EXAMPLE 4

Change 228 centimeters to feet.

SOLUTION $228 \text{ cm} = \dfrac{228 \text{ cm}}{1}\left(\dfrac{1 \text{ in.}}{2.54 \text{ cm}}\right)\left(\dfrac{1 \text{ ft}}{12 \text{ in.}}\right) = \dfrac{228}{2.54 \times 12} \text{ ft} \approx 7.48 \text{ ft}$

This is 1 because ——↑   ↑—— This is 1 because
1 in. = 2.54 cm   1 ft = 12 in.

Therefore, 228 cm ≈ 7.48 ft (rounded to the same accuracy as 228 cm).

The width of a standard paper clip is about 1 cm; 1 in. is about $2\frac{1}{2}$ cm.

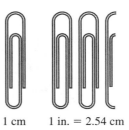

1 cm   1 in. = 2.54 cm

EXAMPLE 5   Change 750 feet per second to kilometers per minute.

SOLUTION $\dfrac{750 \text{ ft}}{\text{sec}} \left(\dfrac{1 \text{ mi}}{5{,}280 \text{ ft}} \right) \left(\dfrac{1.61 \text{ km}}{1 \text{ mi}} \right) \left(\dfrac{60 \text{ sec}}{1 \text{ min}} \right) \approx 14 \text{ km per min}$

This is 1 because —— 1 mi = 5,280 ft

This is 1 because —— 1 mi = 1.61 km

This is 1 because —— 1 min = 60 sec

Therefore, 750 ft per sec $\approx$ 14 km per min (rounded to the same accuracy as 750 fps).

Exercises   B.10

Round off your answers to the allowable accuracy determined by the accuracy of the given numbers and the conversion factors used.

1. 15 in. = _?_ cm
2. 18 in. = _?_ cm
3. 2.12 qt = _?_ L
4. 3.18 qt = _?_ L
5. 0.55 kg = _?_ lb
6. 5.1 kg = _?_ lb
7. 82 km = _?_ mi
8. 140 km = _?_ mi
9. 12 L = _?_ qt
10. 17 L = _?_ qt
11. 33 mi = _?_ km
12. 165 mi = _?_ km
13. 20.6 lb = _?_ kg
14. 28.4 lb = _?_ kg
15. 150 ha = _?_ acres
16. 65 acres = _?_ ha
17. 66 in. = _?_ m
18. 74 in. = _?_ m
19. 2 m = _?_ in.
20. 5 m = _?_ in.
21. 908 g = _?_ lb
22. 1,362 g = _?_ lb
23. 0.75 lb = _?_ g
24. 1.25 lb = _?_ g
25. 1.5 yd = _?_ cm
26. 100 cm = _?_ yd
27. 227 g = _?_ oz
28. 12 oz = _?_ g

29. A road sign in France reads "120 kilometers to Paris." Find this distance in miles.

30. A speed control sign reads "30 kilometers per hour." What is this in miles per hour?

31. Mr. Dubois steps on a scale in Orly Airport. The scale reads 85.7 kilograms. What is his weight in pounds?

32. A crate of transistor radios arrives from Japan marked "67.5 kilograms net weight." Find this weight in pounds.

33. In the Olympics there is a 1,500-meter race, which is about the same distance as our 1-mile race. Which race is longer, and by how much? (Express the difference in feet.)

34. In the Olympics there is a 400-meter race and a quarter-mile race. Which race is longer, and by how much? (Express the difference in feet.)

35. A Howitzer muzzle measures 175 mm. What is this measurement in inches?

36. What is the width in inches of a 35-mm roll of film?

37. How many gallons does a 20-liter container hold?

38. A moon lander is descending at the rate of 1,200 miles per hour. Express this rate of descent in meters per second.

B.II   Converting Units Using a Calculator

Conversion of English system units to metric units (and vice versa) can be easily done using a calculator.

Converting units using a calculator

(These conversion factors have six-significant-digit accuracy.)

English to metric   *Metric to English*

Length

Inches $\boxed{\times}$ 2.54 $\boxed{=}$ Centimeters   Centimeters $\boxed{\times}$ .393701 $\boxed{=}$ Inches

Feet $\boxed{\times}$ 30.48 $\boxed{=}$ Centimeters   Centimeters $\boxed{\times}$ .0328084 $\boxed{=}$ Feet

Yards $\boxed{\times}$ .9144 $\boxed{=}$ Meters   Meters $\boxed{\times}$ 1.09361 $\boxed{=}$ Yards

Miles $\boxed{\times}$ 1.60934 $\boxed{=}$ Kilometers   Kilometers $\boxed{\times}$ .621371 $\boxed{=}$ Miles

Volume (U.S.)   *Volume*

Pints $\boxed{\times}$ .473176 $\boxed{=}$ Liters   Liters $\boxed{\times}$ 2.11338 $\boxed{=}$ Pints

Quarts $\boxed{\times}$ .946353 $\boxed{=}$ Liters   Liters $\boxed{\times}$ 1.05669 $\boxed{=}$ Quarts

Gallons $\boxed{\times}$ 3.78541 $\boxed{=}$ Liters   Liters $\boxed{\times}$ .264172 $\boxed{=}$ Gallons

Weight   *Weight*

Ounces $\boxed{\times}$ 28.3495 $\boxed{=}$ Grams   Grams $\boxed{\times}$ .0352740 $\boxed{=}$ Ounces

Pounds $\boxed{\times}$ 453.592 $\boxed{=}$ Grams   Grams $\boxed{\times}$ .00220462 $\boxed{=}$ Pounds

Temperature   *Temperature*

Changing Fahrenheit to Celsius   Changing Celsius to Fahrenheit

°F $\boxed{-}$ 32 $\boxed{=}$ $\boxed{\times}$ 5 $\boxed{\div}$ 9 $\boxed{=}$ °C   °C $\boxed{\times}$ 9 $\boxed{\div}$ 5 $\boxed{+}$ 32 $\boxed{=}$ °F

EXAMPLE 1   Convert units using a calculator.

a. Change 976 yards to meters.

     S O L U T I O N   976 $\boxed{\times}$ .9144 $\boxed{=}$ 892.4544 ≈ 892 m

b. Change 15.9 gallons to liters.

     S O L U T I O N   15.9 $\boxed{\times}$ 3.78541 $\boxed{=}$ 60.188019 ≈ 60.2 L

c. Change 8.423 ounces to grams.

     S O L U T I O N   8.423 $\boxed{\times}$ 28.3495 $\boxed{=}$ 238.7878385 ≈ 238.8 g

d. Change 52.8 centimeters to inches.

     S O L U T I O N   52.8 $\boxed{\times}$ .393701 $\boxed{=}$ 20.7874128 ≈ 20.8 in.

e. Change 15.76 liters to quarts.

     S O L U T I O N   15.76 $\boxed{\times}$ 1.05669 $\boxed{=}$ 16.6534344 ≈ 16.65 qt

f. Change 19.5°C into °F.

     S O L U T I O N   19.5 $\boxed{\times}$ 9 $\boxed{\div}$ 5 $\boxed{+}$ 32 $\boxed{=}$ 67.1°F

Exercises   *B.II*

Use calculator conversions, and round off the answers to allowable accuracy.

1. 2.65 ft = ? cm   **2.** 34.5 in. = ? cm

3. 419 qt = ? L   **4.** 97 pt = ? L

5. 14.75 oz = ? g   **6.** 5.7 lb = ? g

7. 82°F = ? °C   **8.** 23.5°C = ? °F

9. 1,500 m = ? yd   **10.** 528 cm = ? ft

11. 22 L = ? gal   **12.** 183 L = ? qt

13. 546.7 g = ? lb   **14.** 325.8 g = ? oz

15. 98.6°F = ? °C

Answers to Odd-Numbered Exercises (including Solutions to Selected Exercises), Diagnostic Tests, and Cumulative Review Exercises

Exercises 1.1 (page 4)

1. 4 **3.** 0 **5.** 1 **7.** Cannot be found **9.** 99
11. 100 **13.** No **15.** Yes **17.** 1, 2, 3, 4, 5 **19.** <
21. > **23.** <

Exercises 1.2 (page 6)

1. a. 6 units **b.** 7 tens = 70 units
   c. 5 hundreds = 500 units
3. a. 4 tens **b.** 4 tens = 40 units **c.** 3 hundreds
   d. 30 tens
5. a. 5 **b.** 3 **c.** 2
7. a. 1 **b.** 2 **c.** 6
9. Sixteen thousand, three hundred forty-six
11. One million, seventy thousand, two hundred
13. Forty-nine million, seventy
15. Eight billion, seven hundred twenty-five thousand
17. 16,601   Sixteen thousand, six hundred one
19. 71,040   Seventy-one thousand, forty
21. 710,004,000   Seven hundred ten million, four thousand
23. 5,206,000,000   Five billion, two hundred six million
25. 20,005 **27.** 8,008,808 **29.** 7,000,007
31. 10,000,100,010 **33.** 100,029,006
35. 5,880,000,000,000

Exercises 1.3 (page 9)

1. 4,700 **3.** 930 **5.** 800 **7.** 63,000 **9.** 800
11. 20,000,000 **13.** 52,000,000 **15.** 3,500
17. 45,430,000 **19.** 1,600,000

Exercises 1.4 (page 11)

1. 637 **3.** 1,702 **5.** 1,794 **7.** 1,024 **9.** 1,818
11. 907 **13.** 9,081 **15.** 70,150 **17.** 257
19. 1,014,924,913 **21.** 1,049,125,745 **23.** 892,351,182

25. 392,008,000 **27.**
$$\begin{array}{r} 30,006 \\ 75,000,100 \\ 2,000,000,500 \\ +\ \ 50,100,010 \\ \hline 2,125,130,616 \end{array}$$
29. 45

31. 65,226 ft
33.
$$\begin{array}{lr} \text{Luis} & \$75 \\ \text{Jim} & +\ \ 18 \\ \hline & 93 \quad \text{Amount Luis and Jim have together} \\ & +\ \ 12 \\ \hline \text{Marie} & \$105 \\ \text{Then, Luis} & 75 \\ \text{Jim} & +\ \ 18 \\ \hline & \$198 \quad \text{Total amount the three have together} \end{array}$$

Exercises 1.5 (page 14)

1. 8,205 **3.** 3,005 **5.** 94,200 **7.** 328 **9.** 129
11. 212 **13.** 2,616 **15.** 49,893 **17.** 337,603
19. 230,789 **21.** 2,886 **23.** 92,650,000 mi
25. 3,448 mi **27.** $210
29.
$$\begin{array}{ll} \text{Assim} & 75 \\ \text{Joe} & 57 \\ \text{Jamal} & 92 = 75 + 17 \\ \text{Mike} & \underline{119} = 75 + 57 - 13 \\ & \$343 \end{array}$$

Exercises 1.6 (page 19)

1. 2,552 **3.** 6,734 **5.** 51,976 **7.** 69,318
9. 24,966 **11.** 501,504 **13.** 8,576,220 **15.** 97,760
17. 4,521,606 **19.** 22,594,496 **21.** 72,624
23. 2,697,036 **25.** 7,872,218 **27.** 540,822,290
29. 2,250,795,063 **31.** 940,000 **33.** 29,016,800
35. 1,875,500 **37.** 2,963,400 **39.** 74,983,500
41. 50,176,000 **43.** 713,500 sheets **45.** $160
47. $24,500

Exercises 1.7 (page 21)

1. 5^3 **3.** 3^5 **5.** 8 **7.** 25 **9.** 27 **11.** 36
13. 1 **15.** 100 **17.** 10,000 **19.** 1,000,000 **21.** 0
23. 1 **25.** 16 **27.** 1 **29.** 248,832 **31.** 531,441
33. 14,706,125 **35.** 1

Exercises 1.8   (page 22)

1. 300   **3.** 300   **5.** 160   **7.** 8,000   **9.** 100,000
11. 4   **13.** 2,000   **15.** 80,000   **17.** 9,000
19. 10,300,000

Exercises 1.9   (page 28)

1. 154 R 3   **3.** 1,050 R 4   **5.** 7,306   **7.** 27 R 10
9. 20 R 25   **11.** 53 R 66   **13.** 286   **15.** 207
17. 370 R 5   **19.** 270   **21.** 602 R 2
23. 1,070 R 44   **25.** 1,407 R 8   **27.** 2,084 R 27
29. 351 R 123   **31.** 1,400   **33.** 790 R 240
35. 9,003   **37.** 62 hours   **39. a.** $1,560   **b.** $360
41. 6,025 boxes   **43.** 12 + 4 = 16 boxes
45. All remainders are zero.

Exercises 1.10   (page 29)

1. 5   **3.** 8   **5.** 9   **7.** 4   **9.** 1   **11.** 12   **13.** 10
15. 100   **17.** 23   **19.** 75

Exercises 1.11   (page 32)

1. 25   **3.** 8   **5.** 26   **7.** 17   **9.** 9
11.
$$\begin{aligned}
5^2 - 2^3 + \sqrt{9} &\\
= 25 - 8 + \ \ 3&\\
= \ \ \ \ 17 \ \ + \ \ 3&\\
= 20&
\end{aligned}$$
13. 23   **15.** 42   **17.** 20   **19.** 12   **21.** 5
23.
$$\begin{aligned}
20 - (2 + 4 \cdot 3)&\\
= 20 - (2 + \ 12 \ )&\\
= 20 - \ \ \ \ \ 14&\\
= 6&
\end{aligned}$$
25. 19   **27.** 98
29.
$$\begin{aligned}
5 \cdot 8 + 2\sqrt{100}&\\
= 5 \cdot 8 + 2 \cdot 10&\\
= \ \ 40 \ + \ \ 20&\\
= 60&
\end{aligned}$$
31.
$$\begin{aligned}
24 - 8 \div 4 \cdot 2 + 2^3&\\
= 24 - 8 \div 4 \cdot 2 + 8&\\
= 24 - \ \ \ \ 2 \ \ \cdot 2 + 8&\\
= 24 - \ \ \ \ \ \ 4 \ \ \ \ \ + 8&\\
= \ \ \ \ \ \ \ \ 20 \ \ \ \ \ \ \ + 8&\\
= 28&
\end{aligned}$$
33. 31   **35.** 52   **37.** 205   **39.** 606   **41.** 3,500
43. 684   **45.** 5,909   **47.** 96

Exercises 1.12   (page 34)

1. 6   **3.** 7   **5.** 26   **7.** 79   **9.** 79   **11.** 77 in.
13. 75   **15.** 87   **17.** 9 in.

Exercises 1.13A   (page 35)

1. 40 ft   **3.** 42 in.   **5.** 18 ft
7.

$P = 8 + 8 + 20 + 5 + 12 + 3 = 56$ m
9. 60 in.   **11.** $72

Exercises 1.13B   (page 37)

1. 120 sq. in.   **3.** 36 sq. in.   **5.** 64 sq. ft
7. a. 30 sq. yd   **b.** $450   **9.** $42

Exercises 1.14   (page 39)

1. 19,000   **3.** 70,000   **5.** 2,100,000
7. 200   **9.** 50,000
11. a. 2(400) + 200 + 300 = 1,300 calories   **b.** 1,292 calories
13. a. 9,000 ft   **b.** 8,938 ft

Chapter 1 Review Exercises (page 41)

1. Three billion, seventy-five million, six hundred thousand, eight
2. 5,072,006   **3.** 0, 1, 2, 3, 4   **4.** 10
5. a. 17 > 12   **b.** 0 < 19   **6.** subtraction and division
7. factors or divisors   **8.** subtraction   **9.** trial divisor
10. inverse   **11.**
$$\begin{aligned}
8 &\quad \text{Minuend}\\
-\ 2 &\quad \text{Subtrahend}\\
\overline{6} &\quad \text{Difference}
\end{aligned}$$
12.
$$\text{Divisor} \rightarrow 17\overline{)243} \begin{matrix} 14 \leftarrow \text{Quotient} \\ \ \ \leftarrow \text{Dividend} \end{matrix}$$
$$\begin{aligned}
\underline{17}&\\
73&\\
\underline{68}&\\
5 \leftarrow& \text{Remainder}
\end{aligned}$$
13. a. 280,000   **b.** 9,700   **c.** 43,000,000
14. a. Undefined   **b.** 0   **c.** 4   **d.** Undefined
15. 1,941   **16.** 1,210   **17.** 86,314   **18.** 7,889
19. 5,117   **20.** 679,158   **21.** 27,510   **22.** 6,481,728
23. 3,439,389   **24.** 4,086   **25.** 57 R 14   **26.** 480
27. 25   **28.** 16   **29.** 1   **30.** 6   **31.** 9   **32.** 7,500
33. 80,000   **34.** 30,000
35.
$$\begin{aligned}
10 + 20 \div 2 \cdot 5&\\
= 10 + \ \ \ 10 \ \ \cdot 5&\\
= 10 + \ \ \ \ \ \ 50&\\
= 60&
\end{aligned}$$
36.
$$\begin{aligned}
4\sqrt{100} - 6 + 3 \cdot 2^3&\\
= \ 4 \cdot 10 - 6 + 3 \cdot 8&\\
= \ \ \ \ \ 40 - 6 + \ 24&\\
= \ \ \ \ \ \ \ \ 34 \ \ \ + \ 24&\\
= 58&
\end{aligned}$$
37. 11
38.
$$\begin{aligned}
10 - (8 - 3 \cdot 2) - 6 \div 2&\\
= 10 - (8 - \ \ 6 \ ) - 6 \div 2&\\
= 10 - \ \ \ \ \ \ 2 \ \ \ \ \ - 6 \div 2&\\
= 10 - \ \ \ \ \ \ 2 \ \ \ \ \ - \ \ \ 3&\\
= \ \ \ \ \ \ 8 \ \ \ \ \ \ \ \ \ \ - \ \ \ 3&\\
= 5&
\end{aligned}$$
39. $758   **40.** $1,792   **41.** 2,054,691   **42.** 5 bottles
43. $127   **44.** $48   **45.** 280,000 words
46.
$$\begin{matrix}
\text{Irma} & \$45 &\\
\text{Anita} & 23 &\\
\text{Felisa} & \underline{80} &= 45 + 23 + 12\\
& \$148 &
\end{matrix}$$

47.
$$\begin{array}{r} \$\ 82 \\ \times\ 23 \\ \hline 246 \\ 164 \\ \hline \$1886 \ \ \text{Paid} \end{array}$$
$$\begin{array}{r} \$1,950 \\ -\ 1,886 \\ \hline \$64 \ \ \text{Final payment} \end{array}$$

48. 1,100 people   **49.** $140 a month   **50.** 871,200 times

51.
$$\begin{array}{r} \$\ 67 \\ \times\ 35 \\ \hline 335 \\ 201 \\ \hline \$2345 \ \ \text{Paid} \end{array}$$
$$\begin{array}{r} \$\ 2365 \\ -\ 2345 \\ \hline \$20 \ \ \text{Left} \end{array}$$
52. 352 miles

53. a.
$$\begin{array}{r} \$19 \ \text{a month} \\ 24\overline{)456} \\ \underline{24} \\ 216 \\ \underline{216} \end{array}$$
b.
$$\begin{array}{r} \$19 \\ \times\ 17 \\ \hline 133 \\ 19 \\ \hline \$323 \ \ \text{Paid} \end{array}$$
$$\begin{array}{r} \$456 \\ -\ 323 \\ \hline \$133 \ \ \text{Still owed} \end{array}$$

54.
$$\begin{array}{r} 53,731 \\ -\ 53,408 \\ \hline 323 \ \text{mi} \end{array}$$
$$\begin{array}{r} 17 \ \text{mi per gal} \\ 19\overline{)323} \\ \underline{19} \\ 133 \\ \underline{133} \\ 0 \end{array}$$
55. 22 hr   **56.** 75

57. a. 40 in.   **b.** 96 sq. in.   **58. a.** 36 ft   **b.** 81 sq. ft

59. a. 1,800,000   **b.** 200   **60.** $390

Chapter 1 Diagnostic Test   (page 43)

Following each problem number is the number (in parentheses) of the textbook section in which that kind of problem is discussed.

1. (1.2) Five trillion, eight hundred seventy-nine billion, two hundred million

2. (1.2) 54,007,506,080

3. (1.4)
$$\begin{array}{r} 5,843 \\ 209 \\ +\ 6,027 \\ \hline 12,079 \end{array}$$
4. (1.4)
$$\begin{array}{r} 946 \\ 7,328 \\ 407 \\ +\ 24 \\ \hline 8,705 \end{array}$$

5. (1.5)
$$\begin{array}{r} \overset{2\ \ 1416}{3,564} \\ -\ 782 \\ \hline 2,782 \end{array}$$
6. (1.5)
$$\begin{array}{r} \overset{4\ 10\ 3\ \ \ 9\ \ 16}{50,406} \\ -\ 35,008 \\ \hline 15,398 \end{array}$$

7. (1.6)
$$\begin{array}{r} 576 \\ \times\ 89 \\ \hline 5184 \\ 4608 \\ \hline 51,264 \end{array}$$
8. (1.6)
$$\begin{array}{r} 3,084 \\ \times\ 706 \\ \hline 18504 \\ 21588 \\ \hline 2,177,304 \end{array}$$

9. (1.9)
$$\begin{array}{r} 80 \ \ \text{R}\ 15 \\ 63\overline{)5055} \\ \underline{504} \\ 15 \\ \underline{0} \\ 15 \end{array}$$
10. (1.9)
$$\begin{array}{r} 706 \\ 495\overline{)349,470} \\ \underline{3465} \\ 297 \\ \underline{0} \\ 2970 \\ \underline{2970} \end{array}$$

11. (1.7) **a.** $2^3 = 2 \cdot 2 \cdot 2 = 8$ **b.** $8^2 = 8 \cdot 8 = 64$

12. (1.10) **a.** $\sqrt{16} = 4$ **b.** $\sqrt{100} = 10$

13. (1.8) **a.** $40 \cdot 100 = 4,000$ **b.** $16 \cdot 10^4 = 160,000$

14. (1.3) **a.** 79,000   **b.** 3,700   **15.** (1.5)
$$\begin{array}{r} 71,304 \\ -\ 67,856 \\ \hline 3,448 \ \text{mi} \end{array}$$

16. (1.9)
$$\begin{array}{r} \$56 \\ 36\overline{)2016} \\ \underline{180} \\ 216 \\ \underline{216} \end{array}$$
17. (1.6)
$$\begin{array}{r} \$168 \\ \times\ 52 \\ \hline 336 \\ 840 \\ \hline \$8,736 \end{array}$$

18. (1.6)
$$\begin{array}{r} \$58 \\ \times\ 23 \\ \hline 174 \\ 116 \\ \hline \$1,334 \end{array}$$
$$\begin{array}{r} \$1,350 \\ -\ 1,334 \\ \hline \$16 \ \ \text{Final payment} \end{array}$$

19. (1.13)
$$\begin{array}{l} P = 2l + 2w \\ P = 2 \cdot 18 + 2 \cdot 12 \\ P =\ \ 36\ \ +\ \ 24 \\ P = 60 \ \text{in.} \end{array}$$
20. (1.13)
$$\begin{array}{l} A = s^2 \\ A = 10^2 \\ A = 100 \ \text{sq. cm} \end{array}$$

21. (1.14) $80,000 \cdot 200 = 16,000,000$

22. (1.12)
$$\begin{array}{r} 76 \\ 84 \\ 92 \\ 63 \\ +\ 70 \\ \hline 385 \end{array}$$
$$\begin{array}{r} 77 \ \text{average} \\ 5\overline{)385} \\ \underline{35} \\ 35 \\ \underline{35} \end{array}$$

23. (1.11)
$$\begin{array}{l} 6 + 18 \div 3 \cdot 2 \\ = 6 +\ \ 6\ \ \cdot 2 \\ = 6 +\ \ \ \ 12 \\ = 18 \end{array}$$

24. (1.11)
$$\begin{array}{l} 30 - 4^2 + 4\sqrt{9} \\ = 30 - 16 + 4 \cdot 3 \\ = 30 - 16 +\ \ 12 \\ =\ \ \ \ 14\ \ +\ \ 12 \\ = 26 \end{array}$$

25. (1.11)
$$\begin{array}{l} 5 + 2(10 - 2 \cdot 3) \\ = 5 + 2(10 -\ \ 6\ ) \\ = 5 +\ \ 2 \cdot 4 \\ = 5 +\ \ \ \ 8 \\ = 13 \end{array}$$

Exercises 2.1   (page 47)

1. $\dfrac{3}{8}$ **3.** $\dfrac{1}{6}$ **5.** $\dfrac{3}{7}$ **7.** $\dfrac{5}{11}, \dfrac{17}{22}, \dfrac{1}{2}, \dfrac{98}{107}, \dfrac{1}{31}$ **9.** 5

11. $\dfrac{8}{8}$ and $\dfrac{4}{4}$ **13.** 5   **15.** 4   **17.** 1   **19.** 9

21. 20

Exercises 2.2   (page 49)

1. $\dfrac{10}{21}$ **3.** $\dfrac{21}{32}$ **5.** $\dfrac{12}{25}$ **7.** $\dfrac{8}{27}$ **9.** $\dfrac{9}{16}$ **11.** $\dfrac{25}{96}$

13. $\dfrac{33}{64}$ **15.** $\dfrac{77}{104}$ **17.** $\dfrac{3}{40}$ **19.** $\dfrac{6}{175}$

21. $2 \cdot \dfrac{1}{3} = \dfrac{2}{1} \cdot \dfrac{1}{3} = \dfrac{2 \cdot 1}{1 \cdot 3} = \dfrac{2}{3}$ **23.** $\dfrac{4}{5}$ **25.** $\dfrac{15}{2}$

27. $\dfrac{49}{12}$ **29.** $\dfrac{25}{48}$

Exercises 2.3   (page 51)

1. $1\frac{2}{3}$ **3.** $1\frac{4}{5}$ **5.** $2\frac{3}{5}$ **7.** $3\frac{3}{4}$ **9.** $3\frac{5}{6}$

11. $1\frac{3}{13}$ **13.** $2\frac{6}{7}$ **15.** $\begin{array}{r} 10 \text{ R }17 \\ 19\overline{)207} \\ 19 \\ \hline 17 \end{array} = 10\frac{17}{19}$ **17.** $3\frac{3}{17}$

19. $5\frac{3}{4}$ **21.** $14\frac{2}{5}$

23. *Number of*

students		*Score*		*Points*
3	×	5	=	15
5	×	4	=	20
2	×	2	=	4
10 students				39 points

Average $= \frac{39}{10} = 3\frac{9}{10}$

Exercises 2.4   (page 52)

1. $\frac{3}{2}$ **3.** $\frac{13}{4}$ **5.** $\frac{29}{6}$ **7.** $\frac{37}{10}$ **9.** $\frac{67}{12}$ **11.** $\frac{38}{3}$

13. $\frac{27}{4}$ **15.** $\frac{52}{15}$ **17.** $\frac{63}{4}$ **19.** $\frac{27}{10}$

Exercises 2.5   (page 54)

1. $\frac{5}{10}$ **3.** $\frac{1}{2}$ **5.** $\frac{4}{6}$ **7.** $\frac{3}{5}$ **9.** $\frac{9}{15}$ **11.** $\frac{7}{10}$

13. $\frac{36}{52}$ **15.** $\frac{6}{9}$ **17.** $\frac{40}{55}$ **19.** $\frac{5}{2}$

Exercises 2.6   (page 56)

1. P; 1, 5   **3.** C; 1, 2, 4, 8   **5.** C; 1, 3, 7, 21   **7.** P; 1, 13
9. C; 1, 5, 25   **11.** P; 1, 23   **13.** P; 1, 41
15. C; 1, 2, 4, 5, 10, 20   **17.** $2 \cdot 7$ **19.** $2 \cdot 5 \cdot 7$ **21.** 2^4
23. $2 \cdot 3^2$ **25.** $2^2 \cdot 7$ **27.** 3^4 **29.** $2 \cdot 5^3$ **31.** $2^4 \cdot 3^2$

Exercises 2.7A   (page 59)

1. $\frac{2}{3}$ **3.** $\frac{3}{4}$ **5.** $\frac{3}{4}$ **7.** $\frac{3}{4}$ **9.** $\frac{5}{9}$ **11.** $\frac{4}{5}$

13. $\frac{3}{5}$ **15.** $\frac{4}{5}$ **17.** $\frac{3}{5}$ **19.** $\frac{3}{5}$

21. $\frac{39}{51} = \frac{\overset{1}{\cancel{3}} \cdot 13}{\underset{1}{\cancel{3}} \cdot 17} = \frac{13}{17}$ **23.** $\frac{4}{9}$ **25.** $\frac{5}{7}$

Exercises 2.7B   (page 60)

1. $\frac{4}{7}$ **3.** $\frac{1}{3}$ **5.** 1   **7.** $\frac{5}{32}$ **9.** $\frac{2}{15}$ **11.** $\frac{1}{5}$

13. $\frac{3}{7}$ **15.** $\frac{1}{5}$ **17.** 9   **19.** 240   **21.** 31 lb

23. $\frac{3}{8} \cdot 24 = \frac{3}{\underset{1}{\cancel{8}}} \cdot \frac{\overset{3}{\cancel{24}}}{1} = 9$ gal left
$24 - 9 = 15$ gal used

Exercises 2.8   (page 62)

1. $\frac{2}{3}$ **3.** 1   **5.** $\frac{1}{5}$ **7.** $\frac{1}{3}$ **9.** $1\frac{1}{2}$ **11.** $\frac{5}{8}$

13. $\frac{3}{5}$ **15.** $\frac{17}{27}$ **17.** $1\frac{1}{2}$ **19.** $\frac{2}{3}$

Exercises 2.9   (page 64)

1. 12   **3.** 8   **5.** 15   **7.** 70   **9.** 35   **11.** 90
13. 72   **15.** 600   **17.** 420   **19.** 78   **21.** 132

Exercises 2.10   (page 67)

1. $1\frac{1}{4}$ **3.** $\frac{9}{10}$ **5.** $\frac{1}{2}$ **7.** $\frac{1}{4}$ **9.** $\frac{11}{12}$ **11.** $\frac{5}{6}$

13. $\frac{11}{21}$ **15.** $\frac{1}{5}$ **17.** $\frac{1}{32}$

19. $\begin{aligned} \frac{56}{64} &= \frac{7}{8} = \frac{21}{24} \\ -\frac{14}{24} &= -\frac{7}{12} = -\frac{14}{24} \\ \hline &\quad\;\; \frac{7}{24} \end{aligned}$ $\begin{aligned} 8 &= 2^3 \\ 12 &= 2^2 \cdot 3 \\ \text{LCD} &= 2^3 \cdot 3 \\ &= 24 \end{aligned}$

21. $\frac{11}{40}$ **23.** $1\frac{11}{18}$ **25.** $\frac{17}{60}$ **27.** $\frac{13}{70}$ **29.** 2

31. $1\frac{2}{5}$ **33.** $1\frac{11}{24}$ **35.** $\begin{aligned} \frac{4}{6} &= \frac{2}{3} = \frac{14}{21} \\ \frac{6}{14} &= \frac{3}{7} = \frac{9}{21} \\ +\frac{2}{3} &= \frac{2}{3} = \frac{14}{21} \\ \hline &\quad \frac{37}{21} = 1\frac{16}{21} \end{aligned}$

37. $1\frac{7}{24}$ **39.** $1\frac{11}{28}$

Exercises 2.11   (page 69)

1. $3\frac{7}{10}$ **3.** $5\frac{5}{6}$ **5.** $7\frac{5}{6}$ **7.** $4\frac{1}{8}$ **9.** $11\frac{1}{2}$

11. $5\frac{5}{8}$ **13.** $7\frac{1}{12}$ **15.** $10\frac{5}{8}$ **17.** $37\frac{7}{12}$ **19.** $49\frac{9}{20}$

21. $247\frac{11}{12}$

23. 61   **25.** $120\frac{7}{16}$

27. $\begin{aligned} 117\frac{5}{6} &= 117 + \frac{15}{18} \\ 28\frac{1}{2} &= 28 + \frac{9}{18} \\ +232\frac{7}{9} &= 232 + \frac{14}{18} \\ \hline 377 + \frac{38}{18} &= 377 + 2\frac{2}{18} = 379\frac{2}{18} = 379\frac{1}{9} \end{aligned}$

29. $7\frac{3}{4} = 7 + \frac{60}{80}$ $4 = 2^2$

$5\frac{7}{8} = 5 + \frac{70}{80}$ $8 = 2^3$

$8\frac{5}{16} = 8 + \frac{25}{80}$ $16 = 2^4$

     $5 = 5$

$+ 10\frac{2}{5} = 10 + \frac{32}{80}$ $\text{LCD} = 2^4 \cdot 5 = 80$

$\overline{30 + \frac{187}{80}} = 30 + 2\frac{27}{80} = 32\frac{27}{80}$ oz

31. $10\frac{1}{5} = 10 + \frac{4}{20}$

$5\frac{3}{10} = 5 + \frac{6}{20}$

$12\frac{3}{4} = 12 + \frac{15}{20}$

$+ \ 6\frac{1}{2} = 6 + \frac{10}{20}$

$\overline{33 + \frac{35}{20}} = 33 + 1\frac{15}{20} = 34\frac{3}{4}$ m

Exercises 2.12   (page 71)

I. $4\frac{1}{2}$ **3.** $12\frac{5}{8}$ **5.** $\begin{array}{r} 8 = 7\frac{2}{2} \\ - 4\frac{1}{2} = - 4\frac{1}{2} \\ \hline 3\frac{1}{2} \end{array}$ **7.** $3\frac{2}{5}$

9. $\begin{array}{r} 4\frac{1}{4} = 3\frac{5}{4} \\ - 1\frac{3}{4} = - 1\frac{3}{4} \\ \hline 2\frac{2}{4} = 2\frac{1}{2} \end{array}$ **II.** $5\frac{1}{10}$

13. $\begin{array}{r} 3\frac{1}{12} = 2\frac{13}{12} \\ - 1\frac{1}{6} = - 1\frac{2}{12} \\ \hline 1\frac{11}{12} \end{array}$ **15.** $6\frac{1}{3}$

17. $\begin{array}{r} 68\frac{5}{16} = 67\frac{21}{16} \\ - 53\frac{3}{4} = - 53\frac{12}{16} \\ \hline 14\frac{9}{16} \end{array}$ **19.** $\begin{array}{r} 234\frac{5}{14} = 233\frac{19}{14} \\ - 157\frac{3}{7} = - 157\frac{6}{14} \\ \hline 76\frac{13}{14} \end{array}$

21. $\frac{3}{8}$ lb   **23.** $1\frac{1}{4}$ sq. yd

Exercises 2.13   (page 72)

I. $4\frac{1}{6}$ **3.** 6 **5.** 30 **7.** $10\frac{1}{2}$ **9.** $1\frac{19}{20}$ **II.** $2\frac{2}{3}$

13. $3\frac{3}{10} \cdot \frac{6}{11} \cdot 1\frac{2}{3} = \frac{\overset{3}{\cancel{33}}}{\underset{\underset{1}{\cancel{2}}}{\cancel{10}}} \cdot \frac{6}{\underset{1}{\cancel{11}}} \cdot \frac{\overset{1}{\cancel{5}}}{\underset{1}{\cancel{3}}} = 3$

15. $3\frac{3}{4} \cdot 4\frac{2}{5} \cdot 3\frac{1}{2} = \frac{\overset{3}{\cancel{15}}}{\underset{2}{\cancel{4}}} \cdot \frac{\overset{11}{\cancel{22}}}{\underset{1}{\cancel{5}}} \cdot \frac{7}{2} = \frac{231}{4} = 57\frac{3}{4}$ **17.** $19\frac{1}{2}$ lb

19. $8\frac{2}{5}$ tons   **21.** $10\frac{4}{5}$ m^2

Exercises 2.14   (page 74)

I. $1\frac{1}{2}$ **3.** $\frac{3}{4}$ **5.** $\frac{1}{10}$ **7.** 4 **9.** 2 **II.** 2

13. $\frac{5}{12}$ **15.** 112 **17.** $\frac{35}{16} \div \frac{42}{22} = \frac{35}{16} \cdot \frac{\overset{11}{\cancel{22}}}{\underset{\underset{3}{\cancel{21}}}{\cancel{42}}} = \frac{55}{48} = 1\frac{7}{48}$

19. 45 **21.** $\frac{9}{28}$

23. $\frac{14}{24} \div 210 = \frac{\overset{\frac{1}{2}}{\cancel{14}}}{\underset{12}{\cancel{24}}} \cdot \frac{1}{\underset{30}{\cancel{210}}} = \frac{1}{360}$ **25.** 12

27. $1\frac{7}{9} \div 2\frac{2}{3} = \frac{16}{9} \div \frac{8}{3} = \frac{\overset{2}{\cancel{16}}}{\underset{3}{\cancel{9}}} \cdot \frac{\overset{1}{\cancel{3}}}{\underset{1}{\cancel{8}}} = \frac{2}{3}$ **29.** $2\frac{1}{2}$ **31.** $\frac{1}{5}$

33. $3\frac{1}{2}$ **35.** $1\frac{7}{8}$

37. $4\frac{1}{2} \div 3 = \frac{9}{2} \div \frac{3}{1} = \frac{\overset{3}{\cancel{9}}}{2} \cdot \frac{1}{\underset{1}{\cancel{3}}} = \frac{3}{2} = 1\frac{1}{2}$ tablets

39. $36 \div 1\frac{1}{2} = \frac{36}{1} \div \frac{3}{2} = \frac{\overset{12}{\cancel{36}}}{1} \cdot \frac{2}{\underset{1}{\cancel{3}}} = 24$ books

Exercises 2.15   (page 76)

I. $4\frac{1}{2}$ **3.** $1\frac{1}{3}$ **5.** 2 **7.** $\frac{9}{10}$ **9.** 9 **II.** $1\frac{3}{5}$

13. $\frac{1}{6}$ **15.** $\frac{1}{15}$ **17.** $\frac{2}{5}$ **19.** $3\frac{9}{10}$

21. $\dfrac{\frac{1}{8} + \frac{3}{4}}{\frac{1}{2} - \frac{1}{3}} = \dfrac{\frac{1}{8} + \frac{6}{8}}{\frac{3}{6} - \frac{2}{6}} = \dfrac{\frac{7}{8}}{\frac{1}{6}} = \frac{7}{8} \div \frac{1}{6} = \frac{7}{\underset{4}{\cancel{8}}} \cdot \frac{\overset{3}{\cancel{6}}}{1} = \frac{21}{4} = 5\frac{1}{4}$

23. $\dfrac{\frac{1}{7} + \frac{9}{28}}{\frac{13}{14} - \frac{3}{7}} = \dfrac{\frac{4}{28} + \frac{9}{28}}{\frac{13}{14} - \frac{6}{14}} = \dfrac{\frac{13}{28}}{\frac{7}{14}} = \frac{13}{28} \div \frac{7}{14} = \frac{13}{\underset{2}{\cancel{28}}} \cdot \frac{\overset{1}{\cancel{14}}}{7} = \frac{13}{14}$

25. $\dfrac{2 + 1\frac{3}{4}}{4 - 2\frac{1}{3}} = \dfrac{3\frac{3}{4}}{3\frac{3}{3} - 2\frac{1}{3}} = \dfrac{3\frac{3}{4}}{1\frac{2}{3}} = \dfrac{\frac{15}{4}}{\frac{5}{3}} = \frac{15}{4} \div \frac{5}{3} = \frac{\overset{3}{\cancel{15}}}{4} \cdot \frac{3}{\underset{1}{\cancel{5}}} = \frac{9}{4}$

$= 2\frac{1}{4}$

27. $11\frac{1}{4} + 12\frac{1}{2} + 11\frac{5}{8} + 9\frac{5}{8} = 45$ for four days

If $11 \cdot 5 = 55$ for the week, then $55 - 45 = 10$ for the fifth day.

Exercises 2.16   (page 77)

1. $\dfrac{1}{16}$ **3.** $\dfrac{4}{9}$ **5.** $\dfrac{1}{16}$ **7.** $\dfrac{27}{64}$ **9.** $\left(\dfrac{3}{2}\right)^3 = \dfrac{27}{8} = 3\dfrac{3}{8}$

11. 1   **13.** $\dfrac{64}{81}$ **15.** $\dfrac{3}{10}$ **17.** $\dfrac{4}{9}$

19. $\sqrt{\dfrac{4}{81}} + \left(\dfrac{1}{6}\right)^2$ **21.** $\left(1\dfrac{1}{2}\right)^2 - \sqrt{\dfrac{49}{144}}$

$\qquad = \dfrac{2}{9} + \dfrac{1}{36}$ $\qquad = \left(\dfrac{3}{2}\right)^2 - \sqrt{\dfrac{49}{144}}$

$\qquad = \dfrac{8}{36} + \dfrac{1}{36}$ $\qquad = \dfrac{9}{4} - \dfrac{7}{12}$

$\qquad = \dfrac{9}{36}$ $\qquad = \dfrac{27}{12} - \dfrac{7}{12}$

$\qquad = \dfrac{1}{4}$ $\qquad = \dfrac{20}{12}$

$\qquad\qquad\qquad = \dfrac{5}{3} = 1\dfrac{2}{3}$

23. $A = s^2$

$\qquad A = \left(3\dfrac{1}{3}\right)^2$

$\qquad A = \left(\dfrac{10}{3}\right)^2$

$\qquad A = \dfrac{100}{9}$

$\qquad A = 11\dfrac{1}{9}$ sq. ft

Exercises 2.17   (page 80)

1. $3\dfrac{1}{3}$ **3.** $4\dfrac{5}{6}$ **5.** $1\dfrac{4}{5}$ **7.** $1\dfrac{1}{2}$ **9.** $\dfrac{7}{15}$

11. $\left(\dfrac{2}{3}\right)^2 + 1\dfrac{2}{3} \cdot \dfrac{1}{10}$ **13.** $19\dfrac{3}{4}$ **15.** $4\dfrac{1}{2}$ **17.** 17

$\qquad = \dfrac{2}{3} \cdot \dfrac{2}{3} + \dfrac{\overset{1}{\cancel{5}}}{3} \cdot \dfrac{1}{\underset{2}{\cancel{10}}}$

$\qquad = \dfrac{4}{9} + \dfrac{1}{6}$

$\qquad = \dfrac{8}{18} + \dfrac{3}{18}$

$\qquad = \dfrac{11}{18}$

19. $P = 2l + 2w$

$\qquad P = 2 \cdot 5\dfrac{1}{3} + 2 \cdot 3\dfrac{1}{4}$

$\qquad P = \dfrac{2}{1} \cdot \dfrac{16}{3} + \dfrac{\overset{1}{\cancel{2}}}{1} \cdot \dfrac{13}{\underset{2}{\cancel{4}}}$

$\qquad P = \dfrac{32}{3} + \dfrac{13}{2}$

$\qquad P = \dfrac{64}{6} + \dfrac{39}{6}$

$\qquad P = \dfrac{103}{6} = 17\dfrac{1}{6}$ yd

$\qquad \text{Cost} = 17\dfrac{1}{6} \cdot 12 = \dfrac{103}{\underset{1}{\cancel{6}}} \cdot \dfrac{\overset{2}{\cancel{12}}}{1} = \$206$

Exercises 2.18   (page 81)

1. $\dfrac{5}{6} > \dfrac{3}{4} > \dfrac{2}{3}$ **3.** $\dfrac{3}{4} > \dfrac{11}{16} > \dfrac{5}{8}$ **5.** $\dfrac{3}{4} > \dfrac{5}{7} > \dfrac{9}{14}$

7. $\dfrac{3}{10} > \dfrac{1}{6} > \dfrac{2}{15}$ **9.** $\dfrac{1}{2} = \dfrac{12}{24}$. Because $\dfrac{12}{24} < \dfrac{13}{24}$,

$\qquad\qquad\qquad\qquad\qquad\qquad\qquad \dfrac{1}{2} < \dfrac{13}{24}$

$\qquad\qquad\qquad\qquad\qquad \text{and} \quad 5\dfrac{1}{2} < 5\dfrac{13}{24}$

11. $>$ **13.** $\dfrac{1}{20}$ of 15 ? $\dfrac{5}{16}$ of 2

$\qquad\qquad\quad \dfrac{1}{\underset{4}{\cancel{20}}} \cdot \dfrac{\overset{3}{\cancel{15}}}{1} \ ? \ \dfrac{5}{\underset{8}{\cancel{16}}} \cdot \dfrac{\overset{1}{\cancel{2}}}{1}$

$\qquad\qquad\qquad \dfrac{3}{4} \ ? \ \dfrac{5}{8}$

$\qquad\qquad\qquad \dfrac{6}{8} > \dfrac{5}{8}$

15. Up by $\dfrac{1}{8}$ **17.** 45 in.

Chapter 2 Review Exercises (page 83)

1. $2\dfrac{1}{3}$ **2.** $1\dfrac{3}{8}$ **3.** $5\dfrac{3}{5}$ **4.** $2\dfrac{5}{12}$ **5.** $\dfrac{13}{5}$ **6.** $\dfrac{31}{8}$

7. $\dfrac{101}{11}$ **8.** $\dfrac{83}{6}$ **9.** $\dfrac{7}{9}$ **10.** $\dfrac{3}{7}$ **11.** $\dfrac{28}{57}$ **12.** $\dfrac{4}{5}$

13. $1\dfrac{7}{9}$ **14.** $4\dfrac{4}{15}$ **15.** $4\dfrac{15}{16}$ **16.** $289\dfrac{3}{20}$ **17.** $2\dfrac{1}{10}$

18. $\quad 5\dfrac{1}{3} = \ 5\dfrac{4}{12} = \ 4\dfrac{16}{12}$ **19.** $10\dfrac{1}{8}$ **20.** $78\dfrac{8}{9}$

$\qquad \underline{-2\dfrac{3}{4}} = \underline{-2\dfrac{9}{12}} = \underline{-2\dfrac{9}{12}}$

$\qquad\qquad\qquad\qquad\qquad\qquad 2\dfrac{7}{12}$

21. $10\dfrac{1}{2}$ **22.** $24\dfrac{1}{5}$ **23.** 6   **24.** $12\dfrac{3}{4}$ **25.** $2\dfrac{1}{4}$

26. 26   **27.** $\dfrac{5}{32}$

28. $\quad \dfrac{7}{12} + 2\dfrac{2}{3} \cdot \dfrac{1}{4}$ **29.** $\quad \dfrac{2}{3} \div 1\dfrac{1}{4} \cdot \left(\dfrac{3}{4}\right)^2$

$\qquad = \dfrac{7}{12} + \dfrac{\overset{2}{\cancel{8}}}{3} \cdot \dfrac{1}{\underset{1}{\cancel{4}}}$ $\qquad = \dfrac{2}{3} \div \dfrac{5}{4} \cdot \dfrac{3}{4} \cdot \dfrac{3}{4}$

$\qquad = \dfrac{7}{12} + \dfrac{2}{3}$ $\qquad = \dfrac{\overset{1}{\cancel{2}}}{\underset{1}{\cancel{3}}} \cdot \dfrac{4}{5} \cdot \dfrac{\overset{1}{\cancel{3}}}{\underset{1}{\cancel{4}}} \cdot \dfrac{3}{\underset{2}{\cancel{4}}}$

$\qquad = \dfrac{7}{12} + \dfrac{8}{12}$ $\qquad = \dfrac{3}{10}$

$\qquad = \dfrac{15}{12}$

$\qquad = 1\dfrac{3}{12}$

$\qquad = 1\dfrac{1}{4}$

30. $\left(\dfrac{3}{10} + \dfrac{3}{4}\right) \cdot \sqrt{\dfrac{4}{9}} + \left(\dfrac{5}{6}\right)^0$ **31.** $\dfrac{3}{4}$ **32.** 22

$= \left(\dfrac{6}{20} + \dfrac{15}{20}\right) \cdot \sqrt{\dfrac{4}{9}} + \left(\dfrac{5}{6}\right)^0$

$= \dfrac{21}{20} \quad \cdot \sqrt{\dfrac{4}{9}} + \left(\dfrac{5}{6}\right)^0$

$= \dfrac{\overset{7}{\cancel{21}}}{\underset{10}{\cancel{20}}} \cdot \dfrac{\overset{1}{\cancel{2}}}{\underset{1}{\cancel{3}}} \quad + 1$

$= \dfrac{7}{10} \quad + 1$

$= 1\dfrac{7}{10}$

33. $\dfrac{1\frac{2}{3}}{\frac{3}{4} + \frac{1}{2}} = \dfrac{\frac{5}{3}}{\frac{3}{4} + \frac{2}{4}} = \dfrac{\frac{5}{3}}{\frac{5}{4}} = \dfrac{5}{3} \div \dfrac{5}{4} = \dfrac{\cancel{5}}{3} \cdot \dfrac{4}{\cancel{5}} = \dfrac{4}{3} = 1\dfrac{1}{3}$

34. $\dfrac{\frac{2}{3} + \frac{1}{2}}{\frac{8}{9} - \frac{1}{3}} = \dfrac{\frac{4}{6} + \frac{3}{6}}{\frac{8}{9} - \frac{3}{9}} = \dfrac{\frac{7}{6}}{\frac{5}{9}} = \dfrac{7}{6} \div \dfrac{5}{9} = \dfrac{7}{\cancel{6}_2} \cdot \dfrac{\cancel{9}^3}{5} = \dfrac{21}{10} = 2\dfrac{1}{10}$

35. $\dfrac{2}{5} > \dfrac{3}{10} > \dfrac{1}{4}$ **36.** $\dfrac{5}{6} > \dfrac{2}{3} > \dfrac{5}{8}$ **37.** $\dfrac{11}{16}$ in.

38. 56 books   **39.** $40\dfrac{3}{4}$ hr   **40. a.** $19\dfrac{1}{4}$ sq. yd   **b.** $154

41. 1 week $= 7 \cdot 24 = 168$ hr

$168 \div 2\dfrac{1}{3} = \dfrac{168}{1} \div \dfrac{7}{3} = \dfrac{\cancel{168}^{24}}{1} \cdot \dfrac{3}{\cancel{7}_1} = 72$ trips

42. $24\dfrac{1}{4}$ in.

43. a. $16 \div 3\dfrac{1}{2} = \dfrac{16}{1} \div \dfrac{7}{2} = \dfrac{16}{1} \cdot \dfrac{2}{7} = \dfrac{32}{7} = 4\dfrac{4}{7}$

Therefore, four shelves can be cut from the 16-ft board.

b. $4 \times 3\dfrac{1}{2} = \dfrac{\cancel{4}^2}{1} \times \dfrac{7}{\cancel{2}_1} = 14$ ft for shelves

Four cuts $= 4 \times \dfrac{1}{8} = \dfrac{4}{8} = \dfrac{1}{2}$ in. for saw cuts

Part left $= 16$ ft $- 14$ ft $- \dfrac{1}{2}$ in.

$= 2$ ft $- \dfrac{1}{2}$ in. $= 24$ in. $- \dfrac{1}{2}$ in.

$= 23\dfrac{1}{2}$ in.

44.

$1\dfrac{1}{4} = 1\dfrac{2}{8}$

$+ \dfrac{5}{8} = \dfrac{5}{8}$

$\overline{\phantom{+ \dfrac{5}{8}}}$

$1\dfrac{7}{8}$ yd for each girl

$26 \cdot 1\dfrac{7}{8} = \dfrac{\overset{13}{\cancel{26}}}{1} \cdot \dfrac{15}{\cancel{8}_4} = \dfrac{195}{4}$

$= 48\dfrac{3}{4}$ total yd

45. 8 tons

46.

$3\dfrac{1}{2} = 3\dfrac{4}{8}$

$2\dfrac{3}{4} = 2\dfrac{6}{8}$

$+ 1\dfrac{5}{8} = 1\dfrac{5}{8}$

$\overline{\phantom{+ 1\dfrac{5}{8}}}$

$6\dfrac{15}{8} = 7\dfrac{7}{8}$ lb lost

$200 = 199\dfrac{8}{8}$

$- 7\dfrac{7}{8} = 7\dfrac{7}{8}$

$\overline{\phantom{- 7\dfrac{7}{8}}}$

$192\dfrac{1}{8}$ lb

47. $2\dfrac{1}{8}$ lb

48. Spending $= \left(\dfrac{1}{3} + \dfrac{1}{4} + \dfrac{1}{3}\right)$ of \$18,000

$= \left(\dfrac{4}{12} + \dfrac{3}{12} + \dfrac{4}{12}\right)$ of \$18,000

$= \dfrac{11}{\cancel{12}_2} \cdot \dfrac{\overset{3,000}{\cancel{18,000}}}{1}$

$= \dfrac{33,000}{2}$

$= \$16,500$

Left over $=$
$\begin{array}{r} \$18,000 \\ - \ 16,500 \\ \hline \$1,500 \end{array}$

49. a. 5 ft   **b.** $1\dfrac{9}{16}$ sq. ft   **50. a.** $14\dfrac{1}{2}$ m   **b.** $12\dfrac{3}{8}$ sq. m

Chapter 2 Diagnostic Test   (page 85)

Following each problem number is the number (in parentheses) of the textbook section in which that kind of problem is discussed.

1. (2.3) $\dfrac{69}{8} = 8\overline{)69} \begin{array}{c} 8 \text{ R } 5 \\ 64 \\ \hline 5 \end{array} = 8\dfrac{5}{8}$

2. (2.4) $5\dfrac{7}{12} = \dfrac{5 \cdot 12 + 7}{12} = \dfrac{67}{12}$

3. (2.7A) $\dfrac{\cancel{180}}{\cancel{540}} = \dfrac{\overset{2}{\cancel{18}}}{\underset{6}{\cancel{54}}} = \dfrac{2}{6} = \dfrac{1}{3}$

4. (2.7B) $\dfrac{7}{8}$ of $120 = \dfrac{7}{\cancel{8}_1} \cdot \dfrac{\overset{15}{\cancel{120}}}{1} = 105$

5. (2.10)
$\dfrac{5}{6} = \dfrac{20}{24}$
$+ \dfrac{7}{8} = \dfrac{21}{24}$
$\overline{\phantom{+\dfrac{7}{8}}}$
$\dfrac{41}{24} = 1\dfrac{17}{24}$

6. (2.11)
$4\dfrac{2}{3} = 4 + \dfrac{4}{6}$
$+ 3\dfrac{1}{2} = 3 + \dfrac{3}{6}$
$\overline{\phantom{+3\dfrac{1}{2}}}$
$7 + \dfrac{7}{6} = 7 + 1\dfrac{1}{6} = 8\dfrac{1}{6}$

7. (2.7B) $\dfrac{\overset{1}{\cancel{5}}}{\underset{2}{\cancel{6}}} \cdot \dfrac{\overset{1}{\cancel{3}}}{\underset{4}{\cancel{20}}} = \dfrac{1}{8}$ **8.** (2.13) $1\dfrac{1}{3} \cdot 42 = \dfrac{4}{\cancel{3}_1} \cdot \dfrac{\overset{14}{\cancel{42}}}{1} = 56$

9. (2.14) $\dfrac{3}{8} \div \dfrac{9}{16} = \dfrac{\overset{1}{\cancel{3}}}{\underset{1}{\cancel{8}}} \cdot \dfrac{\overset{2}{\cancel{16}}}{\underset{3}{\cancel{9}}} = \dfrac{2}{3}$

10. (2.10)
$\dfrac{7}{10} = \dfrac{21}{30}$
$- \dfrac{1}{6} = \dfrac{5}{30}$
$\overline{\phantom{-\dfrac{1}{6}}}$
$\dfrac{16}{30} = \dfrac{8}{15}$

11. (2.12)
$8 = 7\dfrac{9}{9}$
$- 3\dfrac{4}{9} = 3\dfrac{4}{9}$
$\overline{\phantom{-3\dfrac{4}{9}}}$
$4\dfrac{5}{9}$

12. (2.14) $2\dfrac{2}{9} \div 3\dfrac{1}{3} = \dfrac{20}{9} \div \dfrac{10}{3} = \dfrac{\overset{2}{\cancel{20}}}{\underset{3}{\cancel{9}}} \cdot \dfrac{\overset{1}{\cancel{3}}}{\underset{1}{\cancel{10}}} = \dfrac{2}{3}$

13. (2.13) $4\dfrac{1}{5} \cdot 2\dfrac{1}{7} = \dfrac{\overset{3}{\cancel{21}}}{\underset{1}{\cancel{5}}} \cdot \dfrac{\overset{3}{\cancel{15}}}{\underset{1}{\cancel{7}}} = 9$

14. (2.12)
$$5\dfrac{1}{4} = 5\dfrac{3}{12} = 4\dfrac{15}{12}$$
$$-2\dfrac{5}{6} = 2\dfrac{10}{12} = 2\dfrac{10}{12}$$
$$\overline{\phantom{-2\dfrac{5}{6} = 2\dfrac{10}{12} = }2\dfrac{5}{12}}$$

15. (2.16) $\left(\dfrac{5}{6}\right)^2 - \sqrt{\dfrac{1}{16}} = \dfrac{25}{36} - \dfrac{1}{4} = \dfrac{25}{36} - \dfrac{9}{36} = \dfrac{16}{36} = \dfrac{4}{9}$

16. (2.10)
$$\dfrac{5}{12} = \dfrac{10}{24}$$
$$\dfrac{3}{8} = \dfrac{9}{24}$$
$$+\dfrac{5}{6} = \dfrac{20}{24}$$
$$\overline{\phantom{+\dfrac{5}{6} = }\dfrac{39}{24} = 1\dfrac{15}{24} = 1\dfrac{5}{8}}$$

17. (2.15) $\dfrac{\frac{5}{8}}{\frac{15}{16}} = \dfrac{5}{8} \div \dfrac{15}{16} = \dfrac{\overset{1}{\cancel{5}}}{\underset{1}{\cancel{8}}} \cdot \dfrac{\overset{2}{\cancel{16}}}{\underset{3}{\cancel{15}}} = \dfrac{2}{3}$

18. (2.15) $\dfrac{\frac{3}{8} + \frac{3}{4}}{7\frac{1}{2}} = \dfrac{\frac{3}{8} + \frac{6}{8}}{\frac{15}{2}} = \dfrac{\frac{9}{8}}{\frac{15}{2}} = \dfrac{9}{8} \div \dfrac{15}{2} = \dfrac{\overset{3}{\cancel{9}}}{\underset{4}{\cancel{8}}} \cdot \dfrac{\overset{1}{\cancel{2}}}{\underset{5}{\cancel{15}}} = \dfrac{3}{20}$

19. (2.17) $\dfrac{4}{5} \div 2\dfrac{2}{3} \cdot \left(\dfrac{2}{3}\right)^2$
$$= \dfrac{4}{5} \div \dfrac{8}{3} \cdot \dfrac{4}{9}$$
$$= \dfrac{\overset{1}{\cancel{4}}}{5} \cdot \dfrac{\overset{1}{\cancel{3}}}{\underset{2}{\cancel{8}}} \cdot \dfrac{\overset{2}{\cancel{4}}}{\underset{3}{\cancel{9}}}$$
$$= \dfrac{2}{15}$$

20. (2.17) $\dfrac{3}{4} + \dfrac{1}{4} \cdot 1\dfrac{3}{5}$
$$= \dfrac{3}{4} + \dfrac{1}{\underset{1}{\cancel{4}}} \cdot \dfrac{\overset{2}{\cancel{8}}}{5}$$
$$= \dfrac{3}{4} + \dfrac{2}{5}$$
$$= \dfrac{15}{20} + \dfrac{8}{20}$$
$$= \dfrac{23}{20}$$
$$= 1\dfrac{3}{20}$$

21. (2.12)
$$1\dfrac{1}{8} = \dfrac{9}{8}$$
$$-\dfrac{3}{4} = -\dfrac{6}{8}$$
$$\overline{\phantom{-\dfrac{3}{4} = -}\dfrac{3}{8}\ \text{lb}}$$

22. (2.14) $7\dfrac{1}{2} \div 3 = \dfrac{15}{2} \div \dfrac{3}{1} = \dfrac{\overset{5}{\cancel{15}}}{2} \cdot \dfrac{1}{\underset{1}{\cancel{3}}} = \dfrac{5}{2} = 2\dfrac{1}{2}$

23. (2.13) $A = lw$
$$A = 5\dfrac{1}{4} \cdot 3\dfrac{1}{3}$$
$$A = \dfrac{\overset{7}{\cancel{21}}}{\underset{2}{\cancel{4}}} \cdot \dfrac{\overset{5}{\cancel{10}}}{\underset{1}{\cancel{3}}}$$
$$A = \dfrac{35}{2}$$
$$A = 17\dfrac{1}{2}\ \text{sq. ft}$$

24. (2.11)
$$23\dfrac{1}{2} = 23\dfrac{4}{8}$$
$$16\dfrac{1}{4} = 16\dfrac{2}{8}$$
$$+37\dfrac{7}{8} = 37\dfrac{7}{8}$$
$$\overline{\phantom{+37\dfrac{7}{8} = }76\dfrac{13}{8} = 76 + 1\dfrac{5}{8} = 77\dfrac{5}{8}\ \text{lb}}$$

25. (2.18) $\dfrac{8}{15} = \dfrac{16}{30}, \dfrac{7}{10} = \dfrac{21}{30}, \dfrac{3}{5} = \dfrac{18}{30}$
$$\dfrac{21}{30} > \dfrac{18}{30} > \dfrac{16}{30}$$
$$\dfrac{7}{10} > \dfrac{3}{5} > \dfrac{8}{15}$$

Exercises 3.1 (page 91)

1. a. 3 **b.** 5 **c.** 6 **3. a.** 2 **b.** 7 **c.** 5
5. Thirty-five hundredths
7. Three and sixteen thousandths **9.** Four ten-thousandths
11. Twenty and nine hundred thousandths
13. Nine thousand and fifty hundredths
15. Five thousand six ten-thousandths **17.** 0.09
19. 2.003 **21.** 0.0400 **23.** 60.08 **25.** 0.720
27. 3,000.55 **29.** 100.0004

Exercises 3.2 (page 94)

1. 7.2 **3.** 9.03 **5.** 400 **7.** 0.0603 **9.** 3.210
11. 107 **13.** 490 **15.** 0.00982 **17.** 0.037
19. 0.10

Exercises 3.3 (page 96)

1. 374.988 **3.** $267.48 **5.** 258.9556 **7.** 11,036.075

9. a. $5 **b.** $27.62 **11.** $\overset{1}{}$3,050.37

$5	
7	5.00002
4	+ 70.0150
6	3,125.38502
+ 5	
$27	

Exercises 3.4 (page 97)

1. 4.41 **3.** 201.34 **5.** 299.855 **7.** 78,499.76
9. 4.8185 **11.** 177.2
13.

Beginning balance plus deposits	Checks written	Computing balance
$254.39	$ 27.15	$750.50
183.50	86.94	− 566.09
233.75	123.47	
78.86	167.66	$184.41 Ending balance
$750.50	122.20	
	38.67	
	$566.09	

Exercises 3.5 (page 99)

1. 0.024 **3.** 0.0024 **5.** 12.6 **7.** 0.136 **9.** 2.7
11. 0.3922 **13.** 395 **15.** 8,170 **17.** $0.9 \times 0.4 = 0.36$

19. $0.7 \times 40 = 28$ **21. a.** $\$6 \times 30 = \$180$ **b.** $159.90
23. a. 90 sq. m **b.** 107.5 sq. m **c.** $2,203.75
25. 51.2 lb **27.**

$$\begin{array}{r} \$875 \\ \times\ .15 \\ \hline 4375 \\ 875 \\ \hline \$131.25 \end{array}$$

$$\begin{array}{r} \$12 \\ \times\ 7 \\ \hline \$84 \end{array}$$

$$\begin{array}{r} \$131.25 \\ + 84.00 \\ \hline \$215.25 \quad \text{Total cost} \end{array}$$

Exercises 3.6A (page 101)

1. 10.87 **3.** 15.6 **5.** 1.05 **7.** 0.175 **9.** 0.079
11. 1.22 **13.** 47.038 **15.** 1.4 **17.** 0.051
19. 0.37 **21.** 8.65 **23.** 2.4 in.
25.

$$\begin{array}{r} 3 \times \$8.50 = \quad \$25.50 \\ + 2 \times \$6.95 = + \quad 13.90 \\ \hline 5 \text{ tapes for} \quad \$39.40 \end{array}$$

$$\text{Average} = \frac{\$39.40}{5} = \$7.88$$

Exercises 3.6B (page 103)

1. 3.56 **3.** 78.4 **5.** 0.064 **7.** 0.014 **9.** 0.906
11. 35.7 **13.** 820 **15.** 2.85 **17.** 2.65 **19.** 30.8
21. 6.406 **23.** 0.0310 **25.** 904.1 **27.** $68.53
29.

$$\begin{array}{r} 1\ 4. \text{ pieces} \\ 3.5_{\wedge}\overline{)5\ 0.0_{\wedge}} \\ \underline{3\ 5} \\ 1\ 5\ 0 \\ \underline{1\ 4\ 0} \\ 1\ 0 \end{array}$$

Exercises 3.7 (page 106)

1. 9.56 **3.** 5.73 **5.** 2,780 **7.** 209.4 **9.** 7.502
11. 984.6 **13.** 0.1 **15.** 274,000
17. $\dfrac{146.35}{10} = 14.635 \approx \$14.64$ **19.** $537 \times 100 = \$53,700$

Exercises 3.8 (page 107)

1. 0.09 **3.** 0.64 **5.** 0.0001 **7.** 1.69 **9.** 0.064
11. 1 **13.** 0.2 **15.** 1.3 **17.** 0.04 **19.** 0.46
21. a. 64 sq. cm **b.** 67.24 sq. cm **23.** 39.0625
25. 2.48832 **27.** 12.4 **29.** 0.78

Exercises 3.9 (page 109)

1. 4.44 **3.** 7.18 **5.** 11.1 **7.** 5.742 **9.** 40.74
11. $2.3 + 1.6 + 1.2 \div 3$ **13.** $4 + 7 - 4 = 7$
$ = 2.3 + 1.6 + \quad 0.4$
$ = 4.3$
15. $2(9 + 5) = 28$ **17.** $\dfrac{2 \times 6}{0.3} = \dfrac{12}{0.3} = 40$ **19.** 10.7802
21. 1,816.71 **23.** 14.76 **25.** $P = 2l + 2w$
$ P = 2(8.65) + 2(5.3)$
$ P = \quad 17.3 \quad + 10.6$
$ P = 27.9 \text{ m}$
27. Fare $= 2 + 0.75 \times 5 = \$5.75$
29. a. $9963\ \boxed{-}\ 9876\ \boxed{=}\ 87$ therms used
$ 87\ \boxed{-}\ 65\ \boxed{=}\ 22$ therms at higher rate
 Cost $= 65\ \boxed{\times}\ .365\ \boxed{+}\ 22\ \boxed{\times}\ .785\ \boxed{=}\ 40.995 \approx \$41.00$
b. $87\ \boxed{\div}\ 30\ \boxed{=}\ 2.9$ therms
c. $41\ \boxed{\div}\ 30\ \boxed{=}\ \$1.37$

Exercises 3.10 (page 112)

1. 0.75 **3.** 2.5 **5.** 0.125 **7.** $0.666\ldots$, or $0.\overline{6}$
9. $0.272727\ldots$, or $0.\overline{27}$ **11.** $3.0666\ldots$, or $3.0\overline{6}$
13. 0.04 **15.** 0.3125 **17.** 6.05 **19.** 0.64
21. 0.22 **23.** 0.087 **25.** 6.467 **27.** 0.5625
29. 2.2917 **31.** 9.3846

Exercises 3.11A (page 113)

1. $\dfrac{3}{5}$ **3.** $\dfrac{1}{20}$ **5.** $\dfrac{3}{40}$ **7.** $\dfrac{33}{50}$ **9.** $2\dfrac{1}{2}$ **11.** $5\dfrac{6}{25}$
13. $\dfrac{1}{16}$ **15.** $37\dfrac{1}{2}$ **17.** $\dfrac{3}{25,000}$ **19.** $\dfrac{7}{8}$ in.

Exercises 3.11B (page 114)

1. $\dfrac{3}{8}$ **3.** $\dfrac{1}{3}$ **5.** $\dfrac{7}{12}$
7.

$$\begin{array}{c} 1.0\dfrac{1}{5} \\ | \\ 1\ 0 \end{array} \longrightarrow \dfrac{10\dfrac{1}{5}}{10} = \dfrac{\dfrac{51}{5}}{10} = \dfrac{51}{5} \div \dfrac{10}{1} = \dfrac{51}{5} \cdot \dfrac{1}{10} = \dfrac{51}{50} = 1\dfrac{1}{50}$$

9.

$$\begin{array}{c} 0.00\dfrac{5}{12} \\ | | | \\ 1\ 0\ 0 \end{array} \longrightarrow \dfrac{\dfrac{5}{12}}{100} = \dfrac{5}{12} \div \dfrac{100}{1} = \dfrac{\overset{1}{\cancel{5}}}{12} \cdot \dfrac{1}{\underset{20}{\cancel{100}}} = \dfrac{1}{240}$$

11.

$$\begin{array}{c} 0.001\dfrac{1}{6} \\ | | | | \\ 1,00\ 0 \end{array} \longrightarrow \dfrac{1\dfrac{1}{6}}{1,000} = \dfrac{\dfrac{7}{6}}{1,000} = \dfrac{7}{6} \div \dfrac{1,000}{1}$$

$$= \dfrac{7}{6} \cdot \dfrac{1}{1,000} = \dfrac{7}{6,000}$$

13. $\dfrac{5}{6}$ **15.** $\dfrac{2}{15}$ **17.** $2\dfrac{3}{28}$

Exercises 3.12 (page 116)

1. 5.32 **3.** 0.55 **5.** 0.375 **7.** 25
9. $0.35 = \dfrac{35}{100} = \dfrac{7}{20}$

$$\dfrac{7}{8} + 0.35 = \dfrac{7}{8} + \dfrac{7}{20} = \begin{array}{r} \dfrac{7}{8} = \dfrac{35}{40} \\ + \dfrac{7}{20} = \dfrac{14}{40} \\ \hline \dfrac{49}{40} = 1\dfrac{9}{40} \end{array}$$

11. $0.75 = \dfrac{75}{100} = \dfrac{3}{4}$

$$6.75 - 2\dfrac{5}{6} = 6\dfrac{3}{4} - 2\dfrac{5}{6} = \begin{array}{r} 6\dfrac{3}{4} = 6\dfrac{9}{12} = 5\dfrac{21}{12} \\ -2\dfrac{5}{6} = 2\dfrac{10}{12} = 2\dfrac{10}{12} \\ \hline 3\dfrac{11}{12} \end{array}$$

13. $\dfrac{7}{15}$ **15.** $3\dfrac{1}{3}$ **17.** $\dfrac{5}{6}$ **19.** $\dfrac{1}{2}$, or 0.5
21. a. $9 + 5 = 14$
b. $8\ \boxed{+}\ 5\ \boxed{\div}\ 6\ \boxed{=}\ \boxed{+}\ 4.797\ \boxed{=}\ \approx 13.630$

3

23. a. $7 \times 0.1 = 0.7$
    b. $7 \boxed{+} 2 \boxed{\div} 11 \boxed{=} \boxed{\times} .1238 \boxed{=} \approx 0.889$
25. $9.38   **27.** $48.38

29. $18{,}000 \div 22\frac{1}{2} = 18{,}000 \div 22.5 =$

$$22.5_\wedge \overline{)18\,000.0_\wedge} \quad \begin{array}{r} 8\,0\,0. \text{ gal} \\ \hline \end{array}$$
$$\begin{array}{r} 1\,800 \\ \hline 000 \end{array}$$

Then, $\begin{array}{r} 1.05 \\ \times \ 8 \ | \ 00 \\ \hline 840 \ | \ 00 \end{array} = \$840$ a year

Exercises 3.13 (page 118)

1. $0.49 > 0.41 > 0.409 > 0.4$
3. $3.1 > 3.075 > 3.05 > 3.009$
5. $0.07501 > 0.075 > 0.0749 > 0.07$
7. $5.5 > 5.0501 > 5.05 > 5.0496 > 5$ **9.** $<$ **11.** $>$
13. $\frac{5}{16} = 0.3125$
    No, 0.325-inch pin > 0.3125-inch hole
15. Star: $\frac{1.02}{7} \approx 14.6$¢ per oz

    Whale: $\frac{1.89}{12} \approx 15.8$¢ per oz

    Therefore, Star Fish is the better buy.
    $15.8 - 14.6 = 1.2$¢ per oz difference.

Chapter 3 Review Exercises (page 119)

1. One hundred forty-five thousandths
2. Two hundred fifty and six hundredths   **3.** 0.0016
4. 500.75   **5.** 0.837   **6.** 0.60   **7.** 24.7   **8.** 190
9. 35,201.8505   **10.** 36.7666   **11.** 4.992   **12.** 0.086
13. 468.255   **14.** 7,850   **15.** 0.0064   **16.** 0.00826
17. 2.05   **18.** 56   **19.** 436.6   **20.** 0.096   **21.** 10.42
22. 3.05
23. $12.5 + 5.6 \times 10^2 + \sqrt{1.44}$ **24.** 59.2
    $= 12.5 + 5.6 \times 10^2 + \quad 1.2$
    $= 12.5 + \quad\quad 560 \quad + \quad 1.2$
    $= 573.7$
25. $72 \div 0.8 \times (0.3)^2$ **26.** 9.2   **27.** 0.875
    $= 72 \div 0.8 \times 0.09$
    $= \quad 90 \quad \times 0.09$
    $= 8.1$
28. $5.0555\ldots$, or $5.0\overline{5}$ **29.** 0.83   **30.** 3.64   **31.** $\frac{17}{25}$

32. $4\frac{1}{40}$ **33.** $0.16\frac{2}{3} = \frac{16\frac{2}{3}}{100} = \frac{50}{3} \div \frac{100}{1} = \frac{\overset{1}{\cancel{50}}}{3} \cdot \frac{1}{\underset{2}{\cancel{100}}} = \frac{1}{6}$

34. $0.4\frac{1}{6} = \frac{4\frac{1}{6}}{10} = \frac{25}{6} \div \frac{10}{1} = \frac{\overset{5}{\cancel{25}}}{6} \cdot \frac{1}{\underset{2}{\cancel{10}}} = \frac{5}{12}$ **35.** 6.435

36. $0.24 = \frac{24}{100} = \frac{6}{25}$

    $\frac{9}{10} + 0.24 = \frac{9}{10} + \frac{6}{25} = \quad \frac{9}{10} = \frac{45}{50}$
    $\quad\quad\quad\quad\quad\quad\quad\quad\quad + \frac{6}{25} = \frac{12}{50}$
    $\quad\quad\quad\quad\quad\quad\quad\quad\quad\quad\quad \frac{57}{50} = 1\frac{7}{50}$

37. $0.025 \times 3\frac{1}{3} = \frac{\overset{1}{\cancel{25}}}{\underset{4}{\cancel{1000}}} \times \frac{10}{3} = \frac{1}{12}$

38. $\frac{4\frac{1}{20}}{0.9} = \frac{\frac{81}{20}}{\frac{9}{10}} = \frac{81}{20} \div \frac{9}{10} = \frac{\overset{9}{\cancel{81}}}{\underset{2}{\cancel{20}}} \cdot \frac{\overset{1}{\cancel{10}}}{\underset{1}{\cancel{9}}} = \frac{9}{2} = 4\frac{1}{2}$, or 4.5

39. $0.3(7 + 4) = 0.3(11) = 3.3$

40. $\frac{0.5 \times 0.5}{8} = \frac{0.25}{8} \approx \frac{0.24}{8} = 0.03$

41. $0.603 > 0.6 > 0.063 > 0.06$
42. $3.908 > 3.9 > 3.89 > 3.098 > 3$ **43.** 7.5   **44.** 0.55
45. a. 21.1 m   **b.** 26.875 sq. m
46. a. 13.6 cm   **b.** 11.56 sq. cm   **47.** 40 pieces
48. Yes, 0.625-inch pin < 0.6875-inch hole   **49.** $121.55

50.

Check written	Deposits made		
$15.98	$250	Beginning balance	$275.38
46.75	+ 350	Deposits	+ 600.00
87.45	$600		875.38
135.46		Checks	− 353.64
+ 68.00		Ending balance	$521.74
$353.64			

51. $19.72   **52.** $4.50
53. $0.065 per mi, which is $6\frac{1}{2}$¢ per mi

54.
$$14\overline{)196} \quad \begin{array}{r} 14 \text{ gal used} \\ \hline \end{array}$$
$$\begin{array}{r} 14 \\ \hline 56 \\ 56 \end{array}$$
$$\begin{array}{r} \$1.3\,2\,9 \\ \times \quad 1\,4 \\ \hline 5\,3\,1\,6 \\ 1\,3\,2\,9 \\ \hline \$1\,8.6\,0\,6 \approx \$18.61 \end{array}$$

Chapter 3 Diagnostic Test (page 121)

Following each problem number is the number (in parentheses) of the textbook section in which that kind of problem is discussed.

1. (3.1) Nine and fifteen thousandths   **2.** (3.1) 420.05
3. (3.3)
$$\begin{array}{r} \overset{1\,1}{7.8} \\ 56. \\ 0.017 \\ + 500.94 \\ \hline 564.757 \end{array}$$
4. (3.4)
$$\begin{array}{r} \overset{3 \ 10 \ \ 5 \ 10}{\cancel{40.60}} \\ - \quad 3.5\,4 \\ \hline 3\,7.0\,6 \end{array}$$
5. (3.5)
$$\begin{array}{r} 3.7 \\ \times 0.0\,5\,8 \\ \hline 2\,9\,6 \\ 1\,8\,5 \\ \hline 0.2\,1\,4\,6 \end{array}$$
6. (3.6B)
$$.0\,7\,8_\wedge \overline{)6\,2.7\,9\,0_\wedge} \quad \begin{array}{r} 8\,0\,5. \\ \hline \end{array}$$
$$\begin{array}{r} 6\,2\,4 \\ \hline 3\,9\,0 \\ 3\,9\,0 \end{array}$$
7. (3.7) $0.46 \times 10^3 = 460$ **8.** (3.7) $6.9 \div 100 = 0.069$
9. (3.9) $75 \div 10 - 0.2 \times 0.06$
    $= \quad 7.5 \quad - 0.2 \times 0.06$
    $= \quad 7.5 \quad - \quad 0.012$
    $= 7.488$
10. (3.8 & 3.9) $(0.8)^2 + 3 \times \sqrt{0.25}$
    $= 0.64 + 3 \times 0.5$
    $= 0.64 + \quad 1.5$
    $= 2.14$

11. (3.6B)

$$0.35_\wedge)\overline{7.2\,0_\wedge 0\,0}\qquad 2\,0.5\,7 \approx 20.6$$

$$\begin{array}{r}7\,0 \\ \hline 2\,0 \\ 0 \\ \hline 2\,0\,0 \\ 1\,7\,5 \\ \hline 2\,5\,0 \\ 2\,4\,5 \\ \hline \end{array}$$

12. (3.10) $\quad 2\dfrac{3}{16} = \dfrac{35}{16} =$

$$16)\overline{3\,5.0\,0\,0\,0}\qquad 2.1\,8\,7\,5$$

$$\begin{array}{r}3\,2 \\ \hline 3\,0 \\ 1\,6 \\ \hline 1\,4\,0 \\ 1\,2\,8 \\ \hline 1\,2\,0 \\ 1\,1\,2 \\ \hline 8\,0 \\ 8\,0 \\ \hline \end{array}$$

13. (3.11A) $\quad 0.78 = \dfrac{78}{100} = \dfrac{39}{50}$

14. (3.11B) $\quad 0.1\dfrac{9}{11} = \dfrac{1\frac{9}{11}}{10} = \dfrac{\frac{20}{11}}{\frac{10}{1}} = \dfrac{\overset{2}{\cancel{20}}}{11} \cdot \dfrac{1}{\cancel{10}} = \dfrac{2}{11}$

15. (3.12) $\quad 1\dfrac{2}{3} \times 0.35 = \dfrac{\overset{1}{\cancel{5}}}{3} \times \dfrac{\overset{7}{\cancel{35}}}{\underset{\underset{4}{20}}{\cancel{100}}} = \dfrac{7}{12}$

16. (3.9) $\quad \dfrac{10 \times 2}{0.4} = \dfrac{20}{0.4} = 50$

17. (3.13) $\quad 7.48\boxed{0} > 7.408 > 7.4\boxed{00} > 7.084$
$\qquad\qquad 7.48 \quad > 7.408 > 7.4 \quad > 7.084$

18. (3.6A) $\quad \dfrac{45.60 + 42.76 + 44.02}{3} = \dfrac{132.38}{3}$

$$3)\overline{1\,3\,2.3\,8\,0}\qquad 4\,4.1\,2\,6 \approx 44.13\text{ sec}$$

$$\begin{array}{r}1\,2 \\ \hline 1\,2 \\ 1\,2 \\ \hline 0\,3 \\ 3 \\ \hline 0\,8 \\ 6 \\ \hline 2\,0 \\ 1\,8 \\ \hline \end{array}$$

19. (3.5)
$$\begin{array}{r}2\,4 \\ \times\,4.8 \\ \hline 1\,9\,2 \\ 9\,6 \\ \hline 1\,1\,5.2\text{ lb} \end{array}$$

20. (3.6B)
$$0.75_\wedge)\overline{1\,2.0\,0_\wedge}\qquad 1\,6.\text{ pieces}$$
$$\begin{array}{r}7\,5 \\ \hline 4\,5\,0 \\ 4\,5\,0 \\ \hline \end{array}$$

Exercises 4.1 (page 127)

1. $\dfrac{2}{3}$ **3.** $\dfrac{9}{4}$ **5.** $\dfrac{5}{1}$ **7.** $\dfrac{1}{5}$ **9.** $\dfrac{7}{20}$ **11.** $\dfrac{3}{1}$

13. $\dfrac{1}{4}$ **15.** $\dfrac{9}{1}$ **17.** $\dfrac{1}{8}$ **19.** $\dfrac{1}{2}$ **21.** $\dfrac{3}{4}$ **23.** $\dfrac{1}{4}$

25. a. $\dfrac{24}{30} = \dfrac{4}{5}$ **b.** $\dfrac{24}{6} = \dfrac{4}{1}$ **c.** $\dfrac{6}{24} = \dfrac{1}{4}$

27. a. $\dfrac{12}{8} = \dfrac{3}{2}$ **b.** $\dfrac{12}{4} = \dfrac{3}{1}$
c. Total pies = 12 + 8 + 4 = 24
$\dfrac{8}{24} = \dfrac{1}{3}$

29. $\dfrac{2\text{ hr }40\text{ min}}{4\text{ hr}} = \dfrac{160\ \cancel{\text{min}}}{240\ \cancel{\text{min}}} = \dfrac{2}{3}$

31. $\dfrac{4\frac{1}{2}\text{-ft shadow}}{15\text{-ft pole}} = 4\dfrac{1}{2} \div 15 = \dfrac{\overset{3}{\cancel{9}}}{2} \cdot \dfrac{1}{\underset{5}{\cancel{15}}} = \dfrac{3}{10}$

33. 48 mi per hr **35.** 7.5 pages per minute

37. a. 26.9¢ per oz **b.** 21.5¢ per slice

39. 4.2¢ per oz

Exercises 4.2 (page 133)

1. a. 8 **b.** 14 **c.** 16 **d.** 28 **e.** 14 and 16
f. 8 and 28

3. Yes **5.** No

7. Yes, $\quad \dfrac{2\frac{1}{4}}{27} = \dfrac{3\frac{1}{3}}{40}$

$$2\dfrac{1}{4} \cdot 40 = 27 \cdot 3\dfrac{1}{3}$$

$$\dfrac{9}{\underset{1}{\cancel{4}}} \cdot \dfrac{\overset{10}{\cancel{40}}}{1} = \dfrac{\overset{9}{\cancel{27}}}{1} \cdot \dfrac{10}{\underset{1}{\cancel{3}}}$$

$$90 = 90$$

9. 12 **11.** 8 **13.** 75 **15.** $2\dfrac{2}{3}$ **17.** 8

19. $\dfrac{7}{20}$, or 0.35 **21.** 2 **23.** 6 **25.** 4

27. $\dfrac{\frac{5}{6}}{1\frac{1}{4}} = \dfrac{24}{x}$ **29.** 0.208 **31.** 320

$$\dfrac{5}{6} \cdot x = \dfrac{5}{\underset{1}{\cancel{4}}} \cdot \dfrac{\overset{6}{\cancel{24}}}{1}$$

$$\dfrac{\overset{1}{\cancel{\dfrac{5}{6}}} \cdot x}{\underset{1}{\cancel{\dfrac{5}{6}}}} = \dfrac{30}{\frac{5}{6}}$$

$$x = 30 \div \dfrac{5}{6}$$

$$x = \dfrac{\overset{6}{\cancel{30}}}{1} \cdot \dfrac{6}{\underset{1}{\cancel{5}}}$$

$$x = 36$$

Exercises 4.3 (page 136)

1. 30 gal **3.** 30 ft **5.** $9,000 **7.** $750 **9.** 25 lb
11. 630 lb **13.** $3.96

4

15. $\dfrac{\frac{1}{2}\text{ in.}}{12\text{ mi}} = \dfrac{2\frac{1}{4}\text{ in.}}{x\text{ mi}}$ **17.** 1,400 mi   **19.** 30 mechanics

$$\frac{1}{2} \cdot x = \frac{\overset{3}{\cancel{12}}}{1} \cdot \frac{9}{\underset{1}{\cancel{4}}}$$

$$\frac{\frac{\overset{1}{\cancel{1}}}{\cancel{2}} \cdot x}{\underset{1}{\frac{\cancel{1}}{\cancel{2}}}} = \frac{27}{\frac{1}{2}}$$

$$x = 27 \div \frac{1}{2}$$

$$x = \frac{27}{1} \cdot \frac{2}{1}$$

$$x = 54\text{ mi}$$

21. Let l = length of room
    w = width of room

$$\frac{1\text{ in.}}{8\text{ ft}} = \frac{3\text{ in.}}{l\text{ ft}} \qquad \frac{1\text{ in.}}{8\text{ ft}} = \frac{2\frac{1}{2}\text{ in.}}{w\text{ ft}}$$

$$1 \cdot l = 8 \cdot 3 \qquad 1 \cdot w = \frac{\overset{4}{\cancel{8}}}{1} \cdot \frac{5}{\underset{1}{\cancel{2}}}$$

$$l = 24\text{ ft} \qquad w = 20\text{ ft}$$

Exercises 4.4A   (page 138)

1. 27%   **3.** 6%   **5.** 140%   **7.** 18.6%   **9.** 7.5%
11. 290%   **13.** 200.5%   **15.** 136%   **17.** 400%

Exercises 4.4B   (page 139)

1. 0.45   **3.** 1.25   **5.** 0.065   **7.** 0.09

9. $2\frac{1}{2}\% = 2.5\% = 0.025$ **11.** $3\frac{1}{4}\% = 3.25\% = 0.0325$

13. 0.1005   **15.** $\frac{3}{4}\% = 0.75\% = 0.0075$

17. $66\frac{2}{3}\% \approx 66.67\% = 0.6667$

Exercises 4.4C   (page 139)

1. 50%   **3.** 40%   **5.** 37.5%   **7.** 90%   **9.** 16%

11. $\frac{1}{3} \approx 0.3333 = 33.33\%$ **13.** $\frac{5}{6} \approx 0.8333 = 83.33\%$

15. 43.75%   **17.** $2\frac{5}{8} = 2.625 = 262.5\%$

Exercises 4.4D   (page 140)

1. $\frac{3}{4}$ **3.** $\frac{1}{10}$ **5.** $\frac{7}{20}$ **7.** $\frac{4}{5}$ **9.** $\frac{1}{20}$

11. $250\% = \frac{250}{100} = \frac{5}{2} = 2\frac{1}{2}$

13. $\frac{1}{2}\% = \dfrac{\frac{1}{2}}{100} = \frac{1}{2} \div 100 = \frac{1}{2} \cdot \frac{1}{100} = \frac{1}{200}$

15. $2\frac{1}{2}\% = \dfrac{2\frac{1}{2}}{100} = 2\frac{1}{2} \div 100 = \frac{\overset{1}{\cancel{5}}}{2} \cdot \frac{1}{\underset{20}{\cancel{100}}} = \frac{1}{40}$

17. $33\frac{1}{3}\% = \dfrac{33\frac{1}{3}}{100} = 33\frac{1}{3} \div 100 = \frac{\overset{1}{\cancel{100}}}{3} \cdot \frac{1}{\underset{1}{\cancel{100}}} = \frac{1}{3}$

19. 0.5; 50%   **21.** $\frac{3}{5}$; 60%   **23.** $\frac{1}{10}$; 0.1

25. 0.75; 75%   **27.** $\frac{3}{50}$; 0.06   **29.** $\frac{12}{25}$; 48%

31. 1.125; 112.5%   **33.** $\frac{11}{25}$; 0.44

35. $\frac{1}{40}$; 2.5%   **37.** $3\frac{1}{2}$; 3.50

39. $6\frac{1}{4}$; 625%   **41.** $\frac{21}{400}$; 0.0525

43. $\frac{3}{400}$; 0.0075

45. a. $\frac{10}{40} = \frac{1}{4}$ **b.** $\frac{1}{4} = 4\overline{)1.00}$ **c.** $0.25_\wedge = 25\%$

$$\begin{array}{r}0.25 \\ 4\overline{)1.00} \\ \underline{8} \\ 20 \\ \underline{20} \\ 0\end{array}$$

47. $100\% - 20\% = 80\% = \frac{80}{100} = \frac{4}{5}$

49. $25\% + 20\% = 45\% = \frac{45}{100} = \frac{9}{20}$

Exercises 4.5   (page 143)

1. 40   **3.** 12   **5.** 9   **7.** 30   **9.** $2\frac{2}{5}$ **11.** 1.5

13. $13\frac{1}{3}\% = \dfrac{13\frac{1}{3}}{100} = 13\frac{1}{3} \div 100 = \frac{\overset{2}{\cancel{40}}}{3} \cdot \frac{1}{\underset{5}{\cancel{100}}} = \frac{2}{15}$

    Then, $13\frac{1}{3}\%$ of $600 = \frac{2}{\underset{1}{\cancel{15}}} \cdot \frac{\overset{40}{\cancel{600}}}{1} = 80$

15. $1\frac{1}{2}$, or 1.5   **17. a.** $101.25   **b.** $273.75

19. 8 days   **21.** $922.50   **23.** 56 hr   **25.** $2,116.50
27. 7.25   **29.** 101.7   **31.** 238.96   **33.** 168.75
35. $45.72

Exercises 4.6   (page 146)

1. 50   **3.** 46%   **5.** 10   **7.** 850   **9.** 225%
11. 31.2   **13.** 600   **15.** 125%   **17.** 24   **19.** 250

21. $P = 66\frac{2}{3}$, $A = 42$, B is unknown.

$$\frac{42}{B} = \frac{66\frac{2}{3}}{100}$$

$$66\frac{2}{3} \cdot B = 4,200$$

$$B = \frac{4,200}{66\frac{2}{3}}$$

$$B = 4,200 \div 66\frac{2}{3}$$

$$B = \frac{4,200}{1} \div \frac{200}{3}$$

$$B = \frac{\overset{21}{\cancel{4,200}}}{1} \cdot \frac{3}{\underset{1}{\cancel{200}}}$$

$$B = 63$$

Exercises 4.7   (page 150)

1. 85 games   **3.** $16\frac{2}{3}\%$ **5.** 25%; 75%   **7.** $6,500

9. $\overset{\text{What}}{\underset{A}{\bigcirc}}$ is $\overset{24}{\underset{P}{\bigcirc}}$ % of $\overset{1,800}{\underset{B}{\bigcirc}}$?

$$\frac{A}{1,800} = \frac{24}{100}$$

$$\frac{100 \cdot A}{100} = \frac{43,2\cancel{00}}{1\cancel{00}}$$

$$A = 432$$

Then, 432

$\underline{-\ 256}$

176 more students

11. 2,303   **13.** Yes; 40% of 50 = 20   **15.** 8.6%

17. 30 − 18 = 12 meets lost

$\overset{12}{\underset{A}{\bigcirc}}$ is $\overset{\text{what}}{\underset{P}{\bigcirc}}$ % of $\overset{30}{\underset{B}{\bigcirc}}$?

$$\frac{12}{30} = \frac{P}{100}$$

$$30 \cdot P = 1200$$

$$\frac{30 \cdot P}{30} = \frac{1200}{30}$$

$$P = 40\% \text{ lost}$$

19. 42 women   **21. a.** $2,240   **b.** $30,240   **23.** 12,220

25. 6%   **27. a.** $34.50   **b.** 75%

29. $12,000

$\underline{-\ \ 5,400}$

$6,600$ Decrease

$\overset{\$6,600}{\underset{A}{\bigcirc}}$ is $\overset{\text{what}}{\underset{P}{\bigcirc}}$ % of $\overset{\$12,000}{\underset{B}{\bigcirc}}$? ⟵ Original cost

$$\frac{6,600}{12,000} = \frac{P}{100}$$

$$12,000 \cdot P = 660,000$$

$$\frac{12,000 \cdot P}{12,000} = \frac{660,000}{12,000}$$

$$P = 55\%$$

Chapter 4 Review Exercises (page 152)

1. $\frac{9}{4}$ **2.** $\frac{1}{9}$ **3.** $\frac{6}{1}$ **4.** Yes   **5.** No   **6.** Yes

7. 36   **8.** 63   **9.** $11\frac{1}{4}$, or 11.25

10. $\dfrac{x}{\frac{2}{3}} = \dfrac{\frac{9}{16}}{\frac{5}{6}}$

$$\frac{5}{6} \cdot x = \frac{\overset{1}{\cancel{2}}}{\underset{1}{\cancel{3}}} \cdot \frac{\overset{3}{\cancel{9}}}{\underset{8}{\cancel{16}}}$$

$$\frac{\frac{5}{6} \cdot x}{\frac{5}{6}} = \frac{\frac{3}{8}}{\frac{5}{6}}$$

$$x = \frac{3}{8} \div \frac{5}{6}$$

$$x = \frac{3}{\underset{4}{\cancel{8}}} \cdot \frac{\overset{3}{\cancel{6}}}{5}$$

$$x = \frac{9}{20}$$

11. 3.5   **12.** $\frac{3}{4}$ **13.** $1.60 per package   **14.** 11.7¢ per oz

15. $1.50   **16.** 8 quarts   **17.** 315 students   **18.** 72 ft

19. 70 mi   **20.** 256 parts   **21. a.** 70%   **b.** 125%

22. a. 0.04   **b.** 0.65   **23. a.** 60%   **b.** 87.5%

24. a. $\frac{6}{25}$ **b.** $\frac{3}{8}$ **25.** 20   **26.** 48   **27.** 135

28. 2.4   **29.** 125%   **30.** 240

31. $\dfrac{A}{3,560} = \dfrac{5}{100}$ **32.** 75%

$$100 \cdot A = 17,800$$

$$\frac{100 \cdot A}{100} = \frac{17,800}{100}$$

$$A = 178$$

Fall enrollment = 3,560 + 178 = 3,738

33. $76   **34.** $450   **35.** 15%   **36.** $1,700

37. $28,090   **38.** Increase = $1,015 − $875 = $140

$$\frac{140}{875} = \frac{P}{100}$$

$$875 \cdot P = 14,000$$

$$\frac{875 \cdot P}{875} = \frac{14,000}{875}$$

$$P = 16\%$$

39. $6.46   **40.**

$$(100\% - 25\%)$$

$$\text{Sale price} = 250 \times 75 \ \%$$

$$= \$187.50$$

$$(100\% - 15\%)$$

$$\text{Employee pays} = 187.5 \times 85 \ \%$$

$$= 159.375 \approx \$159.38$$

Chapter 4 Diagnostic Test   (page 154)

Following each problem number is the number (in parentheses) of the textbook section in which that kind of problem is discussed.

1. (4.1) **a.** $\frac{15}{24} = \frac{5}{8}$ **b.** $\frac{15}{9} = \frac{5}{3}$

2. (4.1) $\dfrac{\overset{52}{\cancel{260}} \text{ words}}{\underset{1}{\cancel{5}} \text{ minutes}} = 52$ words per min

3. (4.2) $\dfrac{x}{12} = \dfrac{\overset{2}{\cancel{40}}}{\underset{3}{\cancel{60}}}$ **4.** (4.2) $\dfrac{\overset{4}{\cancel{20}}}{\underset{15}{\cancel{75}}} = \dfrac{x}{300}$

$\quad\dfrac{x}{12} = \dfrac{2}{3}$ $\quad\dfrac{4}{15} = \dfrac{x}{300}$

$\quad 3 \cdot x = 24$ $\quad 15 \cdot x = 1200$

$\quad\dfrac{\overset{1}{\cancel{3}} \cdot x}{\underset{1}{\cancel{3}}} = \dfrac{24}{3}$ $\quad\dfrac{\overset{1}{\cancel{15}} \cdot x}{\underset{1}{\cancel{15}}} = \dfrac{1200}{15}$

$\quad x = 8$ $\quad x = 80$

5

5. (4.2) $\dfrac{3\frac{1}{2}}{x} = \dfrac{21}{40}$ 　　　　**6.** (4.4A) $0.80_{\wedge} = 80\%$

$21 \cdot x = \dfrac{7}{\cancel{2}_{1}} \cdot \dfrac{\cancel{40}^{20}}{1}$

$\dfrac{\cancel{21}^{1} \cdot x}{\cancel{21}_{1}} = \dfrac{140}{21}$

$x = \dfrac{20}{3} = 6\dfrac{2}{3}$

7. (4.4B) $_{\wedge}12.5\% = 0.125$

8. (4.4C) $\dfrac{3}{40} = 40\overline{)3.000}\quad .075 = 7.5\%$

9. (4.4D) $76\% = \dfrac{76}{100} = \dfrac{19}{25}$

10. (4.5) $\dfrac{5}{8}$ of $72 = \dfrac{5}{\cancel{8}_{1}} \cdot \dfrac{\cancel{72}^{9}}{1} = 45$

11. (4.5) 0.7 of $35 = 0.7 \times 35 = 24.5$

12. (4.6) $\dfrac{15}{B} = \dfrac{\cancel{125}^{5}}{\cancel{100}_{4}}$　　**13.** (4.6) $\dfrac{12}{30} = \dfrac{P}{100}$

$\dfrac{15}{B} = \dfrac{5}{4}$　　　　　　$30 \cdot P = 12 \cdot 100$

$5 \cdot B = 15 \cdot 4$　　　　　$\dfrac{\cancel{30}^{1} \cdot P}{\cancel{30}_{1}} = \dfrac{1200}{30}$

$\dfrac{\cancel{5}^{1} \cdot B}{\cancel{5}_{1}} = \dfrac{60}{5}$　　　　　　$P = 40\%$

$B = 12$

14. (4.3) Let $x = $ cost of $7\frac{1}{2}$ lb

$\dfrac{5 \text{ lb}}{\$4} = \dfrac{7\frac{1}{2}\text{ lb}}{\$x}$

$5 \cdot x = \dfrac{\cancel{4}^{2}}{1} \cdot \dfrac{15}{\cancel{2}_{1}}$

$\dfrac{\cancel{5}^{1} \cdot x}{\cancel{5}_{1}} = \dfrac{30}{5}$

$x = \$6$

15. (4.3) Let $x = $ height of pole

$\dfrac{\cancel{6}^{3}\text{-ft person}}{\cancel{4}_{2}\text{-ft shadow}} = \dfrac{x\text{-ft pole}}{18\text{-ft shadow}}$

$2 \cdot x = 3 \cdot 18$

$\dfrac{\cancel{2}^{1} \cdot x}{\cancel{2}_{1}} = \dfrac{54}{2}$

$x = 27 \text{ ft}$

16. (4.3) Let $x = $ quarts of oil for 6,000 mi

$\dfrac{2 \text{ qt}}{1,500 \text{ mi}} = \dfrac{x \text{ qt}}{6,000 \text{ mi}}$

$1,500 \cdot x = 2 \cdot 6,000$

$\dfrac{\cancel{1,500}^{1} \cdot x}{\cancel{1,500}_{1}} = \dfrac{12,000}{1,500}$

$x = 8 \text{ qt}$

17. (4.7) ⟨34⟩ is ⟨85⟩ % of ⟨what⟩ ?
　　　　　A　　　P　　　　　B

$\dfrac{34}{B} = \dfrac{85}{100}$

$85 \cdot B = 34 \cdot 100$

$\dfrac{\cancel{85}^{1} \cdot B}{\cancel{85}} = \dfrac{3,400}{85}$

$B = 40 \text{ problems}$

18. (4.7) ⟨Discount⟩ is ⟨25⟩ % of ⟨96⟩
　　　　　　A　　　　　P　　　　B

$\dfrac{A}{96} = \dfrac{\cancel{25}^{1}}{\cancel{100}_{4}}$

$4 \cdot A = 96$

$\dfrac{\cancel{4}^{1} \cdot A}{\cancel{4}} = \dfrac{96}{4}$

$A = 24$

Sale price $= \$96 - \$24 = \$72$

19. (4.7) Increase $= 4,600 - 4,000 = 600$

⟨600⟩ is ⟨what⟩ % of ⟨4,000⟩ ?
　A　　　　P　　　　　B

$\dfrac{600}{4,000} = \dfrac{P}{100}$

$4,000 \cdot P = 600 \cdot 100$

$\dfrac{\cancel{4,000}^{1} \cdot P}{\cancel{4,000}} = \dfrac{60,000}{4,000}$

$P = 15\%$

20. (4.7) ⟨What⟩ is ⟨40⟩ % of ⟨80⟩?
　　　　A　　　P　　　　B

$\dfrac{A}{80} = \dfrac{40}{100}$

$100 \cdot A = 3200$

$\dfrac{\cancel{100}^{1} \cdot A}{\cancel{100}} = \dfrac{3200}{100}$

$A = 32 \text{ men}$

Then, $80 - 32 = 48 \text{ women}$

Exercises 5.1　(page 159)

1. 30,000,000 people　　**3.** Texas　　**5.** California

7. 13,000,000 more people　　**9.** 2 times　　**11.** \$40,000

13. February　　**15.** \$15,000　　**17.** January to February

19. $\$30,000 - \$25,000 = \$5,000$ increase

$\dfrac{5,000}{25,000} = \dfrac{P}{100}$

$\dfrac{1}{5} = \dfrac{P}{100}$

$5P = 100$

$P = 20\%$

21. 17,000,000 sq. mi　　**23.**　　9,500,000
　　　　　　　　　　　　　　　　$+\ $7,000,000
　　　　　　　　　　　　　　　16,500,000 sq. mi

25.　　9,500,000　　　　　**27.** $\dfrac{1}{3}$
　　$-\ $7,000,000
　　　2,500,000 sq. mi

Exercises 5.2   (page 161)

1. 4,000,000 farms   **3.** Decrease

5. 500,000 more farms   **7.** 20%   **9.** 3,500 students

11. 3,500 + 3,500 = 7,000 students

13. 3,500 + 6,000 = 9,500 students   **15.** 20,000 students

17. $\dfrac{7,000}{20,000} = \dfrac{P}{100}$ **19.** $\dfrac{7}{13}$

$\dfrac{7}{20} = \dfrac{P}{100}$

$20P = 700$

$P = 35\%$

Exercises 5.3   (page 165)

1. 20,000,000 people   **3.** 1980   **5.** 6,000,000 people

7. 1940 to 1990   **9.** 25%   **11.** $10   **13.** $25 in 1991

15. Stock A   **17.** 1993 to 1994   **19.** $5 more   **21.** $\dfrac{5}{4}$

Exercises 5.4   (page 168)

1. 34%   **3.** 10% + 5% + 10% = 25%

5. 21–25 years old   **7.** Under 21 years old

9. 5,000 students   **11.** 7,500 students   **13.** 40%

15. 20%   **17.** Husband's salary   **19.** 15%

21. $15,000   **23.** $1,000

Exercises 5.5   (page 172)

1. a. 5   **b.** 4   **c.** 4   **d.** 7

3. a. 25   **b.** 24   **c.** None   **d.** 12

5. a. 8.1 mpg   **b.** 19.8 mpg   **c.** 20 mpg

7. a. 173   **b.** 176   **c.** 54

9. a. 7.5   **b.** 7   **c.** 6   **d.** 4

11. a. 740 hr   **b.** 726 hr   **c.** 147 hr

Chapter 5 Review Exercises (page 174)

1. 60°F   **2.** 70°F   **3.** February   **4.** August

5. February to August   **6.** August to February

7. February to March   **8.** September to October

9. 10%   **10.** $33\dfrac{1}{3}\%$ **11.** 250 ft   **12.** 175 ft

13. 20 mph   **14.** 40 mph   **15.** 80 ft   **16.** 130 ft

17. 30%   **18.** 24%   **19.** 96%   **20.** 58%

21. 420 people   **22.** 42,000 people   **23.** $\dfrac{4}{5}$ **24.** $\dfrac{7}{5}$

25. Saturday   **26.** Wednesday   **27.** $8,000   **28.** $7,500

29. Monday   **30.** Wednesday   **31.** $5,500   **32.** $50,000

33. $\dfrac{10,000}{50,000} = \dfrac{P}{100}$ **34.** 9%   **35.** $\dfrac{9}{20}$ **36.** $\dfrac{3}{4}$ **37.** 5

$\dfrac{1}{5} = \dfrac{P}{100}$

$5P = 100$

$P = 20\%$

38. 5.5   **39.** 6   **40.** 6   **41.** 67°F   **42.** 67.5°F

43. 60°F and 80°F   **44.** 40°F

Chapter 5 Diagnostic Test   (page 176)

Following each problem number is the number (in parentheses) of the textbook section in which that kind of problem is discussed.

1. (5.3) 60°F   **2.** (5.3) 2 P.M.

3. (5.3) 70°F − 65°F = 5°F   **4.** (5.3) 6 A.M. to 2 P.M.

5. (5.3) 75° − 60° = 15° increase   **6.** (5.1) Cheerios

$\dfrac{15}{60} = \dfrac{P}{100}$

$60P = 1,500$

$\dfrac{\overset{1}{\cancel{60}}P}{\underset{1}{\cancel{60}}} = \dfrac{1,500}{60}$

$P = 25\%$

7. (5.1) 3 g   **8.** (5.1) Raisin Bran   **9.** (5.1) 4 times

10. (5.1) 11 − 3 = 8 g   **11.** (5.2) 50 cars

12. (5.2) Carson   **13.** (5.2) $\dfrac{60 \text{ cars}}{90 \text{ cars}} = \dfrac{2}{3}$

14. (5.2) 10 × 20 = 200 cars

15. (5.2) $\dfrac{60}{200} = \dfrac{P}{100}$ **16.** (5.4) 10%

$200P = 6,000$

$\dfrac{\overset{1}{\cancel{200}}P}{\underset{1}{\cancel{200}}} = \dfrac{6,000}{200}$

$P = 30\%$

17. (5.4) 35% + 25% + 10% = 70%   **18.** (5.4) C

19. (5.4) $\dfrac{A}{48} = \dfrac{25}{100}$

$\dfrac{A}{48} = \dfrac{1}{4}$

$4A = 48$

$A = 12$ students

20. (5.5) $\dfrac{87 + 72 + 82 + 93 + 71}{5} = \dfrac{405}{5} = 81$

21. (5.5) 71, 72, 82, 87, 93

82 is the median

22. (5.5) 71°F − 53°F = 18°F

23. (5.5) $\dfrac{53 + 66 + 71 + 62}{4} = \dfrac{252}{4} = 63°F$

24. (5.5) 2, 2, 4, 5, 6, 6, 6, 9

$\dfrac{5 + 6}{2} = \dfrac{11}{2} = 5.5$

25. (5.5) 6

Chapters 1–5 Cumulative Review Exercises   (page 178)

1. 5,700.09   **2.** $1\dfrac{5}{9}$ **3.** $12\dfrac{3}{4}$ **4.** $4\dfrac{3}{8}$ **5.** $1\dfrac{5}{9}$

6. 49.12   **7.** 4.608   **8.** 350   **9.** 16   **10.** $\dfrac{5}{6}, \dfrac{3}{4}, \dfrac{7}{12}$

11. 6   **12.** $\dfrac{3}{125}$ **13.** 40%   **14.** $\dfrac{9}{20}$ **15.** 80

16. 60 mi   **17.** 56 cm   **18.** No   **19.** $45

20. 6.1 in.   **21.** 19 yd   **22.** $96   **23.** 15%

24. $3,250   **25.** $1,100

Exercises 6.1 (page 183)

1. Negative seventy-five   **3.** −54   **5.** −62   **7.** −2
9. −10   **11.** −1   **13.** Cannot be found   **15.** *F*
17. *D*   **19.** *A*   **21.** *E*   **23.** >   **25.** <   **27.** <
29. >   **31.** Mt. Mitchell

Exercises 6.2 (page 188)

1. 9   **3.** −7   **5.** −1   **7.** 4   **9.** −11   **11.** 3
13. 11   **15.** −3   **17.** −9   **19.** −7   **21.** 0
23. 7   **25.** −12   **27.** −9   **29.** −6   **31.** 3
33. −9   **35.** 2   **37.** −21   **39.** 7   **41.** 10
43. −40   **45.** 41   **47.** −32

49.
$$\left(-1\frac{1}{2}\right) = \left(-1\frac{5}{10}\right)$$
$$+\left(-3\frac{2}{5}\right) = \left(-3\frac{4}{10}\right)$$
$$\overline{-4\frac{9}{10}}$$

51.
$$\left(4\frac{5}{6}\right) = \left(4\frac{5}{6}\right)$$
$$+\left(-1\frac{1}{3}\right) = \left(-1\frac{2}{6}\right)$$
$$\overline{3\frac{3}{6} = 3\frac{1}{2}}$$

53. −12.78   **55.** 0.084

57. $(-35) + (53) = 18°F$ **59.** −7   **61.** 6   **63.** $-\frac{2}{3}$

65. 8 ⎡+/−⎤ ⎡+⎤ 3 ⎡=⎤ −5
67. 12 ⎡+/−⎤ ⎡+⎤ 26 ⎡+/−⎤ ⎡=⎤ −38

Exercises 6.3 (page 190)

1. 6   **3.** −1   **5.** −8   **7.** 14   **9.** 9   **11.** 2
13. −7   **15.** −4   **17.** 3   **19.** −12   **21.** −7
23. −6   **25.** 14   **27.** 7   **29.** −12   **31.** −3
33. −4   **35.** 5   **37.** −25   **39.** −55   **41.** −59

43. 663   **45.**
$$\left(-4\frac{2}{3}\right) = \left(-4\frac{4}{6}\right)$$
$$-\left(2\frac{1}{6}\right) = +\left(-2\frac{1}{6}\right)$$
$$\overline{-6\frac{5}{6}}$$

47.
$$\left(-3\frac{3}{4}\right) = \left(-3\frac{9}{12}\right)$$
$$-\left(-2\frac{1}{6}\right) = +\left(+2\frac{2}{12}\right)$$
$$\overline{-1\frac{7}{12}}$$

49. $(-10) - (-7) = (-10) + (7) = -3$ **51.** −9.36

53. −47.67   **55.**
$$\left(-5\frac{3}{10}\right) = \left(-5\frac{6}{20}\right)$$
$$-\left(+3\frac{1}{4}\right) = +\left(-3\frac{5}{20}\right)$$
$$\overline{-8\frac{11}{20}}$$

57. $(42) - (-7) = (42) + (7) = 49°F$ rise   **59.** 2,022 ft
61. $(-141) - (68) = (-141) + (-68) = -209$ ft
63. 8 ⎡−⎤ 2 ⎡+/−⎤ ⎡=⎤ 10
65. 17 ⎡+/−⎤ ⎡−⎤ 6 ⎡+/−⎤ ⎡=⎤ −11

Exercises 6.4 (page 194)

1. −6   **3.** −10   **5.** 16   **7.** −32   **9.** −18
11. −54   **13.** 28   **15.** −30   **17.** −63   **19.** 100
21. −56   **23.** −260   **25.** 200   **27.** −1,125
29. −150   **31.** 140   **33.** −40   **35.** 56

37. $\left(2\frac{1}{4}\right)\left(-1\frac{1}{3}\right) = -\left(\dfrac{\overset{3}{\cancel{9}}}{\cancel{4}} \cdot \dfrac{\overset{1}{\cancel{4}}}{\cancel{3}}\right) = -3$

39. $\left(-1\frac{7}{8}\right)\left(-2\frac{4}{5}\right) = +\left(\dfrac{\overset{3}{\cancel{15}}}{\cancel{8}} \cdot \dfrac{\overset{7}{\cancel{14}}}{\cancel{5}}\right) = \dfrac{21}{4} = 5\frac{1}{4}$ **41.** −274

43. 38.18   **45.** $\frac{1}{7}$ **47.** $-\frac{1}{6}$ **49.** $\frac{3}{2}$

51. 9 ⎡+/−⎤ ⎡×⎤ 8 ⎡=⎤ −72
53. 15 ⎡+/−⎤ ⎡×⎤ 25 ⎡+/−⎤ ⎡=⎤ 375

Exercises 6.5 (page 196)

1. 2   **3.** −4   **5.** −5   **7.** 2   **9.** −5   **11.** −4
13. 3   **15.** −3   **17.** 9   **19.** −15   **21.** −3
23. −3   **25.** $\dfrac{-15}{6} = \dfrac{-5}{2} = -2\frac{1}{2}$, or −2.5   **27.** −15
29. 7   **31.** −3.67

33. $\dfrac{2\frac{1}{2}}{-5} = \dfrac{\frac{5}{2}}{-\frac{5}{1}} = \dfrac{5}{2} \div \left(-\dfrac{5}{1}\right) = -\left(\dfrac{\overset{1}{\cancel{5}}}{2} \cdot \dfrac{1}{\cancel{5}}\right) = -\dfrac{1}{2}$

35. $\dfrac{-4\frac{1}{2}}{-1\frac{7}{8}} = \dfrac{-\frac{9}{2}}{-\frac{15}{8}} = \left(-\dfrac{9}{2}\right) \div \left(-\dfrac{15}{8}\right) = +\left(\dfrac{\overset{3}{\cancel{9}}}{\cancel{2}} \cdot \dfrac{\overset{4}{\cancel{8}}}{\cancel{15}}\right) = \dfrac{12}{5}$
$$= 2\frac{2}{5}$$

37. 84 ⎡+/−⎤ ⎡÷⎤ 7 ⎡=⎤ −12
39. 132 ⎡+/−⎤ ⎡÷⎤ 12 ⎡+/−⎤ ⎡=⎤ 11

Exercises 6.6 (page 199)

1. True, because of the commutative property of addition (order of numbers changed)
3. True, because of the associative property of addition (grouping changed)
5. False; the commutative property does not hold for subtraction.
$$6 - 2 \overset{?}{=} 2 - 6$$
$$4 \neq -4$$
7. True, because of the associative property of multiplication
9. False; the commutative property does not hold for division.
$$8 \div 4 \overset{?}{=} 4 \div 8$$
$$2 \neq \frac{1}{2}$$
11. True, because of the commutative property of multiplication
13. False: $(4)(-5) \overset{?}{=} (-5) + (4)$
$$-20 \neq -1$$

15. True, because of the commutative property of addition

17. True, because of the commutative property of multiplication

19. False; division is not associative.

21. False; subtraction is not commutative.

23. True, because of the associative property of multiplication (grouping changed)

25. True, because of the commutative property of addition (order changed)

27. False; subtraction is not associative.

29. True, because of the commutative property of multiplication

Exercises 6.7   (page 200)

1. 0   **3.** 4   **5.** −6   **7.** 0   **9.** 11

11. Undefined   **13.** 0   **15.** Undefined   **17.** −789

19. Undefined

21. $9 \div 0 = E \longleftarrow$ The E in the display indicates that an error has been made. In this case, the error is division by 0, which is undefined.

23. Undefined   **25.** 0

Exercises 6.8   (page 203)

1. 27   **3.** $(-5)^2 = (-5)(-5) = 25$ **5.** 49   **7.** 0

9. −10   **11.** 1,000   **13.** −100,000   **15.** 4

17. $-2^2 = -(2 \cdot 2) = -4$ **19.** 625   **21.** 64,000

23. −16   **25.** −1   **27.** 1   **29.** 1   **31.** −16

33. $\frac{9}{16}$ **35.** $-\frac{1}{1,000}$ **37.** 59,049   **39.** −16,384

41. 5.0625

Exercises 6.9   (page 205)

1. 4 because $4^2 = 16$ **3.** $-\sqrt{4} = -(\sqrt{4}) = -2$

5. 9   **7.** 10   **9.** −7   **11.** 8   **13.** −10   **15.** $\frac{4}{5}$

17. 3.606   **19.** 2.646   **21.** 7.071   **23.** 13.565

Chapter 6 Review Exercises (page 207)

1. 8, 9   **2.** 0, 1, 2   **3.** 9   **4.** 1   **5.** Cannot be done

6. −1   **7. a.** <   **b.** <   **c.** >   **d.** >   **e.** <

8. a. True, because of the associative property of multiplication
b. True, because of the commutative property of addition
c. False: $5 - (-2) = 5 + 2 = 7$
   $(-2) - 5 = (-2) + (-5) = -7$
d. True, because of the associative property of addition
e. True, because of the commutative property of multiplication
f. True, because of the commutative property of addition

9. 3   **10.** $(-5) - (-3) = (-5) + (3) = -2$ **11.** 3

12. 12   **13.** −11   **14.** −72   **15.** 4   **16.** −6

17. −9   **18.** −6   **19.** −42   **20.** −10

21. $-3^2 = -(3 \cdot 3) = -9$ **22.** −4   **23.** 3   **24.** −3

25. 0   **26.** $(-5)^2 = (-5)(-5) = 25$ **27.** −12

28. −24   **29.** 5   **30.** 0   **31.** 6

32. $-2^4 = -(2 \cdot 2 \cdot 2 \cdot 2) = -16$ **33.** 13   **34.** −5

35. 48   **36.** 9   **37.** −7   **38.** 0   **39.** −14   **40.** 5

41. −11   **42.** −9   **43.** 1   **44.** −7   **45.** −126

46. $(-3)^3 = (-3)(-3)(-3) = -27$ **47.** 12   **48.** −5

49. −3   **50.** 5   **51.** 4   **52.** 80   **53.** −17

54. 10,000   **55.** −11   **56.** −16   **57.** Undefined

58. $-\frac{4}{3} = -1\frac{1}{3}$ **59.** $2\frac{1}{6}$ **60.** $-1\frac{13}{20}$ **61.** $-12\frac{1}{2}$

62. Undefined   **63.** $1\frac{5}{22}$ **64.** $-5\frac{11}{12}$

65. $(-7) - (9) = (-7) + (-9) = -16$ **66.** −8

67. 7.280   **68.** 9.592   **69.** 12.369   **70.** 13.601

Chapter 6 Diagnostic Test   (page 208)

Following each problem number is the number (in parentheses) of the textbook section in which that kind of problem is discussed.

1. (6.1) <   **2.** (6.1) >   **3.** (6.1) <   **4.** (6.1) 0, 1, 2

5. (6.1) 1   **6.** (6.3) $(38) - (-20) = (38) + (+20) = 58°F$

7. (6.1) Negative thirty-five

8. (6.6) True, because of the commutative property of addition

9. (6.6) True, because of the associative property of addition

10. (6.6) True, because of the commutative property of multiplication

11. (6.6) False; subtraction is not commutative.
   $7 - 4 \overset{?}{=} 4 - 7$
   $3 \neq -3$

12. (6.2) −4   **13.** (6.3) $20 - (-10) = 20 + (+10) = 30$

14. (6.4) 50   **15.** (6.5) −3   **16.** (6.8) 36   **17.** (6.2) 3

18. (6.4) −36

19. (6.3) $(-10) - (+5) = (-10) + (-5) = -15$

20. (6.7) Undefined   **21.** (6.4) −40   **22.** (6.5) $-\frac{5}{6}$

23. (6.8) −64   **24.** (6.2) 7

25. (6.3) $(-9) - (-7) = (-9) + (7) = -2$ **26.** (6.4) −42

27. (6.3) $(5) - (12) = (5) + (-12) = -7$

28. (6.3) $(-2) - (-6) = (-2) + (6) = 4$ **29.** (6.2) −11

30. (6.8) 1   **31.** (6.4) −54   **32.** (6.5) 4

33. (6.3) $(6) - (10) = (6) + (-10) = -4$ **34.** (6.2) −9

35. (6.4) −24   **36.** (6.2) 5   **37.** (6.8) 0

38. (6.3) $(-4) - (8) = (-4) + (-8) = -12$ **39.** (6.2) −7

40. (6.4) $\left(-1\frac{7}{8}\right)\left(2\frac{2}{5}\right) = -\left(\frac{\overset{3}{\cancel{15}}}{\underset{2}{\cancel{8}}} \cdot \frac{\overset{3}{\cancel{12}}}{\underset{1}{\cancel{5}}}\right) = -\frac{9}{2} = -4\frac{1}{2}$

41. (6.2) $\begin{array}{r} \left(-2\frac{3}{4}\right) = \left(-2\frac{6}{8}\right) \\ + \left(-1\frac{5}{8}\right) = \left(-1\frac{5}{8}\right) \\ \hline -3\frac{11}{8} = -4\frac{3}{8} \end{array}$ **42.** (6.5) −3

43. (6.8) $(-2)^4 = (-2)(-2)(-2)(-2) = 16$ **44.** (6.7) 0

45. (6.7) 0   **46.** (6.3) $(9) - (-5) = (9) + (5) = 14$

47. (6.2) −5   **48.** (6.9) 7   **49.** (6.9) 9   **50.** (6.9) −8

Exercises 7.1   (page 212)

1. −2   **3.** −14   **5.** 15   **7.** 3   **9.** 25   **11.** −1

13. 18   **15.** 36   **17.** 0   **19.** 50   **21.** 20

23. $-3^2 - 4^2 = -9 - 16 = -9 + (-16) = -25$ **25.** −6

27. 36   **29.** $-2(5-3)-5(7-3)$ **31.** 20
$$= -2 \cdot 2 \quad - 5 \cdot 4$$
$$= \quad -4 \quad - \quad 20$$
$$= \quad -4 \quad + (-20)$$
$$= -24$$

33. $-6 - 8 \div 2 \cdot 4$ **35.** -28
$$= -6 - \quad 4 \ \cdot 4$$
$$= -6 - \quad 16$$
$$= -22$$

37. $-4^2 - 2(-5) - 2\sqrt{9}$ **39.** -7 **41.** 11
$$= -16 - 2(-5) - 2 \cdot 3$$
$$= -16 - (-10) - \quad 6$$
$$= -16 + \quad 10 \quad + (-6)$$
$$= \quad -6 \quad + (-6)$$
$$= -12$$

43. $2 \cdot \dfrac{7}{16} + \dfrac{9}{20} \div \dfrac{3}{5}$ **45.** $\left(\dfrac{2}{3}\right)^2 + 3\dfrac{1}{3} \cdot \dfrac{1}{4}$

$$= \dfrac{\overset{1}{\cancel{2}}}{1} \cdot \dfrac{7}{\underset{8}{\cancel{16}}} + \dfrac{\overset{3}{\cancel{9}}}{\underset{4}{\cancel{20}}} \cdot \dfrac{\overset{1}{\cancel{5}}}{\underset{1}{\cancel{3}}} \qquad = \dfrac{4}{9} + \dfrac{\overset{5}{\cancel{10}}}{3} \cdot \dfrac{1}{\underset{2}{\cancel{4}}}$$

$$= \dfrac{7}{8} + \dfrac{3}{4} \qquad\qquad = \dfrac{4}{9} + \dfrac{5}{6}$$

$$= \dfrac{7}{8} + \dfrac{6}{8} \qquad\qquad = \dfrac{8}{18} + \dfrac{15}{18}$$

$$= \qquad \dfrac{13}{8} \qquad\qquad = \qquad \dfrac{23}{18}$$

$$= 1\dfrac{5}{8} \qquad\qquad = 1\dfrac{5}{18}$$

47. $2.3 + 5(3.7) \div 100$ **49.** -98 **51.** -34
$$= 2.3 + \quad 18.5 \div 100$$
$$= 2.3 + \qquad 0.185$$
$$= 2.485$$

53. 43   **55.** -72 **57.** 132

Exercises 7.2   (page 215)

1. -27 **3.** 12   **5.** -6 **7.** 12

9. $9 + 2[3 - (-4)]$ **11.** 5   **13.** -1 **15.** 1
$$= 9 + 2[3 + 4]$$
$$= 9 + 2[7]$$
$$= 9 + \quad 14$$
$$= 23$$

17. 5   **19.** 60   **21.** $-4 - 3[5 + (-7)]$
$$= -4 - 3[-2]$$
$$= -4 - (-6)$$
$$= -4 + \quad 6$$
$$= 2$$

23. $-2^2 + [3 + (-6)]^2$ **25.** -6
$$= -2^2 + [-3]^2$$
$$= -4 + \quad 9$$
$$= 5$$

27. $6 - 3\{8 - [4 - (-1)]\}$ **29.** 20   **31.** -3
$$= 6 - 3\{8 - [4 + 1]\}$$
$$= 6 - 3\{8 - 5\}$$
$$= 6 - 3\{3\}$$
$$= 6 - \quad 9$$
$$= -3$$

33. -2 **35.** $\dfrac{3^2 + 5}{2} - \dfrac{(-4)^2}{8}$
$$= \dfrac{9 + 5}{2} - \dfrac{16}{8}$$

$$= \dfrac{14}{2} - \dfrac{16}{8}$$
$$= \quad 7 \quad - \quad 2$$
$$= 5$$

37. 5   **39.** -2 **41.** $15 - \{4 - [2 - 3(6 - 4)]\}$
$$= 15 - \{4 - [2 - 3(2)]\}$$
$$= 15 - \{4 - [2 - 6]\}$$
$$= 15 - \{4 - [-4]\}$$
$$= 15 - \{8\}$$
$$= 7$$

Exercises 7.3   (page 217)

1. -20 **3.** -4 **5.** 25   **7.** -49 **9.** 147

11. 21   **13.** 21   **15.** $3b - ab + xy$
$$= 3(-5) - (3)(-5) + (4)(-7)$$
$$= -15 + \quad 15 \quad + (-28)$$
$$= -28$$

17. -33 **19.** -5 **21.** 19   **23.** 13

25. $a^2 - 2ab + b^2$ **27.** 64   **29.** -1
$$= (3)^2 - 2(3)(-5) + (-5)^2$$
$$= 9 - 2(3)(-5) + 25$$
$$= 9 + \quad 30 \quad + 25$$
$$= 64$$

31. $\dfrac{1}{2}$ **33.** 0   **35.** 11   **37.** $-\dfrac{2}{3}$

39. $\dfrac{(1 + G)^2 - 1}{H}$ **41.** -1

$$= \dfrac{[1 + (-5)]^2 - 1}{-4}$$

$$= \dfrac{[-4]^2 - 1}{-4}$$

$$= \dfrac{16 - 1}{-4}$$

$$= \dfrac{15}{-4}$$

$$= -3\dfrac{3}{4}$$

43. $G - \sqrt{G - 4EH}$
$$= (-5) - \sqrt{(-5)^2 - 4(-1)(-4)}$$
$$= (-5) - \sqrt{25 - 16}$$
$$= (-5) - \sqrt{9}$$
$$= (-5) - 3$$
$$= -8$$

45. $\dfrac{3x^2}{z} = \dfrac{3\left(\dfrac{1}{2}\right)^2}{\left(\dfrac{3}{8}\right)} = \dfrac{\dfrac{3}{1} \cdot \dfrac{1}{4}}{\dfrac{3}{8}} = \dfrac{\dfrac{3}{4}}{\dfrac{3}{8}} = \dfrac{3}{4} \div \dfrac{3}{8} = \dfrac{\overset{1}{\cancel{3}}}{\underset{1}{\cancel{4}}} \cdot \dfrac{\overset{2}{\cancel{8}}}{\underset{1}{\cancel{3}}} = 2$

47. $(x + y + z)^2$

$$= \left[\left(\dfrac{1}{2}\right) + \left(-\dfrac{3}{4}\right) + \left(\dfrac{3}{8}\right)\right]^2$$

$$= \left[\dfrac{4}{8} + \left(-\dfrac{6}{8}\right) + \dfrac{3}{8}\right]^2$$

$$= \left[\dfrac{1}{8}\right]^2$$

$$= \dfrac{1}{64}$$

Exercises 7.4   (page 219)

1. 105   **3.** 5   **5.** 243   **7.** 77   **9.** 314   **11.** 144
13. 400   **15.** $\sigma = \sqrt{npq}$ **17.** -20
                    $\sigma = \sqrt{(100)(0.9)(0.1)}$
                    $\sigma = \sqrt{9}$
                    $\sigma = 3$

19. $C = \dfrac{a}{a+12} \cdot A$ **21.** 1,628.89   **23.** 9.4

$C = \dfrac{6}{6+12} \cdot 30$

$C = \dfrac{\overset{1}{\cancel{6}}}{\underset{\underset{1}{3}}{\cancel{18}}} \cdot \dfrac{\overset{10}{\cancel{30}}}{1}$

$C = 10$

Chapter 7 Review Exercises (page 220)

1. 9   **2.** $11 - 7 \cdot 3$ **3.** 9   **4.** -12 **5.** 18
                    $= 11 - 21$
                    $= -10$

6. 23   **7.** 7   **8.** 10   **9.** $6 - [8 - (3-4)]$
                                            $= 6 - [8 - (-1)]$
                                            $= 6 - \quad\quad 9$
                                            $= -3$

10. 16   **11.** -5 **12.** 45   **13.** 2   **14.** -17
15. 21   **16.** -1 **17.** $9 - \{6 - [4 - (7-8)]\}$
                                            $= 9 - \{6 - [4 - (-1)]\}$
                                            $= 9 - \{6 - 5\}$
                                            $= 9 - 1$
                                            $= 8$

18. 3   **19.** -2 **20.** $\left(\dfrac{3}{4} - \dfrac{1}{3}\right) \div \left(\dfrac{4}{9} + \dfrac{2}{3}\right)$

$= \left(\dfrac{9}{12} - \dfrac{4}{12}\right) \div \left(\dfrac{4}{9} + \dfrac{6}{9}\right)$

$= \quad\quad \dfrac{5}{12} \quad\quad \div \quad\quad \dfrac{10}{9}$

$= \quad\quad \dfrac{\overset{1}{\cancel{5}}}{\underset{4}{\cancel{12}}} \quad \cdot \quad \dfrac{\overset{3}{\cancel{9}}}{\underset{2}{\cancel{10}}}$

$= \dfrac{3}{8}$

21. 16   **22.** -14 **23.** 2   **24.** 5   **25.** -13
26. $6 - x[y - z]$ **27.** 36
        $= 6 - (-2)[3 - (-4)]$
        $= 6 - (-2)[7]$
        $= 6 - (-14)$
        $= 6 + 14$
        $= 20$

28. $y^2 - 2yz + z^2$
        $= (3)^2 - 2(3)(-4) + (-4)^2$
        $= 9 - 2(3)(-4) + 16$
        $= 9 - 6(-4) + 16$
        $= 9 + 24 + 16$
        $= 49$

29. $x - 2(y - xz)$
        $= -2 - 2[3 - (-2)(-4)]$
        $= -2 - 2[3 - 8]$
        $= -2 - 2[-5]$
        $= -2 - (-10)$
        $= -2 + 10$
        $= 8$

30. $\dfrac{(x+y)^2 - z^2}{x - 2y}$

$= \dfrac{(-2+3)^2 - (-4)^2}{-2 - 2(3)}$

$= \dfrac{1 - 16}{-2 - 6}$

$= \dfrac{-15}{-8}$

$= 1\dfrac{7}{8}$

31. $(x^2 + y^2)(x^2 - y^2)$
        $= [(-2)^2 + (3)^2][(-2)^2 - (3)^2]$
        $= [4 + 9][4 - 9]$
        $= \quad [13] \quad\quad [-5]$
        $= -65$

32. $(x - y)(x^2 + xy + y^2)$ **33.** 14
        $= [(-2) - (3)][(-2)^2 + (-2)(3) + (3)^2]$
        $= \quad [-5] \quad\quad [4 + (-6) + 9]$
        $= \quad [-5] \quad\quad\quad\quad [7]$
        $= -35$

34. 31.5   **35.** $C = \dfrac{5}{9}(F - 32)$ **36.** 5   **37.** 1,300

$C = \dfrac{5}{9}\left(15\dfrac{1}{2} - 32\right)$

$C = \dfrac{5}{9}\left(\dfrac{31}{2} - \dfrac{64}{2}\right)$

$C = \dfrac{5}{\underset{3}{\cancel{9}}}\left(-\dfrac{\overset{11}{\cancel{33}}}{2}\right)$

$C = -\dfrac{55}{6}$

$C = -9\dfrac{1}{6}$

38. $s = \dfrac{1}{2}gt^2$ **39.** 5   **40.** 628

$s = \dfrac{1}{2}(32)\left(\dfrac{3}{2}\right)^2$

$s = \dfrac{1}{\underset{1}{\cancel{2}}} \cdot \dfrac{\overset{\overset{4}{\cancel{8}}}{\cancel{32}}}{1} \cdot \dfrac{9}{\underset{1}{\cancel{4}}}$

$s = 36$

Chapter 7 Diagnostic Test (page 222)

Following each problem number is the number (in parentheses) of the textbook section in which that kind of problem is discussed.

1. (7.1) $17 - 9 - 6 + 11$ **2.** (7.1) $5 + 2 \cdot 3$
                $= \quad 8 \quad - 6 + 11$ $= 5 + 6$
                $= \quad\quad 2 \quad\quad + 11$ $= 11$
                $= 13$

3. (7.1) $-12 \div 2 \cdot 3$ **4.** (7.1) $-5^2 + (-4)^2$
                $= \quad -6 \quad \cdot 3$ $= -25 + 16$
                $= -18$ $= -9$

5. (7.1) $2 \cdot 3^2 - 4$ **6.** (7.1) $3\sqrt{25} - 5(-4)$
                $= 2 \cdot 9 - 4$ $= 3 \cdot 5 - 5(-4)$
                $= \quad 18 - 4$ $= \quad 15 + 20$
                $= 14$ $= 35$

7. (7.2) $\dfrac{8 - 12}{-6 + 2} = \dfrac{-4}{-4} = 1$ **8.** (7.1) $(2^3 - 7)(5^2 + 4^2)$
                                                                                            $= (8 - 7)(25 + 16)$
                                                                                            $= \quad (1) \quad\quad (41)$
                                                                                            $= 41$

8

9. (7.2) $\quad 10 - [6 - (5 - 7)]$
$= 10 - [6 - (-2)]$
$= 10 - [6 + 2]$
$= 10 - \quad 8$
$= 2$

10. (7.1) $\quad 2 + (6 - 3 \cdot 4)$
$= 2 + (6 - \quad 12 \ )$
$= 2 + \quad (-6)$
$= -4$

11. (7.2) $\quad \sqrt{10^2 - 6^2}$
$= \sqrt{100 - 36}$
$= \sqrt{64}$
$= 8$

12. (7.2) $\quad 5 - 2[4 - (6 - 9)]$
$= 5 - 2[4 - (-3)]$
$= 5 - 2[4 + 3]$
$= 5 - 2 \ [7]$
$= 5 - \quad 14$
$= -9$

13. (7.3) $\quad 3a \ + \ bx \ - \quad cy$
$= 3(-2) + 4(5) - (-3)(-6)$
$= \ -6 \ + \ 20 \ - \quad 18$
$= \quad \quad 14 \quad - \quad 18$
$= -4$

14. (7.3) $\quad 4x \ - \quad [a - (3c - b)]$
$= 4(5) - [-2 - \{3(-3) - 4\}]$
$= 4(5) - [-2 - \{-9 \ - 4\}]$
$= 4(5) - [-2 - \quad \{-13\} \ ]$
$= 4(5) - \quad \quad [11]$
$= 20 - \quad \quad 11$
$= 9$

15. (7.3) $\quad x^2 + \quad 2xy \quad - \quad y^2$
$= (5)^2 + 2(5)(-6) - (-6)^2$
$= \ 25 + 2(5)(-6) - \ 36$
$= \ 25 + \ (-60) \ - \ 36$
$= \quad \quad -35 \quad - \ 36$
$= -71$

16. (7.4) $\quad C = \dfrac{5}{9}(F - 32); \quad F = -4$
$C = \dfrac{5}{9}(-4 - 32)$
$C = \dfrac{5}{\underset{1}{\cancel{9}}}\left(-\dfrac{\overset{4}{\cancel{36}}}{1}\right)$
$C = -20$

17. (7.4) $\quad \pi \approx 3.14, \ r = 20$
$A = \pi r^2$
$A \approx (3.14)(20)^2$
$A \approx 3.14(400)$
$A \approx 1{,}256$

18. (7.4) $\quad a = 7, \ A = 38$
$C = \dfrac{a}{a + 12} \cdot A$
$C = \dfrac{7}{7 + 12}(38)$
$C = \dfrac{7}{\underset{1}{\cancel{19}}}\left(\dfrac{\overset{2}{\cancel{38}}}{1}\right)$
$C = 14$

19. (7.4) $\quad P = 600, \ r = 0.10, \ t = 2.5$
$A = P(1 + rt)$
$A = 600(1 + 0.10(2.5))$
$A = 600(1 + 0.25)$
$A = 600(1.25)$
$A = 750$

20. (7.4) $\quad C = 600, \ r = 0.04, \ t = 10$
$V = C - Crt$
$V = 600 - 600(0.04)(10)$
$V = 600 - 240$
$V = 360$

Exercises 8.1 (page 225)

1. a. One term   **b.** No second term
3. a. Three terms   **b.** $-5x$
5. a. One term   **b.** No second term
7. a. Two terms   **b.** $-5(x + 3)$
9. a. Two terms   **b.** $b(x + y)$
11. a. Three terms   **b.** $-\dfrac{x + y}{2}$ **13.** -3 **15.** 2
17. -1 **19.** 1   **21.** $\dfrac{3}{4}$ **23.** $-\dfrac{1}{5}$

Exercises 8.2 (page 229)

1. x^{13} **3.** y^{10} **5.** x^5 **7.** 1   **9.** a^5 **11.** x^{28}
13. a^4 **15.** z **17.** 10,000,000   **19.** 1,000,000
21. 10,000,000,000   **23.** 5
25. $x^2 y^3$ cannot be simplified because the bases are different.
27. $\dfrac{6x^2}{2x} = \dfrac{\overset{3}{\cancel{6}}}{\underset{1}{\cancel{2}}} \cdot \dfrac{x^2}{x} = \dfrac{3}{1} \cdot \dfrac{x}{1} = 3x$
29. $\dfrac{a^3}{b^2}$ cannot be simplified because the bases are different.
31. $\dfrac{10x^4}{5x^3} = \dfrac{\overset{2}{\cancel{10}}}{\underset{}{\cancel{5}}} \cdot \dfrac{x^4}{x^3} = \dfrac{2}{1} \cdot \dfrac{x}{1} = 2x$
33. $\dfrac{12h^4 k^3}{8h^2 k} = \dfrac{12}{8} \cdot \dfrac{h^4}{h^2} \cdot \dfrac{k^3}{k} = \dfrac{3}{2} \cdot \dfrac{h^2}{1} \cdot \dfrac{k^2}{1} = \dfrac{3h^2 k^2}{2}$ **35.** 5^{u+v}
37. Same form. See Word of Caution, page 228.   **39.** y^6
41. a^2 **43.** z^8 **45.** 1   **47.** 729
49. Cannot be simplified   **51.** Cannot be simplified
53. $\dfrac{5a^2 b}{3}$ **55.** $2 \cdot 2^3 \cdot 2^2 = 2^{1+3+2} = 2^6 = 64$
57. $x \cdot x^3 \cdot x^4 = x^{1+3+4} = x^8$ **59.** $x^2 y^3 x^5 = x^{2+5}y^3 = x^7 y^3$
61. $ab^3 a^5 = a^{1+5}b^3 = a^6 b^3$ **63.** $3^{a \cdot b}$ **65.** 1

Exercises 8.3 (page 233)

1. $(-2a)(4a^2) = -(2 \cdot 4)(a \cdot a^2) = -8a^3$ **3.** $-35x^2 y$
5. $64x^6$ **7.** $30x^6$ **9.** $-16m^6 n^5$
11. $(5 \cdot 2)(x^2 \cdot x)(y \cdot y)(z^3 \cdot z) = 10x^3 y^2 z^4$ **13.** $5a + 30$
15. $3m - 12$ **17.** $-2x + 6$ **19.** $-6x^2 + 12x - 15$
21. $12x^3 - 24x$ **23.** $-10x^3 - 6x^2 + 8x$ **25.** $-x^2 + y^2$
27. $-x - 3$ **29.** $-2x^2 - 4x + 7$ **31.** $6x - 24$
33. $7y^2 - 28y + 21$ **35.** $8x^3 - 12x^2 + 20x$ **37.** $x^2 y - 3x$
39. $3a^2 b - 6a^3$ **41.** $6xy - 12x^2 y^2$
43. $\quad -2xy(x^2 y - y^2 x - y - 5)$
$= (-2xy)(x^2 y) + (-2xy)(-y^2 x) + (-2xy)(-y) + (-2xy)(-5)$
$= -2x^3 y^2 + 2x^2 y^3 + 2xy^2 + 10xy$
45. $\quad (3x^3 - 2x^2 y + y^3)(-2xy)$
$= (3x^3)(-2xy) + (-2x^2 y)(-2xy) + (y^3)(-2xy)$
$= -6x^4 y + 4x^3 y^2 - 2xy^4$
47. $12a^5 b^4 - 8a^2 b^3 - 20ab^3$
49. $-12x^3 y^3 z^2 + 6xy^3 z^3 + 8x^2 y^2 z^4$

Exercises 8.4 (page 235)

1. $12x$ **3.** $-7a$ **5.** $4x^2$ **7.** $8x^3$ **9.** $-7xy$
11. $4x^2 + 3x$ **13.** $3a$ **15.** x **17.** $2y$ **19.** 0

21. $-2xy$ **23.** $-2a^2b$ **25.** $3x + 10$ **27.** $5x - 11$

29. $-4x^2 - 7x$ **31.** $3x^2 + 5x - 2$ **33.** $-2x^3 - 2x^2 - 4x$

35. $x^3 - 5x^2 - x + 3$ **37.** $5x^2 + 5xy - 6y^2$

39. $3x^2y - 3xy^2$

41. $\dfrac{2}{3}y^2 - \dfrac{1}{2}y^2 + \dfrac{5}{6}y^2$ **43.** 15.02 sec

$= \left(\dfrac{4}{6} - \dfrac{3}{6} + \dfrac{5}{6}\right)y^2$

$= \dfrac{6}{6}y^2$

$= y^2$

Exercises 8.5A (page 237)

1. a. 2nd   **b.** 2nd degree polynomial

3. a. 4th   **b.** 5th degree polynomial

5. a. 1st   **b.** 2nd degree polynomial

7. a. 3rd   **b.** 5th degree polynomial

9. a. 4th   **b.** 4th degree polynomial   **11.** Not a polynomial

13. $x^3 - 3x^2 + x - 4$ **15.** $8x^5 + 7x^3 - 4x - 5$

17. $xy^3 + 8xy^2 - 4x^2y$

Exercises 8.5B (page 240)

1. $5x + 2$ **3.** $x - 12$ **5.** $8x^2 + 8x$ **7.** $-4x^2 + 9x$

9. $5m^2 - 1$ **11.** $-2x^2 + 7x - 17$ **13.** $13b^2 - 2b - 22$

15. $2x^3 + 4x^2 - x + 3$ **17.** $-a^2 + 3a + 12$

19. $a^2 - 8a + 10$ **21.** $4x^2 - 10x - 1$ **23.** $-2x$

25. $2y^2 - 7y + 4$ **27.** $17a^3 + 8a^2 - 2a$

29. $12x^2y^3 - 9xy^2$ **31.** $7x^3 - 4x^2 - 9x + 17$

33. $3a^3 + 4a^2 + 3a + 3$ **35.** $v^3 + v^2 + 8v + 13$

37. $4x^2y^2 - 10x^2y + 6xy + 3$

39. $(5x - 3) - [(x + 2) + (2x - 6)]$ **41.** $3x^2 + 10$

$= 5x - 3 - [3x - 4]$

$= 5x - 3 - 3x + 4$

$= 2x + 1$

43. $5x^2 + 4x - 7$ **45.** $-x^2 + 8x - 10$

47. $-16a^3 + 11a^2 + 34a - 26$

49. $[(5 + xy^2 + x^3y) + (-6 - 3xy^2 + 4x^3y)]$
  $\quad - [(x^3y + 3xy^2 - 4) + (2x^3y - xy^2 + 5)]$

$= [5 + xy^2 + x^3y - 6 - 3xy^2 + 4x^3y]$
$\quad - [x^3y + 3xy^2 - 4 + 2x^3y - xy^2 + 5]$

$= [5x^3y - 2xy^2 - 1] - [3x^3y + 2xy^2 + 1]$

$= 5x^3y - 2xy^2 - 1 - 3x^3y - 2xy^2 - 1$

$= 2x^3y - 4xy^2 - 2$

51. $4x^2 + 3x + 8$ **53.** $4.4x^2 + 4.1x + 11.8$

55. $-4.78x^2 - 20.85x + 2.46$

Exercises 8.6 (page 243)

1. $-15z^5 + 12z^4 - 6z^3 + 24z^2$ **3.** $6x^3y^2 - 2x^2y^3 + 8xy^4$

5. $15a^3b^2 - 30a^2b^3 + 25ab^4$

7. $3x^5y^4z^4 + 12x^4y^4z^6 - 6x^2y^3z^8$

9. $x^2 + x - 6$ **11.** $2y^2 - 3y - 20$

13. $(x + 4)^2 = (x + 4)(x + 4)$ **15.** $2x^3 + 4x^2 - 5x - 10$

$$\begin{array}{r} x + 4 \\ x + 4 \\ \hline 4x + 16 \\ x^2 + 4x \\ \hline x^2 + 8x + 16 \end{array}$$

17. $x^3 - x^2 - 10x + 6$ **19.** $z^3 - 64$

21. $6x^3 + 17x^2 - 3x - 20$ **23.** $4x^3 + 25x^2 - 45x + 18$

25. $-2x^3 + 11x^2 - 14x + 45$

27. $2x^4 - 8x^3 + 3x^2 - 5x - 28$

29. $2y^4 - 17y^3 + 23y^2 - 11y + 12$

31. $x^4 + 2x^3 - x^2 + 22x - 24$

33. $x^4 + 4x^3 + 10x^2 + 12x + 9$ **35.** $x^3 + 6x^2 + 12x + 8$

37. $(x + y)^2(x - y)^2$

$$\begin{array}{r} (x^2 + 2xy + y^2)(x^2 - 2xy + y^2) \\ x^2 + 2xy + y^2 \\ x^2 - 2xy + y^2 \\ \hline x^2y^2 + 2xy^3 + y^4 \\ -2x^3y - 4x^2y^2 - 2xy^3 \\ x^4 + 2x^3y + x^2y^2 \\ \hline x^4 \qquad - 2x^2y^2 \qquad + y^4 \end{array}$$

Exercises 8.7 (page 246)

1. $x^2 + 5 + 4$ **3.** $a^2 + 7a + 10$ **5.** $m^2 - 2m - 8$

7. $72 + y - y^2$ **9.** $x^2 - 4$ **11.** $h^2 - 36$

13. $a^2 + 8a + 16$ **15.** $x^2 + 6x + 9$ **17.** $16 - 8b + b^2$

19. $6x^2 - 7x - 20$ **21.** $4x^2 - 9y^2$ **23.** $2a^2 + 7ab + 5b^2$

25. $8x^2 + 26xy - 7y^2$ **27.** $49x^2 - 140xy + 100y^2$

29. $(3x + 4)^2 = (3x)^2 + 2(3x)(4) + (4)^2 = 9x^2 + 24x + 16$

31. $25m^2 - 20mn + 4n^2$ **33.** $2x^4 + 5x^2y^2 + 3y^4$

35. $4m^4 - 10m^2n^2 - 6n^4$ **37.** $9x^4 - y^4$

39. $x^4 + 2x^2y^2 + y^4$

41. $\left(\dfrac{2x}{3} + 1\right)\left(\dfrac{2x}{3} - 1\right) = \left(\dfrac{2x}{3}\right)^2 - (1)^2 = \dfrac{4x^2}{9} - 1$

43. $\left(2 - \dfrac{x}{3}\right)^2 = 2^2 + 2(2)\left(-\dfrac{x}{3}\right) + \left(-\dfrac{x}{3}\right)^2 = 4 - \dfrac{4x}{3} + \dfrac{x^2}{9}$

Exercises 8.8A (page 247)

1. $x + 2$ **3.** $1 + 2x$ **5.** $-3x + 4y$ **7.** $2x + 3$

9. $3a - 1$ **11.** $3z^2 - 4z + 2$ **13.** $x - \dfrac{4}{5} + \dfrac{2}{x^2}$

15. $3xyz + 6$ **17.** $x - 2y + 3$ **19.** $5x^4y^3 + 9xy + 1$

Exercises 8.8B (page 251)

1. $x + 3$ **3.** $x - 5$ **5.** $2x + 3$ **7.** $3v + 8$ R 66

9. $4a + b$ **11.** $3a - 2b$ R $7b^2$

13.

$$\require{enclose}\begin{array}{r} x^2 - 4x + 6 \\ x + 4 \enclose{longdiv}{x^3 + 0x^2 - 10x + 24} \\ \ominus \quad \ominus \\ x^3 + 4x^2 \\ \hline -4x^2 - 10x \\ \oplus \quad \oplus \\ -4x^2 - 16x \\ \hline 6x + 24 \\ \ominus \quad \ominus \\ 6x + 24 \\ \hline 0 \end{array}$$

15. $4x^2 + 8x + 8$ R -6 **17.** $a^2 + 2a + 4$

19. $2x^2 - 3x + 4$ **21.** $3x^2 - 2x - 1$

23. $x^2 - 2x + 5$ R 3   **25.** $16x^2 - 4x + 1$

27. $x^2 + x - 1$ R 3

8

Chapter 8 Review Exercises (page 252)

1. a. Three terms   **b.** $2ab$

2. a. Two terms   **b.** $-2(x^2 + y^2)$

3. a. Two terms   **b.** $-4x$ **4.** $3x^4 + 7x^2 - x - 6$

5. a. 2nd   **b.** 3rd degree polynomial   **6.** x^4 **7.** m^9

8. m^{18} **9.** m^3 **10.** $6x^5$ **11.** $4x^6$ **12.** 1

13. y^{12} **14.** 9   **15.** 64   **16.** 1,000,000   **17.** 1

18. $8x^2y + 5xy^2$ **19.** $9a^3 - 3ab^2 + 5$

20. $9x^2y + xy^2 - 12 + 2y^2$ **21.** $4a^2b + 9ab^2 + 4a - 8$

22. $6a^2 - 11a + 2$ **23.**
$$(2x + 9) - [(3x - 5) - (2 - x)]$$
$$= (2x + 9) - [3x - 5 - 2 + x]$$
$$= (2x + 9) - [4x - 7]$$
$$= 2x + 9 - 4x + 7$$
$$= -2x + 16$$

24.
$$(5y^2 - 6y - 4) - (3y^2 - 5y + 1)$$
$$= 5y^2 - 6y - 4 - 3y^2 + 5y - 1$$
$$= 2y^2 - y - 5$$

25. $2a^3 - 5a^2 - 11a - 4$ **26.** $-2x^2y^2 - 5x^2y + 5xy$

27. $-63x^{12}y^8$ **28.** $20a^{10}b^6$ **29.** $6x^3 - 8x^2$

30. $15x^3y^3 + 20x^3y - 10x^2yz$ **31.** $6a^2 - 2a - 20$

32. $9x^2 + 24x + 16$ **33.** $x^2 - 25y^2$ **34.** $4m^2 - 18m + 18$

35. $15u^2 + 11u + 2$ **36.** $49x^2 - 16$ **37.** $x^4 - 12x^2 + 36$

38. $35x^2 - 18xy - 8y^2$ **39.** $2x^4 - x^2y^2 - 3y^4$

40.
$$\left(\frac{2x}{3} + 4\right)^2 = \left(\frac{2x}{3}\right)^2 + 2\left(\frac{2x}{3}\right)(4) + (4)^2$$
$$= \frac{4x^2}{9} + \frac{16x}{3} + 16$$

41. $x^3 - 8$ **42.** $x - 2y$ **43.** $3ab^2 - \dfrac{4b}{5} + 2$

44.
$$\begin{array}{r} 2x - 1 \\ 2x + 1 \overline{)\ 4x^2 + 0x - 1} \\ \ominus \quad \ominus \\ \underline{4x^2 + 2x} \\ -2x - 1 \\ \oplus \quad \oplus \\ \underline{-2x - 1} \\ 0 \end{array}$$

45. $3x + 3$ R 25

46. $2a + 5b$ **47.** $2x^2 + 3x + 1$ R 2

48.
$$\begin{array}{r} a^2 \qquad\quad - 1 \\ 2a^2 - a + 3 \overline{)\ 2a^4 - a^3 + a^2 + a - 3} \\ \ominus \quad \oplus \quad \ominus \\ \underline{2a^4 - a^3 + 3a^2} \\ -2a^2 + a - 3 \\ \oplus \quad \ominus \quad \oplus \\ \underline{-2a^2 + a - 3} \\ 0 \end{array}$$

Chapter 8 Diagnostic Test   (page 254)

Following each problem number is the number (in parentheses) of the textbook section in which that kind of problem is discussed.

1. (8.1), (8.5A) $x^2 - 4xy^2 + 5$

  a. The first term is 2nd degree.

  b. The polynomial is 3rd degree.

  c. The numerical coefficient of the 2nd term is -4.

2. (8.5B)
$$(-4z^2 - 5z + 10) - (10 - z + 2z^2)$$
$$= -4z^2 - 5z + 10 - 10 + z - 2z^2$$
$$= -6z^2 - 4z$$

3. (8.5B)
$$\begin{array}{r} -5x^3 + 2x^2 \qquad\quad - 5 \\ 7x^3 \qquad\quad + 5x - 8 \\ 3x^2 - 6x + 10 \\ \hline 2x^3 + 5x^2 - x - 3 \end{array}$$

4. (8.5B)
$$\begin{array}{ll} 9a^2b - 2ab^2 - 3ab & 9a^2b - 2ab^2 - 3ab \\ \underline{-(4a^2b + 6ab^2 - 7ab)} \longrightarrow \underline{-4a^2b - 6ab^2 + 7ab} \\ & 5a^2b - 8ab^2 + 4ab \end{array}$$

5. (8.5B)
$$(6x^2 - 5 - 11x) + (15x - 4x^2 + 7)$$
$$= 6x^2 - 5 - 11x + 15x - 4x^2 + 7$$
$$= 2x^2 + 4x + 2$$

6. (8.5B)
$$(3xy^2 - 4xy) - (6xy - 3xy^2) + (4xy + x^3)$$
$$= 3xy^2 - 4xy - 6xy + 3xy^2 + 4xy + x^3$$
$$= 6xy^2 - 6xy + x^3$$

7. (8.5B)
$$(2m^2 - 5) - [(7 - m^2) - (4m^2 - 3)]$$
$$= (2m^2 - 5) - [7 - m^2 - 4m^2 + 3]$$
$$= 2m^2 - 5 - 7 + m^2 + 4m^2 - 3$$
$$= 7m^2 - 15$$

8. (8.5B)
$$(2x - 8 - 7x^2) + (12 - 2x^2 + 11x) - (5x^2 - 9x + 6)$$
$$= 2x - 8 - 7x^2 + 12 - 2x^2 + 11x - 5x^2 + 9x - 6$$
$$= -14x^2 + 22x - 2$$

9. (8.2) $10^2 \cdot 10^3 = 10^5 = 100,000$ **10.** (8.2) $(a^3)^5 = a^{15}$

11. (8.2) $\dfrac{12x^6}{2x^2} = 6x^4$ **12.** (8.2) $x^3 \cdot x^5 = x^8$

13. (8.2) $\dfrac{2^8}{2^5} = 2^3 = 8$

14. (8.3) $(-3xy)(5x^3y)(-2xy^4)$
Because two factors are negative, the answer is positive.
$(3 \cdot 5 \cdot 2)(xx^3x)(yyy^4) = 30x^5y^6$

15. (8.3) $4x(3x^2 - 2) = (4x)(3x^2) + (4x)(-2) = 12x^3 - 8x$

16. (8.3)
$$-2ab(5a^2 - 3ab^2 + 4b)$$
$$= (-2ab)(5a^2) + (-2ab)(-3ab^2) + (-2ab)(4b)$$
$$= -10a^3b + 6a^2b^3 - 8ab^2$$

17. (8.7) $(2x - 5)(3x + 4) = 6x^2 - 7x - 20$

18. (8.7)
$$(2y - 3)^2$$
$$= (2y)^2 + 2(2y)(-3) + (-3)^2$$
$$= 4y^2 - 12y + 9$$

19. (8.7) $(m + 3)(2m + 6) = 2m^2 + 12m + 18$

20. (8.7) $(5a + 2)(5a - 2) = (5a)^2 - (2)^2 = 25a^2 - 4$

21. (8.7) $(4x - 2y)(3x - 5y) = 12x^2 - 26xy + 10y^2$

22. (8.6)
$$\begin{array}{r} w^2 + 3w + 8 \\ w - 3 \\ \hline -3w^2 - 9w - 24 \\ w^3 + 3w^2 + 8w \\ \hline w^3 \qquad\quad - w - 24 \end{array}$$

23. (8.8A) $\dfrac{6x^3 - 4x^2 + 8x}{2x} = \dfrac{6x^3}{2x} + \dfrac{-4x^2}{2x} + \dfrac{8x}{2x} = 3x^2 - 2x + 4$

24. (8.8B)
$$\begin{array}{r} 5x + 3 \quad \text{R} -2,\ \text{or}\ 5x + 3 - \dfrac{2}{2x - 1} \\ 2x - 1 \overline{)\ 10x^2 + x - 5} \\ \ominus \quad \oplus \\ \underline{10x^2 - 5x} \\ 6x - 5 \\ \ominus \quad \oplus \\ \underline{6x - 3} \\ -2 \end{array}$$

25. (8.8B)

$$x + 3 \overline{\smash{\big)}\, 2x^3 + 0x^2 - 20x - 6}$$

quotient: $2x^2 - 6x - 2$

$$\underset{\ominus \quad \ominus}{2x^3 + 6x^2}$$

$$- 6x^2 - 20x$$

$$\underset{\oplus \quad \oplus}{- 6x^2 - 18x}$$

$$- 2x - 6$$

$$\underset{\oplus \quad \oplus}{- 2x - 6}$$

$$0$$

Exercises 9.1 (page 260)

1. Yes **3.** No **5.** Yes **7.** 3 **9.** 7 **11.** −7
13. 4 **15.** 3 **17.** −3 **19.** 42 **21.** 4 **23.** 23
25. 12 **27.** −10 **29.** 28
31.

$$\begin{array}{l} -28 = -15 + x \\ \underline{+\ 15 \quad +\ 15} \\ -13 = \qquad x \end{array}$$

Check: $-28 = -15 + x$
$-28 \overset{?}{=} -15 + (-13)$
$-28 = -28$

33. 45
35.

$$\begin{array}{l} 5.6 + x = \quad 2.8 \\ \underline{-\ 5.6 \qquad -\ 5.6} \\ \quad x = -\ 2.8 \end{array}$$

Check: $5.6 + x = 2.8$
$5.6 + (-2.8) \overset{?}{=} 2.8$
$2.8 = 2.8$

37. 7 **39.** −3 **41.** −117 **43.** 31.4 **45.** 8
47. $\dfrac{9}{10}$ **49.** $\dfrac{5}{12}$

Exercises 9.2 (page 262)

1. 2 **3.** 2 **5.** 13 **7.** −4 **9.** 3 **11.** −6
13. −2 **15.** 1 **17.** −2
19.

$$\begin{array}{l} -x - 6 = \quad 4 \\ \underline{\quad +\ 6 \quad +\ 6} \\ -x \quad = \quad 10 \\ \dfrac{-x}{-1} = \dfrac{10}{-1} \\ x = -10 \end{array}$$

Check: $-x - 6 = 4$
$-(-10) - 6 \overset{?}{=} 4$
$10 - 6 \overset{?}{=} 4$
$4 = 4$

21. −6 **23.** 15 **25.** −6 **27.** 4 **29.** $-\dfrac{1}{3}$
31. $\dfrac{2}{3}$ **33.** $\dfrac{4}{3}$ **35.** $-\dfrac{1}{2}$ **37.** 1.3 **39.** −10.8

Exercises 9.3 (page 264)

1. 3 **3.** 3 **5.** −2 **7.** 7 **9.** 5 **11.** 2
13. −3 **15.** 3 **17.** −3 **19.** −1 **21.** 4 **23.** 3
25.

$$\begin{array}{l} 7 - 9x - 12 = 3x + 5 - 8x \\ -9x - 5 = -5x + 5 \\ \underline{+\ 9x \qquad +\ 9x} \\ -\ 5 \qquad 4x + 5 \\ \underline{-\ 5 \qquad -\ 5} \\ -\ 10 = \quad 4x \\ \dfrac{-10}{4} = \dfrac{4x}{4} \\ -\dfrac{5}{2} = x,\ \text{or}\ x = -2\dfrac{1}{2} \end{array}$$

27. 2 **29.** $\dfrac{3}{4}$

Exercises 9.4 (page 266)

1. −5 **3.** 8 **5.** 3 **7.** 8 **9.** 2 **11.** 3
13. 1 **15.** 4 **17.** 1
19.

$$\begin{array}{l} 2(3x - 6) - 3(5x + 4) = 5(7x - 8) \\ 6x - 12 - 15x - 12 = 35x - 40 \\ -9x - 24 = 35x - 40 \\ 16 = 44x \\ \dfrac{16}{44} = \dfrac{44x}{44} \\ \dfrac{4}{11} = x \end{array}$$

21. 5 **23.** 2
25. −6 **27.** 8 **29.** 2
31.

$$\begin{array}{l} 5(3 - 2x) - 10 = 4x + [-(2x - 5) + 15] \\ 15 - 10x - 10 = 4x + [-2x + 5 + 15] \\ 15 - 10x - 10 = 4x - 2x + 20 \\ -15 = 12x \\ -\dfrac{15}{12} = \dfrac{12x}{12} \\ -1\dfrac{1}{4} = x \end{array}$$

33. −3
35.

$$10 - 6\left(\dfrac{1}{2}x - \dfrac{2}{3}\right) = 8\left(\dfrac{3}{4}x + \dfrac{5}{8}\right)$$

$$10 - \dfrac{\overset{3}{\cancel{6}}}{1} \cdot \dfrac{1}{\underset{1}{\cancel{2}}}x + \dfrac{\overset{2}{\cancel{6}}}{1} \cdot \dfrac{2}{\underset{1}{\cancel{3}}} = \dfrac{\overset{2}{\cancel{8}}}{1} \cdot \dfrac{3}{\underset{1}{\cancel{4}}}x + \dfrac{\overset{1}{\cancel{8}}}{1} \cdot \dfrac{5}{\underset{1}{\cancel{8}}}$$

$$\begin{array}{l} 10 - \quad 3x \quad + \quad 4 \quad = \quad 6x \quad + \quad 5 \\ -3x + 14 = \quad 6x + 5 \\ \underline{+3x \qquad\qquad +3x} \\ 14 = \quad 9x + 5 \\ \underline{-5 \qquad\qquad -\ 5} \\ 9 = \quad 9x \\ 1 = x \end{array}$$

Exercises 9.5 (page 268)

1. 5 **3.** No solution **5.** Identity **7.** 2
9. Identity **11.** No solution
13.

$$\begin{array}{l} 7(2 - 5x) - 32 = 10x - 3(6 + 15x) \\ 14 - 35x - 32 = 10x - 18 - 45x \\ -35x - 18 = -35x - 18 \quad \text{Identity} \end{array}$$

15.

$$\begin{array}{l} 2(2x - 5) - 3(4 - x) = 7x - 20 \\ 4x - 10 - 12 + 3x = 7x - 20 \\ -22 \neq -20 \quad \text{No solution} \end{array}$$

17.

$$\begin{array}{l} 2[3 - 4(5 - x)] = 2(3x - 11) \\ 2[3 - 20 + 4x] = 6x - 22 \\ 6 - 40 + 8x = 6x - 22 \\ 2x = 12 \\ x = 6 \end{array}$$

19. Identity

Exercises 9.6 (page 273)

1. $x < 7$

3. $x > -4$

5. $x \le -5$

7. $x \le 3$

9. $6 - x > 2$
$-x > -4$
$\dfrac{-x}{-1} \boxed{<} \dfrac{-4}{-1}$
$x < 4$

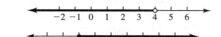

11. $x \geq -2$

13. $x < -1$

15. $2x - 9 > 3(x - 2)$
$2x - 9 > 3x - 6$
$-x - 9 > -6$
$-x > 3$
$\dfrac{-x}{-1} \boxed{<} \dfrac{3}{-1}$
$x < -3$

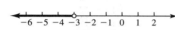

17. $x \geq 2$

19. $x \geq -5$

21. $x \leq 1$

23. $x < 3\dfrac{1}{8}$

25. $2[3 - 5(x - 4)] < 10 - 5x$
$2[3 - 5x + 20] < 10 - 5x$
$2[23 - 5x] < 10 - 5x$
$46 - 10x < 10 - 5x$
$46 - 5x < 10$
$-5x < -36$
$\dfrac{-5x}{-5} \boxed{>} \dfrac{-36}{-5}$
$x > 7\dfrac{1}{5}$

27. $4x - \{6 - [2x - (x + 3)]\} \leq 6$
$4x - \{6 - [2x - x - 3]\} \leq 6$
$4x - \{6 - [x - 3]\} \leq 6$
$4x - \{6 - x + 3\} \leq 6$
$4x - \{9 - x\} \leq 6$
$4x - 9 + x \leq 6$
$5x - 9 \leq 6$
$5x \leq 15$
$x \leq 3$

Chapter 9 Review Exercises (page 275)

1. Yes   **2.** No   **3.** 3   **4.** 3   **5.** 2
6. No solution   **7.** 75   **8.** 196   **9.** 5
10. Identity   **11.** $\dfrac{5}{6}$ **12.** $-\dfrac{1}{2}$ **13.** $\dfrac{2}{3}$ **14.** 14
15. $\dfrac{2}{3}$ **16.** No solution   **17.** -6.9 **18.** Identity
19. $4[-24 - 6(3x - 5) + 22x] = 0$
$4[-24 - 18x + 30 + 22x] = 0$

$4[6 + 4x] = 0$
$24 + 16x = 0$
$16x = -24$
$x = -1\dfrac{1}{2}$

20. $2[-7y - 3(5 - 4y) + 10] = 10y - 12$
$2[-7y - 15 + 12y + 10] = 10y - 12$
$2[5y - 5] = 10y - 12$
$10y - 10 = 10y - 12$
$-10 \neq -12$ No solution

21. $x < 3$

22. $x \geq -6$

23. $x \geq -7$

24. $x < -2$

25. $3[2 - 5(x - 1)] > 3 - 6x$
$3[2 - 5x + 5] > 3 - 6x$
$3[7 - 5x] > 3 - 6x$
$21 - 15x > 3 - 6x$
$21 - 9x > 3$
$-9x > -18$
$\dfrac{-9x}{-9} \boxed{<} \dfrac{-18}{-9}$
$x < 2$

26. $2(x + 4) - 13 \leq 7x - 5(2x - 3)$
$2x + 8 - 13 \leq 7x - 10x + 15$
$2x - 5 \leq -3x + 15$
$5x - 5 \leq 15$
$5x \leq 20$
$x \leq 4$

Chapter 9 Diagnostic Test   (page 276)

Following each problem number is the number (in parentheses) of the textbook section in which that kind of problem is discussed.

1. (9.1) $4x - 5 \overset{?}{=} -7$
$4(-3) - 5 \overset{?}{=} -7$
$-12 - 5 \overset{?}{=} -7$
$-19 \neq -7$
Therefore, -3 is not a solution.

2. (9.1) $\begin{aligned} x - 3 &= 7 \\ +3 \quad & +3 \\ \hline x &= 10 \end{aligned}$

3. (9.1) $\begin{aligned} 10 &= x + 4 \\ -4 \quad & -4 \\ \hline 6 &= x \end{aligned}$

4. (9.2) $\begin{aligned} 3x + 2 &= 14 \\ -2 \quad & -2 \\ \hline 3x &= 12 \\ \dfrac{3x}{3} &= \dfrac{12}{3} \\ x &= 4 \end{aligned}$

5. (9.2) $\begin{aligned} 15 - 2x &= 7 \\ -15 \quad & -15 \\ \hline -2x &= -8 \\ \dfrac{-2x}{-2} &= \dfrac{-8}{-2} \\ x &= 4 \end{aligned}$

6. (9.1) $-5 = \dfrac{x}{7}$
$7(-5) = \dfrac{\overset{1}{7}}{1}\left(\dfrac{x}{\underset{1}{7}}\right)$
$-35 = x$

7. (9.3) $4x + 5 = 17 - 2x$
$6x + 5 = 17$
$\dfrac{6x}{6} = \dfrac{12}{6}$
$x = 2$

8. (9.3) $6x - 3 = 5x + 2$
$x - 3 = 2$
$x = 5$

9. (9.3) $9 + 7x = 5x - 1$
$9 + 2x = -1$
$2x = -10$
$x = -5$

10. (9.3) $2x - 5 = 8x - 3$
$-6x - 5 = -3$
$\dfrac{-6x}{-6} = \dfrac{2}{-6}$
$x = -\dfrac{1}{3}$

11. (9.3) $3z - 21 + 5z = 4 - 6z + 17$
$8z - 21 = 21 - 6z$
$14z = 42$
$\dfrac{14z}{14} = \dfrac{42}{14}$
$z = 3$

12. (9.4) $6k - 3(4 - 5k) = 9$
$6k - 12 + 15k = 9$
$21k - 12 = 9$
$21k = 21$
$\dfrac{21k}{21} = \dfrac{21}{21}$
$k = 1$

13. (9.5) $6x - 2(3x - 5) = 10$
$6x - 6x + 10 = 10$
$10 = 10$ Identity

14. (9.4) $2m - 4(3m - 2) = 5(6 + m) - 7$
$2m - 12m + 8 = 30 + 5m - 7$
$-10m + 8 = 23 + 5m$
$-15m = 15$
$\dfrac{-15m}{-15} = \dfrac{15}{-15}$
$m = -1$

15. (9.5) $2(3x + 5) = 14 + 3(2x - 1)$
$6x + 10 = 14 + 6x - 3$
$6x + 10 = 11 + 6x$
$10 \neq 11$ No solution

16. (9.4) $3[7 - 6(y - 2)] = -3 + 2y$
$3[7 - 6y + 12] = -3 + 2y$
$3[19 - 6y] = -3 + 2y$
$57 - 18y = -3 + 2y$
$60 = 20y$
$\dfrac{60}{20} = \dfrac{20y}{20}$
$3 = y$

17. (9.6)
$5x - 2 > 10 - x$
$\underline{+\ x \qquad\qquad +\ x}$
$6x - 2 > 10$
$\underline{\quad +\ 2 \qquad +\ 2}$
$6x \qquad > \quad 12$
$\dfrac{6x}{6} > \dfrac{12}{6}$
$x > 2$

18. (9.6) $4x + 5 < x - 4$
$3x + 5 < -4$
$\dfrac{3x}{3} < \dfrac{-9}{3}$
$x < -3$

19. (9.6)
$2(x - 4) - 5 \geq 6 + 3(2x - 1)$
$2x - 8 - 5 \geq 6 + 6x - 3$
$2x - 13 \geq 6x + 3$
$\underline{-6x \qquad\qquad -6x}$
$-4x - 13 \geq \qquad 3$
$\underline{+\ 13 \qquad\qquad +\ 13}$
$-4x \qquad \geq \qquad 16$
$\dfrac{-4x}{-4} \leq \dfrac{16}{-4}$
$x \leq -4$

20. (9.6) $7x - 2(5 + 4x) \leq -7$
$7x - 10 - 8x \leq -7$
$-10 - x \leq -7$
$\underline{+10 \qquad\quad +10}$
$-x \leq 3$
$\dfrac{-x}{-1} \geq \dfrac{3}{-1}$
$x \geq -3$

Exercises 10.1A   (page 279)

1. $x + 10$ **3.** $A - 5$ **5.** $6z$ **7.** $C - D$

9. $x + 10$ **11.** $x - UV$ **13.** $5x^2$ **15.** $(A + B)^2$

17. $\dfrac{x + 7}{y}$ **19.** $\dfrac{A}{C + 10}$ **21.** $15 + 2F$ **23.** $3x - 5y$

25. $x(y - 6)$ **27.** $3(A + 5)$ **29.** $2(6 - B)$

31. $6x + 7$

Exercises 10.1B   (page 281)

1. Let S = Goh's salary
Then $S + 75$ is Goh's salary plus seventy-five dollars.

3. Let N = number of children
Then $N - 2$ is two less than the number of children.

5. Let x = Joyce's age
Then $4x$ is four times Joyce's age.

7. Let c = cost of a record
Then $20c + 89$ is twenty times the cost of a record increased by eighty-nine cents.

9. Let c = cost of a hamburger
Then $\dfrac{1}{5}c$, or $\dfrac{c}{5}$, is one-fifth the cost of a hamburger.

11. Let x = speed of car
Then $5x + 100$ is five times the speed of the car, plus one hundred miles per hour.

13. Let x = unknown number
Then x^2 = the square of the unknown, and $5x^2 - 10$ is ten less than five times the square of the unknown.

15. Let x = unknown number
Then $\dfrac{8 + x}{x^2}$ represents the statement.

17. Let C = Centigrade temperature
Then $32 + \dfrac{9}{5}C$ represents the statement.

19. Let x = unknown number
Then $2(5 - x)$ represents the statement.

21. There are two unknowns.
First solution:
Let w = width
Then $w + 12$ = length

Second solution:
Let l = length
Then $l - 12$ = width

23. Let x = one number
Then $60 - x$ = other number

25. Let B = brother's age
Then $B - 5$ = Sanjay's age

27. Let M = Miri's money
Then $M + 139$ = Tom's money

29. Let x = first consecutive integer
Then $x + 1$ = second consecutive integer
Then the product is $x(x + 1)$.

31. Let x = first odd
Then $x + 2$ = second odd
and $x + 4$ = third odd
Then the sum is $x + (x + 2) + (x + 4)$.

33. Let w = width
Then $2w - 4$ = length

35. Let x = unknown number
Then $9 + 3x$ represents the statement.

37. Let x = first even
Then $x + 2$ = second even
Then $2x - (x + 2)$ represents the statement.

Exercises 10.2 (page 285)

1. a. Let x = unknown number
b. $13 + 2x = 25$
$x = 6$
c. The unknown number is 6.

3. a. Let x = unknown number
b. $5x - 8 = 22$
$x = 6$
c. The unknown number is 6.

5. a. Let x = unknown number
b. $\dfrac{x}{12} = 6$
$x = 72$
c. The unknown number is 72.

7. a. Let x = unknown number
b. $7 - x = x + 1$
$x = 3$
c. The unknown number is 3.

9. a. Let x = unknown number
b. $x + 7 = 2x$
$x = 7$
c. The unknown number is 7.

11. a. Let x = unknown number
b. $2(5 + x) = 26$
$x = 8$
c. The unknown number is 8.

13. a. Let x = unknown number
b. $2(4 + x) = x - 10$
$x = -18$
c. The unknown number is -18.

15. a. Let x = first integer
Then $x + 1$ = second integer
and $x + 2$ = third integer
b. $x + (x + 1) + (x + 2) = 63$
$x = 20$
c. The integers are 20, 21, and 22.

17. a. Let x = first even
Then $x + 2$ = second even
and $x + 4$ = third even
b. $x + (x + 2) - (x + 4) = 4$
$x = 6$
c. The integers are 6, 8, and 10.

19. a. Let x = first odd
Then $x + 2$ = second odd
and $x + 4$ = third odd
b. $x + (x + 2) = 4(x + 4)$
$-7 = x$
c. The integers are -7, -5, and -3.

21. a. Let w = width
b. $36 = 2(12) + 2w$
$6 = w$
c. The width is 6 in.

23. a. Let w = width
Then $w + 3$ = length
b. $50 = 2(w + 3) + 2w$
$11 = w$
c. The width is 11 ft, and the length is 14 ft.

25. a. Let h = height
b. $24 = \dfrac{1}{2}(8)(h)$
$6 = h$
c. The height is 6 m.

Exercises 10.3 (page 288)

1. a. Let x = lb of peanuts
Then $12 - x$ = lb of cashews
b. $4x + 8(12 - x) = 5(12)$
$x = 9$
c. 9 lb of peanuts; 3 lb of cashews

3. a. Let x = lb of caramels
b. $1.25(15) + 1.50x = 1.35(x + 15)$
$x = 10$
c. 10 lb of caramels

5. a. Let x = lb of nuts
Then $10 - x$ = lb of raisins
b. $3.40x + 1.90(10 - x) = 25.00$
$x = 4$
c. 4 lb of nuts; 6 lb of raisins

7. a. Let x = dollars per bushel of corn
Note: 100 bu mixture
$\underline{-\quad 30 \text{ bu soybeans}}$
70 bu corn
b. $8.00(30) + x(70) = 4.85(100)$
$x = 3.5$
c. \$3.50 per bushel of corn

9. a. Let x = number of nickels
Then $13 - x$ = number of dimes
b. $5x + 10(13 - x) = 95$
$x = 7$
c. 7 nickels and 6 dimes

11. a. Let x = number of nickels
Then $12 - x$ = number of quarters
b. $5x + 25(12 - x) = 220$
$x = 4$
c. 4 nickels and 8 quarters

13. a. Let $\quad x =$ number of nickels
Then $x + 4 =$ number of quarters
and $\quad 3x =$ number of dimes
b. $5x + 25(x + 4) + 10(3x) = 400$
$\qquad\qquad\qquad\qquad x = 5$
c. 5 nickels, 9 quarters, and 15 dimes

Exercises 10.4 (page 289)

1. a. Let $x =$ cc of 20% solution
b. $0.20x + 0.50(100) = 0.25(x + 100)$
$\qquad\qquad\qquad\qquad x = 500$
c. 500 cc of 20% solution

3. a. Let $\quad x =$ gal of 30% solution
Then $100 - x =$ gal of 90% solution
b. $0.30x + 0.90(100 - x) = 0.75(100)$
$\qquad\qquad\qquad\qquad x = 25$
c. 25 gal of 30% solution; 75 gal of 90% solution

5. a. Let $x =$ mL of water
b. $0.40(500) + 0(x) = 0.25(500 + x)$
$\qquad\qquad\qquad\qquad x = 300$
c. 300 mL of water

7. a. Let $x =$ oz of 10% solution
b. $0.10x + 0.25(20) = 0.16(x + 20)$
$\qquad\qquad\qquad\qquad x = 30$
c. 30 oz of 10% solution

9. a. Let $\quad x =$ mL of 50% solution
Then $50 - x =$ mL of 25% solution
b. $0.50x + 0.25(50 - x) = 0.30(50)$
$\qquad\qquad\qquad\qquad x = 10$
c. 10 mL of 50% solution; 40 mL of 25% solution

Exercises 10.5 (page 293)

1. Let $\quad x =$ speed of Malone's car
Then $x + 9 =$ speed of King's car
$6x = 5(x + 9)$
$\quad x = 45$
a. Malone's car, 45 mph; King's car, 54 mph
b. Distance $= 6x = 6(45) = 270$ mi

3. Let $\quad x =$ time to lake
Then $x - 3 =$ time returning
$2x = 5(x - 3)$
$\quad x = 5$
a. Time to lake $= x = 5$ hr
b. Distance $= 2x = 2(5) = 10$ mi

5. a. Let $\quad x =$ Anita's speed
Then $x + 2 =$ Felix's speed
b. $3x + 3(x + 2) = 54$
$\qquad\qquad\quad x = 8$
c. Anita's speed, 8 mph; Felix's speed, $x + 2 = 10$ mph

7. a. Let $\quad x =$ speed in still water
Then $x + 2 =$ speed downstream
and $\quad x - 2 =$ speed upstream
b. $3(x + 2) = 5(x - 2)$
$\qquad\qquad x = 8$
c. Speed in still water $= 8$ mph
Distance downstream $= 3(x + 2) = 3(8 + 2) = 30$ mi

9. a. Let $\quad x =$ time going
Then $10 - x =$ time returning
b. $80x = 120(10 - x)$
$\quad x = 6$
c. Distance $= 80x = 480$ mi

Exercises 10.6 (page 298)

1. 28 **3.** 12 **5.** 3 **7.** -2 **9.** -125
11. 1,000 **13.** $\dfrac{3}{4}$ **15.** $\dfrac{3}{2}$
17. $\begin{aligned} s &= kt^2 \\ 64 &= k(2)^2 \\ 64 &= 4k \\ 16 &= k \end{aligned}$ $\qquad \begin{aligned} s &= kt^2 \\ s &= 16(3)^2 \\ s &= 16 \cdot 9 \\ s &= 144 \text{ ft} \end{aligned}$

19. $\begin{aligned} I &= \dfrac{k}{d^2} \\ 15 &= \dfrac{k}{(10)^2} \\ 15 &= \dfrac{k}{100} \\ 1{,}500 &= k \end{aligned}$ $\qquad \begin{aligned} I &= \dfrac{k}{d^2} \\ I &= \dfrac{1{,}500}{(15)^2} \\ I &= \dfrac{1{,}500}{225} \\ I &= \dfrac{20}{3} \\ I &= 6\dfrac{2}{3} \text{ cd} \end{aligned}$

21. $1\dfrac{1}{5}$ in. **23.** 1,000 cc **25.** $17\dfrac{7}{9}$ lb

Chapter 10 Review Exercises (page 300)

1. 2 **2.** 9 **3.** $-8, -7,$ and -6 **4.** 11 and 13
5. Length $= 20$ ft; width $= 15$ ft **6.** Height $= 10$ cm
7. 11 nickels; 16 dimes
8. Let $\quad x =$ number of nickels
Then $x + 4 =$ number of quarters
and $\quad 4x =$ number of dimes

Amount of money in nickels	+	amount in quarters	+	amount in dimes	$=$	240¢
$5x$	+	$25(x + 4)$	+	$10(4x)$	$=$	240

$\qquad 5x + 25(x + 4) + 10(4x) = 240$
$\qquad 5x + 25x + 100 + 40x = 240$
$\qquad\qquad\qquad\qquad 70x = 140$
$\qquad\qquad\qquad\qquad\quad x = 2 \quad$ Nickels
$\qquad\qquad\qquad\qquad x + 4 = 6 \quad$ Quarters
$\qquad\qquad\qquad\qquad\quad 4x = 8 \quad$ Dimes
2 nickels, 6 quarters, and 8 dimes

9. 3 lb of almonds; 7 lb of walnuts
10. 30 lb of Colombian coffee; 70 lb of Brazilian coffee
11. 50 cc of 50% solution **12.** 15 mL of water
13. First car's speed $= 40$ mph; second car's speed $= 50$ mph
14. Plane A's speed $= 200$ mph; plane B's speed $= 150$ mph
15. $x = 4$ **16.** $y = -12$ **17.** $y = 36$
18. $y = \dfrac{9}{2}$, or $4\dfrac{1}{2}$

19. $\begin{aligned} R &= \dfrac{k}{d^2} \\ 9 &= \dfrac{k}{(0.02)^2} \\ 9 &= \dfrac{k}{0.0004} \\ 0.0036 &= k \end{aligned}$ $\qquad \begin{aligned} R &= \dfrac{k}{d^2} \\ R &= \dfrac{0.0036}{(0.03)^2} \\ R &= \dfrac{0.0036}{0.0009} \\ R &= 4 \text{ ohms} \end{aligned}$

20.

$$R = kn$$
$$\frac{12,000}{800} = \frac{k(800)}{800}$$
$$15 = k$$

$$R = kn$$
$$15,000 = 15n$$
$$1,000 = n$$

21. 35¢ phone call uses 2 coins (1 quarter and 1 dime).
Therefore, after the phone call he had $10 - 2 = 8$ coins.

Let x = number of dimes
Then $8 - x$ = number of quarters

$$10x + 25(8 - x) = 110$$
$$10x + 200 - 25x = 110$$
$$-15x = -90$$
$$x = 6 \quad \text{Dimes}$$

22. Let c = speed of current
Then $20 + c$ = speed downstream
and $20 - c$ = speed upstream

$$6(20 - c) = 4(20 + c)$$
$$120 - 6c = 80 + 4c$$
$$40 = 10c$$
$$4 = c$$

Speed of current = 4 mph

Chapter 10 Diagnostic Test (page 301)

Following each problem number is the number (in parentheses) of the textbook section in which that kind of problem is discussed.

1. (10.2) Let x = unknown number

When sixteen	is added to	three	times	an unknown number	the sum is	37
16	+	3	.	x	=	37

$$16 + 3x = 37$$
$$3x = 21$$
$$x = 7$$

The unknown number is 7.

2. (10.2) Let x = first even
Then $x + 2$ = second even
and $x + 4$ = third even

$$x + (x + 2) + (x + 4) = 54$$
$$3x + 6 = 54$$
$$3x = 48$$
$$x = 16$$
$$x + 2 = 18$$
$$x + 4 = 20$$

The integers are 16, 18, and 20.

3. (10.3) Let x = number of dimes
Then $15 - x$ = number of quarters

Value of dimes	+	value of quarters	=	300¢

$$10x + 25(15 - x) = 300$$
$$10x + 375 - 25x = 300$$
$$375 - 15x = 300$$
$$-15x = -75$$
$$x = 5$$
$$15 - x = 10$$

5 dimes and 10 quarters

4. (10.2) Let x = width
Then $x + 10$ = length

$$2(x + 10) + 2x = 60$$
$$2x + 20 + 2x = 60$$
$$20 + 4x = 60$$

$$4x = 40$$
$$x = 10 \text{ in.}   \text{Width}$$
$$x + 10 = 20 \text{ in.}   \text{Length}$$

5. (10.2) Let x = unknown number

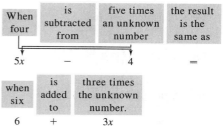

When four	is subtracted from	five times an unknown number	the result is the same as
$5x$	$-$	4	=

when six	is added to	three times the unknown number.
6	+	$3x$

$$5x - 4 = 6 + 3x$$
$$2x = 10$$
$$x = 5$$

The unknown number is 5.

6. (10.3) Let x = lb of Brand A
Then $10 - x$ = lb of Brand B

Cost of Brand A	+	cost of Brand B	=	cost of mixture

$$2.50x + 3.75(10 - x) = 3.00(10)$$
$$250x + 375(10 - x) = 300(10)$$
$$250x + 3750 - 375x = 3000$$
$$3750 - 125x = 3000$$
$$-125x = -750$$
$$x = 6 \text{ lb of Brand A}$$
$$10 - x = 4 \text{ lb of Brand B}$$

7. (10.6)

$$y = \frac{k}{x}$$
$$3 = \frac{k}{-4}$$
$$-12 = k$$

$$y = \frac{k}{x}$$
$$y = \frac{-12}{\frac{4}{5}}$$
$$y = \frac{-12}{1} \div \frac{4}{5}$$
$$y = \frac{-12}{1} \cdot \frac{5}{4}$$
$$y = -15$$

8. (10.4) Let x = pt of 2% solution

Disinfectant in 2% solution	+	disinfectant in 12% solution	=	disinfectant in 4% solution
$0.02x$	+	$0.12(4)$	=	$0.04(x + 4)$

$$2x + 12(4) = 4(x + 4)$$
$$2x + 48 = 4x + 16$$
$$32 = 2x$$
$$16 = x$$

16 pt of the 2% solution

9. (10.5) Let x = speed going home
Then $x + 12$ = speed to mountains

	r	$\cdot$	t	$=$	d
To mountains	$x + 12$		4		$4(x + 12)$
Returning home	x		5		$5x$

Distance to mountains	=	distance home

$$4(x + 12) = 5x$$
$$4x + 48 = 5x$$
$$48 = x$$
$$5x = 5(48) = 240 \text{ mi} \quad \text{Distance}$$

10. (10.6) $\quad p = kd \qquad\qquad p = kd$

$$\frac{17.32}{40} = \frac{k(40)}{40} \qquad p = 0.433(70)$$

$$0.433 = k \qquad\qquad p = 30.31 \text{ psi}$$

Chapters 1–10 Cumulative Review Exercises   (page 301)

1. $17\frac{1}{10}$　　**2.** $3\frac{1}{3}$　　**3.** 16.725　　**4.** 850　　**5.** -4

6. -7　　**7.** $2x^2 - 2x - 7$　　**8.** $4x^2 - 20x + 25$

9. $y^3 - 5y^2 + 10y - 8$　　**10.** $2x - 4$　　**11.** $x = -2$

12. $x = -10$　　**13.** $m = 2$　　**14.** $x \le -5$　　**15.** $\frac{1}{12}$

16. 125%　　**17.** \$3.28　　**18.** 15 qt　　**19.** The number is 8.

20. $y = 75$　　**21.** 4 oz

22. Boat A's speed $= 50$ mph; boat B's speed $= 30$ mph

23. 42　　**24.** 66 in.　　**25.** 65 in.

Exercises 11.1   (page 308)

1. C; {1, 3, 7, 21}　　**3.** P; {1, 41}　　**5.** C; {1, 2, 3, 6, 9, 18}

7. C; {1, 2, 4, 8, 16}　　**9.** $2^2 \cdot 3$　　**11.** 3^3　　**13.** $3 \cdot 5^2$

15. $2^4 \cdot 3^2$　　**17.** 3　　**19.** 6　　**21.** 4　　**23.** $5x^3$

25. $3xy$　　**27.** 2　　**29.** $x^3 y^2$　　**31.** $3a$　　**33.** $2(x - 4)$

35. $5(a + 2)$　　**37.** $3(2y - 1)$

39. $-5(3x + 2)$, or $5(-3x - 2)$　　**41.** $4(2x + 3y)$

43. $3x(3x + 1)$　　**45.** $-2y(2y - 3)$, or $2y(-2y + 3)$

47. $5a^2(2a - 5)$　　**49.** Cannot be factored　　**51.** $5y(3x + 4)$

53. $2ab(a + 2b)$　　**55.** $6c^2 d^2(2c - 3d)$

57. $4x(x^2 - 3 - 6x)$　　**59.** $8(3a^4 + a^2 - 5)$

61. $6x^3 y^2(3y^3 + 4xy^2 - 2x^2)$　　**63.** Cannot be factored

65. $-14xy^3(x^7 y^6 - 3x^4 y + 2)$, or $14xy^3(-x^7 y^6 + 3x^4 y - 2)$

67. $8m^3 n^6(4m^2 n - 3m^5 n^3 - 5)$　　**69.** $(a + b)(m + n)$

71. $(y + 1)(x - 1)$　　**73.** $(a - 2b)(3a + 2)$

Exercises 11.2   (page 311)

1. $(a + b)(m + n)$　　**3.** $(m - n)(x - y)$

5. $(y + 1)(x - 1)$　　**7.** $(a - 2b)(3a + 2)$

9. $(3e - f)(2e - 3)$　　**11.** $(x + y)(a + 2b)$

13. Cannot be factored　　**15.** $\quad 10xy - 15y + 8x - 12$
$$= 5y(2x - 3) + 4(2x - 3)$$
$$= (2x - 3)(5y + 4)$$

17. $\quad x^3 + 3x^2 - 2x - 6$
$$= x^2(x + 3) - 2(x + 3)$$
$$= (x + 3)(x^2 - 2)$$

Exercises 11.3   (page 312)

1. $(x + 3)(x - 3)$　　**3.** $(b + 1)(b - 1)$

5. $(m + n)(m - n)$　　**7.** $(2c + 5)(2c - 5)$

9. $(5a + 2b)(5a - 2b)$　　**11.** $(8a + 7b)(8a - 7b)$

13. $2x^2 - 8 = 2(x^2 - 4) = 2(x + 2)(x - 2)$

15. $5x(y + a)(y - a)$　　**17.** $3x(x + 5)(x - 5)$

19. Cannot be factored　　**21.** $(x^3 + a^2)(x^3 - a^2)$

23. $(7u^2 + 6v^2)(7u^2 - 6v^2)$　　**25.** $(ab + cd)(ab - cd)$

27. $(7 + 5wz)(7 - 5wz)$

29. $3x^4 - 27x^2 = 3x^2(x^2 - 9) = 3x^2(x + 3)(x - 3)$

31. $ab(2a + 3b)(2a - 3b)$

Exercises 11.4   (page 315)

1. $(x + 2)(x + 4)$　　**3.** $(x + 1)(x + 4)$　　**5.** $(k + 1)(k + 6)$

7. $(u + 2)(u + 5)$　　**9.** Cannot be factored

11. $(b - 2)(b - 7)$　　**13.** $(z - 4)(z - 5)$

15. $(x - 2)(x - 9)$　　**17.** $(x - 1)(x + 10)$

19. $(z + 2)(z - 3)$　　**21.** Cannot be factored

23. $(u^2 - 8)(u^2 - 8) = (u^2 - 8)^2$

25. $(v - 4)(v - 4) = (v - 4)^2$　　**27.** $(b + 4d)(b - 15d)$

29. $(r + 3s)(r - 16s)$　　**31.** $(w - 4)(w + 6)$

33. $(n + 2)(n - 12)$　　**35.** $(w + 4z)(w - 12z)$

Exercises 11.5   (page 319)

1. $(x + 2)(3x + 1)$　　**3.** $(x + 1)(5x + 2)$

5. $(x + 1)(4x + 3)$　　**7.** Cannot be factored

9. $(a - 3)(5a - 1)$　　**11.** $(b - 7)(3b - 1)$

13. $(z - 7)(5z - 1)$　　**15.** $(n + 5)(3n - 1)$

17. $(k - 7)(5k + 1)$　　**19.** $(x + 3y)(7x + 2y)$

21. $(h - k)(7h - 4k)$　　**23.** Cannot be factored

25. $(v - 3)(5v - 2)$　　**27.** $(2e^2 - 5)(3e^2 + 4)$

29. $(5k - 1)(7k - 1)$　　**31.** $(7x - 5)(x + 1)$

33. $(2m - 7n)(2m - n)$

35. $\quad 17u - 12 + 5u^2$
$$= 5u^2 + 17u - 12 \quad \text{Descending powers}$$
$$= (5u - 3)(u + 4)$$

Exercises 11.6   (page 321)

1. $2(x + 2y)(x - 2y)$　　**3.** $5(a^2 + 2b)(a^2 - 2b)$

5. $(x^2 + y^2)(x + y)(x - y)$　　**7.** $2(v + 4)(2v - 1)$

9. $4(z - 2)(2z + 1)$　　**11.** $2(2x - 1)(3x + 4)$

13. $a(b - 1)^2$　　**15.** $3(a + 5b)(a - 5b)$

17. $(m^2 + 1)(m + 1)(m - 1)$　　**19.** Cannot be factored

21. Cannot be factored　　**23.** $mn^3(2m + n)(2m - n)$

25. $5w(z + 2w)(z - w)$　　**27.** $(x^2 + 9)(x + 3)(x - 3)$

29. $a^3 b^2(a + 2b)(a - 2b)$　　**31.** $2a(x + 2ay)(x - 2ay)$

33. $2u(u + 3v)(u - 2v)$

35. $\quad 8h^3 - 20h^2 k + 12hk^2$　　**37.** $\quad 12 + 4x - 3x^2 - x^3$
$$= 4h(2h^2 - 5hk + 3k^2) \qquad\quad = 4(3 + x) - x^2(3 + x)$$
$$= 4h(2h - 3k)(h - k) \qquad\quad = (3 + x)(4 - x^2)$$
$$\qquad\qquad\qquad\qquad\qquad\qquad = (3 + x)(2 + x)(2 - x)$$

39. $3(2my - 3nz + 5mz)$　　**41.** $6(a + b)(c - d)$

43. $3(4x - 1)(2x + 3)$　　**45.** $5a(9a - 4)(a - 1)$

47. $\quad x^3 - 4x^2 - 4x + 16$　　**49.** $3(3a + 2b)^2$
$$= x^2(x - 4) - 4(x - 4)$$
$$= (x - 4)(x^2 - 4)$$
$$= (x - 4)(x + 2)(x - 2)$$

51. $\quad x^4 - 2x^2 + 1$
$$= (x^2 - 1)(x^2 - 1)$$
$$= (x + 1)(x - 1)(x + 1)(x - 1)$$
$$= (x + 1)^2(x - 1)^2$$

53. $4xy^2(2x + 3y)(x - y)$

Exercises 11.7 (page 325)

1. 5, −4   **3.** 0, 4   **5.** $\frac{3}{2}$, −10

7. $3(x + 2)(x - 1) = 0$ **9.** 0, $\frac{2}{3}$, $-\frac{9}{5}$

$$3 \neq 0 \quad \begin{array}{c} x + 2 = 0 \\ -2 \quad -2 \\ \hline x \quad = -2 \end{array} \quad \begin{array}{c} x - 1 = 0 \\ +1 \quad +1 \\ \hline x \quad = 1 \end{array}$$

11. −1, −8   **13.** 4, −3   **15.** 9, −2   **17.** 0, $\frac{5}{3}$

19. 0, 6   **21.** 6, −6   **23.** 5, −5   **25.** 3, $\frac{1}{5}$

27. −1, $\frac{5}{3}$ **29.** 3, −6   **31.** 3, $-\frac{2}{3}$ **33.** 0, 3

35. 3 (double root)   **37.** $\frac{1}{2}$, −2

39. $(y - 3)(y - 6) = -2$ **41.** 1, 4   **43.** 6, −2
$y^2 - 9y + 18 = -2$
$y^2 - 9y + 20 = 0$
$(y - 4)(y - 5) = 0$
$\begin{array}{c|c} y - 4 = 0 & y - 5 = 0 \\ y = 4 & y = 5 \end{array}$

45. $2x^3 + x^2 = 3x$ **47.** 0, 5
$2x^3 + x^2 - 3x = 0$
$x(2x^2 + x - 3) = 0$
$x(x - 1)(2x + 3) = 0$
$\begin{array}{c|c|c} x = 0 & x - 1 = 0 & 2x + 3 = 0 \\ & x = 1 & 2x = -3 \\ & & x = -\frac{3}{2} \end{array}$

49. $\dfrac{x - 1}{x + 4} = \dfrac{3}{x}$ **51.** 0, 3
$x(x - 1) = 3(x + 4)$
$x^2 - x = 3x + 12$
$x^2 - 4x - 12 = 0$
$(x - 6)(x + 2) = 0$
$\begin{array}{c|c} x - 6 = 0 & x + 2 = 0 \\ x = 6 & x = -2 \end{array}$

Exercises 11.8 (page 327)

1. a. Let x = smaller number
Then $x + 5$ = larger number
b. $x(x + 5) = 14$
$x = 2$ or $x = -7$
c. Two answers: {2, 7} and {−7, −2}

3. a. Let x = one number
Then $12 - x$ = other number
b. $x(12 - x) = 35$
$x = 5$ or $x = 7$
c. The numbers are 5 and 7.

5. a. Let x = first consecutive odd
Then $x + 2$ = second consecutive odd
b. $x(x + 2) = 63$
$x = 7$ or $x = -9$
c. Two answers: {7, 9} and {−9, −7}

7. a. Let x = first consecutive
Then $x + 1$ = second consecutive

b. $x(x + 1) = 11 + x + x + 1$
$x = 4$ or $x = -3$
c. Two answers: {4, 5} and {−3, −2}

9. a. Let x = first consecutive
Then $x + 1$ = second consecutive
and $x + 2$ = third consecutive
b. $x(x + 1) = 7 + (x + 2)$
$x = 3$ or $x = -3$
c. Two answers: {3, 4, 5} and {−3, −2, −1}

11. a. Let x = first consecutive
Then $x + 1$ = second consecutive
b. $x^2 + (x + 1)^2 = 25$
$x = -4$ or $x = 3$
c. Two answers: {−4, −3} and {3, 4}

13. a. Let x = one number
Then $3x$ = other number
b. $x^2 + (3x)^2 = 40$
$x = 2$ or $x = -2$
c. Two answers: {2, 6} and {−2, −6}

15. a. Let w = width
Then $w + 5$ = length
b. $w(w + 5) = 84$
$w = 7$ or $w \neq -12$
c. Width = 7 ft; length = 12 ft

17. a. Let x = height
Then $x + 3$ = base

$$\text{Area} = \frac{1}{2}bh$$

b.
$$20 = \frac{1}{2}(x + 3)x$$
$$\frac{20}{1} = \frac{(x + 3)x}{x} \quad \text{A proportion}$$
$$1 \cdot (x + 3)x = 20(2)$$
$$x^2 + 3x = 40$$
$$x^2 + 3x - 40 = 0$$
$$(x + 8)(x - 5) = 0$$
$$\begin{array}{c|cc} x = -8 & x = 5 & \text{Height} \\ \text{No meaning} & x + 3 = 8 & \text{Base} \end{array}$$
c. Height = 5 in.; base = 8 in.

19. a. Let l = length
Then $l - 4$ = width
b. $l(l - 4) = 17 + 2l + 2(l - 4)$
$l = 9$ or $l \neq -1$
c. Length = 9 yd; width = 5 yd

21. a. Let x = side of large square
Then $x - 3$ = side of small square

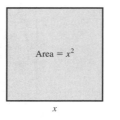

b.

The area of the larger square	is	4 times as great as	the area of the smaller square.
x^2	=	4 ·	$(x - 3)^2$

$$x^2 = 4(x - 3)^2$$
$$x^2 = 4(x^2 - 6x + 9)$$

$$x^2 = 4x^2 - 24x + 36$$
$$0 = 3x^2 - 24x + 36$$
$$0 = 3(x^2 - 8x + 12)$$
$$0 = 3(x - 2)(x - 6)$$

$x = 2$	$x = 6$ Large square
$x - 3 = -1$	$x - 3 = 3$ Small square
Not possible	

c. Side of large square is 6 cm; side of small square is 3 cm

23. a. Let x = first term
Then $x + 7$ = fourth term

b. $\dfrac{x}{3} = \dfrac{6}{x + 7}$
$x = 2$ or $x = -9$

c. The proportions are $\dfrac{2}{3} = \dfrac{6}{9}$, or $\dfrac{-9}{3} = \dfrac{6}{-2}$

Chapter 11 Review Exercises (page 328)

1. $12 = 2^2 \cdot 3$ **2.** $36 = 2^2 \cdot 3^2$ **3.** 31 is prime.

4. $42 = 2 \cdot 3 \cdot 7$ **5.**

2	210
5	105
3	21
	7

$210 = 2 \cdot 3 \cdot 5 \cdot 7$

6. $4(2x - 1)$ **7.** $(m + 2)(m - 2)$ **8.** $(x + 3)(x + 7)$

9. $-2x(x + 3)(x - 3)$ **10.** $(z + 2)(z - 9)$

11. $(4x - 1)(x - 6)$

12. $9k^2 - 144 = 9(k^2 - 16) = 9(k + 4)(k - 4)$

13. $8 - 2a^2 = 2(4 - a^2) = 2(2 + a)(2 - a)$

14. $ab + 2b - a - 2 = b(a + 2) - 1(a + 2) = (a + 2)(b - 1)$

15. $(y - 4)(x + 3)$ **16.** $(3x - 2)(x + 2)$

17. $3uv(5u - 1)$ **18.** $(a^2 + b^2)(a + b)(a - b)$

19. $(x^4 + 1)(x^2 + 1)(x + 1)(x - 1)$

20. $x^4 - 5x^2 + 4 = (x^2 - 4)(x^2 - 1)$
$= (x + 2)(x - 2)(x + 1)(x - 1)$

21. $4xy(x - 2y + 1)$

22. $15a^2 + 15ab - 30b^2 = 15(a^2 + ab - 2b^2)$
$= 15(a + 2b)(a - b)$

23. $3uv(2u^2v - 3v^2 - 4)$ **24.** $5, -3$ **25.** $7, -2$

26. $6, -3$ **27.** $3, -3$ **28.** $0, 9$ **29.** $\dfrac{5}{2}, -\dfrac{1}{3}$

30. $-6, 2$

31.
$$6x^3 = 9x^2 + 6x$$
$$6x^3 - 9x^2 - 6x = 0$$
$$3x(2x^2 - 3x - 2) = 0$$
$$3x(2x + 1)(x - 2) = 0$$

$3x = 0$	$2x + 1 = 0$	$x - 2 = 0$
$x = 0$	$2x = -1$	$x = 2$
	$x = -\dfrac{1}{2}$	

32.
$$\frac{x - 2}{9} = \frac{4}{x + 7}$$
$$(x - 2)(x + 7) = 36$$
$$x^2 + 5x - 14 = 36$$
$$x^2 + 5x - 50 = 0$$
$$(x + 10)(x - 5) = 0$$

$x + 10 = 0$	$x - 5 = 0$
$x = -10$	$x = 5$

33. Let x = smaller number
Then $x + 3$ = large number
$x(x + 3) = 28$
$x = 4$ or $x = -7$
There are two answers: $\{4, 7\}$ and $\{-7, -4\}$

34. Let x = first consecutive
Then $x + 1$ = second consecutive
and $x + 2$ = third consecutive
$$x(x + 1) - (x + 2) = 7$$
$$x^2 + x - x - 2 = 7$$
$$x^2 - 2 = 7$$
$$x^2 - 9 = 0$$
$$(x + 3)(x - 3) = 0$$

$x + 3 = 0$	$x - 3 = 0$
$x = -3$	$x = 3$
$x + 1 = -2$	$x + 1 = 4$
$x + 2 = -1$	$x + 2 = 5$

There are two answers: $\{-3, -2, -1\}$ and $\{3, 4, 5\}$

35. Let l = length
Then $l - 3$ = width
$l(l - 3) = 40$

$l = -5$	$l = 8$
Not possible	$l - 3 = 5$

The length is 8, and the width is 5.

36. Let x = side of small square
Then $x + 6$ = side of large square
$$(x + 6)^2 = 16x^2$$
$$x^2 + 12x + 36 = 16x^2$$
$$0 = 15x^2 - 12x - 36$$
$$0 = 3(5x^2 - 4x - 12)$$
$$0 = 3(5x + 6)(x - 2)$$

$3 \neq 0$	$5x + 6 = 0$	$x - 2 = 0$
	$5x = -6$	$x = 2$
	$x = \dfrac{-6}{5}$	$x + 6 = 8$
	Not possible	

The side of the small square is 2, and the side of the large square is 8.

37. Let x = side of square
$x^2 = 3(4x)$
$x = 0$ or $x = 12$
The side of the square is 12.

38. Let w = width
Then $w + 6$ = length
$w(w + 6) = 12 + 2w + 2(w + 6)$

$w = -6$	$w = 4$
Not possible	$w + 6 = 10$

The width is 4 yd, and the length is 10 yd.

Chapter 11 Diagnostic Test (page 330)

Following each problem number is the number (in parentheses) of the textbook section in which that kind of problem is discussed.

1. (11.1) **a.** $18 = 2 \cdot 9 = 3 \cdot 6$
18 is composite because it has factors other than itself and 1.

b. $21 = 3 \cdot 7$ Composite

c. $31 = 1 \cdot 31$ Prime
31 has no factors other than itself and 1.

2. (11.1)

5	45
3	9
	3

$45 = 3^2 \cdot 5$

3. (11.1)

$$160 = 2^5 \cdot 5$$

2	160
2	80
2	40
2	20
2	10
	5

4. (11.1) $5x + 10 = 5(x + 2)$

5. (11.1) $3x^3 - 6x^2 = 3x^2(x - 2)$

6. (11.3) $16x^2 - 49y^2 = (4x + 7y)(4x - 7y)$

7. (11.6) $2z^2 - 8 = 2(z^2 - 4) = 2(z + 2)(z - 2)$

8. (11.4) $z^2 + 6z + 8 = (z + 2)(z + 4)$

9. (11.4) $m^2 + m - 6 = (m + 3)(m - 2)$

10. (11.5) $5w^2 - 12w + 7 = (5w - 7)(w - 1)$

11. (11.5) $3v^2 + 14v - 5 = (3v - 1)(v + 5)$

12. (11.6) $x^2 + 25$ cannot be factored.

13. (11.5) $4x^2 - 11x + 6 = (4x - 3)(x - 2)$

14. (11.4) $x^2 + 6x - 8$ cannot be factored.

15. (11.2)
$$5n - mn - 5 + m$$
$$= n(5 - m) - 1(5 - m)$$
$$= (5 - m)(n - 1)$$

16. (11.6)
$$6h^2k - 8hk^2 + 2k^3$$
$$= 2k(3h^2 - 4hk + k^2)$$
$$= 2k(3h - k)(h - k)$$

17. (11.6)
$$x^4 - 8x^2 - 9$$
$$= (x^2 + 1)(x^2 - 9)$$
$$= (x^2 + 1)(x + 3)(x - 3)$$

18. (11.6)
$$24a^2 + 2a - 12$$
$$= 2(12a^2 + a - 6)$$
$$= 2(3a - 2)(4a + 3)$$

19. (11.7) $x^2 - 12x + 20 = 0$
$$(x - 2)(x - 10) = 0$$

$x - 2 = 0$	$x - 10 = 0$
$x = 2$	$x = 10$

20. (11.7)
$$3x^2 = 8 - 2x$$
$$3x^2 + 2x - 8 = 0$$
$$(3x - 4)(x + 2) = 0$$

$3x - 4 = 0$	$x + 2 = 0$
$3x = 4$	$x = -2$
$x = \dfrac{4}{3}$	

21. (11.7)
$$3x^2 = 12x$$
$$3x^2 - 12x = 0$$
$$3x(x - 4) = 0$$

$3x = 0$	$x - 4 = 0$
$x = 0$	$x = 4$

22. (11.7) $36y = 18y^2 + 18$
$$0 = 18y^2 - 36y + 18$$
$$0 = 18(y^2 - 2y + 1)$$
$$0 = 18(y - 1)(y - 1)$$

$18 \neq 0$	$y - 1 = 0$	$y - 1 = 0$
	$y = 1$	$y = 1$

23. (11.7) $(x + 1)(x + 3) = 24$
$$x^2 + 4x + 3 = 24$$
$$x^2 + 4x - 21 = 0$$
$$(x + 7)(x - 3) = 0$$

$x + 7 = 0$	$x - 3 = 0$
$x = -7$	$x = 3$

24. (11.8) Let w = width
Then $w + 3$ = length
$$w(w + 3) = 28$$
$$w^2 + 3w = 28$$
$$w^2 + 3w - 28 = 0$$
$$(w + 7)(w - 4) = 0$$

$w + 7 = 0$	$w - 4 = 0$
$w = -7$	$w = 4$
Not possible	$w + 3 = 7$

The width is 4 ft, and the length is 7 ft.

25. (11.8) Let x = first odd
Then $x + 2$ = second odd
$$x(x + 2) = 7 + x + x + 2$$
$$x^2 + 2x = 2x + 9$$
$$x^2 - 9 = 0$$
$$(x + 3)(x - 3) = 0$$

$x + 3 = 0$	$x - 3 = 0$
$x = -3$	$x = 3$
$x + 2 = -1$	$x + 2 = 5$

Two answers: $\{-3, -1\}$ and $\{3, 5\}$

Exercises 12.1   (page 335)

1. $x \neq 2$ **3.** None   **5.** $x \neq 1$ and $x \neq -2$ **7.** None

9. $x \neq 2$ and $x \neq -1$ **11.** $x \neq 0$ and $x \neq -3$ **13.** -5

15. y **17.** -7

19. -5, the sign of the numerator was changed,
$y - 6 = -(6 - y)$. Therefore, the sign of 5 must be changed to -5.

21. $\dfrac{3}{4}$ **23.** $2b$ **25.** $2x$ **27.** 5   **29.** -1

31. $\dfrac{x + 6}{2(x - 4)}$ **33.** $\dfrac{2 + 4}{4} = \dfrac{6}{4} = \dfrac{3}{2}$

35. Cannot be reduced   **37.** $\dfrac{y}{2x + y}$ **39.** $x - 1$

41. $\dfrac{3x - 12}{4 - x} = \dfrac{3\overset{-1}{\cancel{(x - 4)}}}{\underset{1}{\cancel{4 - x}}} = -3$ **43.** $\dfrac{x - 6}{x - 4}$ **45.** $\dfrac{x + 4}{x + 3}$

47. $\dfrac{3x - 2}{5x - 1}$ **49.** $\dfrac{x - y}{x + y}$

51. $\dfrac{(a - 2b)\overset{-1}{\cancel{(b - a)}}}{(2a + b)\underset{1}{\cancel{(a - b)}}} = \dfrac{-1(a - 2b)}{2a + b} = \dfrac{2b - a}{2a + b}$

53. $\dfrac{2y^2 + xy - 6x^2}{3x^2 + xy - 2y^2} = \dfrac{\overset{-1}{\cancel{(2y - 3x)}}(y + 2x)}{\underset{1}{\cancel{(3x - 2y)}}(x + y)}$
$$= \dfrac{-(y + 2x)}{x + y}, \text{ or } -\dfrac{y + 2x}{x + y}$$

55. $\dfrac{8x^2 - 2y^2}{2ax - ay + 2bx - by} = \dfrac{2(4x^2 - y^2)}{a(2x - y) + b(2x - y)}$
$$= \dfrac{2(2x + y)\overset{1}{\cancel{(2x - y)}}}{\underset{1}{\cancel{(2x - y)}}(a + b)}$$
$$= \dfrac{2(2x + y)}{a + b}, \text{ or } \dfrac{4x + 2y}{a + b}$$

57. $\dfrac{xy + 2x + 3y + 6}{xy + x + 3y + 3} = \dfrac{x(y + 2) + 3(y + 2)}{x(y + 1) + 3(y + 1)}$
$$= \dfrac{(y + 2)\overset{1}{\cancel{(x + 3)}}}{(y + 1)\underset{1}{\cancel{(x + 3)}}}$$
$$= \dfrac{y + 2}{y + 1}$$

Exercises 12.2 (page 338)

1. $\dfrac{1}{2}$ **3.** $\dfrac{a}{b}$ **5.** $\dfrac{3x}{2}$ **7.** $\dfrac{5}{2}$ **9.** $\dfrac{x}{y}$ **11.** $\dfrac{5}{x^2}$

13. $\dfrac{b}{2}$ **15.** $\dfrac{a}{a-2}$ **17.** 1 **19.** $\dfrac{5}{z-4}$ **21.** $-\dfrac{3}{2}$

23. 1 **25.** $\dfrac{4a}{3b}$ **27.** $\dfrac{3u^2}{2}$ **29.** $\dfrac{1}{4}$ **31.** $\dfrac{x^2}{x+3}$

33. -1

35. $\dfrac{x-y}{9x+9y} \div \dfrac{x^2-y^2}{3x^2+6xy+3y^2}$

$= \dfrac{\overset{1}{\cancel{(x-y)}}}{\underset{3}{\cancel{9}}\cancel{(x+y)}} \cdot \dfrac{\overset{1}{\cancel{3}}\cancel{(x+y)}\overset{1}{\cancel{(x+y)}}}{\cancel{(x+y)}\cancel{(x-y)}}$

$= \dfrac{1}{3}$

37. $\dfrac{2x^3y+2x^2y^2}{6x} \div \dfrac{x^2y^2-xy^3}{y-x}$

$= \dfrac{\overset{1}{\cancel{2}}x^2y(x+y)}{\underset{3}{\cancel{6}}x} \cdot \dfrac{\overset{-1}{\cancel{y-x}}}{xy^2\cancel{(x-y)}}$

$= \dfrac{-(x+y)}{3y}$, or $-\dfrac{x+y}{3y}$

39. $\dfrac{x^2-8x-9}{x^3-3x^2-x+3} \cdot \dfrac{x^2-9}{x^2-9x}$

$= \dfrac{(x-9)(x+1)}{x^2(x-3)-1(x-3)} \cdot \dfrac{(x+3)(x-3)}{x(x-9)}$

$= \dfrac{(x-9)(x+1)}{(x-3)(x^2-1)} \cdot \dfrac{(x+3)(x-3)}{x(x-9)}$

$= \dfrac{\overset{1}{\cancel{(x-9)}}\overset{1}{\cancel{(x+1)}}}{\cancel{(x-3)}\cancel{(x+1)}(x-1)} \cdot \dfrac{(x+3)\overset{1}{\cancel{(x-3)}}}{x\cancel{(x-9)}}$

$= \dfrac{x+3}{x(x-1)}$

41. $\dfrac{a^2-3ab+2b^2}{2a+2b} \cdot \dfrac{2a^2+3ab+b^2}{a^2-ab-2b^2} \div \dfrac{4a^2+10ab+4b^2}{4a+8b}$

$= \dfrac{(a-b)\overset{1}{\cancel{(a-2b)}}}{\underset{}{\cancel{2}}(a+b)} \cdot \dfrac{\overset{1}{\cancel{(2a+b)}}\overset{1}{\cancel{(a+b)}}}{\cancel{(a-2b)}\cancel{(a+b)}} \cdot \dfrac{\overset{2}{\cancel{4}}\overset{1}{\cancel{(a+2b)}}}{\cancel{2}\cancel{(2a+b)}\cancel{(a+2b)}}$

$= \dfrac{a-b}{a+b}$

Exercises 12.3 (page 340)

1. $\dfrac{9}{a}$ **3.** $\dfrac{2}{a}$ **5.** $\dfrac{4}{x}$ **7.** $\dfrac{4}{x-y}$ **9.** 2 **11.** 1

13. 1 **15.** 3 **17.** -2 **19.** 3 **21.** $\dfrac{-8}{y-2}$, or $\dfrac{8}{2-y}$

23. $\dfrac{a+2}{2a+1} - \dfrac{1-a}{2a+1}$ **25.** -1 **27.** 3

$= \dfrac{a+2-(1-a)}{2a+1}$

$= \dfrac{a+2-1+a}{2a+1}$

$= \dfrac{2a+1}{2a+1} = 1$

29. $\dfrac{7z}{8z-4} + \dfrac{6-5z}{4-8z}$

$= \dfrac{7z}{8z-4} + \dfrac{5z-6}{8z-4}$

$= \dfrac{7z+5z-6}{8z-4}$

$= \dfrac{12z-6}{8z-4}$

$= \dfrac{\overset{3}{\cancel{6}}\cancel{(2z-1)}}{\underset{2}{\cancel{4}}\cancel{(2z-1)}}$

$= \dfrac{3}{2}$

31. $\dfrac{x^2+x}{x+2} - \dfrac{x+4}{x+2}$

$= \dfrac{x^2+x-(x+4)}{x+2}$

$= \dfrac{x^2+x-x-4}{x+2}$

$= \dfrac{x^2-4}{x+2}$

$= \dfrac{\overset{1}{\cancel{(x+2)}}(x-2)}{\underset{1}{\cancel{x+2}}}$

$= x-2$

Exercises 12.4 (page 342)

1. 10 **3.** $2y$ **5.** 4 **7.** $6x$ **9.** x^3 **11.** $12y^2$

13. $36u^3v^3$ **15.** $a(a+3)$

17. (1) $2(x+2)$, 2^2x (denominators in factored form)
(2) 2, x, $(x+2)$ are the different factors.
(3) 2^2, x, $(x+2)$ are the highest powers of each factor.
(4) LCD $= 4x(x+2)$

19. $(x+2)^2$

21. (1) $(x+1)(x+2)$, $(x+1)^2$ **23.** $(x+y)^2(x-y)$
(2) $(x+1)$, $(x+2)$
(3) $(x+1)^2$, $(x+2)$
(4) LCD $= (x+1)^2(x+2)$

25. (1) $2^2 \cdot 3 \cdot x^2(x+2)$, $(x-2)^2$, $(x+2)(x-2)$
(2) 2, 3, x, $(x+2)$, $(x-2)$
(3) 2^2, 3, x^2, $(x+2)$, $(x-2)^2$
(4) LCD $= 12x^2(x+2)(x-2)^2$

Exercises 12.5 (page 346)

1. $\dfrac{3a+2}{a^3}$ **3.** $\dfrac{3x+4}{x^2}$ **5.** $\dfrac{x^2+6x-10}{2x^2}$ **7.** $\dfrac{2-3x}{xy}$

9. $\dfrac{3y+4}{y}$ **11.** $\dfrac{3x^2-3}{x}$ **13.** $\dfrac{5x+2}{6x^2}$ **15.** $\dfrac{10y-21x}{18x^2y^2}$

17. $\dfrac{a^2+3a+9}{a(a+3)}$ **19.** $\dfrac{x^2-x+2}{x(x-2)}$

21. LCD $= 2x(x+2)$ **23.** $\dfrac{x^3-2x^2-x-4}{x(x-2)}$

$\dfrac{3}{2(x+2)} \cdot \dfrac{x}{x} - \dfrac{1}{2x} \cdot \dfrac{x+2}{x+2}$

$= \dfrac{3x}{2x(x+2)} - \dfrac{x+2}{2x(x+2)}$

$= \dfrac{3x-(x+2)}{2x(x+2)}$

$= \dfrac{3x-x-2}{2x(x+2)}$

$= \dfrac{2x-2}{2x(x+2)}$

$= \dfrac{\overset{1}{\cancel{2}}(x-1)}{\underset{1}{\cancel{2}}x(x+2)}$

$= \dfrac{x-1}{x(x+2)}$

12

25. $\dfrac{a^2}{b(a-b)}$ **27.** $\dfrac{5}{6(e-1)}$ **29.** $\dfrac{8}{m-2}$

31. $\dfrac{x+10}{(x-2)(x+2)}$ **33.** $\dfrac{3}{(x+2)(x+1)}$

35. $\dfrac{4x}{(x-1)(x+1)}$

37. LCD $= 4x(x+4)$ **39.** $\dfrac{1}{6}$

$$\dfrac{2x+4}{x(x+4)}\cdot\boxed{\dfrac{4}{4}}+\dfrac{x}{4(x+4)}\cdot\boxed{\dfrac{x}{x}}$$
$$=\dfrac{8x+16}{4x(x+4)}+\dfrac{x^2}{4x(x+4)}$$
$$=\dfrac{x^2+8x+16}{4x(x+4)}$$
$$=\dfrac{(x+4)\overset{1}{\cancel{(x+4)}}}{4x\underset{1}{\cancel{(x+4)}}}$$
$$=\dfrac{x+4}{4x}$$

41. LCD $= xy(x-y)$

$$\dfrac{y}{x(x-y)}\cdot\boxed{\dfrac{y}{y}}+\dfrac{-x}{y(x-y)}\cdot\boxed{\dfrac{x}{x}}$$
$$=\dfrac{y^2-x^2}{xy(x-y)}$$
$$=\dfrac{(y+x)\overset{-1}{\cancel{(y-x)}}}{xy\underset{1}{\cancel{(x-y)}}}$$
$$=\dfrac{-(y+x)}{xy},\text{ or }-\dfrac{y+x}{xy}$$

43. LCD $= (x-3)(x+3)$

$$\dfrac{2x}{x-3}\cdot\boxed{\dfrac{x+3}{x+3}}-\dfrac{2x}{x+3}\cdot\boxed{\dfrac{x-3}{x-3}}+\dfrac{36}{(x+3)(x-3)}$$
$$=\dfrac{2x^2+6x}{(x+3)(x-3)}-\dfrac{2x^2-6x}{(x+3)(x-3)}+\dfrac{36}{(x+3)(x-3)}$$
$$=\dfrac{2x^2+6x-(2x^2-6x)+36}{(x+3)(x-3)}$$
$$=\dfrac{2x^2+6x-2x^2+6x+36}{(x-3)(x+3)}$$
$$=\dfrac{12x+36}{(x-3)(x+3)}$$
$$=\dfrac{12\overset{1}{\cancel{(x+3)}}}{(x-3)\underset{1}{\cancel{(x+3)}}}=\dfrac{12}{x-3}$$

45. $\dfrac{2x+2}{(x+2)^2}$ **47.** $\dfrac{1}{x+1}$ **49.** $\dfrac{x}{x+4}$

51. $\dfrac{3}{(x-1)(x+2)}$ **53.** $\dfrac{a}{5}$

55. $\dfrac{3x}{(x-5)(x+4)}\cdot\boxed{\dfrac{x-2}{x-2}}-\dfrac{5}{(x-5)(x-2)}\cdot\boxed{\dfrac{x+4}{x+4}}$
$$=\dfrac{3x^2-6x}{(x-5)(x+4)(x-2)}-\dfrac{5x+20}{(x-5)(x-2)(x+4)}$$
$$=\dfrac{3x^2-6x-(5x+20)}{(x-5)(x+4)(x-2)}$$
$$=\dfrac{3x^2-6x-5x-20}{(x-5)(x+4)(x-2)}$$
$$=\dfrac{3x^2-11x-20}{(x-5)(x+4)(x-2)}$$

$$=\dfrac{(3x+4)\overset{1}{\cancel{(x-5)}}}{\underset{1}{\cancel{(x-5)}}(x+4)(x-2)}$$
$$=\dfrac{3x+4}{(x+4)(x-2)}$$

57. LCD $= (x+1)^2(x+2)$

$$\dfrac{(x-4)}{(x+1)(x+2)}\cdot\boxed{\dfrac{x+1}{x+1}}+\dfrac{3x+1}{(x+1)^2}\cdot\boxed{\dfrac{x+2}{x+2}}$$
$$=\dfrac{(x-4)(x+1)}{(x+1)^2(x+2)}+\dfrac{(3x+1)(x+2)}{(x+1)^2(x+2)}$$
$$=\dfrac{x^2-3x-4+3x^2+7x+2}{(x+1)^2(x+2)}$$
$$=\dfrac{4x^2+4x-2}{(x+1)^2(x+2)}$$

59. $\dfrac{2x(2x-y)}{4x^2+y^2}$

61. LCD $= (x+2)(x-2)$

$$\dfrac{2x}{x+2}\cdot\boxed{\dfrac{x-2}{x-2}}+\dfrac{x}{x-2}\cdot\boxed{\dfrac{x+2}{x+2}}-\dfrac{3x+2}{(x+2)(x-2)}$$
$$=\dfrac{2x^2-4x}{(x+2)(x-2)}+\dfrac{x^2+2x}{(x-2)(x+2)}-\dfrac{3x+2}{(x+2)(x-2)}$$
$$=\dfrac{2x^2-4x+(x^2+2x)-(3x+2)}{(x+2)(x-2)}$$
$$=\dfrac{2x^2-4x+x^2+2x-3x-2}{(x+2)(x-2)}$$
$$=\dfrac{3x^2-5x-2}{(x+2)(x-2)}$$
$$=\dfrac{(3x+1)\overset{1}{\cancel{(x-2)}}}{(x+2)\underset{1}{\cancel{(x-2)}}}$$
$$=\dfrac{3x+1}{x+2}$$

63. $\dfrac{2}{x-1}$

Exercises 12.6   (page 349)

1. $\dfrac{2}{7}$ **3.** $2\dfrac{2}{25}$ **5.** $\dfrac{3x^2}{2y^3}$

7. $\dfrac{\dfrac{18cd^2}{5a^3b}}{\dfrac{12cd^2}{15ab^2}}=\dfrac{18cd^2}{5a^3b}\div\dfrac{12cd^2}{15ab^2}=\dfrac{\overset{3}{\cancel{18cd^2}}}{\underset{1}{\cancel{5a^3b}}}\cdot\dfrac{\overset{3}{\cancel{15ab^2}}}{\underset{2}{\cancel{12cd^2}}}=\dfrac{9b}{2a^2}$

9. ab **11.** 1 **13.** $\dfrac{a+b}{a-b}$ **15.** $\dfrac{d}{c-2d}$

17. The LCD of the secondary fractions is x.

$$\boxed{\dfrac{x}{x}}\cdot\dfrac{1}{1-\dfrac{1}{x}}=\dfrac{\boxed{x}\,(1)}{\boxed{x}\,(1)-\boxed{\dfrac{x}{1}}\left(\dfrac{1}{x}\right)}=\dfrac{x}{x-1}$$

19. $\dfrac{xy+1}{xy}$ **21.** $\dfrac{1+x^2}{1-x^2}$ **23.** $\dfrac{1}{3}$

25. LCD $= a^2$

$$\boxed{\dfrac{a^2}{a^2}}\cdot\dfrac{1-\dfrac{1}{a^2}}{\dfrac{1}{a}-\dfrac{1}{a^2}}=\dfrac{\boxed{\dfrac{a^2}{1}}\cdot1-\boxed{\dfrac{a^2}{1}}\cdot\dfrac{1}{a^2}}{\boxed{\dfrac{a^2}{1}}\cdot\dfrac{1}{a}-\boxed{\dfrac{a^2}{1}}\cdot\dfrac{1}{a^2}}=\dfrac{a^2-1}{a-1}$$

$$=\dfrac{(a+1)\overset{1}{\cancel{(a-1)}}}{\underset{1}{\cancel{a-1}}}=a+1$$

27. $\dfrac{x - 2y}{y}$

29. LCD $= a^2b^2$

$$\frac{a^2b^2}{a^2b^2} \cdot \frac{\dfrac{1}{a^2} - \dfrac{1}{b^2}}{\dfrac{1}{a} + \dfrac{1}{b}} = \frac{\dfrac{a^2b^2}{1} \cdot \dfrac{1}{a^2} - \dfrac{a^2b^2}{1} \cdot \dfrac{1}{b^2}}{\dfrac{a^2b^2}{1} \cdot \dfrac{1}{a} + \dfrac{a^2b^2}{1} \cdot \dfrac{1}{b}}$$

$$= \frac{b^2 - a^2}{ab^2 + a^2b} = \frac{(\overset{1}{\cancel{b + a}})(b - a)}{ab(\underset{1}{\cancel{b + a}})}$$

$$= \frac{b - a}{ab}$$

31. LCD $= x + 3$

$$\frac{2 + \dfrac{6}{x + 3}}{x - \dfrac{18}{x + 3}} \cdot \frac{x + 3}{x + 3} = \frac{2 \cdot (x + 3) + \dfrac{6}{x + 3} \cdot \dfrac{x + 3}{1}}{x \cdot (x + 3) - \dfrac{18}{x + 3} \cdot \dfrac{x + 3}{1}}$$

$$= \frac{2x + 6 + 6}{x^2 + 3x - 18}$$

$$= \frac{2x + 12}{x^2 + 3x - 18}$$

$$= \frac{2(\overset{1}{\cancel{x + 6}})}{(\underset{1}{\cancel{x + 6}})(x - 3)}$$

$$= \frac{2}{x - 3}$$

33. $x + 1$

35. LCD $= (x + 2)(x - 2)$

$$\frac{\dfrac{2}{x - 2} - \dfrac{1}{x + 2}}{\dfrac{8}{(x + 2)(x - 2)} + \dfrac{2}{x - 2}} \cdot \frac{(x + 2)(x - 2)}{(x + 2)(x - 2)}$$

$$= \frac{\dfrac{2}{x - 2} \cdot \dfrac{(x + 2)(x - 2)}{1} - \dfrac{1}{x + 2} \cdot \dfrac{(x + 2)(x - 2)}{1}}{\dfrac{8}{(x + 2)(x - 2)} \cdot \dfrac{(x + 2)(x - 2)}{1} + \dfrac{2}{x - 2} \cdot \dfrac{(x + 2)(x - 2)}{1}}$$

$$= \frac{2x + 4 - x + 2}{8 + 2x + 4} = \frac{x + 6}{2x + 12} = \frac{\overset{1}{\cancel{x + 6}}}{2(\underset{1}{\cancel{x + 6}})} = \frac{1}{2}$$

Exercises 12.7 (page 355)

1. 12 **3.** 20 **5.** 10 **7.** $\dfrac{3}{2}$ **9.** $-4, 4$ **11.** 2

13. $\dfrac{9}{8}$ **15.** 3 **17.** 5 **19.** 1 **21.** 4, -3

23. 0, -2 **25.** 0, -1 **27.** -2 **29.** 2 **31.** $\dfrac{14}{17}$

33. LCD $= x - 2$ Excluded value: $x \neq 2$

$$\left(\frac{\overset{1}{\cancel{x - 2}}}{1}\right) \frac{x}{\underset{1}{\cancel{x - 2}}} = \left(\frac{\overset{1}{\cancel{x - 2}}}{1}\right) \frac{2}{\underset{1}{\cancel{x - 2}}} + \left(\frac{x - 2}{1}\right)(5)$$

$$x = 2 + 5x - 10$$
$$8 = 4x$$
$$x = 2 \quad \text{(Excluded value)}$$

Therefore, this equation has no solution.

35. LCD $= 5(x + 2)(x - 2)$
Excluded values: $x \neq -2$ and $x \neq 2$

$$5(x + 2)(\overset{1}{\cancel{x - 2}}) \frac{1}{\underset{1}{\cancel{x - 2}}} - 5(\overset{1}{\cancel{x + 2}})(x - 2) \frac{4}{\underset{1}{\cancel{x + 2}}}$$

$$= \cancel{5}(x + 2)(x - 2) \frac{1}{\underset{1}{\cancel{5}}}$$

$$5(x + 2) - 20(x - 2) = (x + 2)(x - 2)$$
$$5x + 10 - 20x + 40 = x^2 - 4$$
$$0 = x^2 + 15x - 54$$
$$0 = (x + 18)(x - 3)$$

$$x + 18 = 0 \qquad\qquad x - 3 = 0$$
$$x = -18 \qquad\qquad x = 3$$

37. LCD $= x(x - 3)$ Excluded values: $x \neq 0$ and $x \neq 3$

$$\frac{2}{\underset{1}{\cancel{x - 3}}} \cdot \frac{\overset{1}{\cancel{x(x - 3)}}}{1} + 2 \cdot x(x - 3) = \frac{6}{\underset{1}{\cancel{x(x - 3)}}} \cdot \frac{\overset{1}{\cancel{x(x - 3)}}}{1}$$

$$2x + 2x^2 - 6x = 6$$
$$2x^2 - 4x - 6 = 0$$
$$2(x^2 - 2x - 3) = 0$$
$$2(x - 3)(x + 1) = 0$$

$$2 \neq 0 \quad\Big|\quad x - 3 = 0 \quad\Big|\quad x + 1 = 0$$
$$\qquad\qquad x = 3 \qquad\qquad x = -1$$
$$\qquad\qquad \text{Excluded value}$$

Therefore, $x = -1$ is the only solution.

39. LCD $= 4(2x + 5)$ Excluded value: $x = -\dfrac{5}{2}$

$$\frac{4(\overset{1}{\cancel{2x + 5}})}{1} \cdot \frac{3}{\underset{1}{\cancel{2x + 5}}} + \frac{4(2x + 5)}{1} \cdot \frac{x}{\underset{1}{\cancel{4}}} = \frac{4(2x + 5)}{1} \cdot \frac{3}{\underset{1}{\cancel{4}}}$$

$$4(3) + x(2x + 5) = 3(2x + 5)$$
$$12 + 2x^2 + 5x = 6x + 15$$
$$2x^2 - x - 3 = 0$$
$$(x + 1)(2x - 3) = 0$$

$$x + 1 = 0 \quad\Big|\quad 2x - 3 = 0$$
$$x = -1 \qquad\quad 2x = 3$$
$$\qquad\qquad\quad x = \dfrac{3}{2}$$

Exercises 12.8 (page 357)

1. $x = \dfrac{4 - y}{2}$ **3.** $z = y + 8$ **5.** $y = 2x + 4$

7. $x = \dfrac{3y + 6}{2}$ **9.** $x = 5 - 2y$ **11.** $x = \dfrac{y - 4}{2}$

13. $x = 6y + 4$ **15.** $V = \dfrac{k}{P}$ **17.** $x = \dfrac{10z}{3y}$

19. $x = \dfrac{y - b}{m}$ **21.** $r = \dfrac{A - P}{Pt}$

23.
$$\frac{S}{1} = \frac{a}{1 - r} \qquad \text{This is a proportion}$$
$$S(1 - r) = a \qquad \text{Product of means} =$$
$$S - Sr = a \qquad \text{product of extremes}$$
$$-Sr = a - S$$
$$\frac{-Sr}{-S} = \frac{a - S}{-S}$$
$$r = \frac{S - a}{S}$$

25. $x = \dfrac{c}{a + b}$ **27.** $R = \dfrac{zr}{r - z}$

12

29. LCD $= Fuv$

$$\frac{\overset{1}{\cancel{Fuv}}}{1} \cdot \frac{1}{\underset{1}{\cancel{F}}} = \frac{\overset{1}{\cancel{F}uv}}{1} \cdot \frac{1}{\underset{1}{\cancel{u}}} + \frac{\overset{1}{Fu\cancel{v}}}{1} \cdot \frac{1}{\underset{1}{\cancel{v}}}$$

$uv = Fv + Fu$

$uv - Fu = Fv$

$u(v - F) = Fv$

$\dfrac{u(v - F)}{(v - F)} = \dfrac{Fv}{(v - F)}$

$u = \dfrac{Fv}{v - F}$

31. $F = \dfrac{9C + 160}{5}$

33. Multiply by LCD $= x$

$$\frac{\overset{1}{\cancel{x}}}{1}\left(\frac{m + n}{\underset{1}{\cancel{x}}}\right) + \frac{x}{1}\left(\frac{-a}{1}\right) = \frac{\overset{1}{\cancel{x}}}{1}\left(\frac{m - n}{\underset{1}{\cancel{x}}}\right) + \frac{x}{1}\left(\frac{c}{1}\right)$$

$m + n - ax = m - n + cx$

$\quad 2n = ax + cx \quad$ Collecting terms

$\quad 2n = x(a + c) \quad$ Factoring x from $ax + cx$

$\dfrac{2n}{a + c} = \dfrac{x\overset{1}{\cancel{(a + c)}}}{\underset{1}{\cancel{a + c}}} \quad$ Dividing both sides by $a + c$

$\dfrac{2n}{a + c} = x$

Exercises 12.9   (page 362)

1. a. Let $x =$ the number

b. $x + \dfrac{1}{x} = \dfrac{25}{12}$

$x = \dfrac{3}{4} \;$ or $\; x = \dfrac{4}{3}$

c. Two answers: $\dfrac{3}{4}$ and $\dfrac{4}{3}$

3. a. Let $x =$ the number

b. $x - \dfrac{1}{x} = \dfrac{8}{3}$

$x = -\dfrac{1}{3} \;$ or $\; x = 3$

c. Two answers: $-\dfrac{1}{3}$ and 3

5. a. Let $\quad x =$ numerator
Then $x + 2 =$ denominator

b. $\dfrac{x - 1}{(x + 2) + 5} = \dfrac{1}{5}$

$x = 3$

c. The fraction is $\dfrac{3}{5}$.

7. a. Let $\quad x =$ numerator
Then $2x =$ denominator

b. $\dfrac{x + 2}{2x + 8} = \dfrac{2}{5}$

$x = 6$

c. The fraction is $\dfrac{6}{12}$.

9. a. Let $x =$ time each works

b. $\dfrac{x}{2} + \dfrac{x}{6} = 1$

$x = \dfrac{3}{2}$

c. $1\dfrac{1}{2}$ hr

11. a. Let $x =$ time all pipes are on

b. $\dfrac{x}{3} + \dfrac{x}{2} + \dfrac{x}{6} = 1$

$x = 1$

c. 1 hr

13. a. Let $x =$ time each works

b. $\dfrac{x}{30} + \dfrac{x}{50} = 4$

$x = 75$

c. 75 min, or 1 hr 15 min

15. a. Let $x =$ time to sort mail

b. $\dfrac{1}{4} + \dfrac{x}{3} = 1$

$x = \dfrac{9}{4}$

c. $2\dfrac{1}{4}$ hr

17. a. Let $\quad x =$ Tonya's speed
$\dfrac{4}{5}x =$ Irma's speed

b. $\qquad 3x + 3\left(\dfrac{4x}{5}\right) = 54$

$5 \cdot 3x + \overset{1}{\cancel{5}} \cdot \dfrac{12x}{\underset{1}{\cancel{5}}} = 54 \cdot 5$

$15x + 12x = 270$

$27x = 270$

$x = 10$

c. Tonya's speed $= x = 10$ mph
Irma's speed $= \dfrac{4}{5}x = \dfrac{4(10)}{5} = 8$ mph

19. a. Let $x =$ score on fourth exam

b. $\dfrac{70 + 85 + 83 + x}{4} = 80$

$\dfrac{238 + x}{4} = 80$

$238 + x = 320$

$x = 82$

c. Shakil must score at least 82.

Chapter 12 Review Exercises (page 364)

1. $x \ne -4$ **2.** $x \ne 0$

3. $x^2 - 3x - 10 = 0$
$(x - 5)(x + 2) = 0$
$x - 5 = 0 \;\big|\; x + 2 = 0$
$\quad x = 5 \;\big|\quad x = -2$
Therefore, $x \ne 5$ and $x \ne -2$.

4. 5   **5.** $x - 2$ **6.** 2   **7.** $\dfrac{2x^2}{y}$ **8.** $1 + 2m$

9. $a - 2$ **10.** $\dfrac{1}{x - 4}$ **11.** $\dfrac{y - x}{y + x}$

12. $\dfrac{a - b}{ax + ay - bx - by} = \dfrac{a - b}{a(x + y) - b(x + y)}$

$= \dfrac{\overset{1}{\cancel{a - b}}}{(x + y)\underset{1}{\cancel{(a - b)}}}$

$= \dfrac{1}{x + y}$

13. $\dfrac{5}{z}$ **14.** $\dfrac{10x - 3}{2x}$ **15.** 1   **16.** $-\dfrac{3ab}{2}$ **17.** $\dfrac{x^2}{4}$

18. LCD $= 15$

$$\frac{3}{3} \cdot \frac{x+4}{5} - \frac{5}{5} \cdot \frac{x-2}{3} = \frac{3x+12}{15} - \frac{5x-10}{15}$$

$$= \frac{3x+12-(5x-10)}{15}$$

$$= \frac{3x+12-5x+10}{15}$$

$$= \frac{-2x+22}{15}$$

19. LCD $= (a-1)(a+2)$

$$\frac{a+2}{a+2} \cdot \frac{a-2}{a-1} + \frac{a-1}{a-1} \cdot \frac{a+1}{a+2} = \frac{a^2-4+a^2-1}{(a+2)(a-1)}$$

$$= \frac{2a^2-5}{(a+2)(a-1)}$$

20. $\dfrac{x^2-y^2}{x^2 y^2}$ **21.** 3 **22.** $\dfrac{-x}{x+3}$

23. LCD $= (x+4)(x+1)(x+2)$

$$\frac{3}{(x+4)(x+1)} \cdot \frac{x+2}{x+2} - \frac{2}{(x+4)(x+2)} \cdot \frac{x+1}{x+1}$$

$$= \frac{3x+6}{(x+4)(x+1)(x+2)} - \frac{2x+2}{(x+4)(x+2)(x+1)}$$

$$= \frac{3x+6-(2x+2)}{(x+4)(x+1)(x+2)} = \frac{3x+6-2x-2}{(x+4)(x+1)(x+2)}$$

$$= \frac{\overset{1}{\cancel{x+4}}}{\underset{1}{\cancel{(x+4)}}(x+1)(x+2)} = \frac{1}{(x+1)(x+2)}$$

24. $\dfrac{3k}{2m}$ **25.** $\dfrac{x+3y}{x-3y}$ **26.** $\dfrac{y}{x}$ **27.** $\dfrac{6}{5}$ **28.** -3

29. 40 **30.** $\dfrac{14}{11}$ **31.** LCD $= 2z$

$$(2z)\frac{4}{2z} + (2z)\frac{2}{z} = (2z)1$$

$$4 + 4 = 2z$$

$$8 = 2z$$

$$z = 4$$

32. LCD $= 4x^2$

$$\frac{4x^2}{1}\left(\frac{3}{x}\right) - \frac{4x^2}{1}\left(\frac{8}{x^2}\right) = \frac{4x^2}{1}\left(\frac{1}{4}\right)$$

$$12x - 32 = x^2$$

$$0 = x^2 - 12x + 32$$

$$0 = (x-4)(x-8)$$

$$x-4 = 0 \ \Big| \ x-8 = 0$$

$$x = 4 \ \Big| \qquad x = 8$$

33. $3x - 4y = 12$

$$3x = 4y + 12$$

$$x = \frac{4y+12}{3}$$

34. LCD $= n$

$$(n)\frac{2m}{n} = (n)P$$

$$2m = nP$$

$$\frac{2m}{P} = n, \ \text{ or } \ n = \frac{2m}{P}$$

35. $H = \dfrac{V}{LW}$ **36.** $m = \dfrac{Egr}{v^2}$

37.
$$\frac{F-32}{C} = \frac{9}{5} \quad \text{This is a proportion}$$

$5(F-32) = 9C \quad$ Product of means $=$ product of extremes

$$5F - 160 = 9C$$

$$\frac{5F-160}{9} = C$$

38. $x = \dfrac{b}{c-a}$ **39.** $\dfrac{29}{49}$ **40.** Two answers: $\dfrac{1}{2}$ and 3

41. 200 min $=$ 3 hr 20 min

42. Yelena's speed $=$ 8 mph; Chun's speed $=$ 6 mph

Chapter 12 Diagnostic Test (page 366)

Following each problem number is the number (in parentheses) of the textbook section in which that kind of problem is discussed.

1. (12.1) **a.** $x - 4 = 0$
$$x = 4$$
Therefore, $x \neq 4$.

(12.1) **b.** $x^2 + 2x = 0$
$$x(x+2) = 0$$
$$x = 0 \ \Big| \ x+2 = 0$$
$$\Big| \qquad x = -2$$
Therefore, $x \neq 0$ and $x \neq -2$.

2. (12.1) **a.** $-\dfrac{-4}{5} = +\dfrac{+4}{⑤}$

b. $\dfrac{-3}{x-y} = \dfrac{+3}{-(x-y)} = \dfrac{③}{y-x}$

3. (12.1) $\dfrac{\overset{2}{\cancel{6}}x^3 y}{\underset{3}{\cancel{9}}x^2 y^2} = \dfrac{2x}{3y}$

4. (12.1) $\dfrac{x^2+8x+16}{x^2-16} = \dfrac{\overset{1}{\cancel{(x+4)}}(x+4)}{\underset{1}{\cancel{(x+4)}}(x-4)} = \dfrac{x+4}{x-4}$

5. (12.1) $\dfrac{6a^2+11ab-10b^2}{6a^2 b-4ab^2} = \dfrac{(2a+5b)\overset{1}{\cancel{(3a-2b)}}}{2ab\underset{1}{\cancel{(3a-2b)}}}$

$$= \frac{2a+5b}{2ab}$$

6. (12.2) $\dfrac{2}{x^2-4} \cdot \dfrac{x^2-2x-8}{2x-8}$

$$= \frac{\overset{1}{\cancel{2}}}{\underset{1}{\cancel{(x+2)}}(x-2)} \cdot \frac{\overset{1}{\cancel{(x-4)}}\overset{1}{\cancel{(x+2)}}}{\underset{1}{\cancel{2}}\underset{1}{\cancel{(x-4)}}}$$

$$= \frac{1}{x-2}$$

7. (12.3) $\dfrac{3y}{y-5} + \dfrac{15}{5-y} = \dfrac{3y}{y-5} + \dfrac{-15}{y-5}$

$$= \frac{3y-15}{y-5}$$

$$= \frac{3\overset{1}{\cancel{(y-5)}}}{\underset{1}{\cancel{y-5}}}$$

$$= 3$$

8. (12.5) LCD $= 12x^2$

$$\frac{1}{6x} + \frac{3}{4x^2} = \frac{1}{6x} \cdot \frac{2x}{2x} + \frac{3}{4x^2} \cdot \frac{3}{3}$$

$$= \frac{2x}{12x^2} + \frac{9}{12x^2}$$

$$= \frac{2x+9}{12x^2}$$

9. (12.5) LCD $= b(b-1)$

$$\frac{b}{b-1} - \frac{b+1}{b} = \frac{b}{b-1} \cdot \frac{b}{b} - \frac{b+1}{b} \cdot \frac{b-1}{b-1}$$

$$= \frac{b^2}{b(b-1)} - \frac{b^2-1}{b(b-1)}$$

$$= \frac{b^2 - (b^2 - 1)}{b(b - 1)}$$
$$= \frac{b^2 - b^2 + 1}{b(b - 1)}$$
$$= \frac{1}{b(b - 1)}$$

10. (12.2) $\dfrac{x^2}{x^2 - 3x} \div \dfrac{3x - 15}{x^2 - 8x + 15} = \dfrac{\overset{x}{\cancel{x^2}}}{\underset{1}{\cancel{x}}\underset{1}{(\cancel{x - 3})}} \cdot \dfrac{\underset{1}{(\cancel{x - 5})}\underset{1}{(\cancel{x - 3})}}{3\underset{1}{(\cancel{x - 5})}}$

$$= \frac{x}{3}$$

11. (12.5) LCD $= (x + 3)(x + 1)(x + 2)$

$$\frac{2}{(x + 3)(x + 1)} \cdot \boxed{\frac{x + 2}{x + 2}} - \frac{1}{(x + 3)(x + 2)} \cdot \boxed{\frac{x + 1}{x + 1}}$$

$$= \frac{2x + 4}{(x + 3)(x + 1)(x + 2)} - \frac{x + 1}{(x + 3)(x + 2)(x + 1)}$$

$$= \frac{2x + 4 - (x + 1)}{(x + 3)(x + 1)(x + 2)} = \frac{2x + 4 - x - 1}{(x + 3)(x + 1)(x + 2)}$$

$$= \frac{\overset{1}{\cancel{x + 3}}}{\underset{1}{(\cancel{x + 3})}(x + 1)(x + 2)} = \frac{1}{(x + 1)(x + 2)}$$

12. (12.6) $\dfrac{\dfrac{9x^5}{10y}}{\dfrac{3x^2}{20y^3}} = \dfrac{9x^5}{10y} \div \dfrac{3x^2}{20y^3} = \dfrac{\overset{3}{\cancel{9}}x^5}{\cancel{10y}} \cdot \dfrac{\overset{2}{\cancel{20}}y^3}{\underset{1}{\cancel{3}}\underset{1}{\cancel{x^2}}} = 6x^3y^2$

13. (12.6) LCD $= a^2$

$$\boxed{\frac{a^2}{a^2}} \cdot \frac{1 + \dfrac{2}{a}}{1 - \dfrac{4}{a^2}} = \frac{\boxed{\dfrac{a^2}{1}} \cdot 1 + \boxed{\dfrac{a^2}{1}} \cdot \dfrac{2}{a}}{\boxed{\dfrac{a^2}{1}} \cdot 1 - \boxed{\dfrac{a^2}{1}} \cdot \dfrac{4}{a^2}}$$

$$= \frac{a^2 + 2a}{a^2 - 4}$$

$$= \frac{a\underset{1}{(\cancel{a + 2})}}{\underset{1}{(\cancel{a + 2})}(a - 2)}$$

$$= \frac{a}{a - 2}$$

14. (12.7) LCD $= 3 \cdot 4 = 12$

$$\boxed{\frac{\overset{4}{\cancel{12}}}{1}} \cdot \left(\frac{y}{\underset{1}{\cancel{3}}}\right) - \boxed{\frac{\overset{3}{\cancel{12}}}{1}} \cdot \left(\frac{y}{\underset{1}{\cancel{4}}}\right) = \boxed{12} (1)$$
$$4y - 3y = 12$$
$$y = 12$$

15. (12.7) LCD $= 2 \cdot 5 = 10$

$$\boxed{\frac{\overset{2}{\cancel{10}}}{1}} \cdot \frac{x - 2}{\underset{1}{\cancel{5}}} = \boxed{\frac{\overset{5}{\cancel{10}}}{1}} \cdot \frac{x + 1}{\underset{1}{\cancel{2}}} + \boxed{\frac{\overset{2}{\cancel{10}}}{1}} \cdot \frac{3}{\underset{1}{\cancel{5}}}$$
$$2(x - 2) = 5(x + 1) + 2 \cdot 3$$
$$2x - 4 = 5x + 5 + 6$$
$$-3x = 15$$
$$x = -5$$

16. (12.7) Excluded values: $a \neq -4$ and $a \neq 0$

$$\frac{3}{a + 4} = \frac{5}{a}$$ This is a proportion
$$5a + 20 = 3a$$ Product of means =
$$2a = -20$$ product of extremes
$$a = -10$$

17. (12.7) LCD $= x^2$ Excluded value: $x \neq 0$

$$\boxed{\frac{x^2}{1}} \cdot \frac{1}{x} + \boxed{\frac{x^2}{1}} \cdot \frac{6}{x^2} = \boxed{x^2} \cdot 1$$

$$x + 6 = x^2$$
$$0 = x^2 - x - 6$$
$$0 = (x - 3)(x + 2)$$
$$x - 3 = 0 \quad | \quad x + 2 = 0$$
$$x = 3 \quad | \quad x = -2$$

18. (12.8) $3x - 4y = 9$
$$3x = 4y + 9$$
$$\frac{3x}{3} = \frac{4y + 9}{3}$$
$$x = \frac{4y + 9}{3}$$

19. (12.8) $PM = Q + PN$
$$PM - PN = Q$$
$$P(M - N) = Q$$
$$\frac{P(\overset{1}{\cancel{M - N}})}{\underset{1}{\cancel{M - N}}} = \frac{Q}{M - N}$$
$$P = \frac{Q}{M - N}$$

20. (12.9) Let $x =$ time each works

	Rate	·	Time	=	Amount of work
Sid	$\dfrac{1}{20}$		x		$\dfrac{x}{20}$
Jorgé	$\dfrac{1}{30}$		x		$\dfrac{x}{30}$

$$\boxed{\begin{array}{c}\text{Amount} \\ \text{Sid unloads}\end{array}} + \boxed{\begin{array}{c}\text{amount} \\ \text{Jorgé unloads}\end{array}} = \boxed{\begin{array}{c}\text{one} \\ \text{truck}\end{array}}$$

$$\frac{x}{20} + \frac{x}{30} = 1$$

$$\frac{\overset{3}{\cancel{60}}}{1} \cdot \frac{x}{\underset{1}{\cancel{20}}} + \frac{\overset{2}{\cancel{60}}}{1} \cdot \frac{x}{\underset{1}{\cancel{30}}} = 60 \cdot 1$$
$$3x + 2x = 60$$
$$5x = 60$$
$$x = 12$$

It will take 12 min.

Exercises 13.1   (page 370)

1.

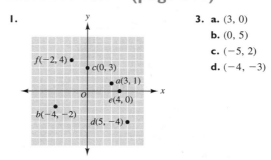

3. **a.** $(3, 0)$
b. $(0, 5)$
c. $(-5, 2)$
d. $(-4, -3)$

5. **a.** 5 because A is 5 units to the right
b. -3 because C is 3 units to the left
c. 3 because E is 3 units to the right
d. 1 because F is 1 unit to the right

13

7. Origin   **9.**

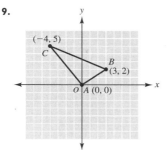

9.

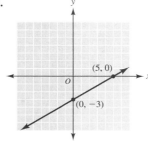

Exercises 13.2 (page 378)

1.

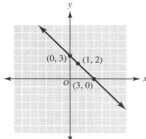

11.

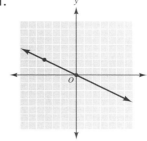

3.

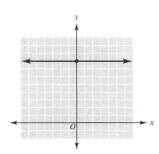

13.

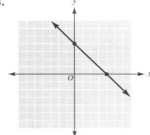

5. x can be any number but y is always 8.

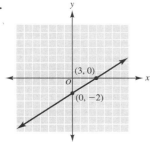

15.

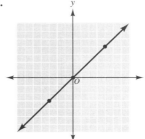

7. x is always -5. $x + 5 = 0$
$$x = -5$$

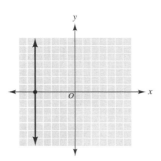

17.

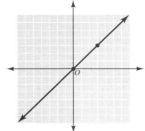

13

19.

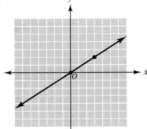

21.

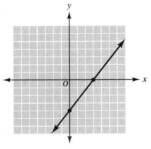

23.

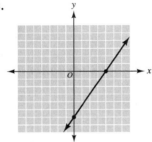

25.

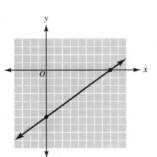

27.

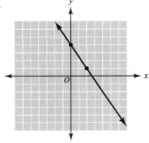

29. a. $x - y = 5$

x	y
0	-5
5	0

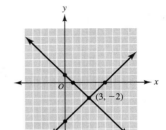

$(3, -2)$

b. $x + y = 1$

x	y
0	1
1	0

Exercises 13.3 (page 384)

1. $-\dfrac{17}{11}$ **3.** $-\dfrac{1}{3}$ **5.** 0   **7.** $-\dfrac{8}{15}$ **9.** 1

11. Undefined   **13.** $-\dfrac{2}{3}$ **15.** $\dfrac{2}{9}$ **17.** $-\dfrac{5}{3}$

19. Slope $= \dfrac{2}{3}$ **21.** Slope $= -\dfrac{1}{2}$

y-intercept $= 4$ y-intercept $= -\dfrac{2}{3}$

23. Slope $= -1$ **25.** Slope $= 0$

y-intercept $= 0$ y-intercept $= 0$

27. $3x + 4y = 12$ **29.** Slope $= \dfrac{2}{3}$

$\quad\quad 4y = -3x + 12$ y-intercept $= -2$

$\quad\quad \dfrac{4y}{4} = \dfrac{-3x}{4} + \dfrac{12}{4}$

$\quad\quad\quad y = -\dfrac{3}{4}x + 3$

Slope $= -\dfrac{3}{4}$

y-intercept $= 3$

31.

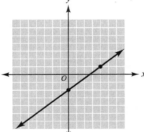

33.

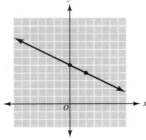

35.

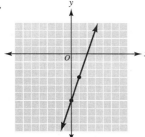

37.

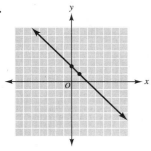

39. Slope $= -\dfrac{3}{2}$

y-intercept $= 6$

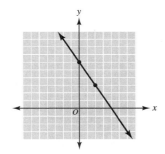

41. Slope $= \dfrac{3}{5}$

y-intercept $= 3$

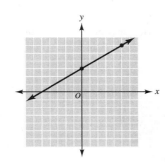

Exercises 13.4 (page 388)

1. $3x - 2y = -4$ **3.** $3x + 4y = -8$ **5.** $4x - 6y = 1$

7. $y - 2 = \dfrac{3}{4}(x + 1)$ **9.** $x - 2y = -5$

   $\boxed{4}\,(y - 2) = \boxed{4}^{\,1} \cdot \dfrac{3}{\underset{1}{4}}\,(x + 1)$

   $4y - 8 = 3x + 3$
   $-3x + 4y = 11$
   $3x - 4y = -11$

11. $2x + 3y = -8$ **13.** $x + 2y = 0$ **15.** $3x - 4y = 12$
17. $2x - 7y = 14$ **19.** $4x + 10y = 5$

21. Use the two points to find the slope, then use m and one point to find the equation of the line.

$m = \dfrac{y_2 - y_1}{x_2 - x_1} = \dfrac{4 - (-1)}{2 - 4} = -\dfrac{5}{2}$

$y - y_1 = m(x - x_1)$

$y - 4 = -\dfrac{5}{2}(x - 2)$

$2y - 8 = -5x + 10$

$5x + 2y = 18$

23. $4x - 3y = 0$ **25.** $x = 4$ **27.** $3x + 4y = 7$
29. $y = 5$ **31.** $x = -6$

Exercises 13.5 (page 390)

1.

x	y
-3	9
-2	4
-1	1
0	0
1	1
2	4
3	9

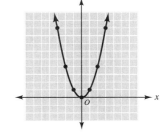

3.

x	y
-3	$4\frac{1}{2}$
-2	2
-1	$\frac{1}{2}$
0	0
1	$\frac{1}{2}$
2	2
3	$4\frac{1}{2}$

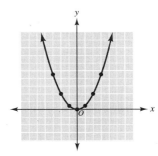

5.

x	y
-2	8
-1	3
0	0
1	-1
2	0
3	3
4	8

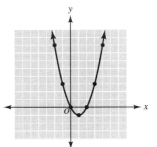

7.

x	y
-2	-8
-1	-3
0	0
1	1
2	0
3	-3
4	-8

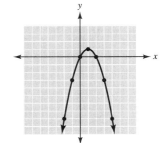

13

9.

x	y
−2	−8
−1	−1
0	0
1	1
2	8

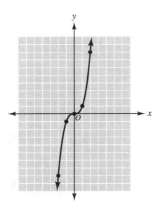

11.

x	y
−2	2
−1	6
0	4
1	2
2	6

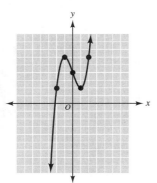

Exercises 13.6   (page 397)

1.

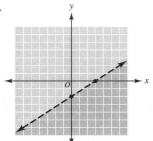

3.

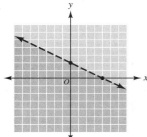

5. $x \geq -2$
   All points to the right
   and including the line
   $x = -2$

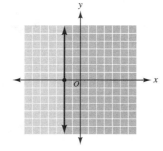

7.

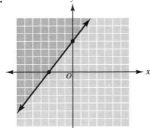

9.

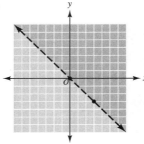

11. $3x - 4y \geq 10$
   Boundary line: $3x - 4y = 10$
   $x = 0, y = \dfrac{10}{-4} = -2\frac{1}{2}$
   $y = 0, x = \frac{10}{3} = 3\frac{1}{3}$

x	y
0	$-2\frac{1}{2}$
$3\frac{1}{3}$	0

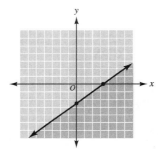

The correct half-plane does not include the origin

13.

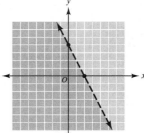

15.

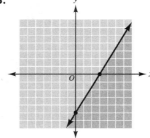

17.

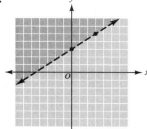

19.

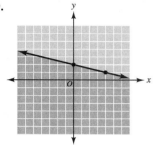

21.

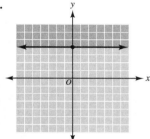

23.

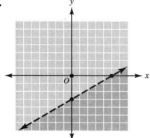

Chapter 13 Review Exercises (page 399)

1.

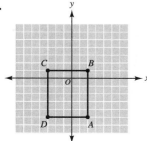

2.

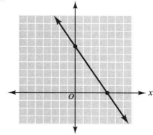

3.

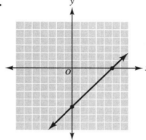

4.

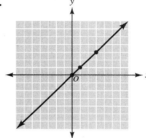

5.

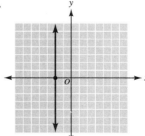

6.

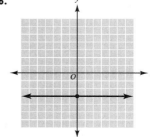

13

7.

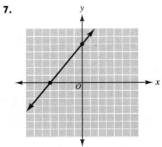

8.

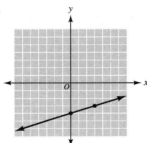

9.

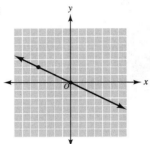

10.

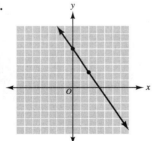

11. $y = \dfrac{x^2}{2}$

$y = \dfrac{(-4)^2}{2} = \dfrac{16}{2} = 8$

$y = \dfrac{(-2)^2}{2} = \dfrac{4}{2} = 2$

$y = \dfrac{(-1)^2}{2} = \dfrac{1}{2}$

$y = \dfrac{0^2}{2} = 0$

$y = \dfrac{1^2}{2} = \dfrac{1}{2}$

$y = \dfrac{2^2}{2} = \dfrac{4}{2} = 2$

$y = \dfrac{4^2}{2} = \dfrac{16}{2} = 8$

x	y
-4	8
-2	2
-1	$\frac{1}{2}$
0	0
1	$\frac{1}{2}$
2	2
4	8

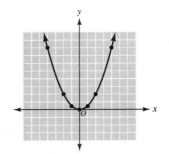

12. $y = x^3 - 3x$

$y = (-3)^3 - (3)(-3)$
$= -27 + 9 = -18$
$y = (-2)^3 - 3(-2)$
$= -8 + 6 = -2$
$y = (-1)^3 - 3(-1)$
$= -1 + 3 = 2$
$y = 0^3 - 3(0) = 0$
$y = 1^3 - 3(1) = -2$
$y = 2^3 - 3(2) = 2$
$y = 3^3 - 3(3) = 18$

x	y
-3	-18
-2	-2
-1	2
0	0
1	-2
2	2
3	18

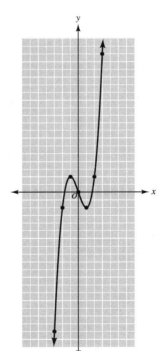

13. $-\dfrac{11}{5}$ **14.** Slope $= \dfrac{4}{15}$

y-intercept $= -2$

15. $x + 2y = 12$ **16.** $2x + 3y = -9$

17. First find m; then use m and one point to find the equation.

$m = \dfrac{y_2 - y_1}{x_2 - x_1} = \dfrac{-2 - 4}{1 - (-3)} = \dfrac{-6}{4} = -\dfrac{3}{2}$

$y - y_1 = m(x - x_1)$

$y - (-2) = -\dfrac{3}{2}(x - 1)$

$2(y + 2) = -3(x - 1)$

$2y + 4 = -3x + 3$

$3x + 2y = -1$

13

18. $x = -7$

19.

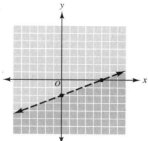

20.

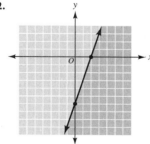

21.

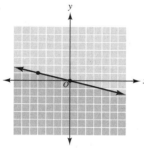

22.

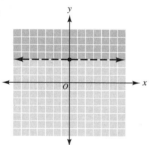

Chapter 13 Diagnostic Test (page 401)

Following each problem number is the number (in parentheses) of the textbook section in which that kind of problem is discussed.

1. (13.1)

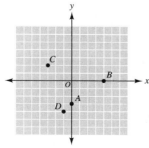

2. (13.3)
$$5x + 2y = 8$$
$$2y = -5x + 8$$
$$\frac{2y}{2} = \frac{-5x}{2} + \frac{8}{2}$$
$$y = -\frac{5}{2}x + 4$$
$$\text{Slope} = -\frac{5}{2}$$
$$y\text{-intercept} = 4$$

3. (13.2)

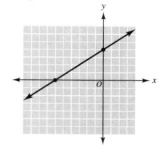

4. (13.3)

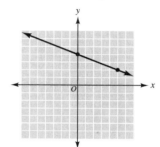

5. (13.2)

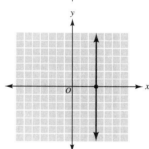

6. (13.5)
$$y = x^2 + x - 6$$
$$y = (-4)^2 + (-4) - 6 = 6$$
$$y = (-3)^2 + (-3) - 6 = 0$$
$$y = (-2)^2 + (-2) - 6 = -4$$
$$y = (-1)^2 + (-1) - 6 = -6$$
$$y = (0)^2 + (0) - 6 = -6$$
$$y = (1)^2 + (1) - 6 = -4$$
$$y = (2)^2 + (2) - 6 = 0$$
$$y = (3)^2 + (3) - 6 = 6$$

x	y
-4	6
-3	0
-2	-4
-1	-6
0	-6
1	-4
2	0
3	6

13

7. (13.6)   Boundary line: $3x - 2y = 6$
If $x = 0$, $3(0) - 2y = 6$
$y = -3$
If $y = 0$, $3x - 2(0) = 6$
$x = 2$

x	y
0	−3
2	0

The *boundary line is solid* because equality included.
The *correct half-plane* includes $(0, 0)$, because
$3x - 2y \leq 6$
$3(0) - 2(0) \leq 6$
$0 \leq 6$ is true

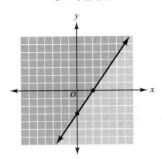

8. (13.3) $m = \dfrac{y_2 - y_1}{x_2 - x_1} = \dfrac{-1 - 2}{1 - (-5)} = \dfrac{-3}{6} = -\dfrac{1}{2}$

9. (13.4)   A horizontal line has an equation in the form $y = b$.
Because the y-coordinate of $(2, 5)$ is 5, the equation is
$y = 5$.

10. (13.4)
$$y = mx + b$$
$$y = -\frac{2}{3}x + (-4)$$
$$3y = -2x - 12$$
$$2x + 3y = -12$$

11. (13.4)
$$y - y_1 = m(x - x_1)$$
$$y - 5 = \frac{1}{4}(x - (-2))$$
$$4(y - 5) = 1(x + 2)$$
$$4y - 20 = x + 2$$
$$-x + 4y = 22$$
$$x - 4y = -22$$

12. (13.4) $P_1(-4, 2)$, $P_2(1, -3)$
$$m = \frac{-3 - 2}{1 - (-4)} = \frac{-5}{5} = -1$$
$$y - y_1 = m(x - x_1)$$
$$y - (-3) = -1(x - 1)$$
$$y + 3 = -x + 1$$
$$x + y = -2$$

Exercises 14.1   (page 407)

1. Is not a solution   **3.** Is a solution

5.

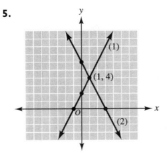

7.

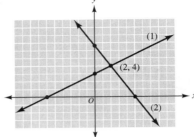

9. (1) $x + 2y = 0$ (2) $x - 2y = -2$

x	y
0	0
2	−1

x	y
0	1
−2	0

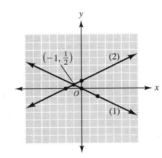

11.

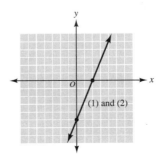

Inconsistent (no solution; parallel lines)

13. (1) $10x - 4y = 20$ has intercepts $(2, 0)$ and $(0, -5)$
(2) $6y - 15x = -30$ has intercepts $(2, 0)$ and $(0, -5)$
Because both lines have the same intercepts, they are the same line. Any point on the line is a solution.

Dependent (many solutions; same line)

15.

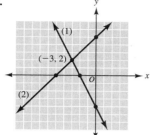

17.

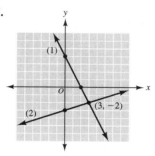

19.

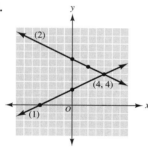

Exercises 14.2 (page 412)

1. $(-2, 0)$ **3.** $(6, -2)$ **5.** $(3, -1)$ **7.** $(-1, 3)$

9. $(0, 0)$

11.
$$\begin{array}{ll} 3] & 4\ x + 3y = 2 \Rightarrow 12x + 9y = 6 \\ -4] & 3\ x + 5y = -4 \Rightarrow \underline{-12x - 20y = 16} \\ & \hspace{4.5cm} -11y = 22 \\ & \hspace{5cm} y = -2 \end{array}$$

Substituting -2 for y in Equation 1:
$$4x + 3(-2) = 2$$
$$4x - 6 = 2$$
$$4x = 8$$
$$x = 2$$
Solution: $(2, -2)$

13.
$$\begin{array}{lll} 9] & 3] & 6\ x - 10y = 6 \Rightarrow 18x - 30y = 18 \\ -6] & -2] & 9\ x - 15y = -4 \Rightarrow \underline{-18x + 30y = 8} \\ & & \hspace{4.5cm} 0 = 26 \quad \text{False} \end{array}$$

Inconsistent (no solution)

15.
$$\begin{array}{ll} 2] & 3x - 5y = -2 \Rightarrow 6x - 10y = -4 \\ 1] & -6x + 10y = 4 \Rightarrow \underline{-6x + 10y = 4} \\ & \hspace{4cm} 0 = 0 \quad \text{True} \end{array}$$

Dependent (many solutions)

17. $(-2, 3)$ **19.** $\left(3, -\dfrac{3}{2}\right)$

21. Dependent (many solutions) **23.** $(-3, -2)$

Exercises 14.3 (page 416)

1. $(5, 3)$ **3.** $(2, -1)$ **5.** $(1, -2)$

7. Dependent (many solutions) **9.** $(2, 1)$ **11.** $(5, -1)$

13. Inconsistent (no solution) **15.** $(-2, -6)$ **17.** $\left(\dfrac{1}{2}, 1\right)$

19. $(1, 1)$ **21.**
$$8\left(\dfrac{x}{2}\right) + 8\left(\dfrac{y}{8}\right) = 8 \ (1)$$
$$(1) \quad 4x + y = 8$$
$$y = 8 - 4x$$
$$10\left(\dfrac{x}{2}\right) + 10\left(\dfrac{y}{5}\right) = 10 \ (4)$$
$$(2) \quad 5x + 2y = 40$$

Substituting $8 - 4x$ for y in Equation 2:
$$5x + 2(\ 8 - 4x\ ) = 40$$
$$5x + 16 - 8x = 40$$
$$-3x = 24$$
$$x = -8$$
Substituting -8 for x in $y = 8 - 4x$:
$$y = 8 - 4(-8)$$
$$y = 8 + 32$$
$$y = 40$$
Therefore, the solution is $(-8, 40)$.

23.
$$\overset{1}{4}\left(\dfrac{x}{\underset{1}{4}}\right) + \overset{2}{4}\left(\dfrac{y+1}{\underset{1}{2}}\right) = 4 \ (2)$$
$$(1) \quad x + 2y + 2 = 8$$
$$x = 6 - 2y$$
$$\overset{3}{6}\left(\dfrac{x}{\underset{1}{2}}\right) + \overset{2}{6}\left(\dfrac{y-2}{\underset{1}{3}}\right) = 6 \ (1)$$
$$(2) \quad 3x + 2y - 4 = 6$$
$$3x + 2y = 10$$

Substituting $6 - 2y$ for x in Equation 2:
$$3\ (6 - 2y)\ + 2y = 10$$
$$18 - 6y + 2y = 10$$
$$-4y = -8$$
$$y = 2$$
Substituting 2 for y in $x = 6 - 2y$:
$$x = 6 - 2(2)$$
$$x = 6 - 4$$
$$x = 2$$
Therefore, the solution is $(2, 2)$.

Exercises 14.4 (page 419)

1. a. Let x = one number
$\phantom{\text{Let }}$ y = other number
b. $x + y = 30$
$\phantom{\text{b.}}$ $x - y = 12$
c. The numbers are 9 and 21.

3. a. Let x = one angle
$\phantom{\text{Let }}$ y = other angle
b. $x + y = 90$
$\phantom{\text{b.}}$ $x - y = 40$
c. The angles are 25° and 65°.

5. a. Let x = smaller number
$\phantom{\text{Let }}$ y = larger number
b. $2x + 3y = 34$
$\phantom{\text{b.}}$ $5x - 2y = 9$
c. 5; 8

7. a. Let x = lb of almonds
$\phantom{\text{Let }}$ y = lb of hazelnuts

14

b. $x + y = 20$
$0.85x + 1.40y = 19.75$
c. 15 lb almonds; 5 lb hazelnuts

9. a. Let l = length
w = width

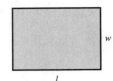

$2w + 2l$ = perimeter
b. (1) $2w + 2l = 19$
(2) $l = \boxed{w + 1.5}$

Substituting $\boxed{w + 1.5}$ for l in Equation 1:
$$2w + 2l = 19$$
$$2w + 2(w + 1.5) = 19$$
$$2w + 2w + 3 = 19$$
$$4w = 16$$
$$w = 4$$

Substituting 4 for w in Equation 2:
$l = \boxed{w + 1.5}$
$l = (4) + 1.5$
$l = 5.5 = 5$ ft 6 in.
c. Width is 4 ft; length is 5 ft 6 in.

11. a. Let x = number of 10¢ stamps
y = number of 25¢ stamps
b. $x + y = 22$
$0.10x + 0.25y = 3.40$
c. The number of 10¢ stamps is 14. The number of 25¢ stamps is 8.

13. a. Let n = numerator
d = denominator
b. $\dfrac{n}{d} = \dfrac{2}{3}$ and $\dfrac{n + 4}{d - 2} = \dfrac{6}{7}$
$2d = 3n$ $6(d - 2) = 7(n + 4)$
$d = \dfrac{3n}{2}$ $6d - 12 = 7n + 28$
                  $6d = 7n + 40$

Substituting $\dfrac{3n}{2}$ for d in $6d = 7n + 40$:
$$6\left(\dfrac{3n}{2}\right) = 7n + 40$$
$$9n = 7n + 40$$
$$2n = 40$$
$$n = 20$$

Substituting 20 for n in $d = \dfrac{3n}{2}$:
$$d = \dfrac{3(20)}{2}$$
$$d = 30$$
c. Therefore, the original fraction is $\dfrac{20}{30}$.

15. a. Let x = number of adult tickets
y = number of children's tickets
b. $x + y = 8$
$1.95x + 0.95y = 10.60$
c. 3 adult tickets; 5 children's tickets

17. a. Let x = number of \$15 rolls
y = number of \$20 rolls
b. $x + y = 40$
$15x + 20y = 725$
c. 15 \$15 rolls; 25 \$20 rolls

19. a. Let x = hr as tutor
y = hr as cashier
b. $x + y = 30$
$4x + 5y = 140$
c. 10 hr as tutor; 20 hr as cashier

21. a. Let x = cost of tie
y = cost of pin
b. $x + y = 1.10$
$x = 1.00 + y$
c. Pin costs \$0.05; tie costs \$1.05.

Chapter 14 Review Exercises
(page 421)

1.

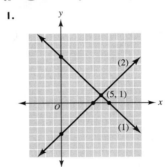

2.

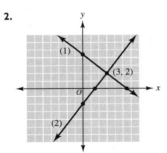

3. (1) $2x - 3y = 3$ has intercepts $\left(1\dfrac{1}{2}, 0\right)$ and $(0, -1)$
(2) $3y - 2x = 6$ has intercepts $(-3, 0)$ and $(0, 2)$

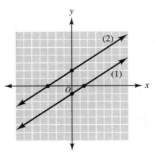

Inconsistent (no solution)

4.

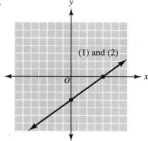

(1) and (2)

Dependent (many solutions)

5. $(1, 2)$ **6.** Dependent (many solutions)

7. Inconsistent (no solution)

8.
$$\boxed{4} \cdot \frac{x}{2} - \boxed{4} \cdot \frac{y}{4} = \boxed{4} \cdot 4$$
(1) $\qquad 2x - y = 16$
$$\boxed{12} \cdot \frac{x}{3} + \boxed{12} \cdot \frac{y}{4} = \boxed{12} \cdot 1$$
(2) $\qquad 4x + 3y = 12$
(1) 3] $2x - y = 16 \Rightarrow 6x - 3y = 48$
(2) $\quad 4x + 3y = 12 \Rightarrow \underline{4x + 3y = 12}$
$$10x \qquad = 60$$
$$x = 6$$
Substituting 6 for x in Equation 1:
$$2x - y = 16$$
$$2(6) - y = 16$$
$$12 - y = 16$$
$$-y = 4$$
$$y = -4$$
Solution: $(6, -4)$

9. $(7, 5)$ **10.** $(-4, -11)$ **11.** $(0, 2)$

12.
$$\boxed{6} \cdot \frac{x}{3} + \boxed{6} \cdot \frac{y}{6} = \boxed{6} \,(-1)$$
(1) $\qquad 2x + y = -6$
$$\boxed{12} \cdot \frac{x}{4} + \boxed{12} \cdot \frac{y}{12} = \boxed{12} \,(-1)$$
(2) $\qquad 3x + y = -12$
$$y = -3x - 12$$
Substituting $\boxed{-3x - 12}$ for y in Equation 1:
$$2x + y = -6$$
$$2x + \boxed{(-3x - 12)} = -6$$
$$-x - 12 = -6$$
$$-x = 6$$
$$x = -6$$
Substituting -6 for x in $y = \boxed{-3x - 12}$:
$$y = -3(-6) - 12$$
$$y = 18 - 12$$
$$y = 6$$
Solution: $(-6, 6)$

13. $(5, -4)$ **14.** Inconsistent **15.** $(-3, -2)$

16.
$$\overset{3}{\boxed{18}} \cdot \frac{x - 1}{\underset{1}{6}} + \overset{2}{\boxed{18}} \cdot \frac{y + 5}{\underset{1}{9}} = \boxed{18} \cdot 2$$
$$3x - 3 + 2y + 10 = 36$$
(1) $\qquad 3x + 2y = 29$
$$\overset{3}{\boxed{12}} \cdot \frac{x + 1}{\underset{1}{4}} - \overset{2}{\boxed{12}} \cdot \frac{y + 2}{\underset{1}{6}} = \boxed{12} \cdot 1$$
$$3x + 3 - 2y - 4 = 12$$
(2) $\qquad 3x - 2y = 13$

(1) $3x + 2y = 29$
(2) $\underline{3x - 2y = 13}$
$$6x \qquad = 42$$
$$x = 7$$
Substituting 7 for x in Equation 1:
$$3x + 2y = 29$$
$$3(7) + 2y = 29$$
$$21 + 2y = 29$$
$$2y = 8$$
$$y = 4$$
Solution: $(7, 4)$

17. 53 and 31 **18.** 8 and 5 **19.** Length is 11 ft; width is 7 ft

20. Let $x = $ hr as tutor
$\qquad y = $ hr as waiter

Brian worked at two jobs	for a total of	32 hours
(1) $\quad x + y \qquad = \qquad 32$
$$x = 32 - y$$

He received	a total of	$216
(2) $\quad 8x + 6y \qquad = \qquad 216$
Substituting $\boxed{32 - y}$ for x in Equation 2:
$$8x + 6y = 216$$
$$8 \,\boxed{(32 - y)} + 6y = 216$$
$$256 - 8y + 6y = 216$$
$$-2y = -40$$
$$y = 20 \text{ hr}$$
Substituting 20 for y in $x = \boxed{32 - y}$:
$$x = 32 - 20 = 12 \text{ hr}$$
Solution: 12 hr as a tutor, 20 hr as a waiter

21. Let $x = $ number of \$10 rolls
$\qquad y = $ number of \$8 rolls

The total number of rolls	is	80
(1) $\qquad x + y \qquad = \qquad 80$
$$x = 80 - y$$

The total amount paid for stamps	is	\$730
(2) $\qquad 10x + 8y \qquad = \qquad 730$
Substituting $\boxed{80 - y}$ for x in Equation 2:
$$10x + 8y = 730$$
$$10 \,\boxed{(80 - y)} + 8y = 730$$
$$800 - 10y + 8y = 730$$
$$-2y = -70$$
$$y = 35 \text{ (\$8 rolls)}$$
Substituting 35 for y in $x = \boxed{80 - y}$:
$$x = 80 - 35 = 45 \text{ (\$10 rolls)}$$
Solution: 45 \$10 rolls and 35 \$8 rolls

22. 4 boxes legal size, 16 boxes letter size

Chapter 14 Diagnostic Test (page 422)

Following each problem number is the number (in parentheses) of the textbook section in which that kind of problem is discussed.

1. (14.1) (1) $3x + 2y = 2$
$\qquad$ (2) $2x - 3y = 10$
$\qquad$ (1) Intercepts: $\left(\dfrac{2}{3}, 0\right)$ $(0, 1)$
$\qquad$ Checkpoint: $(-2, 4)$

(2) Intercepts: $(5, 0)$, $\left(0, -3\frac{1}{3}\right)$

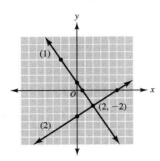

Solution: $(2, -2)$

2. (14.2)

$\boxed{-3}$] $\quad 3x - \boxed{4}\,y = 1 \Rightarrow -9x + 12y = -3$
$\boxed{4}$] $\quad 5x - \boxed{3}\,y = 9 \Rightarrow \underline{20x - 12y = 36}$
$\qquad\qquad\qquad\qquad\qquad 11x \qquad = 33$
$\qquad\qquad\qquad\qquad\qquad\qquad x = 3$

Substituting 3 for x in Equation 1:
$3x - 4y = 1$
$3(3) - 4y = 1$
$9 - 4y = 1$
$\quad -4y = -8$
$\qquad y = 2$
Solution: $(3, 2)$

3. (14.2)

$\boxed{-6}$] $\boxed{-3}$] $\quad 3x - \boxed{4}\,y = 1 \Rightarrow -9x + 12y = -3$
$\boxed{4}$] $\boxed{2}$] $\quad 5x - \boxed{6}\,y = 5 \Rightarrow \underline{10x - 12y = 10}$
$\qquad\qquad\qquad\qquad\qquad x \qquad = 7$

Substituting 7 for x in Equation 1:
$3x - 4y = 1$
$3(7) - 4y = 1$
$21 - 4y = 1$
$\quad -4y = -20$
$\qquad y = 5$
Solution: $(7, 5)$

4. (14.3) (1) $3x - 5y = 14$
$\qquad\quad$ (2) $x = \boxed{y + 2}$

Substituting $\boxed{y + 2}$ for x in Equation 1:
$3x - 5y = 14$
$3(\boxed{y + 2}) - 5y = 14$
$3y + 6 - 5y = 14$
$\quad -2y = 8$
$\qquad y = -4$
Substituting -4 for y in $x = \boxed{y + 2}$:
$x = -4 + 2 = -2$
Solution: $(-2, -4)$

5. (14.3) (1) $4x + \ y = 2 \Rightarrow y = \boxed{2 - 4x}$
$\qquad\quad$ (2) $2x - 3y = 8$

Substituting $\boxed{2 - 4x}$ for y in Equation 2:
$2x - 3y = 8$
$2x - 3\boxed{(2 - 4x)} = 8$
$2x - 6 + 12x = 8$
$\qquad 14x = 14$
$\qquad\quad x = 1$
Substituting 1 for x in $y = \boxed{2 - 4x}$:
$y = 2 - 4(1) = 2 - 4 = -2$
Solution: $(1, -2)$

6. (14.3) (1) $4x - 3y = \ 7$
$\qquad\quad$ (2) $\ x - 2y = -2 \Rightarrow x = \boxed{2y - 2}$

Substituting $\boxed{2y - 2}$ for x in Equation 1:
$4x - 3y = 7$
$4(\boxed{2y - 2}) - 3y = 7$
$8y - 8 - 3y = 7$
$\qquad 5y = 15$
$\qquad\ y = 3$
Substituting 3 for y in $x = \boxed{2y - 2}$:
$x = 2(3) - 2 = 6 - 2 = 4$
Solution: $(4, 3)$

7. (14.2)

$\boxed{10}$] $\boxed{5}$] $\quad \boxed{6}\,x - \ 9y = \ 2 \Rightarrow \ 30x - 45y = \ 10$
$\boxed{6}$] $\boxed{3}$] $\boxed{-10}\,x + 15y = -5 \Rightarrow \underline{-30x + 45y = -15}$
$\qquad\qquad\qquad\qquad\qquad\qquad\qquad 0 = \ -5\ \text{False}$

Inconsistent (no solution)

8. (14.4) Let x = larger number
$\qquad\quad$ and y = smaller number

| The sum of two numbers | is | 18 |

(1) $\qquad x + y \qquad = \ 18$

| Their difference | is | 42 |

(2) $\qquad x - y \qquad = \ 42$
Using addition:
(1) $\ x + y = 18$
(2) $\underline{\ x - y = 42}$
$\quad 2x \quad = 60$
$\qquad\ x = 30$ Larger number
Substituting 30 for x in Equation 1:
$\ x + y = 18$
$30 + y = 18$
$\qquad y = -12$ Smaller number
Solution: 30 is the larger number; -12 is the smaller number.

9. (14.4) Let x = number of classical records
$\qquad\quad$ and y = number of pop records

| Amount of money for classical records | + | amount of money for pop records | = | total amount spent |

(1) $\qquad 2.99x \qquad + \qquad 0.88y \qquad = \quad 19.53$

| Number of classical records | + | number of pop records | = | total records |

(2) $\qquad x \qquad + \qquad y \qquad = \quad 15$
$\qquad\qquad\qquad\qquad\qquad\qquad\qquad y = 15 - x$
Substituting $\boxed{15 - x}$ for y in Equation 1:
$299x + 88(\boxed{15 - x}) = 1953$
$299x + 1320 - 88x = 1953$
$\quad 211x + 1320 = 1953$
$\qquad\qquad 211x = 633$
$\qquad\qquad\quad x = 3$
Solution: 3 classical records; $15 - 3 = 12$ pop records

10. (14.4) Let l = length
$\qquad\quad\ $ and w = width

Length is 3 cm more than width

(1) $l = 3 + w$

$2l + 2w =$ perimeter

(2) $2l + 2w = 102$

Substituting $3 + w$ for l in Equation 2:

$2l + 2w = 102$

$2(3 + w) + 2w = 102$

$6 + 2w + 2w = 102$

$4w = 96$

$w = 24$

Substituting 24 for w in $l = 3 + w$:

$l = 3 + 24$

$l = 27$

Solution: Length is 27 cm; width is 24 cm.

Exercises 15.1   (page 428)

1. $\dfrac{1}{x^4}$ **3.** $\dfrac{4}{a^3}$ **5.** a^4 **7.** $\dfrac{y^3}{x^2}$ **9.** $\dfrac{s}{r^2 t^4}$

11. $\dfrac{1}{x^3 y^2}$ **13.** $h^2 k^4$ **15.** $\dfrac{3}{x^4 y}$ **17.** $\dfrac{a}{b^2}$ **19.** $a^3 c^2$

21. $\dfrac{p^4 t^2}{r}$ **23.** $\dfrac{5y^2}{x}$ **25.** x^{-2} **27.** hk^{-1} **29.** $x^2 y^{-1} z^{-5}$

31. x **33.** $\dfrac{1}{x^3}$ **35.** $\dfrac{1}{x^8}$ **37.** a^6 **39.** x^6 **41.** $\dfrac{1}{y^7}$

43. 1   **45.** $\dfrac{b^2}{a^7}$ **47.** x^7 **49.** $\dfrac{2}{3x^4}$ **51.** $\dfrac{4h^2}{7}$

53. $\dfrac{y^3}{2x^3}$ **55.** $\dfrac{3m^3}{n^6}$ **57.** 1

59. $x^{3m} \cdot x^{-m} = x^{3m-m} = x^{2m}$

61. $(x^{3b})^{-2} = x^{3b(-2)} = x^{-6b} = \dfrac{1}{x^{6b}}$

63. $\dfrac{x^{2a}}{x^{-5a}} = x^{2a-(-5a)} = x^{7a}$ **65.** $\dfrac{x + \dfrac{1}{y}}{y} \cdot \dfrac{y}{y} = \dfrac{xy + 1}{y^2}$

67. $\dfrac{1}{8}$ **69.** $\dfrac{1}{25}$ **71.** 100

73. 16   **75.** $\dfrac{1}{100,000,000}$ **77.** $\dfrac{1}{16}$

79. 10,000   **81.** $\dfrac{1}{32}$ **83.** 25

85. $\dfrac{1}{36}$

87. $\dfrac{10^{-3} \cdot 10^2}{10^5} = 10^{-3+2-5} = 10^{-6} = \dfrac{1}{10^6} = \dfrac{1}{1,000,000}$

89. -1

Exercises 15.2   (page 430)

1. $x^5 y^5$ **3.** $16x^2$ **5.** $\dfrac{1}{5x}$ **7.** $\dfrac{x^2}{25}$ **9.** $a^4 b^6$

11. $4z^6$ **13.** $\dfrac{n^4}{m^8}$ **15.** $\dfrac{x^8}{y^{12}}$ **17.** $\dfrac{a^6}{27}$ **19.** $\dfrac{r^3 s^9}{t^{12}}$

21. $\dfrac{9x^6}{y^2}$ **23.** $\dfrac{1}{M^8 N^{12}}$ **25.** $a^4 b^2$

27. $\left(\dfrac{10^2 \cdot 10^{-1}}{10^{-2}} \right)^2 = (10^{2-1+2})^2 = (10^3)^2 = 10^{3 \cdot 2} = 10^6 = 1,000,000$

29. $m^4 n^2$ **31.** $\dfrac{1}{x^5 y^2}$

33. $\left(\dfrac{\overset{2}{\cancel{8}} s^{-3}}{\underset{1}{\cancel{4}} st^2} \right)^{-2} = (2^1 s^{-3-1} t^{-2})^{-2} = 2^{1(-2)} s^{-4(-2)} t^{-2(-2)}$

$= 2^{-2} s^8 t^4 = \dfrac{s^8 t^4}{2^2} = \dfrac{s^8 t^4}{4}$

35. k^8

37. $(x^3 + y^4)^5$

Cannot be simplified by rules of exponents. Rule 2 or Rule 8 *cannot* be used here because the + sign means that x^3 and y^4 are *not* factors.

39. 1   **41.** $\dfrac{m^2}{6p}$ **43.** 1,000,000   **45.** $\dfrac{1}{x^5 y^2}$

47. $x^4 y^{10}$ **49.** $\dfrac{z}{8y}$

Exercises 15.3   (page 434)

Decimal notation	Scientific notation
1. $7_{\wedge}48.$ $\rightarrow$	7.48×10^2
3. $0.06_{\wedge}3$	6.3×10^{-2}
5. $0.001_{\wedge}732$ $\leftarrow$	1.732×10^{-3}
7. $3_{\wedge}470,000.$	3.47×10^6
9. $0.000001_{\wedge}91$	1.91×10^{-6}
11. $0.00005_{\wedge}63$	5.63×10^{-5}
13. $0.00005_{\wedge}8$	5.8×10^{-5}
15. $0.000000006_{\wedge}547$	6.547×10^{-9}
17. $0.0000000004_{\wedge}77$	4.77×10^{-10}
19. $8_{\wedge}3,600,000.$	8.36×10^7

21. 8.5×10^{-3} **23.** 3×10^5 **25.** 8.12×10

27. 1.5×10^{-2} **29.** 6×10^4 **31.** 3.6×10^{10}

33. 3.75×10^{-12} **35.** 2.28×10^{12}

Exercises 15.4   (page 436)

1. A rational number is any number that can be expressed in the form $\dfrac{a}{b}$, where a and b are integers and b is not zero.

Therefore, $-5 = \dfrac{-5}{1}$; $\dfrac{3}{4}$; $2\dfrac{1}{2} = \dfrac{5}{2}$; $3.5 = \dfrac{7}{2}$; $\sqrt{4} = \dfrac{2}{1}$; and

$3 = \dfrac{3}{1}$ are rational numbers.

3. a. 17   **b.** $x + 1$ **5.** 6   **7.** 1   **9.** 9   **11.** -2

13. $\dfrac{4}{7}$

Exercises 15.5A   (page 440)

1. z **3.** x^2 **5.** $4x^3$ **7.** $2\sqrt{3}$ **9.** $3\sqrt{2}$

11. $2\sqrt{2}$ **13.** $2\sqrt{6}$ **15.** $x\sqrt{x}$ **17.** $m^3\sqrt{m}$

19. $hk\sqrt{h}$ **21.** $6\sqrt{7}$ **23.** $20\sqrt{2}$ **25.** $5x\sqrt{3x}$

27. $2x^2 y^3\sqrt{5}$ **29.** $6x^{18}$ **31.** $2xy\sqrt{5y}$ **33.** $4x^2 y\sqrt{xy}$

35. $3ab^2\sqrt{3a}$ **37.** $5x^2 y\sqrt{10y}$ **39.** 13   **41.** 3

43. x **45.** $5x$ **47.** 5   **49.** $6x$

15

Exercises 15.5B (page 442)

1. $\frac{2}{9}$ **3.** $\frac{3}{7}$ **5.** $\frac{a}{b^3}$ **7.** $\frac{x^3}{v^4}$ **9.** $\frac{2x}{3}$ **11.** $\frac{m}{3}$

13. $\frac{x}{5}$ **15.** $\frac{2\sqrt{3}}{5}$ **17.** $\frac{5\sqrt{2}}{9}$ **19.** $\frac{x\sqrt{x}}{y^2}$ **21.** $\frac{\sqrt{2}}{2}$

23. $\frac{\sqrt{a}}{a}$ **25.** $\frac{2\sqrt{5}}{5}$ **27.** $\frac{8}{\sqrt{2}} \cdot \frac{\sqrt{2}}{\sqrt{2}} = \frac{8\sqrt{2}}{2} = 4\sqrt{2}$

29. $2\sqrt{x}$

Exercises 15.6 (page 443)

1. $7\sqrt{3}$ **3.** $4\sqrt{5}$ **5.** $9\sqrt{x}$ **7.** $6\sqrt{7}$
9. $8\sqrt{2} - 4\sqrt{3}$ **11.** $8\sqrt{x} - 2\sqrt{y}$ **13.** $7\sqrt{2}$ **15.** $\sqrt{2}$
17. $15\sqrt{3}$ **19.** $-4\sqrt{3}$ **21.** $7\sqrt{x}$ **23.** $5 - 2\sqrt{5}$
25. $7\sqrt{3}$ **27.** $5\sqrt{5} - 6\sqrt{3}$ **29.** $8 - 7\sqrt{2}$

Exercises 15.7 (page 445)

1. 6   **3.** $3x$ **5.** $6a^2$ **7.** $2\sqrt{3}$ **9.** $3x\sqrt{2}$
11. $5a\sqrt{2a}$ **13.** 15   **15.** 24
17. $\sqrt{12} \cdot 5\sqrt{2}$
    $= 5\sqrt{12 \cdot 2}$
    $= 5\sqrt{24}$
    $= 5\sqrt{4 \cdot 6}$
    $= 5 \cdot 2\sqrt{6}$
    $= 10\sqrt{6}$
19. $12\sqrt{10}$ **21.** $2 + \sqrt{2}$ **23.** $10 + 3\sqrt{5}$
25. $x - 2\sqrt{x}$ **27.** $6 - 2\sqrt{3}$ **29.** $13 + 5\sqrt{7}$
31. $22 - 11\sqrt{5}$
33. $(5\sqrt{3} - 2)(2\sqrt{3} + 4)$ **35.** 7
    $= 5\sqrt{3} \cdot 2\sqrt{3} + 4 \cdot 5\sqrt{3} - 2 \cdot 2\sqrt{3} - 2 \cdot 4$
    $= \quad 30 \quad + 20\sqrt{3} - 4\sqrt{3} - 8$
    $= 22 + 16\sqrt{3}$
37. 2   **39.** $19 + 8\sqrt{3}$ **41.** $9 - 6\sqrt{x} + x$
43. $8 + 2\sqrt{15}$

Exercises 15.8 (page 447)

1. 2   **3.** $\frac{1}{2}$ **5.** x **7.** $\frac{1}{a^2}$ **9.** $3x^2$ **11.** $2\sqrt{2}$

13. $2\sqrt{3}$ **15.** $\frac{\sqrt{3}}{3}$ **17.** $\frac{\sqrt{15}}{5}$ **19.** $\frac{\sqrt{6}}{3}$

21. $\frac{2}{\sqrt{8}} \cdot \frac{\sqrt{2}}{\sqrt{2}} = \frac{2\sqrt{2}}{\sqrt{16}} = \frac{2\sqrt{2}}{4} = \frac{\sqrt{2}}{2}$

23. $\frac{\sqrt{3}}{2}$ **25.** $\sqrt{2} + 1$ **27.** $\sqrt{5} - \sqrt{2}$ **29.** $3\sqrt{2} + 3$

31. $\frac{6}{\sqrt{5} + \sqrt{2}} \cdot \frac{\sqrt{5} - \sqrt{2}}{\sqrt{5} - \sqrt{2}}$

    $= \frac{6(\sqrt{5} - \sqrt{2})}{5 - 2}$

    $= \frac{\overset{2}{\cancel{6}}(\sqrt{5} - \sqrt{2})}{\underset{1}{\cancel{3}}}$

    $= 2\sqrt{5} - 2\sqrt{2}$
33. $4 - 2\sqrt{2}$ **35.** $4\sqrt{6} + 8$ **37.** $\sqrt{x} - 2$

Exercises 15.9 (page 450)

1. 25   **3.** 64   **5.** 8   **7.** 7   **9.** 15   **11.** 40

---

13. 8   **15.** 6   **17.** 8   **19.** 2, 1   **21.** 5   **23.** $\frac{1}{2}$

25. $\sqrt{x - 3} + 5 = x$
    $(\sqrt{x - 3})^2 = (x - 5)^2$
      $x - 3 = x^2 - 10x + 25$
         $0 = x^2 - 11x + 28$
         $0 = (x - 4)(x - 7)$
$x - 4 = 0 \;\big|\; x - 7 = 0$
  $x = 4 \;\big|\;\;\;\; x = 7$
Check: $x = 7$ *Check:* $x = 4$
$\sqrt{x - 3} + 5 = x$ $\sqrt{x - 3} + 5 = x$
$\sqrt{7 - 3} + 5 \overset{?}{=} 7$ $\sqrt{4 - 3} + 5 \overset{?}{=} 4$
  $\sqrt{4} + 5 \overset{?}{=} 7$ $\sqrt{1} + 5 \overset{?}{=} 4$
    $2 + 5 \overset{?}{=} 7$ $1 + 5 \neq 4$
       $7 = 7$
Therefore, 7 is the only solution.
27. 3   **29.** 6

Chapter 15 Review Exercises (page 451)

1. $\frac{1}{x^{11}}$ **2.** a^2 **3.** $\frac{1}{c^5}$ **4.** p^7 **5.** $\frac{1}{x^9}$ **6.** m^3

7. $\frac{1}{p^{15}}$ **8.** m^8 **9.** 1   **10.** $x^8 y^{12}$ **11.** $\frac{p^2}{r^6}$

12. $\frac{2}{a^3}$ **13.** $9x^8$ **14.** $\frac{1}{16b^6}$ **15.** $\frac{x^{10} y^{15}}{z^{20}}$ **16.** $x^{12} z^8$

17. $\frac{w^{12}}{u^{15} v^6}$ **18.** $(5a^3 b^{-4})^{-2} = 5^{-2} a^{-6} b^8 = \frac{b^8}{25a^6}$

19. $\left(\frac{4h^2}{ij^{-2}}\right)^{-3} = \frac{4^{-3} h^{-6}}{i^{-3} j^6} = \frac{i^3}{4^3 h^6 j^6} = \frac{i^3}{64 h^6 j^6}$

20. $\left(\frac{x^{10} y^5}{x^5 y}\right)^3 = (x^5 y^4)^3 = x^{15} y^{12}$

21. 1   **22.** x^{4d} **23.** $\frac{1}{x^{8a}}$ **24.** 6^{3x} **25.** $x^2 y^{-3}$

26. $m^2 n^3$ **27.** $a^3 b^{-2} c^{-5}$ **28.** $\frac{1}{16}$ **29.** $\frac{1}{27}$ **30.** 125

31. 16   **32.** $\frac{(-8)^2}{-8^2} = \frac{(-8)(-8)}{-(8)(8)} = \frac{64}{-64} = -1$

33. $\left(\frac{10^{-4} \cdot 10}{10^{-2}}\right)^5 = (10^{-4+1+2})^5 = (10^{-1})^5 = 10^{-5} = \frac{1}{10^5} = \frac{1}{100,000}$

34. 2.25×10^{-4} **35.** 9.6×10^5

36. $\frac{1}{200} = 0.005 = 5 \times 10^{-3}$ **37.** 7,800   **38.** 0.0000406

39. 0.01207   **40.** 9.6×10^{-2} **41.** 3×10^6

42. 7×10^3

43. a. $\sqrt{3}, 2\sqrt{5}, \sqrt{5}$, because they cannot be expressed as a fraction
    b. $2\sqrt{5}$ and $\sqrt{5}$

44. a. $9x$ **b.** $\frac{1}{2}$ **c.** $x - 5$ **45.** 9   **46.** $4\sqrt{3}$

47. 8   **48.** a^3 **49.** $x\sqrt{x}$ **50.** $4xy^2$ **51.** $ab^2\sqrt{ab}$

52. $30\sqrt{2}$ **53.** $4\sqrt{5}$ **54.** $\frac{3}{x}$ **55.** $\sqrt{2}$ **56.** $\frac{\sqrt{10}}{5}$

57. $2\sqrt{3}$ **58.** $\sqrt{3}$ **59.** $10 + 3\sqrt{2}$ **60.** -4

61. $7 + 4\sqrt{3}$

62. $(3\sqrt{2} + 1)(2\sqrt{2} - 1)$
    $= 3\sqrt{2} \cdot 2\sqrt{2} - 1 \cdot 3\sqrt{2} + 1 \cdot 2\sqrt{2} - 1 \cdot 1$
    $= \quad 12 \quad - 3\sqrt{2} + 2\sqrt{2} - 1$
    $= 11 - \sqrt{2}$

15

63. $\dfrac{8}{\sqrt{3}-2} \cdot \dfrac{\sqrt{3}+2}{\sqrt{3}+2} = \dfrac{8(\sqrt{3}+2)}{3-4} = -8(\sqrt{3}+2)$

$\qquad\qquad\qquad\qquad\qquad\qquad = -8\sqrt{3}-16$

64. $\dfrac{10}{\sqrt{6}-2} \cdot \dfrac{\sqrt{6}+2}{\sqrt{6}+2} = \dfrac{10(\sqrt{6}+2)}{6-4} = \dfrac{\overset{5}{\cancel{10}}(\sqrt{6}+2)}{\underset{1}{\cancel{2}}}$

$\qquad\qquad\qquad\qquad\qquad\qquad = 5\sqrt{6}+10$

65. $\sqrt{45} - 2\sqrt{27} + \sqrt{20}$
$\quad = \sqrt{9\cdot5} - 2\sqrt{9\cdot3} + \sqrt{4\cdot5}$
$\quad = 3\sqrt{5} - 2\cdot3\sqrt{3} + 2\sqrt{5}$
$\quad = 3\sqrt{5} - 6\sqrt{3} + 2\sqrt{5}$
$\quad = 5\sqrt{5} - 6\sqrt{3}$

66. $x = 16$ **67.** $a = 12$ **68.** $x = 13$ **69.** $a = 3$

70. $(\sqrt{7x-6})^2 = (x)^2$
$\qquad 7x - 6 = x^2$
$\qquad\quad 0 = x^2 - 7x + 6$
$\qquad\quad 0 = (x-6)(x-1)$
$\quad x - 6 = 0 \;\big|\; x - 1 = 0$
$\qquad x = 6 \;\big|\qquad x = 1$

71. $\sqrt{2x-1} + 2 = x$
$\quad (\sqrt{2x-1})^2 = (x-2)^2$
$\qquad 2x - 1 = x^2 - 4x + 4$
$\qquad\quad 0 = x^2 - 6x + 5$
$\qquad\quad 0 = (x-5)(x-1)$
$\quad x - 5 = 0 \;\big|\; x - 1 = 0$
$\qquad x = 5 \;\big|\qquad x = 1$

    Check: $x = 5$ *Check:* $x = 1$
$\quad \sqrt{2(5)-1} + 2 \overset{?}{=} 5 \quad\;\; \sqrt{2(1)-1} + 2 \overset{?}{=} 1$
$\qquad\quad \sqrt{9} + 2 \overset{?}{=} 5 \qquad\qquad \sqrt{1} + 2 \overset{?}{=} 1$
$\qquad\quad\; 3 + 2 = 5 \qquad\qquad\quad 1 + 2 \neq 1$

    Therefore, $x = 5$ is the only solution.

Chapter 15 Diagnostic Test (page 453)

Following each problem number is the number (in parentheses) of the textbook section in which that kind of problem is discussed.

1. (15.1) $a^{-6} \cdot a^2 = a^{-6+2} = a^{-4} = \dfrac{1}{a^4}$

2. (15.2) $(a^{-3}b)^2 = a^{-3\cdot2}b^{1\cdot2} = a^{-6}b^2 = \dfrac{b^2}{a^6}$

3. (15.1) $x^{3a} \cdot x^{2a} = x^{3a+2a} = x^{5a}$

4. (15.1) $\dfrac{x^5}{x^{-2}} = x^{5-(-2)} = x^7$

5. (15.2) $\left(\dfrac{3x^{-4}}{y^2}\right)^2 = \dfrac{3^2 x^{-8}}{y^4} = \dfrac{9}{x^8 y^4}$

6. (15.2) $\left(\dfrac{x^5 y}{x^2 y^3}\right)^3 = (x^{5-2}y^{1-3})^3 = (x^3 y^{-2})^3 = x^9 y^{-6} = \dfrac{x^9}{y^6}$

7. (15.1) $(10^{-3})^2 = 10^{-6} = \dfrac{1}{10^6} = \dfrac{1}{1,000,000}$

8. (15.1) $\dfrac{2^{-2} \cdot 2^3}{2^{-4}} = 2^{-2+3-(-4)} = 2^5 = 32$

9. (15.1) $\dfrac{-4^2}{(-4)^2} = \dfrac{-16}{16} = -1$

10. (15.1) $\dfrac{x^2}{yz^3} = x^2 y^{-1} z^{-3}$

11. (15.3) **a.** $723,000 = 7.23 \times 10^5$
$\qquad\qquad$ **b.** $0.0048 = 4.8 \times 10^{-3}$

12. (15.3) $\dfrac{4200}{0.00014} = \dfrac{4.2 \times 10^3}{1.4 \times 10^{-4}} = 3 \times 10^{3-(-4)} = 3 \times 10^7$

13. (15.7) $\sqrt{2}\sqrt{18x^2} = \sqrt{2\cdot18x^2} = \sqrt{36x^2} = 6x$

14. (15.7) $\sqrt{3}(2\sqrt{3}-5) = (\sqrt{3})(2\sqrt{3}) + (\sqrt{3})(-5)$
$\qquad\qquad\qquad\qquad = 6 - 5\sqrt{3}$

15. (15.5A) $\sqrt{72} = \sqrt{36\cdot2} = 6\sqrt{2}$

16. (15.5B) $\sqrt{\dfrac{18}{2m^2}} = \sqrt{\dfrac{9}{m^2}} = \dfrac{3}{m}$

17. (15.8) $\dfrac{\sqrt{16}}{\sqrt{50}} = \sqrt{\dfrac{16}{50}} = \sqrt{\dfrac{8}{25}} = \dfrac{\sqrt{4\cdot2}}{\sqrt{25}} = \dfrac{2\sqrt{2}}{5}$

18. (15.8) $\dfrac{4}{\sqrt{3}+1} \cdot \dfrac{\sqrt{3}-1}{\sqrt{3}-1} = \dfrac{4(\sqrt{3}-1)}{3-1}$

$\qquad\qquad\qquad\qquad\quad = \dfrac{\overset{2}{\cancel{4}}(\sqrt{3}-1)}{\underset{1}{\cancel{2}}}$

$\qquad\qquad\qquad\qquad\quad = 2\sqrt{3} - 2$

19. (15.5A) $\sqrt{12x^4 y^3} = \sqrt{4\cdot3\cdot x^4 \cdot y^2 \cdot y}$
$\qquad\qquad\qquad\quad = 2x^2 y\sqrt{3y}$

20. (15.6) $2\sqrt{20} + \sqrt{45} = 2\sqrt{4\cdot5} + \sqrt{9\cdot5}$
$\qquad\qquad\qquad\qquad = 2\cdot2\sqrt{5} + 3\sqrt{5}$
$\qquad\qquad\qquad\qquad = 4\sqrt{5} + 3\sqrt{5}$
$\qquad\qquad\qquad\qquad = 7\sqrt{5}$

21. (15.7) $(3\sqrt{2}+5)(2\sqrt{2}+1)$
$\quad = 3\sqrt{2}\cdot2\sqrt{2} + 1\cdot3\sqrt{2} + 5\cdot2\sqrt{2} + 5\cdot1$
$\quad = \quad 12 \quad + \quad 3\sqrt{2} \quad + \quad 10\sqrt{2} \quad + \quad 5$
$\quad = 17 + 13\sqrt{2}$

22. (15.5B) $\sqrt{\dfrac{4}{5}} = \dfrac{\sqrt{4}}{\sqrt{5}} = \dfrac{2}{\sqrt{5}} \cdot \dfrac{\sqrt{5}}{\sqrt{5}} = \dfrac{2\sqrt{5}}{5}$

23. (15.6) $\sqrt{28} + \sqrt{75} - \sqrt{27}$
$\quad = \sqrt{4\cdot7} + \sqrt{25\cdot3} - \sqrt{9\cdot3}$
$\quad = 2\sqrt{7} + 5\sqrt{3} - 3\sqrt{3}$
$\quad = 2\sqrt{7} + 2\sqrt{3}$

24. (15.9) $(\sqrt{4x+5})^2 = (5)^2$ *Check:* $\sqrt{4x+5} = 5$
$\qquad\qquad 4x + 5 = 25 \qquad\qquad\qquad \sqrt{4(5)+5} \overset{?}{=} 5$
$\qquad\qquad\quad\; 4x = 20 \qquad\qquad\qquad\qquad \sqrt{25} \overset{?}{=} 5$
$\qquad\qquad\qquad x = 5 \qquad\qquad\qquad\qquad\quad 5 = 5$

25. (15.9) $(\sqrt{5x-6})^2 = (x)^2$
$\qquad\qquad 5x - 6 = x^2$
$\qquad\quad x^2 - 5x + 6 = 0$
$\qquad (x-3)(x-2) = 0$
$\quad x - 3 = 0 \;\big|\; x - 2 = 0$
$\qquad x = 3 \;\big|\qquad x = 2$

    Check: $x = 3$ *Check:* $x = 2$
$\quad \sqrt{5x-6} = x \qquad\quad \sqrt{5x-6} = x$
$\; \sqrt{5(3)-6} \overset{?}{=} 3 \qquad \sqrt{5(2)-6} \overset{?}{=} 2$
$\qquad\quad \sqrt{9} \overset{?}{=} 3 \qquad\qquad\quad \sqrt{4} \overset{?}{=} 2$
$\qquad\quad\; 3 = 3 \qquad\qquad\qquad\; 2 = 2$

Exercises 16.1   (page 457)

1. $2x^2 - 5x - 3 = 0$
$\quad a = 2,\, b = -5,\, c = -3$

3. $6x^2 - x = 0$
$\quad a = 6,\, b = -1,\, c = 0$

5. $9x^2 - 16 = 0$
$\quad a = 9,\, b = 0,\, c = -16$

7. $\qquad x^2 - \dfrac{5x}{4} = \dfrac{2}{3} \quad$ LCD $= 12$

$\quad 12 \cdot x^2 - \dfrac{\overset{3}{\cancel{12}}}{1} \cdot \dfrac{5x}{\underset{1}{\cancel{4}}} = \dfrac{\overset{4}{\cancel{12}}}{1} \cdot \dfrac{2}{\underset{1}{\cancel{3}}}$

$\qquad 12x^2 - 15x = 8$
$\qquad 12x^2 - 15x - 8 = 0$
$\qquad a = 12,\, b = -15,\, c = -8$

9. $2x^2 - 3x - 10 = 0$
$\quad a = 2,\, b = -3,\, c = -10$

11. $x^2 - 3x - 4 = 0$
$a = 1, b = -3, c = -4$
13. $x^2 + 2x - 11 = 0$
$a = 1, b = 2, c = -11$
15. $(x - 2)(x + 1) = 3x(x + 2)$
$x^2 - x - 2 = 3x^2 + 6x$
$0 = 2x^2 + 7x + 2$
$a = 2, b = 7, c = 2$

Exercises 16.2   (page 459)

1. $2, -3$ **3.** $5, -3$ **5.** $1, -\dfrac{1}{2}$ **7.** $1, -2$

9. $\dfrac{9}{5}, -1$ **11.** $\dfrac{3}{2}, \dfrac{1}{2}$

13. $\dfrac{x}{2} + \dfrac{2}{x} = \dfrac{5}{2}$ LCD $= 2x$

$\dfrac{\overset{1}{\cancel{2x}}}{1} \cdot \dfrac{x}{\underset{1}{\cancel{2}}} + \dfrac{\overset{1}{\cancel{2x}}}{1} \cdot \dfrac{2}{\underset{1}{\cancel{x}}} = \dfrac{\overset{1}{\cancel{2x}}}{1} \cdot \dfrac{5}{\underset{1}{\cancel{2}}}$

$x^2 + 4 = 5x$
$x^2 - 5x + 4 = 0$
$(x - 4)(x - 1) = 0$
$x - 4 = 0 \mid x - 1 = 0$
$x = 4 \mid x = 1$

15. $2, -6$
17. $(x + 2)(x + 3) = x + 3$
$x^2 + 5x + 6 = x + 3$
$x^2 + 4x + 3 = 0$
$(x + 3)(x + 1) = 0$
$x + 3 = 0 \mid x + 1 = 0$
$x = -3 \mid x = -1$

Exercises 16.3   (page 463)

1. $9, 0$ **3.** ±3 **5.** $\pm2\sqrt{2}$ **7.** $\pm\dfrac{2\sqrt{30}}{5}$ **9.** $\dfrac{9}{4}, 0$

11. $x^2 + 9 = 0$
$x^2 = -9$
$\sqrt{x} = \pm\sqrt{-9}$
The solution is not a real number because the radicand is negative.

13. $\dfrac{1}{6}, 0$ **15.** ±5 **17.** $2, 0$ **19.** $\pm\dfrac{\sqrt{6}}{3}$

Exercises 16.4   (page 467)

1. $1, -\dfrac{2}{3}$ **3.** $2 \pm \sqrt{3}$ **5.** $2 \pm \sqrt{2}$ **7.** $\dfrac{4 \pm \sqrt{6}}{2}$

9. $3x^2 + 2x + 1 = 0$
$a = 3, b = 2, c = 1$
$x = \dfrac{-(2) \pm \sqrt{(2)^2 - 4(3)(1)}}{2(3)}$
$x = \dfrac{-2 \pm \sqrt{4 - 12}}{6} = \dfrac{-2 \pm \boxed{\sqrt{-8}}}{6}$
The solution is not a real number because the radicand is negative.

11. $3, -1$ **13.** $4, 1$ **15.** $\dfrac{1}{2}, -3$ **17.** $1 \pm \sqrt{3}$

Exercises 16.5   (page 469)

1. a. Let $x =$ side of square
b. $x^2 = 24$
$x = \pm2\sqrt{6}$
c. Side $= 2\sqrt{6}$ in.

3. a. Let $x =$ width
Then $3x =$ length
b. $3x \cdot x = 54$
$x = \pm3\sqrt{2}$
c. Width $= 3\sqrt{2}$ ft
Length $= 9\sqrt{2}$ ft
5. a. Let $x =$ width
Then $x + 2 =$ length
b. $x(x + 2) = 4$
$x = -1 \pm \sqrt{5}$
c. Width $= -1 + \sqrt{5}$ in.
Length $= 1 + \sqrt{5}$ in.
7. a. Let $x =$ side of original square
Then $x + 1 =$ side of new square
b. $(x + 1)^2 = 2 \cdot x^2$
$x = 1 \pm \sqrt{2}$
c. Side of original square $= 1 + \sqrt{2}$
9. a. Let $x =$ first consecutive integer
Then $x + 1 =$ second consecutive integer
b. $x^2 + (x + 1)^2 = 25$
$x = 3 \mid x = -4$
c. Two answers: $\{3, 4\}$ and $\{-4, -3\}$
11. a. Let $x =$ the number
Then $\dfrac{1}{x} =$ its reciprocal
b. $x - \dfrac{1}{x} = 6$
$x = 3 \pm \sqrt{10}$
c. Two answers: $3 + \sqrt{10}$ and $3 - \sqrt{10}$

Chapter 16 Review Exercises (page 470)

1. $4x^2 + 3x - 7 = 0$
$a = 4, b = 3, c = -7$
2. $x^2 - 10 = 0$
$a = 1, b = 0, c = -10$
3. $2x^2 - 12x - 1 = 0$
$a = 2, b = -12, c = -1$
4. $x^2 + 3x - 12 = 0$
$a = 1, b = 3, c = -12$
5. $2, -3$ **6.** $5, -2$ **7.** $0, 25$ **8.** ±7
9. $x^2 - 2x - 4 = 0$
$a = 1, b = -2, c = -4$
$x = \dfrac{-(-2) \pm \sqrt{(-2)^2 - 4(1)(-4)}}{2(1)}$
$= \dfrac{2 \pm \sqrt{4 + 16}}{2} = \dfrac{2 \pm \sqrt{20}}{2}$
$= \dfrac{2 \pm 2\sqrt{5}}{2} = 1 \pm \sqrt{5}$

10. $2 \pm \sqrt{3}$ **11.** $0, 5$ **12.** $0, 7$ **13.** $\pm\dfrac{3\sqrt{10}}{5}$

14. Not a real number   **15.** $\dfrac{-3 \pm 3\sqrt{5}}{2}$ **16.** $-3 \pm \sqrt{10}$

17. $\dfrac{x + 2}{3} = \dfrac{1}{x - 2} + \dfrac{2}{3}$ LCD $= 3(x - 2)$

$\dfrac{\overset{1}{\cancel{3}(x - 2)}}{1} \cdot \dfrac{(x + 2)}{\underset{1}{\cancel{3}}} = \dfrac{3\cancel{(x - 2)}}{1} \cdot \dfrac{1}{\cancel{(x - 2)}} + \dfrac{\overset{1}{\cancel{3}(x - 2)}}{1} \cdot \dfrac{2}{\underset{1}{\cancel{3}}}$

$(x - 2)(x + 2) = 3 + 2(x - 2)$
$x^2 - 4 = 3 + 2x - 4$

$$x^2 - 2x - 3 = 0$$
$$(x-3)(x+1) = 0$$
$$x - 3 = 0 \mid x + 1 = 0$$
$$x = 3 \mid x = -1$$

18. 2, −4 **19.** ±3√5 **20.** ±2√3

21. Not a real number **22.** Not a real number

23. ±2 **24.** $\pm\dfrac{3\sqrt{5}}{5}$

25. a. Let x = width
Then $2x$ = length
b. $2x \cdot x = 80$
$x = \pm 2\sqrt{10}$
c. Width = $2\sqrt{10}$ cm
Length = $4\sqrt{10}$ cm

26. a. Let x = number
b. $x^2 = 2x + 4$
$x = 1 \pm \sqrt{5}$
c. Two answers: $1 + \sqrt{5}$ and $1 - \sqrt{5}$

27. a. Let x = the number
Then $\dfrac{1}{x}$ = its reciprocal
b. $x + \dfrac{1}{x} = \dfrac{10}{3}$
$x = 3$ or $x = \dfrac{1}{3}$
c. Two answers: 3 and $\dfrac{1}{3}$

Chapter 16 Diagnostic Test (page 471)

Following each problem number is the number (in parentheses) of the textbook section in which that kind of problem is discussed.

1. (16.1) $3x^2 = 9 - 2x$
$3x^2 + 2x - 9 = 0$
$a = 3, b = 2, c = -9$

2. (16.1) $5(x-1) = x(x+2)$
$5x - 5 = x^2 + 2x$
$0 = x^2 - 3x + 5$
$a = 1, b = -3, c = 5$

3. (16.2) $x^2 + 3x = 10$
$x^2 + 3x - 10 = 0$
$(x+5)(x-2) = 0$
$x + 5 = 0 \mid x - 2 = 0$
$x = -5 \mid x = 2$

4. (16.3) $3x^2 - 15x = 0$
$3x(x-5) = 0$
$3x = 0 \mid x - 5 = 0$
$x = 0 \mid x = 5$

5. (16.3) $2x^2 = 9$
$\dfrac{2x^2}{2} = \dfrac{9}{2}$
$\sqrt{x^2} = \pm\sqrt{\dfrac{9}{2}}$
$x = \pm\dfrac{3}{\sqrt{2}} \cdot \dfrac{\sqrt{2}}{\sqrt{2}}$
$x = \pm\dfrac{3\sqrt{2}}{2}$

6. (16.4) $2x^2 - 6x + 1 = 0$
$a = 2, b = -6, c = 1$
$x = \dfrac{-(-6) \pm \sqrt{(-6)^2 - 4(2)(1)}}{2(2)}$
$x = \dfrac{6 \pm \sqrt{36 - 8}}{4} = \dfrac{6 \pm \sqrt{28}}{4}$
$x = \dfrac{6 \pm 2\sqrt{7}}{4} = \dfrac{\overset{1}{2}(3 \pm \sqrt{7})}{\underset{2}{4}}$
$x = \dfrac{3 \pm \sqrt{7}}{2}$

7. (16.2) $6x^2 = 5 - 7x$
$6x^2 + 7x - 5 = 0$
$(2x-1)(3x+5) = 0$
$2x - 1 = 0 \mid 3x + 5 = 0$
$2x = 1 \mid 3x = -5$
$x = \dfrac{1}{2} \mid x = -\dfrac{5}{3}$

8. (16.3) $\dfrac{x^2}{8} = 2x$ LCD = 8
$\dfrac{\overset{1}{8}}{1} \cdot \dfrac{x^2}{\underset{1}{8}} = 8 \cdot 2x$
$x^2 = 16x$
$x^2 - 16x = 0$
$x(x-16) = 0$
$x = 0 \mid x - 16 = 0$
$\mid x = 16$

9. (16.4) $2(x+1) = x^2$
$2x + 2 = x^2$
$0 = x^2 - 2x - 2$
$a = 1, b = -2, c = -2$
$x = \dfrac{-(-2) \pm \sqrt{(-2)^2 - 4(1)(-2)}}{2(1)}$
$x = \dfrac{2 \pm \sqrt{4 + 8}}{2} = \dfrac{2 \pm \sqrt{12}}{2}$
$x = \dfrac{2 \pm 2\sqrt{3}}{2} = 1 \pm \sqrt{3}$

10. (16.5) Let x = side of the square
$x^2 = 48$
$\sqrt{x^2} = \pm\sqrt{48}$
$x = \pm\sqrt{16 \cdot 3}$
$x = \pm 4\sqrt{3}$
Side = $4\sqrt{3}$ in.

Exercises 17.1 (page 479)

1. a. Obtuse **b.** Straight **c.** Right **d.** Acute
3. 65° **5.** 107° **7.** 45° and 135° **9.** 30° and 60°
11. 110° **13.** ∠1 = 65° **15.** ∠CBE = 35°
∠2 = 25° ∠BED = 35°
∠DEF = 145°
17. ∠1 = 60° **19. a.** Quadrilateral
∠3 = 45° **b.** Pentagon
∠4 = 120° **c.** Triangle
∠5 = 60° **d.** Hexagon
∠7 = 135°

Exercises 17.2 (page 485)

1. a. Equilateral **3.** 34° **5.** 99°
b. Isosceles
c. Scalene
d. Isosceles right

7. $x + x + x = 180$
$3x = 180$
$x = 60°$

9. $52°$ **11.** $30°$ and $60°$

13. Let $3x = \angle A$
$x = \angle B$
$x - 10 = \angle C$
$3x + x + x - 10 = 180$
$5x - 10 = 180$
$5x = 190$
$x = 38$
Then $\angle A = 3x = 3(38) = 114°$
$\angle B = x = 38°$
$\angle C = x - 10 = 38 - 10 = 28°$

15. $\angle A = 58°$ **17.** $\angle 2 = 70°$
$\angle ACD = 32°$ $\angle 3 = 110°$
$\angle BCD = 58°$ $\angle 4 = 80°$
                           $\angle 6 = 70°$
                           $\angle 7 = 30°$

19. $5^2 + 12^2 \overset{?}{=} 13^2$ **21.** Is not a right triangle
$25 + 144 \overset{?}{=} 169$
$169 = 169$
Is a right triangle

23. $x = 20$ **25.** $x = 3$ **27.** $x = 3\sqrt{2}$
29. $(x + 1)^2 + (\sqrt{20})^2 = (x + 3)^2$
$x^2 + 2x + 1 + 20 = x^2 + 6x + 9$
$x^2 + 2x + 21 = x^2 + 6x + 9$
$21 = 4x + 9$
$12 = 4x$
$3 = x$

31. 6 and 8   **33.** $x^2 + b^2 = (2x)^2$
$x^2 + b^2 = 4x^2$
$b^2 = 3x^2$
$\sqrt{b^2} = \sqrt{3x^2}$
$b = x\sqrt{3}$

35. a.
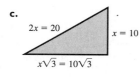
$2x = 8$ $x = 4$
$x\sqrt{3} = 4\sqrt{3}$

b.
$2x = 12$ $x = 6$
$x\sqrt{3} = 6\sqrt{3}$

c.
$2x = 20$ $x = 10$
$x\sqrt{3} = 10\sqrt{3}$

d.
$2x = 2$ $x = 1$
$x\sqrt{3} = \sqrt{3}$

Exercises 17.3   (page 490)

1. $\angle ABD = \angle CDB$ **3.** $\angle A = \angle B$
$\angle ADB = \angle CBD$ $\angle ADF = \angle BEF$
$\angle A = \angle C$ $\angle DFA = \angle EFB$
$DB = DB$ $AF = BF$
$AB = DC$ $AD = BE$
$AD = BC$ $DF = EF$

5. $\angle ACD = \angle CAB$ **7.** $\angle F = 50°$ **9.** $\angle D = 60°$
$\angle ADC = \angle CBA$ $\angle FBE = 40°$ $\angle ECB = 55°$
$\angle DAC = \angle BCA$
$AC = AC$
$AD = CB$
$DC = AB$

11. $x = 6$ **13.** $x = 5$ **15.** $\triangle ACD \cong \triangle ACB$ by SSS
$y = 10$ $y = 6$
17. $\triangle ABC \cong \triangle DCB$ by ASA   **19.** $\triangle ADC \cong \triangle BDC$ by SAS
21. $\triangle AFD \cong \triangle BFE$ by ASA

Exercises 17.4   (page 494)

1. $DF = 9$ **3.** $AC = 16$ **5.** $AE = 14$ **7.** $BC = 12$
$EF = 6$ $EF = 8$
9. $\dfrac{6}{9} = \dfrac{4}{4 + BD}$
$36 = 24 + 6BD$
$12 = 6BD$
$2 = BD$

Exercises 17.5   (page 498)

1. 22 ft   **3.** 50 in.   **5.** 4π in.   **7.** 16 in.   **9.** 6 ft
11. $(6\pi + 12)$ ft   **13.** $(4\pi + 24)$ cm

15.
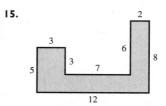

$P = 3 + 3 + 7 + 6 + 2 + 8 + 12 + 5 = 46$ yd
17. 30 cm   **19.** 46 in.

21. Let $x = $ width
Then $2x - 5 = $ length
$2l + 2w = P$
$2(2x - 5) + 2x = 38$
$4x - 10 + 2x = 38$
$6x = 48$
$x = 8$
Width $= x = 8$ yd
Length $= 2x - 5 = 2(8) - 5 = 11$ yd

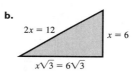
x
$2x - 5$

Exercises 17.6   (page 503)

1. 96 sq. ft   **3.** $7\frac{1}{2}$ sq. in.   **5.** 64π sq. ft
7. 25 sq. in.   **9.** 18π sq. ft   **11.** $(8\pi + 80)$ m^2
13. 15 cm^2 **15.** $(9\pi + 72)$ sq. in.

17. $A = \dfrac{1}{2}bh$
$30 = \dfrac{1}{2}b(12)$
$30 = 6b$
$5 = b$
$a^2 + b^2 = c^2$
$12^2 + 5^2 = c^2$
$144 + 25 = c^2$
$169 = c^2$
$13 = c$
Leg is 5 in.
Hypotenuse is 13 in.

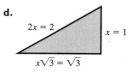
c $a = 12$
b

19. 8 m   **21.** 9 sq. ft   **23.** 196 sq. in.

17

25.
$$C = 2\pi r$$
$$\frac{10\pi}{2\pi} = \frac{2\pi r}{2\pi}$$
$$5 = r$$
$$A = \pi r^2$$
$$A = \pi 5^2$$
$$A = 25\pi \text{ sq. ft}$$

27. 26 in.

Exercises 17.7 (page 506)

1. 900 cu. ft   **3.** 640π cu. in.   **5.** 288π cu. in.
7. 112 cm³   **9.** $(24\pi + 200)$ cu. ft   **11.** 1,728 cu. in.
13. 8 in.   **15. a.** 3600 in.³   **17. a.** $V = \pi r^2 h$
          b. 130 lb $V = \pi(10)^2(50)$
                                  $V = 5000\pi$ cu. in.
                              b. $\dfrac{5000(3.14)}{231} \approx 68$ gal

Chapter 17 Review Exercises (page 510)

1. 15° and 75°   **2.** 70° and 110°
3. $\angle 1 = 85°$ **4.** $\angle 2 = 40°$ **5.** 128°   **6.** 30°
  $\angle 2 = 30°$ $\angle 3 = 105°$
  $\angle 3 = 65°$ $\angle 4 = 75°$
  $\angle 4 = 150°$ $\angle 5 = 75°$
  $\angle 5 = 115°$ $\angle 6 = 105°$
7. 12   **8.** 4   **9.** $2\sqrt{3}$ **10.** $2\sqrt{5}$ cm   **11.** $8\sqrt{2}$ in.
12. $\angle C = 50°$ **13.** $\angle DAE = 37°$
  $\angle DBE = 40°$ $AE = 4$
14. $\triangle ADE \cong \triangle BCE$ by ASA   **15.** $\triangle ADB \cong \triangle BCA$ by SAS
16. $DE = 6$ **17.** Let $x = EC$
$$\frac{8}{6} = \frac{3 + x}{x}$$
$$8x = 18 + 6x$$
$$2x = 18$$
$$x = 9 = EC$$

18. $P = 60$ cm   **19.** $P = 28$ in.
  $A = 150$ cm² $A = 24$ in.²
20. $P = 36$ ft   **21.** $P = (5\pi + 10)$ m
  $A = 60$ ft² $A = 12\frac{1}{2}\pi$ m²
22. $P = 42$ yd   **23.** $P = (4\pi + 24)$ ft
  $A = 57$ sq. yd $A = (8\pi + 64)$ sq. ft

24.

Shaded area	=	area of rectangle	−	area of square
A	=	$l \cdot w$	−	s^2

$$A = 8 \cdot 6 - 4^2$$
$$A = 48 - 16$$
$$A = 32 \text{ sq. in.}$$

25. 5π mm²   **26.** $(144 - 36\pi)$ sq. ft   **27.** 28 m²
28. 24 cm   **29.** 16π ft   **30.** 24 in.   **31.** 4 in.
32. 28 m   **33.** 64 cu. ft   **34.** 3 cm
35. $V = \dfrac{1}{2}\left(\dfrac{4}{3}\pi r^3\right)$ **36.** $\dfrac{3}{4}V = \dfrac{3}{4}(\pi r^2 h)$ Tank is $\dfrac{3}{4}$ full

$V = \dfrac{1}{2} \cdot \dfrac{4}{3} \cdot \pi \cdot 6^3$ $= \dfrac{3}{4}\pi(4)^2(8)$

$V = \dfrac{1}{\overset{}{2}} \cdot \dfrac{4}{\overset{}{3}} \cdot \pi \cdot \overset{36}{\underset{1}{\cancel{216}}}$ $= \dfrac{3}{4} \cdot \pi \cdot \overset{4}{\cancel{16}} \cdot 8$

$V = 144\pi$ m³ $= 96\pi$ cu. ft

Chapter 17 Diagnostic Test (page 513)

Following each problem number is the number (in parentheses) of the textbook section in which that kind of problem is discussed.

1. (17.1) $\angle 2 = 75°$
    $\angle 3 = 80°$
    $\angle 4 = 155°$
    $\angle 5 = 25°$
    $\angle 7 = 100°$

2. (17.2) $\angle A + \angle B + \angle ACB = 180$
    $30 + \angle B + 115 = 180$
    $\angle B = 35°$
    $\angle B + \angle BDC + \angle BCD = 180$
    $35 + 90 + \angle BCD = 180$
    $\angle BCD = 55°$

3. (17.2) $x^2 + 1^2 = 3^2$ **4.** (17.2) $2^2 + (\sqrt{5})^2 = x^2$
    $x^2 + 1 = 9$ $4 + 5 = x^2$
    $x^2 = 8$ $9 = x^2$
    $\sqrt{x^2} = \sqrt{8}$ $3 = x$
    $x = 2\sqrt{2}$

5. (17.1) Let $x =$ one angle
    Then $x + 30 =$ other angle
    $x + x + 30 = 180$
    $2x = 150$
    $x = 75$
    $x + 30 = 105$
    Angles are 75° and 105°.

6. (17.5) $9^2 + 12^2 = c^2$
    $81 + 144 = c^2$
    $225 = c^2$
    $15 = c$
    $P = a + b + c$
    $P = 9 + 12 + 15$
    $P = 36$ in.

7. (17.5) $P = 4s$
    $20 = 4s$
    $5 = s$
    $5^2 + 5^2 = d^2$
    $25 + 25 = d^2$
    $50 = d^2$
    $\sqrt{50} = \sqrt{d^2}$
    $5\sqrt{2}$ cm $= d$

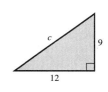

8. (17.6) $A = b \cdot h$
    $A = 8 \cdot 4$
    $A = 32$ sq. ft

9. (17.5)

$$P = 3 + 1 + 2 + 2 + 5 + 3 = 16 \text{ m}$$

10. (17.6) Area of triangle $= A_1 = \dfrac{1}{2}bh$
          $= \dfrac{1}{2}(6)(2)$
          $= 6$ sq. in.
    Area of square $= A_2 = s^2$
          $= 6^2$
          $= 36$ sq. in.

17

Total area $= A_1 + A_2$
$= 6 + 36$
$= 42$ sq. in.

11. (17.6) $A = lw$
$54 = l \cdot 6$
$9 = l$
$P = 2l + 2w$
$P = 2(9) + 2(6)$
$P = 18 + 12$
$P = 30$ ft

12. (17.6) $C = 2\pi r$
$\dfrac{12\pi}{2\pi} = \dfrac{2\pi r}{2\pi}$
$6 = r$
$A = \pi r^2$
$A = \pi 6^2$
$A = 36\pi$ sq. ft

13. (17.7) $V = lwh$
$V = 12 \cdot 3 \cdot 5$
$V = 180$ cu. in.

14. (17.7) $d = 2r$
$6 = 2r$
$3 = r$
$V = \dfrac{4}{3}\pi r^3$
$V = \dfrac{4}{3}\pi(3)^3$
$V = \dfrac{4}{3} \cdot \pi \cdot \overset{9}{\cancel{27}}$
$V = 36\pi$ m^3

15. (17.4) $\dfrac{AB}{4} = \dfrac{12}{6}$
$6AB = 48$
$AB = 8$

16. (17.4) Let $\quad x = AD$
Then $x + 5 = AC$
$\dfrac{4}{8} = \dfrac{5}{x + 5}$
$4x + 20 = 40$
$4x = 20$
$x = 5$

17. (17.3) $\angle DCB = \angle ABC = 90°$
$\angle DCB + \angle D + \angle CBD = 180$
$90 \;+\; 35 + \angle CBD = 180$
$\angle CBD = 55°$

18. (17.3) $AD = AD$ common side
$\triangle ABD \cong \triangle ACD$ by SAS

Chapters 1–17 Cumulative Review Exercises   (page 515)

1. $14\dfrac{11}{18}$ **2.** $2\dfrac{1}{3}$ **3.** 87.14 **4.** 40.803 **5.** -8

6. $3x^2 - 4x - 3$ **7.** $15x^2 - 14xy - 8y^2$

8. $16x^2 - 24x + 9$ **9.** $\dfrac{1}{3x^2}$ **10.** $\dfrac{6x}{x - 6}$ **11.** $x + 3$

12. $\dfrac{x^2 - 5x - 8}{x(x + 4)}$ **13.** $x - 5$ **14.** $x = 3$

15. $x = \dfrac{5}{3}$ or $x = -2$ **16.** $x = 2$ **17.** $x = 5$

18. $(-1, 2)$ **19.** $(5, -2)$

20. $x < -5$

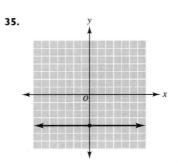

21. $L = \dfrac{P - 2W}{2}$ **22.** $\dfrac{9x^6}{y^8}$ **23.** $\dfrac{b}{a^5}$ **24.** $\dfrac{2y^5}{x^3}$

25. $2x^2 y\sqrt{3y}$ **26.** $2\sqrt{3} + 6\sqrt{5}$ **27.** $-2 + 17\sqrt{2}$

28. ≈ 5.22 **29.** $2x^2 - 3x + 4$ **30.** $\dfrac{19}{250}$ **31.** -31

32. 8.525 **33.** 6×10^{-7}

34.

35.

36. $4x + y = 18$ **37.** Slope $= -\dfrac{3}{2}$ **38.** $4\dfrac{1}{4}$ hr
$\qquad\qquad\qquad$ y-intercept $= 4$

39. $7\dfrac{1}{2}$ lb   **40.** 85%   **41.** 8,000 students   **42.** 45 mph

43. 15 dimes   **44.** $y = 15$ **45.** 45 sq. ft   **46.** 6π

47. $AC = 2\sqrt{5}$ **48.** $\angle CED = 48°$

49. Between 1993 and 1994   **50.** $\$1,500,000$

Exercises B.2A   (page 525)

1. 1,800 m   **3.** 249 g   **5.** 0.0605 kL   **7.** 0.275 kg

9. 780 m   **11.** 72.35 kg   **13.** 0.125 kL

Exercises B.2B   (page 526)

1. 2.79 m   **3.** 0.083 g   **5.** 250 cm   **7.** 390.6 cg

9. 6.32 L   **11.** 2.63 cg   **13.** 182 cm

Exercises B.2C   (page 527)

1. 0.091 m   **3.** 0.47 g   **5.** 2,600 mm   **7.** 108 mg

9. 0.23 L   **11.** 7,040 mg   **13.** 2,000 cc

15. 0.175 L   **17.** 1,500 mg

Exercises B.2D   (page 529)

1. 3,540 m   **3.** 47,000 L   **5.** 2.546 kg   **7.** 0.34 L

9. 516 dm   **11.** 0.895 hL   **13.** 0.784 km

15. 456,000 dg   **17.** 0.3402 dag   **19.** 0.5614 daL

21. 3,750 mm

Exercises B.2E   (page 530)

1. 68°F   **3.** 10°C   **5.** 46°F   **7.** 22°C   **9.** 5°C

Exercises B.4   (page 534)

1. 15 ft   **3.** $\dfrac{3\frac{1}{3}\ \cancel{yd}}{1}\left(\dfrac{3\ ft}{1\ \cancel{yd}}\right)=\dfrac{10}{3}\times\dfrac{3}{1}\ ft=10\ ft$

5. 32 qt   **7.** 14 pt   **9.** 90 min   **11.** 9,240 ft

13. 30 ft, 360 in.   **15.** 2 ft   **17.** $1\frac{1}{2}$ min   **19.** 1 mi

21. 2 yr   **23.** 6 tbsp   **25.** 4 tsp   **27.** 40 oz

29. 5,000 lb   **31.** 6,600 ft

33. $\dfrac{2\ \cancel{wk}}{1}\left(\dfrac{7\ \cancel{da}}{1\ \cancel{wk}}\right)\left(\dfrac{24\ hr}{1\ \cancel{da}}\right)=2\times7\times24\ hr=336\ hr$

35. $\dfrac{2,160\ \cancel{min}}{1}\left(\dfrac{1\ \cancel{hr}}{60\ \cancel{min}}\right)\left(\dfrac{1\ da}{24\ \cancel{hr}}\right)=\dfrac{2,160}{60\times24}\ da=1.5\ da$

37. 40 mph

39. $\dfrac{1,100\ \cancel{ft}}{1\ \cancel{sec}}\left(\dfrac{60\ \cancel{sec}}{1\ \cancel{min}}\right)\left(\dfrac{60\ \cancel{min}}{1\ hr}\right)\left(\dfrac{1\ mi}{5,280\ \cancel{ft}}\right)$

$=\dfrac{1,100\times60\times60\ mi}{5280\ hr}=750\ mph$

Exercises B.5   (page 537)

1. 5 ft 3 in.   **3.** 3 wk 2 da   **5.** 6 gal 3 qt

7. 7 yd 3 in.   **9.** 3 hr 14 min 50 sec   **11.** 5 gal 1 qt 1 pt

13. 3 mi 720 ft   **15.** 3 yr 6 wk 5 da   **17.** 3 tons 1,500 lb

19. 5 lb 4 oz   **21.** 3 da 15 min

Exercises B.6   (page 538)

1. 13 ft 1 in.   **3.** 13 hr 29 min

5. 17 gal 3 qt 1 pt   **7.** 14 yd 1 ft 2 in.

9. 10 da 6 hr 56 min   **11.** 16 tons 1,800 lb

13. 45 lb 13 oz   **15.** 10 yd 1 ft 1 in.

Exercises B.7   (page 539)

1. 5 ft 6 in.   **3.** 6 yd 1 ft   **5.**
$$\begin{array}{r} \overset{7}{\cancel{8}}\ gal\ \overset{\overset{5}{\cancel{4}}}{2}\ qt\ \overset{2}{\cancel{0}}\ pt \\ -\ 3\ gal\ 3\ qt\ 1\ pt \\ \hline 4\ gal\ 2\ qt\ 1\ pt \end{array}$$

7. 2 lb 11 oz   **9.** 1 ton 1,500 lb   **11.** 1 da 14 hr

13. 2 yd 9 in.   **15.** 2 mi 3,580 ft   **17.** 14 min 45 sec

19. 2 da 18 hr 37 min   **21.** 2 hr 55 min 45 sec

Exercises B.8   (page 541)

1. 14 wk 6 da   **3.** 33 mi 1,260 ft   **5.** 12 yd 3 in.

7. 5 hr 40 min 44 sec   **9.**
$$\begin{array}{r} 3\ gal\quad 3\ qt\ 1\ pt \\ \times\qquad\qquad\quad 8 \\ \hline 24\ gal\ 24\ qt\ \cancel{8}\ pt \\ \boxed{4}\quad\ \boxed{0} \\ \cancel{28}\ qt \\ \boxed{7}\quad\quad\ \boxed{0} \\ \hline 31\ gal \end{array}$$

11. 8,448 sq. ft   **13.** 8,100 sq. ft

15. Worker-hours $=17\times12\times8=1,632$
Cost $=\$16.50\times1,632=\$26,928$

Exercises B.9   (page 542)

1. 1 ft 3 in.

3.
$$\begin{array}{r} 0\ qt\ 1\frac{1}{4}\ pt \\ 4\overline{)2\ qt\quad 1\ pt} \\ \ \longrightarrow\ 4 \\ \hline 5\ pt \\ 4 \\ \hline 1\ pt\ \rightarrow\ \frac{1\ pt}{4}=\frac{1}{4}\ pt \end{array}$$

5. 1 hr 50 min   **7.** 1 gal 2 qt 1 pt

9.
$$\begin{array}{r} 1\ yd\qquad 2\ ft\quad 4\frac{2}{5}\ in. \\ 5\overline{)8\ yd\qquad 2\ ft\quad 10\ in.} \\ 5 \\ \hline 3\ yd=\quad 9\ ft \\ \hline 11\ ft \\ 10\ ft \\ \hline 1\ ft=\quad 12\ in. \\ \hline 22\ in. \\ 20 \\ \hline 2\ in.\rightarrow\frac{2\ in.}{5}=\frac{2}{5}\ in. \end{array}$$

11. $12\frac{4}{7}$ oz   **13.** 4 wk 4 da 5 hr   **15.** 1 mi 2,510 ft

17. 8   **19.** 11 postcards, with 10¢ left over   **21.** 45

Exercises B.10   (page 545)

1. 38 cm   **3.** 2.00 L   **5.** 1.2 lb   **7.** 51 mi   **9.** 13 qt

11. 53 km   **13.** 9.36 kg   **15.** 370 acres   **17.** 1.7 m

19. 80 in.   **21.** 2.00 lb   **23.** 340 g

25. $\dfrac{1.5\ \cancel{yd}}{1}\left(\dfrac{36\ \cancel{in.}}{1\ \cancel{yd}}\right)\left(\dfrac{2.54\ cm}{1\ \cancel{in.}}\right)=137.16\ cm\approx140\ cm$

27. $\dfrac{227\ \cancel{g}}{1}\left(\dfrac{1\ \cancel{lb}}{454\ \cancel{g}}\right)\left(\dfrac{16\ oz}{1\ \cancel{lb}}\right)\approx8.00\ oz$ **29.** 75 mi

31. 189 lb

33. $\dfrac{1,500\ \cancel{m}}{1}\left(\dfrac{39.4\ \cancel{in.}}{1\ \cancel{m}}\right)\left(\dfrac{1\ ft}{12\ \cancel{in.}}\right)=4,925\ ft$
$\qquad\qquad\qquad\qquad\qquad\qquad\qquad\begin{array}{r}5,280\\-\ 4,925\\\hline 355\end{array}$
1 mile = 5,280 ft.
Therefore, the 1,500 m race is 355 ft shorter than the mile race.

35. $\dfrac{175\ \cancel{mm}}{1}\left(\dfrac{1\ \cancel{cm}}{10\ \cancel{mm}}\right)\left(\dfrac{1\ in.}{2.54\ \cancel{cm}}\right)=6.88976378\ in.\approx6.89\ in.$

37. $\dfrac{20\ \cancel{L}}{1}\left(\dfrac{1.06\ \cancel{qt}}{1\ \cancel{L}}\right)\left(\dfrac{1\ gal}{4\ \cancel{qt}}\right)=5.30\ gal\approx5\ gal$

Exercises B.11   (page 546)

1. 80.8 cm   **3.** 397 L   **5.** 418.2 g   **7.** 28°C

9. 1,600 yd   **11.** 5.8 gal   **13.** 1.205 lb   **15.** 37°C

B

INDEX